Lecture Notes in Computer Science 13399

Günter Rudolph · Anna V. Kononova ·
Hernán Aguirre · Pascal Kerschke ·
Gabriela Ochoa · Tea Tušar (Eds.)

Parallel Problem Solving from Nature – PPSN XVII

17th International Conference, PPSN 2022
Dortmund, Germany, September 10–14, 2022
Proceedings, Part II

 Springer

Editors
Günter Rudolph (iD)
TU Dortmund
Dortmund, Germany

Anna V. Kononova (iD)
Leiden University
Leiden, The Netherlands

Hernán Aguirre (iD)
Shinshu University
Nagano, Japan

Pascal Kerschke (iD)
Technische Universität Dresden
Dresden, Germany

Gabriela Ochoa (iD)
University of Stirling
Stirling, UK

Tea Tušar (iD)
Jožef Stefan Institute
Ljubljana, Slovenia

ISSN 0302-9743   ISSN 1611-3349 (electronic)
Lecture Notes in Computer Science
ISBN 978-3-031-14720-3   ISBN 978-3-031-14721-0 (eBook)
https://doi.org/10.1007/978-3-031-14721-0

This Springer imprint is published by the registered company Springer Nature Switzerland AG
The registered company address is: Gewerbestrasse 11, 6330 Cham, Switzerland

Preface

The first major gathering of people interested in discussing natural paradigms and their application to solve real-world problems in Europe took place in Dortmund, Germany, in 1990. What was planned originally as a small workshop with about 30 participants finally grew into an international conference named Parallel Problem Solving from Nature (PPSN) with more than 100 participants. The interest in the topics of the conference has increased steadily ever since leading to the pleasant necessity of organizing PPSN conferences biennially within the European region.

In times of a pandemic, it is difficult to find a host for a conference that should be held locally if possible. To ensure the continuation of the conference series, the 17th edition, PPSN 2022, returned to its birthplace in Dortmund. But even at the time of writing this text, it is unclear whether the conference can be held on-site or whether we shall have to switch to virtual mode at short notice.

Therefore, we are pleased that many researchers shared our optimism by submitting their papers for review. We received 185 submissions from which the program chairs have selected the top 85 after an extensive peer-review process. Not all decisions were easy to make but in all cases we benefited greatly from the careful reviews provided by the international Program Committee consisting of 223 scientists. Most of the submissions received four reviews, but all of them got at least three reviews. This led to a total of 693 reviews. Thanks to these reviews we were able to decide about acceptance on a solid basis.

The papers included in these proceedings have been assigned to 12 fuzzy clusters, entitled *Automated Algorithm Selection and Configuration, Bayesian- and Surrogate-Assisted Optimization, Benchmarking and Performance Measures, Combinatorial Optimization, (Evolutionary) Machine Learning and Neuroevolution, Evolvable Hardware and Evolutionary Robotics, Fitness Landscape Modeling and Analysis, Genetic Programming, Multi-Objective Optimization, Numerical Optimizaiton, Real-World Applications, and Theoretical Aspects of Nature-Inspired Optimization*, that can hardly reflect the true variety of research topics presented in the proceedings at hand. Following the tradition and spirit of PPSN, all papers were presented as posters. The 7 poster sessions consisting of about 12 papers each were compiled orthogonally to the fuzzy clusters mentioned above to cover the range of topics as widely as possible. As a consequence, participants with different interests would find some relevant papers in every session and poster presenters were able to discuss related work in sessions other than their own. As usual, the conference also included one day with workshops (Saturday), one day with tutorials (Sunday), and three invited plenary talks (Monday to Wednesday) for free.

Needless to say, the success of such a conference depends on the authors, reviewers, and organizers. We are grateful to all authors for submitting their best and latest work, to all the reviewers for the generous way they spent their time and provided their valuable expertise in preparing the reviews, to the workshop organizers and tutorial presenters for their contributions enhancing the value of the conference, and to the local organizers who helped to make PPSN 2022 happen.

Last but not least, we would like to thank for the donations of the *Gesellschaft der Freunde der Technischen Universität Dortmund e.V. (GdF)* and the *Alumni der Informatik Dortmund e.V. (aido)*. We are grateful for Springer's long-standing support of this conference series. Finally, we thank the *Deutsche Forschungsgemeinschaft (DFG)* for providing financial backing.

July 2022

Günter Rudolph
Anna V. Kononova
Hernán Aguirre
Pascal Kerschke
Gabriela Ochoa
Tea Tušar

Organization

General Chair

Günter Rudolph   TU Dortmund University, Germany

Honorary Chair

Hans-Paul Schwefel   TU Dortmund University, Germany

Program Committee Chairs

Hernán Aguirre	Shinshu University, Japan
Pascal Kerschke	TU Dresden, Germany
Gabriela Ochoa	University of Stirling, UK
Tea Tušar	Jožef Stefan Institute, Slovenia

Proceedings Chair

Anna V. Kononova   Leiden University, The Netherlands

Tutorial Chair

Heike Trautmann   University of Münster, Germany

Workshop Chair

Christian Grimme   University of Münster, Germany

Publicity Chairs

Nicolas Fischöder	TU Dortmund University, Germany
Peter Svoboda	TU Dortmund University, Germany

Social Media Chair

Roman Kalkreuth   TU Dortmund University, Germany

Digital Fallback Chair

Hestia Tamboer	Leiden University, The Netherlands

Steering Committee

Thomas Bäck	Leiden University, The Netherlands
David W. Corne	Heriot-Watt University, UK
Carlos Cotta	Universidad de Málaga, Spain
Kenneth De Jong	George Mason University, USA
Gusz E. Eiben	Vrije Universiteit Amsterdam, The Netherlands
Bogdan Filipič	Jožef Stefan Institute, Slovenia
Emma Hart	Edinburgh Napier University, UK
Juan Julián Merelo Guervós	Universida de Granada, Spain
Günter Rudolph	TU Dortmund University, Germany
Thomas P. Runarsson	University of Iceland, Iceland
Robert Schaefer	University of Krakow, Poland
Marc Schoenauer	Inria, France
Xin Yao	University of Birmingham, UK

Program Committee

Jason Adair	University of Stirling, UK
Michael Affenzeller	University of Applied Sciences Upper Austria, Austria
Hernán Aguirre	Shinshu University, Japan
Brad Alexander	University of Adelaide, Australia
Richard Allmendinger	University of Manchester, UK
Marie Anastacio	Leiden University, The Netherlands
Denis Antipov	ITMO University, Russia
Claus Aranha	University of Tsukuba, Japan
Rolando Armas	Yachay Tech University, Ecuador
Dirk Arnold	Dalhousie University, Canada
Anne Auger	Inria, France
Dogan Aydin	Dumlupinar University, Turkey
Jaume Bacardit	Newcastle University, UK
Thomas Bäck	Leiden University, The Netherlands
Helio Barbosa	Laboratório Nacional de Computação Científica, Brazil
Andreas Beham	University of Applied Sciences Upper Austria, Austria
Heder Bernardino	Universidade Federal de Juiz de Fora, Brazil
Hans-Georg Beyer	Vorarlberg University of Applied Sciences, Austria

Julian Blank	Michigan State University, USA
Aymeric Blot	University College London, UK
Christian Blum	Spanish National Research Council, Spain
Peter Bosman	Centrum Wiskunde & Informatica, The Netherlands
Jakob Bossek	University of Münster, Germany
Jürgen Branke	University of Warwick, UK
Dimo Brockhoff	Inria, France
Alexander Brownlee	University of Stirling, UK
Larry Bull	University of the West of England, UK
Maxim Buzdalov	ITMO University, Russia
Arina Buzdalova	ITMO University, Russia
Stefano Cagnoni	University of Parma, Italy
Fabio Caraffini	De Montfort University, UK
Ying-Ping Chen	National Chiao Tung University, Taiwan
Francisco Chicano	University of Málaga, Spain
Miroslav Chlebik	University of Sussex, UK
Sung-Bae Cho	Yonsei University, South Korea
Tinkle Chugh	University of Exeter, UK
Carlos Coello Coello	CINVESTAV-IPN, Mexico
Ernesto Costa	University of Coimbra, Portugal
Carlos Cotta	Universidad de Málaga, Spain
Nguyen Dang	St Andrews University, UK
Kenneth De Jong	George Mason University, USA
Bilel Derbel	University of Lille, France
André Deutz	Leiden University, The Netherlands
Benjamin Doerr	Ecole Polytechnique, France
Carola Doerr	Sorbonne University, France
John Drake	University of Leicester, UK
Rafal Drezewski	AGH University of Science and Technology, Poland
Paul Dufossé	Inria, France
Gusz Eiben	Vrije Universiteit Amsterdam, The Netherlands
Mohamed El Yafrani	Aalborg University, Denmark
Michael Emmerich	Leiden University, The Netherlands
Andries Engelbrecht	University of Stellenbosch, South Africa
Anton Eremeev	Omsk Branch of Sobolev Institute of Mathematics, Russia
Richard Everson	University of Exeter, UK
Pedro Ferreira	Universidade de Lisboa, Portugal
Jonathan Fieldsend	University of Exeter, UK
Bogdan Filipič	Jožef Stefan Institute, Slovenia

Ke Li	University of Exeter, UK
Arnaud Liefooghe	University of Lille, France
Giosuè Lo Bosco	Università di Palermo, Italy
Fernando Lobo	University of Algarve, Portugal
Daniele Loiacono	Politecnico di Milano, Italy
Nuno Lourenço	University of Coimbra, Portugal
Jose A. Lozano	University of the Basque Country, Spain
Rodica Ioana Lung	Babes-Bolyai University, Romania
Chuan Luo	Peking University, China
Gabriel Luque	University of Málaga, Spain
Evelyne Lutton	INRAE, France
Manuel López-Ibáñez	University of Málaga, Spain
Penousal Machado	University of Coimbra, Portugal
Kaitlin Maile	ISAE-SUPAERO, France
Katherine Malan	University of South Africa, South Africa
Vittorio Maniezzo	University of Bologna, Italy
Elena Marchiori	Radboud University, The Netherlands
Asep Maulana	Tilburg University, The Netherlands
Giancarlo Mauri	University of Milano-Bicocca, Italy
Jacek Mańdziuk	Warsaw University of Technology, Poland
James McDermott	National University of Ireland, Galway, Ireland
Jörn Mehnen	University of Strathclyde, UK
Marjan Mernik	University of Maribor, Slovenia
Olaf Mersmann	TH Köln, Germany
Silja Meyer-Nieberg	Bundeswehr University Munich, Germany
Efrén Mezura-Montes	University of Veracruz, Mexico
Krzysztof Michalak	Wroclaw University of Economics, Poland
Kaisa Miettinen	University of Jyväskylä, Finland
Edmondo Minisci	University of Strathclyde, UK
Gara Miranda	University of La Laguna, Spain
Mustafa Misir	Istinye University, Turkey
Hugo Monzón	RIKEN, Japan
Sanaz Mostaghim	Fraunhofer IWS, Germany
Mario Andres Muñoz Acosta	University of Melbourne, Australia
Boris Naujoks	TH Köln, Germany
Antonio J. Nebro	University of Málaga, Spain
Aneta Neumann	University of Adelaide, Australia
Frank Neumann	University of Adelaide, Australia
Michael O'Neill	University College Dublin, Ireland
Pietro S. Oliveto	University of Sheffield, UK
Una-May O'Reilly	MIT, USA
José Carlos Ortiz-Bayliss	Tecnológico de Monterrey, Mexico

Patryk Orzechowski	University of Pennsylvania, USA
Ender Özcan	University of Nottingham, UK
Gregor Papa	Jožef Stefan Institute, Slovenia
Gisele Pappa	Universidade Federal de Minas Gerais, Brazil
Luis Paquete	University of Coimbra, Portugal
Andrew J. Parkes	University of Nottingham, UK
David Pelta	University of Granada, Spain
Leslie Perez-Caceres	Pontificia Universidad Católica de Valparaíso, Chile
Stjepan Picek	Delft University of Technology, The Netherlands
Martin Pilat	Charles University, Czech Republic
Nelishia Pillay	University of KwaZulu-Natal, South Africa
Petr Pošík	Czech Technical University in Prague, Czech Republic
Raphael Prager	University of Münster, Germany
Michał Przewoźniczek	Wroclaw University of Science and Technology, Poland
Chao Qian	University of Science and Technology of China, China
Xiaoyu Qin	University of Birmingham, UK
Alma Rahat	Swansea University, UK
Khaled Rasheed	University of Georgia, USA
Frederik Rehbach	TH Köln, Germany
Lucas Ribeiro	Universidade Federal de Goiás, Brazil
Eduardo Rodriguez-Tello	CINVESTAV, Tamaulipas, Mexico
Andrea Roli	University of Bologna, Italy
Jonathan Rowe	University of Birmingham, UK
Günter Rudolph	TU Dortmund University, Germany
Thomas A. Runkler	Siemens Corporate Technology, Germany
Conor Ryan	University of Limerick, Ireland
Frédéric Saubion	University of Angers, France
Robert Schaefer	AGH University of Science and Technology, Poland
Andrea Schaerf	University of Udine, Italy
David Schaffer	Binghamton University, USA
Lennart Schäpermeier	TU Dresden, Germany
Marc Schoenauer	Inria Saclay Île-de-France, France
Oliver Schütze	CINVESTAV-IPN, Mexico
Michele Sebag	CNRS, Université Paris-Saclay, France
Moritz Seiler	University of Münster, Germany
Bernhard Sendhoff	Honda Research Institute Europe GmbH, Germany

Marc Sevaux   Université de Bretagne Sud, France
Shinichi Shirakawa   Yokohama National University, Japan
Moshe Sipper   Ben-Gurion University of the Negev, Israel
Jim Smith   University of the West of England, UK
Jorge Alberto Soria-Alcaraz   Universidad de Guanajuato, Mexico
Patrick Spettel   FH Vorarlberg, Austria
Giovanni Squillero   Politecnico di Torino, Italy
Catalin Stoean   University of Craiova, Romania
Thomas Stützle   Université Libre de Bruxelles, Belgium
Mihai Suciu   Babes-Bolyai University, Romania
Dirk Sudholt   University of Sheffield, UK
Andrew Sutton   University of Minnesota, USA
Ricardo H. C. Takahashi   Universidade Federal de Minas Gerais, Brazil
Sara Tari   Université du Littoral Côte d'Opale, France
Daniel Tauritz   Auburn University, USA
Dirk Thierens   Utrecht University, The Netherlands
Sarah Thomson   University of Stirling, UK
Kevin Tierney   Bielefeld University, Germany
Renato Tinós   University of São Paulo, Brazil
Alberto Tonda   INRAE, France
Leonardo Trujillo   Instituto Tecnológico de Tijuana, Mexico
Tea Tušar   Jožef Stefan Institute, Slovenia
Ryan J. Urbanowicz   University of Pennsylvania, USA
Koen van der Blom   Leiden University, The Netherlands
Bas van Stein   Leiden University, The Netherlands
Nadarajen Veerapen   University of Lille, France
Sébastien Verel   Université du Littoral Côte d'Opale, France
Diederick Vermetten   Leiden University, The Netherlands
Marco Virgolin   Centrum Wiskunde & Informatica,
                                  The Netherlands
Aljoša Vodopija   Jožef Stefan Institute, Slovenia
Markus Wagner   University of Adelaide, Australia
Stefan Wagner   University of Applied Sciences Upper Austria,
                                  Austria
Hao Wang   Leiden University, The Netherlands
Hui Wang   Leiden University, The Netherlands
Elizabeth Wanner   CEFET, Brazil
Marcel Wever   LMU Munich, Germany
Dennis Wilson   ISAE-SUPAERO, France
Carsten Witt   Technical University of Denmark, Denmark
Man Leung Wong   Lingnan University, Hong Kong
Bing Xue   Victoria University of Wellington, New Zealand

Kaifeng Yang	University of Applied Sciences Upper Austria, Austria
Shengxiang Yang	De Montfort University, UK
Estefania Yap	University of Melbourne, Australia
Furong Ye	Leiden University, The Netherlands
Martin Zaefferer	TH Köln, Germany
Aleš Zamuda	University of Maribor, Slovenia
Saúl Zapotecas	Instituto Nacional de Astrofísica, Óptica y Electrónica, Mexico
Christine Zarges	Aberystwyth University, UK
Mengjie Zhang	Victoria University of Wellington, New Zealand

Keynote Speakers

Doina Bucur	University of Twente, The Netherlands
Claudio Semini	IIT, Genoa, Italy
Travis Waller	TU Dresden, Germany

Contents – Part II

Numerical Optimizaiton

Contents – Part I

Benchmarking and Performance Measures

Combinatorial Optimization

Evolvable Hardware and Evolutionary Robotics

Fitness Landscape Modeling and Analysis

Genetic Programming

Digging into Semantics: Where Do Search-Based Software Repair Methods Search?

Hammad Ahmad[1(✉)], Padriac Cashin[2], Stephanie Forrest[2], and Westley Weimer[1]

[1] University of Michigan, Ann Arbor, MI 48109, USA
{hammada,weimerw}@umich.edu
[2] Arizona State University, Tempe, AZ 85281, USA
{pcashin,steph}@asu.edu

Abstract. Search-based methods are a popular approach for automatically repairing software bugs, a field known as automated program repair (APR). There is increasing interest in empirical evaluation and comparison of different APR methods, typically measured as the rate of successful repairs on benchmark sets of buggy programs. Such evaluations, however, fail to explain *why* some approaches succeed and others fail. Because these methods typically use syntactic representations, i.e., source code, we know little about how the different methods explore their semantic spaces, which is relevant for assessing repair quality and understanding search dynamics. We propose an automated method based on program semantics, which provides quantitative and qualitative information about different APR search-based techniques. Our approach requires no manual annotation and produces both mathematical and human-understandable insights. In an empirical evaluation of 4 APR tools and 34 defects, we investigate the relationship between search-space exploration, semantic diversity and repair success, examining both the overall picture and how the tools' search unfolds. Our results suggest that population diversity alone is not sufficient for finding repairs, and that searching in the right place is more important than searching broadly, highlighting future directions for the research community.

Keywords: Semantic search spaces · Program repair · Patch diversity

1 Introduction

Early works on automatically repairing defects in software demonstrated that evolutionary computation (EC) and related *search-based* approaches can be surprisingly successful in this domain [1,2,20,35,54]. Since then, there has been an explosion of research into what is now called the *automated program repair* (APR) problem. This research has produced a wide variety of techniques and tools aimed at reducing the manual effort required to repair software bugs or otherwise improve software [23,31]. These tools typically operate on source code

G. Rudolph et al. (Eds.): PPSN 2022, LNCS 13399, pp. 3–18, 2022.
https://doi.org/10.1007/978-3-031-14721-0_1

containing one or more bugs, or *defects*, together with a test suite that encodes required functionality and at least one test that exposes the defect. Multiple candidate patches are often generated, which both repair the defect and pass the test suite [1,20,37,38]. The field has standardized on a small number of benchmark test suites to compare the performance of different tools and techniques, often by measuring the fraction of successful repairs [12,21]. However, we still have little insight into fundamental questions such as: *Why* do some algorithms outperform others? *Which* components of an algorithm are most responsible for its success (or failure)? *How different* are the patches produced by different techniques? *What* kinds of bugs is APR better or worse at solving?

Traditional evaluation approaches are not always helpful for these questions. For example, it can be difficult to determine from a pseudocode description of a new repair algorithm whether it will find a more *diverse* set of repairs than existing ones, or which parts of a search space it will visit [26]. Importantly, today's search-based APR methods use a syntactic representation, i.e., source code, even though repairing bugs involves changing semantics.

Earlier research tackled some of these questions by considering the extent to which proposed repairs are overfit to a test suite [19,30,34,43,46], non-functional properties such as repair readability and maintainability [11,50], and repair diversity [6,18,30,33,36,49,56]. Within the context of diversity, previous studies examined the search space of a single tool to better understand patch construction [14,55] and compared the search spaces explored by different tools with respect to high-level program characteristics [30,32,52]. However, to the best of our knowledge, this previous work considers only program variants that are *test-suite adequate* [38], or *plausible*, meaning that they repair the bug and pass all required tests. This approach ignores how the search process discovers plausible repairs. In this paper, we propose a method for comparing the semantic search spaces of different APR algorithms, and characterize the program variants generated during the search in addition to the end product.

Insight and Approach. Generating candidate variants through syntactic program manipulation is central to search-based APR tools, yet their ultimate value depends on inducing meaningful semantic change. We hypothesize that the effective *search spaces* (the sets of candidate program variants considered or potentially constructed) of different APR tools for a given software defect are distinct but not disjoint. We further hypothesize that lightweight analysis of the run-time semantics of each variant generated, regardless of correctness, can shed light on how different APR tools search for repairs. To analyze the effective search space of a particular tool, we propose to embed its generated variants in a semantic *invariant space*, admitting an approximate notion of *similarity*. Because many individual variants generated during a search are syntactically distinct but semantically equivalent [53], we focus on source-level formal invariants. Since test suites are generally available in this domain, we propose leveraging them for efficient dynamic invariant detection [8], rather than resorting to expensive static or manual approaches. Once each individual variant is characterized by its set of detected invariants, we propose to use a form of weighted vector distance

(*Canberra distance* [17]) to assess differences. Because most programs have many invariants, our vector distance approach has a significant scalability advantage over other approaches, such as checking logical implications between invariant sets with a theorem prover. Ultimately, our approach allows both mathematical (i.e., via principal component analysis) and human-understandable (i.e., two-dimensional visualization) analysis of search spaces.

Contributions. The main contributions of this work are as follows:

- A framework for comparing the effective semantic search spaces of APR algorithms.
- An automated analysis of individual program variants to produce a two-dimensional visualization of their semantic diversity.
- An empirical analysis on four established search-based APR tools.
- A discussion of the relationship between syntactic and semantic diversity and implications for APR algorithm design.

2 Background and Contextual Motivation

Automated Program Repair. Automated program repair (APR) methods seek to locate and repair software defects without introducing side effects. Typically, this involves modifying the program's source code to produce a patched version. Most methods rely on a test suite to certify the repaired program's correctness.

Over the past decade, many search-based methods for APR have been proposed, with some more recognizable as Genetic Programming (GP) solutions than others (see Monperrus [31] or Le Goues et al. [23] for comprehensive reviews). In this paper, we evaluate on four established tools that represent different search-based APR techniques. GenProg implements a form of GP to search for repairs [22,54]. CapGen uses the same mutation operators as GenProg, but allows more granular mutations to sub-elements of statements and mines contextual information to select effective mutations [55]. SimFix mines prior patches, both to construct particular mutations and to guide the selection of operations based on code similarity measurements [14]. TBar is a recent approach that uses 35 different "fix patterns", or templates, to modify the buggy program [25]. Over time these tools have incorporated heuristic information about the software-repair domain to what was originally a pure GP-based approach.

Dynamic Invariant Detection. To capture semantics, we use dynamic invariants (i.e., logical predicates that always hold during the execution of the program) to approximate code functionality. Dynamic invariant detection [8] algorithms trace program state during execution to construct such invariants. These traces contain the state of in-scope variables at specific points in execution, usually before and after function calls. Because they do not rely on program source code to construct invariants (cf. static invariant detection), dynamic approaches are modular and scalable to our problem. However, a finite set of dynamic traces may not capture all possible or relevant future executions and can overfit to the observed traces. Because we are interested in how small regions of code (patches) differ from one another, this issue is less of a concern for our task.

Semantic Search Space. Earlier studies have investigated how well different APR methods explore the search space created by their mutation operators. Typically, the search space is defined as the union of the mutants that can be created by applying n mutations to the original program [18,30,36]. For instance, if an APR tool can only insert one statement before another, then its first-order search space consists of all programs that can be constructed by applying that *insert* operator a single time to the original. This approach has been used to characterize the search space by measuring the density of programs with specific characteristics, such as the number of passing tests or the number of correct patches [36]. By contrast, we define the semantic search space of an APR algorithm in terms of the set of reachable program invariants (via any of its generated mutants) when applying its mutation operations to the original program. Since the goal of the APR process is to construct a semantically correct program, understanding what functionality a given algorithm can construct is crucial to understanding its behavior. Similar to the syntactic search space, the semantic search space is effectively infinite, even with simple operators. Rather than enforcing an n-mutation restriction, as the aforementioned approaches do, we rely on the normal operation of each APR tool, unchanged, to define its semantic search space. This allows us to describe the search spaces of APR tools as they apply in practice.

Contextual Motivation: Does Diversity Lead to More Repairs? Some researchers have suggested that higher population diversity (syntactic or semantic) leads to higher repair rates and better repairs [7,10,39,44], and some tools (e.g., Marriagent [16]) favor high-diversity edits. Other results suggest that high semantic diversity does not necessarily improve repair rates [5,6,51]. If the latter is true, it suggests that researchers should focus less on high diversity mutants, and more on other properties of repair algorithms. If exploring widely in the search space predicts high repair rates, we would expect to observe a correlation between how much of the semantic search space is sampled and an ability to discover repairs. Across the board, however, as we perform quantitative and qualitative analyses to investigate the relationship between semantic diversity and repair rates for APR tools, we find little evidence that this is true (see Sect. 5). This finding challenges the conventional hypothesis that generating diverse mutants is the key to improving repair rates, and supports recent results arguing otherwise.

3   Technical Approach

Even though most of today's search-based APR methods inherit the concept of mutation from evolutionary computation, such tools do not significantly rely on crossover [24,37,38,53]. We thus focus on the mutation operators of APR tools. We begin with a set of mutants for each APR method. These are mutated variants of the original program, which pass all of the positive (regression) tests and may or may not pass the negative (bug inducing) tests. Given a set of mutants, or candidate patches, we next use Daikon [9] (the state-of-the-art for dynamic invariant detection) to generate a set of invariants, one set for each individual variant, regardless of correctness (Sect. 3.1), and then apply an efficient heuristic to measure semantic similarity between invariant sets (Sect. 3.2). Since large invariant sets are challenging to interpret and compare, we also present two visualizations of induced APR search spaces (Sect. 3.3).

3.1 Sampling APR Search Spaces

We aim to reason qualitatively about the search spaces induced by different APR tools and the techniques they employ (e.g., genetic operator-based vs. template-based mutation). Schulte et al. have previously treated the syntactic representation of each variant generated by an APR tool as a sample of the tool's search space [42]. We hypothesize, however, semantic diversity may be a more relevant consideration for understanding tool effectiveness. Our approach is motivated in part by the fact that syntactic variants often leave functionality unchanged (neutral) [41,42,53]. Ultimately, an APR tool's utility relates to its ability to find new functionality that addresses the defect.

We sample the semantic search space in two ways. First, we consider the early phase of a search by selecting the first x variants generated by each tool, reflecting real-world scenarios with scarce computational resources. Our second sampling method provides a broader picture. Some tools might initially search less widely, but focus in later. Thus, we evaluate y mutants selected uniformly at random after each tool completes its search.

We next consider how to capture the behavior of a mutant. Since our benchmarks total 357,000 lines of code and have over 20,000 test cases [15], static analysis methods will not scale for our experiments. Instead, we use dynamic analysis and restrict attention to a subset of the test cases. Because we are interested in repairing bugs, we assume that the greatest variation in mutant functionality will be along faulty execution paths, represented by the failing test. Intuitively, since repair algorithms aim to retain required functionality, they are much more likely to agree semantically on regression (positive) tests. We thus collect only traces associated with negative tests, one set for each distinct mutant. The set of invariants represents the most relevant program behavior. To compare variants, we then compute the difference between each pair of invariant sets across all tools in our study using a computational shortcut, which is surprisingly effective.

3.2 Computing Mutant Similarity

Earlier work defined a metric for computing semantic distance between two programs, based on logical implication between their sets of invariants [4]. This metric reflects the content of individual invariants, and as such quantifies difference precisely. Unfortunately, implementations of this approach have $\mathcal{O}(n!)$ time complexity in the worst case. Invariant detectors (e.g., Daikon) often report thousands of invariants for a single complex program. Thus, implication-based distance approaches are too expensive for use in our setting.

Instead, we use an efficient approximation of the semantic distance between two mutants. By treating invariant sets as bit vectors (one dimension for each invariant), we can compute the Canberra distance [17], a numerical measure of the distance between pairs of points in a vector space, between two invariant sets. To do this, we define a canonical ordering of the union of all invariants found across all mutants, and then associate one bit vector with each mutant, where the nth bit is set if and only if the nth invariant was detected for that program. We then compute the Canberra distance between the bit vectors, and use these distances to embed each mutant in an implied semantic vector space. In our setting, candidate patches are mostly identical except for a small number of mutations, and thus, Canberra distance provides a scalable approach that captures invariant differences between programs effectively.

3.3 Visualizing Search Spaces

Simply presenting a raw set of invariants, or even a string difference between two sets of invariants, is not informative to humans [45]. As such, for each defect, we compute the pairwise distance between invariant sets for every mutant, producing one number per pair, regardless of the APR tool that generated it. We use this information to visualize the semantic subspaces generated by each tool by embedding it in a single two-dimensional plot. Since our metric is relative (i.e., we compute the relative distance between the inferred invariant sets for two mutants), we anchor the measurements to two key points: the invariant set for the original defect, and the invariant set for the human-generated repair. Once the distance measurements are computed, our vector distance metric embeds mutants into a human-friendly two-dimensional visualization.

To complement the distance information, we also consider the number of unique semantic invariants introduced by each new mutant. For each tool, we examine the number of new unique invariants inferred for each mutant produced and evaluated. While the 2D embeddings show where each tool is sampling in semantic space, the rate at which unique invariants accumulate shows how much time the tool spends generating mutants with new semantics (and thus new functionality) compared to rediscovering old functionality with new syntax.

These two visualizations decompose our analysis into a spatial and temporal component, both of which are key to understanding the APR search for solutions.

4   Experimental Setup

We now describe our experimental setup for comparing the search spaces of various APR tools. We also make our replication materials publicly available.

Candidate Patch Collection. We gather candidate variants (mutants) from four established tools: CapGen [55], GenProg [22], SimFix [14], and TBar [25]. All four tools use search-based techniques, but each tool uses different mutation operators and search methods. We ran each tool on 34 representative Java defects from Defects4J [15] that all of the tools we consider operate on (see Table 1).

Table 1. Experimental Benchmarks: 34 Java defects selected from Defects4J. ✓ means that the tool produced a repair. Defects not repaired by any tool (omitted for space) comprise Math 7, 9, 12, 16, 17, 18, 19.

	Chart			Lang				Math																		
	8	11	24	6	26	57	59	1	2	3	4	5	8	11	15	20	30	33	53	57	59	63	65	70	75	80	85
CapGen	✓	✓	✓	✓	✓	✓	✓					✓				✓	✓	✓	✓	✓	✓	✓	✓	✓	✓	✓	✓
GenProg				✓								✓	✓												✓	✓	✓
SimFix				✓								✓	✓		✓		✓	✓	✓	✓	✓			✓	✓	✓	✓
TBar	✓	✓	✓		✓	✓		✓	✓	✓	✓	✓	✓	✓			✓	✓		✓	✓	✓	✓	✓	✓	✓	✓

CapGen reports each generated variant in numeric order, regardless of its correctness. We instrumented the other tools to collect similar information. We note that GenProg caches fitness to increase efficiency, so we record only the variants that would be independently evaluated against the test suite, ignoring duplicates. For all tools, we timestamp and store each variant that is evaluated against the test suite to record how the search proceeds. These modifications account for fewer than 20 lines of code and do not affect search logic.

Invariant Detection. For each program variant in our dataset, we apply the mutations to a clean instance of the Defects4J bug and record a trace of a driver program. Each driver is a small Java program that executes the failing test cases for the mutant. A *trace* is a series of program state observations used to infer program semantics. For each such trace, we then use Daikon to obtain a set of invariants, representing the pre- and post-conditions of executed functions.

We use the invariant sets of the first $x = 600$ mutants generated by each tool to construct a view of the early stages of its search process. We find that the number of semantically unique invariants tapers off at around 300 mutants (Fig. 2a, Sect. 5.1), so we conservatively chose 600 as our cutoff point. We also sample $y = 1000$ mutants uniformly at random from all generated variants (per tool) to provide an overview of the space searched by each tool.

5 Experimental Results

This section presents our results which address the following research questions:

- **RQ1.** Do searches that explore more of semantic space find more repairs?
- **RQ2.** Do different APR tools generate semantically-distinct mutants for a given defect?
- **RQ3.** How does the syntactic diversity of mutants produced by different APR tools relate to their semantic diversity?

5.1 RQ1. APR Search Space Exploration and Repair Rates

We hypothesize that some APR methods sample more widely, that these differences arise from algorithmic decisions, and that these differences lead to differential repair rates for each tool depending on the search budget. We studied

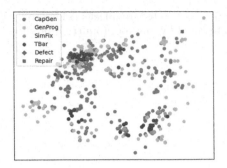
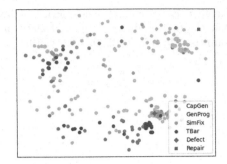

(a) Semantic search space embedding for the first 600 mutants generated by each APR tool for Math 80.

(b) Semantic search space embedding for the randomly sampled 1000 mutants by each APR tool for Math 80.

Fig. 1. Search space visualization of the Math 80 defect. Invariant sets for to generated mutants are embedded in 2D space using multidimensional scaling. Green square is the correct repair, while red diamond is the defect. GenProg and CapGen explore more of the search space than TBar and SimFix. (Color figure online)

each tool's search progress on a representative defect from Defects4J, Math 80 (which relates to integer multiplication and Eigen decomposition).

Figure 1 visualizes our results using the two-dimensional embedding, for both the resource-limited early sampling and the final sampling. In the resource-limited cases (panel (a)), GenProg and CapGen explore more broadly (i.e., enclose the largest area) than either SimFix or TBar, which spend most of their evaluations in localized regions, and rarely test radically-different functionality. We conjecture that the heuristics used to order the mutated programs for testing in CapGen lead to a wider range of functionality being explored with relatively few samples. Panel (b), however, shows substantial differences. GenProg samples more broadly than the others, followed by CapGen, even though both use the same *insert*, *delete*, and *swap* mutation operators. TBar and SimFix, by contrast, are more clustered, with jumps between clusters from different repair templates.

The visualizations in Fig. 1 show the relative scope of each tool's search, but they do not show the search trajectory. To address this issue, we treat the number of unique invariants as a countable proxy for unique functionality and ask how many unique invariants are explored by each additional individual program mutant that the tool evaluates (Fig. 2a). This allows us to visualize both the number of unique invariants that are considered and approximately when they are discovered. The results, shown in Fig. 2, indicate that CapGen and GenProg explore more unique functionality early in the search than TBar and SimFix. Both TBar and SimFix plateau early and remain relatively flat for the remainder of the search. We observed similar trends across the 1000-sample datasets (Fig. 2b) and across all the defects we studied (data not shown).

These results support the hypothesis that APR searches that explore more widely also sample more semantically unique variants. However, the results do

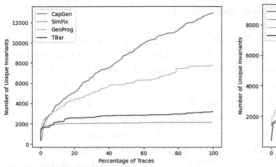 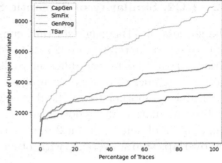

(a) Each tool's unique invariant accumulation for the first 600 mutants (b) Each tool's unique invariant accumulation for 1000 randomly sampled mutants

Fig. 2. Unique invariants from each APR tool for Math 80 over time. x-axis is % of traces evaluated, y-axis is the number of unique invariants. Tools that explore more of the search space also find more unique functionality over time.

Table 2. Semantic overlaps among APR tools. Each row reports the % of mutants that are semantically equivalent to at least one mutant from another tool.

	CapGen	GenProg	SimFix	TBar
CapGen	–	29.0%	25.2%	23.8%
GenProg	31.5%	–	10.8%	37.4%
SimFix	20.2%	86.0%	–	81.9%
TBar	38.0%	59.5%	52.6%	–

not predict relative repair rates. Remarkably, the tools that sample the largest extent of semantic space have lower reported repair rates across the entire Defects4J database, and vice versa. For instance, TBar has the best reported repair rates despite having the lowest exploration reach. Similarly, GenProg, which searched most broadly, reports the lowest repair rate. To summarize, we find that the targeted repair operations used by SimFix and TBar appear to outweigh the advantage of a high-diversity search. This surprising result highlights the key role of *representation*, since the implementation of mutation encodes a choice about representation—although we acknowledge that this result could also be related to the nature of the bug scenarios we studied. What remains unknown is how repairs are distributed throughout the search space: when repairs are close to the original program (e.g., defects in popular APR datasets that can be repaired with only one or two code edits), a thorough search of the nearby region will likely succeed more often than an extensive search of a wider region.

5.2   RQ2. Similarity of Semantic Search Spaces

The success of TBar suggests that combining multiple operators into a single tool increases the repair rate [25]. To test this, we examined the overlap between variants produced by the different tools in our study. We define overlap to be the total number of times each tool generates a mutant that is identical to one generated by another tool. The degree of overlap between two tools is a proxy for their similarity: we hypothesize that tools with high overlap will also repair a similar set of defects. Table 2 reports these results. CapGen and GenProg have low overlap, ≈26% average, with other tools. SimFix and TBar, on the other hand, are much more similar, as expected. TBar uses repair templates taken from several APR tools, often corresponding directly to the mutation operators of other tools in our study. It is thus unsurprising that TBar has the highest minimum overlap (38%). SimFix uses learned templates mined from human-generated repairs, but these also contain *fix patterns* [25] that mirror approaches found in the other tools.

On our dataset, SimFix and TBar have average *repair overlap* comparable to their semantic overlap rates (raw data not shown for brevity): 63% for SimFix and 50% for TBar. GenProg, however, has a much higher repair overlap (83%) compared to its semantic overlap (26%). Of the GenProg repairs, 67% are shared with CapGen and all are shared with TBar. This result can be explained: TBar incorporates all of GenProg's mutation operators. On average, CapGen has 52% repair overlap, ranging from 21% with GenProg to 84% with TBar.

This experiment reveals similarities among tools that may not be evident from their formal descriptions. It also suggests that the strategy of incorporating methods from earlier tools into a composite approach (e.g., TBar [25], Repairnator [47], and ARJA-p [57]) often succeeds. However, each such addition increases system complexity. An ideal combination would maximize performance and minimize cost and complexity. Search space visualizations (such as Fig. 1) support making semantically-guided choices. Finally, focusing only on mutation operators may be misleading, as the tools we studied lack a powerful search heuristic. Even the GenProg family of tools, based on evolutionary computation, searches only in a limited way and relies primarily on mutation.

5.3   RQ3. Syntactic and Semantic Diversity of Mutants

To investigate the relationship between syntactic and semantic diversity for the mutants generated by different APR tools, we compared the rate at which semantically-distinct variants are discovered against the rate of at which unique syntactic variants are discovered. We find that syntactic variants are discovered much more frequently than semantic variants, e.g., between 4 and 20 times greater for Math80, depending on the tool. We observed similar trends for all other defects in our dataset. One explanation for this finding is that many syntactically distinct programs can compile to the same functionality.

Given this disparity, it is natural to ask if a higher semantic discovery ratio (i.e., techniques that find more semantically unique variants per syntactically unique variant) leads to higher overall performance. Our experiments do not support this hypothesis. Instead, we find that high semantic discovery ratios correlate with repair success only 30% of the time. GenProg had the highest ratio (approximately 38%) and the lowest repair rate. Conversely, SimFix had the lowest ratio across 30 of the defects while maintaining a high repair rate. For different defects, TBar and CapGen are typically intermediate between GenProg and Simfix in terms of this ratio, with TBar having the higher ratio of the two.

These results show that repairs are sparse in the search space and that targeting regions of the space where repairs are likely to be found is more effective than randomly sampling a large area of the semantic space. Although each tool finds many more syntactically-unique mutants than semantically-unique ones, it is unclear that this is problematic, given the apparent inverse correlation between semantic reach and repair rates. The success of the search algorithm depends heavily on problem representation, as is well-known in evolutionary computation.

6   Limitations and Threats to Validity

Soundness of Invariant Detection. Despite being the gold standard for dynamic invariant detection, Daikon can infer invariants that may not hold in some parts of the program. To combat this limitation, we consider only invariants marked "high confidence." Additionally, since our approach is based on relative distances between detected invariants, any consistent detection errors are factored out by the difference operation and are unlikely to affect our results.

Syntactically-Invalid Patches. Some mutants produced by APR tools fail syntax or type checks, and cannot be analyzed by our approach. We note that other analysis methods also often fail on ill-formed patches [20], and a majority of the patches produced by the tools we consider are included in our analysis.

Generality. The results from our experimental study may not generalize to other APR tools beyond the four tools we examined, posing a threat to external validity. To mitigate this threat, we chose two tools from each of the main sub-categories of APR tools that fall under the search-based paradigm (i.e., atomic change operators and template-based change operators [12, Sect. 6.1]).

7   Related Work

Earlier APR and Genetic Improvement work also considers the search space, typically characterizing it with respect to a specific characteristic, such as patch correctness, energy efficiency, or neutrality [13,18,27,30,36,40,42,48,49]. Researchers have characterized *neutral* mutations [13,42] (mutations do not discernibly change program behavior—also called *sosies* or *safe*) and developed methods to combine them effectively [40]. Similar to neutral mutation work, Veerapen et al. visualized search spaces by considering local searches of the mutation graph [48,49]. Langdon et al. also completed an exhaustive experiment on the triangle problem [18], concluding that the number of programs that pass all tests is much smaller than the overall search space.

Long et al. [30] characterized the effect on the search space of different configurations of the SPR and Prophet APR tools [28,29], and found that increasing the search space generally increased the number of reachable repairs but also made it harder to find repairs. Similarly, we found that increasing the size of the semantic search space was not sufficient to find more repairs. This trade-off regarding choosing the best representation for a repair problem was explicitly addressed by the Genesis tool, which attempted to manage the size of the search space [27]. This prior work, however, does not consider the semantics of the underlying program beyond measuring how many tests passed. In the end, program behavior determines whether a patch correctly repairs a defect. This motivated us to consider mutant semantic similarity based on invariant set similarity.

Population-based repair tools have used semantics to increase initial population diversity [3] or guide exploration [5,6]. In both cases, the authors failed to find conclusive evidence that increasing population diversity leads to better APR performance. Similarly, we find no correlation between methods that consider a semantically-diverse set of programs and their ability to find repairs. However, our approach enables quantitative and qualitative analysis to investigate this relationship in greater detail than any of the previous works.

8   Conclusion

Many APR algorithms have been proposed, but relatively few ways have been proposed to compare them beyond empirical measurements of success at passing test cases or human assessment of patch quality. We add a new dimension to this work by proposing to assess how these methods explore *semantic search spaces*, extending earlier syntax-based analyses. Our automated, scalable approach leverages dynamic invariant detection and an efficient distance calculation to highlight the semantic differences between program variants. Further, our approach can be easily visualized in 2D space, admitting human interpretability.

Our empirical evaluation of four different search-based tools showed that, contrary to expectation, those methods that search most broadly can experience relatively low repair rates. This surprising result suggests that increasing semantic diversity in the search may not be as helpful as is generally believed. Second, tools that explore semantic mutants that are shared with other tools tend to have higher repair rates, providing an explanation for the success of modern composite tools like TBar or ARJA-p. Finally, tools that search extensively for novel semantics do not necessarily find more repairs, suggesting that tools with targeted repair mechanisms may explore important subsets of the search space. Our results suggest several new research directions. For instance, a deeper understanding of how repairs are distributed throughout syntactic and semantic search spaces would refine our understanding of these results. We hope that results like these will lead to a deeper re-examination of how APR tools are studied and compared, ultimately leading to even more improvements in the future.

Acknowledgements. We gratefully acknowledge the partial support of the NSF (CCF 2211749, 2141300, 1763674, 1908633, and CICI 2115075), DARPA (N6600120C4020, FA8750-19C-0003, HR001119S0089-AMP-FP-029), and AFRL (FA8750-19-1-0501).

References

1. Ackling, T., Alexander, B., Grunert, I.: Evolving patches for software repair. In: GECCO 2011, Dublin, Ireland, pp. 1427–1434. ACM (2011). https://doi.org/10.1145/2001576.2001768
2. Arcuri, A.: Evolutionary repair of faulty software. Appl. Soft Comput. **11**(4), 3494–3514 (2011)
3. Beadle, L., Johnson, C.G.: Semantic analysis of program initialisation in genetic programming. Genet. Program. Evolvable Mach. **10**(3), 307–337 (2009). https://doi.org/10.1007/s10710-009-9082-5. https://link.springer.com/article/10.1007/s10710-009-9082-5
4. Cashin, P., Martinez, C., Weimer, W., Forrest, S.: Understanding automatically-generated patches through symbolic invariant differences. In: ASE 2019, San Diego, USA, pp. 411–414. IEEE (November 2019). https://doi.org/10.1109/ASE.2019.00046
5. Ding, Z.Y.: Patch quality and diversity of invariant-guided search-based program repair. arXiv (March 2020). https://arxiv.org/abs/2003.11667v1
6. Ding, Z.Y., Lyu, Y., Timperley, C., Le Goues, C.: Leveraging program invariants to promote population diversity in search-based automatic program repair. In: 2019 IEEE/ACM International Workshop on Genetic Improvement (GI), pp. 2–9. IEEE (2019)
7. Eiben, A.E., Smith, J.E.: Introduction to Evolutionary Computing. Natural Computing Series, vol. 53. Springer, Heidelberg (2003). https://doi.org/10.1007/978-3-662-05094-1
8. Ernst, M.D., Czeisler, A., Griswold, W.G., Notkin, D.: Quickly detecting relevant program invariants. In: Proceedings of the 22nd International Conference on Software Engineering, pp. 449–458 (2000)
9. Ernst, M.D., et al.: The Daikon system for dynamic detection of likely invariants. Sci. Comput. Program. **69**(1–3), 35–45 (2007). https://doi.org/10.1016/j.scico.2007.01.015
10. Feldt, R.: Generating diverse software versions with genetic programming: an experimental study. IEE Proc. Softw. **145**(6), 228–236 (1998)
11. Fry, Z.P., Landau, B., Weimer, W.: A human study of patch maintainability. In: ISSTA 2012, Minneapolis, USA, p. 177. ACM (2012). https://doi.org/10.1145/2338965.2336775. http://dl.acm.org/citation.cfm?doid=2338965.2336775
12. Gazzola, L., Micucci, D., Mariani, L.: Automatic software repair: a survey. IEEE Trans. Softw. Eng. **45**(1), 34–67 (2017). https://doi.org/10.1109/TSE.2017.2755013
13. Harrand, N., Allier, S., Rodriguez-Cancio, M., Monperrus, M., Baudry, B.: A journey among Java neutral program variants. Genet. Program Evolvable Mach. **20**(4), 531–580 (2019). https://doi.org/10.1007/s10710-019-09355-3

14. Jiang, J., Xiong, Y., Zhang, H., Gao, Q., Chen, X.: Shaping program repair space with existing patches and similar code. In: ISSTA 2018, Amsterdam, Netherlands, vol. 18, pp. 298–309. ACM (July 2018). https://doi.org/10.1145/3213846.3213871. https://dl.acm.org/doi/10.1145/3213846.3213871

15. Just, R., Jalali, D., Ernst, M.D.: Defects4J: a database of existing faults to enable controlled testing studies for Java programs. In: ISSTA 2014, San Jose, USA, pp. 437–440. ACM (July 2014). https://doi.org/10.1145/2610384.2628055. http://dl.acm.org/citation.cfm?doid=2610384.2628055

16. Kou, R., Higo, Y., Kusumoto, S.: A capable crossover technique on automatic program repair. In: IWESEP 2016, Osaka, Japan, pp. 45–50. IEEE (2016). https://doi.org/10.1109/IWESEP.2016.15

17. Lance, G.N., Williams, W.T.: A general theory of classificatory sorting strategies: 1. Hierarchical systems. Comput. J. 9(4), 373–380 (1967)

18. Langdon, W.B., Veerapen, N., Ochoa, G.: Visualising the search landscape of the triangle program. In: McDermott, J., Castelli, M., Sekanina, L., Haasdijk, E., García-Sánchez, P. (eds.) EuroGP 2017. LNCS, vol. 10196, pp. 96–113. Springer, Cham (2017). https://doi.org/10.1007/978-3-319-55696-3_7

19. Le, X.B.D., Thung, F., Lo, D., Goues, C.L.: Overfitting in semantics-based automated program repair. Empir. Softw. Eng. 23(5), 3007–3033 (2018)

20. Le Goues, C., Dewey-Vogt, M., Forrest, S., Weimer, W.: A systematic study of automated program repair: fixing 55 out of 105 bugs for $8 each. In: ICSE 2012, Zürich, Switzerland, pp. 3–13. IEEE (2012). https://doi.org/10.1109/ICSE.2012.6227211

21. Le Goues, C., et al.: The ManyBugs and IntroClass benchmarks for automated repair of C programs. IEEE Trans. Softw. Eng. 41(12), 1236–1256 (2015)

22. Le Goues, C., Nguyen, T., Forrest, S., Weimer, W.: GenProg: a genetic method for automatic software repair. IEEE Trans. Softw. Eng. 38(1), 54–72 (2012). https://doi.org/10.1109/TSE.2011.104

23. Le Goues, C., Pradel, M., Roychoudhury, A.: Automated program repair (December 2019). https://doi.org/10.1145/3318162. https://dl.acm.org/doi/10.1145/3318162

24. Le Goues, C., Weimer, W., Forrest, S.: Representations and operators for improving evolutionary software repair. In: Proceedings of the 14th Annual Conference on Genetic and Evolutionary Computation, pp. 959–966 (2012)

25. Liu, K., Koyuncu, A., Kim, D., Bissyandé, T.F.: TBAR: revisiting template-based automated program repair. In: ISSTA 2019, Beijing, China, pp. 43–54. ACM (July 2019). https://doi.org/10.1145/3293882.3330577. https://dl.acm.org/doi/10.1145/3293882.3330577

26. Liu, K., et al.: A critical review on the evaluation of automated program repair systems. J. Syst. Softw. 171, 110817 (2021)

27. Long, F., Amidon, P., Rinard, M.: Automatic inference of code transforms for patch generation. In: ESEC/FSE 2017, Paderborn, Germany, vol. Part F1301, pp. 727–739. ACM (August 2017). https://doi.org/10.1145/3106237.3106253. https://dl.acm.org/doi/10.1145/3106237.3106253

28. Long, F., Rinard, M.: Prophet: automatic patch generation via learning from successful patches. Technical report, MIT-CSAIL (July 2015). www.csail.mit.edu

29. Long, F., Rinard, M.: Staged program repair with condition synthesis. In: ESEC/FSE 2015, Bergamo, Italy, pp. 166–178. ACM (August 2015). https://doi.org/10.1145/2786805.2786811. https://dl.acm.org/doi/10.1145/2786805.2786811

30. Long, F., Rinard, M.: An analysis of the search spaces for generate and validate patch generation systems. In: ICSE 2016, Austin, Texas, May, vol. 14–22, pp. 702–713. IEEE Computer Society (May 2016). https://doi.org/10.1145/2884781. 2884872

31. Monperrus, M.: Automatic software repair: a bibliography. ACM Comput. Surv. (CSUR) **51**(1), 17 (2018)

32. Motwani, M., Sankaranarayanan, S., Just, R., Brun, Y.: Do automated program repair techniques repair hard and important bugs? Empir. Softw. Eng. **23**(5), 2901–2947 (2018). https://doi.org/10.1007/s10664-017-9550-0. https:// link.springer.com/article/10.1007/s10664-017-9550-0

33. Motwani, M., Soto, M., Brun, Y., Just, R., Le Goues, C.: Quality of automated program repair on real-world defects. IEEE Trans. Softw. Eng. **48**, 637–661 (2020)

34. Nilizadeh, A., Leavens, G.T., Le, X.B.D., Păsăreanu, C.S., Cok, D.R.: Exploring true test overfitting in dynamic automated program repair using formal methods. In: 2021 14th IEEE Conference on Software Testing, Verification and Validation (ICST), pp. 229–240. IEEE (2021)

35. Orlov, M., Sipper, M.: Genetic programming in the wild: evolving unrestricted bytecode. In: Proceedings of the 11th Annual Conference on Genetic and Evolutionary Computation, pp. 1043–1050 (2009)

36. Petke, J., Brownlee, A.E.I., Alexander, B., Wagner, M., Barr, E.T., White, D.R.: A survey of genetic improvement search spaces. In: GECCO 2019, Prague, Czech Republic, pp. 1715–1721. ACM (July 2019). https://doi.org/10.1145/3319619. 3326870. https://dl.acm.org/doi/10.1145/3319619.3326870

37. Qi, Y., Mao, X., Lei, Y., Dai, Z., Wang, C.: The strength of random search on automated program repair. In: ICSE 2014, Hyderabad, India, pp. 254–265. ACM (2014). https://doi.org/10.1145/2568225.2568254

38. Qi, Z., Long, F., Achour, S., Rinard, M.: An analysis of patch plausibility and correctness for generate-and-validate patch generation systems. In: ISSTA 2015, Baltimore, USA, pp. 24–36. ACM (2015). https://doi.org/10.1145/2771783.2771791

39. Renzullo, J., Weimer, W., Forrest, S.: Multiplicative weights algorithms for parallel automated software repair. In: 35th IEEE International Parallel and Distributed Processing Symposium (2021)

40. Renzullo, J., Weimer, W., Moses, M., Forrest, S.: Neutrality and epistasis in program space. In: ICSE 2018, Gothenburg, Sweden, vol. 18, pp. 1–8. IEEE Computer Society (June 2018). https://doi.org/10.1145/3194810.3194812. https://dl. acm.org/doi/10.1145/3194810.3194812

41. Schulte, E., Forrest, S., Weimer, W.: Automated program repair through the evolution of assembly code. In: ASE 2010, Antwerp, Belgium, pp. 313–316. ACM (2010). https://doi.org/10.1145/1858996.1859059. http://portal.acm.org/citation. cfm?doid=1858996.1859059

42. Schulte, E., Fry, Z.P., Fast, E., Weimer, W., Forrest, S.: Software mutational robustness. Genet. Program. Evolvable Mach. **15**(3), 281–312 (2014). https://doi.org/10.1007/s10710-013-9195-8. https://link.springer.com/article/10. 1007/s10710-013-9195-8

43. Smith, E.K., Barr, E.T., Le Goues, C., Brun, Y.: Is the cure worse than the disease? Overfitting in automated program repair. In: ESEC/FSE 2015, Bergamo, Italy, pp. 532–543. ACM (2015). https://doi.org/10.1145/2786805.2786825

44. Soto, M.: Improving patch quality by enhancing key components of automatic program repair. In: 2019 34th IEEE/ACM International Conference on Automated Software Engineering (ASE), pp. 1230–1233. IEEE (2019)

45. Staats, M., Hong, S., Kim, M., Rothermel, G.: Understanding user understanding: determining correctness of generated program invariants. In: ISSTA 2012, Minneapolis, MN, p. 188. ACM (2012). https://doi.org/10.1145/2338965.2336776. http://dl.acm.org/citation.cfm?doid=2338965.2336776
46. Tan, S.H., Yoshida, H., Prasad, M.R., Roychoudhury, A.: Anti-patterns in search-based program repair. In: ESEC/FSE 2016, November, vol. 13–18, pp. 727–738. ACM, New York (November 2016). https://doi.org/10.1145/2950290.2950295. https://dl.acm.org/doi/10.1145/2950290.2950295
47. Urli, S., Yu, Z., Seinturier, L., Monperrus, M., Monperrus, M.: How to design a program repair bot? Insights from the repairnator project. In: ICSE-SEIP 2018, vol. 10 (2018). https://doi.org/10.1145/3183519
48. Veerapen, N., Daolio, F., Ochoa, G.: Modelling genetic improvement landscapes with local optima networks. In: GECCO 2017, vol. 6, pp. 1543–1548. ACM, New York (July 2017). https://doi.org/10.1145/3067695.3082518. https://dl.acm.org/doi/10.1145/3067695.3082518
49. Veerapen, N., Ochoa, G.: Visualising the global structure of search landscapes: genetic improvement as a case study. Genet. Program. Evolvable Mach. **19**(3), 317–349 (September 2018). https://doi.org/10.1007/s10710-018-9328-1
50. Vessey, I., Weber, R.: Some factors affecting program repair maintenance: an empirical study. Commun. ACM **26**(2), 128–134 (1983)
51. Villanueva, O.M., Trujillo, L., Hernandez, D.E.: Novelty search for automatic bug repair. In: GECCO 2020, Cancun, Mexico, pp. 1021–1028. ACM (2020). https://doi.org/10.1145/3377930.3389845. https://dl.acm.org/doi/10.1145/3377930.3389845
52. Wang, S., et al.: Automated patch correctness assessment: how far are we? ASE **2020**, 968–980 (2020). https://doi.org/10.1145/3324884.3416590
53. Weimer, W., Fry, Z.P., Forrest, S.: Leveraging program equivalence for adaptive program repair: models and first results. In: ASE 2013, Silicon Valley, USA, pp. 356–366. IEEE (2013). https://doi.org/10.1109/ASE.2013.6693094
54. Weimer, W., Nguyen, T., Le Goues, C., Forrest, S.: Automatically finding patches using genetic programming. In: ICSE 2009, Vancouver, Canada, pp. 364–367. IEEE (2009). https://doi.org/10.1109/ICSE.2009.5070536
55. Wen, M., Chen, J., Wu, R., Hao, D., Cheung, S.C.: Context-aware patch generation for better automated program repair. In: ICSE 2018, Pittsburgh, Pennsylvania, January, vol. 2018, pp. 1–11. IEEE Computer Society (2018). https://doi.org/10.1145/3180155.3180233
56. Yang, D., Qi, Y., Mao, X.: Evaluating the strategies of statement selection in automated program repair. In: Bu, L., Xiong, Y. (eds.) SATE 2018. LNCS, vol. 11293, pp. 33–48. Springer, Cham (2018). https://doi.org/10.1007/978-3-030-04272-1_3
57. Yuan, Y., Banzhaf, W.: Making better use of repair templates in automated program repair: a multi-objective approach. In: Evolution in Action: Past, Present and Future. GEC, pp. 385–407. Springer, Cham (2020). https://doi.org/10.1007/978-3-030-39831-6_26

Gene-pool Optimal Mixing in Cartesian Genetic Programming

Joe Harrison[1,2(✉)], Tanja Alderliesten[3], and Peter A. N. Bosman[1,2]

[1] Centrum Wiskunde & Informatica, Amsterdam, The Netherlands
{Joe,Peter.Bosman}@cwi.nl
[2] Delft University of Technology, Delft, The Netherlands
[3] Leiden University Medical Center, Leiden, The Netherlands
T.Alderliesten@lumc.nl

Abstract. Genetic Programming (GP) can make an important contribution to explainable artificial intelligence because it can create symbolic expressions as machine learning models. Nevertheless, to be explainable, the expressions must not become too large. This may, however, limit their potential to be accurate. The re-use of subexpressions has the unique potential to mitigate this issue. The Genetic Programming Gene-pool Optimal Mixing Evolutionary Algorithm (GP-GOMEA) is a recent model-based GP approach that has been found particularly capable of evolving small expressions. However, its tree representation offers no explicit mechanisms to re-use subexpressions. By contrast, the graph representation in Cartesian GP (CGP) is natively capable of re-use. For this reason, we introduce CGP-GOMEA, a variant of GP-GOMEA that uses graphs instead of trees. We experimentally compare various configurations of CGP-GOMEA with GP-GOMEA and find that CGP-GOMEA performs on par with GP-GOMEA on three common datasets. Moreover, CGP-GOMEA is found to produce models that re-use subexpressions more often than GP-GOMEA uses duplicate subexpressions. This indicates that CGP-GOMEA has unique added potential, allowing to find even smaller expressions than GP-GOMEA with similar accuracy.

Keywords: Cartesian genetic programming · Gene-pool Optimal Mixing · Subexpression re-use · Evolutionary computation · Symbolic regression

1 Introduction

Automated decision-making using Machine Learning (ML) is becoming more prevalent in domains where interpretability is critical such as medicine or law [9]. Unfortunately, many common ML techniques currently used are based on opaque black-box models. Interpretable models are increasingly desired and sometimes even required by law [15].

Symbolic Regression (SR) is the task of finding an expression of a function that fits the samples of a dataset. Typically, SR techniques are used in the hope of obtaining an interpretable expression. Expressions consists of operators, variables, and constants. Genetic Programming (GP) [7] is a popular tree-based technique used for SR. The resulting expressions from the classic version of GP are, however, often too large to comprehend [18], even when its subexpressions are easy to understand by themselves. This is due to a phenomenon called bloat [8]. Generally, the smaller the expression, the higher the likelihood that it will be interpretable. However, smaller expressions may also be less accurate.

A key reason why classic GP results in large expressions, is because it is easier to represent accurate function estimates with larger trees. One way to combat bloat, is to use a fixed-size tree template. However, enforcing a small tree this way makes the search for high quality solutions more difficult, necessitating more sophisticated evolutionary search. One such approach is the Gene-pool Optimal Mixing Evolutionary Algorithm [2] (GOMEA), of which several variants have been developed for different domains, including tree-based GP (GP-GOMEA) [17,18]. GP-GOMEA is particularly adept at finding small expressions while retaining high accuracy. GOMEA attempts to leverage linkage among problem variables to prevent important building blocks from being disrupted during variation while mixing them well. Linkage information can be prespecified if the optimisation problem is sufficiently understood or can be attempted to be learned during evolution by analysing emerging patterns in the population.

It was suggested that GP-GOMEA may benefit from including repeating subexpressions [18]. In GP-trees subexpression re-use only occurs when the same subexpression is evolved multiple times independently. In Cartesian GP (CGP) [10] expressions are represented by an acyclic feedforward graph rather than a tree. This opens up the opportunity for subexpression re-use. The re-use of subexpressions is interesting because it contributes to the decomposability and interpretability of an expression. Subexpression re-use does not directly decrease the expression length, but rather decreases the number of subexpressions that need to be independently understood. In CGP, these subexpressions can be automatically found during the evolutionary process. However, these subexpressions are not considered Automatically Defined Functions (ADFs) [8], but rather Automatic Re-used Outputs (AROs). AROs require the function in its entirety to remain the same whereas ADFs have dummy arguments where different inputs can be instantiated [12,20]. Nevertheless, the two are closely related. Given that the problem is of sufficient complexity, GP can find smaller expressions using ADFs for some problems [8].

CGP has the ability to produce expressions that re-use subexpressions natively without the need to evolve the same subexpression multiple times. Given the observed advantages brought by GOMEA for GP, it is therefore interesting and potentially of added value to see whether CGP can also benefit from an integration with concepts from GOMEA. Vertices in subexpressions that are re-used can possibly benefit from the simultaneous swaps of genes that happen during linkage-based variation in GOMEA as to not disrupt the salient subexpression.

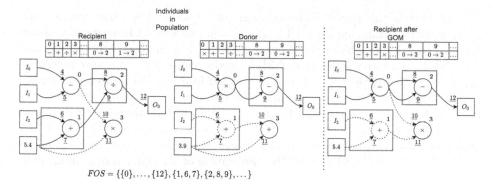

$FOS = \{\{0\}, \ldots, \{12\}, \{1, 6, 7\}, \{2, 8, 9\}, \ldots\}$

Fig. 1. Illustration of how GOM works in CGP-GOMEA. Operator vertices in the 2×2 grid have a problem variable index on the right diagonally above the operator vertex and two underlined problem variable indices to the left representing the location of the incoming vertex. Problem variables in the orange rectangle are an example of variables with high linkage and appear together in the FOS. Problem variables in the blue rectangle are swapped simultaneously from donor to recipient (i.e. clone). Above each graph is a corresponding string representation. Intron vertices and arcs are indicated by dashed lines, and the active graph by filled lines. (Color figure online)

The main contribution of this paper is realising and studying the integration of GOMEA principles in CGP, which we will call CGP-GOMEA[1]. We will compare and contrast CGP-GOMEA with GP-GOMEA and CGP and investigate performance in terms of accuracy, expression length, and subexpression re-use.

2   Methods

Below we outline the relevant details on GOMEA, CGP, and their integration. Special attention is brought to the differences between GP-GOMEA and CGP-GOMEA since these are both GP variants combined with GOMEA. When discussing CGP, the terms vertices and arcs are used, while for GP-trees and GP and CGP in general, the terms nodes and connections are used.

2.1   GOMEA

GOMEA operates on a fixed-length string representation of the problem variables in a genotype. Any mapping from genotype to string can be used as long as the mapping is unique. For instance in GP-GOMEA, nodes in fixed height trees are mapped to a fixed-length string using the pre-order traversal of the tree [17]. Once a mapping is defined, a model describing the linkage between string indices is learned in the form of a Family Of Subsets (FOS), which is a set of subsets of all string indices. Alternatively, the FOS can be provided exogenously.

---

[1] Code and data can be found at https://github.com/matigekunstintelligentie/CGP-GOMEA.

The FOS in this paper is learned each generation and is a hierarchical cluster tree, called a Linkage Tree (LT), where string indices with strong linkage are grouped together in a hierarchical fashion. We used Normalised Mutual Information (NMI) as a proxy for linkage. NMI is used because it is a measure of mutual dependence among variables (in this case string indices). For indices with strong mutual dependence it might be beneficial if the genetic material associated with these indices, is varied in a joint fashion. The algorithm Unweighted Pair Group Method with Arithmetic mean (UPGMA) [5] is used to build the LT. UPGMA only needs the NMI between pairs of problem variables as input, represented by an NMI matrix, to build an LT. The application of UPGMA results in an FOS of size $2l - 1$, where l is the number of string indices. The subset containing all string indices is removed as to not swap entire individuals. The effective FOS size is $\mathbf{2l - 2}$. A randomly initialised population is expected to have no linkage, but due to the NMI matrix being estimated using finite samples some linkage is measured, especially in the case of GP [18]. To combat this, [18] introduced a linear bias correction measured from the initial population such that the NMI matrix is identity at the start of the evolutionary process. This correction is measured once and used throughout the evolutionary process.

Variation in GOMEA happens by means of Gene-pool Optimal Mixing (GOM). Each generation, each individual of the population is first cloned and then undergoes GOM. For each subset in the FOS, a random donor is sampled and then each problem variable instantiation indicated by the subset is copied from the donor to the clone. If the expression of the clone has changed, its fitness is evaluated. If the fitness is equal or better than its original, the change is kept and otherwise it is discarded. The clones replace the entire original population.

2.2 CGP

In CGP, an expression is encoded using a Cartesian grid. Each vertex in the grid has incoming arcs that can potentially come from any preceding column in the grid, making it an acyclic feedforward graph. Note that this makes skip connections and vertex re-use possible (see Fig. 1). By limiting to which preceding column in the grid a vertex can connect, the number of subexpression re-uses can be influenced. This parameter is called Levels-Back (LB). A CGP graph consists of four types of vertices:

1. Ephemeral Random Constants (ERCs) - vertices that output a constant value sampled at the start of the evolutionary process.
2. Inputs (I_i) - vertices that return an input feature of a dataset.
3. Outputs (O_j) - vertices that return the output of an expression.
4. Operators - vertices that apply operations to its incoming arcs.

Only operator vertices are part of the CGP grid. For each operator vertex, the number of incoming arcs is equal to the maximum arity of all operators used. Unary operators only use the first input and ignore other inputs. For the remainder of the paper, the maximum input arity of each operator is two (as in [18]). A vertex in the CGP grid can always connect to an input or ERC vertex regardless of what the value of the LB parameter is. The grid size and number of ERC vertices are (manually) determined a priori and highly depend on the problem and desired shape of the resulting expressions. The number of input vertices depends on the number of inputs in the dataset. Note that even though a vertex appears in the grid it might not be connected to an output vertex, see for example the vertex with string index 3 in Fig. 1. The part of the graph consisting of all vertices and arcs that are connected to a particular output vertex will be referred to as the active graph for that output. Other vertices are considered introns. In CGP it is possible to have multiple outputs or recurrent connections, which enables interesting use-cases. However, in this paper, only feedforward graphs are used for the CGP experiments and only problems with a single output are experimented with in order to compare with GP-GOMEA.

In classic CGP variation happens by means of point mutation [11]. An individual is mutated through point mutation of the operators and arcs until the active graph has changed. A notable difference in our implementation is that ERCs are not mutated in order to be able to fairly compare to the GOMEA algorithms. Originally, selection happens in a $1+\lambda$ scheme [10]. However, tournament selection is also common for larger population sizes [11].

2.3 Adapting GOMEA for CGP

In trees, the location of a problem variable explicitly encodes the location of the incoming child nodes and arcs too, whereas this is not the case in feedforward graphs. To adapt GOMEA for CGP, the incoming arcs in the graph must be added as problem variables in addition to the operator problem variables in the grid. Additionally, a string index is needed for the arc from the grid to the output vertex. When an LT is used, the number of problem variables, and consequently the FOS size, required for a template that can accommodate a similar tree as in GP-GOMEA, is larger. The formula for the number of problem variables in CGP used to build the LT is $3rc + 1$ (for maximum arity of two), where r and c are the number of rows and columns in the CGP grid respectively. An important distinction is that the ERCs and input vertices, as opposed to the original GP-GOMEA implementation [19], are not part of the LT FOS because they are encoded at a fixed position in the grid in CGP-GOMEA. This also means that there is no need for converting continuous ERCs to discrete values (bins) as is needed in GP-GOMEA [18]. To mix ERCs in the population, ERCs are added as unary subsets to the FOS after building the LT. Note that this means that the FOS size increases by the number of ERCs used.

Any unique mapping from vertex and arc to problem variable index can be used. Here, a mapping is used where, starting from the nodes in the first column, each vertex is given three problem variable indices, one for the operator and two for the incoming arcs. The mapping used in this paper is illustrated in Fig. 1.

A larger population size positively impacts the accuracy of the NMI estimation [18]. Typically, there are more inputs and ERCs than operators. This means that there are more possible arcs than operators, especially for the output which can connect to any of the grid vertices. This makes the NMI estimate less accurate for the same population size compared to GP-GOMEA because the cardinality of the variables is higher. Hence, for small population sizes, GP-GOMEA is expected to lead to better results. This, together with the larger FOS size, increases the run-time of GOM as it depends on both factors. GOM is the most costly part for both GP- and CGP-GOMEA due to the many fitness evaluations performed inside GOM. One way to make CGP-GOMEA more efficient is by shrinking the FOS size. We here consider two ways to do this: truncate the FOS or trade expressivity for speed by making the grid smaller.

Table 1. Information about the datasets used in the experiments.

Dataset	#Features	#Samples	Variance y
Boston Housing	13	506	84.59
Yacht Hydrodynamics	6	308	229.84
Tower	25	4999	7703.36

3   Experimental Setup

3.1   General Setup

Each experiment is repeated 30 times using a different random seed for each repetition, but equal random seeds across different experiments to create identical dataset splits for each experiment. Significance is tested using the Wilcoxon signed-rank test using the Pratt tie handling procedure [13] with $\alpha = 0.05/\beta$, where β is the Bonferroni correction coefficient [3, 18].

Initialisation. In GP-GOMEA, ERC and input nodes are sampled with probabilities $\frac{1}{1+\#\text{inputs}}$ and $\frac{\#\text{inputs}}{1+\#\text{inputs}}$ respectively. ERC nodes, therefore, occur much less often as a terminal node, especially when there are many inputs. In CGP-GOMEA and CGP-Classic, the number of ERCs needs to be defined beforehand. The number of ERCs is set to half the number of terminal locations in a full GP tree. For example, a GP tree of height 4 has 16 terminal nodes, in this case, 8 ERCs are instantiated for CGP. The probability of connecting to an ERC or input vertex in CGP is equal. The values for ERCs are sampled uniformly between the minimum and maximum target value in the training set.

In this paper, we focus on small expressions with a total number of symbols smaller than or equal to 32, a limitation posed on the expression length based on findings by [18]. This corresponds with a GP tree of height 4 and arity of 2 with an additional output node. In GP-GOMEA trees are initialised half-and-half as in [16, 18]. For CGP models with a grid with many columns, the full initialisation method [12] often creates large graphs that exceed the 32 node limit. Therefore, only the grow method will be used for all CGP algorithms. Graphs that exceed 32 nodes are penalized in their fitness with a fitness penalty of $10e^6$, severely limiting the chance of selection in the tournament selection of CGP-classic. In GP-GOMEA and CGP-GOMEA, changes due to subset swaps during GOM resulting in a penalty are likely to be discarded.

Operators. The following operators are used: $\{+, -, \div, \times, min, max, exp, pow, log, sqrt, sin, cos, asin, acos\}$. Note that no protected operators are used. This is done to enhance interpretability as protected operators add complexity to each operator. Expressions that return an error on samples in the training set are penalised with a high fitness offset of $10e^6$.

Linear Scaling. To improve performance while keeping an expression small, Linear Scaling (LS) [4,6] is applied to each solution during fitness evaluation unless stated otherwise. LS effectively adds four symbols to each expression. These symbols are however not counted towards the total expression length.

Grid Sizes. For the CGP-GOMEA experiments, four different grid sizes are experimented with: 16×4 (rows $\times$ columns), 8×8, 1×10, and 1×64. The 16×4 grid serves as a comparison to trees of height 4. This grid size is chosen because it is the minimum size that can accommodate any tree of height 4 evolved by GP-GOMEA. An 8×8 grid is used to test what happens if the grid is more flexible in terms of graph depth. A 1×64 grid, which can represent more graph configurations than the 16×4 grid, is tested as a suggestion from literature [11]. A 16×4 grid with $LB = 1$ is also tried. All other experiments have the LB parameter set equal to the number of columns of their respective grid. The 16×4, 8×8, and 1×64 grids all have an FOS size of 384. A grid of 1×10 is tested because it has the same FOS size as a tree of height 4 in GP-GOMEA. Further, truncation of the FOS of a 16×4 CGP grid is investigated. After shuffling the FOS during GOM, only the first k subsets of the FOS are considered, where k is the truncation value. With a truncation of 61, the same FOS size as a tree of height 4 is reached.

Performance Metrics. The training and test coefficients of determination (R^2) and expression length are reported. The expression length is counted as the total number of nodes used in the active graph including the output node. The mean squared error of the training set is optimised instead of optimising the R^2 directly. In particular, we are interested in the re-use of nodes. GP trees can evolve the same subexpression multiple times, whereas CGP has the native ability to re-use vertices. Subexpressions can have the same semantic outcome while differing syntactically. To test whether CGP-Classic and CGP-GOMEA re-use subexpressions more often than that GP-GOMEA evolves duplicate subexpressions, we therefore count the number of re-uses by comparing the output of each connection in the graph or tree with all other connection outputs, except connections to terminal-nodes. Outputs are generated by using the training set augmented with 1000 samples from a normal distribution as input. The re-use count is incremented when two outputs are within a $10e^{-6}$ range of each other.

Computational Budget. The number of evaluations made in GOM is the most time-consuming part of GOMEA [18]. Since CGP-GOMEA has a larger FOS size, the number of evaluations per generation is also much higher. We have therefore opted for a time-based comparison where each run gets a budget of 5000 s. We empirically found that 5000 s leaves enough time for populations of most sizes to converge. A run is terminated when one or more of the following conditions is met: the run reaches 5000 s, the mean fitness and best fitness are equal, the best fitness remains unchanged for 100 generations, or, the mean fitness remains unchanged for 5 generations.

3.2   Setup Main Experiment

Three commonly used datasets will be used in our main experiment: Boston Housing, Yacht Hydrodynamics, and Tower (see Table 1) [1]. The datasets are split into a training and test set of 75% and 25% of the samples respectively. Two sets of experiments are done. One where only inputs are used as terminal nodes and one where both inputs and ERCs can appear as terminal nodes. These sets of experiments are done because there is a difference in how ERCs are handled between CGP- and GP-GOMEA. GP-GOMEA needs to convert continuous ERCs to discrete problem variables. This is done in GP-GOMEA by binning ERCs into 100 bins, the most successful method from [18].

Due to the relatively large size of populations used in this paper, tournament selection is used with a tournament size of 4 for classic CGP to select the parents of the new population that will be mutated to create offspring. The individual with the best fitness is directly copied into the new population.

3.3   Population Size Study

The grid size influences the population size that is needed to ensure the variety of subexpressions in the initial population is large enough, which is important for the success of GOMEA variants. For the main experiments we chose a fixed population size of 1000 as in [15], but this choice is not necessarily optimal. To show the influence of choosing a population size, we do a study to find the optimal population size for GP-GOMEA, CGP-GOMEA 16 × 4, and CGP-Classic on the Boston Housing dataset without ERCs under the time constraint of 5000 s. In Table 4 experimental results are reported using the found optimal population sizes for the Boston Housing dataset on the Yacht and Tower dataset.

3.4   Setup Known Ground Truth Experiment

The optimal formula for the three datasets in Table 1 and the required grid size or tree height is unknown. It is equally unknown whether the datasets have a bias for solutions with less or no subexpression re-use. We want to know whether a known expression with multiple re-used subexpressions is more easily found by CGP-GOMEA compared to GP-GOMEA and CGP-Classic. To this end, we devised a synthetic dataset with a specific known expression that re-uses subexpressions: $I_0^4 - I_1^4 + \frac{I_2^4}{I_3^4}$. To search for this expression, we only allow operators $\{+, -, \div, \times\}$ to be used. The 4th powers in the expression can thus only be created by re-using sub-expressions with the $\times$ operator multiple times. The synthetic dataset has 1000 samples each with 4 input variables, each sampled from a normal distribution with $\sigma = 0.25$ and $\mu = \{0, 1, 2, 3\}$ respectively as to generate slightly overlapping yet mostly distinct input samples. The grid size was chosen so that it is possible to evolve the formula exactly. Only GP-GOMEA, CGP-GOMEA 16×4, 8×8, 16×4 LB $= 1$, and CGP-Classic are tested. In this experiment, LS is not used, because finding the formula rather than optimising for accuracy is what matters here. Nor are ERCs used.

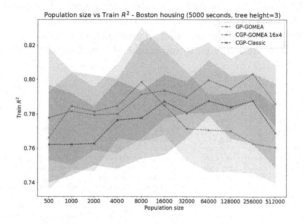

Fig. 2. Results population size study. Shaded area between 10th and 90th percentile.

4   Results

Main Experiment. The results of the main experiment are shown in Table 2. The best training R^2 on the experiments both with and without the use of ERCs is achieved with the CGP-GOMEA 8×8 configuration on Boston Housing and Yacht, and with GP-GOMEA on Tower. The Tower dataset has more variables than available terminal-nodes, which makes it difficult to re-use subexpressions. This is because subexpression re-use means that fewer variables can be used, since re-used subexpressions still count towards the 32 node expression limit. The Yacht Hydrodynamics dataset has a much smaller number of variables and much more re-use is observed for this dataset. A notable difference between the experiments with and without ERCs is that less subexpressions are re-used when ERCs are used, with some CGP configurations even re-using zero subexpressions.

ERCs are used in favour of repeating subexpressions. This could be due to the way ERCs are mixed. In an experiment where ERCs are not added to the FOS and therefore remain unmixed, the re-use of subexpressions was higher for each experiment with similar training R^2.

Truncation, as described earlier, is not a viable method of reducing the FOS size. It consistently ranks among the worst R^2 for all experiments. Trading expressivity for speed is also detrimental to the R^2. A small grid such as 10×1 forces re-use, while as mentioned re-use may not be part of the optimal expression. The configuration from literature, one row with multiple columns, similarly results in low R^2. This is because many individuals in the initial population are penalised for having an expression over 32 nodes, which makes it difficult to create better offspring during GOM without a dedicated constraint handling mechanism in GOMEA, which is currently lacking.

Table 2. Experiment results of various algorithms with and without ERCs as terminal nodes. Median R^2 values are reported due to high variance in test and train R^2. Numbers in bold are best performing for the respective parameter and dataset. Underlined numbers significantly outperform all other algorithms. tr is short for truncation. Test R^2 values filtered from outliers due to unprotected functions are indicated with *.

Algorithm	Without ERCs				With ERCs			
	Median train R^2	Median test R^2	Mean expression length	Mean subexpression re-use	Median train R^2	Median test R^2	Mean expression length	Mean subexpression re-use
Boston Housing								
GP-G	$0.803 \pm 1.71e^{-2}$	$0.761 \pm 5.67e^{-2}$	19.1 ± 3.53	$0.1 \pm 3.00e^{-1}$	$\mathbf{0.83 \pm 1.77e^{-2}}$	$0.758 \pm 1.15e^{-1}$	21.6 ± 3.90	$0.1 \pm 3.00e^{-1}$
CGP-G 16×4	$0.81 \pm 2.24e^{-2}$	$0.756 \pm 5.60e^{-2}*$	18.1 ± 4.13	$0.1 \pm 3.00e^{-1}$	$0.806 \pm 2.67e^{-2}$	$0.783 \pm 4.79e^{-2}$	18.3 ± 5.46	0.0
CGP-G tr	$0.768 \pm 2.29e^{-2}$	$0.729 \pm 5.81e^{-2}$	$\mathbf{11.5 \pm 4.54}$	$0.133 \pm 5.62e^{-1}$	$0.788 \pm 2.23e^{-2}$	$0.733 \pm 5.60e^{-2}$	12.2 ± 3.90	0.0
CGP-G 8×8	$\underline{\mathbf{0.846}} \pm 2.05e^{-2}$	$\mathbf{0.807} \pm 4.53e^{-2}$	27.7 ± 4.64	$0.433 \pm 6.16e^{-1}$	$0.830 \pm 2.94e^{-2}$	$0.787 \pm 7.45e^{-2}$	25.0 ± 6.42	$\mathbf{0.333 \pm 6.50e^{-1}}$
CGP-G 1×10	$0.785 \pm 2.29e^{-2}$	$0.743 \pm 4.48e^{-2}$	15.7 ± 5.63	1.03 ± 2.12	$0.772 \pm 2.96e^{-2}$	$0.75 \pm 6.30e^{-2}$	$\mathbf{11.1 \pm 3.90}$	$0.1 \pm 3.96e^{-1}$
CGP-G LB=1	$0.824 \pm 2.01e^{-2}$	$0.779 \pm 4.43e^{-2}$	21.3 ± 4.08	$0.133 \pm 4.27e^{-1}$	$0.824 \pm 2.25e^{-2}$	$\underline{0.789} \pm 5.56e^{-2}*$	19.6 ± 4.38	$0.0333 \pm 1.80e^{-1}$
CGP-G 1×64	$0.807 \pm 2.01e^{-2}$	$0.78 \pm 6.91e^{-2}$	$15.7 \pm 4.58e^{-1}$	$0.133 \pm 3.40e^{-1}$	$0.810 \pm 1.90e^{-2}$	$0.767 \pm 2.08e^{-1}$	$15.4 \pm 9.12e^{-1}$	$0.0333 \pm 1.80e^{-1}$
CGP-C	$0.789 \pm 2.88e^{-2}$	$0.767 \pm 7.35e^{-2}$	16.9 ± 4.94	$0.367 \pm 7.06e^{-1}$	$0.801 \pm 2.26e^{-2}$	$0.762 \pm 4.50e^{-2}$	18.1 ± 4.29	$0.0333 \pm 1.80e^{-1}$
Yacht Hydrodynamics								
GP-G	$0.995 \pm 7.52e^{-4}$	$0.992 \pm 1.71e^{-3}$	17.7 ± 4.30	0.367 ± 1.28	$0.995 \pm 7.59e^{-4}$	$0.994 \pm 1.78e^{-3}$	17.5 ± 4.30	$0.1 \pm 3.00e^{-1}$
CGP-G 16×4	$0.995 \pm 8.75e^{-4}$	$0.994 \pm 2.08e^{-3}$	21.2 ± 6.05	1.2 ± 1.56	$0.995 \pm 7.61e^{-4}$	$0.993 \pm 2.07e^{-3}$	18.5 ± 4.51	$0.3 \pm 7.37e^{-1}$
CGP-G tr	$0.994 \pm 1.02e^{-3}$	$0.992 \pm 2.20e^{-3}$	18.7 ± 5.05	1.53 ± 2.26	$0.995 \pm 9.51e^{-4}$	$0.993 \pm 1.91e^{-3}$	$\mathbf{15.0 \pm 3.81}$	$0.167 \pm 5.82e^{-1}$
CGP-G 8×8	$\underline{\mathbf{0.996}} \pm 8.89e^{-4}$	$\mathbf{0.994} \pm 1.62e^{-3}$	28.7 ± 3.38	1.83 ± 2.25	$\underline{\mathbf{0.997}} \pm 8.66e^{-4}$	$\mathbf{0.995} \pm 1.71e^{-3}$	26.6 ± 4.92	0.667 ± 1.07
CGP-G 1×10	$0.994 \pm 6.85e^{-4}$	$0.992 \pm 1.80e^{-3}$	26.4 ± 4.07	$\mathbf{12.6 \pm 1.11e^{1}}$	$0.995 \pm 6.21e^{-4}$	$0.993 \pm 2.10e^{-3}$	18.2 ± 3.90	$\mathbf{0.733 \pm 1.46}$
CGP-G LB=1	$0.995 \pm 5.24e^{-4}$	$0.994 \pm 1.81e^{-3}$	25.1 ± 3.72	1.77 ± 1.91	$0.996 \pm 6.38e^{-4}$	$0.994 \pm 2.25e^{-3}$	21.0 ± 4.31	0.467 ± 1.98
CGP-G 1×64	$0.995 \pm 4.92e^{-4}$	$0.994 \pm 1.63e^{-3}$	$\mathbf{15.4 \pm 9.87e^{-1}}$	$0.1 \pm 3.00e^{-1}$	$0.995 \pm 5.79e^{-4}$	$0.994 \pm 1.88e^{-3}$	$15.8 \pm 4.96e^{-1}$	$0.1 \pm 3.00e^{-1}$
CGP-C	$0.994 \pm 1.21e^{-3}$	$0.992 \pm 2.19e^{-3}$	22.7 ± 3.96	0.967 ± 1.43	$0.995 \pm 9.49e^{-4}$	$0.994 \pm 2.08e^{-3}$	19.1 ± 3.87	$0.433 \pm 8.44e^{-1}$
Tower								
GP-G	$\underline{\mathbf{0.873}} \pm 8.08e^{-3}$	$\mathbf{0.878} \pm 1.10e^{-2}$	28.2 ± 3.91	$0.0667 \pm 2.49e^{-1}$	$\underline{\mathbf{0.877}} \pm 6.43e^{-3}$	$\underline{\mathbf{0.874}} \pm 1.09e^{-2}$	29.3 ± 2.58	$0.0667 \pm 2.49e^{-1}$
CGP-G 16×4	$0.846 \pm 1.29e^{-2}$	$0.853 \pm 1.52e^{-2}$	17.2 ± 3.89	$0.0667 \pm 2.49e^{-1}$	$0.84 \pm 3.14e^{-2}$	$0.847 \pm 3.26e^{-2}$	16.0 ± 4.01	$0.0333 \pm 1.80e^{-1}$
CGP-G tr	$0.817 \pm 3.78e^{-2}$	$0.821 \pm 3.99e^{-2}$	$\mathbf{13.0 \pm 4.45}$	$0.1 \pm 3.00e^{-1}$	$0.764 \pm 4.15e^{-2}$	$0.789 \pm 4.55e^{-2}$	11.2 ± 2.86	0.0
CGP-G 8×8	$0.868 \pm 1.09e^{-2}$	$0.872 \pm 1.67e^{-2}$	22.6 ± 6.28	$0.2 \pm 6.00e^{-1}$	$0.864 \pm 1.50e^{-2}$	$0.866 \pm 1.31e^{-2}$	21.7 ± 6.05	$\mathbf{0.233 \pm 6.16e^{-1}}$
CGP-G 1×10	$0.816 \pm 3.34e^{-2}$	$0.827 \pm 3.63e^{-2}$	13.9 ± 3.97	$0.3 \pm 9.00e^{-1}$	$0.769 \pm 3.39e^{-2}$	$0.767 \pm 3.57e^{-2}$	$\mathbf{9.7 \pm 2.79}$	0.0
CGP-G LB=1	$0.861 \pm 1.81e^{-2}$	$0.861 \pm 1.87e^{-2}$	21.0 ± 3.89	$0.167 \pm 3.73e^{-1}$	$0.85 \pm 2.33e^{-2}$	$0.851 \pm 2.51e^{-2}$	17.0 ± 3.62	$0.0667 \pm 3.59e^{-1}$
CGP-G 1×64	$0.851 \pm 1.46e^{-2}$	$0.845 \pm 1.58e^{-2}$	$15.9 \pm 3.00e^{-1}$	$0.1 \pm 3.00e^{-1}$	$0.847 \pm 1.59e^{-2}$	$0.85 \pm 1.70e^{-2}$	$15.9 \pm 3.40e^{-1}$	$0.0333 \pm 1.80e^{-1}$
CGP-C	$0.844 \pm 2.69e^{-2}$	$0.837 \pm 2.84e^{-2}$	19.3 ± 3.70	$\mathbf{0.333 \pm 5.96e^{-1}}$	$0.823 \pm 3.35e^{-2}$	$0.835 \pm 3.38e^{-2}$	16.1 ± 4.75	$0.1 \pm 3.00e^{-1}$

Population Sizing Study. The results of the population sizing study (see Fig. 2) show that all three algorithms initially have a positive trend upwards in terms of R^2 as the population size increases. For CGP-GOMEA 16×4 this trend declines for population sizes above 8000, a smaller population size than observed for the onset of decline in GP-GOMEA and CGP-Classic. This is because although the larger population size positively impacts the quality of the linkage information it also severely limits the number of generations that can be achieved within the maximum time budget, because the run-time of the GOM procedure depends on the FOS size which is larger for CGP-GOMEA.

This exemplifies the importance of using the right parameters in population-based search such as GP. Moreover, what is key to notice from the graph is that the best performance of GP-GOMEA is equal to that of CGP-GOMEA. As these algorithms can represent similar solutions, this was to be expected. However, this search-space-based expectation only holds if the search algorithm is capable of finding high-quality solutions in that space effectively. CGP using classic point-based mutation is not capable of performing equally well. This reconfirms the potential of GOMEA for the GP domain and also confirms that our integration of GOMEA to CGP is essentially successful.

This result also shows that the population size of 1000 used in the main experiment, while congruent with much of literature, can potentially lead to wrong conclusions about the maximum performance of the algorithms tested. Still, the conclusions are valid within the assumed limits. Moreover, the most important comparison, between CGP-GOMEA and GP-GOMEA holds, as from the population sizing experiment we expect these algorithms to perform similarly.

Known Ground Truth Experiment. As mentioned, the expressions found for some datasets have more subexpression re-use than others. If re-use can be found then CGP-GOMEA is a good option. In Table 3 it can be seen that CGP-GOMEA algorithms have a better training R^2 and re-use more subexpressions than GP-GOMEA and CGP-Classic. The CGP-GOMEA 16×4 LB $= 1$ configuration is the only algorithm that can find the exact expression (twice).

Table 3. Results on synthetic dataset. Numbers marked in bold are best performing for the respective parameter. GP-G, CGP-G and CGP-C are short for GP-GOMEA, CGP-GOMEA and CGP-Classic respectively. The value after ± is the standard deviation.

Algorithm	Median train R^2	Median test R^2	Mean expression length	Times expression found
GP-G	$0.995 \pm 3.33e^{-3}$	$0.995 \pm 4.65e^{-3}$	$0.63 \pm 7.52e^{-1}$	0
CGP-G 16×4	$0.997 \pm 2.88e^{-3}$	$0.997 \pm 3.01e^{-3}$	1.77 ± 1.36	0
CGP-G 8×8	$0.999 \pm 2.15e^{-3}$	$\mathbf{0.999 \pm 2.44e^{-3}}$	3.27 ± 2.43	0
CGP-G LB $= 1$	$\mathbf{0.999 \pm 6.98e^{-4}}$	$0.998 \pm 1.30e^{-3}$	$\mathbf{3.87 \pm 3.19}$	**2**
CGP-C	$0.998 \pm 4.27e^{-3}$	0.998 ± 4.08^{-3}	2.0 ± 1.37	0

5 Discussion

In CGP-GOMEA a grid size still needs to be defined a priori. A large enough grid could be defined such that it could accommodate any possible tree with 32 nodes, but this would lead to a very large FOS size and subsequently very long run-times. This effectively means that the expressions in CGP-GOMEA are always bounded by a predefined grid size. Potentially a technique akin to NeuroEvolution of Augmenting Topologies (NEAT) [14] could be used to evolve unbounded graphs while still being able to swap homologous blocks using a GOMEA-like approach.

Table 4. Training R^2 of various algorithms trained on Yacht and Tower dataset with population size found in population sizing study for the Boston Housing dataset. Median R^2 values are reported due to the high variance in test and train R^2 values. Numbers marked in bold are best performing for the respective parameter and dataset. Underlined numbers significantly outperform all other algorithms.

	Boston Housing	Yacht Hydrodynamics	Tower
GP-G	**0.803** $\pm$ 1.27e^{-2}	0.994 $\pm$ 5.85e^{-4}	0.769 $\pm$ 1.60e^{-2}
CGP-G	0.791 $\pm$ 2.80e^{-2}	**0.994** $\pm$ 3.93e^{-4}	**<u>0.844</u>** $\pm$ 2.66e^{-2}
CGP-C	0.788 $\pm$ 1.70e^{-2}	0.993 $\pm$ 9.49e^{-4}	0.780 $\pm$ 1.98e^{-2}

The R^2, expression length, and potentially the number of re-used subexpressions are of interest to optimise. In this paper, however, only the training R^2 is optimised. No pressure is applied on evolving short expressions or expressions with subexpression re-use. Instead, these are just attributes that resulted from single-objective training. As a result, less re-use may have been observed than what is possible. A multi-objective setting may overcome this as well as give more insight into just how much re-use is possible.

Further, a potentially interesting line of research is using the subgraphs with multiple re-uses found by CGP-GOMEA as building blocks for other algorithms. Since these re-uses clearly have value [7,17]. Re-used subexpressions are easily found in CGP graphs without using exogenous processes.

Limitations of this work are the use of only one population size in the main experiment, a restricted number of datasets, and a fixed runtime. Ideally, the optimal population size is used for each configuration and dataset. This would however quickly exceed our computational budget. Of high interest are approaches that adaptively set the population size, increasing resources over time so that an anytime algorithm is obtained. More research is needed to identify in more detail what datasets can benefit from models with native re-use. Finally, only one configuration of CGP-Classic is compared against. While we believe the comparison was fair, showcasing the potential of GOMEA within a basic representation space of CGP, more versions of CGP exist [11] and should be compared against in future work, with similar augmentations on the GOMEA side.

6 Conclusion

In this paper, we showed how GOMEA principles can be applied to CGP and we thereby introduced CGP-GOMEA. We find that CGP-GOMEA with a grid-size of 8×8 strikes a good balance between CGP grid depth and breath and obtains similar training R^2 compared to GP-GOMEA and superior training R^2 compared CGP-Classic on three common datasets while re-using more subexpressions. On a synthetic dataset, where the expression to regress to is known, that has multiple subexpression re-uses, CGP-GOMEA is better able to find expressions that are close to optimal compared to GP-GOMEA and CGP-Classic. We therefore conclude that CGP-GOMEA can successfully leverage the advantageous properties of GOMEA within the CGP representation, enabling re-use integrated within the search procedure, opening up interesting avenues of research.

Acknowledgement. This research is part of the research programme Open Competition Domain Science-KLEIN with project number OCENW.KLEIN.111, which is financed by the Dutch Research Council (NWO). We further thank the Maurits en Anna de Kock Foundation for financing a high-performance computing system. We also thank Marco Virgolin aiding in implementing GP-GOMEA, and Dazhuang Liu and Evi Sijben for their fruitful discussions and reviews.

References

1. Asuncion, A., Newman, D.: UCI machine learning repository (2007)
2. Bosman, P.A.N., Thierens, D.: On measures to build linkage trees in LTGA. In: Coello, C.A.C., Cutello, V., Deb, K., Forrest, S., Nicosia, G., Pavone, M. (eds.) PPSN 2012. LNCS, vol. 7491, pp. 276–285. Springer, Heidelberg (2012). https://doi.org/10.1007/978-3-642-32937-1_28
3. Demšar, J.: Statistical comparisons of classifiers over multiple data sets. J. Mach. Learn. Res. **7**, 1–30 (2006)
4. Dick, G., Owen, C.A., Whigham, P.A.: Feature standardisation and coefficient optimisation for effective symbolic regression. In: Proceedings of the Genetic and Evolutionary Computation Conference, pp. 306–314 (2020)
5. Gronau, I., Moran, S.: Optimal implementations of UPGMA and other common clustering algorithms. Inf. Process. Lett. **104**(6), 205–210 (2007)
6. Keijzer, M.: Improving symbolic regression with interval arithmetic and linear scaling. In: Ryan, C., Soule, T., Keijzer, M., Tsang, E., Poli, R., Costa, E. (eds.) EuroGP 2003. LNCS, vol. 2610, pp. 70–82. Springer, Heidelberg (2003). https://doi.org/10.1007/3-540-36599-0_7
7. Koza, J.R.: Genetic Programming: On the Programming of Computers by Means of Natural Selection, vol. 1. MIT Press, Cambridge (1992)

8. Koza, J.R.: Genetic Programming II: Automatic Discovery of Reusable Programs, vol. 17. MIT Press, Cambridge (1994)
9. Lipton, Z.C.: The mythos of model interpretability: in machine learning, the concept of interpretability is both important and slippery. Queue **16**(3), 31–57 (2018)
10. Miller, J.F., et al.: An empirical study of the efficiency of learning boolean functions using a cartesian genetic programming approach. In: Proceedings of the Genetic and Evolutionary Computation Conference, vol. 2, pp. 1135–1142 (1999)
11. Miller, J.F.: Cartesian genetic programming: its status and future. Genet. Program Evolvable Mach. **21**, 1–40 (2019). https://doi.org/10.1007/s10710-019-09360-6
12. Poli, R., Banzhaf, W., Langdon, W.B., Miller, J.F., Nordin, P., Fogarty, T.C.: Genetic Programming. Springer (2004)
13. Pratt, J.W.: Remarks on zeros and ties in the Wilcoxon signed rank procedures. J. Am. Stat. Assoc. **54**(287), 655–667 (1959)
14. Stanley, K.O., Miikkulainen, R.: Evolving neural networks through augmenting topologies. Evol. Comput. **10**(2), 99–127 (2002)
15. Vilone, G., Longo, L.: Explainable artificial intelligence: a systematic review. arXiv preprint arXiv:2006.00093 (2020)
16. Virgolin, M., Alderliesten, T., Bel, A., Witteveen, C., Bosman, P.A.: Symbolic regression and feature construction with GP-GOMEA applied to radiotherapy dose reconstruction of childhood cancer survivors. In: Proceedings of the Genetic and Evolutionary Computation Conference, pp. 1395–1402 (2018)
17. Virgolin, M., Alderliesten, T., Witteveen, C., Bosman, P.A.: Scalable genetic programming by gene-pool optimal mixing and input-space entropy-based building-block learning. In: Proceedings of the Genetic and Evolutionary Computation Conference, pp. 1041–1048 (2017)
18. Virgolin, M., Alderliesten, T., Witteveen, C., Bosman, P.A.: Improving model-based genetic programming for symbolic regression of small expressions. Evol. Comput. **29**(2), 211–237 (2021)
19. Virgolin, M., De Lorenzo, A., Medvet, E., Randone, F.: Learning a formula of interpretability to learn interpretable formulas. In: Bäck, T., et al. (eds.) PPSN 2020. LNCS, vol. 12270, pp. 79–93. Springer, Cham (2020). https://doi.org/10.1007/978-3-030-58115-2_6
20. Woodward, J.R.: Complexity and cartesian genetic programming. In: Collet, P., Tomassini, M., Ebner, M., Gustafson, S., Ekárt, A. (eds.) EuroGP 2006. LNCS, vol. 3905, pp. 260–269. Springer, Heidelberg (2006). https://doi.org/10.1007/11729976_23

Genetic Programming for Combining Directional Changes Indicators in International Stock Markets

Xinpeng Long$^{(\boxtimes)}$ ⓘ, Michael Kampouridis ⓘ, and Panagiotis Kanellopoulos ⓘ

School of Computer Science and Electronic Engineering,
University of Essex, Wivenhoe Park, UK
{xl19586,mkampo,panagiotis.kanellopoulos}@essex.ac.uk

Abstract. The majority of algorithmic trading studies use data under fixed physical time intervals, such as daily closing prices, which makes the flow of time discontinuous. An alternative approach, namely directional changes (DC), is able to convert physical time interval series into event-based series and allows traders to analyse price movement in a novel way. Previous work on DC has focused on proposing new DC-based indicators, similar to indicators derived from technical analysis. However, very little work has been done in combining these indicators under a trading strategy. Meanwhile, genetic programming (GP) has also demonstrated competitiveness in algorithmic trading, but the performance of GP under the DC framework remains largely unexplored.

In this paper, we present a novel GP that uses DC-based indicators to form trading strategies, namely GP-DC. We evaluate the cumulative return, rate of return, risk, and Sharpe ratio of the GP-DC trading strategies under 33 datasets from 3 international stock markets, and we compare the GP's performance to strategies derived under physical time, namely GP-PT, and also to a buy and hold trading strategy. Our results show that the GP-DC is able to outperform both GP-PT and the buy and hold strategy, making DC-based trading strategies a powerful complementary approach for algorithmic trading.

Keywords: Directional changes · Genetic programming · Algorithmic trading

1 Introduction

Algorithmic trading has always been a vibrant research topic of paramount importance within the finance domain [8]. The majority of algorithmic trading research takes place on physical time scale, e.g. using hourly, daily, and weekly data. However, using such fixed time scales has the drawback of making data discontinuous and omitting important information between two data points, e.g. daily data would not have captured the flash crash that occurred across US stock indices on 6 May 2010 from 2:32 pm to 3:08 pm, as prices rebounded shortly afterwards [2].

© The Author(s), under exclusive license to Springer Nature Switzerland AG 2022
G. Rudolph et al. (Eds.): PPSN 2022, LNCS 13399, pp. 33–47, 2022.
https://doi.org/10.1007/978-3-031-14721-0_3

An alternative approach is to summarise prices as events. The rationale is to record key events in the market representing significant movements in price, such as a change of, for instance, 5%. Directional changes (DC) is a relatively recent event-based technique, which relies on a threshold θ to detect significant price movements. It was first proposed in [12] and formally defined in [19]. In the DC framework, a physical time series is divided into upward and downward trends, where each such trend marks a DC event at the moment the price change exceeds θ; the DC event is usually followed by an overshoot (OS) event representing the time interval of price movement along the trend beyond the DC event.

In this work, we are interested in using DC-based indicators to perform algorithmic trading. Indicators are mathematical patterns derived from past data and are used to predict future price trends. They are commonly used in technical analysis, e.g., in the form of moving averages, and trade breakout rules. With the evolution of DC research, new DC-based indicators have been proposed, see e.g. [3,20,21]. Therefore, in this paper we will combine 28 different DC indicators under a genetic programming (GP) algorithm [18], namely GP-DC. We apply the derived trading strategies to 33 different datasets from three international markets, namely the DAX performance index, Nikkei 225, and the Russell 2000 index. Our goal is to show that the DC paradigm is not only competitive compared to the physical time paradigm, but has even the potential to outperform it. To achieve this goal, we benchmark GP-DC with another GP-based physical time trading strategy, namely GP-PT, that uses technical analysis indicators under physical time. We compare the GP-DC's results to results obtained by GP-PT. We compare the two GPs' performance on different financial metrics, such as cumulative returns, average rate of return per trade, risk, and Sharpe ratio. We also compare the GPs' performance against the buy-and-hold strategy, which is a common financial benchmark.

The remainder of this paper is organized as follows. In Sect. 2, we present background information and discuss the DC-related literature. Section 3 introduces the methodology of our experiments, and then Sect. 4 presents the experimental set up, as well as the datasets, benchmarks, and parameter tuning process. Section 5 presents the experimental results and finally, Sect. 6 concludes the paper and discusses future work.

2   Background and Literature Review

2.1   Overview of Directional Changes

Directional Changes form an event-based approach for summarising market price movements, as opposed to a fixed-interval-based approach. A DC event is identified only when the price movement of the objective financial instrument exceeds a threshold predefined by the trader. Depending on the direction of price movement, such DC events could be either upturn events or downturn events. Frequently, after the confirmation of a DC event, an overshoot (OS) event follows; the OS event ends when a new price movement starts in an opposite trend,

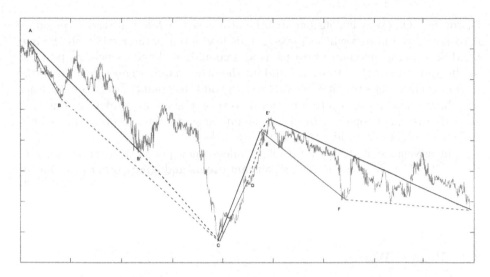

Fig. 1. An example for DC. The grey line indicates the physical time series, the red line denotes a series of DC and OS events as defined by a threshold of 0.01%, while the blue line denotes a series of DC and OS events as defined by a threshold of 0.018%. DC events are depicted with solid lines, while dotted lines denote the OS events. (Color figure online)

eventually leading to a new DC event. Recent studies, however, have pointed out that a DC event is not necessarily followed by an OS one [1].

Figure 1 presents an example of how to convert physical time series to DC and OS events using two different thresholds (see the red and blue lines). Note that thresholds may in principle vary, as traders need not necessarily agree on which price movement constitutes a significant event; each such threshold leads to a different event series. A smaller threshold leads to the identification of more events and increases the opportunity of trade, while a larger threshold leads to fewer events with greater price movement. Thus, selecting an appropriate threshold is a key challenge.

By looking at the historical daily price movement (grey line) and the events created by the threshold of 0.01% (red lines), there are plenty of price movements that are not classified as events under the DC framework, as these do not exceed the threshold. Only when a price change is larger than the threshold is the time series divided into DC events (solid lines) and OS events (dotted lines). For example, the solid red line from A to B is considered a DC event on a downturn, while an OS event follows (from B to C). Then, a new DC event (in the opposite direction) is detected from C to D and this is followed by an OS event from D to E in an upturn, and so on.

It is worth noting that the change of trend can be confirmed only when the price movement exceeds the threshold. In other words, we do not know when the OS event ends until the next DC event (in the opposite direction) is confirmed. For example, in Fig. 1, the point D is a DC event confirmation point. Before point

D, the last OS event is considered to be still active, while the trader considers it to have been in a downward event. This leads to a paradox that on the one hand, in order to maximise returns, trades should be closed as near as possible to the endpoint of the OS event, and on the other hand, when the endpoint of the OS event is detected, it is already well beyond that point. Therefore, figuring out the extreme point where direction is reversed, such as point C in Fig. 1, is an active research topic on the DC domain. In particular, several scaling laws have been suggested to identify the OS event length.

The advantage of DC is that it offer traders a new perspective on price movements; it allows them to focus on significant events and ignore other price movements that could be considered as noise. Therefore, DC leads to new research directions and challenges that are not relevant under physical time periods; in the following section, we present existing work on DC.

2.2 Related Work

DC was first proposed by Guillaume et al. [12] and was formally defined by Tsang [19] as an alternative, event-based method to the traditional physical time model. Since DC is appropriate to handle non-fixed time intervals and high-frequency data, a series of papers applies it on tick data from the Forex market, see e.g. [7,14,15]. There exist two key issues in DC. The first is when do the OS events end; clearly, this has impact on profit maximisation. In other words, we are interested in figuring out the relationship between DC events and OS events. In this direction, Glattfelder et al. [11] introduced 12 new empirical scaling laws to establish quantitative relationships between price movements and transactions in the foreign exchange market. Following along this path, Aloud and Fasli [5] considered four new scaling laws under the DC framework and concluded that these perform successfully on the foreign exchange market. To name an example, one of the most prominent scaling laws states that OS takes, on average, twice as long to reach the same amount of price change as the DC event length. Recently, Adegboye and Kampouridis [1] proposed a novel DC trading strategy which does not assume that a DC event is always followed by an OS event; their results suggest that this strategy outperforms other DC-based trading strategies, as well as the buy and hold strategy, when tested on 20 Forex currency pairs.

The second key issue is the application of technical analysis under a DC framework; technical analysis has been frequently used on physical time by capturing features of markets, namely technical indicators. Aloud [3] converted physical time data into event-based data and introduced a first set of indicators tailored for the DC framework. Further DC indicators were suggested in [20] and [21]. These DC indicators were applied to summarise price changes in the Saudi Stock Market with the aim to help investors discover and capture valuable information. Furthermore, Ao and Tsang [6] proposed two DC-based trading strategies, namely TA1 and TA2, derived from the Average Overshoot Length scaling law. Their results indicated a positive return for most cases in FTSE 100, Hang Seng, NASDAQ 100, Nikkei 225, and S&P 500 stock market indices. Very recently, a combination of DC with reinforcement learning, trained by the Q-learning algorithm, was proposed by Aloud and Alkhamees [4] on S&P 500, NASDAQ, and Dow Jones stock market. Their results showcase substantial return and an increase in the Sharpe ratio.

The above discussion reveals a relative scarcity of DC studies on the stock market. Moreover, using DC-based indicators to derive trading strategies is still in its infancy compared to the, well-established, technical analysis under physical time. We remark that GP has been very effective in the past in combining different (technical) indicators to derive profitable trading strategies, see e.g. [9,10,13]. This naturally begs the question of how effective GP would be when combined with DC-based indicators, and, hence, motivates us to compare such an approach with a physical time model. Next, we introduce the GP methodology while also presenting the GP-DC trading strategy we used.

3   Methodology

This section presents GP-DC, a genetic programming approach using indicators suggested for the DC framework.

3.1   Genetic Programming Model

Terminal Set. After obtaining the daily closing prices for a dataset, we apply the DC framework to summarise the prices as events. Then, from the event series, we calculate the values of 28 indicators specific to the DC framework, much alike technical indicators being derived from technical analysis in physical time [16]. These 28 DC indicators have been introduced and discussed in [3] and, together with an Ephemeral Random Constant (ERC), form the terminal set. Whenever ERC is called, it returns a random number following the uniform distribution and ranging between -1 and 1. In order to fit the range of ERC, the DC indicators have been normalised.

Table 1 lists the DC indicators. In particular, there is a collection of 11 indicators, some of which are calculated over a certain period (e.g. the total number of DC events N_{DC} can be calculated over a period of 10, 20, 30, 40, or 50 days), thus leading to a total of 28 indicators. The third column in Table 1 takes the value N/A for indicators not requiring a period length (namely OSV, TMV, T_{DC}, and R_{DC}).

Table 1. DC indicators

Indicator	Description	Periods (days)
TMV	TMV is the price movement between the extreme point at the beginning and end of a trend, normalised by the threshold θ	N/A
OSV	OSV is calculated by the percentage difference between the current price with the last directional change confirmation price divided by the threshold θ	N/A
Average OSV	This is the average value of the OSV over the selected period	3, 5, 10
R_{DC}	R_{DC} represents the time-adjusted return of DC. It could be calculated by the TMV times threshold θ divided by the time intervals between each extreme point	N/A
Average R_{DC}	This is the average value of the R_{DC} over the selected period	3, 5, 10
T_{DC}	This is the time spent on a trend	N/A
Average T_{DC}	This is the average value of T_{DC} over the selected period	3, 5, 10
N_{DC}	N_{DC} is the total number of DC events over the selected period	10, 20, 30, 40, 50
C_{DC}	C_{DC} is defined as the sum of the absolute value of the TMV over the selected period	10, 20, 30, 40, 50
A_T	A_T represents the difference between the time DC spends on the up trends and down trends over the selected period	10, 20, 30, 40, 50

Function Set. The function set includes two logical operators, namely AND and OR, and two logical expressions, namely less than ($<$) and greater than ($>$).

Model Representation. The GP evolves logical expressions, where the root is one of AND, OR, $<$, or $>$. These expressions are then integrated as the first branch of an If-Then-Else (ITE) statement; see Part 1 of Fig. 2. The rest of the ITE tree contains a 'Then' and an 'Else' branch; the former represents a buy action, and always returns a leaf node with a value of 1. The latter represents a hold action, and always returns a leaf node with a value of 0. Note that there is no sell action during this structure; we will discuss the part of sell action in Sect. 3.2. We did not include Part 2 in the GP is as its values are constants, either 0 or 1; there was thus no need to evolve them.

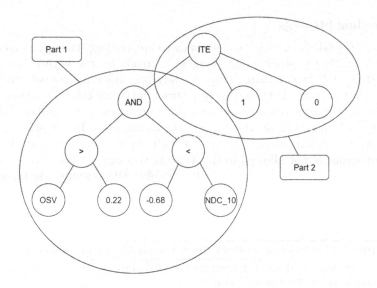

Fig. 2. An example of the GP tree and the If-Then-Else structures. If OSV is greater than 0.22 and N_{DC} for 10 days is greater than −0.68, then we get a signal for a buy action; otherwise, we hold.

Fitness Function. We use the Sharpe ratio as the fitness function of the GP trading strategies. The advantage of using the Sharpe ratio is that it takes into account both returns and risk:

$$
\text{SharpeRatio} = \frac{\text{E}(R) - R_f}{\sqrt{\text{Var}(R)}}, \tag{1}
$$

where E and Var stand for the sample mean value and the sample variance, R stands for the rate of returns and R_f is the risk-free rate. The data used, i.e., the returns, for computing the Sharpe ratio were obtained by the trading algorithm outlined in Sect. 3.2, which indicates when the selling of the stocks will take place.

Selection Method and Operators. We use elitism, sub-tree crossover and point mutation. We also use tournament selection to choose individuals as parents for the above operators.

A summary of the GP configuration is presented in Table 2.

Table 2. Configuration of the GP algorithm

Configuration	Value
Function set	AND, OR, >, <
Terminal set	28 DC indicators and ERC
Genetic operators	Elitism, subtree crossover and point mutation
Selection	Tournament

3.2 Trading Strategy

The goal of the GP tree, which corresponds to our trading strategy, is to answer the question: "Is the stock price going to increase by $r\%$ within the next n days?". If the GP tree returns True, we buy one amount of stock, unless we already own the stock. If the GP tree returns False, we take no action (hold). When we already own a stock, and the price increases by $r\%$ within the next n days, we sell the stock on the given day this happens. If the price does not increase by $r\%$ within the next n days, we sell the stock on the n-th day. Note that short-selling is not allowed in this trading strategy. At the end of each sell action, we calculate and record the resulting profit. All positions take transaction costs into account; the transaction cost is 0.025% per trade. The above trading strategy is summarised in Algorithm 1.

Algorithm 1. Our trading strategy given threshold $r\%$ and duration n days

---

Require: Initialise variables (O represents the prediction of the GP tree, while *index* indicates whether the stock is held)

1: **if** $O = 1$ and $index = 0$ **then**
2:     Buy one amount of stock
3:     $index \leftarrow 1$
4:     $N \leftarrow i$ //Starting time for trade: i is always the current time
5:     $K \leftarrow p$ //Stock price when buying: p is always the current price
6: **else**
7:     **if** ($index = 1$ and $p > (1 + r/100) \times K$) OR $(i - K) > n$ **then**
8:         Sell the stock
9:         $index \leftarrow 0$
10:        Calculate and record profit
11:    **end if**
12: **end if**

The rate of return from each trade is computed based on the price P_b we bought and the price P we sold the stock; see Eq. (2). These returns are saved as a list and, eventually, we compute the sample mean of that list, which gives the overall rate of return; this is the input to Eq. (1) to determine the Sharpe ratio. The *risk*, as seen in Eq. 3, is the standard deviation of that list.

$$R = \left\{ \frac{0.99975 \cdot P - 1.00025 \cdot P_b}{1.00025 \cdot P_b} \right\} \cdot 100\% \tag{2}$$

$$\text{Risk} = \sqrt{\text{Var}(R)} \tag{3}$$

4 Experimental Set up

4.1 Data

Recall that, as discussed in Sect. 1, our goal is to evaluate GP-DC algorithm on the stock market. We use data from three international markets, namely the

DAX performance index, Nikkei 225, and the Russell 2000 index. From each index, we downloaded 10 stocks from Yahoo! Finance, as well as the data for the index themselves. Therefore in total, we use 33 datasets (3 markets × 10 stocks + 3 indices). Each dataset consists of daily closing prices for the period 2015 to 2020 and was split into three parts, namely training (2015 to 2018), validation (2019), and test (2020), as follows: 60%:20%:20%. All data were then converted into DC indicators (see Table 1), and normalised, as explained in Sect. 3.1.

4.2   Benchmarks

We compare the performance of the GP-DC trading strategy against GP-PT as well as buy-and-hold, a typical financial benchmark. For the GP-PT algorithm, we use the same GP as the one described above in Sect. 3. The only difference is that its terminals are now based on technical analysis (physical time), rather than directional changes. To make the comparison fairer, the number of technical indicators in the GP-PT algorithm is equal to that of the DC indicators in the GP-DC algorithm. We select the indicators which are prevalent in the finance field [17]. These indicators are: each of Moving Average, Commodity Channel Index, Relative Strength Index, and William's %R with periods of 10, 20, 30, 40, and 50 days, each of Average True Range, and Exponential Moving Average with periods of 3, 5, 10 days, and finally, On Balance Volume and parabolic SAR without periods; hence, we obtain 28 technical indicators.

4.3   Parameter Tuning for GP

We performed a grid search to decide on the optimal GP parameters for both the GP-DC and GP-PT algorithms, and tuning took place by using the validation set. Based on [18], we adopted the most common values for each parameter, namely 4, 6, 8 (max depth); 100, 300, 500 (population size); 0.75, 0.85, 0.95 (crossover probability); 2, 4, 6 (tournament size); and 25, 35, 50 (number of generations). Mutation probability is equal to (1-crossover probability), so we did not need to separately tune this parameter. Table 3 shows the selected parameters and their value after tuning.

Table 3. Parameters of the GP algorithm

Parameters	Value
Max depth	6
Population size	500
Crossover probability	0.95
Tournament size	2
Numbers of generation	50

4.4 Parameter Tuning for Trading Strategy

Recall that there are 3 parameters on our trading strategy, 2 parameters derived for the question "whether the stock price will increase by $r\%$ during the next n days?" and one parameter is the threshold on DC. Rather than tuning the above parameters and then selecting the best set across all datasets (which is what we did for the GP), we decided to allow for tailored values for each dataset. The configuration space for these three parameters is presented in Table 4.

Buy and hold is also a useful benchmark, as it compares the GPs' performance against the market performance. We will thus also report the buy and hold performance of each dataset.

Table 4. Configuration space for the trading strategy

Parameters	Configuration space
n (days-ahead of prediction)	1, 5, 15
r (percentage of price movement)	1%, 5%, 10%, 20%
Threshold of DC	0.001, 0.002, 0.005, 0.01, 0.02

5 Result and Analysis

In this section, we present our results for the DC model, the physical time model and the traditional benchmark of buy and hold. Our aim is to study the competitiveness of the DC-based indicators and whether the resulting trading strategies can outperform the traditional technical analysis (GP-PT) trading strategies.

5.1 Comparison Between GP-DC and GP-PT

Table 5 presents summary statistics across all 33 datasets under rate of return (ROR), risk, and Sharpe ratio (SR). As we can observe, the GP-DC algorithm outperformed GP-PT algorithm in terms of average, median, and maximum results for ROR and SR. On the other hand, GP-PT algorithm did better in terms of average, median, and maximum risk.

Figure 3 presents the box plots of the above results, and we can reach similar conclusions as from Table 5. Furthermore, not only the values but also the overall box plot of GP-DC algorithm is higher in terms of ROR and SR, when compared to the GP-PT algorithm. When arguing about risk, the GP-DC's plot is higher than the GP-PT's one, indicating more risky behavior by GP-DC. Furthermore, the ROR for each trade of DC is concentrated above zero. In contrast, the results of the GP-PT algorithm have many negative values, which indicate that GP-DC algorithm is more competitive than the GP-PT algorithm in terms of rate of return.

To confirm the above results, we performed the non-parametric Kolmogorov-Smirnov test between the GP-DC and GP-PT results distributions. We ran the test for each metric (ROR, risk, and SR). The p-value for each test was 0.0082, 0.8107, and 0.6015, respectively. As the p-value for ROR was below 0.05, it denotes that the null hypothesis is rejected at the 5% significance level, thus making the differences in rate of return between GP-DC and GP-PT statistically significant. On the other hand, even though GP-PT algorithm had a lower risk, the differences were not statistically significant. Similarly, even though GP-DC algorithm outperformed GP-PT algorithm in terms of SR, their difference was not statistically significant.

Table 5. Summary statistics of the GP-DC and GP-PT algorithm. The best values per metric appear in boldface.

Measurement	Rate of return		Risk		Sharpe ratio (SR)	
Algorithms	GP-DC	GP-PT	GP-DC	GP-PT	GP-DC	GP-PT
Average	**1.4949%**	−0.0566%	0.1062	**0.0898**	**0.3403**	0.2919
Median	**1.7943%**	−0.2495%	0.0814	**0.0757**	**0.2985**	0.1207
Maximum	**9.7798%**	7.9318%	0.3273	**0.2340**	**1.3688**	1.3382
Minimum	−7.5580%	**−4.7622%**	0.0280	0.0280	−0.5037	**−0.2604**

These results show the potential of the DC approach to act as a complementary approach to the physical time one, as it can yield statistically higher returns than physical time technical analysis indicators. However, it should also be noted that this happened at the expense of a slightly higher risk. Therefore, it deserves further study whether more fine-tuned DC strategies can also lead to lower risk, or, perhaps, whether a mix of DC and physical time strategies is to be suggested.

5.2   Buy and Hold

We now compare the performance of the GP-DC and GP-PT algorithms with the buy-and-hold strategy, where one unit of stock is bought on the first day of trading and sold on the last day. Because of the nature of buy-and-hold, the standard deviation cannot be calculated since there is only a single buy-sell action and thus a single profit value; similarly, we cannot calculate risk and SR. Besides, rate of return is not a very meaningful metric for comparison, as both GP-DC and GP-PT algorithms have a high number of trades, while buy-and-hold has a single trade. To make a fairer comparison, we instead use the cumulative returns over the test set.

As we can observe in Table 6, the GP-DC algorithm has a significantly higher average and median values compared to the GP-PT algorithm and buy-and-hold

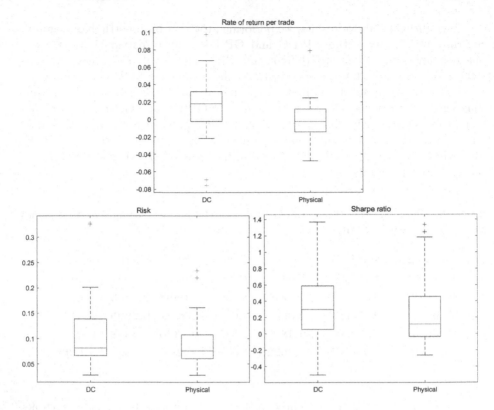

Fig. 3. Box plot of DC and physical time

(GP-DC average: 13.85%; median: 11.94%. GP-PT average: −1.53%; median: −2.73%. Buy-and-hold average: −4.08%; median: −10.81%). On the other hand, the highest cumulative returns is observed for buy-and-hold (around 135%), and the lowest for GP-DC (around −33%). It is also worth noting that the markets tested in this article are predominately bear markets, as it is also evident by the negative average and median cumulative returns of the buy and hold strategy. Since we use 2020 data as our test data, the occurrence of COVID 19 in 2020 significantly affects the stock market explaining the negative cumulative returns. Therefore the fact that GP-DC algorithm has achieved strong average and median cumulative return performance indicates its high potential as a profitable trading paradigm.

The above results are also confirmed by looking at the distribution of results presented in Fig. 4. The majority of the values presented in the box plot for GP-DC algorithm have higher values (i.e. cumulative returns) than the other two approaches. These results are supported by the Kolmogorov-Smirnov tests, which returned a p-value of 0.0082 in the comparison of GP-DC and GP-PT algorithms, and a p-value of 4.83E−04 for the comparison of DC and buy-and-hold. It should be noted that statistical significance in this case at the 5% level

Table 6. Cumulative returns of GP-DC, GP-PT, and buy and hold. Best values denoted in boldface.

Model	Average	Median	Maximum	Minimum
GP-DC	**13.8498%**	**11.9383%**	83.2537%	**−33.0880%**
GP-PT	−1.5341%	−2.7340%	59.4906%	−33.5400%
Buy and hold	−4.0821%	−10.8100%	**135.9218%**	−42.7290%

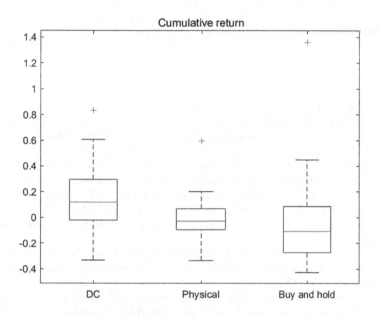

Fig. 4. Box plot of cumulative returns for GP-DC, GP-PT, and buy-and-hold

is for p-values below 0.025, after taking into account the Bonferroni correction for the (two) multiple comparisons.

6 Conclusion

We have explored the benefit of combining genetic programming with indicators tailored for a directional changes framework. Our main contribution is to provide evidence for the effectiveness of this approach in the stock market. To do so, we conducted experiments on 33 datasets from 3 different international stock markets. Over these datasets, our approach (GP-DC) statistically outperformed the GP-PT algorithm, that combines genetic programming with technical indicators based on physical time, as well as the buy and hold strategy, in terms of cumulative return, rate of return, and Sharpe ratio. On the other hand, GP-PT algorithm had lower risk than GP-DC, although this finding is not statistically significant. The above results demonstrate that GP-DC is competitive against these two benchmarks in the stock market and can also be considered as a complementary technique to physical time.

Future work will thus focus on creating new trading strategies that combine technical analysis (physical time) and DC indicators. We believe that such strategies have the potential to bring in further improvements in profitability and risk and outperform the standalone strategies from technical analysis and directional changes.

References

1. Adegboye, A., Kampouridis, M.: Machine learning classification and regression models for predicting directional changes trend reversal in FX markets. Exp. Syst. Appl. **173**, 114645 (2021)
2. Adegboye, A., Kampouridis, M., Johnson, C.G.: Regression genetic programming for estimating trend end in foreign exchange market. In: 2017 IEEE Symposium Series on Computational Intelligence (SSCI), pp. 1–8. IEEE (2017)
3. Aloud, M.E.: Time series analysis indicators under directional changes: the case of Saudi stock market. Int. J. Econ. Financ. Issues **6**(1), 55–64 (2016)
4. Aloud, M.E., Alkhamees, N.: Intelligent algorithmic trading strategy using reinforcement learning and directional change. IEEE Access **9**, 114659–114671 (2021)
5. Aloud, M., Fasli, M.: Exploring trading strategies and their effects in the foreign exchange market: exploring trading strategies. Comput. Intell. **33**(2), 280–307 (2016)
6. Ao, H., Tsang, E.: Trading algorithms built with directional changes. In: 2019 IEEE Conference on Computational Intelligence for Financial Engineering & Economics (CIFEr), pp. 1–7. IEEE (2019)
7. Bakhach, A., Tsang, E., Ng, W.L., Chinthalapati, V.R.: Backlash agent: a trading strategy based on directional change. In: 2016 IEEE Symposium Series on Computational Intelligence (SSCI), pp. 1–9. IEEE (2016)
8. Brabazon, A., Kampouridis, M., O'Neill, M.: Applications of genetic programming to finance and economics: past, present, future. Genet. Program Evolvable Mach. **21**(1), 33–53 (2020)
9. Christodoulaki, E., Kampouridis, M., Kanellopoulos, P.: Technical and sentiment analysis in financial forecasting with genetic programming. In: IEEE Symposium on Computational Intelligence for Financial Engineering and Economics (CIFEr) (2022)
10. Claveria, O., Monte, E., Torra, S.: Evolutionary computation for macroeconomic forecasting. Comput. Econ. **53**(2), 833–849 (2019)
11. Glattfelder, J.B., Dupuis, A., Olsen, R.B.: Patterns in high-frequency FX data: discovery of 12 empirical scaling laws. Quant. Financ. **11**(4), 599–614 (2011)
12. Guillaume, D.M., Dacorogna, M.M., Davé, R.R., Müller, U.A., Olsen, R.B., Pictet, O.V.: From the bird's eye to the microscope: a survey of new stylized facts of the intra-daily foreign exchange markets. Financ. Stochast. **1**(2), 95–129 (1997)
13. Hamida, S.B., Abdelmalek, W., Abid, F.: Applying dynamic training-subset selection methods using genetic programming for forecasting implied volatility. arXiv preprint arXiv:2007.07207 (2020)
14. Hussein, S.M.: Event-based microscopic analysis of the FX market. Ph.D. thesis, University of Essex (2013)

15. Kampouridis, M., Adegboye, A., Johnson, C.: Evolving directional changes trading strategies with a new event-based indicator. In: Shi, Y., et al. (eds.) SEAL 2017. LNCS, vol. 10593, pp. 727–738. Springer, Cham (2017). https://doi.org/10.1007/978-3-319-68759-9_59

16. Kampouridis, M., Tsang, E.: Investment opportunities forecasting: extending the grammar of a GP-based tool. Int. J. Comput. Intell. Syst. 5(3), 530–541 (2012)

17. Kelotra, A., Pandey, P.: Stock market prediction using optimized Deep-ConvLSTM model. Big Data 8(1), 5–24 (2020)

18. Poli, R., Langdon, W.B., McPhee, N.F.: A field guide to genetic programming (2008). Published via http://lulu.com and freely available at http://www.gp-field-guide.org.uk. (With contributions by JR Koza)

19. Tsang, E.: Directional changes, definitions. Working Paper WP050-10 Centre for Computational Finance and Economic Agents (CCFEA), University of Essex Revised 1, Technical report (2010)

20. Tsang, E. P. K., Tao, R., Ma, S.: Profiling financial market dynamics under directional changes. Quantit. Finan. (2016). https://doi.org/10.1080/14697688.2016.1164887

21. Tsang, E.P., Tao, R., Serguieva, A., Ma, S.: Profiling high-frequency equity price movements in directional changes. Quant. Financ. 17(2), 217–225 (2017)

Importance-Aware Genetic Programming for Automated Scheduling Heuristics Learning in Dynamic Flexible Job Shop Scheduling

Fangfang Zhang[1]([⊠])(iD), Yi Mei[1](iD), Su Nguyen[2](iD), and Mengjie Zhang[1](iD)

[1] School of Engineering and Computer Science, Victoria University of Wellington,
PO BOX 600, Wellington 6140, New Zealand
{fangfang.zhang,yi.mei,mengjie.zhang}@ecs.vuw.ac.nz
[2] Centre for Data Analytics and Cognition, La Trobe University, Bundoora, Australia
P.Nguyen4@latrobe.edu.au

Abstract. Dynamic flexible job shop scheduling (DFJSS) is a critical and challenging problem in production scheduling such as order picking in the warehouse. Given a set of machines and a number of jobs with a sequence of operations, DFJSS aims to generate schedules for completing jobs to minimise total costs while reacting effectively to dynamic changes. Genetic programming, as a hyper-heuristic approach, has been widely used to learn scheduling heuristics for DFJSS automatically. A scheduling heuristic in DFJSS includes a routing rule for machine assignment and a sequencing rule for operation sequencing. However, existing studies assume that the routing and sequencing are equally important, which may not be true in real-world applications. This paper aims to propose an importance-aware GP algorithm for automated scheduling heuristics learning in DFJSS. Specifically, we first design a rule importance measure based on the fitness improvement achieved by the routing rule and the sequencing rule across generations. Then, we develop an adaptive resource allocation strategy to give more resources for learning the more important rules. The results show that the proposed importance-aware GP algorithm can learn significantly better scheduling heuristics than the compared algorithms. The effectiveness of the proposed algorithm is realised by the proposed strategies for detecting rule importance and allocating resources. Particularly, the routing rules play a more important role than the sequencing rules in the examined DFJSS scenarios.

Keywords: Importance-aware scheduling heuristics learning · Genetic programming · Hyper-heuristic · Dynamic flexible job shop scheduling

1 Introduction

Dynamic flexible job shop scheduling (DFJSS) [1,2] is an important combinatorial optimisation problem which is valuable in real-world applications such as production scheduling in manufacturing and processing industries [3,4].

G. Rudolph et al. (Eds.): PPSN 2022, LNCS 13399, pp. 48–62, 2022.
https://doi.org/10.1007/978-3-031-14721-0_4

The goal of DFJSS is to find effective schedules to process a number of jobs by a set of machines [5]. In DFJSS, each job consists of a number of operations, and each operation can be processed by more than one machine. Two decisions, i.e., machine assignment to allocate operations to machines and operation sequencing to select an operation to be processed next by an idle machine, need to be made simultaneously. In addition, the decision marking has to be made under dynamic environments such as continuously job arrival [6,7].

Genetic programming (GP) [8], as a hyper-heuristic approach [9–12], has been successfully used to learn scheduling heuristics for DFJSS [13,14]. For GP in DFJSS, a scheduling heuristic consists of a routing rule and a sequencing rule which are used to make decisions on machine assignment and operation sequencing, respectively. The quality of schedules depends on the interaction of the routing rule and the sequencing rule. These two rules in previous studies are normally regarded as being equally important, and are given the same amount of computational resources to evolve. However, this is not necessarily the case in real world applications, and giving too many computational resources to less important rules may lead to a waste of resources and negatively affect the quality of schedules.

To this end, this paper aims to propose an effective importance-aware scheduling heuristics learning GP approach for DFJSS. The developed rule importance measure reflects the significance of the routing rule and the sequencing rule in DFJSS. Inspired by the computation resource allocation strategy that is widely used to allocate resources to sub-problems [15–19], an adaptive computational resource allocation strategy is designed to give more resources for learning the more important rules. In this paper, we use the number of individuals to represent the magnitude of the resources. The proposed algorithm aims to help GP find better scheduling heuristics by allocating proper number of individuals between learning the routing and sequencing rules in DFJSS. Specifically, this paper has the following research objectives.

- Develop an effective strategy to measure the importance of the routing rule and the sequencing rule in the decision making of DFJSS.
- Propose an adaptive computational resource allocation strategy based on the rule importance.
- Analyse the effectiveness of the proposed algorithm in terms of the performance of learned rules.
- Analyse how the proposed algorithm affects GP's behaviour in terms of the number and the ratio of individuals assigned, and the reward for each rule.
- Analyse the effect of the proposed algorithm on the sizes of the learned scheduling heuristics.

2 Background

2.1 Dynamic Flexible Job Shop Scheduling

In DFJSS, m machines $\mathcal{M} = \{M_1, M_2, ..., M_m\}$ are required to process n jobs $\mathcal{J} = \{J_1, J_2, ..., J_n\}$. Each job J_j has a sequence of operations $\mathcal{O}_j = \{O_{j1}, O_{j2}, ..., O_{jl_j}\}$ that need to be processed one by one, where l_j is the number of operations of job J_j.

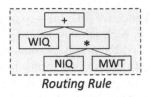

Routing Rule

Fig. 1. An example of the routing rule learned by GP.

Each operation O_{ji} can be processed on more than one machine $M(O_{ji}) \subseteq \pi(O_{ji})$ [20]. Thus, the machine that processes an operation determines its processing time $\delta(O_{ji}, M(O_{ji}))$. This paper focuses on one of the most common dynamic events in real life, i.e., new jobs arrive dynamically [21,22]. The information about a new job is unknown until it arrives on the shop floor. Below are the main constraints of the DFJSS problems.

- A machine can only process one operation at a time.
- Each operation can be handled by only one of its candidate machines.
- An operation cannot be handled until its precedents have been processed.
- Once started, the processing of an operation cannot be stopped.

We consider three commonly used objectives in this paper. The calculations of them are shown as follows.

- Mean-flowtime (Fmean): $\frac{\sum_{j=1}^{n}(C_j - r_j)}{n}$
- Mean-tardiness (Tmean): $\frac{\sum_{j=1}^{n} max\{0, C_j - d_j\}}{n}$
- Mean-weighted-tardiness (WTmean): $\frac{\sum_{j=1}^{n} w_j * max\{0, C_j - d_j\}}{n}$

where C_j represents the completion time of a job J_j, r_j represents the release time of J_j, d_j represents the due date of J_j, and n represents the number of jobs.

2.2 GP for DFJSS

GP starts with a randomly initialised population that contains a number of individuals (i.e., *initialisation*). GP programs consist of terminals and functions, which are naturally priority functions for prioritising machines and operations in DFJSS. Figure 1 shows an example of the routing rule which is a priority function of WIQ + NIQ * MWT where WIQ and NIQ are the workload and the number of operations in the queue of a machine, MWT is the needed time of a machine to finish its current processing operation. The quality of individuals is evaluated by measuring the decision marking performance of applying individuals on DFJSS simulations (i.e., *evaluation*). New offspring are generated by genetic operators, i.e., crossover, mutation, and reproduction, with selected parents (i.e., *evolution*). New offspring will be put into and evaluated in the next generation. GP improves the quality of individuals generation by generation until the stopping criterion is met. The best learned scheduling heuristic at the last generation is reported as the output of the GP algorithm [23,24].

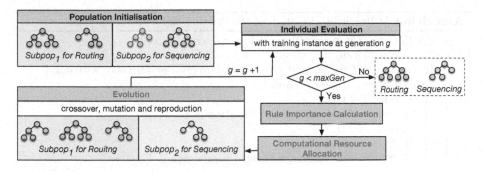

Fig. 2. The flowchart of the proposed algorithm.

3 Importance-Aware Scheduling Heuristic Learning

3.1 An Overview of the Proposed Algorithm

Figure 2 shows the flowchart of the proposed algorithm, and the newly developed components are highlighted in red. We use cooperative coevolution strategy to learn the routing rule and the sequencing rule simultaneously [25,26]. The population consists of two subpopulations, and the first (second) subpopulation $Subpop_1$ ($Subpop_2$) is used to learn the routing (sequencing) rule. The evolutionary processes of the two subpopulations are independent except for the individual evaluation. Since a routing rule and a sequencing rule have to work together to make decisions in DFJSS, for individual evaluation, the individuals in $Subpop_1$ ($Subpop_2$) at the current generation are evaluated with the best individual in $Subpop_2$ ($Subpop_1$) at the previous generation. The best scheduling heuristic obtained from the whole population is reported as the best at the current generation, i.e., can either be from $Subpop_1$ or from $Subpop_2$. Since there is no previous generation for the first generation, for individual evaluation in $Subpop_1$ ($Subpop_2$), we randomly select one individual in $Subpop_2$ ($Subpop_1$).

Before evolution, we first measure the importance of the routing rule and the sequencing rule. Then, we use the rule importance information to allocate computational resources, i.e., individuals, for learning different rules. More computational resources will be allocated to the important rule, which is expected to improve the overall scheduling effectiveness in DFJSS. The number of individuals in $Subpop_1$ and $Subpop_2$ is adaptive. As the example shown in Fig. 2, at the beginning, there are two individuals in each subpopulation for learning each rule. After the computational resources allocation, three individuals are used to learn the routing rule and one individual is utilised for the sequencing rule. The details of the developed new components are shown in the following subsections.

3.2 Measure the Importance of the Routing and Sequencing Rules

This paper measures the importance of rules based on their contributions to the fitness improvement which is calculated according to consecutive generations.

Algorithm 1: Reward Calculation of the Routing and Sequencing Rule

---

1: $rewardRouting = 0$, $rewardSequencing = 0$, $counter = 3$
2: **while** $counter \leqslant g$ **do**
3:    **if** $fitness_1 < fitness'_1$ and $fitness_2 < fitness'_2$, or $fitness_1 > fitness'_1$ and $fitness_2 > fitness'_2$ **then**
4:       **if** $\triangle 1 < \triangle 2$ **then**
5:          | $rewardRouting = rewardRouting + 1$
6:       **end**
7:       **if** $\triangle 1 > \triangle 2$ **then**
8:          | $rewardSequencing = rewardSequencing + 1$
9:       **end**
10:      **if** $\triangle 1 = \triangle 2$ **then**
11:         $rewardRouting = rewardRouting + 0$
12:         $rewardSequencing = rewardSequencing + 0$
13:      **end**
14:    **else**
15:       $rewardRouting = rewardRouting + 0$
16:       $rewardSequencing = rewardSequencing + 0$
17:    **end**
18:    $counter = counter + 1$
19: **end**
20: **return** $rewardRouting$, $rewardSequencing$

We assume the best fitness of $Subpop_1$ and $Subpop_2$ are $fitness_1$ and $fitness_2$ at the current generation, and $fitness'_1$ and $fitness'_2$ at the previous generation. We calculate the fitness improvement of $Subpop_1$ for learning routing rules and $Subpop_2$ for learning sequencing rules as $\triangle 1 = (fitness_1 - fitness'_1)/fitness'_1$ and $\triangle 2 = (fitness_2 - fitness'_2)/fitness'_2$, respectively. In the general minimisation problems, we can compare $\triangle 1$ and $\triangle 2$ directly (i.e., $\triangle 1 \leqslant 0$ and $\triangle 2 \leqslant 0$), and treat the one with smaller $\triangle$ (i.e., larger $|\triangle|$) as the important one. However, it is not always the case in this paper due to the used instance rotation strategy, i.e., different generations use different training instances, which has been successfully used to train scheduling heuristics with GP [27,28]. This indicates that the fitness scales are different across generations due to the difference of training instances, and we are not sure whether the fitness will increase or decrease across consecutive generations. Thus, this paper defines that the routing rule will be more important than the sequencing rule when $\triangle 1 < \triangle 2$ under the conditions of either $fitness_1 > fitness'_1$ and $fitness_2 > fitness'_2$ or $fitness_1 < fitness'_1$ and $fitness_2 < fitness'_2$ (lines 3–13). The rewards for the routing rule $rewardRouting$ and the sequencing rule $rewardSequencing$ are calculated as shown in Algorithm 1, where g is the current generation number. It is noted that we do not measure the rule importance at the generations in either of the following two cases, i.e., If $fitness_1 > fitness'_1$ and $fitness_2 < fitness'_2$, or If $fitness_1 < fitness'_1$ and $fitness_2 > fitness'_2$ (lines 14–17), due to the unknown fitness change information. How to measure rule importance in these two cases will be studied in our future work.

3.3   Adaptive Computational Resource Allocation Strategy

We start to measure the rule importance from generation three (i.e., $g \geqslant 3$, population at generation 1 is randomly initialised, and we do not consider it for avoiding randomness). At a generation, we use the reward obtained by rules so far to decide the number of individuals for learning each rule. The ratios for deciding the number of individuals for the routing rule is shown as below:

$$ratioRouting = \frac{rewardRouting}{rewardRouting + rewardSequencing} \tag{1}$$

Thus, the number of offspring generated per generation for learning the routing rule and sequencing rule is $popsize * ratioRouting$ and $popsize * (1 - ratioRouting)$, respectively. The number of individuals for learning the routing and sequencing rule is adaptive over generations, which are highly related to the rule importance.

4   Experiment Design

Simulation Model: This paper considers to process 6000 jobs including 1000 warm-up jobs with ten machines. The importance of jobs varies which are represented by weight, i.e., 20%, 60%, 20% jobs are with weights 1, 2, and 4, respectively [29]. Each job has a certain number of operations which follows a uniform discrete distribution between one and ten. Each operation can be processed by more than one machine, where the number of options follows a uniform discrete distribution between one and ten. The processing time of each operation follows a uniform discrete distribution with the range [1, 99]. Utilisation level (P) is a factor to simulate different DFJSS scenarios, and a higher utilisation level indicates a busier DFJSS. The utilisation is calculated as $P = \mu * P_M / \lambda$, where μ is the average processing time of machines, P_M is the probability of a job visiting a machine, λ is the rate of the Poisson process for simulating job arrival.

Design of Comparisons: GP, which has an equal number of individuals for learning the routing and sequencing rules, is selected as a baseline for comparison. The algorithm that gives the *important* rule more individuals is named IGP. To measure the performance of IGP, IGP will be compared with GP. To further verify the effectiveness of IGP, we compare with a reverse algorithm named UNIGP that gives *unimportant* rule more individuals by swapping the number of individuals for routing and sequencing rules obtained by the proposed individual allocation strategy. The scenarios with utilisation levels of 0.75, 0.85 and 0.95 are used to measure the performance of algorithms. The scenarios are represented as <objective, utilisation level> such as <Fmean, 0.75>.

Parameter Settings: All the algorithms have 1000 individuals with two subpopulations. Each subpopulation is 500 individuals. IGP and UNIGP have an adaptive number of individuals across generations. Each individual of the algorithm consists of terminals, i.e., shown in Table 1 [21], and functions, i.e., +,

Table 1. The terminal and function sets.

	Terminals	Description
Machine-related	NIQ	The number of operations in the queue
	WIQ	Current work in the queue
	MWT	Waiting time of a machine
Operation-related	PT	Processing time of an operation
	NPT	Median processing time for next operation
	OWT	Waiting time of an operation
Job-related	WKR	Median amount of work remaining of a job
	NOR	The number of operations remaining of a job
	W	Weight of a job
	TIS	Time in system

Table 2. The mean (standard deviation) of objective values on test instances of GP, IGP, and UNIGP according to 30 independent runs in nine scenarios.

Scenarios	GP	IGP	UNIGP
<Fmean, 0.75>	336.23(1.26)	335.63(1.07)(↑)	335.94(1.19)(≈)(≈)
<Fmean, 0.85>	384.69(1.63)	383.79(1.50)(↑)	386.97(4.06)(↓)(↓)
<Fmean, 0.95>	550.94(5.79)	549.69(2.95)(≈)	558.08(9.64)(↓)(↓)
<Tmean, 0.75>	13.28(0.40)	13.09(0.29)(↑)	13.76(0.77)(↓)(↓)
<Tmean, 0.85>	40.27(1.85)	39.56(0.82)(≈)	42.15(2.92)(↓)(↓)
<Tmean, 0.95>	175.49(2.85)	174.25(2.43)(↑)	182.88(6.94)(↓)(↓)
<WTmean, 0.75>	27.04(1.05)	26.66(1.02)(↑)	27.71(2.22)(≈)(↓)
<WTmean, 0.85>	75.82(3.83)	74.46(1.90)(↑)	76.57(4.37)(≈)(↓)
<WTmean, 0.95>	294.58(9.65)	290.45(6.10)(↑)	303.93(15.40)(↓)(↓)
Average rank	2	**1.51**	2.49

* An algorithm is compared with its left algorithm(s) one by one if has.

$-$, $*$, protected $/$, max, min. The initialised GP programs are generated by the ramp-half-and-half method with a minimal (maximal) depth of 2 (6). The depths of all programs are no more than 8. Tournament selection with size 7 is used to select parents for producing offspring. The new offspring are generated by elites of a value of 10, and crossover, mutation and reproduction with rates 80%, 15%, and 5%, respectively. The maximal number of generations of algorithms is 51.

5 Results and Discussions

We use the results from 30 independent runs to verify the performance of the proposed algorithm. We apply the Friedman test to see whether there is a significant difference among algorithms. If yes, then Wilcoxon test with a significance level of 0.05. is used to compare two algorithms, and "↑", "↓", "≈" indicate an algorithm is better, worse or similar with the compared algorithm.

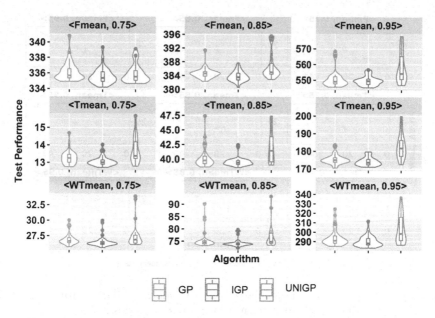

Fig. 3. Violin plots of the obtained objective values on test instances of GP, IGP and UNIGP according to 30 independent runs in nine scenarios.

Quality of Learned Scheduling Heuristics: Table 2 shows the mean and standard deviations of objective values on unseen instances of GP, IGP and UNIGP over 30 independent runs in nine scenarios. The results show that IGP is significantly better than GP in most of the examined scenarios. This verifies the effectiveness of the proposed algorithm with adaptive computational resources allocation strategy. In addition, UNIGP is much worse than baseline GP and IGP which is as expected, since UNIGP applies the opposite idea from IGP. This verifies the proposed algorithm from a reverse point of view. Overall, we can see that IGP is the best algorithm among them with the smallest rank value of 1.51. Figure 3 shows the violin plots of obtained test objective values of GP, IGP, and UNIGP based on 30 independent runs in nine scenarios. We can see that the proposed algorithm IGP shows its superiority and achieves the best performance with a lower objective distribution.

Accumulated Rewards for Routing and Sequencing Rules: Figure 4 shows the curves of average accumulated reward values of IGP-Routing and IGP-Sequencing based on 30 independent runs in nine scenarios. It is clear that the reward values for the routing rule are increasing steadily along with the generations. However, there is only a small increase on the reward values of the sequencing rules in most of the scenarios. The results show that the routing rule plays a more important role than the sequencing rule in DFJSS. The proposed algorithm is expected to give more individuals for learning the routing rules.

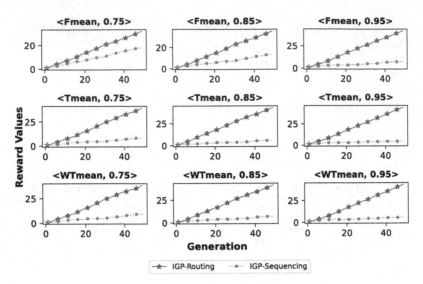

Fig. 4. Curves of average accumulated reward values of IGP-Routing and IGP-Sequencing according to 30 independent runs in nine scenarios.

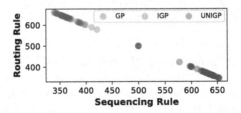

Fig. 5. Scatter plots of the number of individuals for learning the routing rule and the sequencing rule across all generations of GP, IGP, and UNIGP.

The Number of Individuals for Learning Rules: Figure 5 shows the scatter plots of individuals for learning the routing and sequencing rule across all generations of GP, IGP and UNIGP in scenario <Fmean, 0.75>. For GP, a fixed number of individuals are equally set for learning the routing rule (i.e., 500 individuals) and the sequencing rule (i.e., 500 individuals). IGP gives more individuals for learning the routing rule, which UNIGP biases more on the learning on the sequencing rule. The results show that the routing rule is more important than the sequencing rule in DFJSS. Furthermore, the proposed algorithm IGP can adaptively allocate more individuals for learning the routing rule. Similar pattern is also found in other scenarios.

Ratios of Number of Individuals for Learning the Routing Rule: Figure 6 shows the curves of average ratios of the number of individuals for learning the routing rule along with generations of IGP and UNIGP. At the first two generations, the ratios of the number of individuals for learning routing

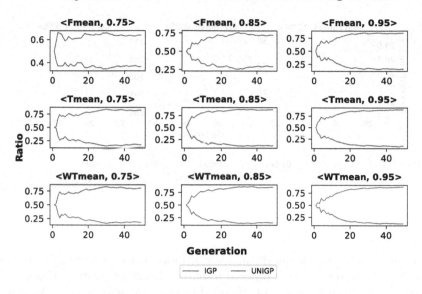

Fig. 6. Curves of the average ratios of the number of individuals for learning the routing rule along with generations of IGP and UNIGP according to 30 independent runs.

and sequencing rules are the same, which are 0.5 in all scenarios (computational resource allocation starts at generation three). From generation three, IGP starts to increase the ratios of the number of individuals for learning routing rules. After generation 20, the ratios arrive at a relatively steady state, which are around 0.85 in most scenarios. UNIGP has shown the opposite trend, where the ratios of the number of individuals for learning the routing rule keep decreasing to about 0.15 at generation 20 and stay at a relatively constant number after that.

Comparison with Algorithms with Fixed Number of Individuals: Based on the discussion in the previous section, we can see that the found ratios of individuals for the routing rule are around 0.85. In other words, about 850 (i.e., 1000 * 0.85) individuals are used by IGP for learning the routing rule. It is interesting to know whether fixing the number of individuals for learning rules can get the same performance as IGP or not. To investigate this, we compare IGP with GP500, GP650 and GP850, where 500 (500), 650 (350), and 850 (150) are the number of individuals for learning the routing (sequencing) rule. We choose the most complex scenarios (i.e., <WTmean, 0.75>, <WTmean, 0.85>, and <WTmean, 0.95>) that consider the job importance for this investigation. Figure 7 shows the curves of the average objective values of GP500, GP650, GP850 and IGP on test instances in mean-weighted-tardiness related scenarios over 30 independent runs. The results show that GP850 performs better than GP500 and GP650 in most cases. This indicates that finding a good threshold for the number of individuals for learning rules can improve the performance. In addition, the results also show that IGP shows its superiority compared with all other algorithms in terms of the convergence speeds and final performance.

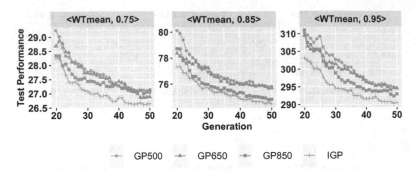

Fig. 7. Curves of average objective values of GP500, GP650, GP850 and IGP on test instances in mean-weighted-tardiness related scenarios based on 30 independent runs.

We can see that simply fixing the number of individuals for learning rules is not effective as an adaptive computational resource allocation strategy. One possible reason is that the importance of the routing rule and sequencing rules' importance may differ on different instances at different generations. Another possible reason is that although the routing rule and the sequencing rule differ in importance to the schedule, it is still necessary to allocate enough resources to learn the less important rule to get a good enough rule before generation 20 as shown in Fig. 6 (i.e., the schedule quality in DFJSS depends on two rules). The superior performance of IGP verifies the effectiveness of the proposed algorithm to detect rule importance and allocate computational resources automatically and adaptively.

Sizes of the Learned Scheduling Heuristics: To verify the effect of the proposed IGP on learned scheduling heuristics, this section investigates rule size. We use the number of nodes for measuring the rule sizes [30]. Since the routing rule and the sequencing rule work together in DFJSS, it is reasonable to look at the sum of their rule sizes. We find that there is no significant difference between the rule sizes (routing rule plus sequencing rule) of GP and IGP, such as with the mean and standard deviation of 98.53(23.93) and 101.87(21.83)($\approx$) in <Fmean, 0.75>. Figure 8 shows the violin plots of the average sizes of routing rules and sequencing rules over population in nine scenarios. Overall, the results show that the routing rules are larger than the sequencing rules for both GP and IGP. This also demonstrates that the routing rule is more important than the sequencing rule. We can also see that there is an increase on the routing rule sizes of IGP compared with GP, especially in the three scenarios with Tmean (i.e., <Tmean, 0.75>, <Tmean, 0.85> and <Tmean, 0.95>). This trend is clearer in the scenarios with higher utilisation levels. In contrast with the increase of the routing rule size, the sequencing rule size becomes smaller in most of the scenarios. This indicates that using more (less) resources on learning the rule can increase (decrease) the corresponding rule size.

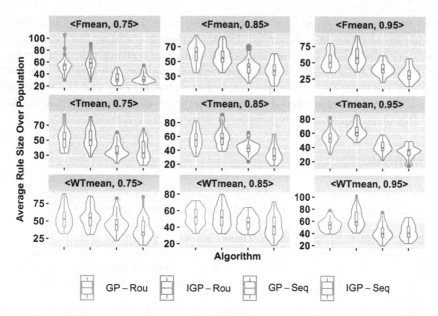

Fig. 8. Violin plots of the sizes of average routing rules and sequencing rules over population of GP and IGP in nine scenarios.

6 Conclusions and Future Work

The goal of this paper is to develop an effective importance-aware scheduling heuristics learning GP approach to automatically learn the routing and sequencing rules for DFJSS. The goal has been achieved by proposing a novel rule importance measure, and an adaptive strategy to allocate computational resources, i.e., GP individuals, for learning the routing rule and the sequencing rule.

The results show that the importance of the routing rule and the sequencing rule differs, and the routing rule is more important than the sequencing rule in the examined DFJSS scenarios. The proposed rule importance strategy based on the improvement of fitness across generations can detect the rule importance in DFJSS properly. Furthermore, the developed adaptive computational resources allocation strategy based on the rule importance measure has successfully optimised the learning process for the routing rule and the sequencing rule. The effectiveness of the proposed IGP has also been verified by the analyses in terms of the exact number and the ratios of allocated individuals for rules, the accumulated reward for rules, and the comparison with the algorithms with a fixed number of individuals for learning rules. Further analyses show that there is no significant difference between the rule size of the pairs of routing rule and sequencing rule, however, the routing (sequencing) rule obtained by the proposed algorithm is larger (smaller) than compared algorithms. In addition, we observe that the routing rule is normally larger than the sequencing rule learned by GP algorithms, which can also be an indicator of the importance of the routing rule.

Some interesting directions can be further investigated in the near future. The rule importance in different DFJSS scenarios may differ. For example, the sequencing rule might be more important than the routing rule if there are a small number of machines. More comprehensive analyses are needed. Moreover, this paper confirms that the importance of the routing rule and the sequencing rule can differ. A more advanced strategy to improve the overall decision marking in DFJSS by recognising such differences is worth investigating.

References

1. Nie, L., Gao, L., Li, P., Li, X.: A GEP-based reactive scheduling policies constructing approach for dynamic flexible job shop scheduling problem with job release dates. J. Intell. Manuf. **24**(4), 763–774 (2013)
2. Zhang, F., Mei, Y., Nguyen, S., Zhang, M.: Guided subtree selection for genetic operators in genetic programming for dynamic flexible job shop scheduling. In: Hu, T., Lourenço, N., Medvet, E., Divina, F. (eds.) EuroGP 2020. LNCS, vol. 12101, pp. 262–278. Springer, Cham (2020). https://doi.org/10.1007/978-3-030-44094-7_17
3. Zhang, F., Nguyen, S., Mei, Y., Zhang, M.: Genetic Programming for Production Scheduling. MLFMA, Springer, Singapore (2021). https://doi.org/10.1007/978-981-16-4859-5
4. Nguyen, S., Zhang, M., Johnston, M., Chen Tan, K.: Hybrid evolutionary computation methods for quay crane scheduling problems. Comput. Oper. Res. **40**(8), 2083–2093 (2013)
5. Hart, E., Ross, P., Corne, D.: Evolutionary scheduling: a review. Genet. Program Evolvable Mach. **6**(2), 191–220 (2005)
6. Jaklinović, K., Durasević, M., Jakobović, D.: Designing dispatching rules with genetic programming for the unrelated machines environment with constraints. Exp. Syst. Appl. **172**, 114548 (2021)
7. Zhang, F., Mei, Y., Nguyen, S., Zhang, M.: Correlation coefficient-based recombinative guidance for genetic programming hyperheuristics in dynamic flexible job shop scheduling. IEEE Trans. Evol. Comput. **25**(3), 552–566 (2021). https://doi.org/10.1109/TEVC.2021.3056143
8. Koza, J.R.: Genetic programming as a means for programming computers by natural selection. Stat. Comput. **4**(2), 87–112 (1994)
9. Burke, E.K., et al.: Hyper-heuristics: a survey of the state of the art. J. Oper. Res. Soc. **64**(12), 1695–1724 (2013)
10. Braune, R., Benda, F., Doerner, K.F., Hartl, R.F.: A genetic programming learning approach to generate dispatching rules for flexible shop scheduling problems. Int. J. Prod. Econ. **243**, 108342 (2022)
11. Pillay, N., Qu, R.: Hyper-Heuristics: Theory and Applications. Springer, Cham (2018). https://doi.org/10.1007/978-3-319-96514-7
12. Zhang, F., Mei, Y., Nguyen, S., Zhang, M.: Collaborative multifidelity-based surrogate models for genetic programming in dynamic flexible job shop scheduling. IEEE Trans. Cybern. **52**(8), 8142–8156 (2022). https://doi.org/10.1109/TCYB.2021.3050141

13. Zhang, F., Mei, Y., Nguyen, S., Zhang, M., Tan, K.C.: Surrogate-assisted evolutionary multitask genetic programming for dynamic flexible job shop scheduling. IEEE Trans. Evol. Comput. **25**(4), 651–665 (2021)
14. Zhang, F., Mei, Y., Nguyen, S., Tan, K.C., Zhang, M.: Multitask genetic programming-based generative hyper-heuristics: a case study in dynamic scheduling. IEEE Trans. Cybern. (2021). https://doi.org/10.1109/TCYB.2021.3065340
15. Shen, X., Guo, Y., Li, A.: Cooperative coevolution with an improved resource allocation for large-scale multi-objective software project scheduling. Appl. Soft Comput. **88**, 106059 (2020)
16. Ren, Z., Liang, Y., Zhang, A., Yang, Y., Feng, Z., Wang, L.: Boosting cooperative coevolution for large scale optimization with a fine-grained computation resource allocation strategy. IEEE Trans. Cybern. **49**(12), 4180–4193 (2018)
17. Yang, M., et al.: Efficient resource allocation in cooperative co-evolution for large-scale global optimization. IEEE Trans. Evol. Comput. **21**(4), 493–505 (2017). https://doi.org/10.1109/TEVC.2016.2627581
18. Jia, Y.-H., Mei, Y., Zhang, M.: Contribution-based cooperative co-evolution for nonseparable large-scale problems with overlapping subcomponents. IEEE Trans. Cybern. **52**(6), 4246–4259 (2020). https://doi.org/10.1109/TCYB.2020.3025577
19. Zhang, X.-Y., Gong, Y.-J., Lin, Y., Zhang, J., Kwong, S., Zhang, J.: Dynamic cooperative coevolution for large scale optimization. IEEE Trans. Evol. Comput. **23**(6), 935–948 (2019)
20. Brucker, P., Schlie, R.: Job-shop scheduling with multi-purpose machines. Computing **45**(4), 369–375 (1990)
21. Zhang, F., Mei, Y., Nguyen, S., Zhang, M.: A preliminary approach to evolutionary multitasking for dynamic flexible job shop scheduling via genetic programming. In: Proceedings of the Genetic and Evolutionary Computation Conference, pp. 107–108. ACM (2020)
22. Durasevic, M., Jakobovic, D.: A survey of dispatching rules for the dynamic unrelated machines environment. Exp. Syst. Appl. **113**, 555–569 (2018)
23. Hart, E., Sim, K.: A hyper-heuristic ensemble method for static job-shop scheduling. Evol. Comput. **24**(4), 609–635 (2016)
24. Zhang, F., Mei, Y., Nguyen, S., Zhang, M.: Genetic programming with adaptive search based on the frequency of features for dynamic flexible job shop scheduling. In: Paquete, L., Zarges, C. (eds.) EvoCOP 2020. LNCS, vol. 12102, pp. 214–230. Springer, Cham (2020). https://doi.org/10.1007/978-3-030-43680-3_14
25. Yska, D., Mei, Y., Zhang, M.: Genetic programming hyper-heuristic with cooperative coevolution for dynamic flexible job shop scheduling. In: Castelli, M., Sekanina, L., Zhang, M., Cagnoni, S., García-Sánchez, P. (eds.) EuroGP 2018. LNCS, vol. 10781, pp. 306–321. Springer, Cham (2018). https://doi.org/10.1007/978-3-319-77553-1_19
26. Zhang, F., Mei, Y., Zhang, M.: A two-stage genetic programming hyper-heuristic approach with feature selection for dynamic flexible job shop scheduling. In: Proceedings of the Genetic and Evolutionary Computation Conference, pp. 347–355. ACM (2019)
27. Hildebrandt, T., Heger, J., Reiter, B.S.: Towards improved dispatching rules for complex shop floor scenarios: a genetic programming approach. In: Proceedings of the Conference on Genetic and Evolutionary Computation, pp. 257–264. ACM (2010)

28. Zhang, F., Mei, Y., Nguyen, S., Tan, K.C., Zhang, M.: Instance rotation based surrogate in genetic programming with brood recombination for dynamic job shop scheduling. IEEE Trans. Evol. Comput. (2022). https://doi.org/10.1109/TEVC.2022.3180693
29. Hildebrandt, T., Branke, J.: On using surrogates with genetic programming. Evol. Comput. **23**(3), 343–367 (2015)
30. Zhang, F., Mei, Y., Nguyen, S., Zhang, M.: Evolving scheduling heuristics via genetic programming with feature selection in dynamic flexible job-shop scheduling. IEEE Trans. Cybern. **51**(4), 1797–1811 (2021)

Towards Discrete Phenotypic Recombination in Cartesian Genetic Programming

Roman Kalkreuth[⊠][iD]

Computational Intelligence Research Group, Chair XI Algorithm Engineering,
Department of Computer Science, TU Dortmund University,
Dortmund, North Rhine-Westphalia, Germany
roman.kalkreuth@tu-dortmund.de
https://ls11-www.cs.tu-dortmund.de/

Abstract. The tree-based representation model of Genetic Programming (GP) is largely used with subtree crossover for genetic variation. Unlike Cartesian Genetic Programming (CGP) which is commonly used merely with mutation. Compared to comprehensive knowledge about recombination in the field of tree-based GP, the state of knowledge in CGP appears to be comparatively poor. Even if CGP was officially introduced over twenty years ago, the role of recombination in CGP has been recently considered an open issue. Several promising steps have been taken in recent years, but more research is needed to develop towards a more comprehensive and holistic perspective on crossover in CGP. In this work, we propose a phenotypic variation method for discrete recombination in CGP. We compare our method to the traditional mutation-only CGP approach on a set of well-known symbolic regression problems. The initial results presented in this work demonstrate that the use of our proposed discrete recombination method performs significantly better than the traditional mutation-only approach.

Keywords: Cartesian Genetic Programming · Crossover · Phenotypic variation

1 Introduction

Cartesian Genetic Programming can be considered a well-established graph-based GP variant. Initial work towards CGP was done by Miller, Thompson, Kalganova, and Fogarty [8,14,15] by the introduction of a two-dimensional graph encoding model of functional nodes. CGP can be seen as an extension to the traditional tree-based GP representation model since its representation allows many graph-based applications such as digital circuit design [26], evolution of neural network topologies [16,28] and synthesis of cryptographic Boolean functions [5,7]. CGP has introduced over two decades ago but is still predominantly used only with a probabilistic point mutation operator. The reason for this is that various standard

© The Author(s), under exclusive license to Springer Nature Switzerland AG 2022
G. Rudolph et al. (Eds.): PPSN 2022, LNCS 13399, pp. 63–77, 2022.
https://doi.org/10.1007/978-3-031-14721-0_5

genotypic crossover operators failed to improve the search performance of standard CGP in the past [3, 15]. Overall, the state of knowledge about recombination in CGP appears to be weak when compared to the number of publications in tree-based GP. The role of recombination in CGP was recently surveyed by Miller [17] and is still considered to be an open issue. Even if some progress has been made in recent years, comprehensive and advanced knowledge about recombination in CGP is still missing [17]. In the field of evolutionary computation (EC), discrete recombination is a well-established form of recombination in various subfields. Discrete recombination typically selects each gene from one of the two parents with equal probability. According to Rudolph [22], this method can be therefore considered as a dynamic n-point crossover since each gene for the chromosome of the offspring is selected from the first or second parent with equal probability. In this work, we take a step forward on the issue of crossover and introduce a method for the adaption of discrete recombination in CGP. We initially evaluate our method on a set of well-known symbolic regression benchmarks. Our results demonstrate the effectiveness of our approach for these problems.

Section 2 of this work describes CGP. Related work on crossover in CGP is surveyed in Sect. 3. This section also gives a brief historical overview of discrete recombination in the field of EC. In Sect. 4, we introduce our new method. Section 5 is devoted to the description of our experiments and the presentation of our results. Our findings are discussed in Sect. 6. Finally, Sect. 7 gives a conclusion and outlines our future work.

2   Cartesian Genetic Programming

In contrast to tree-based GP, CGP represents a genetic program via genotype-phenotype mapping as an indexed, acyclic, and directed graph. In this way, CGP can be seen as an extension of the traditional tree-based GP approach. The CGP representation model is based on a rectangular grid or row of nodes. Each genetic program is encoded in the genotype of an individual and is decoded to its corresponding phenotype. A definition of a cartesian genetic program $\mathcal{P}$ is given in Definition 1. Let $\phi : \mathcal{P} \mapsto \Psi$ be a decode function which maps $\mathcal{P}$ to a phenotype Ψ. Originally, the structure of the graph was represented by a rectangular grid of n_r rows and n_c columns, but later work focused on a representation with one row. The CGP decoding procedure processes groups of genes, and each group refers to a node of the graph, except the last one, which represents the outputs of the phenotype. Each node is represented by two types of genes that index the function number in the GP function set and the node inputs. These nodes are called *function nodes* and execute functions on the input values. The number of input genes depends on the maximum arity n_a of the function set.

Definition 1 (Cartesian Genetic Program). *A cartesian genetic program $\mathcal{P}$ is an element of the Cartesian product $\mathcal{N}_i \times \mathcal{N}_f \times \mathcal{N}_o \times \mathcal{F}$:*

- *$\mathcal{N}_i$ is a finite non-empty set of input nodes*
- *$\mathcal{N}_f$ is a finite set of function nodes*

- $\mathcal{N}_o$ is a finite non-empty set of output nodes
- $\mathcal{F}$ is a finite non-empty set of functions

A backward search is conducted to decode the corresponding phenotype. The decoding itself starts at the output nodes and continues until the inputs nodes are reached. The decoding procedure is done for all output genes. The result of the decoding procedure can be described as a set of directed paths Ω. Given the input set I and the output set O, let $\omega = I \times \Omega \mapsto O$ be an output function. An example of the backward search of the most popular one-row integer representation is illustrated in Fig. 1. The backward search starts from the program output and processes all nodes which are linked in the genotype. In this way, only active nodes are processed during evaluation. The genotype in Fig. 1 is grouped by its function nodes. The first (underlined) gene of each group refers to the function number in the corresponding function set. The non-underlined genes represent the input connections of the node. Inactive function nodes are shown in gray color and with dashed lines.

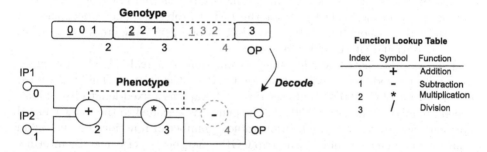

Fig. 1. Example of the decoding procedure of a CGP genotype to its corresponding phenotype. The identifiers IP1 and IP2 stand for the two input nodes with node index 0 and 1. The identifier OP stands for the output node of the graph.

The number of inputs n_i, outputs n_o, and the length of the genotype is fixed. Every candidate program is represented with $n_r * n_c * (n_a + 1) + n_o$ integers. Even if the length of the genotype is fixed for each candidate program, the length of the corresponding phenotype in CGP is variable, which can be considered as an advantage of the CGP representation. CGP is traditionally used with a $(1 + \lambda)$ evolutionary algorithm (EA). The $(1 + \lambda)$-EA is often used with a selection strategy called *neutrality*, which is based on the idea that genetic drift yields to diverse individuals having equal fitness. The genetic drift is implemented into the selection mechanism in a way that individuals which have the same fitness as the normally selected parent are determined, and one of these same-fitness individuals is returned uniformly at random. The new population in each generation consists of the best individual of the previous population and the λ created offspring. The breeding procedure is mostly done by a point mutation that swaps genes in the genotype of an individual in the valid range by chance.

Another point mutation is the flip of the functional gene, which changes the functional behavior of the corresponding function node.

3   Related Work

3.1   Recombination in CGP

According to the reports of Clegg et al. [3], the first attempts of recombination in standard CGP included testing of various genotypic crossover techniques. For instance, the genetic material was recombined by swapping parts of the genotypes of the parent individuals or randomly exchanging selected nodes. Clegg et al. reported that all techniques failed to improve the convergence of CGP and that merely swapping the integers disrupts the search performance. In comparison to mutation only CGP, the addition of genotypic crossover techniques hindered the performance. In one of the first empirical studies about CGP, Miller [15] analyzed its computational efficiency on Boolean function problems. More precisely, Miller analyzed and studied the influence of population size on the efficiency of CGP. The key finding of his study was that extremely low populations perform most effectively for the tested problems. The experiments of this study also demonstrated that the addition of a genotypic crossover reduces the computational effort only marginally.

This was the motivation for the introduction of a real-valued representation and intermediate recombination for CGP by Clegg et al. The real-valued representation of CGP represents the directed graph as a fixed-length list of real-valued numbers in the interval [0, 1]. The genes are decoded to the integer-based representation by their normalization values (number of functions or maximum input range). The recombination of two CGP genotypes is performed by intermediate recombination with a random weighting factor. Clegg et al. demonstrated that the new representation in combination with crossover improves the convergence behavior of CGP on one of the two tested symbolic regression problems. However, for the later generations, Clegg et al. found that the use of crossover in real-valued CGP disrupts the convergence on one problem. Later work by Turner [30] presented results with intermediate recombination on three additional classes of computational problems, digital circuit synthesis, function optimization, and agent-based wall avoidance. On these problems, it was found that the real-valued representation together with the crossover operation performed worse than mutation-only CGP.

Kalkreuth et al. [10] introduced and investigated subgraph crossover in CGP which exchanges and links subgraphs of active function nodes between two selected parents and the block crossover exchanges blocks of active function genes. In recent comparative studies, its use has been found beneficial for several symbolic regression benchmarks since it led to a significant decrease in the number of fitness evaluations needed to find the ideal solution [9,11]. Contrarily, the gain of the search performance was considerably lower for the tested Boolean function problems [9,11]. Moreover, the results of the experiments clearly showed that the subgraph crossover failed to improve the search performance on some of the tested Boolean benchmarks when compared to the results of the traditional $1 + \lambda$ selection strategy.

Husa and Kalkreuth [6] proposed block crossover which selects active function nodes by chance in accordance with a predefined block size but without any order. The function genes of the selected active nodes are then swapped. The block crossover has been compared to mutation-only CGP on a suite of Boolean functions and symbolic regression problems. The outcome of the study gave significant evidence that the $(1+\lambda)$-CGP cannot be considered the most efficient CGP algorithm in the Boolean function domain, although it seems to be often a good choice. The outcome of the study gave the first evidence, that it is possible for crossover operators to outperform the standard $1 + \lambda$ selection strategy.

Sivla et al. [27] introduced a form of crossover for multiple output problems. The proposed method combines the subgraphs of the best outputs of the parent individuals produce an offspring. The proposed crossover technique was applied to the synthesis of combinational logic circuits with multiple outputs. The so-called X-CGP obtained the best results when compared to single chromosome CGP representations and performed better than the multi-chromosome representation for some of the tested problems. The experiments of Siliva et al. indicate that the proposed method is promising. On the other hand, the authors concluded that more studies are needed since X-CGP performed no better than the mutation-only multi-chromosome techniques on the majority of the tested problems.

3.2   Historical Background of Discrete Recombination

Discrete recombination in EC was first described by Rechenberg [20,21] for the simulation of the first type of a multimembered evolutionary strategy (ES) called $(\mu + 1)$ or steady-state ES. Rechenberg demonstrated that recombination can improve the speed of the evolutionary process if the measure is taken per generation rather than per function evaluation [2]. Schwefel [23,24] later utilized discrete recombination among five types of recombination for two further versions of the multimembered ES, called $(\mu + \lambda)$- and (μ, λ)-ES [1]. Schwefel [24] performed an empirical study with 50 uni- and multimodal test functions and compared ESs to the most traditional direct optimization strategies and the outcome showed good results for ESs. According to Bäck et al. [1], the best results were achieved with the use of several types of recombination. In the field of GAs, discrete recombination is commonly referred to as *uniform crossover* and has been found to be a useful search operator [4]. Uniform crossover was first proposed for the binary encoding model of GA by Syswerda [29] and its search performance was found superior to the one- and two-point crossover in the most cases. Uniform recombination in GA inspired the adaption in tree-based GP [18,19] where function nodes and subtrees are exchanged between two parent individuals in accordance with a uniform rate. If the uniform rate is set to 50%, this method represents the tree-based GP equivalent of the uniform crossover for binary strings.

4 The Proposed Method

We adapt discrete recombination in CGP by means of phenotypic functional variation which is performed through the exchange of function genes of active function nodes. The phenotype of a CGP individual is represented by its active function nodes which are determined before the crossover procedure. After selecting two individuals, the minimum and a maximum number of active function nodes of the two individuals is determined. The reason for this is that the size of the phenotype in CGP is not fixed and can vary among individuals. To perform the exchange of active function genes, the crossover procedure iterates over the minimum number of active nodes. A binary decision is made by chance in each iteration whether the function genes are swapped or kept. In the case that both phenotypes differ in size, our method performs a special step in the last iteration called *boundary extension* which extends the selection of active function genes. The idea behind this step is to include active function genes of the larger phenotype into the selection which would not be considered if the lists of active function nodes are merely interated in order. Just like the uniform crossover in GA, our method produces two offspring. The algorithmic implementation of our method is described in Algorithm 1. Exemplifications of the procedure on genotypic and phenotypic level are illustrated in Fig. 2 and 3. An implementation for the CGP extension package of the Java Evolutionary Computation Research System (ECJ) [25] is provided in the ECJ GitHub repository[1].

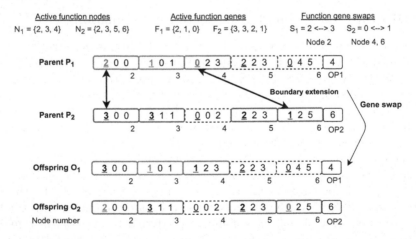

Fig. 2. Exemplification of discrete recombination in CGP: Active function genes of two CGP genotypes are recombined by means of discrete recombination. Function genes, which have been randomly selected for the exchange, are connected with a double-sided arrow in the figure. The active function nodes and genes of the respective parent and offspring individuals are highlighted in red and blue color. (Color figure online)

---

[1] https://github.com/GMUEClab/ecj.

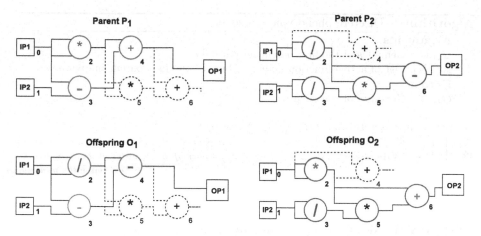

Fig. 3. Illustration of discrete recombination in CGP on the phenotypic level based on the genotypic exemplification presented in Fig. 2. Active function nodes of the respective parent and offspring individuals are highlighted in red and blue color. (Color figure online)

5   Experiments

5.1   Experimental Setup

We performed experiments with symbolic regression problems. We compared the traditional $(1 + \lambda)$-CGP to a canonical EA equipped with our proposed discrete recombination and tournament selection. The algorithms which we used in our experiments are listed in Table 1. To evaluate the search performance, we measured the number of fitness evaluations until the CGP algorithm terminated successfully as recommended by McDermott et al. [13]. Termination was triggered when an ideal solution was found or a predefined budget of fitness evaluation was exceeded. We defined a maximum number of 10^8 fitness evaluations for our experiments and calculated the success rate (SR). In addition to the mean values of the measurements, we also calculated the standard deviation (SD), median (Q2) as well as lower and upper quartile (Q1 and Q3). Meta-optimization experiments have been performed to compare the algorithms fairly and are described in more detail in the following subsection. All tested algorithms were compared on the same number of function nodes to exclude conditions, which can distort the search performance comparison. Our method was tested against the traditional $(1 + \lambda)$-CGP which we declared as the baseline algorithm for our experiments. In our experiments, we exclusively used the single-row standard integer-based representation of CGP. Since we cannot guarantee normally distributed values in our samples, we used the nonparametric two-tailed Mann-Whitney U test to evaluate statistical significance. More precisely, we tested the null hypothesis that two samples come from the same population (i.e. have the same median). We performed 100 independent runs with different random seeds. The levels back parameter l was set to ∞.

Algorithm 1. Discrete phenotypic crossover

Arguments
G_1, G_2: Genomes of the first parent individuals
N_1, N_2: List of active function node numbers of the first parent inidividuals
Return
$\widetilde{G}_1$, $\widetilde{G}_2$: Genomes of the offspring

```
 1: function DiscreteCrossover(G₁, G₂, N₁, N₂)
 2:     l₁ ← |N₁|                              ▷ Number of active nodes of the first parent
 3:     l₂ ← |N₂|                              ▷ Number of active nodes of the second parent
 4:     min ← Min(l₁, l₂)                                    ▷ Determine the minimum
 5:     max ← Max(l₁, l₂)                                    ▷ Determine the maximum
 6:     i ← 0
 7:     while i < min do                  ▷ Iterate over the minimum number of active nodes
 8:         if RandomBoolean() = true then ▷ Decision by chance to keep or swap genes
 9:             ▷ Check if conditions for boundary extension are satisfied
10:             if i = min − 1 and l₁ ≠ l₂ then
11:                 r ← RandomInteger(0, max - i)           ▷ Determine a random offset
12:                 if l₁ < l₂ then     ▷ If the first parent has the minimum of active nodes
13:                     n₁ ← N₁[i]
14:                     ▷ Extend node selection for the second, phenotypically larger, parent
15:                     n₂ ← N₂[i + r]
16:                 else                ▷ Otherwise, extend the selection for the first parent
17:                     n₁ ← N₁[i + r]
18:                     n₂ ← N₂[i]
19:                 end if
20:             else               ▷ Without boundary extension, just select the nodes in order
21:                 n₁ ← N₁[i]
22:                 n₂ ← N₂[i]
23:             end if
24:             p₁ ← PositionFromNodeNumber(n₁)              ▷ Function gene position of n₁
25:             p₂ ← PositionFromNodeNumber(n₂)              ▷ Function gene position of n₂
26:             G̃₁, G̃₂ ← SwapGenes(G₁, G₂, p₁, p₂)             ▷ Swap the function genes
27:         end if
28:         i ← i + 1                                          ▷ Loop counter increment
29:     end while
30:     return G̃₁, G̃₂
31: end function
```

Table 1. Identifiers for the tested CGP algorithms.

Identifier	Description
$1 + \lambda$	$1 + \lambda$ selection strategy with neutral genetic drift
Canonical	Canonical EA with phenotypic uniform crossover and tournament selection

5.2   Benchmarks

We chose eleven symbolic regression problems from the work of McDermott et al. [13] for better GP benchmarks. The reason for our choice of these problems is the fact that we can find an ideal solution more likely on average and evaluate the search performance of the whole evolutionary process. Our set of benchmarks covers uni- as well as bivariate polynomial, trigonometric, logarithmic, and power functions. The functions of the problems are shown in Table 2. A training data set $U[a, b, c]$ refers to c uniform random samples drawn from a to b inclusive. We used the extended Koza function set as recommended by McDermott et al. The function set is shown in Table 3. The fitness of the individuals was represented by a cost function value. The cost function was defined by the sum of the absolute difference between the real function values and the values of an evaluated individual. Let $T = \{x_p\}_{p=1}^{\mathcal{P}}$ be a training dataset of $\mathcal{P}$ random points and $f_{\text{ind}}(x_p)$ the value of an evaluated individual and $f_{\text{ref}}(x_p)$ the true function value. Let

$$C := \sum_{p=1}^{\mathcal{P}} |f_{\text{ind}}(x_p) - f_{\text{ref}}(x_p)|$$

be the cost function. When the difference of all absolute values becomes less than 0.01, the algorithm is classified as converged.

5.3   Meta-optimization

We tuned relevant parameters for all tested CGP algorithms on the set of benchmark problems. Moreover, we used the meta-optimization toolkit of ECJ. The parameter space for the respective algorithms, explored by meta-optimization, is presented in Table 4. For the meta-level, we used a canonical GA equipped with intermediate recombination and point mutation. Since GP benchmark problems can be very noisy in terms of finding the ideal solution, we oriented the meta-optimization with a common approach that has been used in previous studies [6,11,12]. The meta-evolution process at the base level was repeated multiple times for each candidate setting and the most effective settings were compared to find the best setting. For the problems Koza 1–3 and Nguyen 4–7, we selected effective settings of certain parameters for the $(1 + \lambda)$-CGP from previous parametrization studies [11,12].

Table 2. List of symbolic regression benchmarks.

Problem	Objective function	Vars	Training set	Function set
Koza-1	$x^4 + x^3 + x^2 + x$	1	U[−1, 1, 20]	Koza
Koza-2	$x^5 - 2x^3 + x$	1	U[−1, 1, 20]	Koza
Koza-3	$x^6 - 2x^4 + x^2$	1	U[−1, 1, 20]	Koza
Nguyen-4	$x^6 + x^5 + x^4 + x^3 + x^2 + x$	1	U[−1, 1, 20]	Koza
Nguyen-5	$\sin(x^2)\cos(x) - 1$	1	U[−1, 1, 20]	Koza
Nguyen-6	$\sin(x) + \sin(x + x^2)$	1	U[−1, 1, 20]	Koza
Nguyen-7	$\ln(x + 1) + \ln(x^2 + 1)$	1	U[0, 2, 20]	Koza
Nguyen-8	$\sqrt{x}$	1	U[0, 4, 20]	Koza
Nguyen-9	$\sin(x^2) + \sin(y^2)$	2	U[0, 2, 20]	Koza
Nguyen-10	$2 * \sin(x) * \cos(x)$	2	U[0, 2, 20]	Koza
Nguyen-11	x^y	2	U[0, 2, 20]	Koza

Table 3. Function set used for the experiments.

Name	Functions	Constants
Koza	$+$ $-$ $*$ $/$ $\sin$ $\cos$ e^n $\ln(\lvert n \rvert)$	Constant input with a value of 1

Table 4. Parameter space explored by meta-optimization for the $1 + \lambda$ and canonical CGP algorithm.

Algorithm	Parameter	Description	Range
$1 + \lambda$	λ	Number of offspring	[1, 1024]
	N	Number of function nodes	[10, 1000]
	M	Point mutation rate [%]	[1.0, 30.0]
Canonical	N	Number of function nodes	[10, 1000]
	M	Point mutation rate [%]	[1.0, 30.0]
	C	Crossover rate [%]	[10, 100]
	P	Population size	[10, 500]
	T	Tournament size	[2, 20]

5.4 Results

The results of our meta-optimization and search performance evaluation are presented in Table 5 and it is clearly visible that the Canonical-CGP with discrete recombination reduces the number of fitness evaluations to termination significantly on all tested problems. Moreover, on the more complex problems, the Canonical-CGP achieves higher success rates. Violin plots are provided in Fig. 4.

Table 5. Results of the meta-optimization and search performance evaluation.

Problem	Algorithm	Parametrization						Search performance evaluation						p
		N	λ	M [%]	C [%]	P	T	MFE	SD	1Q	2Q	3Q	SR	
Koza-1	$1+\lambda$	10	4	20	–	–	–	3,285,238	8,974,193	518,516	1,408,326	3,460,391	1.0	
	Canonical	10	–	20	70	50	4	**532,957**	**652,332**	**76,868**	**311,983**	**724,563**	1.0	10^{-9}
Koza-2	$1+\lambda$	10	4	20	–	–	–	2,325,581	7,830,950	340,608	1,260,496	2,463,527	1.0	
	Canonical	10	–	10	50	50	4	**733,925**	**934,455**	**67,387**	**394,075**	**982,325**	1.0	10^{-6}
Koza-3	$1+\lambda$	10	4	20	–	–	–	428,778	663,576	26,527	159,290	502,686	1.0	
	Canonical	10	–	20	50	50	4	**122,629**	**264,791**	**17,113**	**41,282**	**113,324**	1.0	10^{-4}
Nguyen-4	$1+\lambda$	100	16	10	–	–	–	91,228,744	24,303,588	100,000,000	100,000,000	100,000,000	0.16	
	Canonical	100	–	8	50	50	4	**59,767,376**	**38,075,889**	**23,060,887**	**59,816,675**	100,000,000	**0.62**	10^{-10}
Nguyen-5	$1+\lambda$	60	16	7	–	–	–	64,092,121	42,126,017	14,078,020	96,894,232	100,000,000	0.50	
	Canonical	60	–	7	70	50	4	**9,758,166**	**23,157,856**	**190,312**	**833,400**	**6,072,437**	**0.96**	10^{-13}
Nguyen-6	$1+\lambda$	100	16	10	–	–	–	16,757,903	18,877,924	2,980,764	10,508,376	23,852,124	0.95	
	Canonical	100	–	8	70	50	4	**1,634,090**	**4,399,397**	**21,962**	**132,575**	**888,900**	1.0	10^{-15}
Nguyen-7	$1+\lambda$	200	16	7	–	–	–	64,033,983	35,411,800	30,458,912	67,583,400	100,000,000	0.67	
	Canonical	200	–	7	50	50	7	**23,424,276**	**32,155,768**	**2,622,975**	**7,966,750**	**26,935,237**	**0.93**	10^{-13}
Nguyen-8	$1+\lambda$	150	16	15	–	–	–	1,554,341	1,745,877	93,096	911,720	2,386,644	1.0	
	Canonical	150	–	15	50	50	7	**764,404**	**890,860**	**149,262**	**415,800**	**1,028,275**	1.0	0.02
Nguyen-9	$1+\lambda$	150	16	15	–	–	–	1,141,109	1,681,517	32,288	560,416	1,572,280	1.0	
	Canonical	150	–	15	50	50	4	**291,008**	**613,343**	**14,350**	**50,975**	**255,450**	1.0	10^{-5}
Nguyen-10	$1+\lambda$	60	128	20	–	–	–	905,799	1,659,653	26,144	130,176	1,201,152	1.0	
	Canonical	60	–	15	70	50	4	**139,754**	**178,352**	**22,837**	**76,375**	**185,050**	1.0	0.002
Nguyen-11	$1+\lambda$	50	64	10	–	–	–	155,608	165,428	14,944	111,488	224,784	1.0	
	Canonical	50	–	10	70	50	4	**56,685**	**65,100**	**111,62**	**37,175**	**75,225**	1.0	10^{-4}

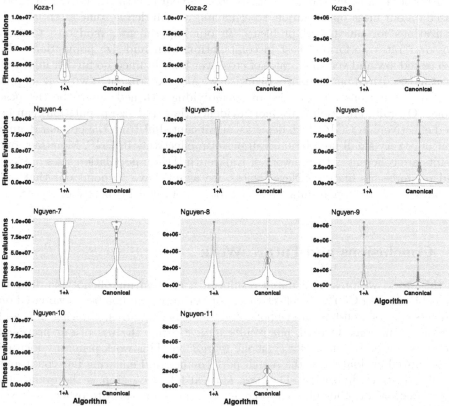

Fig. 4. Violin plots for all tested problems and algorithms of our experiments.

6   Discussion

The experiments presented in this work allow certain points that are worthy of discussion. Even if the initial results of our proposed method are promising we have to emphasize that more experiments are needed to achieve insight into how our method performs in other problem domains. Since former work [10] on recombination in CGP presented promising results with symbolic regression problems, we initially tested our proposed method in this problem domain. However, we have to evaluate our method in problem domains where the search space differs from the continuous search spaces of our tested symbolic regression problems. Recent work [9,11] led to more insight into the antagonism between continuous and discrete search spaces and its implications for the success of crossover-based algorithms in CGP. For our experiments, we also did not include the comparison to other crossover operators that have been proposed for CGP. For our initial evaluation and generally as a first step we concentrated on comparisons to the most commonly used algorithm in CGP and ensuring fair conditions with meta-optimization. But since several crossover operators have been proposed in recent years, more comparative studies are needed in the field of CGP and should be addressed by future work.

Another point that should be discussed is the parametrization of our method. Based on our meta-optimization experiments, we can derive some essential generalizations for our tested problems. In our experiments, moderate to high crossover rates performed best in combination with mid-size populations. We also tested low and very high rates of crossover but obtained no further improvement in the search performance. Likewise, we also experimented with bigger and smaller populations but the size of 50 individuals turned out to be the best choice. Overall, our results give more evidence that mid-size populations can be used effectively in CGP which depicts a significant shift from the popular dogma that only very small populations can perform effectively in CGP. Moreover, our results are coherent with the work of Kalkreuth [11] on population sizes in CGP and reinforce his findings. Nevertheless, we again, have to point out that our findings are based on results that have been obtained in merely one problem category.

7   Conclusions and Future Work

In this work, we presented initial results of a method for phenotypic discrete recombination in CGP. The effectiveness of our approach has been evaluated on a diverse set of well-known symbolic regression benchmarks, covering uni- and bivariate functions. Overall, our results indicate that the use of our proposed methods can be beneficial for symbolic regression. This work primarily focused on an initial evaluation of the search performance and ensuring fair conditions through meta-optimization. The next natural following step is the evaluation of our method in other problem domains and in comparison to other crossover operators. Therefore, our future work will primarily focus on comparative studies. Another part of our future work will be devoted to analytical experiments to study the effects caused by the phenotypic discrete crossover.

References

1. Bäck, T., Hoffmeister, F., Schwefel, H.: A survey of evolution strategies. In: Belew, R.K., Booker, L.B. (eds.) Proceedings of the 4th International Conference on Genetic Algorithms, San Diego, CA, USA, July 1991, pp. 2–9. Morgan Kaufmann (1991)
2. Beyer, H., Schwefel, H.: Evolution strategies - a comprehensive introduction. Nat. Comput. 1(1), 3–52 (2002). https://doi.org/10.1023/A:1015059928466
3. Clegg, J., Walker, J.A., Miller, J.F.: A new crossover technique for cartesian genetic programming. In: Thierens, D., et al. (eds.) Proceedings of the 9th Annual Conference on Genetic and Evolutionary Computation, GECCO 2007, London, 7–11 July 2007, vol. 2, pp. 1580–1587. ACM Press (2017). https://doi.org/10.1145/1276958. 1277276. http://www.cs.bham.ac.uk/~wbl/biblio/gecco2007/docs/p1580.pdf
4. De Jong, K., Spears, W.: On the virtues of parameterized uniform crossover. In: Proceedings of the 4th International Conference on Genetic Algorithms, pp. 230–236. Morgan Kaufmann Publishers, San Mateo (1991)
5. Hrbacek, R., Dvorak, V.: Bent function synthesis by means of cartesian genetic programming. In: Bartz-Beielstein, T., Branke, J., Filipič, B., Smith, J. (eds.) PPSN 2014. LNCS, vol. 8672, pp. 414–423. Springer, Cham (2014). https://doi. org/10.1007/978-3-319-10762-2_41
6. Husa, J., Kalkreuth, R.: A comparative study on crossover in cartesian genetic programming. In: Castelli, M., Sekanina, L., Zhang, M., Cagnoni, S., García-Sánchez, P. (eds.) EuroGP 2018. LNCS, vol. 10781, pp. 203–219. Springer, Cham (2018). https://doi.org/10.1007/978-3-319-77553-1_13
7. Husa, J., Sekanina, L.: Evolving cryptographic boolean functions with minimal multiplicative complexity. In: IEEE Congress on Evolutionary Computation, CEC 2020, Glasgow, United Kingdom, 19–24 July 2020, pp. 1–8. IEEE (2020). https:// doi.org/10.1109/CEC48606.2020.9185517.
8. Kalganova, T.: Evolutionary approach to design multiple-valued combinational circuits. In: Proceedings of the 4th International Conference on Applications of Computer Systems, ACS 1997, Szczecin, Poland, pp. 333–339 (1997)
9. Kalkreuth, R.: A comprehensive study on subgraph crossover in cartesian genetic programming. In: Guervós, J.J.M., Garibaldi, J.M., Wagner, C., Bäck, T., Madani, K., Warwick, K. (eds.) Proceedings of the 12th International Joint Conference on Computational Intelligence, IJCCI 2020, Budapest, Hungary, 2–4 November 2020, pp. 59–70. SCITEPRESS (2020). https://doi.org/10.5220/0010110700590070.
10. Kalkreuth, R., Rudolph, G., Droschinsky, A.: A new subgraph crossover for cartesian genetic programming. In: McDermott, J., Castelli, M., Sekanina, L., Haasdijk, E., García-Sánchez, P. (eds.) EuroGP 2017. LNCS, vol. 10196, pp. 294–310. Springer, Cham (2017). https://doi.org/10.1007/978-3-319-55696-3_19
11. Kalkreuth, R.T.: Reconsideration and Extension of Cartesian Genetic Programming. Ph.D. thesis (2021). https://doi.org/10.17877/DE290R-22504. http://dx. doi.org/10.17877/DE290R-22504

12. Kaufmann, P., Kalkreuth, R.: An empirical study on the parametrization of cartesian genetic programming. In: Proceedings of the Genetic and Evolutionary Computation Conference Companion, GECCO 2017, pp. 231–232. ACM, New York (2017). https://doi.org/10.1145/3067695.3075980. http://doi.acm.org/10.1145/3067695.3075980

13. McDermott, J., et al.: Genetic programming needs better benchmarks. In: Proceedings of the 14th International Conference on Genetic and Evolutionary Computation Conference, GECCO 2012, Philadelphia, Pennsylvania, USA, 7–11 July 2012, pp. 791–798. ACM (2012). https://doi.org/10.1145/2330163.2330273

14. Miller, J.F., Thomson, P., Fogarty, T.: Designing electronic circuits using evolutionary algorithms. arithmetic circuits: a case study. In: Genetic Algorithms and Evolution Strategies in Engineering and Computer Science, pp. 105–131. Wiley (1997)

15. Miller, J.F.: An empirical study of the efficiency of learning boolean functions using a cartesian genetic programming approach. In: Banzhaf, W., et al. (eds.) Proceedings of the Genetic and Evolutionary Computation Conference, Orlando, Florida, USA, 13–17 July 1999, vol. 2, pp. 1135–1142. Morgan Kaufmann (1999). http://citeseer.ist.psu.edu/153431.html

16. Miller, J.F., Wilson, D.G., Cussat-Blanc, S.: Evolving programs to build artificial neural networks. In: Adamatzky, A., Kendon, V. (eds.) From Astrophysics to Unconventional Computation. ECC, vol. 35, pp. 23–71. Springer, Cham (2020). https://doi.org/10.1007/978-3-030-15792-0_2

17. Miller, J.F.: Cartesian genetic programming: its status and future. Genet. Program. Evolvable Mach. 21(1), 129–168 (2020). https://doi.org/10.1007/s10710-019-09360-6

18. Poli, R., Langdon, W.B.: On the ability to search the space of programs of standard, one-point and uniform crossover in genetic programming. Technical report CSRP-98-7, University of Birmingham, School of Computer Science (January 1998). ftp://ftp.cs.bham.ac.uk/pub/tech-reports/1998/CSRP-98-07.ps.gz. Presented at GP-98

19. Poli, R., Langdon, W.B.: On the search properties of different crossover operators in genetic programming. In: Koza, J.R., et al. (eds.) Genetic Programming 1998: Proceedings of the 3rd Annual Conference, University of Wisconsin, Madison, Wisconsin, USA, 22–25 July 1998, pp. 293–301. Morgan Kaufmann (1998). http://www.cs.essex.ac.uk/staff/poli/papers/Poli-GP1998.pdf

20. Rechenberg, I.: Evolutionsstrategie: Optimierung technischer Systeme nach Prinzipien der biologischen Evolution. Dr.-Ing. Ph.D. thesis, Thesis, Technical University of Berlin, Department of Process Engineering (1971)

21. Rechenberg, I.: Evolutionsstrategie Optimierung technischer Systeme nach Prinzipien der biologishen Evolution. Frommann Holzboog Verlag, Stuttgart (1973)

22. Rudolph, G.: Global optimization by means of distributed evolution strategies. In: Schwefel, H.-P., Männer, R. (eds.) PPSN 1990. LNCS, vol. 496, pp. 209–213. Springer, Heidelberg (1991). https://doi.org/10.1007/BFb0029754

23. Schwefel, H.P.: Evolutionsstrategien für die numerische Optimierung, pp. 123–176. Birkhäuser Basel, Basel (1977). https://doi.org/10.1007/978-3-0348-5927-1_5

24. Schwefel, H.P.: Numerical Optimization of Computer Models. Wiley, USA (1981)

25. Scott, E.O., Luke, S.: ECJ at 20: toward a general metaheuristics toolkit. In: López-Ibáñez, M., Auger, A., Stützle, T. (eds.) Proceedings of the Genetic and Evolutionary Computation Conference Companion, GECCO 2019, Prague, Czech Republic, 13–17 July 2019, pp. 1391–1398. ACM (2019). https://doi.org/10.1145/3319619.3326865

26. Sekanina, L., Walker, J.A., Kaufmann, P., Platzner, M.: Evolution of electronic circuits. In: Miller, J.F. (ed.) Cartesian Genetic Programming. Natural Computing Series, pp. 125–179. Springer, Heidelberg (2011). https://doi.org/10.1007/978-3-642-17310-3_5

27. da Silva, J.E.H., Bernardino, H.: Cartesian genetic programming with crossover for designing combinational logic circuits. In: 7th Brazilian Conference on Intelligent Systems, BRACIS 2018, São Paulo, Brazil, 22–25 October 2018, pp. 145–150. IEEE Computer Society (2018). https://doi.org/10.1109/BRACIS.2018.00033

28. Suganuma, M., Kobayashi, M., Shirakawa, S., Nagao, T.: Evolution of deep convolutional neural networks using cartesian genetic programming. Evol. Comput. **28**(1), 141–163 (2020). https://doi.org/10.1162/evco_a_00253

29. Syswerda, G.: Uniform crossover in genetic algorithms. In: Schaffer, J.D. (ed.) Proceedings of the 3rd International Conference on Genetic Algorithms, George Mason University, Fairfax, Virginia, USA, June 1989, pp. 2–9. Morgan Kaufmann (1989)

30. Turner, A.J.: Improving crossover techniques in a genetic program. Master's thesis, Department of Electronics, University of York (2012)

Multi-Objective Optimization

A General Architecture for Generating Interactive Decomposition-Based MOEAs

Giomara Lárraga$^{(\boxtimes)}$ and Kaisa Miettinen

Faculty of Information Technology, University of Jyvaskyla, 40014 Jyvaskyla, Finland
{giomara.g.larraga-maldonado,kaisa.miettinen}@jyu.fi

Abstract. Evolutionary algorithms have been widely applied for solving multiobjective optimization problems. Such methods can approximate many Pareto optimal solutions in a population. However, when solving real-world problems, a decision maker is usually involved, who may only be interested in a subset of solutions that meet their preferences. Several methods have been proposed to consider preference information during the solution process. Among them, interactive methods support the decision maker in learning about the trade-offs among objectives and the feasibility of solutions. Also, such methods allow the decision maker to provide preference information iteratively. Typically, interactive multiobjective evolutionary algorithms are modifications of existing *a priori* or *a posteriori* algorithms. However, they mainly focus on finding a region of interest and do not support the decision maker finding the most preferred solution. In addition, the cognitive load imposed on the decision maker is usually not considered. This article proposes an architecture for developing interactive decomposition-based evolutionary algorithms that can support the decision maker during the solution process. The proposed architecture aims to improve the applicability of interactive methods in solving real-world problems by considering the needs of a decision maker. We apply our proposal to generate an interactive decomposition-based algorithm utilizing a reference vector re-arrangement procedure and MOEA/D. We demonstrate the performance of our proposal with a real-world problem and multiple benchmark problems.

Keywords: Multiobjective optimization · Evolutionary algorithms · Preference information · Decision making · Interactive methods · Interactive preference incorporation

1 Introduction

Multiobjective optimization problems involve multiple conflicting objective functions that must be optimized simultaneously. Because of the conflict among the objective functions, these problems do not have a single optimal solution, but a set of trade-off solutions named a Pareto optimal set. The goal of solving a multiobjective optimization problem is to help a decision maker (DM) find the most preferred trade-offs among objectives. A DM is a person with expertise

G. Rudolph et al. (Eds.): PPSN 2022, LNCS 13399, pp. 81–95, 2022.
https://doi.org/10.1007/978-3-031-14721-0_6

about the problem and is usually interested in a subset of solutions that meets their preferences, known as a region of interest.

Methods for solving multiobjective optimization problems can be classified according to the role of the DM in the solution process into no preference, *a priori*, interactive, and *a posteriori* methods [22]. No preference methods are utilized when no DM is available and the problem is solved without considering any preference information. *A priori* methods ask for preference information once at the beginning of the solution process. On the other hand, *a posteriori* methods generate multiple solutions representing Pareto optimal ones and consider the preference information afterward. In interactive methods, the DM can provide preference information iteratively, allowing them to direct the solution process progressively. When studying interactive solution processes of DMs, one can often observe two phases: learning and decision phases, as stated in [21]. The DM explores different solutions during the learning phase until they find a region of interest. Then, in the decision phase, the DM fine-tunes the search to find the most preferred solution in that region.

Several scalarization-based methods [22] and evolutionary algorithms [7] have been proposed to solve multiobjective optimization problems. Multiobjective evolutionary algorithms (MOEAs) are population-based metaheuristics capable of representing the Pareto optimal set with approximated solutions. MOEAs can be divided into three main classes [29]: dominance-based, indicator-based, and decomposition-based algorithms. Decomposition-based MOEAs [14] have recently gained researchers' attention because of their scalability in terms of the number of objectives. These MOEAs decompose the original multiobjective optimization problem into multiple single-objective optimization problems or simpler multiobjective optimization problems to be solved collaboratively with the use of a scalarizing function and a set of so-called reference vectors. Decomposition-based MOEAs are suitable for preference incorporation as they can easily focus on certain parts of the Pareto optimal set by modifying the decomposition. MOEAs have been typically utilized as *a posteriori* methods. Although some interactive decomposition-based MOEAs are available in the literature (e.g. [3,11,12]), most of them focus only on the learning phase and identifying the region of interest. In other words, they do not consider a decision phase to help the DM find the most preferred solution.

In this article, we propose a general architecture for developing interactive decomposition-based MOEAs that address the needs of a decision maker. Our proposal consists of multiple modules that can be utilized to convert *a priori* and *a posteriori* methods into interactive ones. Each module contains different procedures that some interactive MOEAs have employed in the literature. In addition, new procedures can be incorporated into each one of the modules. The rest of the article is structured as follows. Section 2 presents background information on the main concepts used in the article. Then, a brief review of the existing interactive decomposition-based MOEAs is presented in Sect. 3. Section 4 describes some desirable properties of an interactive solution process. Then, we present the proposed architecture to meet the desirable properties of interactive decomposition-based MOEAs in Sect. 5. As a proof of concept, we present some results and an algorithmic comparison in Sect. 6. We conclude the article in Sect. 7.

2 Background

A multiobjective optimization problem minimizing k (with $k \geq 2$) conflicting objective functions f_i ($i = 1, \ldots, k$) can be mathematically formulated as follows:

$$
\begin{aligned}
\text{minimize} \quad & \mathbf{F}(\mathbf{x}) = (f_1(\mathbf{x}), \ldots, f_k(\mathbf{x})) \\
\text{subject to} \quad & \mathbf{x} \in S,
\end{aligned}
\tag{1}
$$

where $S \subset \mathbb{R}^n$ is the feasible set of decision vectors $\mathbf{x} = (x_1, ..., x_n)^T$ with n decision variables. There is a corresponding objective vector $\mathbf{F}(\mathbf{x})$ for every feasible decision vector $\mathbf{x}$. The problem can involve equality and inequality constraints that must be satisfied by the feasible decision vectors. Because of the conflict among the objective functions in (1), not all of them can achieve their optimal values simultaneously. A solution $\mathbf{x}^1 \in S$ dominates a solution $\mathbf{x}^2 \in S$ if and only if $f_i(\mathbf{x}^1) \leq f_i(\mathbf{x}^2)$ for all $i = 1, \ldots k$, and $f_j(\mathbf{x}^1) < f_j(\mathbf{x}^2)$ for at least one index $j = 1, \ldots, k$. Then, a solution $\mathbf{x}^* \in S$ is Pareto optimal if and only if there is no solution $\mathbf{x} \in S$ that dominates it. A Pareto optimal set is then formed by all Pareto optimal solutions, and the corresponding objective vectors compose a Pareto front.

An ideal $\mathbf{z}^*$ and a nadir $\mathbf{z}^{nad}$ point represent the best and worst objective function values in the Pareto front, respectively. The ideal point can be calculated by minimizing each objective function separately. Calculating the nadir point is usually difficult since it requires computing the entire Pareto optimal set. However, it can be approximated using a pay-off table [22] or other means [9].

Decomposition-based MOEAs [14] utilize a set of reference vectors (which are also known as reference points or weight vectors)[1] to decompose the original multiobjective optimization problem into a set of single-objective optimization problems or simpler multiobjective optimization problems to be solved collaboratively. Usually, in the initialization of decomposition-based MOEAs, a set of reference vectors uniformly distributed in the objective space is generated utilizing e.g. a simplex lattice design [6]. This method requires a parameter p to control the density of the reference vectors. Then, the total amount of reference vectors is given by $\binom{p+k-1}{k-1}$. A scalarizing function is utilized to evaluate the solutions belonging to a part of the objective space. The solutions then evolve in the direction of the reference vector associated with such a part. Scalarizing functions map an objective vector to a real-valued scalar. Examples of decomposition-based MOEAs are MOEA/D [30], RVEA [5], and NSGA-III [8] which utilize dominance in combination with decomposition.

---

[1] For simplicity, we will utilize the term reference vectors throughout this article.

In interactive methods, the DM provides preference information iteratively. Iterations are intervals during which MOEAs ask for preference information from the DM. They typically occur every GEN generations, where GEN is a parameter set before the method start. It is worth noting that a DM can provide preference information in multiple ways [4,20]. Reference points are a common way of representing preference information in MOEAs [4]. A reference point $\mathbf{z}^{ref}$ is a k-dimensional vector consisting of a desirable value for each objective function.

3   Related Works

Some interactive decomposition-based MOEAs have been proposed in the literature. As stated in [4], most of the preference-based MOEAs are modifications of an existing a posteriori MOEA. We can classify interactive decomposition-based MOEAs according to how they accomplish interactivity. Although different types of preference information have been utilized in these methods, we consider here only the ones employing reference points.

The simplest way of imitating interactivity in MOEAs is by performing a series of a priori steps. However, the applicability of such methods in real-world problems is often not considered in the papers where they have been proposed, as some of their properties would significantly increase the DM's workload. For example, they usually do not let the DM decide when to interact with the method. In addition, they typically display an extensive set of solutions to be compared at each iteration. The interactive version of R-MOEA/D [25] is an example of an algorithm utilizing this structure.

Some methods modify the decomposition without altering the structure of the decomposition-based MOEA. Each iteration uses the preferences to update the decomposition and guide the search toward the region of interest. The most common modification to the decomposition involves rearranging the reference vectors according to the preference information [3,12,15,19]. Some other methods utilize the preference information to modify the approximation of the ideal point required by the decomposition-based MOEA [23,24]. The IOPIS framework [27] is another example in this category, as it creates a new (typically lower-dimensional) preference incorporated space (consisting of a set of scalarization functions) to reformulate the problem. It is worth noting that IOPIS can also be applied to other types of MOEAs (e.g., dominance-based and indicator-based); however, it has only been tested with decomposition-based methods. Although the structure of the methods in this category is similar to the methods in the previous category, these methods typically include mechanisms to ensure their applicability to real-world problems (e.g., considering a limited number of solutions to be shown to the DM at each iteration, controlling the frequency of iterations and the size of the region of interest).

Finally, some methods add additional steps for each generation of the decomposition-based algorithm for managing the preference information. Such steps are commonly intended to update the reference vectors inside the evolutionary process (and not before running the method as in the previous category).

MOEA/D-a [31], MOEA/D-b [31], and MOEA/D-c [31] are examples of methods in this category. Interactive WASF-GA [26] is another example, as it replaces the dominance relation of NSGA-II by utilizing an achievement scalarizing function, which directs the search toward the region of interest.

4 Properties of an Interactive Solution Process

In an interactive solution process, a DM iterates by providing preference information to the method and studying the received solutions until the most preferred one is found. A DM learns about the trade-offs among the objective functions as well as the feasibility of the preferences after each iteration. As a result, the DM may change the preference information during the solution process. To ensure the practical usability of the method, it should limit the level of cognitive burden and provide solutions that help the DM gain insight into the problem. Thus, we can summarize the main desirable properties of an interactive method as follows [2, 28]:

1. The method provides accurate information about possible solutions.
2. The DM and the method can communicate quite easily.
3. The method identifies and produces Pareto optimal solutions.
4. The method provides the DM with a clear overview of the Pareto optimal set/Pareto front.
5. The method enables the DM to find a region of interest in a learning phase.
6. The method has a decision phase to enable the DM to fine-tune the solutions in the region of interest.
7. The method gives the DM confidence that the final solution is the most preferred one, or at least close enough to it.

These properties are directly applicable to scalarization-based methods. Although MOEAs have to meet these properties, they have somewhat different needs and characteristics. Instead of producing Pareto optimal solutions, MOEAs can provide a set of non-dominated solutions, as they are metaheuristics and cannot guarantee optimality. In addition, most interactive MOEAs focus only on the learning phase, representing a region of interest without helping the DM select the most preferred solution. In the next section, we present an architecture for developing interactive decomposition-based MOEAs that meet the above-mentioned properties.

5 Proposed Architecture

We propose an architecture consisting of multiple modules that can be utilized to generate interactive decomposition-based MOEAs that meet the properties discussed in the previous section. Also, *a posteriori* or *a priori* decomposition-based algorithms can be converted to interactive ones with the help of the architecture.

The architecture has two types of modules: static and dynamic. Static modules consist of multiple steps that must be considered during the solution process. On the other hand, dynamic modules allow us to personalize the method according to our needs. Such modules present multiple alternatives from which we can select one or multiple. This architecture aims to provide a guideline for developing new interactive decomposition-based MOEAs that consider the structure and properties of an interactive solution process. The alternatives presented in the dynamic modules have been selected after analyzing the structure of multiple interactive MOEAs. This means that the interested user can incorporate new options that accomplish the main aim of each module. The architecture is illustrated in Fig. 1. The static modules have a red marker in the upper right corner of the corresponding box, while the dynamic modules have a green marker. The architecture has seven modules: initialization, preference elicitation, component adaptation, optimization, spread adjustment, selection of solutions, and iteration. Below, we give details of each module.

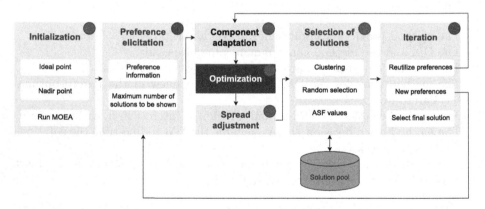

Fig. 1. Proposed architecture for developing interactive decomposition-based MOEAs.

Initialization Module: This module provides a DM information for learning about feasible solutions to the problem before starting the solution process (property 1). The alternatives in this module are: computing the ideal point, estimating the nadir point, and running an MOEA for a fixed number of generations. Usually, showing the ideal and nadir points to the DM may help them provide more realistic preference information within the lower and upper bounds of the objective functions. If we want to show some feasible solutions to the DM before starting the solution process, running an *a posteriori* MOEA would be a good alternative. However, only a representative set of solutions should be displayed to the DM. E.g., a clustering method can be utilized to limit the number of solutions to display.

Preference Elicitation Module: This module retrieves the DM's preference information and the maximum number of solutions (Ns) they want to see at each iteration (property 2). Here, we utilize reference points to represent the DM's preferences. However, we can extend the architecture to support more types of preference information. If an *a priori* method is employed in the optimization module, then the type of preference information is the same utilized by such a method. However, if the optimization module uses an *a posteriori* method, the preference information is selected according to the mechanism employed in the component adaptation module. The preference information provided during all the iterations is stored for further use.

Component Adaptation Module: This module is needed only when an *a posteriori* MOEA is utilized in the optimization module. It aims to modify some elements used by the decomposition-based MOEA to consider preference information. For example, modifying the distribution of the reference vectors, changing the problem formulation, or using the preferences information to replace the approximation of the ideal point required by some MOEAs (e.g., MOEA/D).

Optimization Module: In this module, an MOEA is utilized to solve a multiobjective optimization problem considering the preference information provided by the DM (property 3). There are two alternatives for this module: utilizing an *a priori* MOEA or an *a posteriori* one. In both cases, the methods do not need to be modified.

Spread Adjustment Module: This module controls the size of the region of interest depending on the phase of the interactive solution process (properties 5 and 6). A higher spread value is utilized during the learning phase, as the aim is to learn and eventually find the region of interest. The value will be reduced during the decision phase to help the DM find the most preferred solution in the region of interest. The preference information stored in the preference elicitation module is used to identify the decision phase. If the preference information starts to be similar among multiple iterations, the decision phase has begun. Some *a priori* methods utilize a spread parameter. When using some of those methods in the optimization module, no additional procedures are needed to control the size of the region of interest, and the parameter is updated iteratively. On the other hand, if the optimization module employs an *a posteriori* method or *a priori* method that does not consider a spread parameter, a mechanism to select a subset of solutions from the region of interest is needed (e.g., the trimming procedure of the R-metric [13]).

Solution Selection Module: This module filters the solutions on the region of interest to show only a reduced set of Ns representative solutions to the DM (property 4). The solutions can be selected in multiple ways: randomly, dividing the solution set into Ns clusters and selecting the solutions closest to each centroid, or selecting the Ns solutions with the best values of a scalarizing function. In addition, this module stores the best solutions in an archive to avoid losing them. A scalarizing function can be utilized to determine which solutions to keep. These solutions can be used when the DM provides a reference point close to another one from a previous iteration.

Iteration Module: In this module, the DM can decide whether or not to provide new preference information. When no new preference information is available, the same values as in the previous iteration are utilized. The DM can also select the final solution among the ones displayed, in which case the solution process is finished (property 7).

We create an interactive decomposition-based MOEA in the following section to demonstrate how the proposed architecture can be used. In addition, we compare it with a method consisting of multiple *a priori* steps.

6   Example Method and Experiments

In this section, as a proof of concept, we demonstrate how a method created with the proposed architecture can support a DM during an interactive solution process. The method considered has the following configuration of modules:

Initialization: MOEA/D is run 200 generations. Then, the ideal and nadir points are estimated from the resulting population.

Preference Elicitation: Reference points are utilized to represent the preferences of the DM. All the reference points provided during the solution process will be stored. A maximum of five solutions ($Ns = 5$) will be shown to the DM during each iteration.

Component Adaptation: The reference vectors are rearranged utilizing the NUMS procedure [15]. It is worth noting that the adaptation is performed based on the initial values of the reference vectors (which are generated utilizing a simplex lattice design [6]).

Optimization: MOEA/D is utilized with 200 generations per each iteration.

Spread Adjustment: We adapt the spread parameter of the NUMS procedure based on the differences among the reference points. Initially, the spread parameter is set as 0.5. If the DM utilizes a reference point close to another one provided in a previous iteration, the spread parameter is divided by two.

Selection of Solutions: The five most representative solutions are shown to the DM. These are obtained by clustering the solution set utilizing the k-means method [18]. These solutions will be stored in the solution pool if other solutions do not dominate them. In addition, the dominated solutions are removed from the pool every time a new solution is incorporated.

Iteration: When a new reference point is provided, the solutions in the pool are included in the initial population of MOEA/D. If the DM does not provide new preferences, the optimization module is run again without modifying the population or the reference point. The process only stops if the DM has found a preferred solution.

6.1   Interactive Solution Process

In what follows, we refer to the method created with our architecture as MOEA/D-NUMS+, while the one consisting of multiple *a priori* steps is called MOEA/D-NUMS. We conduct an experiment with the crash-worthiness design of vehicles problem [17]. It is a real-world engineering problem whose goal is to make vehicles crash-safe. During a collision, the frontal structure of the vehicle absorbs energy, which increases passenger safety. Increasing the mass of the vehicle generally improves its energy absorption capacity. By contrast, lightweight materials are necessary for a vehicle's mass to be reduced and, therefore, its fuel consumption. To achieve a proper design, we must find a compromise between higher energy absorption and lightweight construction. In this problem, five decision variables are used to represent the thickness of five reinforced components surrounding the frontal structure of the vehicle. Specifically, three objective functions need to be minimized: 1) the mass of a vehicle; 2) deceleration during full-frontal crashes (which influence passenger injuries); and 3) toe board intrusion during offset-frontal crashes (which affect the structural integrity of the vehicle). It is worth noting that we will take the role of the DM in this experiment, as its main aim is to exemplify how the method works. Experiments with real DMs will be considered as future work.

Now, we can describe the iterations of the solution process. At the beginning, the ideal and nadir points were shown to the DM, whose values are $\mathbf{z}^* = (1661.708, 6.986, 0.0708)$ and $\mathbf{z}^{nad} = (1666.211, 8.304, 0.106)$.

Iteration 1. First, the DM set the ideal point as the reference point to see how difficult it is to achieve these promising values. The five solutions (obtained after applying the clustering method) displayed to the DM are shown in Table 1. It is worth noting that the solutions are sorted in increasing order of f_1. However, the DM should be able to decide how to see the solutions displayed by the method (e.g., in an increasing or decreasing order of some of the objectives).

Table 1. Results of the first iteration

	Mass (kg)	Deceleration (m/s)	Intrusion (m)
1	1662.032	8.209	0.078
2	1662.420	8.095	0.086
3	1663.010	7.923	0.096
4	1663.839	7.680	0.104
5	1665.229	7.273	0.104

Table 2. Results of the second iteration

	Mass (kg)	Deceleration (m/s)	Intrusion (m)
1	1662.007	8.216	0.078
2	1662.077	8.196	0.079
3	1662.152	8.174	0.081
4	1662.233	8.150	0.082
5	1662.342	8.118	0.085

Iteration 2. Since the reference point had been too optimistic, the DM adjusted its components to be more realistic but focused on improving the value of the first objective. The new reference point was $(1661.9, 8.0, 1.0)$, and the obtained five solutions are shown in Table 2.

Iteration 3. Based on the solutions shown, the DM realized the trade-off between f_1 and f_2 and provided a new reference point $(1665, 7.4, 0.1)$ with the aim of obtaining better values for f_2. The results obtained are shown in Table 3. There was a good improvement on f_2, but f_1 and f_3 impaired.

Table 3. Results of the third iteration

	Mass (kg)	Deceleration (m/s)	Intrusion (m)
1	1663.181	7.873	0.098
2	1663.584	7.755	0.102
3	1664.070	7.612	0.105
4	1664.581	7.463	0.107
5	1664.999	7.340	0.106

Table 4. Results of the fourth iteration

	Mass (kg)	Deceleration (m/s)	Intrusion (m)
1	1662.016	8.214	0.078
2	1662.060	8.201	0.079
3	1662.106	8.187	0.080
4	1662.163	8.171	0.081
5	1662.221	8.154	0.082

Iteration 4. After noticing the impairment in f_1 and f_3, the DM decided to get similar results to the ones of Iteration 2. For this, the DM did not need to provide a new reference point, but it was taken from the list of reference points utilized during the solution process. In addition, the spread of the region of interest was reduced automatically. The results obtained are shown in Table 4.

Iteration 5. The DM was satisfied with the improvement on f_1 and f_3 and wanted to find more solutions in the same region. Thus, the DM kept the same reference point as the previous iteration, and the algorithm automatically reduced the spread of the region of interest. The results are shown in Table 5.

Iteration 6. The DM noticed that the solutions were refined and was satisfied with the fifth one. The DM selected it as the final solution, as its value on f_2 was better than the rest without losing too much on f_1 and f_3.

Table 5. Results of the fifth iteration

	Mass (kg)	Deceleration (m/s)	Intrusion (m)
1	1662.034	8.208	0.078
2	1662.054	8.203	0.079
3	1662.074	8.197	0.079
4	1662.097	8.190	0.080
5	**1662.123**	**8.182**	**0.080**

Now we summarize the advantages of our method compared to another one consisting of multiple *a priori* steps. Providing the ideal and nadir points before starting the solution process can help the DM give a reference point when the DM does not have a clear idea of the feasibility of the solutions, as was shown in iteration 1. The solutions provided in each iteration can be easily compared to identify trade-offs between objectives, as there is only a reduced number of representative solutions. We took advantage of this property on iterations 3 and 4. This reduces the cognitive burden of the DM compared with a method consisting of multiple *a priori* steps. As we stored the preferences provided during the solution process, the DM can easily re-utilize a previous reference point, as was shown in iteration 4. Then, the spread adjustment allowed us to refine the solutions when reaching the decision phase in iterations 5 and 6. If we had used a method consisting of *a priori* steps, the support during the solution process would not be enough for allowing the DM to learn about the problem trade-offs and the feasibility of solutions.

6.2 Algorithmic Comparison

In the previous section, we showed how MOEA/D-NUMS+ could support a DM during an interactive solution process. In this section, we compare the performance of MOEA/D-NUMS+ and MOEA/D-NUMS utilizing the artificial decision maker (ADM) proposed in [1] (to replace the DM), and the R-IGD performance indicator [16]. The ADM adjusts the preference information according to the insight gained during each iteration, producing reference points differently depending on the phase of the solution process. The generated reference points simulate the exploration in the Pareto optimal set during the learning phase. On the other hand, the reference points mimic a progressive convergence on the region of interest obtained from the learning phase during the decision phase. We considered 4 iterations for the learning phase ($L = 4$) and 3 for the decision phase ($D = 3$) in this experiment. For each iteration, ADM computes the R-IGD for the results obtained by each method. After the run, the cumulative R-IGD for the learning phase is obtained by adding the R-IGD values of the first L iterations. The cumulative R-IGD for the decision phase is obtained by adding the R-IGD values of the last D iterations. The methods were tested utilizing the same parameters for both of them: 50 generations per iteration, a lattice resolution of 5, 0.7 as a spread parameter during the learning phase, and 0.3 during the decision phase. We did not employ the spread adjustment procedure in this experiment, as the ADM controls the value of the spread parameter according to the phase of the solution process. We considered two benchmark problems: DTLZ1 and DTLZ3 [10] with 4, 7, and 9 objectives, resulting in six different problems. The number of variables was set as $10 + k - 1$ [10]. We made ten independent runs for each problem. The median R-IGD values of MOEA/D-NUMS and MOEA/D-NUMS+ are shown in Table 6. The best results are highlighted in **boldface**.

For DTLZ1, MOEA/D-NUMS+ outperformed MOEA/D-NUMS in most of the cases, except during the decision phase when four objectives were considered. In the DTLZ3 problem, MOEA/D-NUMS had a better performance than the proposed method only in the learning phase with seven objectives and the decision phase with four objectives. This experiment showed us that the proposed architecture can help in improving the quality of the solutions obtained during the learning and decision phase. However, more extensive experimentation considering different types of problems is needed.

Table 6. R-IGD values for DTLZ1 and DTLZ3 with 7, 5, and 9 objectives.

Problem	k	Phase	MOEA/D-NUMS		MOEA/D-NUMS+	
			Median	Std. dev.	Median	Std. dev.
DTLZ1	4	Learning	2.804056	0.174749	**2.671504**	0.236763
		Decision	**2.722246**	0.283701	2.74031	0.269508
	7	Learning	4.663774	0.126492	**3.523776**	0.16947
		Decision	3.317874	0.210076	**3.046718**	0.515918
	9	Learning	4.332031	0.195637	**3.011431**	0.268385
		Decision	3.36567	0.21437	**3.113235**	0.658471
DTLZ3	4	Learning	0.451812	0.039745	**0.377324**	0.031323
		Decision	**1.472565**	0.019187	1.497671	0.000519
	7	Learning	**0.803064**	0.922071	1.958682	1.522243
		Decision	0.383067	0.429625	**0.234077**	0.122247
	9	Learning	1.182198	0.448281	**0.754516**	0.908085
		Decision	0.81725	0.299627	**0.377474**	0.111261

6.3 Discussion

We utilized the proposed architecture to create an interactive version of MOEA/D. To this aim, we employed a procedure for re-arranging the reference vectors utilized in the optimization method. However, our proposal can also be applied to *a priori* methods. We do not need to use the component adaptation module in such a case, as the *a priori* method internally modifies the decomposition-based MOEA to handle the preference information. It is worth noting that although we only utilized reference points in this article, our proposal can be extended to other types of preference information. The type of preference information in the current architecture is related to the one required by the component adaptation and/or optimization modules. To handle more types of preferences, we would need an additional module for preference unification. Such a module should allow the DM to use any kind of preference information and then transform it into the one required by the rest of the modules. In addition, there are some methods that do not consider a spread parameter to control the size of the region of interest. In this case, we can utilize an external method like the one considered in the R-metric [16]. In this article, we did not consider the case where the same value of the spread parameter can be utilized in multiple iterations. Such behavior can be useful when the MOEA needs more generations to converge to the Pareto optimal set. However, the DM is usually not aware of the technical details of the method. Finally, the selection of the solutions to be shown to the DM can be performed in multiple ways, for example, through different clustering techniques or scalarizing functions.

7   Conclusions

Multiple interactive versions of decomposition-based MOEAs have been proposed in the literature, but they typically do not consider the DM's needs and cognitive load. We have introduced an architecture to create interactive decomposition-based MOEAs by integrating multiple modules with existing *a priori* or *a posteriori* methods. To demonstrate how our architecture can be employed, we created an interactive MOEA utilizing NUMS, a method for rearranging the reference vectors which has been mainly to convert an *a posteriori* method into an *a priori* one. We solved a real-world problem to demonstrate the advantages of using our proposal for improving the applicability of the methods and reducing the cognitive load of the DM. In addition, we compared the proposed method with another one (which does not include the properties of the architecture) consisting of multiple *a priori* steps. According to the results, utilizing the proposed architecture improves the performance in most test problems utilized. This is the first step toward improving the performance of interactive decomposition-based MOEAs. The proposed architecture can be improved in multiple directions, for example, by including a preference unification module to consider different types of preference information and also developing different methods for adapting the spread parameter. In addition, it can be extended to other types of MOEAs (e.g., dominance- and indicator-based MOEAs).

References

1. Afsar, B., Miettinen, K., Ruiz, A.B.: An artificial decision maker for comparing reference point based interactive evolutionary multiobjective optimization methods. In: Ishibuchi, H., et al. (eds.) EMO 2021. LNCS, vol. 12654, pp. 619–631. Springer, Cham (2021). https://doi.org/10.1007/978-3-030-72062-9_49
2. Afsar, B., Miettinen, K., Ruiz, F.: Assessing the performance of interactive multiobjective optimization methods: a survey. ACM Comput. Surv. **54**(4), 1–27 (2021). https://doi.org/10.1145/3448301
3. Aghaei Pour, P., Rodemann, T., Hakanen, J., Miettinen, K.: Surrogate assisted interactive multiobjective optimization in energy system design of buildings. Optim. Eng. **23**, 303–327 (2022). https://doi.org/10.1007/s11081-020-09587-8
4. Bechikh, S., Kessentini, M., Said, L.B., Ghédira, K.: Chapter four - preference incorporation in evolutionary multiobjective optimization: a survey of the state-of-the-art. Adv. Comput. **98**, 141–207 (2015). https://doi.org/10.1016/bs.adcom.2015.03.001. http://www.sciencedirect.com/science/article/pii/S0065245815000273
5. Cheng, R., Jin, Y., Olhofer, M., Sendhoff, B.: A reference vector guided evolutionary algorithm for many-objective optimization. IEEE Trans. Evol. Comput. **20**(5), 773–791 (2016). https://doi.org/10.1109/TEVC.2016.2519378
6. Cornell, J.A.: Experiments with Mixtures: Designs, Models, and the Analysis of Mixture Data. Wiley (2011)
7. Deb, K.: Multi-Objective Optimization using Evolutionary Algorithms. Wiley, Chichester (2001)

8. Deb, K., Jain, H.: An evolutionary many-objective optimization algorithm using reference-point-based nondominated sorting approach, part I: solving problems with box constraints. IEEE Trans. Evol. Comput. **18**(4), 577–601 (2014). https://doi.org/10.1109/TEVC.2013.2281535

9. Deb, K., Miettinen, K., Chaudhuri, S.: Towards an estimation of nadir objective vector using a hybrid of evolutionary and local search approaches. IEEE Trans. Evol. Comput. **14**(6), 821–841 (2010)

10. Deb, K., Thiele, L., Laumanns, M., Zitzler, E.: Scalable test problems for evolutionary multiobjective optimization. In: Abraham, A., Jain, L., Goldberg, R. (eds.) Evolutionary Multiobjective Optimization: Theoretical Advances and Applications, pp. 105–145. Springer, London (2005). https://doi.org/10.1007/1-84628-137-7_6

11. Gong, M., Liu, F., Zhang, W., Jiao, L., Zhang, Q.: Interactive MOEA/D for multi-objective decision making. In: Proceedings of the 13th Annual Conference on Genetic and Evolutionary computation, GECCO 2011. ACM, New York (2011)

12. Hakanen, J., Chugh, T., Sindhya, K., Jin, Y., Miettinen, K.: Connections of reference vectors and different types of preference information in interactive multiobjective evolutionary algorithms. In: 2016 IEEE Symposium Series on Computational Intelligence, SSCI 2016, pp. 1–8. Institute of Electrical and Electronics Engineers Inc. (2017). https://doi.org/10.1109/SSCI.2016.7850220

13. Li, K., Deb, K., Yao, X.: R-metric: evaluating the performance of preference-based evolutionary multiobjective optimization using reference points. IEEE Trans. Evol. Comput. **22**(6), 821–835 (2018)

14. Li, K.: Decomposition multi-objective evolutionary optimization: from state-of-the-art to future opportunities. CoRR abs/2108.09588 (2021). https://arxiv.org/abs/2108.09588

15. Li, K., Chen, R., Min, G., Yao, X.: Integration of preferences in decomposition multiobjective optimization. IEEE Trans. Cybern. **48**(12), 3359–3370 (2018). https://doi.org/10.1109/TCYB.2018.2859363

16. Li, K., Deb, K., Yao, X.: R-metric: evaluating the performance of preference-based evolutionary multiobjective optimization using reference points. IEEE Trans. Evol. Comput. **22**(6), 821–835 (2018). https://doi.org/10.1109/TEVC.2017.2737781

17. Liao, X., Li, Q., Yang, X., Zhang, W., Li, W.: Multiobjective optimization for crash safety design of vehicles using stepwise regression model. Struct. Multi. Optim. **35**(6), 561–569 (2008). https://doi.org/10.1007/s00158-007-0163-x

18. MacQueen, J., et al.: Some methods for classification and analysis of multivariate observations. In: Proceedings of the 5th Berkeley Symposium on Mathematical Statistics and Probability, Oakland, CA, USA, vol. 1, pp. 281–297 (1967)

19. Mazumdar, A., Chugh, T., Hakanen, J., Miettinen, K.: An interactive framework for offline data-driven multiobjective optimization. In: Filipič, B., Minisci, E., Vasile, M. (eds.) BIOMA 2020. LNCS, vol. 12438, pp. 97–109. Springer, Cham (2020). https://doi.org/10.1007/978-3-030-63710-1_8

20. Miettinen, K., Hakanen, J., Podkopaev, D.: Interactive nonlinear multiobjective optimization methods. In: Greco, S., Ehrgott, M., Figueira, J. (eds.) Multiple Criteria Decision Analysis. ISORMS, vol. 233, pp. 927–976. Springer, New York (2016). https://doi.org/10.1007/978-1-4939-3094-4_22

21. Miettinen, K., Ruiz, F., Wierzbicki, A.P.: Introduction to multiobjective optimization: interactive approaches. In: Branke, J., Deb, K., Miettinen, K., Słowiński, R. (eds.) Multiobjective Optimization. LNCS, vol. 5252, pp. 27–57. Springer, Heidelberg (2008). https://doi.org/10.1007/978-3-540-88908-3_2

22. Miettinen, K.: Nonlinear Multiobjective Optimization. Kluwer Academic Publishers, Boston (1999)
23. Nguyen, L., Bui, L.T.: A multi-point interactive method for multi-objective evolutionary algorithms. In: Proceedings of the 4th International Conference on Knowledge and Systems Engineering, KSE 2012, pp. 107–112 (2012). https://doi.org/10.1109/KSE.2012.30
24. Nguyen, L., Duc, D.N., Thanh, H.N.: An enhanced multi-point interactive method for multi-objective evolutionary algorithms. In: Satapathy, S.C., Bhateja, V., Nguyen, B.L., Nguyen, N.G., Le, D.-N. (eds.) Frontiers in Intelligent Computing: Theory and Applications. AISC, vol. 1013, pp. 42–49. Springer, Singapore (2020). https://doi.org/10.1007/978-981-32-9186-7_5
25. Qi, Y., Li, X., Yu, J., Miao, Q.: User-preference based decomposition in MOEA/D without using an ideal point. Swarm Evol. Comput. 44, 597–611 (2019). https://doi.org/10.1016/j.swevo.2018.08.002
26. Ruiz, A.B., Luque, M., Miettinen, K., Saborido, R.: An interactive evolutionary multiobjective optimization method: interactive WASF-GA. In: Gaspar-Cunha, A., Henggeler Antunes, C., Coello, C.C. (eds.) EMO 2015. LNCS, vol. 9019, pp. 249–263. Springer, Cham (2015). https://doi.org/10.1007/978-3-319-15892-1_17
27. Saini, B.S., Hakanen, J., Miettinen, K.: A new paradigm in interactive evolutionary multiobjective optimization. In: Bäck, T., et al. (eds.) PPSN 2020. LNCS, vol. 12270, pp. 243–256. Springer, Cham (2020). https://doi.org/10.1007/978-3-030-58115-2_17
28. Thiele, L., Miettinen, K., Korhonen, P.J., Molina, J.: A preference-based evolutionary algorithm for multi-objective optimization. Evol. Comput. 17(3), 411–436 (2009). https://doi.org/10.1162/evco.2009.17.3.411
29. Zhang, J., Xing, L.: A survey of multiobjective evolutionary algorithms. In: 2017 IEEE International Conference on Computational Science and Engineering (CSE) and IEEE International Conference on Embedded and Ubiquitous Computing (EUC), vol. 1, pp. 93–100 (2017). https://doi.org/10.1109/CSE-EUC.2017.27
30. Zhang, Q., Li, H.: MOEA/D: a multiobjective evolutionary algorithm based on decomposition. IEEE Trans. Evol. Comput. 11(6), 712–731 (2007). https://doi.org/10.1109/TEVC.2007.892759
31. Zheng, J., Yu, G., Zhu, Q., Li, X., Zou, J.: On decomposition methods in interactive user-preference based optimization. Appl. Soft Comput. J. 52, 952–973 (2017). https://doi.org/10.1016/j.asoc.2016.09.032

An Exact Inverted Generational Distance for Continuous Pareto Front

Zihan Wang⑩, Chunyun Xiao$^{(\boxtimes)}$, and Aimin Zhou$^{(\boxtimes)}$⑩

Shanghai Institute of AI for Education, School of Computer Science and Technology,
East China Normal University, Shanghai 200062, China
zhwang@stu.ecnu.edu.cn, {cyxiao,amzhou}@cs.ecnu.edu.cn

Abstract. So far, many performance indicators have been proposed to compare different evolutionary multiobjective optimization algorithms (MOEAs). Among them, the inverted generational distance (IGD) is one of the most commonly used, mainly because it can measure a population's convergence, diversity, and evenness. However, the effectiveness of IGD highly depends on the quality of the reference set. That is to say, all the reference points should be as close to the Pareto front (PF) as possible and evenly distributed to become ready for a fair performance evaluation. Currently, it is still challenging to generate well-configured reference sets, even if the PF can be given analytically. Therefore, biased reference sets might be a significant source of systematic error. However, in most MOEA literature, biased reference sets are utilized in experiments without an error estimation, which may make the experimental results unconvincing. In this paper, we propose an exact IGD (eIGD) for continuous PF, which is derived from the original IGD under an additional assumption that the reference set is perfect, i.e., the PF itself is directly utilized as an infinite-sized reference set. Therefore, the IGD values produced by biased reference sets can be compared with eIGD so that systematic error can be quantitatively evaluated and analyzed.

Keywords: Multiobjective optimization · Evolutionary computation · Performance indicator · Reference set · Differential geometry

1 Introduction

In recent years, many evolutionary multiobjective optimization algorithms (MOEAs) have established themselves in a leading position in dealing with multiobjective optimization problems (MOPs) [17]. In order to compare the performance of MOEAs, several carefully constructed test suites containing various MOPs have been proposed, such as ZDT [19], DTLZ [5], and WFG [7], to name a few.

An MOP usually contains several conflict objectives, which can not be minimized simultaneously. Consequently, a set of optimal solutions, called Pareto optimal solutions, exist such that each solution is equally preferable. The set of all Pareto optimal solutions is called Pareto front (PF) and Pareto set (PS) in

© The Author(s), under exclusive license to Springer Nature Switzerland AG 2022
G. Rudolph et al. (Eds.): PPSN 2022, LNCS 13399, pp. 96–109, 2022.
https://doi.org/10.1007/978-3-031-14721-0_7

the objective space and the decision space respectively. In practice, an MOEA maintains a population with limited size to approximate the PF of an MOP. Convergence and diversity are two important aspects regarding the quality of PF approximations. A good PF approximation should contain solutions as close to the PF as possible and as evenly distributed as possible. Therefore, some performance metrics have been proposed to quantitatively evaluate the quality of the PF approximations produced by MOEAs from specific perspectives, such as generational distance (GD) [15], inverted generational distance (IGD) [2,16], hypervolume (HV) [13], and averaged Hausdorff distance (Δ_p) [12].

Some performance metrics require a set of reference points taken from the PF to evaluate the PF approximations. Inverted generational distance (IGD) is such kind of metric. It is one of the most commonly used metrics, mainly because it can measure both convergence, diversity, and evenness of an approximation. However, there are still some difficulties in applying it in practice, although it has been widely adopted.

1. It is challenging to generate proper reference sets. The accuracy of IGD highly depends on the quality of the reference set as studied in [8,10,11]. It is suggested that the size of a reference set should be large enough to represent PF very well in [18]. It further highlighted the necessity of many uniformly sampled reference points on the entire PF through empirical studies in [8,9]. So far, however, there has been little discussion about how to construct such a large enough and uniform enough reference set, even if the PF is given analytically. Most related work focuses only on some simple situations like linear PF [1,3,4,8]. Although some recent research turned to a little more complicated PF [6,14], there are still very few effective methods toward even distribution.
2. Very few objective approaches are available to evaluate the quality of reference sets. Currently, there some intuitive standards such as
   - the size of the reference set should be big enough,
   - the reference points should be close enough to the PF, better if they are accurately located on PF, and
   - the reference points should be evenly distributed.
   But how large should the reference set be to give a fair comparison? And how to measure evenness? There are only some empirical conventions on these issues, and it is not enough for more precise MOEA studies.
3. The effect of a biased reference set is not clear. Usually, the currently widely-used reference sets may not be so perfect. For a lower computational cost, the reference sets can not be oversized. For PF with complicated shapes, the points may not be uniformly distributed. If a biased reference set is used, are the results still convincing? In most existing work, no error estimation is conducted. More seriously, it is possible to construct reference sets to make some results more satisfactory.

To solve the difficulties mentioned above, we will derive the exact value of IGD, which is defined on a real PF. With the exact value, the quality of reference sets can be evaluated, and the systematic error caused by biased reference sets can be quantitatively estimated. This paper aims to present such an exact IGD, and the major contributions are summarized as follows.

1. The exact IGD value, named as eIGD, is derived from the original IGD using a real PF rather than discrete reference sets.
2. The effect of biased reference sets is empirically studied. By using eIGD, the relative errors caused by different methods can be calculated.

The rest of the paper is organized as follows. Section 2 presents some related work. In Sect. 3, the exact IGD is introduced, and in Sect. 4, an experiment is conducted to study the effect of biased reference sets. Finally, Sect. 5 concludes the paper with some remarks for future work.

2   Related Work

This paper considers the following multiobjective optimization problem (MOP).

$$\text{minimize} \quad \boldsymbol{F}(\boldsymbol{x}) = (f_1(\boldsymbol{x}), f_2(\boldsymbol{x}), \ldots, f_m(\boldsymbol{x})) \tag{1}$$

$$\text{s.t.} \quad \boldsymbol{x} \in \Omega \tag{2}$$

where $\boldsymbol{x} = (x_1, \ldots, x_n)$ is an n-D decision variable vector, $\Omega \subset \mathbb{R}^n$ defines the feasible region of the search space, and $\boldsymbol{F}(\boldsymbol{x})$ is a vector containing $m \geq 2$ objective functions such that $f_i : \Omega \to \mathbb{R}$ for $i \in \{1, 2, \ldots, m\}$.

For assessing algorithm performance, many metrics or indicators have been proposed, among which GD and IGD are representative ones. In the following, we present some definitions and notations required to analyze GD and IGD.

Let $\boldsymbol{u}, \boldsymbol{v} \in \mathbb{R}^n$, $A \subset \mathbb{R}^n$, and $\| \cdot \|$ be L2-norm, the distance between two vectors $\boldsymbol{u}, \boldsymbol{v}$ is defined as

$$\text{dis}(\boldsymbol{u}, \boldsymbol{v}) = \| \boldsymbol{u} - \boldsymbol{v} \|, \tag{3}$$

and the distance between vector $\boldsymbol{u}$ and vector set A is defined as

$$\text{dis}(\boldsymbol{u}, A) = \inf_{\boldsymbol{a} \in A} \| \boldsymbol{u} - \boldsymbol{a} \|. \tag{4}$$

For an objective vector set $A = \{\boldsymbol{a}_1, \boldsymbol{a}_2, \ldots, \boldsymbol{a}_{|A|}\}$ and a reference point set $Z = \{\boldsymbol{z}_1, \boldsymbol{z}_2, \ldots, \boldsymbol{z}_{|Z|}\}$, we define the generational distance (GD) and inverted generational distance (IGD) as

$$\text{GD}(A, Z) = \frac{1}{|A|} \sum_{a \in A} \text{dis}(\boldsymbol{a}, Z) \tag{5}$$

$$\text{IGD}(A, Z) = \frac{1}{|Z|} \sum_{z \in Z} \text{dis}(\boldsymbol{z}, A). \tag{6}$$

Consequently, it is clear that

$$\text{GD}(A, Z) = \text{IGD}(Z, A). \tag{7}$$

In the literature, there exist different definitions for GD and IGD such as

$$GD'_p(A, Z) = \frac{1}{|A|} \sqrt[p]{\sum_{a \in A} \text{dis}(a, Z)^p} \tag{8}$$

$$IGD'_p(A, Z) = \frac{1}{|Z|} \sqrt[p]{\sum_{z \in Z} \text{dis}(z, A)^p}. \tag{9}$$

Schütze et al. [12] reported that such a definition may lead to misleading results. Considering such a situation that A consists n copies of the same element a, $\text{dis}(a, Z) = 1$, and $p > 1$, then we have

$$\lim_{n \to \infty} GD(A, Z) = \lim_{n \to \infty} \frac{\sqrt[p]{n}}{n} = 0. \tag{10}$$

Such a result is counterintuitive since the indicator value becomes better with the increase of n, while the convergence of A remains the same. Therefore, they proposed a modified definition as

$$GD''_p(A, Z) = \sqrt[p]{\frac{1}{|A|} \sum_{a \in A} \text{dis}(a, Z)^p} \tag{11}$$

$$IGD''_p(A, Z) = \sqrt[p]{\frac{1}{|Z|} \sum_{z \in Z} \text{dis}(z, A)^p} \tag{12}$$

This definition can be seen as the power mean of $\text{dis}(a, Z)$ or $\text{dis}(z, A)$ with power p. When $p = 1$, the above definition becomes the same as Eqs. (5) and (6), i.e., the arithmetic mean of $\text{dis}(\cdot, \cdot)$. Moreover, if $p \leq q$, we have

$$\min_{z \in Z} \text{dis}(z, A) \leq IGD''_p(A, Z) \leq IGD''_q(A, Z) \leq \max_{z \in Z} \text{dis}(z, A), \tag{13}$$

and this idea also works for GD. This property makes Eqs. (11) and (12) a more reasonable choice than Eqs. (8) and (9). However, in this paper, we use the most simple form, i.e., $p = 1$, because (i) it makes the definition more clear along with better mathematical properties, (ii) no reported evidence suggests that some $p > 1$ works more reasonably, and (iii) such a definition is more widely used in the literature.

Although the mathematical expressions of GD and IGD are similar, their behaviors are quite different. GD measures the mean distance from the obtained solutions to PF. So as long as the solutions are close enough to PF, the GD value could converge to 0, regardless of the distribution. IGD measures the mean distance from reference points to the closest obtained solutions. In order to have a lower IGD, the solutions must be very close to the PF and cannot miss any part of the whole PF. Consequently, GD is only a convergence indicator, while IGD is able to measure both convergence and diversity. This is why IGD is more commonly used than GD in the literature.

3   Exact Inverted Generational Distance

IGD intends to evaluate how well the solutions approximate a PF. A reference set is fundamentally a discrete approximation of a PF. Based on reference sets, indicators can be calculated more efficiently. However, such a reference-set-based performance indicator, e.g., IGD, is a kind of indirect indicator, which means that it actually measures the approximation to reference points rather than to the real PF. If a biased reference set is used, inaccurate results may be produced. Thus, to guide the generation of reference sets, it is reasonable to derive the exact value of IGD, which is defined on the real PF rather than another discrete approximation, as a standard.

The exact IGD value can be defined by extending the original discrete model to a continuous one, technically by replacing sum with integral. Here we assume that the PF is continuous and smooth. For better understanding, we first consider a bi-objective case, where the PF is a curve L and can be expressed[1] as $y = f(x), x \in [x_0, x_1]$. Due to optimality of PF, $f(x)$ is a monotone decreasing function, so its inverse function $x = f^{-1}(y), y \in [y_0, y_1]$ also exists. Schütze et al. [12] suggested that the exact value should be

$$\text{eIGD}'(A, L) \equiv \frac{1}{x_1 - x_0} \int_{x_0}^{x_1} \text{dis}[(x, y), A] \, dx. \tag{14}$$

However, this calculation has a critical defect. Apparently, the exact value is not related to which integral variable we choose. That is to say, since the PF is a monotone function, its inverse function also exists. So we can both use x and y to calculate eIGD, and the result should be the same. Consider an MOP, in which the PF is part of a hyperbola, writing $y = (x - 2)^{-1} + 1, x \in [0, 1]$, and a solution located on $(2, 1)$. Now we calculate eIGD using Eq. (14). First, we choose x as the integral variable, so we have

$$\text{eIGD}'(A, L) = \frac{1}{x_1 - x_0} \int_{x_0}^{x_1} \text{dis}\left[(x, y), A\right] dx \tag{15}$$

$$= \frac{1}{1 - 0} \int_0^1 \text{dis}\left[(x, y), (2, 1)\right] dx \tag{16}$$

$$= \int_0^1 \sqrt{(x - 2)^2 + (y - 1)^2} \, dx \tag{17}$$

$$= \int_1^2 \sqrt{u^2 + \frac{1}{u^2}} \, du \tag{18}$$

$$= \left. \frac{u\sqrt{u^2 + u^{-2}} \left(\sqrt{u^4 + 1} - \tanh^{-1} \sqrt{u^4 + 1} \right)}{2\sqrt{u^4 + 1}} \right|_{u=1}^{2} \tag{19}$$

$$\approx 1.6714. \tag{20}$$

---

[1] In Sect. 3 and below, the variable x is not related to the x mentioned in Eq. (1), i.e., the solutions in the decision space. Here, since we are discussing performance evaluation, all the points or coordinates are in the objective space.

Then we choose y as the integral variable, and we have

$$\text{eIGD}'(A, L) = \frac{1}{y_1 - y_0} \int_{y_0}^{y_1} \text{dis}\left[(x, y), A\right] dy \tag{21}$$

$$= \frac{1}{1/2 - 0} \int_0^{\frac{1}{2}} \text{dis}\left[(x, y), (2, 1)\right] dy \tag{22}$$

$$= 2 \int_0^{\frac{1}{2}} \sqrt{(x - 2)^2 + (y - 1)^2} \, dy \tag{23}$$

$$= 2 \int_{\frac{1}{2}}^1 \sqrt{\frac{1}{u^2} + u^2} \, du \tag{24}$$

$$= \left. \frac{u\sqrt{u^2 + u^{-2}} \left(\sqrt{u^4 + 1} - \tanh^{-1}\sqrt{u^4 + 1}\right)}{\sqrt{u^4 + 1}} \right|_{u=\frac{1}{2}}^1 \tag{25}$$

$$\approx 1.5968. \tag{26}$$

In this case, the $\text{eIGD}'(A, L)$ values are different by using y and x as the integral variable. The reason is that in Eq. (14), the calculus is not proper. More specifically, in a bi-objective situation, the integral on a PF is conducted on a 1-D curve in a 2-D space. Therefore, we should select the curve element ds, rather than dx. Thus, in this paper, the exact IGD, denoted as eIGD, is defined as

$$\text{eIGD}(A, L) \equiv \frac{\int_L \text{dis}[(x, y), A] \, ds}{\int_L 1 \, ds}. \tag{27}$$

Consider a generalized form where the curve L is expressed in a parametric equation form, i.e.,

$$L : \begin{cases} x = x(t) \\ y = y(t) \end{cases}, t_0 \leq t \leq t_1. \tag{28}$$

Then we have

$$ds = \sqrt{\left(\frac{dx}{dt}\right)^2 + \left(\frac{dy}{dt}\right)^2}. \tag{29}$$

Consequently, Eq. (27) can be written as

$$\text{eIGD}(A, L) = \frac{\int_L \text{dis}\left[(x, y), A\right] ds}{\int_L 1 \, ds} \tag{30}$$

$$= \frac{\int_{t_0}^{t_1} \text{dis}\left[(x(t), y(t)), A\right] \sqrt{\left(\frac{dx}{dt}\right)^2 + \left(\frac{dy}{dt}\right)^2} \, dt}{\int_{t_0}^{t_1} \sqrt{\left(\frac{dx}{dt}\right)^2 + \left(\frac{dy}{dt}\right)^2} \, dt}. \tag{31}$$

Apparently, a specific curve L might have different parametric expressions, and it is easy to prove that different parametric expressions will all result in the same eIGD value. Consider the above example. If x is chosen to be the parameter, the PF can be written as $L : y = (x - 2)^{-1} + 1, \ x \in [0, 1]$, and we have

$$ds = \sqrt{1 + \left(\frac{dy}{dx}\right)^2} \, dx \tag{32}$$

$$= \sqrt{1 + \frac{1}{(x-2)^4}} \, dx. \tag{33}$$

Therefore,

$$\text{eIGD}(A, L) = \frac{\int_L \text{dis}\left[(x, y), A\right] ds}{\int_L 1 \, ds} \tag{34}$$

$$= \frac{\int_0^1 \text{dis}\left[\left(x, \frac{1}{x-2} + 1\right), A\right] \sqrt{1 + \frac{1}{(x-2)^4}} \, dx}{\int_0^1 \sqrt{1 + \frac{1}{(x-2)^4}} \, dx} \tag{35}$$

$$= \frac{\int_1^2 \sqrt{x^2 + \frac{1}{x^2}} \sqrt{1 + \frac{1}{x^4}} \, dx}{\int_1^2 \sqrt{1 + \frac{1}{x^4}} \, dx} \tag{36}$$

$$= \frac{\left. \frac{x^4 - 1}{2x^2} \right|_{x=1}^{2}}{\int_1^2 \sqrt{1 + \frac{1}{x^4}} \, dx} \tag{37}$$

$$\approx 1.6562, \tag{38}$$

and if y is chosen to be the parameter, the PF can be written as $L : x = (y - 1)^{-1} + 2, \ y \in [0, 1/2]$, and we have

$$\text{eIGD}(A, L) = \frac{\int_L \text{dis}\left[(x, y), A\right] ds}{\int_L 1 \, ds} \tag{39}$$

$$= \frac{\int_0^{1/2} \text{dis}\left[\left((y - 1)^{-1} + 2, y\right), (2, 1)\right] \sqrt{1 + (y - 1)^{-4}} \, dy}{\int_0^{1/2} \sqrt{1 + (y - 1)^{-4}} \, dy} \tag{40}$$

$$= \frac{\int_{1/2}^1 y + y^{-3} \, dy}{\int_{1/2}^1 \sqrt{1 + y^{-4}} \, dy} \tag{41}$$

$$\approx 1.6562. \tag{42}$$

The two eIGD values are the same when we choose different parametric expressions.

Moreover, our conclusion can be extended to higher dimensional situations. Consider a 2-D PF in a 3-D space, i.e., $\Sigma : T \subset \mathbb{R}^2 \to S \subset \mathbb{R}^3$, which exists in MOPs with 3 objectives. Due to the optimality of PF, it can be expressed by $\Sigma : z = z(x, y), \ (x, y) \in T \subset \mathbb{R}^2$. Therefore, the eIGD is

$$\text{eIGD}(A, \Sigma) \equiv \frac{\iint_S \text{dis}\,[(x, y, z), A]\,\mathrm{d}\sigma}{\iint_S 1\,\mathrm{d}\sigma}. \tag{43}$$

Since

$$\mathrm{d}\sigma = \sqrt{1 + \left(\frac{\partial z}{\partial x}\right)^2 + \left(\frac{\partial z}{\partial y}\right)^2}\,\mathrm{d}x\,\mathrm{d}y, \tag{44}$$

we have

$$\text{eIGD}(A, \Sigma) = \frac{\iint_S \text{dis}\,[(x, y, z), A]\,\mathrm{d}\sigma}{\iint_S 1\,\mathrm{d}\sigma} \tag{45}$$

$$= \frac{\iint_T \text{dis}\,[(x, y, z), A]\,\sqrt{1 + \left(\frac{\partial z}{\partial x}\right)^2 + \left(\frac{\partial z}{\partial y}\right)^2}\,\mathrm{d}x\,\mathrm{d}y}{\iint_T \sqrt{1 + \left(\frac{\partial z}{\partial x}\right)^2 + \left(\frac{\partial z}{\partial y}\right)^2}\,\mathrm{d}x\,\mathrm{d}y}. \tag{46}$$

Similarly, both x and y can be used as the major variable as well, and the eIGD shall remain the same. More generally, if the PF is parameterized as

$$\Sigma : \begin{cases} x = x(u, v) \\ y = y(u, v) \ , \ (u, v) \in T \subset \mathbb{R}^2, \\ z = z(u, v) \end{cases} \tag{47}$$

then according to first fundamental form, we have

$$\text{eIGD}(A, \Sigma) \tag{48}$$

$$= \frac{\iint_S \text{dis}[(x, y, z), A]\,\mathrm{d}\sigma}{\iint_S 1\,\mathrm{d}\sigma} \tag{49}$$

$$= \frac{\iint_T \text{dis}[(x, y, z), A]\,\sqrt{\left[\frac{\partial(y, z)}{\partial(u, v)}\right]^2 + \left[\frac{\partial(z, x)}{\partial(u, v)}\right]^2 + \left[\frac{\partial(x, y)}{\partial(u, v)}\right]^2}\,\mathrm{d}u\,\mathrm{d}v}{\iint_T \sqrt{\left[\frac{\partial(y, z)}{\partial(u, v)}\right]^2 + \left[\frac{\partial(z, x)}{\partial(u, v)}\right]^2 + \left[\frac{\partial(x, y)}{\partial(u, v)}\right]^2}\,\mathrm{d}u\,\mathrm{d}v}, \tag{50}$$

where

$$\frac{\partial(x, y)}{\partial(u, v)} = \begin{vmatrix} x_u & x_v \\ y_u & y_v \end{vmatrix} = x_u y_v - x_v y_u. \tag{51}$$

In a more generalized situation, consider an m-D PF in an n-D space, i.e., $\Sigma : T \subset \mathbb{R}^m \rightarrow S \subset \mathbb{R}^n$ $(m < n)$ with a parametrization form of

$$\Sigma : \begin{cases} x_1 = x_1(t_1, t_2, \ldots, t_m) \\ x_2 = x_2(t_1, t_2, \ldots, t_m) \\ \quad \cdots \\ x_n = x_n(t_1, t_2, \ldots, t_m) \end{cases} \tag{52}$$

$$\begin{aligned} \boldsymbol{x} &= (x_1, x_2, \ldots, x_n) \in S \subset \mathbb{R}^n, \\ \boldsymbol{t} &= (t_1, t_2, \ldots, t_m) \in T \subset \mathbb{R}^m, \end{aligned} \tag{53}$$

we have

$$\mathrm{eIGD}(A, \Sigma) \equiv \frac{\displaystyle\int_S \mathrm{dis}(\boldsymbol{x}, A) \, \mathrm{d}\sigma}{\displaystyle\int_S 1 \, \mathrm{d}\sigma} \tag{54}$$

$$= \frac{\displaystyle\int_T \mathrm{dis}[\boldsymbol{x}(t), A]\sqrt{\det(J^\top J)} \, \mathrm{d}t_1 \, \mathrm{d}t_2 \ldots \mathrm{d}t_m}{\displaystyle\int_T \sqrt{\det(J^\top J)} \, \mathrm{d}t_1 \, \mathrm{d}t_2 \ldots \mathrm{d}t_m}, \tag{55}$$

where J denotes the Jacobian matrix, i.e.,

$$J = \begin{bmatrix} \dfrac{\partial x_1}{\partial t_1} & \dfrac{\partial x_1}{\partial t_2} & \cdots & \dfrac{\partial x_1}{\partial t_m} \\ \dfrac{\partial x_2}{\partial t_1} & \dfrac{\partial x_2}{\partial t_2} & \cdots & \dfrac{\partial x_2}{\partial t_m} \\ \vdots & \vdots & \ddots & \vdots \\ \dfrac{\partial x_n}{\partial t_1} & \dfrac{\partial x_n}{\partial t_2} & \cdots & \dfrac{\partial x_n}{\partial t_m} \end{bmatrix}. \tag{56}$$

4 Evaluating Discretization Error Using eIGD

This part is devoted to evaluating discretization error, i.e., the systematic error caused by a biased reference set, using eIGD. In this section, we will generate reference sets using four different methods and compare the IGD value with eIGD. In this way, the discretization error can be evaluated and analyzed quantitatively.

Let us consider a bi-objective optimization problem in which the PF is a 1/4 circle, and one solution is located at $(2, 1)$. Formally,

$$L = \{(x, y) \mid x^2 + y^2 = 1, \ 0 \leq x, y \leq 1\}, \tag{57}$$

$$A = \{(2, 1)\}. \tag{58}$$

Then, we can calculate eIGD as

$$\text{eIGD}(A, L) = \frac{\int_L \text{dis}\left[(x, y), A\right] ds}{\int_L 1 \, ds} \tag{59}$$

$$= \frac{2}{\pi} \int_0^1 \frac{\sqrt{(x - 2)^2 + (\sqrt{1 - x^2} - 1)^2}}{\sqrt{1 - x^2}} dx \tag{60}$$

$$\approx 1.459083233376. \tag{61}$$

We consider the following four reference generation methods.

x-uniform. Points are uniformly sampled from the x-axis and projected to the PF, i.e.,

$$Z = \begin{cases} x = i/k \\ y = \sqrt{1 - i^2/k^2} \end{cases}, \ i \in \{0, 1, \ldots, k\}. \tag{62}$$

y-uniform. Points are uniformly sampled from the y-axis and projected to the PF, i.e.,

$$Z = \begin{cases} x = \sqrt{1 - i^2/k^2} \\ y = i/k \end{cases}, \ i \in \{0, 1, \ldots, k\}. \tag{63}$$

Simplex-uniform. Points are uniformly sampled from a unit simplex, i.e., $y = 1 - x$ $(0 \leq x \leq 1)$ for bi-objective situation, and projected to the curve. Usually, the Das-Dennis algorithm [3] is used for uniform sampling on a unit simplex, i.e.,

$$Z = \begin{cases} x = \dfrac{i/k}{\sqrt{i^2/k^2 + (1 - i/k)^2}} \\ y = \dfrac{1 - i/k}{\sqrt{i^2/k^2 + (1 - i/k)^2}} \end{cases}, \ i \in \{0, 1, \ldots, k\}. \tag{64}$$

Curve-uniform. Since the PF is part of a circle in this experiment, we can divide the angle into equal parts and then project them to the PF, i.e.,

$$Z = \begin{cases} x = \cos(\dfrac{i\pi}{2k}) \\ y = \sin(\dfrac{i\pi}{2k}) \end{cases}, \ i \in \{0, 1, \ldots, k\}. \tag{65}$$

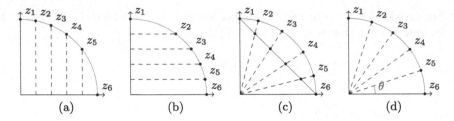

Fig. 1. Illustrations of reference points generated by (a) x-uniform, (b) y-uniform, (c) simplex-uniform, (d) curve-uniform.

The reference points generated by the four approaches are plotted in Fig. 1. Apparently, both x-uniform, y-uniform, and simplex-uniform can not generate evenly distributed reference points. For x-uniform, more points are located on the top of the curve and fewer on the bottom; and on the contrary, for y-uniform, more points are located on the bottom. For simplex-uniform, more points are on the two sides of the curve and fewer in the mid. Because the PF is precisely part of a circle, curve-uniform can generate evenly distributed points, and equal parts of the curve will always contain the same number of points.

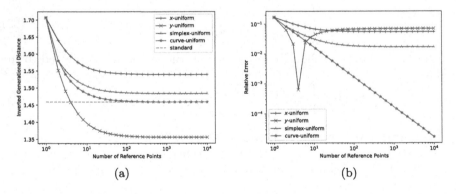

Fig. 2. Statistical results for four reference set generation methods, (a) IGD value versus set size, (b) relative error versus set size.

The four approaches are used to generate reference sets of different sizes, and the IGD values are calculated and plotted in Fig. 2(a). Then, the results are compared with the exact value, and the relative error, which is defined in Eq. (66), is plotted in Fig. 2(b).

$$\text{Relative Error} = \frac{|\text{IGD}(A, Z) - \text{eIGD}(A, L)|}{\text{eIGD}(A, L)} \tag{66}$$

From the results, we can draw some conclusions.

1. With the increase of the set size, the IGD values produced by four methods converge to four different values. This evidence indicates that simply increasing the number of reference points will not necessarily make the IGD more accurate. Not only the set size but also the distribution is a critical factor.
2. The reference sets produced by y-uniform result in the lowest IGD value. The main reason is that, in such reference sets, more points are located on the bottom, and the solution is located in $(2, 1)$. Consequently, the solution set in this experiment may benefit from the distribution from y-uniform. This phenomenon may imply that biased reference sets could have specific preferences, which may result in tendentious experimental results.
3. The relative errors of the first three methods are larger than 10^{-2}, which shows that biased reference sets may significantly influence the experimental results. Thus, this issue deserves to be further studied.
4. The relative error of curve-uniform is nearly in inverse proportion to the size, indicating that discretization error could be limited with properly configured reference sets.
5. Convergence of the relative error of the curve-uniform approach also reflects that our derivation of eIGD is correct.

5   Conclusion and Future Work

In this paper, an exact IGD, named eIGD, is proposed. Exact IGD is derived from the original IGD under an additional assumption that there are infinite reference points evenly distributed on PF. The analytical form of eIGD from 2-D to higher dimensional continuous situations is presented. Moreover, with eIGD, it is possible to evaluate the discretization error, i.e., the error caused by reference sets. An experiment is conducted in which four reference generation methods are studied. The result indicates that the discretization error could be limited with a proper reference set. In contrast, with a biased reference set, the error might be out of control and may cause unconvincing experimental results.

As for future work, there are some possible issues. (i) In this paper, the discretization error is calculated using a fixed population. For practical use, an estimation for the upper bound of the discretization error for a specific reference set is expected. (ii) Since we have derived the integral form of eIGD in detail, it can be used as a practical indicator because the accurate eIGD value can be calculated using numerical integral algorithms. (iii) Similar derivation can be easily extended to other performance indicators, such as GD, IGD+, and Δ_p.

Acknowledgements. This work is supported by the Scientific and Technological Innovation 2030 Major Projects under Grant No. 2018AAA0100902, the Science and Technology Commission of Shanghai Municipality under Grant No. 19511120601, the National Natural Science Foundation of China under Grant No. 61731009 and 61907015, and the Fundamental Research Funds for the Central Universities.

References

1. Blank, J., Deb, K., Dhebar, Y.D., Bandaru, S., Seada, H.: Generating well-spaced points on a unit simplex for evolutionary many-objective optimization. IEEE Trans. Evol. Comput. **25**(1), 48–60 (2021). https://doi.org/10.1109/TEVC.2020.2992387
2. Coello Coello, C.A., Reyes Sierra, M.: A study of the parallelization of a coevolutionary multi-objective evolutionary algorithm. In: Monroy, R., Arroyo-Figueroa, G., Sucar, L.E., Sossa, H. (eds.) MICAI 2004. LNCS (LNAI), vol. 2972, pp. 688–697. Springer, Heidelberg (2004). https://doi.org/10.1007/978-3-540-24694-7_71
3. Das, I., Dennis, J.E.: Normal-boundary intersection: a new method for generating the pareto surface in nonlinear multicriteria optimization problems. SIAM J. Optim. **8**(3), 631–657 (1998). https://doi.org/10.1137/S1052623496307510
4. Deb, K., Jain, H.: An evolutionary many-objective optimization algorithm using reference-point-based nondominated sorting approach, part I: solving problems with box constraints. IEEE Trans. Evol. Comput. **18**(4), 577–601 (2014). https://doi.org/10.1109/TEVC.2013.2281535
5. Deb, K., Thiele, L., Laumanns, M., Zitzler, E.: Scalable test problems for evolutionary multiobjective optimization. In: Abraham, A., Jain, L.C., Goldberg, R.R. (eds.) Evolutionary Multiobjective Optimization. Advanced Information and Knowledge Processing, pp. 105–145. Springer, Heidelberg (2005). https://doi.org/10.1007/1-84628-137-7_6
6. He, C., Pan, L., Xu, H., Tian, Y., Zhang, X.: An improved reference point sampling method on pareto optimal front. In: IEEE Congress on Evolutionary Computation, CEC 2016, Vancouver, BC, Canada, 24–29 July 2016, pp. 5230–5237. IEEE (2016). https://doi.org/10.1109/CEC.2016.7748353
7. Huband, S., Hingston, P., Barone, L., While, L.: A review of multiobjective test problems and a scalable test problem toolkit. IEEE Trans. Evol. Comput. **10**(5), 477–506 (2006). https://doi.org/10.1109/TEVC.2005.861417
8. Ishibuchi, H., Imada, R., Setoguchi, Y., Nojima, Y.: Reference point specification in inverted generational distance for triangular linear pareto front. IEEE Trans. Evol. Comput. **22**(6), 961–975 (2018). https://doi.org/10.1109/TEVC.2017.2776226
9. Ishibuchi, H., Masuda, H., Nojima, Y.: Sensitivity of performance evaluation results by inverted generational distance to reference points. In: IEEE Congress on Evolutionary Computation, CEC 2016, Vancouver, BC, Canada, 24–29 July 2016, pp. 1107–1114. IEEE (2016). https://doi.org/10.1109/CEC.2016.7743912
10. Ishibuchi, H., Masuda, H., Tanigaki, Y., Nojima, Y.: Modified distance calculation in generational distance and inverted generational distance. In: Gaspar-Cunha, A., Henggeler Antunes, C., Coello, C.C. (eds.) EMO 2015. LNCS, vol. 9019, pp. 110–125. Springer, Cham (2015). https://doi.org/10.1007/978-3-319-15892-1_8
11. Li, M., Yao, X.: Quality evaluation of solution sets in multiobjective optimisation: a survey. ACM Comput. Surv. **52**(2), 26:1–26:38 (2019). https://doi.org/10.1145/3300148
12. Schütze, O., Esquivel, X., Lara, A., Coello, C.A.C.: Using the averaged Hausdorff distance as a performance measure in evolutionary multiobjective optimization. IEEE Trans. Evol. Comput. **16**(4), 504–522 (2012). https://doi.org/10.1109/TEVC.2011.2161872
13. Shang, K., Ishibuchi, H., He, L., Pang, L.M.: A survey on the hypervolume indicator in evolutionary multiobjective optimization. IEEE Trans. Evol. Comput. **25**(1), 1–20 (2021). https://doi.org/10.1109/TEVC.2020.3013290

14. Tian, Y., Xiang, X., Zhang, X., Cheng, R., Jin, Y.: Sampling reference points on the pareto fronts of benchmark multi-objective optimization problems. In: 2018 IEEE Congress on Evolutionary Computation, CEC 2018, Rio de Janeiro, Brazil, 8–13 July 2018, pp. 1–6. IEEE (2018). https://doi.org/10.1109/CEC.2018.8477730
15. Valenzuela-Rendón, M., Uresti-Charre, E.: A non-generational genetic algorithm for multiobjective optimization. In: Bäck, T. (ed.) Proceedings of the 7th International Conference on Genetic Algorithms, East Lansing, MI, USA, 19–23 July 1997, pp. 658–665. Morgan Kaufmann (1997)
16. Zhou, A., Jin, Y., Zhang, Q., Sendhoff, B., Tsang, E.P.K.: Combining model-based and genetics-based offspring generation for multi-objective optimization using a convergence criterion. In: IEEE International Conference on Evolutionary Computation, CEC 2006, part of WCCI 2006, Vancouver, BC, Canada, 16–21 July 2006, pp. 892–899. IEEE (2006). https://doi.org/10.1109/CEC.2006.1688406
17. Zhou, A., Qu, B., Li, H., Zhao, S., Suganthan, P.N., Zhang, Q.: Multiobjective evolutionary algorithms: a survey of the state of the art. Swarm Evol. Comput. 1(1), 32–49 (2011). https://doi.org/10.1016/j.swevo.2011.03.001
18. Zhou, A., Zhang, Q., Jin, Y., Sendhoff, B.: Adaptive modelling strategy for continuous multi-objective optimization. In: Proceedings of the IEEE Congress on Evolutionary Computation, CEC 2007, Singapore, 25–28 September 2007, pp. 431–437. IEEE (2007). https://doi.org/10.1109/CEC.2007.4424503
19. Zitzler, E., Deb, K., Thiele, L.: Comparison of multiobjective evolutionary algorithms: empirical results. Evol. Comput. 8(2), 173–195 (2000). https://doi.org/10.1162/106365600568202

Direction Vector Selection for R2-Based Hypervolume Contribution Approximation

Tianye Shu, Ke Shang$^{(\boxtimes)}$, Yang Nan, and Hisao Ishibuchi$^{(\boxtimes)}$

Guangdong Provincial Key Laboratory of Brain-Inspired Intelligent Computation, Department of Computer Science and Engineering, Southern University of Science and Technology, Shenzhen 518055, China
{12132356,shangk,nany,hisao}@sustech.edu.cn, kshang@foxmail.com

Abstract. Recently, an R2-based hypervolume contribution approximation (i.e., R_2^{HVC} indicator) has been proposed and applied to evolutionary multi-objective algorithms and subset selection. The R_2^{HVC} indicator approximates the hypervolume contribution using a set of line segments determined by a direction vector set. Although the R_2^{HVC} indicator is computationally efficient compared with the exact hypervolume contribution calculation, its approximation error is large if an inappropriate direction vector set is used. In this paper, we propose a method to generate a direction vector set for reducing the approximation error of the R_2^{HVC} indicator. The method generates a set of direction vectors by selecting a small direction vector set from a large candidate direction vector set in a greedy manner. Experimental results show that the proposed method outperforms six existing direction vector set generation methods. The direction vector set generated by the proposed method can be further used to improve the performance of hypervolume-based algorithms which rely on the R_2^{HVC} indicator.

Keywords: Evolutionary multi-objective optimization · Hypervolume contribution · Hypervolume contribution approximation

1 Introduction

In evolutionary multi-objective optimization (EMO), convergence and diversity are two desired properties of a solution set. To address the conflicting nature of the two properties, many indicators are proposed such as hypervolume (HV) [25, 31], generational distance (GD) [28], inverted generational distance (IGD) [5], and R2 [13]. These indicators are used not only for evaluating the performance of evolutionary multi-objective optimization algorithms (EMOAs) but also for designing EMOAs.

Hypervolume is one of the most widely used indicators in EMO since hypervolume is Pareto compliant [30]. Many EMOAs are based on the hypervolume indicator such as SMS-EMOA [1,11] and FV-EMOA [18]. In these algorithms,

G. Rudolph et al. (Eds.): PPSN 2022, LNCS 13399, pp. 110–123, 2022.
https://doi.org/10.1007/978-3-031-14721-0_8

hypervolume contribution plays an important role, which is the increment (or decrement) of the hypervolume of a solution set when a solution is added (or removed). For example, SMS-EMOA discards the solution with the least hypervolume contribution from the population in each generation, so that the hypervolume of the remaining population is maximized. Hypervolume contribution is also crucial in hypervolume subset selection (HSS), which aims to select a subset with the maximum hypervolume from a candidate solution set. Greedy HSS methods [2,3,12] usually select (or remove) the solution with the largest (or least) hypervolume contribution iteratively.

One drawback of hypervolume-based algorithms is their expensive computational cost, especially in high-dimensional spaces. This is because the exact calculation of hypervolume and hypervolume contribution is #P-hard [4]. To decrease the computational cost, an R2-based hypervolume contribution approximation method (i.e., R_2^{HVC} indicator) was proposed in [26]. Benefiting from the R_2^{HVC} indicator, an efficient hypervolume-based algorithm R2HCA-EMOA was proposed, which outperforms many state-of-the-art EMOAs on many-objective problems [23]. The R_2^{HVC} indicator was also applied in a greedy approximate HSS algorithm (i.e., GAHSS) whose computational cost is much lower than that of greedy exact HSS algorithms [24].

The basic idea of the R_2^{HVC} indicator is to use different line segments to approximate the hypervolume contribution. Therefore, a set of vectors is needed to determine the directions of these line segments. Nan et al. [21] reported that the performance of the R_2^{HVC} indicator highly depends on the distribution of the used direction vectors, and uniformly distributed direction vectors are not useful for the R_2^{HVC} indicator. However, currently available methods for direction vector set generation are not specially designed for the R_2^{HVC} indicator, and some of them aim to obtain uniformly distributed direction vectors. As a result, these methods are not suitable for the R_2^{HVC} indicator.

In this paper, we propose a direction vector set generation method called the greedy approximation error selection (GAES). Specifically, we formulate the direction vector set generation for the R_2^{HVC} indicator as a subset selection problem. The target is to minimize the approximation error of the R_2^{HVC} indicator, which is defined by the average distance between the ranking of solutions based on the hypervolume contribution and their ranking based on the R_2^{HVC} indicator in a set of training solution sets. The proposed algorithm selects direction vectors one by one from a large candidate direction vector set in a greedy manner. Our experimental results show that the proposed method can achieve the smallest approximation error and the highest correct identification rate among seven direction vector set generation methods. The direction vector set generated by the proposed method can be further used to improve the performance of hypervolume-based algorithms (e.g., GAHSS) which rely on the R_2^{HVC} indicator.

The rest of the paper is organized as follows: In Sect. 2, we briefly review the hypervolume, the hypervolume contribution, the R2-based hypervolume contribution, six direction vector set generation methods, and subset selection. We propose a direction vector set generation method for the R_2^{HVC} indicator in Sect. 3. The performance of the proposed method is tested in Sect. 4. Finally, in Sect. 5, the conclusion is given.

2 Background

2.1 Hypervolume and Hypervolume Contribution

Mathematically, the hypervolume indicator is defined as follows.

Definition 1. *In the objective space R^m with a reference point $r \in R^m$, the hypervolume of a solution set $S \subset R^m$ is defined as*

$$HV(S, r) = \mathcal{L}\left(\bigcup_{s \in S}\{a | s \succeq a \succeq r\}\right), \tag{1}$$

where $\mathcal{L}(\cdot)$ is the Lebesgue measure of a set, and $s \succeq a$ denotes s dominates a (i.e., $s_i \leq a_i \; \forall i \in \{1, 2, .., m\}$ and $s_j < a_j \; \exists j \in \{1, 2, ..., m\}$ in the minimization case).

Based on the definition of the hypervolume, the hypervolume contribution is defined as follows.

Definition 2. *In the objective space R^m with a reference point $r \in R^m$, the hypervolume contribution of a solution $s \in R^m$ to a solution set $S \subset R^m$ is defined as*

$$HVC(s, S, r) = \begin{cases} HV(S, r) - HV(S \setminus \{s\}, r), & \text{if } s \in S, \\ HV(S \cup \{s\}, r) - HV(S, r), & \text{if } s \notin S. \end{cases} \tag{2}$$

Figure 1(a) and (b) illustrate the hypervolume and the hypervolume contribution in the two-objective space.

2.2 R2-Based Hypervolume Contribution Approximation

In [26], an R2-based indicator (i.e., R_2^{HVC} indicator) was proposed to approximate the hypervolume contribution. Suppose we have a solution set S in the m-dimensional objective space. To approximate the hypervolume contribution of the solution s to S with the reference point r, the lengths of a set of line segments are used. The length L of each line segment is determined by each direction vector λ in a given direction vector set V, the solution set $S \setminus \{s\}$ and the reference point r. Mathematically,

$$R_2^{HVC}(s, S, r, V) = \frac{1}{|V|}\sum_{\lambda \in V} L(s, S \setminus \{s\}, r, \lambda)^m$$

$$= \frac{1}{|V|}\sum_{\lambda \in V} min\left\{\min_{s' \in S \setminus \{s\}}\{g^{*2tch}(s'|\lambda, s)\}, g^{mtch}(r|\lambda, s)\right\}^m. \tag{3}$$

For minimization problems, the $g^{*2tch}(s'|\lambda, s)$ function in Eq. (3) is defined as $g^{*2tch}(s'|\lambda, s) = \max\limits_{j \in \{1,2,...,m\}}\left\{\frac{s'_j - s_j}{\lambda_j}\right\}$, and the $g^{mtch}(r|\lambda, s)$ function in Eq. (3) is defined as $g^{mtch}(r|\lambda, s) = \min\limits_{j \in \{1,2,...,m\}}\left\{\frac{|r_j - s_j|}{\lambda_j}\right\}$. The mechanism of the R2-based hypervolume contribution approximation is illustrated in Fig. 1(c).

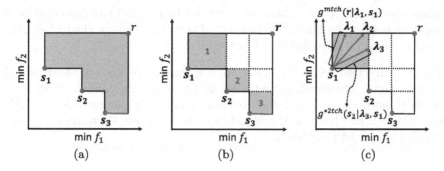

Fig. 1. Illustration of the hypervolume, the hypervolume contribution and the R2-based hypervolume contribution approximation. The grey area in (a) is the hypervolume of the solution set $\{s_1, s_2, s_3\}$. Each grey area (1, 2 and 3) in (b) is the hypervolume contribution of the corresponding solution (s_1, s_2 and s_3) to the solution set $\{s_1, s_2, s_3\}$, respectively. The red lines in (c) illustrate the mechanism of the R2-based hypervolume contribution approximation. (Color figure online)

2.3   Direction Vector Set Generation Methods

Six existing direction vector set generation methods are briefly explained. The first three methods are space filling methods. Every generated weight vector w is normalized to obtain the direction vector λ (i.e., $\lambda = \frac{w}{||w||_2}$).

- **Das and Dennis (DAS) method** [6]: In the m-dimensional space, the DAS method generates a weight vector $w = (w_1, w_2, ..., w_m)$ by uniformly dividing each dimension into H parts. The value of w_j is selected from $\{0, \frac{1}{H}, \frac{2}{H}, ..., 1 - \sum_{i=1}^{j-1} w_i\}$. Totally, C_{H+m-1}^{m-1} weight vectors are generated.
- **Unit normal vector (UNV) method** [10]: In the UNV method, a set of weight vectors is randomly sampled from the m-dimensional normal distribution (i.e., $w \sim N_m(0, I_m)$).
- **JAS method** [17]: The JAS method randomly generates a weight vector $w = (w_1, w_2, ..., w_m)$ with the uniform probability distribution [17]. For $k < m$, w_k is sampled by $w_k = (1 - \sum_{j=1}^{k-1} w_j)(1 - \sqrt[m-k]{\mu})$ where μ is randomly sampled from $[0, 1]$. For the last dimension, $w_m = 1 - \sum_{j=1}^{m-1} w_j$.

The other three methods select the desired direction vector set from a large candidate direction vector set generated by one of the first three methods.

- **Maximally spare selection method with DAS (MSS-D** [7]): A large candidate direction vector set U is generated by the DAS method. Then, m extreme direction vectors $(1, 0, ..., 0)$, $(0, 1, ..., 0)$, ..., $(0, 0, ..., 1)$ are selected as the initial direction vector set V. The direction vector $\lambda \in U$ with the largest distance to the vector set V is selected (i.e., moved from U to V). This step repeats until V reaches the desired size.
- **Maximally spare selection method with UNV (MSS-U** [7]): The only difference between MSS-U and MSS-D is that the candidate direction vector set U in MSS-U is generated by UNV instead of DAS.

– **Kmeans-U** [15]: The method starts with a large candidate direction vector set U generated by the UNV method. Then the k-means clustering [20] is used to obtain a direction vector subset V from U.

In Fig. 2, a direction vector set of size 91 is generated by each method, and the size of the candidate direction vector set is 49,770 for the MSS-D, MSS-U and Kmeans-U methods. In Fig. 2, the direction vector sets generated by the DAS, MSS-D, MSS-U and Kmeans-U methods are more uniform than those generated by the UNV and JAS methods.

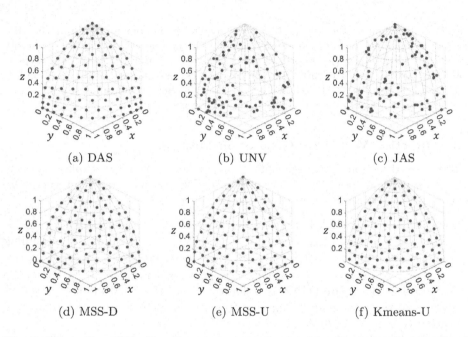

Fig. 2. Illustration of the direction vector sets generated by the six methods outlined in Sect. 2.3. Each direction vector set contains 91 direction vectors.

2.4   Subset Selection

Subset selection is to select a subset from a large candidate set to optimize a given metric [22]. Formally, given a set $U = \{e_1, e_2, ..., e_N\}$, a metric f (to be maximized) and a positive integer k, subset selection aims to find a subset $V \subseteq U$ with $|V| = k$ for maximizing $f(V)$. That is, $V^* = argmax_{V \subseteq U, |V|=k} f(V)$. When the target is to minimize $f(V)$, the problem can be written as $V^* = argmin_{V \subseteq U, |V|=k} f(V)$. To solve the subset selection problem, the greedy inclusion is a simple yet widely used method. For example, hypervolume subset selection (HSS) aims to maximize the hypervolume of the selected solution subset. The greedy inclusion for the HSS [3,12] selects the solution with the largest hypervolume contribution iteratively.

3 Proposed Method for Selecting Direction Vector Set

Since the performance of the R_2^{HVC} indicator strongly depends on the used direction vector set, we propose a simple greedy inclusion algorithm called the greedy approximation error selection (GAES) for obtaining a high-quality direction vector set for the R_2^{HVC} indicator, and analyze its time complexity.

3.1 Approximation Error

Equipped with a good direction vector set, the R_2^{HVC} indicator is supposed to be consistent with the hypervolume contribution. That is, the approximation error should be small. Usually, we are interested in the ranking of solutions based on the hypervolume contribution values in hypervolume-based algorithms. Therefore, we define the approximation error between the ranking based on the hypervolume contribution and the ranking based on the R_2^{HVC} indicator. The hypervolume contribution is calculated by the WFG algorithm [29], and the R_2^{HVC} indicator is calculated by Eq. (3). Suppose we have a solution set $S_i = \{s_1, s_2, ..., s_t\}$. We denote the ranking based on the hypervolume contribution by $\sigma_H(s_1), \sigma_H(s_2), ..., \sigma_H(s_t)$, where $\sigma_H(s_i)$ is the rank of the solution s_i among the t solutions in S_i. In the same manner, we denote the ranking based on the R_2^{HVC} indicator by $\sigma_R(s_1), \sigma_R(s_2), ..., \sigma_R(s_t)$. Spearman's footrule is one of the most well-known distances between rankings [19], which can be described as follows:

$$D(S_i, \sigma_H, \sigma_R) = \sum_{j=1}^{t} |\sigma_H(s_j) - \sigma_R(s_j)|. \tag{4}$$

Small approximation error means that the distance between the two rankings is small. Based on the distance in Eq. (4), we can measure the approximation error. For a set of solution sets $S = \{S_1, S_2, ...S_T\}$, the approximation error is defined as follows:

$$AE(S, V, r) = \frac{1}{T} \sum_{i=1}^{T} D(S_i, \sigma_H, \sigma_R), \tag{5}$$

where r is the reference point, σ_H is the ranking of the solutions in S_i based on the hypervolume contribution, and σ_R is the ranking of these solutions based on the R_2^{HVC} indicator with the direction vector set V.

3.2 Problem Formulation

Now we can formulate the problem of generating a good direction vector set for the R_2^{HVC} indicator as a subset selection problem. Given a large candidate direction vector set $U = \{\lambda_1, \lambda_2, ..., \lambda_N\}$, a set of training solution sets S, and a reference point r, the problem is to find a direction vector subset $V \subset U$ with $|V| = n$ $(n < N)$ to minimize the approximation error $AE(S, V, r)$ in Eq. (5).

3.3 Greedy Inclusion Algorithm

To solve the above problem, a simple greedy inclusion algorithm called the greedy approximation error selection (GAES) is proposed (Algorithm 1). Firstly, a set of training solution sets S should be prepared in advance. Then, a large candidate direction vector set U is generated by some methods. The direction vector set V is empty initially. Iteratively, the direction vector $\boldsymbol{\lambda}^* \in U$ which minimizes $AE(S, V \cup \{\boldsymbol{\lambda}\}, \boldsymbol{r})$ is selected (i.e., moved from U to V) until V reaches the desired size.

We analyze the time complexity of Algorithm 1. The most time-consuming step is to calculate the approximation error AE. Let us consider a single training solution set S_i with size t. In Eq. (4) and (5), we have to obtain two rankings of the solutions in S_i: One is based on the hypervolume contribution and the other is based on the R_2^{HVC} indicator. The hypervolume contribution can be calculated in advance. Therefore, we only need to calculate the R_2^{HVC} indicator with the direction vector set $V \cup \{\boldsymbol{\lambda}\}$ for every $\boldsymbol{\lambda} \in U$. It is worth noting in Eq. (3) that the R_2^{HVC} indicator with a direction vector set V is basically the average length of the line segment determined by the direction vector $\boldsymbol{\lambda}$ for every $\boldsymbol{\lambda} \in V$. Thus, we can calculate the length L for each solution in S_i with each direction vector in U in advance, which requires $O(Nt^2m)$ time. With these lengths, we can update the R_2^{HVC} indicator for each solution in S_i in $O(t)$ time. Sorting these solutions requires $O(t \log t)$ time, which is performed for each $\boldsymbol{\lambda}$ in U in each iteration in Algorithm 1. The total time complexity is $O(T(nNt \log t + Nt^2m))$.

Algorithm 1. Greedy Approximation Error Selection

Input: S (a set of training solution sets), N (size of a candidate direction vector set), n (size of a desired direction vector set), $\boldsymbol{r}$ (reference point)
Output: V (desired direction vector set)
Generate a candidate direction vector set U of size N by some methods.
$V \leftarrow \emptyset$
while $|V| < n$
   $\boldsymbol{\lambda}^* = \underset{\boldsymbol{\lambda} \in U}{argmin} \; AE(S, V \cup \{\boldsymbol{\lambda}\}, \boldsymbol{r})$
   Move $\boldsymbol{\lambda}^*$ from U to V
end while

---

4 Experiments and Discussions

4.1 Direction Vector Selection

Experimental Settings. The first experiment is to illustrate the direction vector selection process of the GAES algorithm (Algorithm 1). We generate 100 training solution sets of size 100, and the hypervolume contribution of each solution in each training solution set is calculated in advance. More specifically, to generate each training solution set S_i with size 100, we first determine the shape (triangular or inverted triangular) and the curvature (linear, convex or concave) of the Pareto front. Then, 100 solutions in S_i are randomly sampled from this Pareto front. The triangular Pareto front $\sum_{i=1}^{m} f_i^p = 1, f_i \geq 0$ for $i = 1, 2, ..., m$ is used in the first 50 training solution sets. The remaining 50 training solution sets use the inverted triangular Pareto front $\sum_{i=1}^{m} (1 - f_i)^p = 1, 0 \leq f_i \leq 1$ for $i = 1, 2, ..., m$. The p value in the two formulas controls the curvature. To make the curvature more diverse, the p value is determined by $p = 2^x$ where x is uniformly sampled from $[-1, 1]$ (i.e., the range of p value is $[0.5, 2]$). The candidate direction vector set of size 10,000 is generated by the UNV method in Sect. 2.3. The size of the desired direction vector set is set as 91, 105 and 120 for 3, 5 and 8-objective cases, respectively. The reference point r is set as $(1.2, 1.2, ..., 1.2)$. The proposed method with UNV is denoted as GAES-U in our experiments.

The six direction vector set generation methods in Sect. 2.3 are used as the baselines. For the MSS-U, MSS-D, and Kmeans-U methods, the size of the candidate direction vector set is set as 49,770, 46,376 and 31,824 for 3, 5 and 8-objective cases, respectively. For each of the six methods and the GAES-U method, 21 direction vector sets are generated from 21 independent runs. We conduct the experiments on a virtual machine equipped with two ADM EPYC 7702 64-Core CUP@2.4 GHz, 256 GB RAM and Ubuntu Operating System. All codes are implemented in MATLAB R2021b and available from https://github.com/HisaoLabSUSTC/GAES.

Results and Discussions. The performance of the selected direction vectors by GEAS-U is shown in Fig. 3 at each iteration (i.e., after selecting a single direction vector, two direction vectors, ..., n direction vectors). The blue curve shows that the approximation error of the R_2^{HVC} indicator on the training solution sets decreases monotonically as more direction vectors are selected by the GAES-U method. The GAES-U method (i.e., the rightmost point of the blue curve) has a better approximation error than the other six methods. With the same number of direction vectors, the approximation error by each method increases as the number of objectives increases. This is because more direction vectors are needed for the R_2^{HVC} indicator to approximate the hypervolume contribution precisely in a higher-dimensional space. The advantage of the GAES-U method is clear in the 8-objective case. Only one direction vector selected by the GAES-U method (i.e., the leftmost point of the blue curve) has a similar approximation error as 120 direction vectors generated by the DAS method (i.e., the top dash line). The 40 direction vectors selected by the GAES-U method have a better approximation error than 120 direction vectors generated by the other methods.

The direction vector sets generated by the GAES-U method are shown in Fig. 4. In the 3-objective case, the direction vector set generated by the GAES-U method in Fig. 4(a) is less uniform than those generated by the DAS, MSS-D, MSS-U and Kmean-U methods in Fig. 2(a), (d), (e) and (f), and is more uniform than those generated by the UNV and JAS methods in Fig. 2(b) and (c).

Figure 5(a) shows the computational time for the training solution sets generation including the hypervolume contribution calculation, which increases severely as the number of the objectives increases. However, this part only needs to be performed once, and the generated training solution sets and the calculated hypervolume contribution can be used for multiple runs of the GAES-U method. The runtime of the GAES-U method increases slightly as the number of the objectives increases as shown in Fig. 5(b).

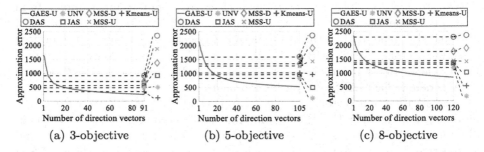

(a) 3-objective   (b) 5-objective   (c) 8-objective

Fig. 3. Approximation errors of the R_2^{HVC} indicator with different direction vector set generation methods on the training solution sets (average results over 21 runs). (Color figure online)

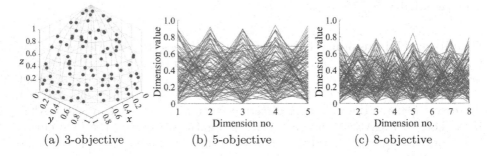

(a) 3-objective   (b) 5-objective   (c) 8-objective

Fig. 4. The direction vector sets generated by the GAES-U method.

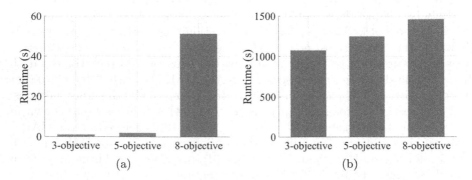

(a)   (b)

Fig. 5. Runtime (a) for generating the training solution sets and pre-calculating the hypervolume contribution of each solution in each training solution set. Runtime (b) for selecting direction vectors by the GAES-U method averaged over 21 runs.

4.2   Test on Six Regular Pareto Fronts

Experimental Settings. In the previous experiment, we have obtained different direction vector sets. Then, in this experiment, we compare the performance of the R_2^{HVC} indicator with these direction vector sets on testing solution sets. To generate testing solution sets, six regular Pareto fronts are considered, which are linear triangular, concave triangular and convex triangular Pareto fronts and their corresponding inverted Pareto fronts. For each type of the front, 100 testing solution sets of size 100 are randomly sampled from the front. Firstly, we calculate the approximation error defined in Eq. (4) on the testing solution sets using different direction vector sets. Then, the correct identification rate (CIR) of the R_2^{HVC} indicator with different direction vector set generation methods is calculated. For a testing solution set, the correct identification means that the solution with the least hypervolume contribution is correctly identified by the R_2^{HVC} indicator, and the CIR implies how many testing solution sets are correctly handled.

Results and Discussions. Figure 6 shows the approximation error of the R_2^{HVC} indicator with different direction vector set generation methods on the testing solution sets. The GAES-U method (i.e., the rightmost point of the blue curve in each figure in Fig. 6) has the smallest approximation error on the testing solution sets. This observation shows the good generalization ability of the GAES-U method. The CIR of the R_2^{HVC} indicator is shown in Table 1. The best CIR (i.e., 58.5%) is obtained by the GAES-U method in Table 1. The UNV and JAS method obtain 51.1% and 50.4% CIR, respectively. The worst one is the Kmeans-U method with the CIR of 33.7%. From these CIR results of the R_2^{HVC} indicator, we can see that the proposed GAES-U method is clearly better than the other six methods.

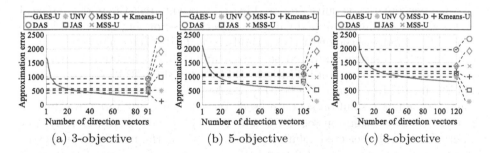

(a) 3-objective   (b) 5-objective   (c) 8-objective

Fig. 6. Approximation errors of the R_2^{HVC} indicator with different direction vector set generation methods on the testing solution sets (average results over 21 runs). (Color figure online)

4.3   Application: GAHSS

Experimental Settings. The direction vector sets generated by the GAES-U method are tested by the greedy approximate hypervolume subset selection

Table 1. Correct identification rate (CIR) of the R_2^{HVC} indicator with different direction vector set generation methods on different solution sets. The rank of each method is in the parenthesis, and a small value means a better rank.

Solution set		GAES-U	DAS	UNV	JAS	MSS-D	MSS-U	Kmeans-U
Linear triangular	3	**76.4%(1)**	59.0%(7)	66.8%(4)	67.0%(3)	68.0%(2)	66.2%(5)	63.4%(6)
	5	**55.4%(1)**	15.0%(7)	46.5%(2)	46.0%(3)	17.0%(6)	18.7%(5)	27.6%(4)
	8	29.6%(5)	23.0%(6)	36.4%(2)	34.9%(3)	**42.0%(1)**	30.0%(4)	14.3%(7)
Linear inverted triangular	3	**73.4%(1)**	53.0%(6)	64.0%(3)	63.2%(5)	67.0%(2)	63.5%(4)	50.0%(7)
	5	**62.1%(1)**	14.0%(7)	51.8%(2)	49.2%(3)	27.0%(6)	30.9%(4)	28.6%(5)
	8	**51.0%(1)**	0.0%(6.5)	37.0%(2)	24.4%(4)	0.0%(6.5)	6.9%(5)	28.1%(3)
Concave triangular	3	**67.4%(1)**	32.0%(7)	57.0%(3)	58.8%(2)	38.0%(5)	36.8%(6)	49.4%(4)
	5	**61.4%(1)**	16.0%(6)	50.7%(2)	49.8%(3)	21.0%(5)	26.7%(4)	13.3%(7)
	8	**65.9%(1)**	53.0%(6)	64.8%(3)	61.7%(4)	56.0%(5)	65.3%(2)	42.3%(7)
Concave inverted triangular	3	**73.8%(1)**	41.0%(7)	59.6%(3)	60.7%(2)	50.0%(5)	50.7%(4)	46.9%(6)
	5	40.1%(3)	26.0%(4)	**42.7%(1)**	42.1%(2)	23.0%(5)	15.9%(6)	12.1%(7)
	8	**55.5%(1)**	0.0%(7)	38.6%(3)	36.4%(4)	10.0%(6)	43.9%(2)	33.0%(5)
Convex triangular	3	**66.5%(1)**	55.0%(6)	55.1%(5)	57.2%(4)	60.0%(3)	60.3%(2)	36.0%(7)
	5	26.2%(2)	17.0%(6)	25.0%(3)	**30.4%(1)**	19.0%(4)	18.8%(5)	4.5%(7)
	8	7.0%(5)	**33.0%(1)**	14.2%(4)	18.8%(3)	30.0%(2)	4.0%(6)	1.3%(7)
Convex inverted triangular	3	**65.5%(1)**	32.0%(6)	50.0%(2)	49.5%(3)	38.0%(4)	35.1%(5)	27.9%(7)
	5	88.5%(2)	**90.0%(1)**	76.6%(5)	79.3%(4)	66.0%(6)	80.8%(3)	42.6%(7)
	8	87.7%(2)	**100.0%(1)**	83.1%(4)	77.6%(6)	61.0%(7)	81.7%(5)	84.8%(3)
Avg. rank		**1.72**	5.42	2.94	3.28	4.47	4.28	5.89
Avg. CIR		**58.5%**	36.6%	51.1%	50.4%	38.5%	40.9%	33.7%

(GAHSS) algorithm [24]. The difference between the GAHSS algorithm and the greedy HSS algorithm mentioned in Sect. 2.4 is that the R_2^{HVC} indicator is used to approximate the hypervolume contribution in GAHSS. A part of the subset selection benchmark test suite proposed in [27] is used to test the performance of the GAHSS algorithm equipped with the direction vector sets generated by the GAES-U method and the other six methods. Specifically, the candidate solution sets consist of the nondominated solutions after 100,000 function evaluations when NSGA-III [8] is run on DTLZ1 [9], DTLZ2 [9], Minus-DTLZ1 [16], Minus-DTLZ2 [16], DTLZ7 [9] and WFG3 [14] problems for 3, 5 and 8-objective cases. Thus, 18 candidate solution sets are used. We select 91, 210 and 156 solutions from the candidate solution sets for 3, 5 and 8-objective cases, respectively. The hypervolume of the selected solution subset is used as the performance metric. The reference point is set as 1.2 times the nadir point of the true Pareto front for each candidate solution set. GAHSS is performed 21 runs with each direction vector set generation method, and the Wilcoxon rank sum test is used to compare the hypervolume performance.

Results and Discussions. Table 2 shows the hypervolume of the solution subset selected by the GAHSS algorithm with different direction vector set generation methods. The best result is obtained by the proposed GAES-U method. One

Table 2. Hypervolume of the solution subset selected by GAHSS with different direction vector set generation methods on different candidate solution sets. The rank of each method is in the parenthesis, and a small value means a better rank. The Wilcoxon rank sum test is used to compare the performance. The symbols "+", "−" and "≈" mean the GAES-U method "is significantly better than", "is significantly worse than" and "has no significant difference with" the corresponding method, respectively.

Candidate solution set		GAES-U	DAS	UNV	JAS	MSS-D	MSS-U	Kmeans-U
DTLZ1	3	**1.90E−1(1)**	1.89E-1(7,+)	1.90E-1(4,+)	1.89E-1(6,+)	1.90E-1(3,+)	1.90E-1(5,+)	1.90E−1(2,≈)
	5	**7.68E−2(1)**	7.56E−2(5,+)	7.67E−2(3,+)	7.67E−2(4,+)	7.54E−2(6,+)	7.53E−2(7,+)	7.67E−2(2,+)
	8	**1.67E−2(1)**	1.20E−2(7,+)	1.67E−2(2,+)	1.67E−2(4,+)	1.46E−2(6,+)	1.67E−2(5,+)	1.67E−2(3,+)
DTLZ2	3	**1.15E+0(1)**	9.30E−1(7,+)	1.15E+0(3,+)	1.15E+0(4,+)	1.01E+0(6,+)	1.05E+0(5,+)	1.15E+0(2,≈)
	5	**2.21E+0(1)**	1.59E+0(7,+)	2.20E+0(3,+)	2.20E+0(4,+)	1.62E+0(6,+)	2.09E+0(5,+)	2.21E+0(2,≈)
	8	4.16E+0(2)	2.21E+0(6,+)	4.15E+0(3,+)	4.14E+0(4,+)	2.08E+0(7,+)	4.12E+0(5,+)	**4.16E + 0(1,−)**
Minus-DTLZ1	3	8.87E+7(2)	8.74E+7(7,+)	8.85E+7(3,+)	8.84E+7(5,+)	8.81E+7(6,+)	8.84E+7(4,+)	**8.88E+7(1,−)**
	5	4.48E+12(2)	4.07E+12(7,+)	4.46E+12(3,+)	4.40E+12(4,+)	4.11E+12(6,+)	4.23E+12(5,+)	**4.48E+12(1,≈)**
	8	**1.36E+19(1)**	8.23E+18(7,+)	1.27E+19(3,+)	1.11E+19(4,+)	8.41E+18(6,+)	9.05E+18(5,+)	1.35E+19(2,+)
Minus-DTLZ2	3	4.47E+1(2)	4.00E+1(7,+)	4.45E+1(3,+)	4.44E+1(4,+)	4.31E+1(5,+)	4.31E+1(6,+)	**4.47E+1(1,≈)**
	5	**2.51E+2(1)**	2.31E+2(7,+)	2.47E+2(3,+)	2.45E+2(4,+)	2.31E+2(6,+)	2.40E+2(5,+)	2.50E+2(2,+)
	8	**1.15E+3(1)**	6.55E+2(7,+)	1.10E+3(3,+)	1.08E+3(4,+)	8.17E+2(6,+)	1.04E+3(5,+)	1.14E+3(2,+)
DTLZ7	3	2.81E+0(2)	2.54E+0(7,+)	2.80E+0(3,+)	2.79E+0(4,+)	2.72E+0(6,+)	2.73E+0(5,+)	**2.81E+0(1,≈)**
	5	**5.08E+0(1)**	4.56E+0(7,+)	5.04E+0(3,+)	5.03E+0(4,+)	4.79E+0(6,+)	4.87E+0(5,+)	5.07E+0(2,+)
	8	**7.56E+0(1)**	6.05E+0(6,+)	7.40E+0(5,+)	7.41E+0(3,+)	5.50E+0(7,+)	7.40E+0(4,+)	7.55E+0(2,≈)
WFG3	3	3.85E+1(2)	3.55E+1(5,+)	3.84E+1(3,≈)	3.83E+1(4,≈)	3.55E+1(6,+)	3.54E+1(7,+)	**3.85E+1(1,−)**
	5	**1.47E+4(1)**	1.37E+4(7,+)	1.46E+4(3,+)	1.47E+4(2,+)	1.37E+4(6,+)	1.46E+4(4,+)	1.46E+4(5,+)
	8	1.01E+8(3)	6.42E+7(7,+)	1.01E+8(4,≈)	**1.01E + 8(1,−)**	9.09E+7(6,+)	1.01E+8(2,−)	9.90E+7(5,+)
Avg. rank		**1.44**	6.67	3.17	3.83	5.89	4.94	2.06
+/−/≈			18/0/0	16/0/2	16/1/1	18/0/0	17/1/0	8/3/7

interesting observation is that the Kmeans-U method, which has poor performance in the CIR experiment in Table 1, shows competitive performance in the GAHSS experiment in Table 2. Future examinations on this interesting observation is needed.

5 Conclusion

In this paper, we formulated the problem of generating a good direction vector set for the R_2^{HVC} indicator as a subset selection problem to minimize the approximation error. A greedy inclusion method called the greedy approximation error selection (GAES) was proposed to solve this problem. Experimental results showed that the GAES method outperforms other available methods for direction vector set generation for the R_2^{HVC} indicator. The direction vector set generated by the GAES method was applied to the greedy approximate hypervolume subset selection, and good performance was demonstrated in comparison with the other direction vector set generation methods. One future research topic is to examine the performance of the GAES method in hypervolme-based evolutionary multi-objective algorithms.

Acknowledgements. This work was supported by National Natural Science Foundation of China (Grant No. 62002152, 61876075), Guangdong Provincial Key Laboratory (Grant No. 2020B121201001), the Program for Guangdong Introducing Innovative and Enterpreneurial Teams (Grant No. 2017ZT07X386), The Stable Support Plan Program of Shenzhen Natural Science Fund (Grant No. 20200925174447003), Shenzhen Science and Technology Program (Grant No. KQTD2016112514355531).

References

1. Beume, N., Naujoks, B., Emmerich, M.: SMS-EMOA: multiobjective selection based on dominated hypervolume. Eur. J. Oper. Res. **181**(3), 1653–1669 (2007)
2. Bradstreet, L., Barone, L., While, L.: Maximising hypervolume for selection in multi-objective evolutionary algorithms. In: Proceedings of IEEE Congress on Evolutionary Computation (CEC), pp. 1744–1751 (2006)
3. Bradstreet, L., While, L., Barone, L.: Incrementally maximising hypervolume for selection in multi-objective evolutionary algorithms. In: Proceedings of IEEE Congress on Evolutionary Computation (CEC), pp. 3203–3210 (2007)
4. Bringmann, K., Friedrich, T.: Approximating the volume of unions and intersections of high-dimensional geometric objects. Comput. Geom. Theor. Appl. **43**(6), 601–610 (2010)
5. Coello Coello, C.A., Reyes Sierra, M.: A study of the parallelization of a coevolutionary multi-objective evolutionary algorithm. In: Monroy, R., Arroyo-Figueroa, G., Sucar, L.E., Sossa, H. (eds.) MICAI 2004. LNCS (LNAI), vol. 2972, pp. 688–697. Springer, Heidelberg (2004). https://doi.org/10.1007/978-3-540-24694-7_71
6. Das, I., Dennis, J.E.: Normal-boundary intersection: a new method for generating the Pareto surface in nonlinear multicriteria optimization problems. SIAM J. Optim. **8**(3), 631–657 (1998)
7. Deb, K., Bandaru, S., Seada, H.: Generating uniformly distributed points on a unit simplex for evolutionary many-objective optimization. In: Deb, K., et al. (eds.) EMO 2019. LNCS, vol. 11411, pp. 179–190. Springer, Cham (2019). https://doi.org/10.1007/978-3-030-12598-1_15
8. Deb, K., Jain, H.: An evolutionary many-objective optimization algorithm using reference-point-based nondominated sorting approach, part I: solving problems with box constraints. IEEE Trans. Evol. Comput. **18**(4), 577–601 (2014)
9. Deb, K., Thiele, L., Laumanns, M., Zitzler, E.: Scalable test problems for evolutionary multiobjective optimization. In: Abraham, A., Jain, L., Goldberg, R. (eds.) EMO 2005, pp. 105–145. Springer, London (2005). https://doi.org/10.1007/1-84628-137-7_6
10. Deng, J., Zhang, Q.: Approximating hypervolume and hypervolume contributions using polar coordinate. IEEE Trans. Evol. Comput. **23**(5), 913–918 (2019)
11. Emmerich, M., Beume, N., Naujoks, B.: An EMO algorithm using the hypervolume measure as selection criterion. In: Coello Coello, C.A., Hernández Aguirre, A., Zitzler, E. (eds.) EMO 2005. LNCS, vol. 3410, pp. 62–76. Springer, Heidelberg (2005). https://doi.org/10.1007/978-3-540-31880-4_5
12. Guerreiro, A.P., Fonseca, C.M., Paquete, L.: Greedy hypervolume subset selection in low dimensions. Evol. Comput. **24**(3), 521–544 (2016)
13. Hansen, M.P., Jaszkiewicz, A.: Evaluating the quality of approximations to the non-dominated set. IMM Technical report, Institute of Mathematical Modelling, Technical University of Denmark (1998)
14. Huband, S., Hingston, P., Barone, L., While, L.: A review of multiobjective test problems and a scalable test problem toolkit. IEEE Trans. Evol. Comput. **10**(5), 477–506 (2006)

15. Ishibuchi, H., Imada, R., Setoguchi, Y., Nojima, Y.: Reference point specification in inverted generational distance for triangular linear Pareto front. IEEE Trans. Evol. Comput. **22**(6), 961–975 (2018)
16. Ishibuchi, H., Setoguchi, Y., Masuda, H., Nojima, Y.: Performance of decomposition-based many-objective algorithms strongly depends on Pareto front shapes. IEEE Trans. Evol. Comput. **21**(2), 169–190 (2017)
17. Jaszkiewicz, A.: On the performance of multiple-objective genetic local search on the 0/1 knapsack problem - a comparative experiment. IEEE Trans. Evol. Comput. **6**(4), 402–412 (2002)
18. Jiang, S., Zhang, J., Ong, Y.S., Zhang, A.N., Tan, P.S.: A simple and fast hypervolume indicator-based multiobjective evolutionary algorithm. IEEE Trans. Cybern. **45**(10), 2202–2213 (2015)
19. Kumar, R., Vassilvitskii, S.: Generalized distances between rankings. In: Proceedings of the 19th International Conference on World Wide Web, pp. 571–580 (2010)
20. MacQueen, J., et al.: Some methods for classification and analysis of multivariate observations. In: Proceedings of the 5th Berkeley Symposium on Mathematical Statistics and Probability, vol. 1, pp. 281–297 (1967)
21. Nan, Y., Shang, K., Ishibuchi, H.: What is a good direction vector set for the R2-based hypervolume contribution approximation. In: Proceedings of the Genetic and Evolutionary Computation Conference (GECCO), pp. 524–532 (2020)
22. Qian, C.: Distributed Pareto optimization for large-scale noisy subset selection. IEEE Trans. Evol. Comput. **24**(4), 694–707 (2020)
23. Shang, K., Ishibuchi, H.: A new hypervolume-based evolutionary algorithm for many-objective optimization. IEEE Trans. Evol. Comput. **24**(5), 839–852 (2020)
24. Shang, K., Ishibuchi, H., Chen, W.: Greedy approximated hypervolume subset selection for many-objective optimization. In: Proceedings of the Genetic and Evolutionary Computation Conference (GECCO), pp. 448–456 (2021)
25. Shang, K., Ishibuchi, H., He, L., Pang, L.M.: A survey on the hypervolume indicator in evolutionary multiobjective optimization. IEEE Trans. Evol. Comput. **25**(1), 1–20 (2021)
26. Shang, K., Ishibuchi, H., Ni, X.: R2-based hypervolume contribution approximation. IEEE Trans. Evol. Comput. **24**(1), 185–192 (2020)
27. Shang, K., Shu, T., Ishibuchi, H., Nan, Y., Pang, L.M.: Benchmarking subset selection from large candidate solution sets in evolutionary multi-objective optimization. arXiv preprint arXiv:2201.06700 (2022)
28. Van Veldhuizen, D.A.: Multiobjective evolutionary algorithms: classifications, analyses, and new innovations. Ph.D. Dissertation, Air Force Institute of Technology (1999)
29. While, L., Bradstreet, L., Barone, L.: A fast way of calculating exact hypervolumes. IEEE Trans. Evol. Comput. **16**(1), 86–95 (2012)
30. Zitzler, E., Brockhoff, D., Thiele, L.: The hypervolume indicator revisited: on the design of pareto-compliant indicators via weighted integration. In: Obayashi, S., Deb, K., Poloni, C., Hiroyasu, T., Murata, T. (eds.) EMO 2007. LNCS, vol. 4403, pp. 862–876. Springer, Heidelberg (2007). https://doi.org/10.1007/978-3-540-70928-2_64
31. Zitzler, E., Thiele, L., Laumanns, M., Fonseca, C.M., Da Fonseca, V.G.: Performance assessment of multiobjective optimizers: an analysis and review. IEEE Trans. Evol. Comput. **7**(2), 117–132 (2003)

Do We Really Need to Use Constraint Violation in Constrained Evolutionary Multi-objective Optimization?

Shuang Li[1] , Ke Li[2] , and Wei Li[1]($\boxtimes$)

[1] Control and Simulation Center, Harbin Institute of Technology, Harbin, China
fleehit@163.com
[2] Department of Computer Science, University of Exeter, Exeter EX4 5DS, UK
k.li@exeter.ac.uk

Abstract. Constraint violation has been a building block to design evolutionary multi-objective optimization algorithms for solving constrained multi-objective optimization problems. However, it is not uncommon that the constraint violation is hardly approachable in real-world blackbox optimization scenarios. It is unclear that whether the existing constrained evolutionary multi-objective optimization algorithms, whose environmental selection mechanism are built upon the constraint violation, can still work or not when the formulations of the constraint functions are unknown. Bearing this consideration in mind, this paper picks up four widely used constrained evolutionary multi-objective optimization algorithms as the baseline and develop the corresponding variants that replace the constraint violation by a crisp value. From our experiments on both synthetic and real-world benchmark test problems, we find that the performance of the selected algorithms have not been significantly influenced when the constraint violation is not used to guide the environmental selection. The supplementary material of this paper can be found in https://tinyurl.com/23dtdne8.

Keywords: Constrained multi-objective optimization · Constraint handling techniques · Evolutionary multi-objective optimization

1 Introduction

Real-world optimization problems in science [36], engineering [1] and economics [30] usually involve multiple conflicting objectives under a number of equality and inequality constraints, a.k.a. constrained multi-objective optimization problems (CMOPs). In this paper, we consider the CMOP defined as follows:

$$\begin{array}{ll} \text{minimize} & \mathbf{F}(\mathbf{x}) = (f_1(\mathbf{x}), \cdots, f_m(\mathbf{x}))^T \\ \text{subject to} & \mathbf{g}(\mathbf{x}) \leq 0 \\ & \mathbf{h}(\mathbf{x}) = 0 \\ & \mathbf{x} = (x_1, \cdots, x_n)^T \in \Omega \end{array}, \qquad (1)$$

This work was supported by UKRI Future Leaders Fellowship (MR/S017062/1), EPSRC (2404317), NSFC (62076056), Royal Society (IES/R2/212077) and Amazon Research Award.

G. Rudolph et al. (Eds.): PPSN 2022, LNCS 13399, pp. 124–137, 2022.
https://doi.org/10.1007/978-3-031-14721-0_9

where $\Omega = [x_i^L, x_i^U]_{i=1}^n \subseteq \mathbb{R}^n$ defines the search (or decision variable) space and $\mathbf{x}$ is an n-dimensional vector therein. $\mathbf{F} : \Omega \rightarrow \mathbb{R}^m$ constitutes m conflicting objective functions, and $\mathbb{R}^m$ is the objective space. $\mathbf{g}(\mathbf{x}) = (g_1(\mathbf{x}), \cdots, g_p(\mathbf{x}))^T$ and $\mathbf{h}(\mathbf{x}) = (h_1(\mathbf{x}), \cdots, h_q(\mathbf{x}))^T$ are vectors of inequality and equality constraints respectively. Given a CMOP, the degree of constraint violation of a solution $\mathbf{x}$ at the i-th constraint is calculated as:

$$c_i(\mathbf{x}) = \begin{cases} \langle 1 - g_j(\mathbf{x})/a_j \rangle, & j = 1, \cdots, q, \ i = j, \\ \langle \epsilon - |h_k(\mathbf{x})/b_k - 1| \rangle, & k = 1, \cdots, p, \ i = k + q, \end{cases} \quad (2)$$

where ϵ is a small tolerance term (e.g., $\epsilon = 10^{-6}$) that relaxes the equality constraints to the inequality constraints. a_j and b_k where $j \in \{1, \cdots, q\}$ and $k \in \{1, \cdots, p\}$ are normalization factors of the corresponding constraints. $\langle \alpha \rangle$ returns 0 if $\alpha \geq 0$ otherwise it returns the negative of α. Given a CMOP, the constraint violation (CV) value of a solution $\mathbf{x}$ is calculated as:

$$CV(\mathbf{x}) = \sum_{i=1}^{\ell} c_i(\mathbf{x}), \quad (3)$$

where $\ell = p + q$. $\mathbf{x}$ is feasible in case $CV(\mathbf{x}) = 0$; otherwise $\mathbf{x}$ is infeasible. Given two feasible solutions $\mathbf{x}^1$ and $\mathbf{x}^2$, $\mathbf{x}^1$ is said to *Pareto dominate* $\mathbf{x}^2$ (denoted as $\mathbf{x}^1 \preceq \mathbf{x}^2$) if and only if $f_i(\mathbf{x}^1) \leq f_i(\mathbf{x}^2)$, $\forall i \in \{1, \cdots, m\}$ and $\exists j \in \{1, \cdots, m\}$ such that $f_j(\mathbf{x}^1) < f_j(\mathbf{x}^2)$. A solution $\mathbf{x}^* \in \Omega$ is *Pareto-optimal* with respect to (1) if $\nexists \mathbf{x} \in \Omega$ such that $\mathbf{x} \preceq \mathbf{x}^*$. The set of all Pareto-optimal solutions is called the *Pareto-optimal set* (PS). Accordingly, $PF = \{\mathbf{F}(\mathbf{x})|\mathbf{x} \in PS\}$ is called the *Pareto-optimal front* (PF).

Due to the population-based property, evolutionary algorithms (EAs) have been widely recognized as an effective approach for multi-objective optimization. Over the past three decades, much effort have been devoted to developing evolutionary multi-objective optimization (EMO) algorithms, e.g. elitist non-dominated sorting genetic algorithm (NSGA-II) [8], indicator-based EA (IBEA) [44] and multi-objective EA based on decomposition (MOEA/D) [43]. However, they cannot be directly applied to CMOPs without the assistance of a constraint handling technique (CHT), which can be seen as a selection mechanism to deal with constraints. In the 90s, some early endeavors to the development of EAs for solving CMOPs (e.g., [11] and [7]) are simply driven by a prioritization of the search for feasible solutions over the 'optimal' one. However, such methods are notorious for the loss of selection pressure in the case where the population is filled with infeasible solutions.

After the development of the constrained dominance relation [8], most, if not all, prevalent CHTs in the EMO community directly or indirectly depend on the CV defined in Eq. (3). Specifically, a solution $\mathbf{x}^1$ is said to constraint-dominate $\mathbf{x}^2$, if: 1) $\mathbf{x}^1$ is feasible while $\mathbf{x}^2$ is not; 2) both of them are infeasible and $CV(\mathbf{x}^1) < CV(\mathbf{x}^2)$; or 3) both of them are feasible and $\mathbf{x}^1 \prec \mathbf{x}^2$. By replacing the Pareto dominance relation with this constrained dominance relation, the state-of-the-art NSGA-II and NSGA-III [16] can be readily used to tackle CMOPs.

Borrowing this idea, several MOEA/D variants (e.g., [6,16,17,24]) use the CV as an alternative criterion in the subproblem update procedure. Moreover, the constrained dominance relation is augmented with terms such as the number of violated constraints [28], ϵ-constraint [4,25,35] and angle between each other [10] to provide an additional selection pressure to infeasible solutions whose CV values have a marginal difference.

In addition to the above feasibility-driven CHTs, the second category aims at balancing the trade-off between convergence and feasibility during the search process. For example, Jiménez et al. [19] proposed a min-max formulation that drives feasible and infeasible solutions to evolve towards optimality and feasibility, respectively. In [31], a Ray-Tai-Seow algorithm was proposed to simultaneously take the objective values, the CV along with the combination of them into consideration to compare and rank non-dominated solutions. Based on the similar rigour, some modified ranking mechanisms (e.g., [2,41,42]) were developed by leveraging the information from both the objective and constraint spaces. Instead of prioritizing feasible solutions, some researchers (e.g., [22,29,34]) proposed to exploit information from infeasible solutions in case they can provide additional diversity to the current evolutionary population.

As a step further, the third category seeks to strike a balance among convergence, diversity and feasibility simultaneously. As a pioneer along this line, Li et al. proposed a two-archive EA that maintains two co-evolving and complementary populations to solve CMOPs [21]. Specifically, one archive, denoted as the convergence-oriented archive (CA), pushes the population towards the PF; while the other one, denoted as the diversity-oriented archive, provides additional diversity. To complement the behavior of the CA, the DA explores the areas under-exploited by the CA including the infeasible region(s). In addition, to take advantage of the complementary effects of both CA and DA, a restricted mating selection mechanism was proposed to adaptively choose appropriate mating parents according to the evolution status of the CA and DA respectively. After [21], there has been a spike of efforts on the development of multi-population strategies (e.g., [23,26,27,32,37,39]) to leverage some complementary effects of both feasible and infeasible solutions simultaneously for solving CMOPs.

Instead of the environmental selection, the last category tries to repair the infeasible solutions in order to drive them towards the feasible region(s). For example, a so-called Pareto descent repair operator [13] was proposed to explore possible feasible solutions along the gradient information around infeasible solutions in the constraint space. In [18], a feasible-guided strategy was developed to guide infeasible solutions towards the feasible region along the 'feasible direction', i.e., a vector starting from an infeasible solution and ending up with its nearest feasible solution. In [33], a simulated annealing algorithm was applied to accelerate the progress of movements from infeasible solutions toward feasible ones.

Remark 1. As discussed at the outset of this subsection, all these prevalent CHTs require the access of the CV. This applies to the last category, since it needs to access the gradient information of the CV. The implicit assumption behind the prevalent CHTs is the access of the closed form of the constraint function(s). However, this is not practical in the real world, such problems are usually a black box (e.g., [3,12,14]). In other words, we can only know whether a solution is feasible or not.

Bearing this consideration in mind, we come up with the overarching research question of this paper: *do the prevalent CHTs in the EMO literature still work when we do not have an access to the CV?*

The rest of this paper is organized as follows. The experimental settings are summarized in Sect. 2 and the results are presented and analyzed in Sect. 3. Finally, Sect. 4 concludes this paper and sheds some light on future directions.

2   Experimental Settings

In this section, we introduce the experimental settings of our empirical study including the benchmark test problems, the peer algorithms, the performance metrics and statistical tests.

2.1   Benchmark Test Problems

In our empirical study, we use 45 benchmark test problems widely studied in the literature to constitute our benchmark suite. More specifically, it consists of C1-DTLZ1, C1-DTLZ3, C2-DTLZ2 and C3-DTLZ4 from the C-DTLZ benchmark suite [16]; DC1-DTLZ1, DC1-DTLZ3, DC2-DTLZ1, DC2-DTLZ3, DC3-DTLZ1, DC3-DTLZ3 chosen from the DC-DTLZ benchmark suite [21]; and other 35 problems picked up from the real-world constrained multi-objective problems (RWCMOPs) benchmark suite [20]. In particular, the RWCMOPs are derived from the mechanical design problems (denoted as RCM1 to RCM21), the chemical engineering problems (denoted as RCM22 to RCM24), the process design and synthesis problems (denoted as RCM25 to RCM29), and the power electronics problems (denoted as RCM30 to RCM35), respectively. All these benchmark test problems are scalable to any number of objectives while we consider $m \in \{2, 3, 5, 10\}$ for C-DTLZ, DC-DTLZ problems and $m \in \{2, 3, 4, 5\}$ for RWC-MOPs in our experiments. The mathematical definitions of these benchmark test problems along with settings of the number of variables and the number of constraints can be found in the supplemental document of this paper.[1]

2.2   Peer Algorithms and Parameter Settings

In our empirical study, we choose to investigate the performance of four widely studied EMO algorithms for CMOPs, including C-NSGA-II [8], C-NSGA-III [16], C-MOEA/D [16], and C-TAEA [21]. To address our overarching research question stated at the end of Sect. 1, we design a variant for each of these peer algorithms (dubbed vC-NSGA-II, vC-NSGA-III, vC-MOEA/D, and vC-TAEA, respectively) by

---

[1] The supplemental document can be downloaded from here.

replacing the CV with a crisp value. Specifically, if a solution $\mathbf{x}$ is feasible, we have $CV(\mathbf{x}) = 1$; otherwise we set $CV(\mathbf{x}) = -1$. The settings of population size and the maximum number of function evaluations are detailed in the supplemental document of this paper.

2.3   Performance Metrics and Statistical Tests

This paper applies the widely used inverted generational distance (IGD) [5], IGD^+ [15], and hypervolume (HV) [45] as the performance metrics to evaluate the performance of different peer algorithms. In our empirical study, each experiment is independently repeated 31 times with a different random seed. To have a statistical interpretation of the significance of comparison results, we use the following two statistical measures in our empirical study.

- Wilcoxon signed-rank test [40]: This is a non-parametric statistical test that makes no assumption about the underlying distribution of the data and has been recommended in many empirical studies in the EA community [9]. In particular, the significance level is set to $p = 0.05$ in our experiments.
- A_{12} effect size [38]: To ensure the resulted differences are not generated from a trivial effect, we apply A_{12} as the effect size measure to evaluate the probability that one algorithm is better than another. Specifically, given a pair of peer algorithms, $A_{12} = 0.5$ means they are *equal*. $A_{12} > 0.5$ denotes that one is better for more than 50% of the times. $0.56 \leq A_{12} < 0.64$ indicates a *small* effect size while $0.64 \leq A_{12} < 0.71$ and $A_{12} \geq 0.71$ mean a *medium* and a *large* effect size, respectively.

3   Experimental Results

The PFs and the feasible regions of the synthetic problems are relatively simple whereas those of RWCMOPs are complex. In this section, we plan to separate the discussion on the synthetic problems (i.e., C-DTLZ and DC-DTLZ) from the RWCMOPs in view of their distinctive characteristics.

3.1   Performance Analysis on Synthetic Benchmark Test Problems

Due to page limit, we leave the complete comparison results of IGD, IGD^+, and HV in Table 3 to Table 8 of supplemental document. Instead, we summarize the Wilcoxon signed-rank test results in the Table 1. From this table, it is clear to see that most comparison results (at least 62.5% and it even goes to 100% for the HV comparisons between C-TAEA and vC-TAEA) do not have any statistical significance. In other words, replacing the CV with a crisp value does not significantly influence the performance on C-DTLZ and DC-DTLZ problems. In addition to the pairwise comparisons, we apply the A_{12} effect size to have a better understanding of the performance difference between the selected EMO algorithm and its corresponding variant. From the collected comparison results

Table 1. Summary of the Wilcoxon signed-rank test results of four selected EMO algorithms against their corresponding variants on IGD, IGD$^+$, and HV.

Problems	Metrics	C-NSGA-II +/−/=	C-NSGA-III +/−/=	C-MOEA/D +/−/=	C-TAEA +/−/=
C-DTLZ and DC-DTLZ	IGD	1/7/32	1/13/26	3/1/36	2/0/38
	IGD$^+$	1/7/32	0/15/25	3/7/36	1/0/39
	HV	0/8/32	0/15/25	4/1/35	0/0/40
RWCMOPs	HV	0/8/27	0/11/24	4/7/24	1/4/30

+, −, and = denote the performance of the selected algorithm is significantly better, worse, and equivalent to the corresponding variant, respectively.

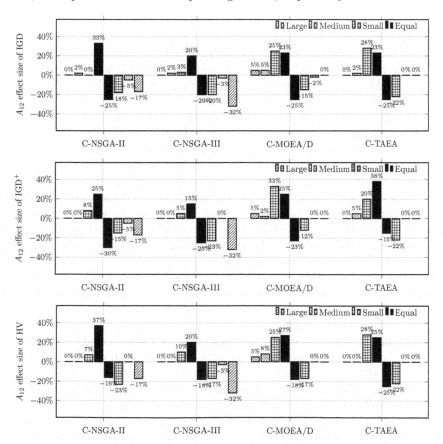

Fig. 1. Percentage of the large, medium, small, and equal A12 effect size of metrics for C-DTLZ and DC-DTLZ problems. + means that the variant that replaces the CV with a crisp value can obtain a better result; − means the opposite case.

$(50 \times 2 = 100$ in total) shown in Fig. 1, we can see that most of the comparison results are classified as *equal* (ranging from 38% to 58%). As reflected in Table 1, it is surprising to see that the corresponding variants (i.e., without using the CV to guide the evolutionary search process) have achieved better performance in many cases. In particular, up to 32% comparison results are classified to be *large*. In the following paragraphs, we plan to analyse some remarkable findings collected from the results.

- Let us first look into the performance of C-NSGA-II and C-NSGA-III w.r.t. their variants vC-NSGA-II and vC-NSGA-III. As shown in Table 1, the performance of C-NSGA-II and C-NSGA-III have been deteriorated (ranging from 17.5% to 37.5%) when replacing the CV with a crisp value in their corresponding CHTs, especially on C1-DTLZ1 and C2-DTLZ2.
    - As the illustrative example shown in Fig. 2, the feasible region of C1-DTLZ1 is a narrow wedge arrow right above the PF. Without the guidance of the CV, both C-NSGA-II and C-NSGA-III struggle in the large infeasible region. In particular, there is no sufficient selection pressure to guide the population to move forward.
    - C2-DTLZ2 has several disparately distributed feasible regions as the illustrative example shown in Fig. 3. Since the CHTs of both C-NSGA-II and C-NSGA-III do not have a dedicated diversity maintenance mechanism, the evolutionary population can be guided towards some, but not all, local feasible region(s) as the examples shown in Fig. 3(c).
    - As for the other test problems, we find that the replacement of CV with a crisp value dose not make a significant impact to the performance of both C-NSGA-II and C-NSGA-III. This can be explained as a large feasible region that makes the Pareto dominance alone can provide sufficient selection pressure towards the PF.
- It is interesting to note that C-MOEA/D uses the same CHT as C-NSGA-II and C-NSGA-III, but its performance does not deteriorate significantly when replacing the CV with a crisp value as shown in Table 1. This can be understood as the baseline MOEA/D that provides a better mechanism to preserve the population diversity during the environmental selection. Thus, the evolutionary population can overcome the infeasible regions towards the PF.
- As for C-TAEA, it is surprising to note that the consideration of the CV does not pose any impact to its performance as evidenced in Table 1 (nearly all comparison results have no statistical significance). This can be explained as the use of the diversity-oriented archive in C-TAEA which does not consider the CV but just relies on the Pareto dominance alone to drive the evolutionary population.

3.2   Performance Analysis on Real-World Benchmark Test Problems

Since the PFs of the RWCMOPs are unknown a priori, we only apply the HV as the performance metric in this study. As in Sect. 3.1, the complete comparison

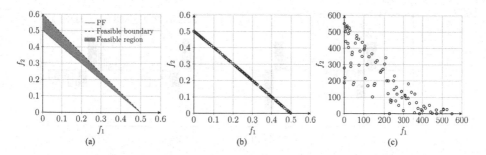

Fig. 2. (a) The illustration of the feasible region of C1-DTLZ1; (b) and (c) are the scatter plots of the non-dominated solutions (with the median IGD value) obtained by C-NSGA-II and vC-NSGA-II, respectively.

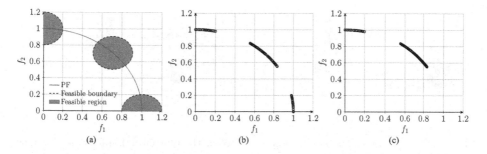

Fig. 3. (a) The illustration of the feasible region of C2-DTLZ2; (b) and (c) are the scatter plots of the non-dominated solutions (with the median IGD value) obtained by C-NSGA-II and vC-NSGA-II, respectively.

results of HV are given in Tables 9 and 10 of the supplemental document while the Wilcoxon signed-rank test results are summarized in Table 1. From these results, we can see that most of the comparisons (around 68.5% to 85.7%) do not have statistical significance. In other words, there is a marginal difference when replacing the CV with a crisp value. To have a better understanding of the difference, we again apply the A_{12} effect size to complement the results of the Wilcoxon signed-rank test. From the bar charts shown in Fig. 4, it is clear to see that most comparison results (ranging from 46% to 69%) are classified to be equal while only up to 14% comparison results are classified to have a large difference. In the following paragraphs, we plan to elaborate some selected results on problems with an equal and a large effect size, respectively.

As for the RWCMOPs whose A_{12} effect size comparison results are classified as equal, we consider the following four test problems in our analysis.

– Let us start from the RCM5 problem. As shown in Fig. 5(a), the feasible and infeasible regions have almost the same size while the PF is located in the intersection between them. In this case, it is natural that the environmental selection can provide necessary selection pressure without using the CV.

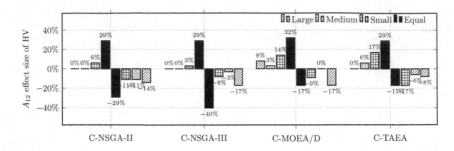

Fig. 4. Percentage of the large, medium, small, and equal A_{12} effect size of metrics for RWCMOPs. + means that the variant that replaces the CV with a crisp value can obtain a better result; − means the opposite case.

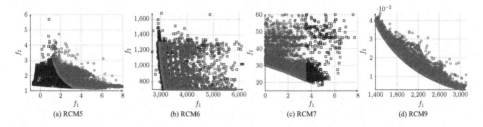

Fig. 5. Distribution of feasible solutions (denoted as the blue circle), infeasible solutions (denoted as the black square), and non-dominated solutions (denoted as red triangle) obtained by C-NSGA-II on RCM5, RCM6, RCM7 and RCM9. (Color figure online)

- As for the RCM6 problem shown in Fig. 5(b), the feasible and infeasible regions are intertwined with each other. Therefore, the infeasible region does not really provide an obstacle to the evolutionary population. Accordingly, the CV plays a marginal role for constraint handling.
- Similar to the RCM5 problem, the RCM7 problem has a large and opening feasible region as shown in Fig. 5(c). In addition, the feasible and infeasible regions are hardly overlapped with each. In this case, the evolutionary population can have a large chance to explore in the feasible region without any interference from the infeasible solutions.
- At the end, as shown in Fig. 5(d), the RCM9 problem can hardly be treated as a CMOP since the feasible region is overtaking the infeasible region. In other words, the feasible region is too large to find an infeasible solution. Accordingly, it is not difficult to understand that the CV becomes useless.

As for the other RWCMOPs, of which the comparison results are classified to be large according to the A_{12} effect size, we pick up two remarkable cases and make some analysis as follows.

- Let us first consider the RCM30 problem. As shown in Fig. 6(a), the feasible region of the RCM30 problem is very narrow and is squeezed towards the

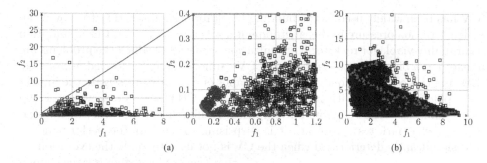

Fig. 6. (a) Distribution of feasible solutions (denoted as the blue circle), infeasible solutions (denoted as the black square), and non-dominated solutions (denoted as red triangle) obtained by C-NSGA-II on RCM30. (b) Distribution of infeasible solutions (denoted as the black square) and non-dominated solutions (denoted as red triangle) obtained by vC-NSGA-II on RCM30. (Color figure online)

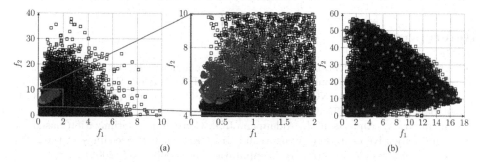

Fig. 7. (a) Distribution of feasible solutions (denoted as the blue circle), infeasible solutions (denoted as the black square), and non-dominated solutions (denoted as red triangle) obtained by C-NSGA-II on RCM35. (b) Distribution of infeasible solutions (denoted as the black square) and non-dominated solutions (denoted as red triangle) obtained by vC-NSGA-II on RCM35. (Color figure online)

PF. Therefore, it is not difficult to understand that the evolutionary population can hardly be navigated without the guidance of the CV. As shown in Fig. 6(b), the solutions obtained by vC-NSGA-II are far away from the PF.

- As shown in Fig. 7(a), comparing to the RCM30 problem, the size of the feasible region of the RCM35 problem is much wider. However, it is still largely surrounded by the infeasible region. In this case, as shown in Fig. 7(b), without the guidance of the CV, the evolutionary population can not only have sufficient selection pressure to move forward, but also can be mislead to the infeasible region that dominates the feasible region.

4 Conclusion

Most, if not all, existing CHT in EMO implicitly assume that the formulation of the constraint function(s) of a CMOP is well defined a priori. Therefore, the

CV has been widely used as the building block for designing CHTs to provide an extra selection pressure in the environmental selection. However, this assumption is arguably viable for real-world optimization scenarios of which the problems are treated as a black box. In this case, the CV cannot usually be derived in practice. Bearing this consideration in mind, this paper empirically investigate the impact of replacing the CV with a crisp value in the CHTs of four prevalent EMO algorithms for CMOPs. From our empirical results on both synthetic and real-world benchmark test problems, it is surprising to see that the performance is not significantly deteriorated when the CV is not used to guide the evolutionary population. One of the potential reasons is that the feasible region is large enough to attract the evolutionary population thus leading to a marginal obstacle for an EMO algorithm to overcome the infeasible region. This directly comes up to the requirement of new benchmark test problems with more challenging infeasible regions. In addition, this also inspires new research opportunity to develop new CHT(s) to handle the CMOP with unknown constraint in the near future.

References

1. Andersson, J.: Applications of a multi-objective genetic algorithm to engineering design problems. In: Fonseca, C.M., Fleming, P.J., Zitzler, E., Thiele, L., Deb, K. (eds.) EMO 2003. LNCS, vol. 2632, pp. 737–751. Springer, Heidelberg (2003). https://doi.org/10.1007/3-540-36970-8_52
2. Angantyr, A., Andersson, J., Aidanpaa, J.O.: Constrained optimization based on a multiobjective evolutionary algorithm. In: CEC 2003: Proceedings of the 2003 IEEE Congress on Evolutionary Computation, pp. 1560–1567 (2003)
3. Ariafar, S., Coll-Font, J., Brooks, D.H., Dy, J.G.: ADMMBO: Bayesian optimization with unknown constraints using ADMM. J. Mach. Learn. Res. **20**, 123:1–123:26 (2019)
4. Asafuddoula, M., Ray, T., Sarker, R.A.: A decomposition-based evolutionary algorithm for many objective optimization. IEEE Trans. Evol. Comput. **19**(3), 445–460 (2015)
5. Bosman, P.A.N., Thierens, D.: The balance between proximity and diversity in multiobjective evolutionary algorithms. IEEE Trans. Evol. Comput. **7**(2), 174–188 (2003)
6. Cheng, R., Jin, Y., Olhofer, M., Sendhoff, B.: A reference vector guided evolutionary algorithm for many-objective optimization. IEEE Trans. Evol. Comput. **20**(5), 773–791 (2016)
7. Coello, C.A.C., Christiansen, A.D.: MOSES: a multiobjective optimization tool for engineering design. Eng. Opt. **31**(3), 337–368 (1999)
8. Deb, K., Agrawal, S., Pratap, A., Meyarivan, T.: A fast and elitist multiobjective genetic algorithm: NSGA-II. IEEE Trans. Evol. Comput. **6**(2), 182–197 (2002)
9. Derrac, J., García, S., Molina, D., Herrera, F.: A practical tutorial on the use of nonparametric statistical tests as a methodology for comparing evolutionary and swarm intelligence algorithms. Swarm Evol. Comput. **1**(1), 3–18 (2011)
10. Fan, Z., Li, W., Cai, X., Hu, K., Lin, H., Li, H.: Angle-based constrained dominance principle in MOEA/D for constrained multi-objective optimization problems. In: CEC 2016: Proceedings of the 2016 IEEE Congress on Evolutionary Computation, pp. 460–467. IEEE (2016)

11. Fonseca, C.M., Fleming, P.J.: Multiobjective optimization and multiple constraint handling with evolutionary algorithms. I. A unified formulation. IEEE Trans. Syst. Man Cybern., Part A **28**(1), 26–37 (1998)
12. Gelbart, M.A., Snoek, J., Adams, R.P.: Bayesian optimization with unknown constraints. In: Proceedings of the Thirtieth Conference on Uncertainty in Artificial Intelligence, UAI 2014, Quebec City, Quebec, Canada, 23–27 July 2014, pp. 250–259. AUAI (2014)
13. Harada, K., Sakuma, J., Ono, I., Kobayashi, S.: Constraint-handling method for multi-objective function optimization: Pareto descent repair operator. In: Obayashi, S., Deb, K., Poloni, C., Hiroyasu, T., Murata, T. (eds.) EMO 2007. LNCS, vol. 4403, pp. 156–170. Springer, Heidelberg (2007). https://doi.org/10. 1007/978-3-540-70928-2_15
14. Hernández-Lobato, J.M., Gelbart, M.A., Hoffman, M.W., Adams, R.P., Ghahramani, Z.: Predictive entropy search for Bayesian optimization with unknown constraints. In: Proceedings of the 32nd International Conference on Machine Learning, ICML 2015, Lille, France, 6–11 July 2015, vol. 37, pp. 1699–1707. JMLR (2015)
15. Ishibuchi, H., Masuda, H., Tanigaki, Y., Nojima, Y.: Modified distance calculation in generational distance and inverted generational distance. In: Gaspar-Cunha, A., Henggeler Antunes, C., Coello, C.C. (eds.) EMO 2015. LNCS, vol. 9019, pp. 110–125. Springer, Cham (2015). https://doi.org/10.1007/978-3-319-15892-1_8
16. Jain, H., Deb, K.: An evolutionary many-objective optimization algorithm using reference-point based nondominated sorting approach, part II: handling constraints and extending to an adaptive approach. IEEE Trans. Evol. Comput. **18**(4), 602–622 (2014)
17. Jan, M.A., Zhang, Q.: MOEA/D for constrained multiobjective optimization: some preliminary experimental results. In: UKCI 2010: Proceedings of the 2010 UK Workshop on Computational Intelligence, pp. 1–6 (2010)
18. Jiao, L., Luo, J., Shang, R., Liu, F.: A modified objective function method with feasible-guiding strategy to solve constrained multi-objective optimization problems. Appl. Soft Comput. **14**, 363–380 (2014)
19. Jiménez, F., Gómez-Skarmeta, A.F., Sánchez, G., Deb, K.: An evolutionary algorithm for constrained multi-objective optimization. In: CEC 2002: Proceedings of the 2002 IEEE Congress on Evolutionary Computation, pp. 1133–1138 (2002)
20. Kumar, A., et al.: A benchmark-suite of real-world constrained multi-objective optimization problems and some baseline results. Swarm Evol. Comput. **67**, 100961 (2021)
21. Li, K., Chen, R., Fu, G., Yao, X.: Two-archive evolutionary algorithm for constrained multiobjective optimization. IEEE Trans. Evol. Comput. **23**(2), 303–315 (2019)
22. Li, K., Deb, K., Zhang, Q., Kwong, S.: An evolutionary many-objective optimization algorithm based on dominance and decomposition. IEEE Trans. Evol. Comput. **19**(5), 694–716 (2015)
23. Liu, Z.Z., Wang, B.C., Tang, K.: Handling constrained multiobjective optimization problems via bidirectional coevolution. IEEE Trans. Cybern., 1–14 (2021, early access)
24. Liu, Z., Wang, Y., Huang, P.: AnD: a many-objective evolutionary algorithm with angle-based selection and shift-based density estimation. Inf. Sci. **509**, 400–419 (2020)

25. Martínez, S.Z., Coello, C.A.C.: A multi-objective evolutionary algorithm based on decomposition for constrained multi-objective optimization. In: CEC 2014: Proceedings of the 2014 IEEE Congress on Evolutionary Computation, pp. 429–436 (2014)

26. Ming, F., Gong, W., Wang, L., Gao, L.: A constrained many-objective optimization evolutionary algorithm with enhanced mating and environmental selections. IEEE Trans. Cybern., 1–13 (2022, early access)

27. Ming, F., Gong, W., Wang, L., Lu, C.: A tri-population based co-evolutionary framework for constrained multi-objective optimization problems. Swarm Evol. Comput. **70**, 101055 (2022)

28. Oyama, A., Shimoyama, K., Fujii, K.: New constraint-handling method for multi-objective and multi-constraint evolutionary optimization. Jpn. Soc. Aeronaut. Space Sci. Trans. **50**, 56–62 (2007)

29. Peng, C., Liu, H., Gu, F.: An evolutionary algorithm with directed weights for constrained multi-objective optimization. Appl. Soft Comput. **60**, 613–622 (2017)

30. Ponsich, A., Jaimes, A.L., Coello, C.A.C.: A survey on multiobjective evolutionary algorithms for the solution of the portfolio optimization problem and other finance and economics applications. IEEE Trans. Evol. Comput. **17**(3), 321–344 (2013)

31. Ray, T., Tai, K., Seow, K.C.: Multiobjective design optimization by an evolutionary algorithm. Eng. Opt. **33**(4), 399–424 (2001)

32. Shan, X., Li, K.: An improved two-archive evolutionary algorithm for constrained multi-objective optimization. In: Ishibuchi, H., et al. (eds.) EMO 2021. LNCS, vol. 12654, pp. 235–247. Springer, Cham (2021). https://doi.org/10.1007/978-3-030-72062-9_19

33. Singh, H.K., Ray, T., Smith, W.: C-PSA: Constrained Pareto simulated annealing for constrained multi-objective optimization. Inf. Sci. **180**(13), 2499–2513 (2010)

34. Ebrahim Sorkhabi, A., Deljavan Amiri, M., Khanteymoori, A.R.: Duality evolution: an efficient approach to constraint handling in multi-objective particle swarm optimization. Soft. Comput. **21**(24), 7251–7267 (2016). https://doi.org/10.1007/s00500-016-2422-5

35. Takahama, T., Sakai, S.: Efficient constrained optimization by the ϵ constrained rank-based differential evolution. In: CEC 2012: Proceedings of the 2012 IEEE Congress on Evolutionary Computation, pp. 1–8 (2012)

36. Thurston, D.L., Srinivasan, S.: Constrained optimization for green engineering decision-making. Environ. Sci. Technol. **37**(23), 5389–5397 (2003)

37. Tian, Y., Zhang, T., Xiao, J., Zhang, X., Jin, Y.: A coevolutionary framework for constrained multiobjective optimization problems. IEEE Trans. Evol. Comput. **25**(1), 102–116 (2021)

38. Vargha, A., Delaney, H.D.: A critique and improvement of the CL common language effect size statistics of McGraw and Wong. J. Educ. Behav. Stat. **25**(2), 101–132 (2000)

39. Wang, J., Li, Y., Zhang, Q., Zhang, Z., Gao, S.: Cooperative multiobjective evolutionary algorithm with propulsive population for constrained multiobjective optimization. IEEE Trans. Syst. Man Cybern.: Syst. **52**, 3476–3491 (2021)

40. Wilcoxon, F.: Individual comparisons by ranking methods. In: Kotz, S., Johnson, N.L. (eds.) Breakthroughs in Statistics. Springer, New York (1945). https://doi.org/10.1007/978-1-4612-4380-9_16

41. Woldesenbet, Y.G., Yen, G.G., Tessema, B.G.: Constraint handling in multiobjective evolutionary optimization. IEEE Trans. Evol. Comput. **13**(3), 514–525 (2009)

42. Young, N.: Blended ranking to cross infeasible regions in constrained multi-objective problems. In: CIMCA 2005: Proceedings of the 2005 International Conference on Computational Intelligence Modeling, Control and Automation, pp. 191–196 (2005)
43. Zhang, Q., Li, H.: MOEA/D: a multiobjective evolutionary algorithm based on decomposition. IEEE Trans. Evol. Comput. **11**(6), 712–731 (2007)
44. Zitzler, E., Künzli, S.: Indicator-based selection in multiobjective search. In: Yao, X., et al. (eds.) PPSN 2004. LNCS, vol. 3242, pp. 832–842. Springer, Heidelberg (2004). https://doi.org/10.1007/978-3-540-30217-9_84
45. Zitzler, E., Thiele, L.: Multiobjective evolutionary algorithms: a comparative case study and the strength Pareto approach. IEEE Trans. Evol. Comput. **3**(4), 257–271 (1999)

Dynamic Multi-modal Multi-objective Optimization: A Preliminary Study

Yiming Peng and Hisao Ishibuchi[(✉)]

Guangdong Provincial Key Laboratory of Brain-Inspired Intelligent Computation,
Department of Computer Science and Engineering, Southern University of Science
and Technology, Shenzhen 518055, China
`11510035@mail.sustech.edu.cn, hisao@sustech.edu.cn`

Abstract. Many real-world multi-modal multi-objective optimization problems are subject to continuously changing environments, which requires the optimizer to track multiple equivalent Pareto sets in the decision space. To the best of our knowledge, this type of optimization problems has not been studied in the literature. To fill the research gap in this area, we provide a preliminary study on dynamic multi-modal multi-objective optimization. We give a formal definition of dynamic multi-modal multi-objective optimization problems and point out some key challenges in solving them. To facilitate algorithm development, we suggest a systematic approach to construct benchmark problems. Furthermore, we provide a feature-rich test suite containing 10 novel dynamic multi-modal multi-objective test problems.

Keywords: Evolutionary multi-objective optimization · Multi-modal multi-objective optimization · Dynamic multi-objective optimization · Benchmark problems

1 Introduction

Over the past few years, multi-modal multi-objective optimization problems (MMOPs) have received increasing attention from researchers and rapidly become a popular research area. This special class of multi-objective optimization problems is characterized by having multiple equivalent Pareto sets in the decision space. As pointed out in [4], equivalent Pareto sets are useful in practical applications since they can provide extra flexibility in the decision-making procedure. Thus, in addition to ensuring a good solution distribution over the Pareto front, a multi-modal multi-objective optimization algorithm is also required to ensure the diversity in the decision space to cover as many equivalent Pareto sets as possible. Various real-world optimization problems such as the rocket

Supplementary Information The online version contains supplementary material available at https://doi.org/10.1007/978-3-031-14721-0_10.

engine design problems [24], the neural architecture search problems [26], and the multi-objective knapsack problems [13] can be formulated as MMOPs.

Recently, many efficient algorithms have emerged to solve MMOPs efficiently, e.g., algorithms in [7,16,18,27]. However, up to now, the algorithm research on multi-modal multi-objective optimization has tended to focus on solving MMOPs in static environments. This greatly limits the value of these algorithms in real-world applications, where the environment in which the optimization problem is posed is often dynamically changing. Due to the dynamic environment, the Pareto front and/or the Pareto set of an MMOP may change over time. For example, it is not uncommon that a rocket engine design obtained by the above-mentioned approach [24] is no longer viable due to the change of some physical constraints. In this case, instead of simply restarting the algorithm to search for a new solution, it would be more efficient to utilize the original Pareto optimal solutions (i.e., the Pareto optimal solutions obtained before the change of the environment). In this paper, we refer to this type of optimization problems as dynamic MMOPs (dMMOPs). From the previous example, we can see that dMMOPs are essentially equivalent to solving a series of MMOPs, which can be viewed as an extension of dynamic multi-objective optimization problems (dMOPs) [8]. Therefore, existing techniques for handling dMOPs are also helpful for dMMOPs.

This paper provides a preliminary study on dMMOPs. We first give a formal definition of dMMOPs and analyze some key challenges in solving them. Being a novel type of optimization problems, dMMOPs pose unprecedented challenges to algorithm designers. To facilitate algorithm development, we provide a systematic approach for constructing dMMOPs for benchmarking the algorithm performance. Furthermore, we suggest an easy-to-use test suite containing 10 novel test problems based on the proposed approach.

2   Related Work

2.1   Multi-modal Multi-objective Optimization

In Sect. 1, we explained that MMOPs have multiple equivalent Pareto sets in the decision space. In our previous work [21], we provided a more precise definition for MMOPs. Furthermore, we pointed out that the main challenge in solving MMOPs comes from the need for the algorithm to maintain the diversity of populations in both the decision and objective spaces. One strategy is to select solutions with good diversity in both the decision and objective spaces in environmental selection. For example, both Omni-optimizer [7] and MO_Ring_PSO_SCD [27] use modified crowding distance [5] metrics which consider the diversity in both spaces. Another popular approach is to use niching [25] mechanisms to "divide" the population into several niches, each of which evolves independently. In this manner, solutions in different niches can converge to different Pareto sets. For example, MMOEA-DC [16] partitions the population into several clusters, DNEA [17] adopts the fitness sharing strategy [10], MOEA/D-MM [20] uses the clearing strategy [22].

2.2   Dynamic Multi-objective Optimization

A basic dMOP can be defined by introducing a time variable t into a standard multi-objective optimization problem as follows:

$$\min \boldsymbol{F}(\boldsymbol{x}) = (f_1(\boldsymbol{x}, t), f_2(\boldsymbol{x}, t), \ldots, f_M(\boldsymbol{x}, t))^T,$$
$$\text{s. t. } \boldsymbol{g}(\boldsymbol{x}, t) \leq 0, \boldsymbol{h}(\boldsymbol{x}, t) = 0, \tag{1}$$

where $\boldsymbol{F}$, $\boldsymbol{g}$, and $\boldsymbol{h}$ are dynamic (i.e., time-dependent) objective functions, inequality constraints, and equality constraints, respectively.

As shown in (1), both the objective functions and the constraints may change over time, which can lead to certain Pareto optimal solutions becoming suboptimal or infeasible. Therefore, the key to efficiently solving dMOPs is to sensitively detect changes in the environment and quickly converge to the new Pareto set.

As suggested by Raquel et al. [23], dynamic multi-objective optimization evolutionary algorithms (dMOEAs) for solving dMOPs can be broadly classified into the following categories:

1. **Diversity-based dMOEAs.** This approach attempts to maintain and/or enhance the diversity of the population in order to quickly detect and react to environmental changes. In [6], Deb et al. proposed two NSGA-II [5] variants called DNSGA-II-A and DNSGA-II-B based on this approach.
2. **Memory-based dMOEAs.** This approach attempts to store (i.e., memorize) historical Pareto optimal solutions for reusing them in the future. This type of algorithms is particularly efficient for dMOPs with periodical changes. Representatives in this category include the algorithms proposed in [1,14].
3. **Prediction-based dMOEAs.** This approach aims to train a model to predict the movement of the Pareto set of a dMOP based on historical data. The main advantage of this approach is that the algorithm can react to the change proactively (i.e., it can take action in advance). This enables the algorithm to swiftly converge to the new Pareto set. However, the performance of prediction-based algorithms largely depends on the accuracy of prediction models. State-of-the-art prediction-based algorithms include MOEA/D-SVR [2] and PPS [15].
4. **Multi-population dMOEAs.** Multi-population-based algorithms explore the search space with multiple subpopulations. Ideally, multiple subpopulations can explore different regions of the search space simultaneously. In this manner, the algorithm can be more sensitive to environmental changes and locate new promising regions in the search space quickly. dCOEA [9] and VEPSO [11] are two well-known multiple-population algorithms for solving dMOPs.

3   Dynamic Multi-modal Multi-objective Optimization

In order to define dMMOPs, we first need to introduce the multi-modal property. Suppose that an objective function vector $\boldsymbol{F}$ defines a multi-objective optimization problem whose Pareto set is S, we say that $\boldsymbol{F}$ is a multi-modal function if and only if the following condition is met:

$$\exists x_1^*, x_2^* \in S, \text{ s. t. } x_1^* \neq x_2^* \text{ and } F(x_1^*) = F(x_2^*), \tag{2}$$

where x_1^* and x_2^* are called equivalent Pareto optimal solutions [20].

Now we can formally define dMMOPs as follows:

Definition 1 (dMMOP). *A dMMOP is a dMOP with multi-modal objective functions.*

Due to the multi-modal property, the optimization goal of dMMOPs is to track all the equivalent Pareto sets in the decision space. Compared to dMOPs, which require the optimizer to track only one Pareto set, dMMOPs pose more difficult challenges to algorithm designers.

First, dMMOPs require the optimizer to manage multiple "subpopulations" for tracking multiple equivalent Pareto sets which may locate in different regions in the decision space. This strategy improves the diversification ability of the optimizer at the cost of reducing its convergence ability. Thus, balancing this trade-off is essential for solving dMMOPs. Second, the time series of multiple equivalent Pareto sets may interfere with each other, making it very difficult for the optimizer to identify them correctly. Figure 1 gives an example showing a dMMOP with two equivalent Pareto sets whose centers are denoted by A and B, respectively. Suppose that at time t, A and B move to A' and B', respectively. As shown in Fig. 1 (a), the actual time series from $t - 1$ to t are $A \rightarrow A'$ and $B \rightarrow B'$. However, as shown in Fig. 1 (b), an optimizer (e.g., a prediction-based dMOEA) may obtain incorrect time series, i.e., $A \rightarrow B'$ and $B \rightarrow A'$. In this case, the algorithm may have a very poor performance.

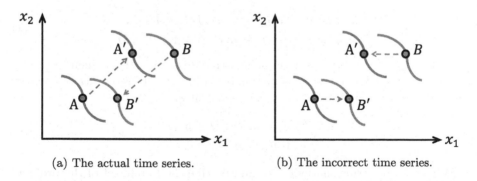

(a) The actual time series.   (b) The incorrect time series.

Fig. 1. Illustration of the difficulty of time series identification when handling dMMOPs.

Furthermore, detecting environmental changes when handling dMMOPs can also be a difficult task. Most existing dynamic multi-objective optimization algorithms detect changes by re-evaluating solutions. However, this strategy may fail when handling dMMOPs. For example, suppose that there is a dMMOP with three equivalent Pareto sets. If a population obtained by an algorithm is only distributed on two of the three Pareto sets, then the algorithm cannot detect the change of the remaining Pareto set by simply re-evaluating the population.

Based on the above discussions, we can see that dMMOPs are very different from standard the dMOPs. Aside from addressing the above-mentioned issues, we also need novel test problems with various characteristics to facilitate the development of efficient algorithms for solving dMMOPs. Thus, in the next section, we suggest a systematic approach for constructing benchmark dMMOPs.

4   A Systematic Approach for Constructing dMMOPs

In this section, we propose a general approach for constructing dMMOPs. Our proposed approach is capable of constructing scalable and flexible test problems with various dynamics.

To construct a dMMOP, we first define a basic dMOP denoted by G as follows:

$$\text{Minimize } \boldsymbol{G}(\boldsymbol{x}, t) = \{g_1(\boldsymbol{x}, t), g_2(\boldsymbol{x}, t), \ldots, g_M(\boldsymbol{x}, t)\}, \tag{3}$$

where M is the number of objectives, $\boldsymbol{x} = (x_1, x_2, \ldots, x_p)^T$ is the decision variable vector with p dimensions, and $g_j \geq 0$ $(j = 1, 2, \ldots, M)$ are M objective functions to be minimized.

By carefully specifying the objective functions, we can ensure that the Pareto front of the problem G changes dynamically whereas its Pareto set is stationary. The reason for this is explained later. We denote the Pareto front and Pareto set $PF(G, t)$ and $PS(G)$, respectively.

Now we construct the desired dMMOP denoted by F as follows:

$$\text{Minimize } \boldsymbol{F}(\boldsymbol{x}, \boldsymbol{y}, t) = (f_1, f_2, \ldots, f_M)^T,$$
$$f_j = g_j(\boldsymbol{x}, t) \cdot [1 + h(\boldsymbol{x}, \boldsymbol{y}, t)], j = 1, 2, \ldots, M, \tag{4}$$

where $\boldsymbol{y} = (y_1, y_2, \ldots, y_q)^T$ is a decision variable vector with q dimensions, and $h(\boldsymbol{x}, \boldsymbol{y}, t) \geq 0$ is a dynamic scalar function regarding two decision variable vectors $\boldsymbol{x}$ and $\boldsymbol{y}$ which satisfy the following constraint at any time t:

$$\forall \boldsymbol{x}^* \in PS(G), \exists \boldsymbol{y} = \boldsymbol{y}^*,$$
$$\text{s. t. } h(\boldsymbol{x}^*, \boldsymbol{y}^*, t) = 0. \tag{5}$$

With the above formulations, we can describe the Pareto set of the problem F as follows:

$$PS(F, t) = \{\boldsymbol{x} = \boldsymbol{x}^*, \boldsymbol{y} = \boldsymbol{y}^* \mid \boldsymbol{x}^* \in PS(G), h(\boldsymbol{x}^*, \boldsymbol{y}^*, t) = 0\}. \tag{6}$$

Notice that when $h = 0$, the problem F is equivalent to G, i.e., their Pareto fronts are the same. This means that the geometry and dynamics of the Pareto front of the problem F only depend on g_j $(j = 1, 2, \ldots, M)$. Recall that when we constructed G, we purposefully made its Pareto set stationary over time. From (6), we can see that the dynamics of the Pareto set can only be controlled by h. Thus, the dynamics for the Pareto front and the Pareto set of the constructed dMMOP can be controlled independently (i.e., by specifying g_j and h functions, respectively). This enables researchers to construct various new test problems by composing different dynamics for the Pareto front and Pareto set. Furthermore, according to (6), F has multiple equivalent Pareto sets when h is a multi-modal function at time t (i.e., $h(\boldsymbol{x}, \boldsymbol{y}, t) = 0$ holds for different values of $\boldsymbol{x}$ and $\boldsymbol{y}$). Thus, by altering the number and positions of the global and local optima of h, we can easily specify the number and distribution of the global and local Parcto sets of F, respectively.

To conclude, the proposed approach can construct scalable dMMOPs with an arbitrary number of decision variables and objective functions. The number of equivalent Pareto sets is also scalable. The proposed approach also allows us to set different dynamics for the Pareto front and Pareto set, thus making it possible to build flexible and sophisticated test problems according to the needs of algorithm designers.

4.1   Case Study on an Example Test Problem

In this section, we use a simple example with only two decision variables x and y to demonstrate how to create a novel dMMOP using our proposed framework. The first step is to define the base dMOP denoted by G_e. We use the

$$\boldsymbol{x} = (x), x \in [0.1, 1],$$
$$\begin{cases} \quad g_1(\boldsymbol{x}, t) = x, \\[2mm] g_2(\boldsymbol{x}, t) = \dfrac{1}{x} + 5\cos^2(0.5\pi t). \end{cases} \tag{7}$$

Then we can construct a dMMOP denoted by dMMOP1 based on G. In our current example, we use the following function h to control the dynamics and geometry of the Pareto set:

$$\boldsymbol{y} = (y), y \in [0, 10],$$
$$h(\boldsymbol{x}, \boldsymbol{y}, t) = \sqrt{|y - 1| \cdot |y - D(t)|}, \tag{8}$$
$$D(t) = 1 + 2\sin^2(0.2\pi t).$$

Since h has two optima (i.e., $y = 1$ and $y = D(t)$), dMMOP1 has two equivalent Pareto sets, one of which varies dynamically over time while the other remains stationary. Notice that when $D(t) = 1$, the two equivalent Pareto sets are overlapping. The Pareto set and Pareto front of dMMOP1 are described in (9) and illustrated in Fig. 2.

$$PS : x \in [0.1, 1.1], y \in \{1, D(t)\},$$
$$PF : g_2 = \frac{1}{g_1} + 5\cos^2(0.5\pi t), g_1 \in [0.1, 1.1]. \tag{9}$$

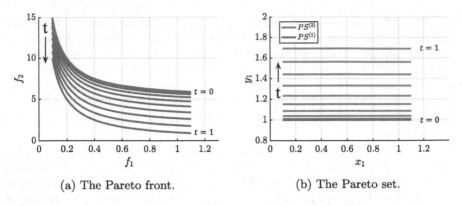

(a) The Pareto front.   (b) The Pareto set.

Fig. 2. The Pareto front and Pareto set of dMMOP1. (a) shows the Pareto front sampled from ten t values vary from 0 to 1. (b) shows the corresponding Pareto sets, where $PS^{(1)}$ and $PS^{(2)}$ denote the first and second equivalent Pareto sets, respectively.

To the best of our knowledge, there are no test problems similar to dMMOP1 in the literature. We further investigate the performance of two algorithms, namely, DNSGA-II-A [6] and MMO-MOES [29] on this test problem. DNSGA-II-A is a diversity-based algorithm for solving dMOPs. It randomly selects and re-evaluates 10% of the population in each generation. If the objective values of any of the solutions have changed, 30% of the population are randomly re-initialized. However, since DNSGA-II-A does not take into account the multi-modal property, we expect it to obtain only one of the two equivalent Pareto sets of dMMOP1. In contrast, MMO-MOES is an algorithm designed for solving MMOPs in static environments. To make it possible to handle dMMOPs, we incorporate the same change detection and change response mechanisms used in DNSGA-II-A into MMO-MOES. We call the resulting algorithm dMMO-MOES.

We use the mean values of the IGD [3] and IGDX [30] indicators (denoted by MIGD and MIGDX, respectively) to measure the performance in tracking the moving Pareto front and Pareto set, respectively. Smaller IGD and IGDX values mean that the obtained solutions can better approximate the Pareto front and the Pareto set, respectively.

The time unit t for dMMOP1 can be calculated with (10), which is modified from [19].

$$t = \max\left\{ \frac{1}{n_t} \lfloor 1 + \frac{\tau - \tau_0}{\tau_t} \rfloor, 0 \right\},$$ (10)

where:

- τ is the current generation counter,
- τ_0 is the number of generations that the optimization problem remains stationary before the first change,
- n_t is the number of distinct time steps in one time unit, which controls the severity of the dynamic change, and

– τ_t is the number of generations where t remains unchanged, which controls the frequency of the dynamic change.

For each algorithm, the population size is 200, and other parameters are set as the suggested values in the corresponding papers [6,29]. For dMMOP1, τ_0, τ_t, and n_t are specified as 50, 20, and 10, respectively. Each algorithm is tested on dMMOP1 for 31 runs with the maximum number of generations being set to $\tau_0 + 100\tau_t$ (i.e., each run comprises 100 environmental changes).

Figures 3 and 4 report the results obtained from a single run with the median MIGD value among 31 runs of each algorithm. From Fig. 3 (a), we can observe that both DNSGA-II-A and dMMO-MOES can track the moving Pareto front. Although these two algorithms use exactly the same change detection and change response mechanisms, DNSGA-II-A clearly outperforms dMMO-MOES regarding the IGD indicator (i.e., the IGD values obtained by DNSGA-II are smaller). DNSGA-II-A not only has more stable performance but also converges faster to the new Pareto front when environmental changes occur.

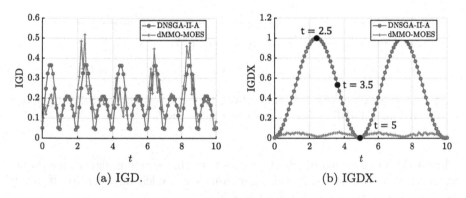

Fig. 3. The change of IGD and IGDX values obtained by DNSGA-II-A and dMMO-MOES on dMMOP1 over 100 environmental changes.

However, in Fig. 3 (b), dMMO-MOES significantly outperforms DNSGA-II-A in terms of IGDX. This is because DNSGA-II can obtain solutions only in one of the two equivalent Pareto subsets. Therefore, as the distance between the two equivalent Pareto sets increases (e.g., t increases from 0 to 2.5), the IGDX value obtained by DNSGA-II-A also increases. Similarly, when $t = 0, 5, 10$, the IGDX values obtained by DNSGA-II-A are the best since the two equivalent Pareto sets of dMMOP1 are overlapping. As shown in Fig. 3 (b), the IGDX values of dMMO-MOES are much smaller than that of DNSGA-II-A over the 100 environmental changes. These observations indicate that dMMO-MOES is more capable of tracking multiple Pareto sets than DNSGA-II-A.

Figure 4 shows the populations obtained by DNSGA-II-A and dMMO-MOES when t equals to 2.5, 3.5 and 5 in the decision space. From this figure, we can verify that DNSGA-II-A obtained solutions only in one of the two equivalent

Pareto sets, whereas dMMO-MOES can track both of them. It is worth noting that Fig. 4 (e) and Fig. 4 (f) also reveal that the convergence ability of dMMO-MOES is noticeably weaker than DNSGA-II-A since many solutions are not on the Pareto sets.

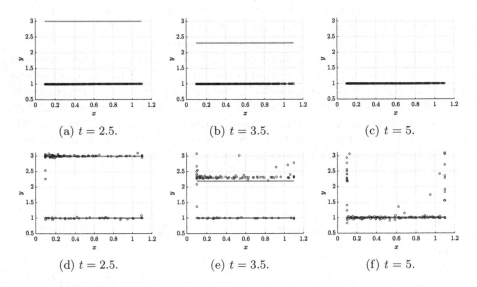

(a) $t = 2.5$.   (b) $t = 3.5$.   (c) $t = 5$.

(d) $t = 2.5$.   (e) $t = 3.5$.   (f) $t = 5$.

Fig. 4. The populations obtained by DNSGA-II-A (i.e., (a)–(c)) and dMMO-MOES (i.e., (d)–(f)) on dMMOP1 when t equals to 2.5, 3.5 and 5.

From these experimental results, we can see that existing algorithms do not perform well on dMMOPs. Novel algorithms are needed in order to efficiently solve this novel type of optimization problems.

5   A Suggested Test Suite

In this section, we provide a novel test suite 10 dMMOPs with various dynamics. All test problems are constructed based on the proposed approach presented in Sect. 4. Definitions of the proposed dMMOPs are summarized in Table 1. Notice that we omit the definition of dMMOP1 in Table 1 since it has already been defined in Sect. 4.1.

Here we briefly describe some main characteristics of each test problem. Details of each test problem are shown in the supplementary file[1]. First, dMMOP2 features a Pareto front whose shape changes periodically from convex to linear and then to concave. It has two stationary equivalent Pareto sets in the decision space. dMMOP3 is the same as dMMOP2 except that its h function is

---

[1] The supplementary file can be found at https://github.com/Yiming-Peng-Official/dMMOP.

Table 1. The proposed dMMOP test suite.

Problem	Definition	Pareto Set
dMMOP1	–	–
dMMOP2	$\boldsymbol{x} = (x_1, x_2)^T, x_{1:2} \in [0,1],$ $\begin{cases} g_1(\boldsymbol{x}, t) = [\cos(\pi/2x_1)\cos(\pi/2x_2)]^{1/D(t)}, \\ g_2(\boldsymbol{x}, t) = [\cos(\pi/2x_1)\sin(\pi/2x_2)]^{1/D(t)}, \\ g_3(\boldsymbol{x}, t) = [\sin(\pi/2x_1)]^{1/D(t)}, \end{cases}$ $\boldsymbol{y} = (y), y_1 \in [0,1],$ $h(\boldsymbol{x}, \boldsymbol{y}, t) = 1 - \sin^2(2\pi y),$ $D(t) = 0.25 + 0.75\sin^2(\pi/12t).$	$x_{1:2} \in [0,1],$ $y = \{\frac{1}{4}, \frac{3}{4}\}.$
dMMOP3	Same as dMMOP2, except that h is defined as follows: $h(\boldsymbol{x}, \boldsymbol{y}, t) = 1 - \exp\left[-(\frac{y-0.2}{0.03})^2\right] - 0.8\exp\left[-(\frac{y-0.6}{0.4})^2\right].$	$x_{1:2} \in [0,1],$ $y = 0.2.$
dMMOP4	Add a minus sign to all objectives of dMMOP2.	Same as dMMOP2.
dMMOP5	$\boldsymbol{x} = (x), x \in [0,1],$ $\begin{cases} g_1(\boldsymbol{x}, t) = x, \\ g_2(\boldsymbol{x}, t) = 1 - x\cos^2(D(t)x\pi), \end{cases}$ $\boldsymbol{y} = (y), y \in [-1,1],$ $h(\boldsymbol{x}, \boldsymbol{y}, t) = 1 - \exp\left[-\left(\frac{(y+x)\cdot(y-x)}{0.4}\right)^2\right],$ $D(t) = 5\sin^2(0.2\pi t).$	All non-dominated solutions satisfying: $g_2 = 1 - g_1\cos^2(D(t)g_1\pi),$ $g_1 \in [0,1].$
dMMOP6	$\boldsymbol{x} = (x), x \in [0,1],$ $\begin{cases} g_1(\boldsymbol{x}, t) = x, \\ g_2(\boldsymbol{x}, t) = 1 - x - \frac{\cos(2D(t)\pi x + \pi/2)}{2D(t)\pi}, \end{cases}$ $\boldsymbol{y} = (y), y \in [0,9],$ $h(\boldsymbol{x}, \boldsymbol{y}, t) = \begin{cases} 2(y - \sin|2\pi x - \pi| + |2\pi x - \pi|)^2, & y \in [0,4] \\ 2(y - 4 - \sin|2\pi x - \pi| + |2\pi x - \pi|), & y \in (4,9] \end{cases}$ $D(t) = 0.1 + 5\sin^2(0.2\pi t).$	$x \in [0,1], y \in [0,4],$ $y = \sin|2\pi x - \pi| + |2\pi x - \pi|.$ and $x \in [0,1], y \in (4,9],$ $y = \sin|2\pi x - \pi| + |2\pi x - \pi| + 4.$
dMMOP7	$\boldsymbol{x} = (x_1, x_2)^T, x_{1:2} \in [0,1],$ $\begin{cases} g_1 = 0.5 + D(t)\cdot(x_1 - 0.5), \\ g_2 = (1 - x_2)(1 - g_1), \\ g_3 = x_2(1 - g_1), \end{cases}$ $D(t) = \cos^2(0.2\pi t),$ $\boldsymbol{y} = (y), y \in [0,1],$ $h(\boldsymbol{x}, \boldsymbol{y}, t) = 1 - \sin^2\left(2\pi(y - S(t)\sin(\pi x_1) + \frac{1}{4})\right),$ $S(t) = 0.5\cos^2(0.2\pi t).$	$y = S(t)\sin(\pi x_1), x_{1:2} \in [0,1].$ and $y = S(t)\sin(\pi x_1) + \frac{1}{2}, x_{1:2} \in [0,1].$
dMMOP8	$\boldsymbol{x} = (x), x \in [0,1],$ $\begin{cases} g_1(\boldsymbol{x}, t) = x, \\ g_2(\boldsymbol{x}, t) = 1 - \sqrt{x}, \end{cases}$ $\boldsymbol{y} = (y), y \in [0,1],$ $h(\boldsymbol{x}, \boldsymbol{y}, t) = 1 - \sin(D(t)\pi y),$ $D(t) = 1 + 9\sin^2(0.2\pi t).$	$x \in [0,1], y = \frac{0.5 + 2i}{D(t)},$ $i = 0, 1, 2, \ldots$
dMMOP9	$\boldsymbol{x} = (x), x \in [1,3],$ $\begin{cases} g_1(\boldsymbol{x}, t) = |x - 2|, \\ g_2(\boldsymbol{x}, t) = 1 - \sqrt{|x - 2|}, \end{cases}$ $\boldsymbol{y} = (y), y \in [-1,1].$ $h(\boldsymbol{x}, \boldsymbol{y}, t) = \begin{cases} 2(y - \sin(2D(t)\pi|y - 2| + \pi))^2, & y \in [1,2) \\ 2(y - \sin(2\pi|y - 2| + \pi))^2, & y \in [2,3] \end{cases}$ $D(t) = 1 + 4\sin^2(\pi/2t)$	$y \in [1,2),$ $y = \sin(2D(t)\pi|x - 2| + \pi)$ and $y \in [2,3],$ $y = \sin(2\pi|x - 2| + \pi).$
dMMOP10	$\boldsymbol{x} = (x), x \in [-0.5, 0.5],$ $\boldsymbol{y} = (y), y \in [-0.5, 0.5],$ $\begin{bmatrix} x_r \\ y_r \end{bmatrix} = \begin{bmatrix} x \\ y \end{bmatrix} \begin{bmatrix} \cos\theta(t) & -\sin\theta(t) \\ \sin\theta(t) & \cos\theta(t) \end{bmatrix}$ $\begin{cases} g_1 = x_r, \\ g_2 = 1/x_r, \end{cases}$ $h(\boldsymbol{x}, \boldsymbol{y}, t) = 1 - \cos^6(2\pi y_r),$ $\theta(t) = 2\pi\sin^2(0.2\pi t).$	$x_r \in [-0.5, 0.5],$ $y_r = \{-\frac{1}{2}, \frac{1}{2}\}.$

modified to have a locally optimal Pareto front. dMMOP4 is another variant of dMMOP2 which is constructed based on the idea proposed in [12]. By adding a minus sign to all objective functions of dMMOP2, the Pareto front of dMMOP4 changes from a triangular shape to an inverted triangular shape. As pointed out in [12], such a Pareto front shape is difficult for decomposition-based multi-objective optimization algorithms. dMMOP5 is a test problem whose Pareto set and Pareto front change from a continuous curve to multiple disconnected segments over time. Furthermore, the number of disconnected Pareto set and Pareto front segments also change dynamically. The dMMOP6 test problem has two equivalent Pareto sets which are the same as the MMF8 test problem in [28]. However, its Pareto front changes dynamically from a regular curve to a mixed convex/concave curve. The number of knee points on the Pareto front also changes dynamically over time. dMMOP7 has two equivalent Pareto sets, each of which is a time-varying manifold with two dimensions. Its Pareto front is a 2-dimensional plane that can degenerate into a line over time. The dMMOP8 test problem has a time-varying number of equivalent Pareto sets. dMMOP9 has two equivalent Pareto sets, one of which changes its geometry over time, while the other is always stationary. In dMMOP10, the equivalent Pareto sets rotate clockwise around the origin as time changes. Since the centroid of the Pareto set is always the origin, dMMOP10 is challenging for some dMOEAs that rely on centroid-based prediction models (e.g., [15]).

In conclusion, the proposed dMMOP test suite provides test problems with various characteristics, thus allowing researchers to evaluate the performance of an algorithm with respect to a wide variety of aspects.

6   Concluding Remarks

In this paper, we introduced a novel type of optimization problem, namely, dMMOPs by extending MMOPs into dynamic environments. Furthermore, we gave a formal definition for dMMOPs and analyzed some key challenges in solving them. Since test problems are essential for algorithm development, we proposed a general approach for constructing dMMOPs for benchmarks. In addition, we provided a novel test suite containing 10 novel dMMOPs. We believe that these test problems can help researchers to develop more efficient algorithms for solving real-world dMMOPs.

This paper only provides a preliminary study on dynamic multi-modal multi-objective optimization, many potential research topics are left for future work. For example, experimental results in Sect. 4.1 show that simply incorporating an existing change response mechanism to multi-modal multi-objective optimization algorithms (e.g., MMO-MOES in our experiments) does not yield satisfactory results, and we still need more efficient algorithms to handle dMMOPs in the future.

Acknowledgements. This work was supported by National Natural Science Foundation of China (Grant No. 61876075), Guangdong Provincial Key Laboratory (Grant No. 2020B121201001), the Program for Guangdong Introducing Innovative and Entrepreneurial Teams (Grant No. 2017ZT07X386), The Stable Support Plan Program of Shenzhen Natural Science Fund (Grant No. 20200925174447003), Shenzhen Science and Technology Program (Grant No. KQTD2016112514355531).

References

1. Azzouz, R., Bechikh, S., Said, L.B.: A dynamic multi-objective evolutionary algorithm using a change severity-based adaptive population management strategy. Soft Comput. **21**(4), 885–906 (2015). https://doi.org/10.1007/s00500-015-1820-4
2. Cao, L., Xu, L., Goodman, E.D., Bao, C., Zhu, S.: Evolutionary dynamic multiobjective optimization assisted by a support vector regression predictor. IEEE Trans. Evol. Comput. **24**(2), 305–309 (2020)
3. Coello Coello, C.A., Reyes Sierra, M.: A study of the parallelization of a coevolutionary multi-objective evolutionary algorithm. In: Monroy, R., Arroyo-Figueroa, G., Sucar, L.E., Sossa, H. (eds.) MICAI 2004. LNCS (LNAI), vol. 2972, pp. 688–697. Springer, Heidelberg (2004). https://doi.org/10.1007/978-3-540-24694-7_71
4. Deb, K.: Multi-objective Optimization Using Evolutionary Algorithms. Wiley, New York (2001)
5. Deb, K., Pratap, A., Agarwal, S., Meyarivan, T.: A fast and elitist multiobjective genetic algorithm: NSGA-II. IEEE Trans. Evol. Comput. **6**(2), 182–197 (2002)
6. Deb, K., Rao N., U.B., Karthik, S.: Dynamic multi-objective optimization and decision-making using modified NSGA-II: a case study on hydro-thermal power scheduling. In: Obayashi, S., Deb, K., Poloni, C., Hiroyasu, T., Murata, T. (eds.) EMO 2007. LNCS, vol. 4403, pp. 803–817. Springer, Heidelberg (2007). https://doi.org/10.1007/978-3-540-70928-2_60
7. Deb, K., Tiwari, S.: Omni-optimizer: a generic evolutionary algorithm for single and multi-objective optimization. Eur. J. Oper. Res. **185**(3), 1062–1087 (2008)
8. Farina, M., Deb, K., Amato, P.: Dynamic multiobjective optimization problems: test cases, approximations, and applications. IEEE Trans. Evol. Comput. **5**(8), 425–442 (2004)
9. Goh, C., Tan, K.C.: A competitive-cooperative coevolutionary paradigm for dynamic multiobjective optimization. IEEE Trans. Evol. Comput. **13**(1), 103–127 (2008)
10. Goldberg, D.E., Richardson, J.: Genetic algorithms with sharing for multimodal function optimization. In: Proceedings of the Second International Conference on Genetic Algorithms and Their Application, pp. 41–49 (1987)
11. Greeff, M., Engelbrecht, A.P.: Solving dynamic multi-objective problems with vector evaluated particle swarm optimisation. In: IEEE Congress on Evolutionary Computation, pp. 2917–2924 (2008)
12. Ishibuchi, H., Matsumoto, T., Masuyama, N., Nojima, Y.: Many-objective problems are not always difficult for Pareto dominance-based evolutionary algorithms. In: Proceedings of the 24th European Conference on Artificial Intelligence (2020)
13. Jaszkiewicz, A.: On the performance of multiple-objective genetic local search on the 0/1 knapsack problem - a comparative experiment. IEEE Trans. Evol. Comput. **6**(4), 402–412 (2002)

14. Jiang, S., Yang, S.: A steady-state and generational evolutionary algorithm for dynamic multiobjective optimization. IEEE Trans. Evol. Comput. **21**(1), 65–82 (2016)
15. Li, Q., Zou, J., Yang, S., Zheng, J., Ruan, G.: A predictive strategy based on special points for evolutionary dynamic multi-objective optimization. Soft Comput. **23**(11), 3723–3739 (2018). https://doi.org/10.1007/s00500-018-3033-0
16. Lin, Q., Lin, W., Zhu, Z., Gong, M., Li, J., Coello, C.A.C.: Multimodal multi-objective evolutionary optimization with dual clustering in decision and objective spaces. IEEE Trans. Evol. Comput. **25**(1), 130–144 (2021)
17. Liu, Y., Ishibuchi, H., Nojima, Y., Masuyama, N., Shang, K.: A double-niched evolutionary algorithm and its behavior on polygon-based problems. In: Proceedings of the Parallel Problem Solving from Nature - PPSN XV, pp. 262–273 (2018)
18. Liu, Y., Yen, G.G., Gong, D.: A multimodal multiobjective evolutionary algorithm using two-archive and recombination strategies. IEEE Trans. Evol. Comput. **23**(4), 660–674 (2019)
19. Nguyen, T.T.: Continuous Dynamic Optimization Using Evolutionary Algorithms. Ph.D. thesis, The University of Birmingham (2010)
20. Peng, Y., Ishibuchi, H.: A decomposition-based multi-modal multi-objective optimization algorithm. In: Proceedings of the 2020 IEEE Congress on Evolutionary Computation, pp. 1–8 (2020)
21. Peng, Y., Ishibuchi, H., Shang, K.: Multi-modal multi-objective optimization: problem analysis and case studies. In: Proceedings of the IEEE Symposium Series on Computational Intelligence, pp. 1865–1872 (2019)
22. Petrowski, A.: A clearing procedure as a niching method for genetic algorithms. In: Proceedings of the IEEE International Conference on Evolutionary Computation, pp. 798–803 (1996)
23. Raquel, C., Yao, X.: Dynamic multi-objective optimization: a survey of the state-of-the-art. In: Evolutionary Computation for Dynamic Optimization Problems, pp. 85–106. Springer, Heidelberg(2013). https://doi.org/10.1007/978-3-642-38416-5_4
24. Schütze, O., Vasile, M., Coello, C.A.C.: Computing the set of epsilon-efficient solutions in multiobjective space mission design. J. Aerosp. Comput. Inf. Commun. **8**(3), 53–70 (2011)
25. Shir, O.M.: Niching in evolutionary algorithms. In: Handbook of Natural Computing, pp. 1035–1069. Springer, Heidelberg (2012). https://doi.org/10.1007/978-3-540-92910-9_32
26. Tian, Y., Liu, R., Zhang, X., Ma, H., Tan, K.C., Jin, Y.: A multi-population evolutionary algorithm for solving large-scale multi-modal multi-objective optimization problems. IEEE Tran. Evol. Comput. **25**(3), 405–418 (2020)
27. Yue, C., Qu, B., Liang, J.: A multiobjective particle swarm optimizer using ring topology for solving multimodal multiobjective problems. IEEE Trans. Evol. Comput. **22**(5), 805–817 (2018)
28. Yue, C., Qu, B., Yu, K., Liang, J., Li, X.: A novel scalable test problem suite for multimodal multiobjective optimization. Swarm Evol. Comput. **48**, 62–71 (2019)
29. Zhang, K., Chen, M., Xu, X., Yen, G.G.: Multi-objective evolution strategy for multi-modal multi-objective optimization. Appl. Soft Comput. **101**, 107004 (2021)
30. Zhou, A., Zhang, Q., Jin, Y.: Approximating the set of Pareto-optimal solutions in both the decision and objective spaces by an estimation of distribution algorithm. IEEE Trans. Evol. Comput. **13**(5), 1167–1189 (2009)

Fair Feature Selection
with a Lexicographic Multi-objective
Genetic Algorithm

James Brookhouse$^{(\boxtimes)}$ and Alex Freitas

School of Computing, University of Kent, Canterbury, UK
james@brookhou.se, A.A.Freitas@kent.ac.uk

Abstract. There is growing interest in learning from data classifiers whose predictions are both accurate and fair, avoiding discrimination against sub-groups of people based e.g. on gender or race. This paper proposes a new Lexicographic multi-objective Genetic Algorithm for Fair Feature Selection (LGAFFS). LGAFFS selects a subset of relevant features which is optimised for a given classification algorithm, by simultaneously optimising one measure of accuracy and four measures of fairness. This is achieved by using a lexicographic multi-objective optimisation approach where the objective of optimising accuracy has higher priority over the objective of optimising the four fairness measures. LGAFFS was used to select features in a pre-processing phase for a random forest algorithm. The experiments compared LGAFFS' performance against two feature selection approaches: (a) the baseline approach of letting the random forest algorithm use all features, i.e. no feature selection in a pre-processing phase; and (b) a Sequential Forward Selection method. The results showed that LGAFFS significantly improved fairness measures in several cases, with no significant difference regarding predictive accuracy, across all experiments.

1   Introduction

Recently, there has been an increased focus on the fairness of the decisions made by automated processes [1,17]; since algorithms that learn from biased data often produce biased predictive models. We address fairness in the classification task of machine learning, where a predictive feature (e.g. gender or race) is set as a sensitive feature. The values of a sensitive feature are used to split individuals (instances in a dataset) into protected and unprotected groups. The protected group contains individuals likely to be victims of discrimination, who are more likely to obtain a negative outcome (class label) than the unprotected group.

A large number of fairness measures have been proposed to capture some notion of fairness in a model learned from data [14,21]. These fairness measures can be categorised into group-level and individual-level fairness measures.

An example of a group-level fairness metrics is the discrimination score [2], which measures the difference between the predicted positive-class probabilities

© The Author(s), under exclusive license to Springer Nature Switzerland AG 2022
G. Rudolph et al. (Eds.): PPSN 2022, LNCS 13399, pp. 151–163, 2022.
https://doi.org/10.1007/978-3-031-14721-0_11

of the protected and unprotected groups. Some group-level metrics of fairness measure the difference between the false positive error rate and/or the false negative error rate between the protected and unprotected groups [3]. Group-level fairness measures have the limitation of not considering fairness at the individual level; i.e., they do not penalise models where two very similar individuals within the same group unfairly receive different outcomes (class labels).

An individual-level fairness metric avoids this limitation, by measuring similarities among individuals. Consistency is an individual fairness metric which compares an individual to its k-nearest neighbours; if all of an individual's neighbours have the same class as the current individual, this test is considered maximally satisfied for that individual, this is then repeated for each individual and an average taken [22]. However, as the number of features grows, the notion of "nearest neighbours" become increasingly meaningless, as the distances between individuals tend to increase, leading to comparisons being made between increasingly different individuals.

In practice, no single fairness measure can be deemed the best in general, and it has also been proved that there is a clear trade-off among some fairness measures, which cannot be simultaneously optimised [3,10].

Hence, intuitively it makes sense to use multiple fairness measures, with different pros and cons, and try to optimise those multiple measures at the same time, in order to achieve more robust fairness results. This is precisely the focus of this paper, where we propose a new multi-objective Genetic Algorithm (GA) for fair feature selection, The GA uses the lexicographic approach to optimise two objectives in decreasing priority order: predictive accuracy and fairness. The accuracy objective involves one measure, but the fairness objective is more complex and involves four measures. Hence, we propose a new procedure for aggregating four fairness measures into a single fairness objective by systematically considering all permutations of lexicographic ordering of those four measures, as described in detail later.

The GA selects a subset of relevant features for a given classification algorithm in a data pre-processing phase [13]. This is a difficult task for two reasons. First, the search space's size is exponential in the number of features, with $2^n - 1$ candidate solutions (feature subsets), where n is the number of features in the dataset (the "$- 1$" discounts the empty feature subset). Second, intuitively the search space is rugged (highly non-convex) with many local optima, even in a single-objective scenario, with the problem being aggravated in the multi-objective scenario.

We focus on GAs for two mains reasons. First, they are robust global search methods, being less likely to get trapped into local optima in the search space, by comparison with conventional local search methods [7,18], and so they tend to cope better with feature interaction (a key issue in feature selection). Second, the fact that they evolve a population of candidate solutions facilitates multi-objective optimisation [5,19], as proposed in this work.

This paper is organised as follows. Section 2 describes the proposed multi-objective genetic algorithm for fair feature selection. Section 3 describes the datasets used in the experiments and the experimental setup. Section 4 reports experimental results and Sect. 5 presents the conclusions and future work.

Algorithm 1: Ramped Population Initialisation

Data: population_size, MIN_P, MAX_P
Result: Population of Individuals
1 **Function** *initialise_population()*:
2   step_size = (MAX_P − MIN_P) / population_size
3   **for** i **to** *population_size* **do**
4   ⌊ p = MIN_P + (i * step_size)
5   population += Individual.initialise(p)
6   **return** *population*

---

2 A Lexicographic-Optimisation Genetic Algorithm for Fair Feature Selection

This section describes our new Lexicographic-optimisation Genetic Algorithm for Fair Feature Selection (LGAFFS), which selects a subset of relevant features for a classification algorithm in a data pre-processing phase. LGAFFS selects individuals for reproduction based on the principle of lexicographic optimisation to combine predictive accuracy and fairness measures, as described later.

In LGAFFS, each individual of the population represents a candidate feature subset. More precisely, each individual consists of a string of N bits (genes), where N is the number of features in the dataset, and the i-th gene takes the value 1 or 0 to indicate whether or not (respectively) the i-th feature is selected.

LGAFFS follows a wrapper approach to feature selection [13], where a base classification algorithm is used to learn a classification model based on the feature subset selected by an individual, and that model's quality (in terms of accuracy and fairness) is used to compute that individual's fitness. Hence, the GA aims at finding the best subset of features for the base classification algorithm. Fitness computation is performed by using a well-known internal cross-validation procedure, which uses only the training set (i.e. not using the test set).

LGAFFS uses uniform crossover and bit-flip mutation as genetic operators to generate new individuals in each generation. However, the population initialisation, tournament selection and elitism selection are non-standard procedures, and hence these are described in detail in the next subsections.

2.1 Population Initialisation

When creating the initial population, each individual has a different probability that each gene (feature) will be selected or not. This ramping initialisation is described in Algorithm 1. As shown in line 4, each individual has the probability (p) that a gene (feature) will be switched on increased by `step_size` compared to the previous individual, where `step_size` is defined in line 2 as a function of the maximum and minimum probabilities for a feature to be selected – denoted `MAX_P` and `MIN_P`, which are input arguments for Algorithm 1.

The motivation for this ramped population initialisation procedure is to promote diversity in the population. If all individuals had the same probability p

---

Algorithm 2: Pseudo-code of Lexicographic Tournament selection.

    Data: Instances, Population, ϵ, fair_win_ϵ

1 **Function** *tournament_selection()*:

2   i1, i2 = select_random_individuals()

3   **if** *not $|i1.accuracy - i2.accuracy| > \epsilon$* **then**

4   i1_win, i2_win = fairness_aggregation(i1,i2) `// See Algorithm 3`

5   **if** *$|i1\_win - i2\_win| > fair\_win\_\epsilon$* **then**

6   **return** *fairest_individual*

7   **return** *best_accuracy_individual*

---

that a single gene is switched on or off, then, as the number of genes (features) in an individual increases, the number of features selected (switched on genes) in each individual would tend to converge to $p \times N_{genes}$, the mean of a binomial distribution, where p is the probability of each gene being switched on and N_{genes} is the number of genes. Hence, all individuals would tend to have a similar number of selected features, and individuals with low or high numbers of selected features in the initial population would be rare, limiting the search of these areas. The ramped population initialisation avoids this problem, giving each individual a different probability p of switching a gene, sweeping from MIN_P to MAX_P.

2.2 Lexicographic Tournament Selection

Lexicographic tournament selection, with tournament size of two, is used to select individuals for reproduction. In the lexicographic-optimisation approach [8], we compare the two individuals in a tournament considering the objectives in decreasing order of priority. Let V_1 and V_2 denote the values of the current objective for individuals 1 and 2. When those two individuals are compared based on the first objective, if $|V_1 - V_2| > \epsilon$ (where ϵ is a very small value), then the best individual is the tournament winner. Otherwise, the two individuals' objective values are deemed equivalent (negligible difference) based on that objective; then the next objective is considered in the same way, and so on. This is repeated until a significant (greater than ϵ) difference is observed and a best individual selected. If there is no significant difference between two individuals for all objectives, then the individual with the best value of the highest priority objective is selected.

The pseudo-code for this is shown in Algorithm 2. When comparing two individuals, the lexicographic approach requires the objectives to be ordered. LGAFFS considers accuracy as the highest-priority objective to be optimised, followed by a lower-priority set of fairness measures which are aggregated into a single fairness objective to be optimised as described in Sect. 2.4.

Note that the lexicographic method avoids the specification of ad-hoc weights to each objective, which would be the case if using a weighted sum of objectives [8]. The lexicographic approach simply requires that an order of priority for the objectives be defined; and intuitively it is easier for users to specify a priority order of objectives than ad-hoc numerical weights. The lexicographic approach requires a small threshold parameter (ϵ); but again, it is intuitively easier for users to specify this parameter than to specify ad-hoc weights for each objective.

Note that an alternative to the lexicographic approach would be the well-known Pareto dominance approach [5]. However, the Pareto approach is not suitable for our feature selection task where the objective of accuracy has higher priority than the objective of fairness, since the Pareto approach ignores this objective prioritisation. In particular, if we used the Pareto approach, once the fairest model is found it would tend to be preserved by the selection operator and remain in the Pareto front along the GA run even if its accuracy was very low; but that model would be a bad solution, given the objectives' priority order. In this case, the GA would waste computational resources searching on areas of the Pareto front around bad solutions (like areas with maximal fairness but low accuracy). In contrast, a lexicographic approach would never select that fairest model due to its very low accuracy (as the highest-priority objective).

2.3   The Four Fairness Measures and the Accuracy Measure

No single fairness measure captures all nuances of a fair model, so LGAFFS optimises four fairness measures to get more robust fairness results. To define these measures we use the following nomenclature:

- S: Protected/sensitive feature: $0 \rightarrow$ unprotected group, $1 \rightarrow$ protected group
- $\hat{Y}$: the predicted class; Y: the actual class; taking class labels 1 or 0
- TP, FP, TN, FN: Number of True Positives, False Positives, True Negatives and False Negatives, respectively

The first measure is the discrimination score (DS) [2], which is defined as:

$$DS = 1 - \left| P(\hat{Y} = 1 | S = 0) - P(\hat{Y} = 1 | S = 1) \right| \tag{1}$$

DS is a group-level fairness measure that takes the optimal value of 1 if both protected and unprotected groups have an equal probability of being assigned to the positive class by the classifier. If DS is used on unbalanced datasets, those where the data shows a large difference between the probability of a positive outcome for both groups, to satisfy DS will require a reduction in accuracy. In this case the lexicographic approach is robust to such selective pressures as the ordering of the objectives prioritises accuracy over the fairness measures.

The second measure used is consistency [22], defined as:

$$C = 1 - \frac{1}{Nk} \sum_{i} \sum_{j \in kNN(x_n)} |\hat{y}_i - \hat{y}_j| \tag{2}$$

Consistency is an individual-level similarity metric that compares the class predicted by a classifier to each instance in the dataset to the class predicted by the classifier to that instance's k nearest instances (neighbours) in the dataset. If all these neighbours have the same predicted class as the current instance, then that instance is considered consistent. The measure computes the average degree of consistency over all instances in the dataset. A fully consistent model has a consistency of 1 and an inconsistent model has a value of 0.

Thirdly, the False Positive Error Rate Balance Score (FPERBS) [3,4] is:

$$FPERBS = 1 - \left| \frac{FP_{S=0}}{FP_{S=0} + TN_{S=0}} - \frac{FP_{S=1}}{FP_{S=1} + TN_{S=1}} \right| \qquad (3)$$

FPERBS measures the difference in the probability that a truly negative instance is incorrectly predicted as positive between protected and unprotected groups. Fourthly, the False Negative Error Rate Balance Score (FNERBS) [3,9,11] measures the difference in the probability that a truly positive instance is incorrectly predicted as negative between protected and unprotected groups:

$$FNERBS = 1 - \left| \frac{FN_{S=0}}{FN_{S=0} + TP_{S=0}} - \frac{FN_{S=1}}{FN_{S=1} + TP_{S=1}} \right| \qquad (4)$$

A score of 1 indicates an optimally fair result for both FPERBS and FNERBS.

As the accuracy measure to be optimised, LGAFFS uses the geometric mean of Sensitivity and Specificity (Eq. 5). This measure was chosen because it incentivises the correct classification of both positive-class and negative-class instances, to counteract pressure from the fairness measures to produce maximally fair models that trivially predict the same class for all instances.

$$Sensitivity = \frac{TP}{TP + FN}, \qquad Specificity = \frac{TN}{TN + FP},$$
$$GM_{Sen \times Spec} = \sqrt{Sensitivity \cdot Specificity} \qquad (5)$$

2.4 Aggregating Fairness Measures

As discussed earlier, LGAFFS optimises one accuracy measure and four fairness measures. We consider accuracy as the highest-priority objective (as usual in machine learning), and the four fairness measures as lower-priority objectives. Among those fairness measures, there is no consensus in the literature about what is the best one, and so it would be "unfair" to prioritise one fairness measure over the others. Hence, we aggregate the four fairness measures into a single objective to be optimised by the GA (in addition to the accuracy objective), by computing all possible 24 (4!) permutations of the four fairness measures.

Algorithm 3 shows how two individuals are compared regarding fairness. Each permutation defines a lexicographic order for the fairness measures which can be evaluated to find the first significant difference between the individuals, at which point the best individual is given a win. A significant difference is one greater than the very small ϵ. After all permutations have been evaluated, the individual with the higher number of wins is declared the best individual overall.

Algorithm 3: Aggregating fairness measures.

Data: Ind_1, Ind_2, ϵ
Result: Number of wins for each individual

```
1  Function fairness_aggregation():
2      i1_win = i2_win = 0
3      permutations = generate_permutations(measures)
4      forall permutations do
5          forall permutation.measures do
6              i1, i2 = compute_fairness_measure(measure, Ind_1, Ind_2)
7              if |i1 - i2| > ε then
8                  if i1 > i2 then
9                      i1_win++
10                     break // Exit inner forall
11                 else
12                     i2_win++
13                     break // Exit inner forall

14     return i1_win, i2_win
```

2.5 Lexicographic Elitism

Recall that the lexicographic approach requires the ranking of objectives, and our GA prioritises the accuracy objective (the geometric mean of Sensitivity and Specificity) over the four fairness measures. The fairness measures are aggregated into a single objective (see Sect. 2.4). To find the **best individual** the procedure in Algorithm 4 is used for implementing elitism.

First the population is sorted by accuracy (line 2 of Algorithm 4), where any individuals with accuracy within ϵ of the most accurate individual are shortlisted for fairness comparison (line 4). These shortlisted individuals have their average

---

Algorithm 4: Lexicographic Selection of the Best Individual

Data: population, ϵ, fair_rank_ϵ
Result: Best individual

```
1  Function get_best_individual():
2      accuracy_rank = sort_population_by_accuracy(population)
3      best_accur_indiv = accuracy_rank.head()
4      individuals = select_all_individuals_within_accuracy_ε(accuracy_rank)
5      avg_fair_rank = average_rank_of_fairness_permutations(individuals)
6      if (avg_fair_rank.head().average_rank – best_accur_indiv.average_rank) >
          fair_rank_ε then
7          return avg_fair_rank.head()
8      else
9          return best_accur_indiv
```

Table 1. Datasets used in all experiments, detailing the number of instances, features and the sensitive features for each dataset.

Data set	Instances	Features	Sensitive features
Adult Income (US Census)	48842	14	Race, Gender, Age
German Credit	1000	20	Age, Gender
Credit Card Default	30000	24	Gender
Communities and Crime	1994	128	Race
Student Performance (Portuguese)	650	30	Age, Gender, Relationship
Student Performance (Maths)	396	30	Age, Gender, Relationship
ProPublica recidivism	6167	52	Race, Gender

rank of fairness computed across all 24 permutations generated as described in Sect. 2.4. For each permutation of the four fairness measures, the set of shortlisted individuals is arranged by its lexicographic order, where the first (last) measure in the permutation is considered the most (least) important. Fairness values within the threshold ϵ are considered equivalent and the less important metrics are considered until a significant difference is found.

If the fairest shortlisted individual (with the lowest average rank) has a significantly better rank than the most accurate shortlisted individual (i.e. the difference between their average ranks is greater than fair_rank_ϵ), the former is selected by elitism as the best individual in preference over the most accurate individual – since the difference in accuracy between those two shortlisted individuals is considered non-significant, i.e., within ϵ. Otherwise, the most accurate shortlisted individual is selected by elitism.

LGAFFS' Python code is available at https://github.com/bunu/LGAFFS.

2.6   Related Work

Quadrianto et al. [16] and Valvidia et al. [20] proposed a GA for fair classification. Both GAs were designed for optimising (hyper)-parameters of a classification algorithm, rather than feature selection; and both GAs use Pareto dominance rather than the lexicographic approach used here. The Pareto approach is sound in general, but as noted earlier, it is not suitable for our feature selection task prioritising accuracy over fairness. La Cava and Moore [12] proposed genetic programming (GP) for feature construction, which can implicitly perform feature selection, but feature construction has a much larger search space than feature selection. Their GP uses lexicase selection, a broadly lexicographic approach. However, instead of ordering the objectives based on user-defined priorities like in LGAFFS; their GP uses *randomised* lexicographic orderings of different subgroups of instances (with different sensitive feature values). The GP evaluates multiple fairness-violation events, each for a different subgroup of instances; but each event is evaluated by the same fairness formula: the difference of error rates (either FP or FN error rates) between all instances and a sub-group of instances. In addition, unlike those three algorithms, LGAFFS combines group-level and individual-level fairness measures, increasing fairness robustness.

Table 2. Results for Research Question 1: Comparing the performance of random forest trained with the features selected by LGAFFS in a pre-processing phase against the performance of random forest trained with all features. Showing the values for all five measures being optimised by LGAFFS.

Dataset	Sensitive feature	$GM_{Sen \times Spec}$		Discrimination score		Consistency		FPERBS		FNERBS	
		LGAFFS	All feats	LGAFFS	All feats	LGAFFS	All feats	LGAFFS	All feats	LGAFFS	All feats
Adult	Age	0.6475	**0.7602**	**0.8485**	0.7557	**0.8656**	0.7887	**0.9522**	0.9084	**0.9189**	0.6416
Adult	Race	0.7409	**0.7632**	**0.9356**	0.8955	**0.8201**	0.7889	**0.9862**	0.9589	**0.9902**	0.9004
Adult	Sex	0.7420	**0.7623**	**0.8498**	0.8152	**0.8163**	0.7877	**0.9490**	0.9217	**0.9416**	0.9110
German Credit	Age	**0.5901**	0.5774	**0.9361**	0.8394	0.7642	**0.8032**	**0.8633**	0.7626	0.8911	**0.8968**
German Credit	Gender	**0.6036**	0.5637	**0.9399**	0.9087	0.7510	**0.8114**	**0.8993**	0.8489	0.9230	**0.9349**
Student Maths	Age	**0.9208**	0.9046	0.7964	**0.8169**	0.8444	**0.8483**	**0.9022**	0.8781	0.8951	**0.8974**
Student Maths	Dalc	0.8951	**0.9072**	**0.7563**	0.7214	0.8377	**0.8447**	0.8051	**0.8312**	0.8157	**0.8160**
Student Maths	Famrel	**0.9052**	0.8914	0.7039	**0.7148**	0.8361	**0.8412**	**0.8760**	0.8480	**0.9222**	0.9189
Student Maths	Romantic	0.8984	**0.9008**	**0.9076**	0.9055	**0.8468**	0.8443	**0.9151**	0.9090	**0.9106**	0.8989
Student Maths	Sex	**0.9027**	0.8977	0.8397	**0.8652**	0.8387	**0.8488**	0.7646	**0.8233**	0.9012	**0.9251**
Student Maths	Walc	0.9000	**0.9012**	**0.8364**	0.8023	0.8376	**0.8503**	0.8206	**0.8362**	**0.9196**	0.8741
Student Portuguese	Age	**0.8196**	0.7864	**0.8638**	0.8536	0.9106	**0.9192**	**0.7038**	0.6686	**0.9578**	0.9428
Student Portuguese	Dalc	**0.7867**	0.7834	0.8470	**0.8758**	0.9128	**0.9251**	**0.6230**	0.5511	0.9088	**0.9376**
Student Portuguese	Famrel	0.8031	**0.8035**	0.8019	**0.8305**	0.9097	**0.9177**	**0.6450**	0.6286	0.9205	**0.9583**
Student Portuguese	Romantic	**0.7846**	0.7825	0.9370	**0.9452**	**0.9211**	0.9189	0.7608	**0.7775**	0.9686	**0.9796**
Student Portuguese	Sex	**0.8110**	0.7994	0.9200	**0.9313**	0.9134	**0.9239**	**0.7646**	0.7029	**0.9708**	0.9650
Student Portuguese	Walc	**0.8035**	0.7831	0.9271	**0.9380**	0.9186	**0.9205**	0.7214	**0.7248**	**0.9679**	0.9671
Communities and Crime	Race	0.8303	**0.8419**	**0.6623**	0.5957	0.6056	**0.6126**	**0.9182**	0.8553	**0.8313**	0.7762
Default of Credit	Sex	0.5896	**0.5910**	**0.9755**	0.9724	0.8354	**0.8395**	0.9739	**0.9776**	**0.9852**	0.9832
Propublica Recidivisim	Race	0.7306	**0.7515**	**0.8324**	0.7744	**0.6849**	0.6836	**0.9022**	0.8169	**0.9105**	0.8834
Propublica Recidivisim	Sex	0.7113	**0.7518**	**0.9408**	0.9189	0.6756	**0.6876**	**0.9254**	0.9125	**0.9451**	0.9400
Number of wins		10	11	13	8	6	15	15	6	13	8
Wilcoxon signed-rank test		0.9442		0.06288		0.07346		▲ 0.0088		0.18024	

3   Datasets and Experimental Setup

Table 1 describes the 7 binary classification datasets used. When a dataset has multiple sensitive features – a sensitive feature is one which represents a protected characteristic and/or a group that is unfairly treated – the algorithm is ran multiple times using a different sensitive feature each time. 6 datasets are from the UCI Machine Learning repository [6]. The 7th dataset is from ProPublica, investigating biases in predicting if criminals would re-offend [1].

For all datasets, except Adult Income, the experiments use a well-known 10-fold cross-validation procedure. Adult Income is already partitioned into a training and test set, so this partition is used instead of cross validation. LGAFFS' parameters were not optimised and were set as follows: ϵ: 0.01 (threshold for significant differences in Algorithms 2, 3 and 4), fair_rank_ϵ: 1 (threshold for significant fairness-rank differences in Algorithm 4), fair_win_ϵ: 1 (threshold for significant difference in the number of wins among 24 permutations of fairness measures in Algorithm 2); population_size: 100, MAX_P: 0.5 and MIN_P: 0.1 (for population initialisation in Algorithm 1), internal cross validation folds: 3, max_iterations: 50, tournament_size: 2, crossover_probability: 0.9, mutation_probability: 0.05.

4   Experimental Results

We addresses two research questions. First, we compare the use of LGAFFS to select features in a pre-processing phase against the baseline of no feature selection in that phase. Second, we compare LGAFFS to the popular Sequential Forward Selection (SFS) method. Both LGAFFS and SFS use the wrapper approach to feature selection; i.e., they repeatedly use a base classification algorithm to evaluate feature subsets. The base algorithm was Random Forest from scikit-learn [15], with default parameter settings; which was chosen because it is a very popular and powerful classification algorithm. Note that the Random Forest algorithm performs embedded feature selection (during its run), but that feature selection considers only accuracy; whilst using LGAFFS to perform feature selection in a pre-processing phase we optimise both accuracy and fairness.

We also calculated the Pearson's linear correlation coefficient for each of the 6 pairs of fairness measures for LGAFFS, the coefficients were: 0.71 for (DS,FNERBS), −0.59 for (C,FPERBS), 0.42 for (C,FNERBS), 0.24 for (DS,C), 0.08 for (DS,FPERBS) and 0.02 for (FPERBS,FNERBS). So, 4 pairs of fairness measures have an absolute value of correlation smaller than 0.5.

4.1   RQ1: Does LGAFFS Select a Better Subset than the Full Set?

The first research question asks whether using LGAFFS to select features in a pre-processing phase leads to better results than the baseline approach of not performing any feature selection. That is, does the random forest algorithm perform better (regarding accuracy and fairness) when it is trained with the features selected by LGAFFS or when it is trained with the full feature set?

Table 3. Results for Research Question 2: Comparing the performance of random forest trained with the features selected by LGAFFS vs. Sequential Forward Selection (both selecting features in a pre-processing phase). Showing the values for all five measures being optimised by LGAFFS.

Dataset	Sensitive feature	$GM_{Sen \times Spec}$		Discrimination score		Consistency		FPERBS		FNERBS	
		LGAFFS	SFS	LGAFFS	SFS	LGAFFS	SFS	LGAFFS	SFS	LGAFFS	SFS
Adult	Age	0.6475	**0.7482**	**0.8485**	0.7968	**0.8656**	0.8142	**0.9522**	0.9437	**0.9189**	0.8306
Adult	Race	0.7409	**0.7465**	0.9356	**0.9358**	**0.8201**	0.8149	0.9862	**0.9887**	**0.9902**	0.9444
Adult	Sex	0.7420	**0.7480**	**0.8498**	0.8331	**0.8163**	0.8145	**0.9490**	0.9396	**0.9416**	0.9057
German Credit	Age	**0.5901**	0.4653	**0.9361**	0.8872	0.7642	**0.8110**	**0.8633**	0.8095	**0.8911**	0.8890
German Credit	Gender	**0.6036**	0.4851	**0.9399**	0.9237	0.7510	**0.8212**	**0.8993**	0.8457	**0.9230**	0.9202
Student Maths	Age	**0.9208**	0.9041	**0.7964**	0.7954	**0.8444**	0.8407	**0.9022**	0.8468	**0.8951**	0.8619
Student Maths	Dalc	**0.8951**	0.8921	**0.7563**	0.6893	0.8377	**0.8392**	**0.8051**	0.7940	0.8157	**0.8177**
Student Maths	Famrel	0.9052	**0.9069**	**0.7039**	0.6707	0.8361	**0.8367**	**0.8760**	0.7993	**0.9222**	0.9096
Student Maths	Romantic	0.8984	**0.9033**	**0.9076**	0.9008	**0.8468**	0.8311	**0.9151**	0.8906	0.9106	**0.9170**
Student Maths	Sex	0.9027	**0.9265**	0.8397	**0.8468**	**0.8387**	0.8361	0.7646	**0.8602**	0.9012	**0.9521**
Student Maths	Walc	0.9000	**0.9084**	**0.8364**	0.8151	**0.8376**	0.8347	0.8206	**0.8796**	**0.9196**	0.9026
Student Portuguese	Age	**0.8196**	0.7812	0.8638	**0.8909**	**0.9106**	0.9050	0.7038	**0.7445**	**0.9578**	0.9576
Student Portuguese	Dalc	0.7867	**0.7917**	0.8470	**0.8580**	**0.9128**	0.9090	**0.6230**	0.6138	0.9088	**0.9460**
Student Portuguese	Famrel	**0.8031**	0.7946	**0.8019**	0.7392	**0.9097**	0.9050	**0.6450**	0.5911	**0.9205**	0.8538
Student Portuguese	Romantic	**0.7846**	0.7833	0.9370	**0.9378**	**0.9211**	0.9062	**0.7608**	0.7273	0.9686	**0.9716**
Student Portuguese	Sex	**0.8110**	0.8017	**0.9200**	0.9148	**0.9134**	0.9032	**0.7646**	0.7504	**0.9708**	0.9554
Student Portuguese	Walc	**0.8035**	0.7845	**0.9271**	0.9179	**0.9186**	0.9087	0.7214	**0.7393**	**0.9679**	0.9541
Communities and Crime	Race	**0.8303**	0.7923	**0.6623**	0.5740	**0.6056**	0.6049	**0.9182**	0.7935	**0.8313**	0.6992
Default of Credit	Sex	**0.5896**	0.5639	0.9755	**0.9816**	0.8354	**0.8580**	0.9739	**0.9807**	0.9852	**0.9889**
Propublica Recidivisim	Race	**0.7306**	0.7200	**0.8324**	0.7763	**0.6849**	0.6617	**0.9022**	0.8026	**0.9105**	0.8769
Propublica Recidivisim	Sex	0.7113	**0.7199**	**0.9408**	0.8660	**0.6756**	0.6661	**0.9254**	0.8492	**0.9451**	0.9058
Number of wins		12	9	15	6	16	5	15	6	15	6
Wilcoxon signed-rank test		0.17068		▲ 0.00634		0.05876		▲ 0.04236		▲ 0.030	

Table 2 show the experimental results for this research question. In this table, the first two columns show the dataset and the sensitive feature. The following ten columns show the accuracy and fairness results of training the Random Forest algorithm with features selected by LGAFFS or with the full feature set.

In each row of this table (i.e. for each pair of a dataset and a sensitive feature), for each pair of columns comparing the accuracy or fairness of LGAFFS against the full feature set, the best result is shown in boldface. The last but one row of the table shows the number of wins for each approach for each of the five measures of performance, whilst the last row shows the p-value obtained by the Wilcoxon signed-rank statistical significance test. Statistically significant results, at the conventional significance level of $\alpha = 0.05$, are marked with a red triangle. In Table 2, there was no substantial difference in the number of wins regarding accuracy. LGAFFS achieved substantially more wins in three of the four fairness measures, with statistical significance in one measure: FPERBS.

4.2   RQ2: Does LGAFFS Perform Better than SFS?

The second question involves the comparison of LGAFFS to a popular local search-based feature selection method, viz. Sequential Forward Selection (SFS). The SFS method is not aware of the 4 fairness measures; it is just optimising the accuracy measure, i.e., the geometric mean of sensitivity and specificity.

Table 3 presents the results for the Random Forest algorithm when using LGAFFS or SFS to select features. LGAFFS achieved more wins in all 5 measures, with statistical significance shown in 3 of the 4 fairness measures.

5   Conclusions

We have proposed a new lexicographic-optimisation Genetic Algorithm for fair feature selection, which selects a feature subset optimised for a classification algorithm based on both predictive accuracy and 4 fairness measures capturing different aspects of fairness, including both group-level and individual-level fairness. No single fairness measure reflects all the nuances of fairness; LGAFFS optimises multiple fairness measures to obtain more robust fairness results.

LGAFFS was compared with 2 other feature selection approaches (no feature selection and Sequential Forward Selection) using Random Forest as the classification algorithm. There was no significant difference in the predictive accuracies of models learned when using LGAFFS versus the 2 other approaches. Regarding fairness, when comparing LGAFFS against the 2 other approaches across the 4 fairness measures, LGAFFS achieved significantly better results in 4 of the 8 comparisons, and there was no significant differences between LGAFFS and the 2 other approaches in the other 4 comparisons.

Future work could include extending SFS to make it a fairness-aware method.

Acknowledgements. This work was funded by a research grant from The Leverhulme Trust, UK, reference number RPG-2020-145.

References

1. Angwin, J., Larson, J., Mattu, S., Kirchner, L.: Machine bias: there's software used across the country to predict future criminals, and it's biased against blacks (2016). https://www.propublica.org/article/machine-bias-risk-assessments-in-criminal-sentencing
2. Calders, T., Verwer, S.: Three Naive Bayes approaches for discrimination-free classification. Data Min. Knowl. Discov. **21**(2), 277–292 (2010)
3. Chouldechova, A.: Fair prediction with disparate impact: a study of bias in recidivism prediction instruments. Big Data **5**(2), 153–163 (2017)
4. Corbett-Davies, S., Goel, S.: The measure and mismeasure of fairness: a critical review of fair machine learning. arXiv preprint arXiv:1808.00023 (2018)
5. Deb, K.: Multi-objective Optimization Using Evolutionary Algorithms. Wiley, New York (2002)
6. Dua, D., Graff, C.: UCI machine learning repository (2017). http://archive.ics.uci.edu/ml
7. Freitas, A.: Data Mining and Knowledge Discovery with Evolutionary Algorithms. Springer, Heidelberg (2002). https://doi.org/10.1007/978-3-662-04923-5
8. Freitas, A.A.: A critical review of multi-objective optimization in data mining: a position paper. ACM SIGKDD Explorat. Newslett. **6**(2), 77–86 (2004)
9. Hardt, M., Price, E., Srebro, N.: Equality of opportunity in supervised learning. In: Advances in Neural Information Processing Systems, pp. 3315–3323 (2016)
10. Kleinberg, J., Mullainathan, S., Raghavan, M.: Inherent trade-offs in the fair determination of risk scores. arXiv preprint arXiv:1609.05807 (2016)
11. Kusner, M.J., Loftus, J., Russell, C., Silva, R.: Counterfactual fairness. In: Advances in Neural Information Processing Systems, pp. 4066–4076 (2017)
12. La Cava, W., Moore, J.: Genetic programming approaches to learning fair classifiers. In: Proceedings of the Genetic and Evolutionary Computation Conference (GECCO-2020), pp. 967–975 (2020)
13. Li, J., et al.: Feature selection: a data perspective. ACM Comput. Surv. **50**(6), 94:1–94:45 (2017)
14. Mehrabi, N., Morstatter, F., Saxena, N., Lerman, K., Galstyan, A.: A survey on bias and fairness in machine learning. arXiv preprint arXiv:1908.09635 (2019)
15. Pedregosa, F., et al.: Scikit-learn: machine learning in python. J. Mach. Learn. Res. **12**, 2825–2830 (2011)
16. Quadrianto, N., Sharmanska, V.: Recycling privileged learning and distributed matching for fairness. In: Proceedings of the 31st Conference on Neural Information Processing Systems (NIPS 2017), pp. 677–688 (2017)
17. Skeem, J.L., Lowenkamp, C.T.: Risk, race, & recidivism: predictive bias and disparate impact. Criminology **54**, 680 (2016)
18. Telikani, A., Tahmassebi, A., Banzhaf, W., Gandomi, A.: Evolutionary machine learning: a survey. ACM Comput. Surv. **54**(8), 161:1–161:35 (2021)
19. Tian, Y., et al.: Evolutionary large-scale multi-objective optimization: a survey. ACM Comput. Surv. **54**(8), 174:1–174:34 (2021)
20. Valdivia, A., Sanchez-Monedero, J., Casillas, J.: How fair can we go in machine learning? Assessing the boundaries of accuracy and fairness. Int. J. Intell. Syst. **36**(4), 1619–1643 (2021)
21. Verma, S., Rubin, J.: Fairness definitions explained. In: 2018 IEEE/ACM International Workshop on Software Fairness (FairWare), pp. 1–7. IEEE (2018)
22. Zemel, R., Wu, Y., Swersky, K., Pitassi, T., Dwork, C.: Learning fair representations. In: International Conference on Machine Learning, pp. 325–333 (2013)

Greedy Decremental Quick Hypervolume Subset Selection Algorithms

Andrzej Jaszkiewicz[ID] and Piotr Zielniewicz[(✉)][ID]

Faculty of Computing and Telecommunications, Poznan University of Technology,
Piotrowo 3, 60-965 Poznan, Poland
jaszkiewicz@cs.put.poznan.pl, piotr.zielniewicz@put.poznan.pl
https://cat.put.poznan.pl/

Abstract. The contribution of this paper is fourfold. First, we present an updated implementation of the Improved Quick Hypervolume algorithm which is several times faster than the original implementation and according to the presented computational experiment it is at least competitive to other state-of-the-art codes for hypervolume computation. Second, we present a Greedy Decremental Lazy Quick Hypervolume Subset Selection algorithm. Third, we propose a modified Quick Hypervolume Extreme Contributor/Contribution algorithm using bounds from previous iterations of a greedy hypervolume subset selection algorithm. According to our experiments these two methods perform the best for greedy decremental hypervolume subset selection. Finally, we systematically compare performance of the fastest algorithms for greedy incremental and decremental hypervolume subset selection using two criteria: CPU time and the quality of the selected subset.

Keywords: Multiobjective optimization · Hypervolume · Greedy algorithms

1 Introduction

Hypervolume is one of the most often used set-quality indicators in multiobjective optimization [24,36]. This indicator measures the hypervolume of the region dominated by a set of points in the objective space. This set of points may for example correspond to solutions generated by a multiobjective evolutionary algorithm. Hypervolume indicator has the advantage of being compatible with the Pareto dominance relation [38]. In the context of evolutionary multiobjective optimization (EMO) hypervolume may be used for evaluation of EMO Algorithms [9], for bounding Pareto archives [20], or for fitness assignment in, so called, indicator-based Algorithms [2,5,19,35,37], which constitute one of the main classes of EMO Algorithms [23].

The potential drawback of the hypervolume indicator is high computational complexity of exact algorithms. Efficient algorithms exist only for the number

This research was funded by the Polish Ministry of Education and Science.

G. Rudolph et al. (Eds.): PPSN 2022, LNCS 13399, pp. 164–178, 2022.
https://doi.org/10.1007/978-3-031-14721-0_12

of objectives, $m = 2$, $m = 3$ [4], and $m = 4$ [14]. For the general case, the best known algorithm has $\mathcal{O}(n^{m/3} \text{polylog } n)$ time complexity [10].

Because of the wide use of the hypervolume indicator in the context of EMO, beside the basic problem of hypervolume calculation, a number of other hypervolume-related problems has been defined in the literature [16,24,30]. One of them is the hypervolume subset selection problem (HSSP), the goal of which is to select a predefined number of points from a larger set to maximize the hypervolume. HSSP may be used to guide EMO algorithms [1,2], to bound a Pareto archive [20], or to select a reduced set of the most representative solutions of a multiobjective problem for further analysis by the decision maker [8]. HSSP is NP-hard for three and more objectives [7], however, since HSSP consists of maximizing a submodular function subject to a cardinality constraint [13], HSSP may be approximated using a greedy approach with approximation guarantee [26]. Thus, several authors proposed greedy algorithms for HSSP [6,14,15]. Some of these approaches are limited, however, to two and three [15], or four objectives [14].

HSSP may be solved in a greedy manner either starting with an empty subset and adding the point with the maximum contribution in each step (incremental approach), or starting with the whole set and removing the point with the minimum contribution in each step (decremental approach), until the desired size of the subset is obtained [3,6]. The relative efficiency of incremental and decremental aproaches may depend on the required size of the subsset. Intuitively, if a small subset is to be selected from a large set, the incremental approach may perform better, while the decremental approach may be better if only relatively few points need to be removed from the original set.

In this paper:

- We report an updated implementation of the improved quick hypervolume algorithm (QHV-II) which is several times faster than the original implementation reported in [17] and according to the presented computational experiment it is at least competitive to other state-of-the-art codes for hypervolume computation.
- We present the Greedy Decremental Lazy Quick HSS algorithm motivated by previously proposed Greedy Incremental Lazy HSS Algorithm [11].
- We propose a modified Quick Hypervolume Extreme Contributor/Contribution (QEHC) Algorithm [18] and use it within greedy hypervolume subset selection algorithms. The modified version of QEHC (QEHC-B) uses bounds from previous iterations to improve its efficiency.
- We present a computational experiment indicating that the two proposed methods perform the best for greedy decremental HSS.
- We systematically compare performance of the fastest algorithms for greedy incremental and decremental HSS using two criteria: CPU time and the quality of the selected subset. According to our knowledge it is one of the first such systematic study reported in the literature.

The paper is organized in the following way. In the next section we provide basic definition. In Sect. 3, we shortly describe the QHV-II Algorithm used as the basic algorithm for hypervolume calculation. Then, we describe the Greedy

Decremental Lazy Quick HSS algorithm introduced in this paper. In Sect. 5, we describe the modified Quick Hypervolume Extreme Contributor/Contribution algorithm. Results of computational experiment are presented in Sect. 6. Finally we present conclusions and directions for further research.

2   Basic Definitions

Definition 1. *(Weak Pareto dominance) Consider two points $r, p \in \mathbb{R}^m$, where $\mathbb{R}^m$ is the space of m objectives to be maximized. We say that p weakly dominates r, also written as $r \preceq p$, if and only if $\forall j \ p_j \geq r_j$.*

Definition 2. *(Hypercuboid) Hypercuboid delimited by points $r, p \in \mathbb{R}^m$, $r \preceq p$ is:*

$$[r,p] = \{q \in \mathbb{R}^m \mid q \preceq p \wedge r \preceq q\} \tag{1}$$

Definition 3. *(Hypervolume indicator) Given a set of points S in the objective space $\mathbb{R}^m$ and a reference point r_* such that $\forall_{p \in S} r_* \preceq p$, the hypervolume indicator of S is the measure of the region weakly dominated by S and dominating r_*, i.e.:*

$$H(S, r_*) = \Lambda(\{q \in \mathbb{R}^m \mid r_* \preceq q \wedge \exists p \in S : q \preceq p\}) \tag{2}$$

where $\Lambda(.)$ denotes the Lebesgue measure. Alternatively, it may be interpreted as the measure of the union of hypercuboids:

$$H(S, r_*) = \Lambda\left(\bigcup_{p \in S} [r_*, p]\right) \tag{3}$$

Definition 4. *(Hypercuboid-bounded hypervolume indicator) Given a set of points S in the objective space $\mathbb{R}^m$ and a hypercuboid $[r_*, r^*]$ such that $\forall_{p \in S} r_* \preceq p$, the hypercuboid-bounded hypervolume indicator of S is the measure of the region weakly dominated by S within $[r_*, r^*]$, i.e.:*

$$H(S, [r_*, r^*]) = \Lambda(\{q \in [r_*, r^*] \mid \exists p \in S : q \preceq p\}) \tag{4}$$

where $\Lambda(.)$ denotes the Lebesgue measure.
    Note that

$$H(S, [r_*, r^*]) = H(nd\text{-}worse(S, r^*), r_*) \tag{5}$$

where $nd\text{-}worse(S, r^*) = \{q \in \mathbb{R}^m \mid \exists q' \in S : \forall j \ q_j = \min(q'_j, r^*_j)\}$ may be interpreted as projection of S onto $[r_*, r^*]$.

Definition 5. *(Hypervolume contribution) Hypervolume contribution of a point s to $H(S \cup \{s\}, r_*)$ (allowing both $s \in S$ or $s \notin S$) is the difference between hypervolume of $S \cup \{s\}$ and hypervolume of $S \setminus \{s\}$, i.e.:*

$$HC(s, S, r_*) = H(S \cup \{s\}, r_*) - H(S \setminus \{s\}, r_*) \tag{6}$$

Hypervolume contribution of a point s defined by Eq. (6) could alternatively be calculated as the difference of hypervolume of $\{s\}$ and hypercuboid-bounded hypervolume of $S \setminus \{s\}$ within $[r_*, s]$, i.e.:

$$HC(s, S, r_*) = \Lambda([r_*, s]) - H(S \setminus \{s\}, [r_*, s]) \tag{7}$$

where $\Lambda([r_*, s])$ is the hypervolume of hypercuboid $[r_*, s]$. In practice, the use of Eq. (7) allows for a faster calculation of hypervolume contribution than Eq. (6), since hypervolume is calculated just once (the time of calculation of $\Lambda([r_*, s])$ is negligible) and many points in S may become dominated after projection onto $[r_*, s]$.

3 Improved Quick Hypervolume Algorithm Scheme

In this paper we propose to use the Improved Quick Hypervolume (QHV-II) Algorithm [17] as the basic algorithm for hypervolume calculation. It is one of the fastest exact algorithms improving previously proposed Quick Hypervolume Algorithm [27,28]. QHV-II calculates the hypervolume in a divide and conquer manner. In each iterations, it selects the point with the maximum individual hypervolume as the pivot point, adds the hypervolume of the region dominated by the pivot point, and then splits the remaining problem (corresponding to the remaining region and the remaining points) into m sub-problems corresponding to non-overlapping hypercuboids. If the number of points is sufficiently small, it uses simple geometric properties to calculate the hypervolume. For further details the reader is referred to [17].

4 Greedy Decremental Lazy Quick HSS Algorithm

HSSP may be approximated in a greedy way using either incremental or decremental approach [6]. The incremental approach starts with an empty subset and then the point with the highest contribution is added to the selected subset in each iteration (see Algorithm 1). The decremental approach starts with the original set of points and then the point with the lowest contribution is removed from the selected subset in each iteration (see Algorithm 2). Both, incremental and decremental approaches provide some approximation guaranties [22,26,29].

In [11] Chen et al. proposed the Greedy Incremental Lazy HSS algorithm which exploits the fact that hypervolume is non-decreasing submodular function, i.e. it is monotone in adding points [32]:

$$H(S \cup \{p\}, r_*) \geq H(S, r_*), \forall S, \forall p \notin S \tag{8}$$

The same obviously applies to hypercuboid-bounded hypervolume. As a direct consequence of (7), hypervolume contribution is non-increasing submodular function:

$$HC(s, S \cup \{p\}, r_*) \leq HC(s, S, r_*), \forall S, \forall p \notin S \tag{9}$$

Algorithm 1. Greedy Incremental HSS algorithm

Input: S_{all} - the original set of points, $k \leq |S_{all}|$ - the number of points to be selected
Output: S the subset of points selected from S_{all}
$S = \emptyset$
while $|S| < k$ **do**
    for all $s \in S_{all}$ **do**
        calculate $HC(s, S, r_*)$
    end for
    select $s^* \in S_{all}$ with the highest $HC(s^*, S, r_*)$
    $S = S \cup \{s^*\}$, $S_{all} = S_{all} \setminus \{s^*\}$
end while

Algorithm 2. Greedy Decremental HSS algorithm

Input: S_{all} - the original set of points, $k \leq |S_{all}|$ - the number of points to be selected
Output: S the subset of points selected from S_{all}
$S = S_{all}$
while $|S| > k$ **do**
    for all $s \in S$ **do**
        calculate $HC(s, S, r_*)$
    end for
    select $s^* \in S$ with the lowest $HC(s^*, S, r_*)$
    $S = S \setminus \{s^*\}$
end while

Thus, the hypervolume contribution of s calculated in a previous iteration could be treated as the upper bound for the contribution in the current iteration of the greedy incremental algorithm, denoted by $HC_{UB}(s, S, r_*)$. If this upper bound for point s is lower than the hypervolume contribution for another points p, then there is no need to recalculate $HC(s, S, r_*)$ in the current iteration [25]. In many cases, the recalculation of the hypervolume contribution of a point results in the same value or only a slightly smaller value than the current upper bound since the inclusion of a single point changes the hypervolume contributions of only its neighbors in the objective space. Thus, the point with the largest hypervolume contribution is often found after examining very few points. The full description of this algorithm may be found in [11]. The algorithm returns exactly the same subset as the original incremental approach with the same tie-breaking mechanism. According to the experiments reported in [11] it is the fastest greedy incremental HSS algorithm for $m \geq 5$ and relatively small subset sizes.

Chen et al. [11] considered only greedy incremental lazy approach, however, the same reasoning may be applied to the decremental approach. In this case, Eq. (9) means that the hypervolume contribution of a point $s \in S$ to the selected subset S never decreases when a point is removed from S. Thus, the hypervolume contribution of s to S calculated in a previous iteration could be treated as the lower bound for the contribution in the current iteration of the greedy decremental algorithm. If the lower bound for point s, denoted by $HC_{LB}(s, S, r_*)$, is greater than the hypervolume contribution for another points p, then there is no need to recalculate $HC(s, S, r_*)$ in the current iteration. The Greedy Decremental Lazy HSS algorithm introduced in this paper is summarized in Algorithm 3.

Algorithm 3. Greedy Decremental Lazy HSS algorithm

Input: S_{all} - the original set of points, $k \leq |S_{all}|$ - the number of points to be selected
Output: S the subset of points selected from S_{all}
$S = S_{all}$, $\mathbb{HC}_{LB} = \emptyset$
while $|S| > k$ do
    if the first iteration then
        for all $s \in S$ do
            calculate $HC(s, S, r_*)$ and add it to $\mathbb{HC}_{LB}$
        end for
        select $s^* \in S$ with the lowest $HC(s^*, S, r_*)$
        $S = S \setminus \{s^*\}$
    else
        while $\mathbb{HC}_{LB} \neq \emptyset$ do
            $s^* =$ point with the lowest upper bound in $\mathbb{HC}_{LB}$
            calculate of $HC(s^*, S, r_*)$
            update $\mathbb{HC}_{LB}$ with $HC(s^*, S, r_*)$
            if s^* has the largest upper bound in $\mathbb{HC}_{UB}$ then
                $S = S \setminus \{s^*\}$
                $\mathbb{HC}_{LB} = \mathbb{HC}_{LB} \setminus HC_{UB}(s^*, S, r_*)$
                break
            end if
        end while
    end if
end while

---

5   The Modified Quick Hypervolume Extreme Contributor/Contribution Algorithm

Greedy HSS algorithms select a point with either the lowest or the highest hypervolume contribution in each iteration. In Algorithms 1 and 2 we assumed that this selection is made by calculating each contribution. There are, however, dedicated methods for selection of the point with the extreme contribution. In [18] we proposed Quick Extreme Hypervolume Contributor/Contribution (QEHC) algorithm that could be used to this end. The idea of this algorithm is to run concurrently processes calculating contributions of each point using an algorithm that provides lower and upper bounds for the contribution in each step. These bounds are then used to stop processes that cannot yield the extreme contributor. Within a greedy HSS algorithm we can further improve QEHC by exploiting contribution bounds from previous iterations.

Let $HC^{i-1}(s, S, r_*)$ and $HC^i(s, S, r_*)$ be the hypervolume contributions of point s in the consecutive iterations $i-1$ and i of the incremental greedy HSS algorithm and let $HC_{UB}^{i-1}(s, S, r_*)$ be the upper bound for contribution obtained in the previous iteration $i-1$. Since a point was added to S after iteration $i-1$, exploiting (9) we have:

$$HC_{UB}^{i-1}(s, S, r_*) \geq HC^{i-1}(s, S, r_*) \geq HC^i(s, S, r_*) \tag{10}$$

In other words the upper bound obtained in the previous iteration remains valid in the subsequent iteration.

Analogously, the lower bound obtained in the previous iteration remains valid in the subsequent iteration for the greedy decremental algorithm:

$$HC_{LB}^{i-1}(s, S, r_*) \leq HC^{i-1}(s, S, r_*) \leq HC^i(s, S, r_*) \tag{11}$$

Thus, the effective lower bound in the greedy decremental algorithm in i-th iteration is:

$$HC_{LB}^{i_e}(s, S, r_*) = \max\{HC_{LB}^{i-1_e}(s, S, r_*), HC_{LB}^i(s, S, r_*)\} \tag{12}$$

and the effective upper bound in the greedy incremental algorithm in i-th iteration is:

$$HC_{UB}^{i_e}(s, S, r_*) = \min\{HC_{UB}^{i-1_e}(s, S, r_*), HC_{UB}^i(s, S, r_*)\} \tag{13}$$

The proposed modified QEHC (QEHC-B) algorithm takes advantage of the effective bounds to obtain a better speed-up. Within the greedy decremental approach the process $P(s)$ of computation of contribution of points s could be stopped if:

$$HC_{LB}^{i_e}(s, S, r_*) > \min_{p \in S} HC_{UB}^i(p, S, r_*) \tag{14}$$

Within the greedy incremental approach the process $P(s)$ could be stopped if:

$$HC_{UB}^{i_e}(s, S, r_*) < \max_{p \in S_{all}} HC_{LB}^i(p, S, r_*) \tag{15}$$

Note that this use of contribution bounds is in fact very similar to that of greedy lazy algorithms.

For further details about QEHC the reader is referred to [18].

6   Computational Experiment

In the computational experiment we use the data sets proposed in [21] (concave, convex, linear) with 5 to 10 objectives and 1000 points[1]. We did not include data sets $m \leq 4$ since dedicated methods exist for such case [14,15]. Of course,

---

[1] All data sets used in this experiment, source code and the detailed results are available at https://chmura.put.poznan.pl/s/DxsmP72OS65Glce.

formally speaking, linear data sets are also convex, but we use the original terminology of the authors of these sets.

To calculate the hypervolume contribution, which according to Eq. (7) boils down to calculating hypervolume, we use either WFG [34] (using the code obtained from the authors of this method) or QHV-II algorithm [17]. Both codes were compiled under Visual Studio C++ with the same settings.

For the purpose of this experiment we have improved the C++ implementation of QHV-II. The new implementation is several times faster than the original one reported in [17] (see Fig. 1). All improvements are technical, like the use of more efficient data structures, improved memory allocation, removing redundant code that was used only for the purpose of the computational experiment reported in [17], and they do not modify the algorithm of QHV-II. To show efficiency of this new implementation we compare it to WFG implementation, which is considered to be among state-of-the-art codes for this task. In Fig. 1 we present running times of full hypervolume computation averaged over 10 data sets of a given type and number of objectives for randomly selected sets of 100, 200, ..., 1000 points, and 6, 8, and 10 objectives. These results indicate that QHV-II was on average faster in all cases except of linear data sets with 10 objectives for which the results were very similar, but with advantage of WFG. We note, however, that the relative performance of WFG improves with the growing number of objectives. To confirm these observations we used the Wilcoxon signed rank tests with the significance level $\alpha = 0.05$. QHV-II was significantly faster in most cases. The main exception are linear data sets with 10 objectives where WFG was significantly faster up to 800 points and then the two methods were not significantly different. Few other exceptions where observed for the smallest subset size (100 and 200). In addition, the new implementation of QHV-II was significantly faster than the old implementation in all cases except one case with the smallest subset size.

As it was mentioned above the greedy decremental HSS algorithm selects a point with the lowest hypervolume contribution in each iteration and there exist dedicated methods for selection of such points like the described above QEHC. Another algorithm that could be used to this end is IWFG [12,33]. Thus, we test also versions of Greedy Decremental HSS algorithm with the use of IWFG, QEHC and QEHC-B. QEHC and QEHC-B may also be used for selection of the point with the maximum contribution, so they may also be used in the incremental approach.

We compare the following greedy decremental methods:

- Greedy Decremental HSS algorithm with the use of QHV-II – GD_QHV-II
- Greedy Decremental HSS algorithm with the use of WFG – GD_WFG
- Update-based Greedy Decremental HSS algorithm with the use of QHV-II – UGD_QHV-II. This algorithm has been proposed in [14] and it takes advantage of the fact that the hypervolume contribution could be obtained by efficiently updating the contribution from the previous iteration after a single new point has been removed from the selected subset.
- Greedy Decremental HSS algorithm with the use of QEHC – GD_QEHC

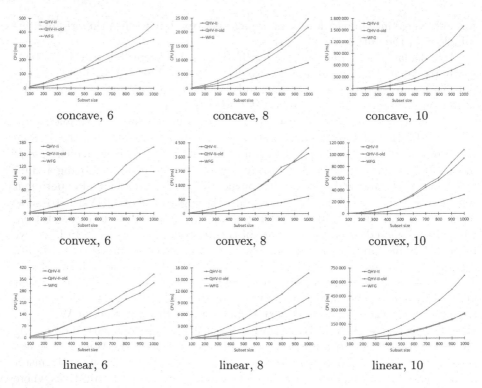

Fig. 1. Full hypervolume computation with the new QHV-II implementation, old QHV-II implementation, and WFG

- Greedy Decremental HSS algorithm with the use of QEHC-B – GD_QEHC-B
- Greedy Decremental HSS algorithm with the use of IWFG – GD_IWFG
- Greedy Decremental Lazy HSS algorithm with the use of QHV-II – GDL_QHV-II
- Greedy Decremental Lazy HSS algorithm with the use of WFG – GDL_WFG

We compare also the following greedy incremental methods:

- Greedy Incremental HSS algorithm with the use of QEHC – GI_QEHC
- Greedy Incremental HSS algorithm with the use of QEHC-B – GI_QEHC-B
- Greedy Incremental Lazy HSS algorithm with the use of QHV-II – GIL_QHV-II
- Greedy Incremental Lazy HSS algorithm with the use of WFG – GIL_WFG

In this case, we do not include other non-lazy algorithms, since they were already evaluated and outperformed by the lazy Algorithm in [11].

In Fig. 2 we present exemplary running times of greedy HSS methods needed to select a subset with a given number of points out of 1000 points for $m = 5, 7, 9$. Since such experiments are time consuming we were not able to complete these calculations for $m = 10$ before the submission of this paper. These results will be

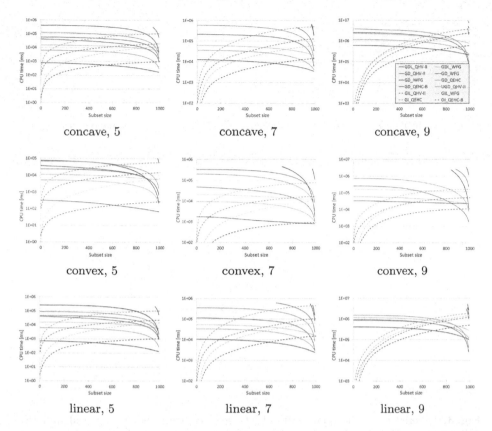

Fig. 2. CPU times of Greedy Decremental and Incremental methods (the legend placed in the top right corner is common to all charts)

presented at the conference and made available at the web page. Note, that some methods were stopped when their running time became much larger than the maximum running time of the best methods. The results, in general, confirm the intuition that the incremental approach is faster when a lower number of points needs to be selected, while the decremental approach is faster when relatively few points needs to be removed. Depending on the number of objectives and the number of points the fastest method is either GIL_QHV-II, GDL_QHV-II, or GD_QEHC-B. The general pattern is that GIL_QHV-II is the fastest method up to a given number of points to be selected (e.g. up to 690 points for linear data sets with $m = 9$) and then, GDL_QHV-II becomes the fastest. For data sets with $m = 9$, GD_QEHC-B becomes the fastest method for the highest number of points to be selected (e.g. from 960 points for linear data sets with $m = 9$). We confirmed the statistical significance of these observations comparing the best and the second best method with the Wilcoxon signed rank tests with the significance level $\alpha = 0.05$.

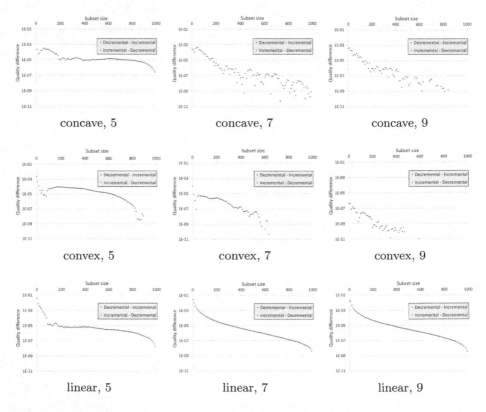

Fig. 3. Differences of the quality of subsets selected with the incremental and the decremental approaches

GD_QEHC-B outperforms GDL_QHV-II only for data sets with the highest number of objectives and when a relatively small number of points need to be removed. This is because GDL_QHV-II needs to calculate all contributions in the first iteration, however then further iterations are very short due to the use of the bounds in the lazy approach. In GD_QEHC-B the first iteration is shorter, however, because many processes calculating contributions are stopped before the final contribution has been obtained, the bounds from the previous iterations are, in general, worse than the bounds used in the lazy approach. Another observation is that methods based on QHV-II perform better than methods based on WFG. Furthermore, the Greedy Decremental HSS algorithm with the use of IWFG – GD_IWFG – performs relatively poorly. Already in [18] we observed that IWFG performs poorly for linear and concave data sets (similar observations have been made for linear data sets in [31]). Furthermore, IWFG does not use any information from the previous iterations of the greedy algorithm like the bounds used in the QEHC-B.

Finally, in Fig. 3 we present the difference of the quality of the subsets selected with the incremental and the decremental approach. Since we use logarithmic scale in this figure, it is not possible to show both positive and negative differences in a single series. Thus, we present two separate series for cases when either the decremental or the incremental approach was better. In some cases the hypervolume of the selected subsets was exactly the same and these points are not shown in this figure since value 0 cannot be shown with logarithmic scale. The decremental approach is almost always better for linear data sets and other types of data sets with $m = 5$. To verify statistical significance of the observed differences we again used the Wilcoxon signed rank tests with the significance level $\alpha = 0.05$. The decremental approach was significantly better in majority of cases for linear data sets and convex/concave data sets with $m = 5$ (for example, for concave data sets with $m = 5$ the decremental approach was significantly better in $64, 8\%$ cases while the incremental approach never was significantly better). Thus, for such data sets the decremental approach may be preferred over the incremental approach even when a relatively small number of points is to be selected, if the quality of the selected subset is more important than the CPU time. For convex/concave data sets with $m = 7, 9$ in majority of cases no significant differences were observed.

7   Conclusions

In this paper we have shown that both the incremental and the decremental greedy hypervolume selection methods are more efficient with the use of the updated implementation of QHV-II than with WFG which is often considered to be the state-of-the-art code for hypervolume computation. We have also proposed two new methods for the decremental greedy hypervolume selection, i.e. Greedy Decremental Lazy HSS algorithm and Greedy Decremental HSS algorithm with the use of QEHC and bounds from the previous iterations. We have also compared the best greedy incremental and decremental methods in terms of the efficiency and the quality of the selected subset. This comparison shows that different methods are the best choices depending on the number of points to be selected, the type of data sets, and the relative importance of the subset quality and running time.

In both lazy algorithms and algorithms using QEHC we regularly recalculate hypervolume contribution of a given point with some other points added or removed from set S. In the future we would like to investigate if, instead of re-running such processes from the scratch, we could save some intermediate states of such processes and then update/continue computations resulting from the added/removed points. In fact already in QEHC we explicitly use a stack of subproblems remaining to be processed and we envision that a similar data structure could be used to store the status of already performed computations. It is not clear, however, if the overhead related to the management of such data structure will not annihilate potential savings.

References

1. Bader, J., Deb, K., Zitzler, E.: Faster hypervolume-based search using monte carlo sampling. In: Ehrgott, M., Naujoks, B., Stewart, T.J., Wallenius, J. (eds.) Multiple Criteria Decision Making for Sustainable Energy and Transportation Systems, pp. 313–326. Springer, Berlin (2010). https://doi.org/10.1007/978-3-642-04045-0_27
2. Bader, J., Zitzler, E.: HypE: an algorithm for fast hypervolume-based many-objective optimization. Evol. Comput. **19**(1), 45–76 (2011)
3. Basseur, M., Derbel, B., Goëffon, A., Liefooghe, A.: Experiments on greedy and local search heuristics for dimensional hypervolume subset selection. In: Proceedings of the Genetic and Evolutionary Computation Conference, GECCO 2016, pp. 541–548. Association for Computing Machinery, New York (2016). https://doi.org/10.1145/2908812.2908949
4. Beume, N., Fonseca, C.M., Lopez-Ibanez, M., Paquete, L., Vahrenhold, J.: On the complexity of computing the hypervolume indicator. IEEE Trans. Evol. Comput. **13**(5), 1075–1082 (2009)
5. Beume, N., Naujoks, B., Emmerich, M.: Sms-emoa: multiobjective selection based on dominated hypervolume. Euro. J. Operat. Res. **181**, 1653–1669 (2007). https://doi.org/10.1016/j.ejor.2006.08.008
6. Bradstreet, L., While, L., Barone, L.: Incrementally maximising hypervolume for selection in multi-objective evolutionary algorithms. In: 2007 IEEE Congress on Evolutionary Computation, pp. 3203–3210. IEEE (2007)
7. Bringmann, K., Cabello, S., Emmerich, M.T.M.: Maximum Volume Subset Selection for Anchored Boxes. In: Aronov, B., Katz, M.J. (eds.) 33rd International Symposium on Computational Geometry (SoCG 2017). Leibniz International Proceedings in Informatics (LIPIcs), vol. 77, pp. 22:1–22:15. Schloss Dagstuhl-Leibniz-Zentrum fuer Informatik, Dagstuhl (2017)
8. Bringmann, K., Friedrich, T., Klitzke, P.: Generic postprocessing via subset selection for hypervolume and epsilon-indicator. In: Bartz-Beielstein, T., Branke, J., Filipič, B., Smith, J. (eds.) PPSN 2014. LNCS, vol. 8672, pp. 518–527. Springer, Cham (2014). https://doi.org/10.1007/978-3-319-10762-2_51
9. Brockhoff, D., Tran, T., Hansen, N.: Benchmarking numerical multiobjective optimizers revisited. In: Proceedings of the 2015 Annual Conference on Genetic and Evolutionary Computation, pp. 639–646. GECCO 2015. Association for Computing Machinery, New York (2015)
10. Chan, T.M.: Klee's measure problem made easy. In: 2013 IEEE 54th Annual Symposium on Foundations of Computer Science, pp. 410–419 (2013)
11. Chen, W., Ishibuchi, H., Shang, K.: Lazy greedy hypervolume subset selection from large candidate solution sets. In: 2020 IEEE Congress on Evolutionary Computation (CEC), pp. 1–8 (2020)
12. Cox, W., While, L.: Improving the iwfg algorithm for calculating incremental hypervolume. In: 2016 IEEE Congress on Evolutionary Computation (CEC), pp. 3969–3976 (2016)
13. Friedrich, T., Neumann, F.: Maximizing submodular functions under matroid constraints by multi-objective evolutionary algorithms. In: Bartz-Beielstein, T., Branke, J., Filipič, B., Smith, J. (eds.) PPSN 2014. LNCS, vol. 8672, pp. 922–931. Springer, Cham (2014). https://doi.org/10.1007/978-3-319-10762-2_91
14. Guerreiro, A.P., Fonseca, C.M.: Computing and updating hypervolume contributions in up to four dimensions. IEEE Trans. Evol. Comput. **22**(3), 449–463 (2018)

15. Guerreiro, A.P., Fonseca, C.M., Paquete, L.: Greedy hypervolume subset selection in low dimensions. Evol. Comput. **24**(3), 521–544 (2016)
16. Guerreiro, A.P., Fonseca, C.M., Paquete, L.: The hypervolume indicator: Problems and algorithms (2020)
17. Jaszkiewicz, A.: Improved quick hypervolume algorithm. Comput. Oper. Res. **90**, 72–83 (2018)
18. Jaszkiewicz, A., Zielniewicz, P.: Quick Extreme Hypervolume Contribution Algorithm, pp. 412–420. Association for Computing Machinery, New York (2021). https://doi.org/10.1145/3449639.3459394
19. Jiang, S., Zhang, J., Ong, Y., Zhang, A.N., Tan, P.S.: A simple and fast hypervolume indicator-based multiobjective evolutionary algorithm. IEEE Trans. Cybern. **45**(10), 2202–2213 (2015). https://doi.org/10.1109/TCYB.2014.2367526
20. Knowles, J.D., Corne, D.W., Fleischer, M.: Bounded archiving using the lebesgue measure. In: The 2003 Congress on Evolutionary Computation, CEC 2003, vol. 4, pp. 2490–2497 (2003)
21. Lacour, R., Klamroth, K., Fonseca, C.M.: A box decomposition algorithm to compute the hypervolume indicator. Comput. Oper. Res. **79**, 347–360 (2017)
22. Laitila, J., Moilanen, A.: New performance guarantees for the greedy maximization of submodular set functions. Optimization Letters **11**(4), 655–665 (2016). https://doi.org/10.1007/s11590-016-1039-z
23. Li, B., Li, J., Tang, K., Yao, X.: Many-objective evolutionary algorithms: a survey. ACM Comput. Surv. **48**(1), 1–35 (2015)
24. Li, M., Yao, X.: Quality evaluation of solution sets in multiobjective optimisation: a survey. ACM Comput. Surv. **52**(2), 1–38 (2019)
25. Minoux, M.: Accelerated greedy algorithms for maximizing submodular set functions. In: Stoer, J. (ed.) Optimization Techniques, pp. 234–243. Springer, Berlin (1978). https://doi.org/10.1007/BFb0006528
26. Nemhauser, G.L., Wolsey, L.A., Fisher, M.L.: An analysis of approximations for maximizing submodular set functions-i. Math. Program. **14**(1), 265–294 (1978)
27. Russo, L.M.S., Francisco, A.P.: Quick Hypervolume. IEEE Trans. Evol. Comput. **18**(4), 481–502 (2014)
28. Russo, L.M.S., Francisco, A.P.: Extending quick hypervolume. J. Heuristics **22**(3), 245–271 (2016). https://doi.org/10.1007/s10732-016-9309-6
29. Seo, M.G., Shin, H.S.: Greedily excluding algorithm for submodular maximization. In: 2018 IEEE Conference on Control Technology and Applications (CCTA), pp. 1680–1685 (2018). https://doi.org/10.1109/CCTA.2018.8511628
30. Shang, K., Ishibuchi, H., He, L., Pang, L.M.: A survey on the hypervolume indicator in evolutionary multiobjective optimization. IEEE Trans. Evol. Comput. **25**(1), 1–20 (2021)
31. Shang, K., Ishibuchi, H., Ni, X.: R2-based hypervolume contribution approximation. IEEE Trans. Evol. Comput. **24**(1), 185–192 (2020)
32. Ulrich, T., Thiele, L.: Bounding the effectiveness of hypervolume-based $(\mu + \lambda)$-archiving algorithms. In: Proceedings of the 6th International Conference on Learning and Intelligent Optimization, LION 2012, pp. 235–249. Springer, Berlin (2012)
33. While, L., Bradstreet, L.: Applying the wfg algorithm to calculate incremental hypervolumes. In: 2012 IEEE Congress on Evolutionary Computation, pp. 1–8 (2012)
34. While, L., Bradstreet, L., Barone, L.: A fast way of calculating exact hypervolumes. IEEE Trans. Evol. Comput. **16**(1), 86–95 (2012)

35. Zitzler, E., Künzli, S.: Indicator-based selection in multiobjective search. In: Yao, X., et al. (eds.) PPSN 2004. LNCS, vol. 3242, pp. 832–842. Springer, Heidelberg (2004). https://doi.org/10.1007/978-3-540-30217-9_84
36. Zitzler, E., Thiele, L.: Multiobjective evolutionary algorithms: a comparative case study and the strength pareto approach. IEEE Trans. Evol. Comput. 3(4), 257–271 (1999)
37. Zitzler, E., Thiele, L., Bader, J.: On set-based multiobjective optimization. IEEE Trans. Evol. Comput. 14(1), 58–79 (2010)
38. Zitzler, E., Thiele, L., Laumanns, M., Fonseca, C.M., da Fonseca, V.G.: Performance assessment of multiobjective optimizers: an analysis and review. IEEE Trans. Evol. Comput. 7(2), 117–132 (2003)

Hybridizing Hypervolume-Based Evolutionary Algorithms and Gradient Descent by Dynamic Resource Allocation

Damy M. F. Ha[1,2](✉) ⬤, Timo M. Deist[2] ⬤, and Peter A. N. Bosman[1,2] ⬤

[1] Delft University of Technology, Delft, The Netherlands
d.m.f.ha@student.tudelft.nl, P.A.N.Bosman@tudelft.nl
[2] Centrum Wiskunde and Informatica, Life Sciences and Health Research Group, Amsterdam, The Netherlands
{dmfh,timo.deist,peter.bosman}@cwi.nl

Abstract. Evolutionary algorithms (EAs) are well-known to be well suited for multi-objective (MO) optimization. However, especially in the case of real-valued variables, classic domination-based approaches are known to lose selection pressure when approaching the Pareto set. Indicator-based approaches, such as optimizing the uncrowded hypervolume (UHV), can overcome this issue and ensure that individual solutions converge to the Pareto set. Recently, a gradient-based UHV algorithm, known as UHV-ADAM, was shown to be more efficient than (UHV-based) EAs if few local optima are present. Combining the two techniques could exploit synergies, i.e., the EA could be leveraged to avoid local optima while the efficiency of gradient algorithms could speed up convergence to the Pareto set. It is a priori however not clear what would be the best way to make such a combination. In this work, therefore, we study the use of a dynamic resource allocation scheme to create hybrid UHV-based algorithms. On several bi-objective benchmarks, we find that the hybrid algorithms produce similar or better results than the EA or gradient-based algorithm alone, even when finite differences are used to approximate gradients. The implementation of the hybrid algorithm is available at https://github.com/damyha/uncrowded-hypervolume.

Keywords: Real-valued optimization · Multi-objective · Hybrid algorithm

1 Introduction

In real-valued multi-objective (MO) optimization, multiple conflicting objectives need to be optimized. The goal of MO optimization often is to find a diverse set

---

Supported by Open Technology Programme (nr. 15586) financed by Dutch Research Council (NWO), Elekta, and Xomnia. Cofunding by Ministry of Economic Affairs: public-private partnership allowance for top consortia for knowledge and innovation (TKIs).

G. Rudolph et al. (Eds.): PPSN 2022, LNCS 13399, pp. 179–192, 2022.
https://doi.org/10.1007/978-3-031-14721-0_13

of (near-)Pareto optimal solutions, and usually to do so as efficiently as possible. Evolutionary algorithms (EAs) (e.g. [6,9]) are known to be well suited for MO optimization [8]. However, in real-valued MO optimization, classic domination-based approaches lose selection pressure when approaching the Pareto set [2]. Indicator-based approaches, such as optimizing the hypervolume (HV) [21] or the uncrowded hypervolume (UHV) [12,17,19] can overcome this issue and ensure that individual solutions converge to the Pareto set. Recently, a gradient-based UHV algorithm known as UHV-ADAM [10] was shown to be more efficient than (UHV-based) EAs if few local optima are present. EAs generally remain more efficient if many local optima are present. Combining the two techniques could exploit synergies, especially in problems with many local optima, i.e., the EA could be leveraged to avoid local optima while the gradient algorithms could be leveraged to efficiently converge to the Pareto set. It is however unknown a priori, how the techniques should be combined to get the best results.

Attempts in the literature have been successful at creating efficient MO hybrid algorithms (also known as memetic algorithms). In [18] a hybrid algorithm was proposed that probabilistically executes different variation operators of EAs. Gradient algorithms however have not been integrated into their work. In [3] a domination-based EA was combined with gradient-based algorithms that exploit either the gradient of a single-objective or a combination thereof that corresponds to maximum improvement in a multi-objective sense. In [3], resources are further-more dynamically assigned to the gradient algorithms via a resource allocation scheme (RAS). A HV-based hybrid algorithm was introduced in [13], which combines both an EA and gradient algorithm that aim to maximize the HV. In contrast to [3] however, [13] executes the gradient algorithm after the EA is finished. Supplementing the EA during evolution however might be of key value.

In this work, we study the potential of unifying the convergence properties of UHV-based MO algorithms with a hybrid interleaving optimization scheme. Specifically, we formulate a new UHV-based hybrid algorithm and show that the hybrid algorithm is capable of performing better than the worst of the original algorithms or in some cases better than both algorithms. For this, we combine a UHV-based EA called UHV-GOMEA [17] with UHV-ADAM [10] by extending the RAS of [3]. The resulting hybrid algorithm is consequently UHV-based. The UHV distinguishes itself in that a set of solutions is optimized instead of individual solutions. Concretely, this means that the UHV-based hybrid (and EA) employ a population of solution sets, not a population of individual solutions. Each solution set is optimized towards the Pareto set. In this work, we empirically determine the hybrid's architecture using a similar set of benchmarks as in [17] and [10]. We then compare the final algorithm with its component algorithms, UHV-GOMEA and UHV-ADAM, and another UHV-based gradient algorithm on the Walking Fish Group (WFG) benchmark set [15]. The remainder of this document is organized as follows: In Sects. 2 and 3 we introduce the UHV indicator and existing UHV algorithms. In Sect. 4 we introduce the hybrid algorithm with its RAS. The experiments follow in Sect. 5, with a discussion and conclusion in Sects. 6 and 7 respectively.

2   Uncrowded Hypervolume Optimization

We consider an MO optimization problem to be a problem where m objective functions need to be minimized. Let $\mathbf{f} : \mathcal{X} \to \mathbb{R}^m$, with $\mathbf{f} = [f_0, ..., f_{m-1}]$, be an m-dimensional vector of objective functions, where $\mathcal{X} \subseteq \mathbb{R}^n$ is an n-dimensional search space. In this work we focus on bi-objective problems ($m = 2$). A solution $\mathbf{x} \in \mathcal{X}$, where $\mathbf{x} = [x_0, ..., x_{n-1}]$, will be called an MO-solution. The goal of MO optimization is to find a set $\mathbb{S}$ of diverse and (near-)Pareto-optimal MO-solutions. To achieve this, we assess the quality of $\mathbb{S}$ via the UHV indicator function in Eq. 1. The UHV measures the hypervolume (HV), i.e. the area in objective space enclosed by the non-dominated solutions of $\mathbb{S}$ and reference point $\mathbf{r} = (r_0, r_1)$, and penalizes the dominated solutions of $\mathbb{S}$ via the uncrowded distance (ud). We refer the reader to [17] for the reasons behind scaling and exponentiation operations on the ud. To calculate the HV, let $\mathcal{A}$ be the approximation set that contains all non-dominated solutions of $\mathbb{S}$. $\mathcal{A}$ then forms an approximation boundary $\partial \mathbf{f}(\mathbb{S})$ in objective space. The reader is referred to [21] on how $\partial \mathbf{f}(\mathbb{S})$ is calculated. The HV is the region encapsulated between approximation boundary $\partial \mathbf{f}(\mathbb{S})$ and reference point $\mathbf{r}$, as shown in Fig. 1. The aforementioned uncrowded distance $\mathrm{ud}(\mathbf{x}, \mathbb{S})$ is the closest Euclidean distance between MO-solution $\mathbf{x}$'s objective values $\mathbf{f}(\mathbf{x})$ and the approximation boundary $\partial \mathbf{f}(\mathbb{S})$. By definition, $\mathrm{ud}(\mathbf{x}, \mathbb{S})$ is zero for a non-dominated solution. Using the UHV indicator, an MO problem is effectively reformulated as a single-objective problem. The goal of UHV-based algorithms is to maximize the UHV, as maximization leads directly to the minimization of the original objective functions as well improving the diversity [1].

$$\mathrm{UHV}(\mathbb{S}) = \mathrm{HV}(\mathbb{S}) - \frac{1}{|\mathbb{S}|} \sum_{\mathbf{x} \in \mathbb{S}} \mathrm{ud}(\mathbf{x}, \mathbb{S})^m \tag{1}$$

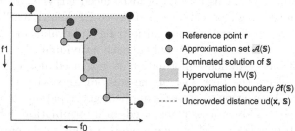

Fig. 1. Illustration of the UHV of $\mathbb{S}$ for an arbitrary bi-objective problem.

3   UHV-Based Algorithms

3.1   UHV-ADAM

UHV-ADAM [10] is based on the single objective stochastic gradient algorithm ADAM [16]. UHV-ADAM parameterizes a single solution set $\mathbb{S}$ of p number of

MO-solutions as ϕ^0, such that $\phi^0 = [\mathbf{x_0}, ..., \mathbf{x_{p-1}}] \in \mathbb{R}^{p \cdot n}$. Let $\mathbf{F}(\phi^0)$ be the operator that assesses the objective functions for every MO-solution in ϕ^0, as displayed in Eq. 2. UHV-ADAM starts by randomly initializing the MO-solutions of ϕ^0 and evaluates the objective values $(f_0(\mathbf{x_i}), f_1(\mathbf{x_i}))$ and objective gradients $(\nabla f_0(\mathbf{x_i}), \nabla f_1(\mathbf{x_i}))$ for every MO-solution $\mathbf{x_i}$ of solution set ϕ^0. Using the objective values and objective function gradients, the gradient of the UHV indicator $\nabla \text{UHV}(\phi^0)$ is calculated. $\nabla \text{UHV}(\phi^0)$ indicates how MO-solutions in the search space must move to (locally) obtain the most UHV gain. The reader is referred to [10,11] on how $\nabla \text{UHV}(\phi^0)$ is exactly calculated. UHV-ADAM then determines the direction in which solutions are moved in the next iteration via a variance-corrected weighted average of $\nabla \text{UHV}(\phi^0)$. How far the solutions are moved is determined by step size factor γ and the variance correction. γ is determined by a shrinking scheme which reduces γ by 1% if no UHV improvement is found. The initial γ is computed by taking 1% of the average initialization range. This initialization method will be used later to reinitialize UHV-ADAM within the hybrid algorithm. UHV-ADAM repeats the process of calculating the UHV gradient and moving the solutions until all computation resources, e.g., a time or function evaluation budget, have been spent or a desired UHV value has been reached.

$$\phi^0 = \begin{bmatrix} \mathbf{x_0} \\ ... \\ \mathbf{x_{p-1}} \end{bmatrix} \to \mathbf{F}(\phi^0) = \begin{bmatrix} \mathbf{f}(\mathbf{x_0}) \\ ... \\ \mathbf{f}(\mathbf{x_{p-1}}) \end{bmatrix} = \begin{bmatrix} f_0(\mathbf{x_0}) & \cdots & f_{m-1}(\mathbf{x_0}) \\ \vdots & \ddots & \vdots \\ f_0(\mathbf{x_{p-1}}) & \cdots & f_{m-1}(\mathbf{x_{p-1}}) \end{bmatrix} \quad (2)$$

3.2 UHV-GOMEA

The Uncrowded Hypervolume Gene-pool Optimal Mixing Evolutionary Algorithm (UHV-GOMEA) [17] is a recently introduced UHV-based EA that leverages strengths of the single-objective model-based EA known as RV-GOMEA [6]. UHV-GOMEA starts off by randomly initializing and evaluating a population of N solution sets: $\phi = [\phi^0, \cdots, \phi^{N-1}]$, where each individual ϕ^i ($i = 0, \cdots, N-1$) has p MO-solutions. Gradient information is not used nor calculated. UHV-GOMEA then selects the best 35% of the solution sets with the highest UHV value as parents. A variation operator is applied on the parents to create new offspring solution sets. This process is repeated until termination. UHV-GOMEA's variation operator makes use of linkage models. In this work, only the marginal product linkage model (Lm) is used. Lm greedily rearranges the MO-solutions of each solution set such that all i'th MO-solution $\mathbf{x_i}$ ($i = 0, \ldots, p-1$) of each solution set is in the same region of the approximation front. It then groups all variables pertaining to $\mathbf{x_i}$ into sets. These sets together compromise a FOS (Family Of Subsets) denoted as $\mathcal{F}$. For each $\mathcal{F}$, a Gaussian distribution is estimated. These Gaussians are used to create offspring by sampling MO-solutions from this marginal product distribution and to inject the new MO-solutions into each individual of the population. If the UHV improves, changes are kept. Otherwise, they are rejected. For more details, including how the Gaussians are estimated and adapted during evolution, the reader is referred to [6,17].

4 Hybridization

4.1 Changes Made to UHV-ADAM

In this work UHV-ADAM has been extended such that the single-solution set solving algorithm is compatible with population-based UHV-GOMEA. To this end, UHV-ADAM steps are applied to population members after a run of UHV-GOMEA. UHV-ADAM instances are assigned to each solution set of the population, allowing the weighted moving average and γ to be tuned accurately and differently to the environment of each solution set in the population. UHV-ADAM instances are reset every time the variation operator of UHV-GOMEA is applied to prevent γ and the moving averages of UHV-ADAM instances to become inaccurate if UHV-GOMEA makes big leaps in the search space. Resetting the UHV-ADAM instances comes at the cost of warming up the moving averages again as well as redetermining γ. γ is re-estimated by creating the tightest box that contains all MO-solutions of the population and to take 1% of the average box width. Finally, a RAS will be used to adaptively determine which algorithm (ADAM or GOMEA) should be used more during a run. After determining the resource distribution, the resources assigned to UHV-ADAM must be distributed over the population members. Early experiments have shown that distributing among the 3 solutions with the highest UHV works the best, but this will be further investigated in Sect. 5.3.

4.2 Resource Allocation Scheme

The hybrid created in this work is based on [3], where a resource allocation scheme (RAS) is used. In this work, only UHV-GOMEA and the modified UHV-ADAM are hybridized. The hybrid algorithm executes UHV-GOMEA and UHV-ADAM sequentially. UHV-GOMEA is always executed once per generation, while the RAS determines the number of UHV-ADAM steps. The RAS of [3] is extended to accommodate the modified UHV-ADAM and works as follows: let the actual number of evaluations and improvements found in generation t by optimizer $o \in \{\text{GOMEA}, \text{ADAM}\}$ be $E_o(t)$ and $I_o(t)$ respectively. An evaluation occurs when one MO-solution $\mathbf{x_i}$ is evaluated. What entails an improvement will be discussed later in Experiment 1. Let the number of evaluations and improvements to be considered for redistribution be $\mathcal{E}_o(t)$ and $\mathcal{I}_o(t)$ respectively. For UHV-ADAM, only the values of the current generation are of interest, that is: $\mathcal{E}_{\text{ADAM}}(t) = E_{\text{ADAM}}(t)$ and $\mathcal{I}_{\text{ADAM}}(t) = I_{\text{ADAM}}(t)$. The number of evaluations and improvements to be considered for UHV-GOMEA is a sum of values of previous generations, that is: $\mathcal{E}_{\text{GOMEA}}(t) = \sum_{t'=t_{min}}^{t} E_{\text{GOMEA}}(t)$ and $\mathcal{I}_{\text{GOMEA}}(t) = \sum_{t'=t_{min}}^{t} I_{\text{GOMEA}}(t)$, where $t_{min} \geq 0$ and t_{min} is chosen as large as possible such that $\mathcal{E}_{\text{GOMEA}}(t) \geq \mathcal{E}_{\text{ADAM}}(t)$ still holds. UHV-GOMEA includes past values for two reasons: it makes the comparison between the gradient algorithm and EA fairer and also allows the number of gradient algorithm calls to grow [4]. Following [4], the EA's variation operator is executed once per generation while the number of executions of the gradient algorithms are related to the respective reward they receive. The reward, displayed in Eq. 3, is the efficiency of finding improvements. The reward is 0 if $\mathcal{E}_o(t) = 0$.

$$\mathcal{R}_o(t) = \frac{\mathcal{I}_o(t)}{\mathcal{E}_o(t)} \tag{3}$$

Let the evaluations to be redistributed to UHV-ADAM be $\mathcal{E}_{\text{ADAM}}^{\text{Red}}(t)$. $\mathcal{E}_{\text{ADAM}}^{\text{Red}}(t)$ is the ratio of UHV-ADAM's contribution to the total reward times the total sum of evaluations to be considered in generation t as shown in Eq. 4.

$$\mathcal{E}_{\text{ADAM}}^{\text{Red}}(t) = \frac{\mathcal{R}_{\text{ADAM}}(t)}{\sum_{o'} \mathcal{R}_{o'}(t)} \sum_{o'} \mathcal{E}_{o'}(t) \tag{4}$$

To calculate the number of iterations UHV-ADAM can execute with budget $\mathcal{E}_{\text{ADAM}}^{\text{Red}}(t)$, let the number of calls be $\mathcal{C}_{\text{ADAM}}^{\text{Red}}(t)$, where $\mathcal{C}_{\text{ADAM}}^{\text{Red}}(t)$ can be calculated by dividing the resources assigned to UHV-ADAM by the average number of evaluations required per call. The average evaluations per call are estimated using the resources and calls of generation t, resulting in Eq. 5.

$$\mathcal{C}_{\text{ADAM}}^{\text{Red}}(t) = \frac{\mathcal{E}_{\text{ADAM}}^{\text{Red}}(t)}{\frac{\mathcal{E}_{\text{ADAM}}(t)}{\mathcal{C}_{\text{ADAM}}(t)}} = \frac{\mathcal{C}_{\text{ADAM}}(t)}{\mathcal{E}_{\text{ADAM}}(t)} \mathcal{E}_{\text{ADAM}}^{\text{Red}}(t) \tag{5}$$

To ensure a smooth decrease in the number of gradient calls, memory decay is implemented in Eq. 6. If the number of calls after redistribution is smaller than the number of calls executed in the current generation, a running average is used to decrease the number of calls. If the number of calls increases, memory decay is not applied in order to stimulate the use of gradient Algorithms [4]. The (memory) decay factor η is kept at the original value of 0.75 [4].

$$\mathcal{C}_{\text{ADAM}}^{\text{Run}}(t+1) = \begin{cases} \mathcal{C}_{\text{ADAM}}^{\text{Red}}(t), & \text{if } \mathcal{C}_{\text{ADAM}}^{\text{Red}}(t) \geq \mathcal{C}_{\text{ADAM}}^{\text{Run}}(t) \\ \eta \mathcal{C}_{\text{ADAM}}^{\text{Run}}(t) + (1-\eta)\mathcal{C}_{\text{ADAM}}^{\text{Red}}(t), & \text{otherwise} \end{cases} \tag{6}$$

The number of UHV-ADAM calls to execute next generation could be set to $\mathcal{C}_{\text{ADAM}}(t+1) = \lfloor \mathcal{C}_{\text{ADAM}}^{\text{Run}}(t+1) \rfloor$, However, if at some point $\mathcal{C}_{\text{ADAM}}(t) = 0$ holds, UHV-ADAM cannot be activated any more. As UHV-ADAM could become useful again in the future, a waiting scheme is used that makes UHV-ADAM wait $\mathcal{W}_{\text{ADAM}}(t)$ generations. In [4], gradient algorithms are only allowed to be executed at most once per individual per generation. Furthermore, at most (population size) N number of total calls can be executed per generation. Early experiments have shown that executing one UHV-ADAM call per individual does not substantially affect convergence. For this reason, multiple gradient calls can be applied to the same individual. Furthermore, a lower bound is introduced such that if UHV-ADAM is to be executed, it executes at least $\mathcal{C}_{\text{ADAM}}^{\text{min}}$ calls.

This ensures that the performance of UHV-ADAM is assessed after it warms up its internal parameters. $\mathcal{C}_{\text{ADAM}}^{\min}$ is set to 10 and has not been further optimized. The cap on total gradient calls is kept and set to N. The modified waiting scheme is shown in Eq. 7. UHV-ADAM is forced to wait for some generations when $\mathcal{C}_{\text{ADAM}}^{\text{Run}}(t+1) \leq \mathcal{C}_{\text{ADAM}}^{\min}$. Because the extended UHV-ADAM executes a minimum number of calls, $\mathcal{C}_{\text{ADAM}}^{\min}$ has been added to prevent the waiting scheme from triggering too early. The actual number of calls to be executed is shown in Eq. 8, where $\mathcal{C}_{\text{ADAM}}^{\min}$ has also been added to the original Equation.

$$
\mathcal{W}_{\text{ADAM}}(t+1) = \begin{cases} \left\lfloor \frac{\mathcal{C}_{\text{ADAM}}^{\min}}{\mathcal{C}_{\text{ADAM}}^{\text{Run}}(t+1)} \right\rfloor, & \text{if } \mathcal{W}_{\text{ADAM}}(t) = 0 \\ \mathcal{W}_{\text{ADAM}}(t) - 1, & \text{otherwise} \end{cases} \tag{7}
$$

$$
\mathcal{C}_{\text{ADAM}}(t) = \begin{cases} \mathcal{C}_{\text{ADAM}}^{\min}, & \text{if } \mathcal{W}_{\text{ADAM}}(t-1) = 1 \\ \min(\lfloor \mathcal{C}_{\text{ADAM}}^{\text{Run}}(t) \rfloor, N), & \text{otherwise} \end{cases} \tag{8}
$$

5 Experiments

5.1 Experimental Setup

The problems used in the experiments are given in Table 1, where n is the problem dimensionality. Problem 0 is uni-modal, objective-wise decomposable [17] and can be quickly solved with gradient Algorithms [10]. Problem 1 is a low multi-modal problem based on the Rosenbrock function which has pair-wise dependencies [5]. It is known for pulling algorithms towards the optimum of the more easily solvable Sphere function while potentially getting solutions stuck in a local optimum of the Rosenbrock function. Problem 2 contains the multi-modal Rastrigin [14] problem, where many local optima are evenly scattered around the solution space. Problem 3 is multi-modal in both objectives where the Pareto set is enveloped by basins. The Pareto sets of all problems lie on a line between the respective optima.

Table 1. The bi-objective benchmark problems selected for the experiments.

#	Problem name	Objectives	Properties
0	Convex bi-sphere	$f_0 = f_{\text{sphere}}(\mathbf{x})$, with $f_{\text{sphere}}(\mathbf{x}) = \sum_{i=0}^{n-1}(x_i)^2$ $f_1 = f_{\text{sphere}}(\mathbf{x} - \mathbf{c}_0)$ $\mathbf{c}_0 = [1, 0, \cdots, 0]$	Uni-modal, decomposable
1	Convex sphere Rosenbrock	$f_0 = \frac{1}{n}f_{\text{sphere}}(\mathbf{x})$ $f_1 = \frac{1}{n-1}f_{\text{ros}}(\mathbf{x})$, with $f_{\text{ros}}(\mathbf{x}) = \sum_{i=0}^{n-1}(100(x_i - x_{i-1}^2)^2 + (1 - x_{i-1})^2)$	Multi-modal, attraction to f_0
2	Convex sphere Rastrigin	$f_0 = f_{\text{sphere}}(\mathbf{x})$ $f_1 = f_{\text{rast}}(\mathbf{x} - \mathbf{c}_2)$, with $f_{\text{rast}}(\mathbf{x}) = An + \sum_{i=0}^{n-1} x_i^2 - A\cos(2\pi x_i)$ $A = 10, \mathbf{c}_2 = [0.5, 0, \cdots, 0]$	Multi-modal
3	Bi-cosine sphere	$f_0 = f_{\cos}(\mathbf{x})$, with $f_{\cos}(\mathbf{x}) = f_{\text{sphere}}(\mathbf{x})(1 - \beta\cos(2\pi f |\mathbf{x}|))$ $f_1 = f_{\cos}(\mathbf{x} - \mathbf{c}_0)$ $\beta = 0.6, f = 0.1$	Multi-modal in f_0 and f_1

5.2   Experiment 1: The Effect of the Improvement Metric

In experiment 1, the problems from Table 1 are used to assess the effects of different improvement metrics $I_o(t)$. Problem 0 is excluded from this experiment as tuning the hybrid algorithm on this easily solvable problem is undesirable. Metrics ΔBestUHV and ΔAverageUHV are the difference between the best found UHV and average population UHV respectively in subsequent generations. CountUHVImproved and CountBestUHVImproved count the number of times the UHV of a solution and that of the best solution have improved respectively. In [3], the number of MO-solutions added to an elitist archive is counted. We will identify that metric with Bosman2012. Here, we use an infinitely large elitist archive to encourage counting MO-solutions that improve the UHV, which otherwise are potentially rejected by a (nearly) full, finite sized elitist archive. For clarity, the elitist archive is not used for anything but the improvement metric.

Gradient calls are applied to the best 3 solutions of the population that have the highest UHV. The solution set size is set to $p = 9$. As we do not know the HV of Pareto set $\mathcal{A}^\star$ analytically, HV($\mathcal{A}^\star$) is set to the maximum HV obtained from running all algorithms 30 times, while initializing the algorithms near the Pareto set. In experiment 1 we run each improvement metric on problems $P = [1, 2, 3]$ for the following problem dimensionalities $D = [2, 5, 10, 20, 40, 80]$. For each dimension, we determine the best population N by running the following population sizes $N = [40, 80, 160, 320, 640, 1280]$ 30 times and select the most efficient population size that reaches a target HV accuracy of ΔHV$_p < 10^{-6}$ with a success rate of at least 29 out of 30 runs. We consider runs that need more than 10^7 MO-evaluations to have failed in finding the target HV. We then compute a performance score, which sums the relative performance of improvement metric imp with respect to the best performing improvement metric amongst all improvement metrics I, over all problems P and problem dimensionalities D in Eq. 9.

$$\text{score(imp)} = \sum_{pr \in P} \sum_{d \in D} \frac{\text{median(MO-Evaluations(pr, d, imp))}}{\min_{imp' \in I}(\text{median(MO-Evaluations(pr, d, imp')))}} \quad (9)$$

Table 2 shows the results. Using ΔBestUHV obtains the best score in all problems except Problem 1. ΔAverageUHV is consistently performing the worst. ΔAverageUHV is generally biased towards rewarding UHV-GOMEA as UHV-ADAM is not designed to efficiently optimize an entire population. Experiment 1 shows that it is not trivial to select an improvement metric that is superior for all problems. Instead, improvement metrics appear to be problem specific. However, as ΔBestUHV has the best average score, it will be used in further experiments.

5.3   Experiment 2: The Effect of the Choice of Method to Distribute Gradient Resources

The effect of the choice of method to distribute the resources assigned to the modified UHV-ADAM, on the required number of MO-evaluations to reach a target HV accuracy of ΔHV$_p < 10^{-6}$ and the corresponding success rate (SR)

Table 2. The scores assigned to each improvement metric. The lower the score, the better. The numbers in bold are the lowest scores of a category.

Problem	ΔBestUHV	ΔAverage UHV	Count UHVImproved	CountBestUHV improved	Bosman 2012
Convex sphere & Rosenbrock (1)	9.0	9.0	**8.3**	9.2	9.0
Convex sphere & Rastrigin (2)	**6.1**	9.8	7.9	6.4	7.1
Bi-cosine sphere (3)	**6.1**	8.6	7.1	6.4	6.4
Average	**7.1**	9.1	7.8	7.3	7.5

is shown in Fig. 2. UHV-GOMEA and UHV-ADAM have also been added as a reference. Table 3 shows the scores obtained by the distribution methods using Eq. 9. Following experiment 1, Problem 0 is excluded from Table 3. Distribution methods that are unable to find a population size that meets the SR threshold of 29 out of 30 runs are disqualified and denoted as "DQ". The evaluation budget, population optimization method and solution set size are the same as in experiment 1. The hybrid uses the ΔBestUHV improvement metric. Among the distribution methods, Best-m-Solutions and Best-m%Population apply gradient calls on the best solutions of the population. The former applies calls to a fixed number of solutions sets and the latter to a percentage of the population. ALL applies calls on all solution sets, starting from $\phi^0, \phi^1, \cdots$ until all calls have been distributed. RANDOM applies calls randomly with replacement. In Fig. 2, UHV-GOMEA is generally amongst the worst performing implementations along with UHV-ADAM, which fails to reach the target SR threshold in all problems except Problem 0. In Problem 1 of Fig. 2, the statistics of the successful runs of UHV-ADAM have been displayed despite not meeting the SR threshold. Interestingly, in [10], UHV-ADAM is able to solve Problem 1 when initialized near the global optima ($[0, 2]^n$). In this experiment however, UHV-ADAM gets stuck on local optima due to a larger initialization range. Among the distribution methods, Best3Solutions and BestSolution are on average among the best performing distribution methods according to Table 3. Distribution methods: ALL, RANDOM, Best5%Population and Best10%Population are disqualified for not reaching the target SR. Interestingly, analysis shows that for Problem 3 the improvement metric chosen generally remains in the waiting state until most local optima are no longer within the scope of the population, after which it maximizes the number of UHV-ADAM calls, resulting in similar performance amongst the Best-m-Solutions and Best-m%Population distribution methods. Table 3 clearly shows that concentrating gradient calls on the best solutions is more efficient than diluting gradient calls over the population.

5.4   Experiment 3: The WFG Benchmark

We use the WFG suite [15] as an independent method to benchmark the results of the hybrid algorithm. For detailed characteristics of these 9 benchmark functions,

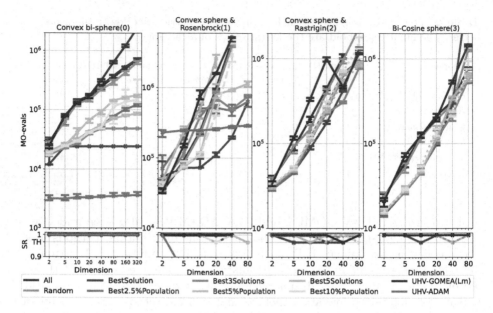

Fig. 2. The effect of different distribution methods on the required number of MO-evaluations to reach a target accuracy of $\Delta HV_p < 10^{-6}$ for various problems. The success rate (SR) measures the fraction of runs that reach the target accuracy out of 30 runs, where the target threshold (TH) is 29/30 runs.

Table 3. The scores assigned to each distribution method. The lower the score, the better. Numbers in bold are the lowest scores of a category. "DQ" denotes distribution methods that fail one or more success rate thresholds.

Problem	All	Random	Best solution	Best3 solutions	Best5 solutions	Best2.5% population	Best5% population	Best10% population
Convex sphere & Rosenbrock(1)	DQ	DQ	**6.6**	18.3	17.4	19.7	DQ	DQ
Convex sphere & Rastrigin(2)	DQ	13.4	7.1	**6.7**	7.7	8.1	8.2	9.4
Bi-Cosine sphere(3)	11.6	11.9	17.8	**6.3**	7.1	6.7	7.1	7.2
Average	DQ	DQ	10.5	**10.4**	10.7	11.5	DQ	DQ

the reader is referred to [15]. WFG1 is a separable problem, with a flat region which can stagnate the search. WFG2 has a uni-modal disconnected convex front. WFG3 is multi-modal and has a linear front. WFG4-9 all have concave fronts, where WFG4 and WFG9 are multi-modal. Following [17], the benchmark is used in a bi-objective setting with $k_{WFG} = 4$ position variables and $l_{WFG} = 20$ distance variables, resulting in $n = 24$ decision variables. The HV reference point $\mathbf{r}$ is set to $r = (11, 11)$. The computation budget is set to 10^7 MO-evaluations for each algorithm. For the solution set size, we use $p = 9$.

The algorithms we consider in this experiment include the base algorithms: UHV-GOMEA(Lm), UHV-ADAM, the constructed hybrid algorithms: ΔBestUHV with distribution methods BestSolution and Best3Solutions, as well as another UHV-based gradient algorithm called UHV-GA-MO [10]. UHV-GA-MO is based on the GA-MO scheme [20]. We refer the reader to [10] for the exact details of UHV-GA-MO. Each algorithm is executed 30 times. Algorithms that use populations have their population sizes set to 200 following [10, 17]. Gradient-based algorithms use finite difference gradient approximations (indicated by the suffix "-FD" in Table 4). Finite difference approximations come at the cost of $(1 + n) \cdot p$ MO-evaluations [10]. Per problem, outcomes are compared to the result with the highest mean value and tested for statistical significance up to 4 decimals by a Wilcoxon two-sided rank-sum test where the initial $\alpha' = 0.05$. α' is Bonferroni corrected by a factor of 36, making the final α to be $\alpha = 0.05/36$.

Table 4 shows that, on average, the best results were obtained with Hybrid-Best3Solutions-FD, followed by Hybrid-BestSolution-FD. Interestingly, the hybrids never obtain a rank worse than 2, indicating that in this experiment, the hybrids on average perform better than the original component algorithms. Furthermore, for problems: WFG1, WFG2, WFG4, WFG6 and WFG8, at least one of the hybrids obtains statistically better HVs than the original component algorithms. Interestingly, in WFG 4, Hybrid-Best3Solutions-FD preforms better than the UHV-GOMEA (Lm) despite WFG 4 being a multi-modal problem.

Table 4. The WFG benchmark for 10^7 MO-evaluations. Hypervolume values are shown (mean, ± standard deviation(rank)). Finite differences (FD) are used for the gradient-based algorithms. Scores in bold are the best or not statistically different from the other bold scores, indicated per problem.

Problem	UHV-GOMEA(Lm)	UHV-ADAM-FD	UHV-GA-MO-FD	Hybrid-BestSolution-FD	Hybrid-Best3Solutions-FD
WFG1	94.63±1.73(5)	97.32±0.60(3)	96.74±0.60(4)	98.90±0.29(2)	**101.57±0.49(1)**
WFG2	110.13±0.03(3)	106.26±5.09(5)	**109.60±6.68(4)**	**110.36±1.20(2)**	110.84±2.04(1)
WFG3	**116.50±0.00(4)**	**116.50±0.00 (1)**	114.78±0.33(5)	**116.50±0.00(1)**	**116.50±0.00(1)**
WFG4	**112.75±0.58(3)**	103.34±3.61(5)	107.21±0.97(4)	113.46±0.35(2)	**114.02±0.13(1)**
WFG5	**112.19±0.10(4)**	112.21±0.03(3)	111.32±0.68(5)	**112.22±0.00(1)**	**112.22±0.00(2)**
WFG6	114.38±0.03(3)	113.79±0.10(4)	110.52±2.00(5)	**114.40±0.00(1)**	114.40±0.00(2)
WFG7	**114.40±0.01(3)**	114.37±0.03(4)	113.88±0.16(5)	**114.40±0.00(2)**	**114.40±0.00(1)**
WFG8	111.43±0.28(3)	110.57±0.81(4)	109.48±1.06(5)	111.70±0.23(2)	**111.82±0.01(1)**
WFG9	**111.46±0.16(3)**	107.54±1.10(4)	103.18±5.31(5)	**111.49±0.03(2)**	**111.51±0.02(1)**
Rank	3.44(3)	3.67(4)	4.67(5)	1.67(2)	1.22(1)

6   Discussion

A real-valued multi-objective (MO) hybrid algorithm was created by combining two uncrowded hypervolume (UHV) indicator-based algorithms via a dynamic resource allocation scheme. In Experiment 1 it was shown that for UHV optimization, picking an improvement metric is not trivial, as problem dependency has been observed. The results of experiment 1 however, also showed that if the hybrid is tasked to do UHV optimization, on average it benefits most from using the ΔBestUHV improvement metric, followed by CountBestUHVImproved. Both metrics quantify the improvement of the best UHV, while the remaining metrics (Bosman2012, CountUHVImproved, ΔAverageUHV) measure the improvement over all solution sets. This opens the question why resource allocation towards the algorithms which improve fewer solution sets with higher UHV is preferable over the full runtime of the hybrid.

Experiment 2 has shown that concentrating gradient calls on a select number of solutions is preferred over diluting calls over the entire population. Distributing resources to the solutions with the top 3 highest UHV performed the best on average. Analysis on this distribution method however, has shown that during convergence, the hybrid frequently stalls due to an inaccurately estimated γ. Substantial improvement in convergence could be obtained by improving γ estimates at reinitialization of UHV-ADAM after executing UHV-GOMEA.

Experiment 2 also provided additional insight in the properties of UHV-ADAM. Figure 2 confirms that problems with few local optima (e.g. Convex sphere & Rosenbrock) can be solved by UHV-ADAM, while problems with many local optima (e.g. Convex sphere & Rastrigin) are not solvable.

One of the limitations of this work is that the problems that were used to tune the hybrid, all share the commonality of having a connected Pareto set. A connected Pareto set simplifies finding all other Pareto optimal solutions as soon as one solution has been determined. If one of the objectives then happens to be easily solvable (e.g. Sphere), it potentially creates situations where even algorithms that are not suited to solve multi-modal problems can still find the Pareto set by first solving the easy objective before moving over to the other objective, bypassing any local optimum. Future work should thus consider disconnected Pareto sets. Another limitation of this work, is that only a single EA, i.e. UHV-GOMEA, has been selected for hybridization. In [17], it was already observed that domination-based EA MO-RV-GOMEA [7] initially performs better than UHV-GOMEA. An even more efficient hybrid algorithm could potentially be created with MO-RV-GOMEA. However, as MO-RV-GOMEA is a domination-based EA, compatibility issues are likely to occur with UHV-based algorithms. Introducing a different EA could furthermore test the robustness of the RAS.

7   Conclusion

In this work, for the first time, a multi-objective optimization algorithm was introduced that hybridizes an uncrowded hypervolume-based (UHV) evolutionary algorithm with a UHV-based gradient algorithm via a dynamic resource allocation scheme (RAS). Experiments used to study the RAS showed that selecting a reward metric for the RAS is not trivial as it was observed that the best metric is problem-dependent. Experiments also showed that concentrating gradient steps on a select number of solutions of the population, outweighs dispersing gradient steps over the entire population. Implementations of the hybrid algorithm have also been compared to other UHV-based algorithms. It was shown that even if finite difference approximations are used to calculate gradients, it is still able to obtain competitive or better results than the original component algorithms as well as other UHV-based algorithms. We conclude that the resulting hybrid is therefore a promising addition to the existing spectrum of evolutionary algorithms for multi-objective optimization.

References

1. Auger, A., Bader, J., Brockhoff, D., Zitzler, E.: Theory of the hypervolume indicator: optimal μ-distributions and the choice of the reference point. In: Proceedings of the Tenth ACM SIGEVO Workshop on Foundations of Genetic Algorithms, pp. 87–102 (2009)
2. Berghammer, R., Friedrich, T., Neumann, F.: Convergence of set-based multiobjective optimization, indicators and deteriorative cycles. Theoret. Comput. Sci. **456**, 2–17 (2012)
3. Bosman, P.A.: On gradients and hybrid evolutionary algorithms for real-valued multiobjective optimization. IEEE Trans. Evol. Comput. **16**(1), 51–69 (2011)
4. Bosman, P.A., De Jong, E.D.: Combining gradient techniques for numerical multiobjective evolutionary optimization. In: Proceedings of the 8th Annual Conference on Genetic and Evolutionary Computation, pp. 627–634 (2006)
5. Bosman, P.A., Grahl, J., Thierens, D.: Benchmarking parameter-free amalgam on functions with and without noise. Evol. Comput. **21**(3), 445–469 (2013)
6. Bouter, A., Alderliesten, T., Witteveen, C., Bosman, P.A.: Exploiting linkage information in real-valued optimization with the real-valued gene-pool optimal mixing evolutionary algorithm. In: Proceedings of the Genetic and Evolutionary Computation Conference, pp. 705–712 (2017)
7. Bouter, A., Luong, N.H., Witteveen, C., Alderliesten, T., Bosman, P.A.: The multiobjective real-valued gene-pool optimal mixing evolutionary algorithm. In: Proceedings of the Genetic and Evolutionary Computation Conference, pp. 537–544 (2017)
8. Deb, K., Kalyanmoy, D.: Multi-Objective Optimization Using Evolutionary Algorithms. John Wiley & Sons Inc., USA (2001)
9. Deb, K., Pratap, A., Agarwal, S., Meyarivan, T.: A fast and elitist multiobjective genetic algorithm: Nsga-ii. IEEE Trans. Evol. Comput. **6**(2), 182–197 (2002)
10. Deist, T.M., Maree, S.C., Alderliesten, T., Bosman, P.A.N.: Multi-objective optimization by uncrowded hypervolume gradient ascent. In: Bäck, T., et al. (eds.) PPSN 2020. LNCS, vol. 12270, pp. 186–200. Springer, Cham (2020). https://doi.org/10.1007/978-3-030-58115-2_13
11. Emmerich, M., Deutz, A.: Time complexity and zeros of the hypervolume indicator gradient field. In: EVOLVE-a Bridge Between Probability, Set Oriented Numerics, And Evolutionary Computation III, pp. 169–193. Springer (2014). https://doi.org/10.1007/978-3-319-01460-9_8
12. Emmerich, M., Deutz, A., Beume, N.: Gradient-based/evolutionary relay hybrid for computing pareto front approximations maximizing the S-metric. In: Bartz-Beielstein, T., et al. (eds.) HM 2007. LNCS, vol. 4771, pp. 140–156. Springer, Heidelberg (2007). https://doi.org/10.1007/978-3-540-75514-2_11
13. Hernández, V.A.S., Schütze, O., Wang, H., Deutz, A., Emmerich, M.: The set-based hypervolume newton method for bi-objective optimization. IEEE Trans. Cybern. **50**(5), 2186–2196 (2018)
14. Hoffmeister, F., Bäck, T.: Genetic algorithms and evolution strategies: similarities and differences. In: Schwefel, H.-P., Männer, R. (eds.) PPSN 1990. LNCS, vol. 496, pp. 455–469. Springer, Heidelberg (1991). https://doi.org/10.1007/BFb0029787

15. Huband, S., Barone, L., While, L., Hingston, P.: A scalable multi-objective test problem toolkit. In: Coello Coello, C.A., Hernández Aguirre, A., Zitzler, E. (eds.) EMO 2005. LNCS, vol. 3410, pp. 280–295. Springer, Heidelberg (2005). https://doi.org/10.1007/978-3-540-31880-4_20

16. Kingma, D.P., Ba, J.: Adam: A method for stochastic optimization. arXiv preprint arXiv:1412.6980 (2014)

17. Maree, S.C., Alderliesten, T., Bosman, P.A.: Uncrowded hypervolume-based multi-objective optimization with gene-pool optimal mixing. Evolutionary Comput. 1–24 (2021)

18. Sharma, S., Blank, J., Deb, K., Panigrahi, B.K.: Ensembled crossover based evolutionary algorithm for single and multi-objective optimization. In: 2021 IEEE Congress on Evolutionary Computation (CEC), pp. 1439–1446. IEEE (2021)

19. Touré, C., Hansen, N., Auger, A., Brockhoff, D.: Uncrowded hypervolume improvement: Como-cma-es and the sofomore framework. In: Proceedings of the Genetic and Evolutionary Computation Conference, pp. 638–646 (2019)

20. Wang, H., Deutz, A., Bäck, T., Emmerich, M.: Hypervolume indicator gradient ascent multi-objective optimization. In: Trautmann, H., et al. (eds.) EMO 2017. LNCS, vol. 10173, pp. 654–669. Springer, Cham (2017). https://doi.org/10.1007/978-3-319-54157-0_44

21. Zitzler, E., Thiele, L.: Multiobjective evolutionary algorithms: a comparative case study and the strength pareto approach. IEEE Trans. Evol. Comput. **3**(4), 257–271 (1999)

Identifying Stochastically Non-dominated Solutions Using Evolutionary Computation

Hemant Kumar Singh[1][(✉)] and Juergen Branke[2]

[1] The University of New South Wales, Canberra, Australia
h.singh@adfa.edu.au
[2] University of Warwick, Coventry, UK
juergen.branke@wbs.ac.uk

Abstract. We consider the problem of finding a solution robust to disturbances of its decision variables, and explain why this should be framed as problem to identify all stochastically non-dominated solutions. Then we show how this can be formulated as an unconventional multi-objective optimization problem and solved using evolutionary computation. Because evaluating stochastic dominance in a black-box setting is computationally very expensive, we also propose more efficient algorithm variants that utilize surrogate models and re-use historical data. Empirical results on several test problems demonstrate that the algorithm indeed finds the stochastically non-dominated solutions, and that the proposed efficiency enhancements are able to drastically cut the number of required function evaluations while maintaining good solution quality.

Keywords: Robust optimization · Stochastic dominance ·
Evolutionary algorithm

1 Background and Motivation

In some real-world environments, the decision variables are subject to disturbances before implementation, e.g., due to manufacturing tolerances [2]. In such cases, it is desirable that the solution is not only good, but also robust. Different definitions of robustness have been proposed in the previous literature:

1. the solution with the best expected performance despite the disturbances. This corresponds to a risk neutral decision maker [13].
2. the solution with the best worst-case performance given the possible range of disturbances. This corresponds to a highly risk sensitive decision maker, willing to sacrifice expected performance for protection from risk [2,12].
3. the solution with the best weighted combination of expected performance plus w times the standard deviation σ. The larger the weight w on the standard deviation, the more risk averse this choice becomes. It has also been suggested to treat this as a multi-objective problem [1,10,14].

G. Rudolph et al. (Eds.): PPSN 2022, LNCS 13399, pp. 193–206, 2022.
https://doi.org/10.1007/978-3-031-14721-0_14

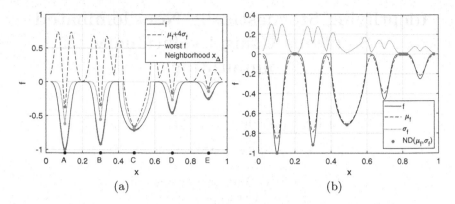

Fig. 1. Illustration of the existing robustness measures. The image of the points $A-E$ in the fitness landscape of deterministic, 4σ robustness and worst-case robustness are marked in red, magenta and green dots, respectively. (Color figure online)

The above definitions can be intuitively understood by the illustration of a single-variable function shown in Fig. 1(a). The objective function is that of the TP3 problem in [13]. The deterministic function is shown with a solid black line. It is assumed that any given design x has an uncertainty uniformly distributed in $x_\Delta = [x - \Delta, x + \Delta]$ with $\Delta = 0.025$. Five points of interest (local optima) A, B, C, D and E have been marked in the design space. For each of these points, we show the region x_Δ by 51 uniformly sampled points, shown as blue dots. The resulting landscape of the robust formulation based on the worst case is shown as dotted line, the landscape for a mean plus 4σ robust formulation is shown as dashed line. Both robust formulations result in the solution C being identified as the robust optimum design. However, it can be seen that the distribution of objective values around design A yields (significantly) better performance under the given variations for some values of $x \in x_\Delta$. Even though the $\mu+4\sigma$ value and the worst value obtained by the design C is better than that of A, design A yields a better or equal performance compared to C with an 88.23% probability (based on the uniform sampling shown)[1].

Moreover, the formulation based on mean plus variance may distort the fitness landscape in undesirable ways. If the uncertain region is slightly larger, say $\Delta = 0.05$, the fitness of solution A becomes even worse than the design $x = 0.2$, whereas the objective value around design A is *never* worse than the latter. Increasing the value of w would magnify the penalty associated with the standard deviation and a solution with extremely poor value but very low standard deviation (e.g. $x = 0.2, 0.8$) is considered equivalent to a solution with much better expected values but higher standard deviation (e.g., A, B, C). To remove the sensitivity of the results to the choice of w, some works have suggested optimizing the expected value and standard deviation as a bi-objective problem [10].

---

[1] Note that these probabilities will change if the uncertainty does not follow a uniform random distribution; a scenario excluded from the scope of this work.

However, as shown in Fig. 1(b), the non-domination sorting based on μ and σ (of 951 uniform samples in $[x_{min}+\Delta, x_{max}-\Delta]$) would also yield several undesirable solutions that have poor objective value, on account of their low/zero variations. Also to note is that some of these solutions (e.g., again $x = 0.2, 0.8$) which have the worst possible objective value of $f = 0$, are preferred over the local minima D, E since the latter get dominated by another point (C) in the search space. The worst case formulation also masks the information regarding the better performance achieved within the variable uncertainties, as seen between the designs A and C. It also renders many of the designs indistinguishable in terms of their fitness (flat regions in Fig. 1). Optimizing the worst case performance is also a *bilevel optimization* which entails other characteristic challenges [9].

In order to overcome some of the shortcomings above, we propose a new way of defining robustness that does not depend only on expected or extreme values, but rather takes into consideration the distribution of the design performance more comprehensively. In particular, we propose to identify *all* solutions that are stochastically non-dominated. The concept of stochastic dominance is often used to compare or rank probability distributions [11]. For two probability distributions $g_A(x)$ and $g_B(x)$, the corresponding cumulative distribution $G_A(x)$ is said to first-order stochastically dominate $G_B(x)$ ($G_A(x) \preceq_{sd} G_B(x)$) if and only if the following inequality holds:

$$G_A(x) \leq G_B(x) \quad \forall x. \tag{1}$$

For any utility function $u(x)$ that is strictly increasing and piece-wise differentiable (which should be true for any rational DM), if $G_A(x) \preceq_{sd} G_B(x)$

$$G_A(x) \preceq_{sd} G_B(x) \Leftrightarrow \mathbb{E}_A(u(X)) \leq \mathbb{E}_B(u(X), \tag{2}$$

where $\mathbb{E}_A$ and $\mathbb{E}_B$ are the expectations over the probability distributions g_A and g_b, respectively. In other words, if we are able to identify all first-order stochastically non-dominated solutions, then we would be sure that among the identified solutions would be the most preferred solution for any rational decision maker, irrespective of their risk preferences.

Our paper is structured as follows. After formulating the problem in Sect. 2, we explain our baseline algorithm and strategies to reduce the number of function evaluations in Sect. 3. Empirical results are reported in Sect. 4. The paper concludes with a summary and some ideas for future work.

2   Proposed Problem Formulation

The proposed definition for robustness is based on the *quantile function (QF)* of the objective computed within the given uncertain region $\mathbf{x_\Delta}$. This function defines, for each possible probability $p \in [0, 1]$ the fitness value that is obtained at least with that probability. More formally,

$$QF(x, p) = \inf\{y \in \mathbb{R} : p \leq G(f(x))\} \tag{3}$$

where $G(f(x))$ is the cumulative probability density function of the fitness value $f(x)$ of solution x given the uncertainty of the disturbance.

To identify all first-order stochastically non-dominated solutions, we are then solving the following optimization problem.

$$\min \qquad QF(x,p) \quad \forall p \tag{4}$$
$$\text{s.t.} \qquad x_i^L \leq x_i \leq x_i^U, \quad i = 1, \ldots n_x. \tag{5}$$

Under the proposed definition, a solution x_A is considered better than another solution x_B if $QF(x_A)$ yields a lower or equal value than $QF(x_B)$ (for minimization) for all values of $p \in [0,1]$. This is equivalent to x_A first-order stochastically dominating x_B.

To understand the proposed measure intuitively, let us consider the QF functions of the solutions $A - E$ previously discussed, as shown in Fig. 2. A given point on the curve, say $(0.5, -0.4043)$ of curve D can be interpreted as: 50% of the designs within the $\mathbf{x_\Delta}$ region of solution D have a better (lower) performance value than -0.4043. From the observed QF curves, it can be inferred that A dominates B, D, E, which means that for any quantile of fitness values A yields a lower fitness than either B, D or E. On the other hand, (A, C) and (B, C) are first order stochastically non-dominated pairs, implying that for each of the pair, there exists a monotonic utility function that would lead to this being the preferred solution. Thus, the set of first-order stochastically non-dominated solutions out of these points is identified as A and C.

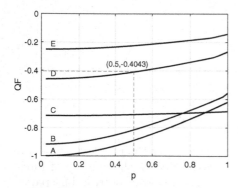
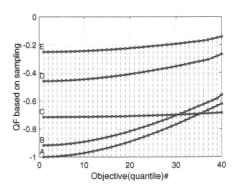

Fig. 2. QF function of solutions $A, B,$ C, D, E in Fig. 1

Fig. 3. Evaluating the quantiles (objectives) (Color figure online)

Interestingly, the above formulation can be regarded as a multi-objective problem with *infinite* number of objectives. Regardless of the nature of the objective function $f(\mathbf{x})$, QF is mathematically a continuous function. Thus, first-order stochastic non-dominance is simply the non-domination criterion applied to compare two continuous functions. For practical implementation of the idea, we approximate QF by a finite (but large) set of objectives M, see below for more details. At the same time, it should also be noted that QF is a strictly non-decreasing function. This characteristic can be used to circumvent some of the scalability issues normally associated with non-domination based sorting for problems with large number of objectives [8].

3   Solution Using an Evolutionary Algorithm

The basic framework of our proposed algorithm is quite similar to a canonical EA used to solve deterministic problems, but its components have been customized to deal with the proposed robust problem formulation. The algorithm assumes no prior information about the nature of the function, considering it as a black-box. For brevity, we refer to the first-order stochastic domination as FOS-domination in the following.

3.1   Discretization and Evaluation of Objectives

In many practical problems, the analytical form of the objective function is unknown. A viable method to *approximate* the quantile function would then be by sampling a finite number (say N_s) of designs within the uncertain region. Furthermore, to practically compare between different solutions and to represent them in a way that can be handled by EAs, a discretization of the quantile function itself is needed. We propose to do so by using M uniformly sampled values of p between 0 and 1. In order to evaluate a solution's performance, the quantile function value corresponding to the ith value of p is assigned as its ith quantile, where $i \in [1, M]$. This is illustrated in Fig. 3, where we chose $M = 40$ objectives and $N_s = 1000$ samples to construct the quantile function. Each vertical dotted line in the figure corresponds to an objective (denoted on the x-axis), and the red dots represent the corresponding robust objective values for a solution (read from the y-axis).

3.2   Parent Selection and Evolution Operators

For evolving offspring, the widely used crossover and mutation operators, simulated binary crossover (SBX) and polynomial mutation (PM) [5] are used. Parents are selected from the current population by pairwise tournament selection. These mechanisms have been selected due to their widespread use in literature, but can be easily substituted with other evolutionary operators.

3.3   Dominance Calculation and Ranking

The process of FOS-domination ranking for a given set S containing N solutions and M objectives (quantiles) is outlined in Algorithm 1 and the key steps are briefly described below.

Firstly, a distance matrix **d** is computed. Each element of the matrix d_{ij} denotes the minimum amount that needs to be added to *all* QF values of the solution i for it to be dominated by solution j (Line 3 in Algorithm 1).

$$d_{ij} = \max\{\max_q\{f_q(j) - f_q(i)\}, 0\} \qquad (6)$$

This quantity will correspond to the quantile in which solution i is better than j by the maximum amount. For example when comparing solution A with B in Fig. 3, $d_{AB} = f_1(B) - f_1(A)$. When comparing C and D, the maximum difference occurs in the 40th objective, so $d_{CD} = f_{40}(D) - f_{40}(C)$. Note that this measure is structurally similar to additive ϵ indicator [16], but applied in quantile space instead of objective space.

Next, for each solution, $dMin$, the minimum of its distance values w.r.t. all other solutions is identified (Line 5). This is the minimum value that needs to be added to each objective of this solution to get dominated by *any* other solution in the set S, i.e., $dMin(i) = \min_{j \in S} d_{ij}$. Thus, in the example above, $dMin(A) = d_{AB}$ and $dMin(C) = d_{CA}$. Note that $dMin$ will be 0 for any solution that is dominated by another solution (B, D, E in this case).

The sequence of elimination is then determined in the Lines 7–17. The solution with the lowest $dMin$ represents the solution that can be dominated most easily, and is therefore added first to the elimination set. Then, the solution is removed, and all corresponding d values (row and column) are set to ∞. Thereafter, $dMin$ is updated, based on the updated $\mathbf{d}$ matrix. The solution with the lowest $dMin$ is again identified as the next solution to be added to the elimination list, and so on. Once all solutions have been added to the list, the order is reversed (Line 18), so as to rank the solutions from best to worst.

Note in the above ranking process that the dominated solutions are indistinguishable from each other, since all of them will have a $dMin = 0$. In order to obtain a full ordering, the FOS-domination ranking can be repeated only on solutions that achieved $dMin = 0$ in the first pass. The solutions that get $dMin = 0$ in the second pass can then be further segregated and ranked; until all solutions have obtained a distinct ranking. Equivalently, one can first do a non-domination sorting of the given solution set, and then apply FOS-domination ranking front-by-front.

3.4   Strategies to Reduce Computational Effort

For the above algorithm, an adequate number of samples needs to be sampled in $\mathbf{x_\Delta}$ to replicate the quantile function accurately. If the population size is N, the number of generations N_G and the number of samples evaluated in the vicinity of each solution $\mathbf{x}$ is N_s, then the total number of function evaluations (calls to the original function $f(\mathbf{x})$) can be calculated as $NFE = N \times N_G \times N_s$. In order to reduce the NFE, we propose two strategies below.

Use of Approximation Models: The use of surrogate models is prevalent in the literature for solving computationally expensive problems with stringent limits on NFE [15]. The basic idea is that based on a few available or prudently sampled designs, a surrogate model can be built and used to partially guide the search in lieu of true evaluations. The true evaluation is then evoked only for

Algorithm 1. FOS-domination ranking

Input: Solution set $S = N \times M$ matrix, where N = No. of solutions to be ranked, M = No. of quantiles considered

1: **for** $i = 1$ **to** N **do**
2:    **for** $j = 1$ **to** N **do**
3:       Compute d_{ij} according to Eq. 6
4:    **end for**
5:    $dMin_i = \min(d_{i,j}; j = 1 : N)$
6: **end for**
7: Initialize $ranklist = \emptyset$;
8: **for** $i = 1$ **to** N **do**
9:    **if** $i \neq N$ **then**
10:       $j = \text{argmin}(dMin_j)$
11:    **else**
12:       $j = 1 : N - ranklist$ {Set difference}
13:    **end if**
14:    $ranklist = [ranklist\ j]$;
15:    $d(:, j) = \infty; d(j, :) = infty$
16:    $dMin_i = \min(d_{ij}; j = 1 : N)$
17: **end for**
18: Return final ranks $R = \text{reverse}(ranklist)$

relatively few solutions during the search that have been identified as promising based on the predictions from the surrogate model.

We use the Kriging model [4] to approximate the function $f(\mathbf{x})$, and by extension, the quantile function and associated quantiles in the neighborhood of any candidate solution $\mathbf{x}$. Instead of using a large sample size, say $N_s = 100$ points in $\mathbf{x_\Delta}$, we use much fewer samples, say $N_{ss} = 10$. A Kriging model is built using the set of data $(\mathbf{x}, f(\mathbf{x}))$ such that for any unknown $\mathbf{x}$, the value of $f(\mathbf{x})$ can be predicted. The required number of samples ($N_s = 100$) are then extracted using this surrogate model to construct the quantile function based on *predicted* $f(\mathbf{x})$ values.

Re-using Samples from Neighboring Solutions: Another way to reduce the computation is to reuse the previously evaluated samples that fall under the $\mathbf{x_\Delta}$ of the solution currently under consideration. The sample and its fitness value can be inherited in such cases in lieu of evaluating a new sample. However, the number of available solutions could be unevenly distributed, and have larger or smaller size than the required number of samples N_s. This would adversely affect the quality of the surrogate built in the region, as consequently the quantile function and objectives. To counter this, we propose a simple strategy that augments the existing points (if any), with new samples required to reach the required number N_s, while maintaining relative uniformity between the samples. The process is illustrated in Fig. 4. Suppose that the point currently under consideration is $x = 5.5$, let $N_s = 10$, and $\Delta = 0.5$. This implies that 11 points (including $x = 5.5$) need to be sampled uniformly in [5.0,6.0] to estimate the quantile values of the point. These are labeled as 'Ideal' points, shown with black dots. If two other points, $x = 5.18$ and 6.354 have previously undergone robustness evaluation, this means that 11 uniformly sampled solutions (each) are available in $x \in [4.68, 5.68]$ and $x \in [5.854, 6.854]$, respectively, shown as blue dots. We examine each of the uniformly distributed samples (black dots) and check if its closest existing sample (blue dot) is $\leq \frac{2\Delta}{N_s}$ away. If so, this original sample and its f value are directly used. If not, then the sample is evaluated instead. Moreover, the point under the robustness evaluation, i.e., $x = 5.5$ is evaluated unless an exact copy of it exists already. Thus, in this case only 2 samples needed to be evaluated (shown in red circles), whereas the remaining 9 samples are picked from an archive. It is also possible to use more sophisticated mechanisms to select the new sample locations, e.g., the one proposed in [6].

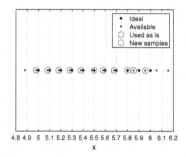

Fig. 4. Re-using the samples to reduce NFE (Color figure online)

4   Numerical Experiments

In this section, we evaluate the proposed approach on a range of benchmark problems. Please note that due to the space limitations, the results for the full set of problems are included in supplementary online material (SOM), which is available at http://www.mdolab.net/Hemant/Research-Data/ppsn22sup.zip; while only a few representative figures are included in this main manuscript.

4.1   Test Problems

We demonstrate the proposed approach on the set of problems (TP1-9) formulated in [13]. Moreover, one problem, TP10, is additionally created for this study, and defined as $f = x \sin(2\pi x - \pi)$ with $x \in [0, 10]$ and $\Delta = 0.5$. The interesting feature of the problem is that (by design) the set of stochastically non-dominated solutions can be readily inferred from observation as $x = \{1, 2, \ldots 9\}$.

4.2   Experimental Setup

The algorithmic parameters used for solving the problems considered are given in Table 1. Four versions of the algorithm are used to solve each problem, configured by setting the use of surrogates and re-use of the previous samples as ON/OFF.

- V1: This is the baseline version, where both the surrogates and re-use of previous points is set to OFF.
- V2: Surrogates ON, re-use previous points OFF
- V3: Surrogates OFF, re-use previous points ON
- V4: Surrogates ON, re-use previous points ON

For each problem, 21 independent runs are conducted using each algorithm variant. The quality of the resulting solutions are assessed visually as well as via unary metrics (discussed in next sub-section). In addition to the quality of solutions, the savings incurred in the cheaper versions (V2–V4) compared to the baseline (V1) version are also observed.

Table 1. Parameters used for the EA

Parameter	Value
Number of quantiles (M)	11
Population size (N)	20
No. of generations (G)	50
Crossover probability (p_c)	0.9
Mutation probability (p_m)	0.1
SBX crossover index (η_c)	10
Polynomial mutation index (η_m)	20
Neighborhood sampling points (N_s)	100 (1000 for TP10)
Reduced sampling size for surrogate-based versions (V3/V4) (N_{ss})	10

4.3   Performance Measurement

In order to quantify the performance of the proposed algorithm and its variants, we resort to the inverted generational distance (IGD) metric [3]. IGD is commonly used in evolutionary multi-objective optimization for benchmarking the performance of algorithms. IGD compares the Pareto front (PF) approximation P obtained by an algorithm with a given *reference set* Q, which is the best estimate of the PF. Both sets P and Q refer to a set of points in the objective space. To compute the IGD, for each point in Q, the nearest point in P is identified and the corresponding Euclidean distance is recorded. Then, IGD is calculated as the mean of these distances; with a lower IGD indicating better performance.

For many of the standard benchmark problems, the true PF is known analytically, so a given number of points can be sampled on it to generate the reference set. If the true optimum of the problem under consideration may not be exactly known, a reference set is constructed e.g. by accumulating a large set of non-dominated solutions by combining solutions examined in multiple runs of all compared algorithms. Among the problems considered in this study, the theoretical optimum can be readily inferred only for three problems - TP1, TP7, and TP10. For TP1 and TP7, the function is monotonically decreasing in the range of $x = [2, 8]$. The only deterministic (global) optimum lies at $x = 8$, and the function value then steps up to 0 (its maximum value) thereafter. Therefore, in terms of stochastic non-dominance, $x = 8 - \Delta = 7.5$ is the true optimum solution for the problem. As for TP10, it is defined in a way as to have multiple peaks with the same periodicity but different, monotonically increasing, amplitudes. The Δ value chosen for the problem is 0.5, which is half the cycle of the function, thereby making 2Δ the full cycle. By observation, the points at the middle of the cycles, i.e., $x = \{1, 2, 3, \dots 9\}$ therefore form the true optimum (stochastically non-dominated) solutions to the problem.

For the remainder of the problems, approximate reference sets have been generated by considering a set of uniformly sampled 1001 solutions within $\pm 0.5\Delta$ of their local and global optimum solutions. Then, the stochastically non-dominated solutions among these are considered to be the reference set.

4.4   Results

The median IGD values obtained using all variants of the proposed algorithms (V1–V4) are listed in Table 2, while the corresponding median function evaluations across 21 runs are listed in Table 3. Moreover, the convergence plots for the median runs for some representative problems are visualized in Fig. 5. Shown in Fig. 6 are the solutions obtained for TP3 in both x and quantile space; with the full set of problems included in the SOM Figs. 2, 3, 4 and 5.

Table 2. Median IGD values obtained by the proposed algorithm. The numbers in parenthesis denote the ratio of IGD compared to baseline (V*/V1), with ↑ or ↓ indicating the ratio to be higher or lower than 1, respectively.

Problem	V1	V2	V3	V4
TP1	0.0002	0.0006 (2.45× ↑)	0.0021 (9.23× ↑)	0.0054 (23.59× ↑)
TP2	0.0002	0.0004 (2.82× ↑)	0.0006 (3.93× ↑)	0.0006 (3.54× ↑)
TP3	0.0012	0.0012 (1.05× ↑)	0.0019 (1.62× ↑)	0.0012 (1.05× ↑)
TP4	0.0004	0.0004 (1.21× ↑)	0.0007 (1.90× ↑)	0.0004 (1.16× ↑)
TP5	0.0012	0.0022 (1.78× ↑)	0.0019 (1.61× ↑)	0.0023 (1.89× ↑)
TP6	0.0072	0.0065 (0.91× ↓)	0.0083 (1.15× ↑)	0.0061 (0.85× ↓)
TP7	0.1421	0.5141 (3.62× ↑)	1.3860 (9.75× ↑)	6.6902 (47.09× ↑)
TP8	0.0384	0.0395 (1.03× ↑)	0.0450 (1.17× ↑)	0.0436 (1.13× ↑)
TP9	0.0051	0.0051 (1.01× ↑)	0.0088 (1.73× ↑)	0.0055 (1.08× ↑)
TP10	0.0117	0.0124 (1.06× ↑)	0.0137 (1.17× ↑)	0.0124 (1.06× ↑)

Table 3. Median function evaluations used by the proposed algorithm. The numbers in parenthesis denote the ratio of evaluations compared to baseline (V1/V*).

Problem	V1	V2	V3	V4
TP1	1.01e+05	13020 (7.76× ↓)	3721 (27.14× ↓)	2824 (35.76× ↓)
TP2	1.01e+05	13020 (7.76× ↓)	3958 (25.52× ↓)	2845 (35.5× ↓)
TP3	1.01e+05	13020 (7.76× ↓)	4000 (25.25× ↓)	2845 (35.5× ↓)
TP4	1.01e+05	13020 (7.76× ↓)	3241 (31.16× ↓)	2341 (43.14× ↓)
TP5	1.01e+05	13020 (7.76× ↓)	3984 (25.35× ↓)	2892 (34.92× ↓)
TP6	1.01e+05	13020 (7.76× ↓)	4869 (20.74× ↓)	3180 (31.76× ↓)
TP7	1.01e+05	13020 (7.76× ↓)	3742 (26.99× ↓)	2776 (36.38× ↓)
TP8	1.01e+05	13020 (7.76× ↓)	5315 (19.00× ↓)	3217 (31.4× ↓)
TP9	1.01e+05	13020 (7.76× ↓)	3927 (25.72× ↓)	2949 (34.25× ↓)
TP10	1.001e+06	31020 (32.27× ↓)	30849 (32.45× ↓)	21031 (47.6× ↓)

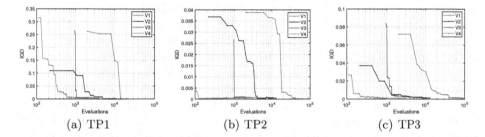

(a) TP1   (b) TP2   (c) TP3

Fig. 5. Convergence plots corresponding to the median IGD run (Color figure online)

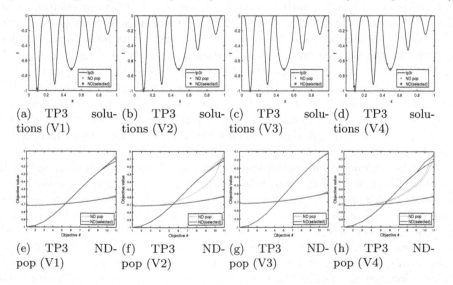

(a) TP3 solutions (V1)   (b) TP3 solutions (V2)   (c) TP3 solutions (V3)   (d) TP3 solutions (V4)

(e) TP3 ND-pop (V1)   (f) TP3 ND-pop (V2)   (g) TP3 ND-pop (V3)   (h) TP3 ND-pop (V4)

Fig. 6. Representative median IGD results obtained for TP3 using all versions (V1–V4) of the proposed algorithm. Results for all problems included in the SOM. (Color figure online)

To provide context to Fig. 6, please note that there are two types of solutions shown. Those in blue represent the non-dominated solutions (based on quantiles) in the final population obtained after executing the EA run. However, some of these solutions are what is referred to in the literature as *dominance resistant solutions* (DRS) [7]. DRS are those that are significantly poor on one/some objective(s), but are non-dominated in the population due to a marginal improvement over another solution in one/some of the objectives. This particularly comes into play when the number of objectives is high, such as is the case here. To eliminate the DRS, we first normalise all objective values between the maximum and minimum values obtained among all objectives. Then, any differences between the normalized objective values that are less than 1% of the range are eliminated by rounding the values to two digits. These subsets of non-dominated solutions are shown in red color in the figures, and used for computation of the metrics.

From Table 2, it can be observed that the median IGD values are generally small, indicating that all four versions of the proposed algorithm were able to locate the correct regions of stochastically non-dominated solutions. The overall accuracy decreases successively when moving from V1 to V4. The % increase (V*/V1) in the IGD value is listed alongside the median IGD for each of the variants. The factors lie in the range of $\approx [1, 4]$. The notable exception to this are TP1 and TP7, for the versions V3 and V4, i.e. those that operate with surrogate-assistance. These two functions have their optimum exactly at $x = 7.5$, and at the edge of x_Δ, i.e., at $x = 8$ there is a significant discontinuity, stepping from the lowest to the highest value of the function instantaneously. However, given that the surrogate models assume a continuous function, the predicted step by the model at $x = 8.0$ will not be exactly vertical, leading to an overestimation of some quantile values. This also implies that any solution right of $x = 7.5$, even slightly, i.e., $x = 7.5 + \Delta; \Delta \to 0$ will have at least one quantile value as 0 (the highest value taken by the objective function). Note this, for example, for the population members (marked blue) in SOM Figs. 2, 3, 4 and 5 for TP7. The results for V3 and V4 of TP7 are also affected by the fact that the range of function values is very large ($[-216,0]$), so small errors will lead to large IGD values. A closer look at SOM Figs. 2, 3, 4 and 5 reveal that the solutions from the median run obtained by V3 and V4 are quite close to those obtained using V1 and V2. The same observations apply to TP1, as evident from SOM Figs. 2, 3, 4 and 5.

For other problems with multiple stochastically nondominated solutions, such as TP3, TP6 shown in Fig. 6 and SOM, respectively. The algorithm shows commendable performance by identifying solutions in all the relevant regions. The same extends to other problems (shown in the SOM), with possible exception of TP8 where the solutions were found typically in 4 out of 5 regions in the median run. Reflecting back on Figs. 2, 3, 4 and 5, it can be seen that the algorithm converged to the two correct regions near points A and C - those with the non-dominated quantile functions among the multiple optima. Notably, the above solutions were obtained with significantly reduced number of evaluations compared to the baseline algorithm (V1/V*). The reduction in function evaluations is typically about 8-fold for V2 in the range of 20–40 folds for V3 and V4 to obtain solutions that are only marginally worse in quality compared to V1. Figure 5 further provides a visualization of how quickly the computationally efficient variants of the proposed algorithm are able to converge relative to the baseline version.

5   Conclusions and Future Work

We proposed a new paradigm for black-box robust optimization, providing first order stochastically non-dominated solutions to a decision-maker. Towards this end, we formulated an underlying multi-objective optimization problem with discretized quantile functions and proposed an evolutionary approach to solve the problem. Since the process is computationally expensive in terms of NFEs consumed, strategies to reduce the NFEs substantially were also proposed, including the use of surrogate approximation and re-use of historical data. The results are encouraging and demonstrate the capability of the proposed algorithm in achieving the targeted solutions, as well as reducing the computational effort in doing so with relatively small compromise in solution quality.

In the future, we would like to make the proposed technique scalable for higher numbers of variables by using more efficient sampling methods, and extend the approach to deal with second order stochastic dominance. Also, the impact of the number of quantiles used for discretization and the density of samples used for performance estimation also needs further investigation to assess the proposed approach more comprehensively.

Acknowledgments. The first author would like to acknowledge the support from Discovery Project DP190102591 from Australian Research Council.

References

1. Asafuddoula, M., Singh, H.K., Ray, T.: Six-sigma robust design optimization using a many-objective decomposition-based evolutionary algorithm. IEEE Trans. Evol. Comput. **19**(4), 490–507 (2014)
2. Beyer, H.G., Sendhoff, B.: Robust optimization - a comprehensive survey. Comput. Methods Appl. Mech. Eng. **196**(33–34), 3190–3218 (2007)
3. Coello Coello, C.A., Reyes Sierra, M.: A study of the parallelization of a coevolutionary multi-objective evolutionary algorithm. In: Monroy, R., Arroyo-Figueroa, G., Sucar, L.E., Sossa, H. (eds.) MICAI 2004. LNCS (LNAI), vol. 2972, pp. 688–697. Springer, Heidelberg (2004). https://doi.org/10.1007/978-3-540-24694-7_71
4. Couckuyt, I., Dhaene, T., Demeester, P.: ooDACE toolbox: a flexible object-oriented kriging implementation. J. Mach. Learn. Res. **15**, 3183–3186 (2014)
5. Deb, K., Pratap, A., Agarwal, S., Meyarivan, T.: A fast and elitist multiobjective genetic algorithm: NSGA-II. IEEE Trans. Evol. Comput. **6**(2), 182–197 (2002)
6. Fei, X., Branke, J., Gülpınar, N.: New sampling strategies when searching for robust solutions. IEEE Trans. Evol. Comput. **23**(2), 273–287 (2018)

7. Ishibuchi, H., Matsumoto, T., Masuyama, N., Nojima, Y.: Effects of dominance resistant solutions on the performance of evolutionary multi-objective and many-objective algorithms. In: Proceedings of the 2020 Genetic and Evolutionary Computation Conference, pp. 507–515 (2020)

8. Ishibuchi, H., Tsukamoto, N., Nojima, Y.: Evolutionary many-objective optimization: a short review. In: 2008 IEEE Congress on Evolutionary Computation (IEEE World Congress on Computational Intelligence), pp. 2419–2426. IEEE (2008)

9. Islam, M.M., Singh, H.K., Ray, T.: A surrogate assisted approach for single-objective bilevel optimization. IEEE Trans. Evol. Comput. 21(5), 681–696 (2017)

10. Jin, Y., Sendhoff, B.: Trade-off between performance and robustness: an evolutionary multiobjective approach. In: Fonseca, C.M., Fleming, P.J., Zitzler, E., Thiele, L., Deb, K. (eds.) EMO 2003. LNCS, vol. 2632, pp. 237–251. Springer, Heidelberg (2003). https://doi.org/10.1007/3-540-36970-8_17

11. Levy, H.: Stochastic dominance and expected utility: survey and analysis. Manage. Sci. 38(4), 555–593 (1992)

12. Lu, K., Branke, J., Ray, T.: Improving efficiency of bi-level worst case optimization. In: Handl, J., Hart, E., Lewis, P.R., López-Ibáñez, M., Ochoa, G., Paechter, B. (eds.) PPSN 2016. LNCS, vol. 9921, pp. 410–420. Springer, Cham (2016). https://doi.org/10.1007/978-3-319-45823-6_38

13. Paenke, I., Branke, J., Jin, Y.: Efficient search for robust solutions by means of evolutionary algorithms and fitness approximation. IEEE Trans. Evol. Comput. 10(4), 405–420 (2006)

14. Sun, G., Li, G., Zhou, S., Li, H., Hou, S., Li, Q.: Crashworthiness design of vehicle by using multiobjective robust optimization. Struct. Multidiscip. Optim. 44(1), 99–110 (2011)

15. Wang, G.G., Shan, S.: Review of metamodeling techniques in support of engineering design optimization. ASME J. Mech. Des. 129(4), 370–380 (2006)

16. Zitzler, E., Künzli, S.: Indicator-based selection in multiobjective search. In: Yao, X., et al. (eds.) PPSN 2004. LNCS, vol. 3242, pp. 832–842. Springer, Heidelberg (2004). https://doi.org/10.1007/978-3-540-30217-9_84

Large-Scale Multi-objective Influence Maximisation with Network Downscaling

Elia Cunegatti[1,2] (ID), Giovanni Iacca[1](✉)(ID), and Doina Bucur[2](ID)

[1] University of Trento, Trento, Italy
elia.cunegatti@studenti.unitn.it, giovanni.iacca@unitn.it
[2] University of Twente, Enschede, The Netherlands
d.bucur@utwente.nl

Abstract. Finding the most influential nodes in a network is a computationally hard problem with several possible applications in various kinds of network-based problems. While several methods have been proposed for tackling the influence maximisation (IM) problem, their runtime typically scales poorly when the network size increases. Here, we propose an original method, based on network downscaling, that allows a multi-objective evolutionary algorithm (MOEA) to solve the IM problem on a reduced scale network, while preserving the relevant properties of the original network. The downscaled solution is then upscaled to the original network, using a mechanism based on centrality metrics such as PageRank. Our results on eight large networks (including two with ∼50k nodes) demonstrate the effectiveness of the proposed method with a more than 10-fold runtime gain compared to the time needed on the original network, and an up to 82% time reduction compared to CELF.

Keywords: Social network · Influence maximisation · Complex network · Genetic algorithm · Multi-objective optimisation

1 Introduction

Given a social network for which the network structure is known and the process of influence propagation can be modelled, the problem of influence maximisation (IM) [29] in its simplest form aims to select a certain number of participants (nodes) in the network, such that their combined influence upon the network is maximal. This is a combinatorial optimisation task, NP-hard for most propagation models [18]. Various metaheuristics have been proposed to solve this problem, including (but not limited to) simulated annealing [16], genetic algorithms [4,24], memetic algorithms [11], particle swarm optimisation [12], and, more recently, evolutionary deep reinforcement learning [25] and multi-transformation evolutionary frameworks [33]. Multi-objective formulations of the IM problems have also been tackled, for instance in [5,6]. In all cases, the drawback of these methods is their long runtime, which makes them infeasible to use on large social networks with more than 10^5 nodes.

G. Rudolph et al. (Eds.): PPSN 2022, LNCS 13399, pp. 207–220, 2022.
https://doi.org/10.1007/978-3-031-14721-0_15

Here, aim to address precisely this computational issue. As in [5,6], we consider a multi-objective formulation of the IM problem, where the two competing objectives are the number of selected nodes (to be minimised), and the combined influence (to be maximised). With respect to the previous literature, we contribute a scalable method for the multi-objective IM problem, which allows to tackle the problem for social networks orders of magnitude larger than before. The method is built around a multi-objective evolutionary algorithm (MOEA), but additionally employs a first step of *graph summarisation* [23] which downscales the network (by a configurable factor) while preserving its important structural properties, and a last step which upscales the solutions from the downscaled network to the original one. This approach allows the MOEA to run in feasible time, since it is executed on a smaller instance of the problem.

The runtime has always been a key issue in the literature on the IM problem. Some previous works have tried to overcome this limitation by improving directly the effectiveness of the algorithm, see [22] for a survey on this topic. Some recent works have tried to minimise the runtime in billion-scale networks [13,21,35]. Yet, to the best of our knowledge no previous work has attempted to tackle the problem by focusing on the input (i.e., the network), instead of the algorithm.

We tested our method on six different networks, using two propagation models. For the downscaling process, we used three different values of scaling factor, to analyse how this affects the performance of our method. For the upscaling process, we evaluated different centrality metrics. Finally, we tested our method on two large networks with high scaling factor values and compared the results with a classical heuristic algorithm. The results show that our method can achieve near-optimal or even better results, compared to the MOEA on the corresponding unscaled networks, while drastically reduce the runtime required.

The rest of the paper is structured as follows. In the next section, we describe our method. In Sect. 3, we present the numerical results. Finally, we give the conclusions in Sect. 4.

2   Method

We provide a first overview of the method in Fig. 1. For the MOEA, we use Non-Dominated Sorting Genetic Algorithm-II (NSGA-II) [8], which has shown good results on this problem in prior work [5,6,15], but whose runtime made it prohibitive on large networks. This computationally heavy method is marked in Fig. 1 on the left with a heavy red arrow. Instead of attempting to further improve the algorithm (with likely minor gains in efficiency), here we design an alternative which fundamentally changes the way we treat the problem input.

Given a large social network (top left in Fig. 1), in step (1) we synthesise a downscaled version of the network, for which the scaling factor is configurable. A network scaled by a factor $s = 2$ would contain half of the nodes of the original, but otherwise preserve all the important properties of the original network: the number of communities is maintained constant, and the node degree distribution is scaled proportionally with the network size. In step (2), we apply the MOEA

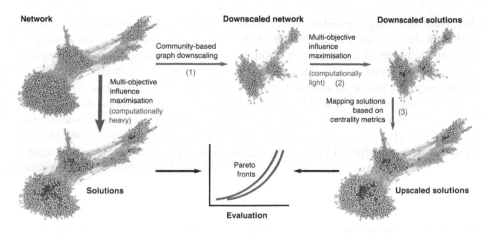

Fig. 1. The method. The larger the original network (top left), the less feasible IM is (bottom left). Our efficient three-step method scales down the network, does IM on the scaled network, then upscales the solutions (right). (Color figure online)

on this downscaled network, obtaining a number of non-dominated solutions (where each solution is a *seed set*, i.e., the set of nodes from which influence is propagated). Only one such seed set is shown in Fig. 1, with the corresponding nodes marked in red. Finally, for step (3) we devise a method based on node centrality metrics to select the seed set in the original network, such that its nodes correspond (in terms of position in the network) to the ones found in step (2). Steps (1–3) output a Pareto front (PF) of solutions. To evaluate how accurate this method is, we also execute the MOEA on the original network, and compare the two PFs. In the rest of this section, we detail the three aforementioned steps[1].

2.1   Step (1): Community-Based Downscaling

The output of the downscaling process is a completely synthetic network, not identical to any part of the original network, but which appears very similar, yet much smaller. The *scaling factor s* is the key parameter when downscaling: we experiment with values in a geometric sequence, $s \in \{2, 4, 8\}$. The downscaling process satisfies the following requirements:

1. preserves the *number of communities* in the network;
2. downscales the *number of nodes and edges* by a factor s;
3. preserves the *node degree distribution*.

We pose the last requirement because dynamic phenomena on a network (such as information propagation) have outcomes which depend on the node degrees. We use the Leiden algorithm [32], a state-of-the-art, scalable method for community detection. The algorithm is stochastic, so variations in the number and size of the communities are possible; because of this, we select one solution,

---

[1] Code available at: https://github.com/eliacunegatti/Influence-Maximization.

then filter out from the network any communities that are too small (i.e., those that contain a number of nodes lower than the scaling factor s). We then preserve this number of communities obtained on the original network, downscaling (proportionally with s) the number of nodes and edges in each community. Per community, the shape of the node degree distribution is also preserved, as follows. Take the number of nodes in some original community to be N. We take a number of samples of size N/s from the original node degrees of each community, then retain only the samples with the smallest Euclidean distance (computed in terms of mean degree and std. dev. per community) compared to the original ones. We thus have a desired node degree distribution for a downscaled network.

The downscaled network is then generated by the Stochastic Block Model (SBM), a random generative network model originally proposed in [14]. This generates random networks with a configured number of communities, number of nodes per community, and edge density per community. Here, we used a more fine-grained, recent SBM method [17,28] which takes also into account node degrees, and is implemented in the Python `graph-tool` library[2].

2.2   Step (2): MOEA on Two Objectives (cascade Size and Seed Set Size)

Single-objective IM is the problem of finding those k nodes with maximum collective influence upon the network. As mentioned above, a candidate solution for this is a *seed set* (a set of node identifiers) of size $k > 0$. The multi-objective formulation in this study aims to maximise the collective influence, while minimising the size of the seed set k. In other words, the fitness of a candidate solution is a tuple of: (a) the **estimated collective influence** of the seed set, and (b) the **size of the seed set**. Both values are normalised with respect to the network size, to allow for a fair comparison between scaled and unscaled networks, as well as between networks of different sizes. We set the maximum possible value of k to be 2.5% of the network size.

As for the estimated collective influence, we model influence cascade using two classic, discrete-time propagation models for social networks [18]. They simulate the dynamics of information adoption in a network modelled by graph G, in which a set S of "seed" nodes are the initial sources of information. At any given time, the nodes in G are in one of two states: "activated" (if they received the information and may propagate it further) or not. Initially, only the nodes in S are activated. The information propagates via network links probabilistically: a probability p models the likelihood of a source node activating a neighbouring destination node via their common link. The important quantity is the number of nodes eventually activated, also called the *cascade size*—for IM, this is typically the main objective.

Algorithm 1 gives a general view of cascade propagation models: set A consists of all the activated nodes, and is initially equal to the seed set S. At each time step, recently activated source nodes try to activate their neighbours independently. If an activation fails, it is never retried (a destination node is assumed

---

[2] https://graph-tool.skewed.de.

to have made its decision). If it succeeds, the propagation may continue; the process stops when no new nodes were activated in a time step.

We use two model variants: **Independent Cascade** (IC) and **Weighted Cascade** (WC). IC was first introduced in marketing, to model the complex effects of word-of-mouth communication [10]. In IC, the probability p is equal across all links (when a node has more than one neighbour, their activations are tried in arbitrary order). WC further models the fact that a node's attention is limited: the probabilities of activation on links leading to a destination node m are not uniform, but inversely proportional to the number of such links in G (in other words, the degree of m), $p = 1/deg(m)$ [18], where $deg(m)$ is the in-degree of node m, i.e., the number of edges incoming to m.

Algorithm 1 Cascade propagation models. G is the graph, S the seed set, and p the probability that a link will be activated.

Input: G, S, p

1: $A \leftarrow S$ ▷ The complete set of activated nodes
2: $B \leftarrow S$ ▷ Nodes activated in the previous iteration
3: **while** B not empty **do**
4:     $C \leftarrow \emptyset$ ▷ Nodes activated in the current iteration
5:     **for** n in B **do**
6:         **for** m in $neighbours(n) \setminus A$ **do**
7:             $C \leftarrow m$ with probability p ▷ Activation attempt
8:         **end for**
9:     **end for**
10:    $B \leftarrow C$ and $A \leftarrow A \cup B$
11: **end while**
12: **return** $|A|$ ▷ The final size of the cascade

Although a single execution of Algorithm 1 is polynomial in the size of the network, the model is stochastic, and computing the expected cascade size exactly for a given seed set S is #P-complete [34]. However, good estimates of $|A|$ can be obtained by Monte Carlo simulations: in our experiments, we run 100 repetitions of Algorithm 1 for each estimation.

Concerning the MOEA, we used the implementation and parameterisation of NSGA-II adopted in prior work [5,15]. In short, the parent solutions are selected by tournament selection; the child solutions are generated by one-point crossover and random mutation. An archive keeps all the non-dominated solutions found, i.e., the PF. The replacement mechanism selects non-dominated solutions by their dominance levels, and then sorts them by crowding distance to prefer isolated solutions and obtain a better coverage of the PF.

To improve the convergence of the MOEA, we apply a *smart initialisation* of its initial population, as proposed in [19]. First, we apply node filtering, which computes the influence of each node in the network separately, and then keeps the 50% most influential nodes. Then, each of these nodes is added to a candidate solution with a probability proportional to its degree. We summarise the parameters of the method in Table 1.

Table 1. Parameters of the method.

Network parameters		NSGA-II parameters	
Scaling factor s	$\{2, 4, 8\}$	Population size	100
Max. seed set size k	2.5% · network size	Generations	1000
IC probability p	0.05	Elites	2
No. simulations	100	Crossover rate	1.0
		Mutation rate	0.1
		Tournament size	5

2.3 Step (3): Upscaling

Once the MOEA has been run on the downscaled network, the last step is to map the solutions back to the original network. This step takes in input two graphs (the original G and the downscaled G_s) and a set seed on G_s, denoted as S_s. The task is to translate S_s into an seed set S on G.

We achieve this by *matching* nodes between the two graphs, based on their *node centrality indicators*, namely node statistics which capture the position of the node in the network. We test the following classical centrality indicators, based on them being shown to be predictive of the node's influence [3]: each node's *degree, eigenvector centrality* and its variants *PageRank* with a 0.85 damping factor and *Katz centrality, closeness, betweenness,* and *core number* (see [27] for their definitions).

For a seed set S_s in G_s, we find a matching seed set S of $|S_s| \times s$ nodes in G. We do this per community. Each node in S_s has a *rank* in its community, based on the centrality values of all nodes in that community. We then search in G (among the nodes in the corresponding community) for s nodes with the most similar ranks. These nodes form S.

Evaluation. We evaluate the PFs obtained, particularly to compare between the MOEA results on the original network and those obtained with our new method. We use the *hypervolume* (HV) indicator (also known as Lebesgue measure, or S metric) proposed in [36]. This is calculated as the volume (of the fitness space) dominated by each solution in the PF with respect to a reference point. The *hyperarea* (HR) [36] is the ratio of two HVs, and is used here in the final evaluation step (bottom center in Fig. 1).

3 Results

Network Data. We test our method on six real-world social network topologies (listed in Table 2). These range between 4 039 and 28 281 nodes, with variable average degrees (and thus network densities), and variable number and size of communities. **Ego Fb.** denotes data merged from many ego networks on Facebook, collected from survey participants at a large university. **Fb. Pol.** is a network of mutually liked, verified politicians' pages on Facebook. **Fb. Pag.** is similar, but with Facebook pages from various categories. **Fb. Org.** is a network of friendships among Facebook users who indicated employment at one corporation. **PGP** is the largest connected component in the network of PGP encryption users. **Deezer** represents online friendships between users of the Deezer music platform. All graphs are undirected and connected.

Table 2. Networks considered in the experimentation.

Network	Nodes	Edges	Communities			Node degrees		
			Num.	Min.	Max.	Avg.	Std.	Max.
Ego Fb. [26]	4 039	88 234	17	19	548	43.90	52.41	1045
Fb. Pol. [30]	5 908	41 729	31	8	562	14.12	20.09	323
Fb. Org. [9]	5 524	94 219	13	35	1045	34.11	31.80	417
Fb. Pag. [30]	11 565	67 114	31	8	1916	11.60	21.28	326
PGP [1]	10 680	24 316	91	8	668	4.55	8.07	205
Deezer [31]	28 281	92 752	71	8	4106	6.55	7.94	172

In the remainder of this section, we experiment with and evaluate our method. We present results for the three distinct steps of the method (as per Fig. 1).

3.1   Community-Based Downscaling of Large Networks

This step obtains synthetic scaled networks, with the scaling factor s. These networks have the same number of communities, a scaled number of nodes and edges, and the same shape of the degree distribution. For $s \in \{2, 4, 8\}$, we show in Fig. 2 the degree distributions (in log-log scale) for the six networks: that of the original (unscaled) network, and that of synthetic, scaled versions. Figure 1 included plots of Fb. Org. before and after downscaling ($s = 4$).

The networks plausibly fit typical power-law degree distributions, in which the fraction of nodes with a certain degree d is proportional to $d^{-\alpha}$ where α is positive (so, decreases with d), after a cutoff point $d_{\min}$. In real-world networks from various domains, the power-law parameter α is often measured between 1.5 and 3 [2], which is mostly the case also for these six social networks. This parameter can be seen in Fig. 2, in the linear slope of all distributions (for high enough degrees). Our downscaling step preserves the power-law parameter α between the original and downscaled networks, naturally while scaling down the degree frequencies with s. We show the fitted values for α in Table 3, from which it is clear that the downscaling method introduced little error in the shape of the degree distribution.

In the next step, we run the MOEA on the downscaled networks (and, for evaluation, also on the original networks).

3.2   MOEA and Solution Upscaling: The Optimality of Solutions

The optimisation process obtains a two-dimensional PF of solutions, where as said each solution is a seed set of k nodes of the network. We show an example

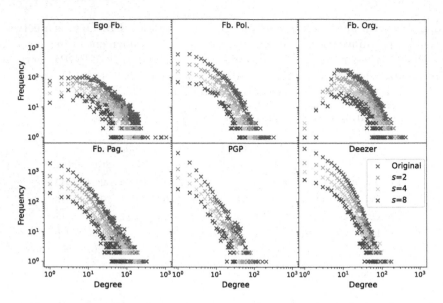

Fig. 2. Degree distributions, before and after downscaling.

Table 3. Degree distribution preservation, power-law coefficients.

	Original	$s = 2$	$s = 4$	$s = 8$
Ego Fb	1.32	1.35	1.38	1.42
Fb. Pol	1.50	1.52	1.54	1.57
Fb. Org	1.40	1.32	1.33	1.35
Fb. Pag	1.60	1.58	1.60	1.62
PGP	2.11	1.94	1.97	1.93
Deezer	1.73	1.66	1.68	1.68

run (randomly selected out of the 10 performed) of PFs in Fig. 3, for the PGP case and for both propagation models (IC and WC), obtained by the MOEA on the original network and on the downscaled networks (for three values of s). The PFs for the original and downscaled networks (top in Fig. 3) show that running the optimisation on the downscaled networks preserves the shape of the PF, but slightly lowers the values reached for the main objective, namely the percentage of influenced nodes (i.e., the cascade size). The more downscaled the network is, the more pronounced this effect appears to be. When upscaling the solutions obtained on the downscaled networks, here using the PageRank centrality in the upscaling process (bottom in Fig. 3), this gap closes for IC, but not for WC propagation.

Of note, the PGP case shown in Fig. 3 is actually one of the cases where our method performs worst (see below). One of the cases where it performs best is Fb. Pag., whose PFs are shown in Fig. 4 (also in this case, for one run of the

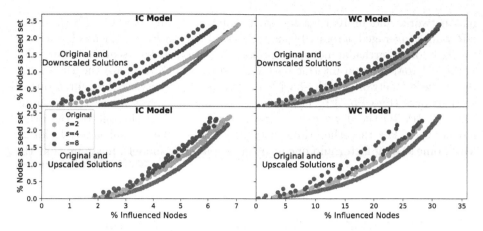

Fig. 3. PGP Pareto fronts: (top) for the original and downscaled networks, and (bottom) for the original network and the upscaled solutions.

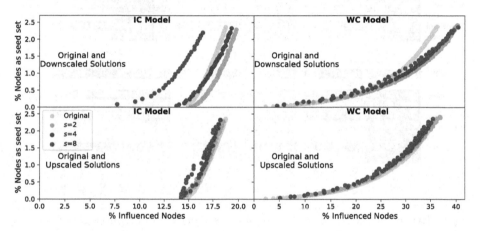

Fig. 4. Fb. Pag. Pareto fronts: (top) for the original and downscaled networks, and (bottom) for the original network and the upscaled solutions.

MOEA on the original network and on the downscaled networks, using PageRank for upscaling). On this network, the MOEA on the downscaled networks often produces a better PF than on the original (top in Fig. 4), and the final, upscaled PFs are comparable to the original one (bottom in Fig. 4). Thus, the method largely preserved the quality of the solutions.

In general and quantitatively, we observe great variation among the networks under test in terms of HR, i.e., on the ratio between (1) the HV subtended by the PF found by the MOEA on the original network and (2) the HV subtended by the PF obtained from the upscaled solutions.

To provide a more robust estimate of the HR values, for each network and propagation model we executed the MOEA in 10 independent runs on the original network, and in 10 runs for each value of scaling factor[3]. We show the HR values (averaged across 10 runs) comparatively in Fig. 5. Each cell contains the HR for a particular network, propagation model, and scaling factor. The "MOEA" rows contain intermediate HR values, which compare the PF on the **downscaled** networks with the PF on the original network. For three out of six networks (Ego Fb., Fb. Pol., and PGP) the HR never reaches 1, meaning that the HV on the downscaled networks is *lower* than the one on the original network. For the other three networks, HR reaches or surpasses 1, for at least some scaling factor and one or both propagation models, meaning that the downscaling step by itself preserved or even raised the optimality of the PF.

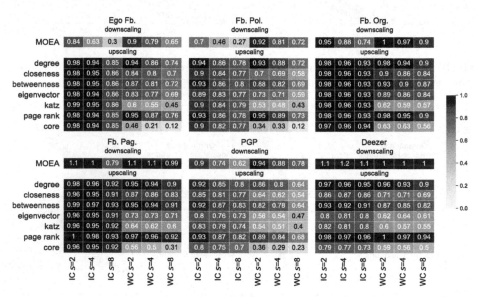

Fig. 5. Hyperarea (averaged across 10 runs) for each network, propagation model, scaling factor, and centrality metric.

The HR values in the rows labelled with the various centrality names in Fig. 5 compare instead the PF of the **upscaled** solutions with the PF on the original network, so serve as evaluation metrics for our method. These show how accurate each centrality is at the upscaling task, and thus help choose the most suitable centrality for any future work. We observe that it is generally easier for all centralities to obtain a good PF of upscaled solutions when using IC. However, three of the centralities consistently yield good upscaling results for both IC and WC: PageRank, betweenness, and degree centralities. PageRank

---

[3] We also compared directly the HV values. We applied the Wilcoxon Rank-Sum test (with $\alpha = 0.05$) to analyse whether the HV values calculated on the downscaled solutions and the upscaled ones (with upscaling based on PageRank) were significantly different from the HV values obtained on the original network, all on 10 runs. All the pairwise comparisons resulted statistically significant, excluding the one related to the downscaled solutions found on Fb. Pag. with $s = 2$ and WC model.

is the best option overall (this is why we used it in Fig. 3 and Fig. 4). Across networks with IC, the HR obtained by PageRank is in the interval $[0.93, 1]$ when $s = 2$, in $[0.86, 0.97]$ when $s = 4$, and in $[0.78, 0.96]$ when $s = 8$. Fairly similar numbers are obtained with WC.

3.3   Runtime Analysis

The proposed method not only can preserve the quality of the solutions, but gains drastically in runtime due to our design based on downscaling the input data. We compute the runtime in terms of number of *activation attempts* (line 7 in Algorithm 1). This is a proxy metric for the actual runtime, as it counts how many times that activation step is executed during all the simulations needed by a MOEA run (either on the original network, or on the downscaled one). We show these measurements in Fig. 6, from which we can see (note the log scale on the y-axis) that the runtime needed by our method decreases by a factor from two to five when the scaling factor is doubled.

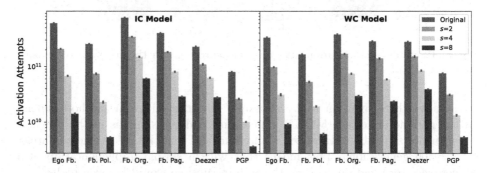

Fig. 6. Runtime (no. activation attempts) in the comparative experiments with MOEA on the original network (average across 10 runs, error bars indicate std. dev.).

3.4   Comparison with Heuristic Algorithm

To further prove the applicability of the proposed method, we conducted a final set of experiments on two different networks with ~50k nodes: soc-gemsec [30] and soc-brightkite [7]. The goal of these experiments was to test our method with higher scaling factors on larger networks.

Due to the large runtime required, we executed only one run of the proposed method with the WC model for $s = 16$ and $s = 32$, and compared the results with those obtained by the deterministic greedy CELF algorithm [20] computed on the unscaled network.

Despite the computational limitations, the comparison is still informative. We see in Fig. 7 that our proposed method is able to achieve better results than CELF, even for these high scaling factors. Not only that: Fig. 8 shows that we obtain these results with a much lower number of activation attempts compared to CELF (with a reduction of up to 82%).

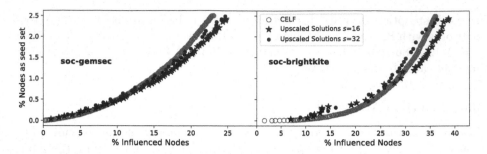

Fig. 7. PFs obtained with CELF and with our method for $s = 16$ and $s = 32$.

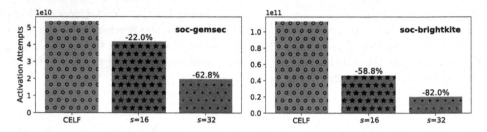

Fig. 8. Runtime (no. activation attempts) measurements in the comparative experiment with CELF.

It is worth to mention that the results obtained with our method depend on the parameters of the MOEA, which remained the same as in the previous experiments, see Table 1. However, we have noticed that the HV reaches a plateau after ∼300 generations. This means that with a simple implementation of a convergence termination criteria, the gain in runtime would be even higher.

4   Discussion and Conclusions

In this paper, we proposed a novel approach to tackle the IM problem where the focus is on the input network instead of the algorithm itself. For this reason, while here we tested the approach on a MOEA, the same method could be applied, in principle, to any other IM algorithm, such as the ones described in Sect. 1.

Our method has been proven to work correctly, although with some differences in the results, with both IC and WC propagation models on all the tested networks, regardless their sizes or properties.

The results demonstrate the effectiveness of our method in terms of quality of the solutions obtained. Furthermore, our method is able to drastically decrease, even for large networks, the runtime of the MOEA. The latter has the additional advantage of providing a whole set of diverse solutions (i.e., seed sets), unlike heuristic methods that are usually designed to return just one seed set.

Our method has proven to work properly with different scaling values. Nevertheless, the results show a lower quality of the solutions as the scaling factor increases. This result is in line with expectations, as it shows a clear trade-off between solution quality and runtime gain.

To the best of our knowledge, this is the first work where a downscaling and upscaling process is proposed on networks to solve the IM problem. Regarding the centrality measure used in the upscaling process, we can state the best one to be PageRank, given the high quality of the upscaling results obtained and the low complexity required to compute it.

References

1. Boguñá, M., Pastor-Satorras, R., Díaz-Guilera, A., Arenas, A.: Models of social networks based on social distance attachment. Phys. Rev. E **70**(5), 056122 (2004)
2. Broido, A.D., Clauset, A.: Scale-free networks are rare. Nat. Commun. **10**(1), 1–10 (2019)
3. Bucur, D.: Top influencers can be identified universally by combining classical centralities. Sci. Rep. **10**(1), 1–14 (2020)
4. Bucur, D., Iacca, G.: Influence maximization in social networks with genetic algorithms. In: EvoApplications (2016)
5. Bucur, D., Iacca, G., Marcelli, A., Squillero, G., Tonda, A.: Improving multi-objective evolutionary influence maximization in social networks. In: EvoApplications, pp. 117–124 (2018)
6. Bucur, D., Iacca, G., Marcelli, A., Squillero, G., Tonda, A.P.: Multi-objective evolutionary algorithms for influence maximization in social networks. In: EvoApplications (2017)
7. Cho, E., Myers, S.A., Leskovec, J.: Friendship and mobility: user movement in location-based social networks. In: Proceedings of the 17th ACM SIGKDD International Conference on Knowledge Discovery and Data Mining, pp. 1082–1090. ACM (2011)
8. Deb, K., Agrawal, S., Pratap, A., Meyarivan, T.: A fast and elitist multiobjective genetic algorithm: NSGA-II. IEEE Trans. Evol. Comput. **6**, 182–197 (2002)
9. Fire, M., Puzis, R.: Organization mining using online social networks. Netw. Spat. Econ. **16**(2), 545–578 (2015). https://doi.org/10.1007/s11067-015-9288-4
10. Goldenberg, J., Libai, B., Muller, E.: Talk of the network: a complex systems look at the underlying process of word-of-mouth. Mark. Lett. **12**, 211–223 (2001)
11. Gong, M., Song, C., Duan, C., Ma, L., Shen, B.: An efficient memetic algorithm for influence maximization in social networks. IEEE Comput. Intell. Mag. **11**, 22–33 (2016)
12. Gong, M., Yan, J., Shen, B., Ma, L., Cai, Q.: Influence maximization in social networks based on discrete particle swarm optimization. Inf. Sci. **367–368**, 600–614 (2016)
13. Güney, E., Leitner, M., Ruthmair, M., Sinnl, M.: Large-scale influence maximization via maximal covering location. Eur. J. Oper. Res. **289**, 144–164 (2021)
14. Holland, P., Laskey, K.B., Leinhardt, S.: Stochastic blockmodels: first steps. Soc. Netw. **5**, 109–137 (1983)
15. Iacca, G., Konotopska, K., Bucur, D., Tonda, A.: An evolutionary framework for maximizing influence propagation in social networks. Softw. Impacts **9**, 100107 (2021)
16. Jiang, Q., Song, G., Cong, G., Wang, Y., Si, W., Xie, K.: Simulated annealing based influence maximization in social networks. In: AAAI (2011)

17. Karrer, B., Newman, M.E.J.: Stochastic blockmodels and community structure in networks. Phys. Rev. E **83**(1Pt2), 016107 (2011)
18. Kempe, D., Kleinberg, J., Tardos, É.: Maximizing the spread of influence through a social network. In: KDD, pp. 137–146 (2003)
19. Konotopska, K., Iacca, G.: Graph-aware evolutionary algorithms for influence maximization. In: GECCO Companion (2021)
20. Leskovec, J., Krause, A., Guestrin, C., Faloutsos, C., Vanbriesen, J.M., Glance, N.S.: Cost-effective outbreak detection in networks. In: KDD (2007)
21. Li, X., Smith, J.D., Dinh, T.N., Thai, M.T.: Tiptop: (almost) exact solutions for influence maximization in billion-scale networks. IEEE/ACM Trans. Networking **27**, 649–661 (2019)
22. Li, Y., Fan, J., Wang, Y., Tan, K.L.: Influence maximization on social graphs: a survey. IEEE Trans. Knowl. Data Eng. **30**, 1852–1872 (2018)
23. Liu, Y., Safavi, T., Dighe, A., Koutra, D.: Graph summarization methods and applications: a survey. ACM Comput. Surv. (CSUR). **51**(3), 1–34 (2018). https://dl.acm.org/doi/abs/10.1145/3186727
24. Lotf, J.J., Azgomi, M.A., Dishabi, M.R.E.: An improved influence maximization method for social networks based on genetic algorithm. Physica A **586**, 126480 (2022)
25. Ma, L., et al.: Influence maximization in complex networks by using evolutionary deep reinforcement learning. IEEE Trans. Emerg. Topics Comput. Intell., 1–15 (2022). https://ieeexplore.ieee.org/document/9679820
26. McAuley, J., Leskovec, J.: Learning to discover social circles in ego networks. In: NIPS (2012)
27. Newman, M.: Networks. Oxford University Press, New York (2018)
28. Peixoto, T.P.: Nonparametric Bayesian inference of the microcanonical stochastic block model. Phys. Rev. E **95**(1), 012317 (2017)
29. Richardson, M., Agrawal, R., Domingos, P.M.: Trust management for the semantic web. In: SEMWEB (2003)
30. Rozemberczki, B., Davies, R., Sarkar, R., Sutton, C.: GEMSEC: graph embedding with self clustering. In: ASONAM, pp. 65–72 (2019)
31. Rozemberczki, B., Sarkar, R.: Characteristic functions on graphs: birds of a feather, from statistical descriptors to parametric models. In: CIKM, pp. 1325–1334 (2020)
32. Traag, V.A., Waltman, L., van Eck, N.J.: From Louvain to Leiden: guaranteeing well-connected communities. Sci. Rep. **9**, 5233 (2019)
33. Wang, C., Zhao, J., Li, L., Jiao, L., Liu, J., Wu, K.: A multi-transformation evolutionary framework for influence maximization in social networks. arXiv preprint arXiv:2204.03297 (2022)
34. Wang, C., Chen, W., Wang, Y.: Scalable influence maximization for independent cascade model in large-scale social networks. Data Min. Knowl. Disc. **25**, 545–576 (2012)
35. Wu, H.H., Küçükyavuz, S.: A two-stage stochastic programming approach for influence maximization in social networks. Comput. Optim. Appl. **69**(3), 563–595 (2017). https://doi.org/10.1007/s10589-017-9958-x
36. Zitzler, E., Thiele, L.: Multiobjective optimization using evolutionary algorithms—a comparative case study. In: Eiben, A.E., Bäck, T., Schoenauer, M., Schwefel, H.-P. (eds.) PPSN 1998. LNCS, vol. 1498, pp. 292–301. Springer, Heidelberg (1998). https://doi.org/10.1007/BFb0056872

Multi-Objective Evolutionary Algorithm Based on the Linear Assignment Problem and the Hypervolume Approximation Using Polar Coordinates (MOEA-LAPCO)

Diana Cristina Valencia-Rodríguez$^{(\boxtimes)}$ ⓘ and Carlos Artemio Coello Coello ⓘ

Department of Computer Science, CINVESTAV-IPN, Av. IPN 2508, San Pedro Zacatenco, 07360 Mexico City, Mexico
diana.valencia@cinvestav.mx, ccoello@cs.cinvestav.mx

Abstract. Hungarian Differential Evolution (HDE) is a Multi-Objective Evolutionary Algorithm that transforms its selection process into a Linear Assignment Problem (LAP). In a LAP, we want to assign n agents to n tasks, where assigning an agent to a task corresponds to a cost. Thus, the aim is to minimize the overall assignment cost. It has been shown that HDE is competitive with respect to state-of-the-art algorithms. However, in this work, we identify two drawbacks in its selection process: it sometimes selects duplicated solutions and occasionally prefers weakly-dominated solutions over non-dominated ones. In this work, we propose an algorithm that tries to fix these drawbacks using the hypervolume indicator. However, since the computation of the hypervolume indicator is expensive, we adopted an approximation that uses a polar coordinates transformation. The resulting algorithm is called "Multi-Objective Evolutionary Algorithm Based on the Linear Assignment Problem and the Hypervolume Approximation using Polar Coordinates (MOEA-LAPCO)." Our experimental results show that our proposed MOEA-LAPCO outperforms the original HDE, and it is competitive with state-of-the-art algorithms.

Keywords: Multi-objective optimization · Linear assignment problem · Hypervolume approximation

1 Introduction

Multi-objective Optimization Problems (MOPs) are types of problems where we want to optimize two or more objectives, usually in conflict (i.e., the improvement of one objective causes the deterioration of another objective). Its formal definition is the following (assuming a minimization problem):

© The Author(s), under exclusive license to Springer Nature Switzerland AG 2022
G. Rudolph et al. (Eds.): PPSN 2022, LNCS 13399, pp. 221–233, 2022.
https://doi.org/10.1007/978-3-031-14721-0_16

$$\text{minimize } \boldsymbol{f}(\boldsymbol{x}) := [f_1(\boldsymbol{x}), f_2(\boldsymbol{x}), \dots, f_k(\boldsymbol{x})] \tag{1}$$

$$s.t. \quad g_i(\boldsymbol{x}) \leq 0 \quad i = 1, 2, \dots, m; \quad h_i(\boldsymbol{x}) = 0 \quad i = 1, 2, \dots, p \tag{2}$$

where $\boldsymbol{x} = [x_1, x_2, \dots, x_n]^T$ is the vector of decision variables, $f_i : \mathbb{R}^n \to \mathbb{R}$, $i = 1, \dots, k$ are the objective functions and $g_i, h_j : \mathbb{R}^n \to \mathbb{R}$, $i = 1, \dots, m$, $j = 1, \dots, p$ are the constraint functions. In MOPs, we usually adopt the concept of *Pareto Dominance* to give a partial order to the solutions. It is said that a vector $\boldsymbol{x} \in \mathbb{R}^n$ *dominates* a vector $\boldsymbol{y} \in \mathbb{R}^n$ (denoted as $\boldsymbol{x} \prec \boldsymbol{y}$, if $f_i(\boldsymbol{x}) \leq f_i(\boldsymbol{y})$ for all $i = 1, \dots, k$, and $\exists j$ such that $f_j(\boldsymbol{x}) < f_j(\boldsymbol{y})$. Moreover, a vector $x \in \mathbb{R}^n$ is called *Pareto optimal*, if there is no vector $y \in \mathbb{R}^n$ such that $y \prec x$. Therefore, we aim to find the set of Pareto optimal solutions (called *Pareto Optimal set*) and its corresponding image (called *Pareto Optimal Front*).

An algorithm designed to solve MOPs is the so-called Hungarian Differential Evolution (HDE). The core idea of this algorithm is to transform its selection process into a Linear Assignment Problem (LAP). It has been shown that HDE is very competitive with state-of-the-art algorithms [8,9]. However, we identified two main drawbacks in its selection process: it may select duplicated solutions and occasionally prefers weakly-dominated solutions[1] over non-dominated ones.

On the other hand, the hypervolume indicator is a popular choice within evolutionary multi-objective optimization due to its compatibility with Pareto dominance. Nevertheless, its computation becomes expensive as the number of objectives increases. For this purpose, Deng and Zhang [4] proposed a novel way to approximate the hypervolume indicator using a polar coordinates transformation to reduce the computational cost.

In this work, we propose a new algorithm that tries to overcome de disadvantages of the HDE's selection process using the approximation of the hypervolume indicator based on polar coordinates. This gives rise to the "Multi-Objective Evolutionary Algorithm Based on the Linear Assignment Problem and the Hypervolume Approximation using Polar Coordinates" (MOEA-LAPCO). Our experimental analysis shows that this algorithm outperforms the original HDE and is competitive with respect to state-of-the-art algorithms.

The remainder of this paper is organized in the following way. First, we explain the preliminary information in Sects. 2 and 3. Then, in Sect. 4, we introduce our proposed approach. After that, we show our experimental analysis in Sect. 5. Finally, we present our conclusions and some possible paths for future research in Sect. 6.

2   Approximating the Hypervolume Contribution Using Polar Coordinates

The hypervolume indicator (denoted by I_H) measures the size of the objective space covered by a set given a reference point. Let $A \subset \mathbb{R}^k$ and $\boldsymbol{z}^u \in \mathbb{R}^k$ be a

---

[1] A solution $\boldsymbol{x}$ is said to *weakly dominate* $\boldsymbol{y}$ (denoted as $\boldsymbol{x} \preceq \boldsymbol{y}$ if $f_i(\boldsymbol{x}) \leq f_i(\boldsymbol{y})$ for all $i = 1, \dots, k$.

reference point dominated by every point in A. Therefore, the I_H of A and z^u can be written as [4]:

$$I_H(A, z^u) = \int_D I_\Omega(z)dz \tag{3}$$

where $z^l = (z_1^l, ..., z_k^l)^T$ s.t. $z_i^l = \min\{y_i \mid y = (y_1, ..., y_k)^T \in A\}$, $D = \{z \in \mathbb{R}^k \mid z^l \prec z \prec z^u\}$, $\Omega = \{z \in \mathbb{R}^k \mid \exists y \in A \text{ such that } y \prec z \prec z^u\}$ and $I_\Omega(z)$ is the characteristic function of Ω. Moreover, the hypervolume contribution of a vector $y \in A$ considering A and z^u is defined as $V(y, A, z^u) = I_H(A, z^u) - I_H(A\backslash\{y\}, z^u)$.

The computational cost of the hypervolume is prohibitive when the number of objectives is bigger than six. To deal with this problem, Deng and Zhang [4] proposed a new method to approximate the hypervolume using polar coordinates. Their idea is to express the hypervolume (displayed in (3)) as a $(k-1) - D$ integral using the polar coordinate system. Deng and Zhang proposed different methods to approximate the hypervolume contribution using the polar coordinates transformation. In this work, we selected the most stable method according to the experimental results in [4].

Let $A \subset \mathbb{R}^k$ and $z^u \in \mathbb{R}^k$ be a reference point dominated by every point in A. To compute the hypervolume contribution, this method first generates n uniformly distributed points $\{\theta^{(1)}, \ldots, \theta^{(n)}\}$ on the $(k-1) - D$ unit sphere in R_+^m using the Unit Normal Vector Approach. Therefore, each point θ^i is generated as follows:

$$\theta^{(i)} = \frac{|x|}{||x||_2} \text{ where } x \sim \mathcal{N}(0, I_k). \tag{4}$$

In addition, a matrix M is constructed, whose (i,j)-entry is the jth largest value in $\{l_{\bar{y}}(\theta^{(i)}) \mid \bar{y} \in A\}$ where

$$l_{\bar{y}}(\theta) = \min_{1 \leq m \leq k}(1/\theta_m)(z_m^u - \bar{y}_m). \tag{5}$$

Finally, the contribution $V(y, A, z^u)$ is approximated using the following expression:

$$\tilde{V}(y, A, z^u) = \frac{\Phi}{2^k} \frac{1}{kn} \sum_{i=1}^{n} \begin{cases} M_{i1}^k - M_{i2}^k & \text{if } l_y(\theta^{(i)}) = M_{i1} \\ 0, & \text{otherwise,} \end{cases} \tag{6}$$

where $\Phi = [2\pi^{(k/2)}/\Gamma(k/2)]$ is the area of the $(k-1) - D$ unit sphere and $\Gamma(x) = \int_0^\infty z^{x-1}e^{-z}dz$ is the analytic continuation of the factorial function.

3   Hungarian Differential Evolution

The core idea of HDE is to transform the selection process of a Multi-Objective Evolutionary Algorithm into a LAP. In a LAP, we have to assign n agents to n

tasks. Assigning an agent i to a task j implies a cost c_{ij}. Therefore, the aim is to find the assignment with minimal cost. Formally, a LAP can be modeled as [1]:

$$\min_{x \in \chi} \sum_{i=1}^{n} \sum_{j=1}^{n} c_{ij} X_{ij} \tag{7}$$

$$s.t. \quad \sum_{j=1}^{n} X_{i,j} = 1 \quad (i = 1, 2, \ldots, n), \tag{8}$$

$$\sum_{i=1}^{n} X_{ij} = 1 \quad (j = 1, 2, \ldots, n), \tag{9}$$

$$X_{ij} \in \{0, 1\} \quad (i, j = 1, 2, \ldots, n) \tag{10}$$

where $\chi = (X_{ij})$ is a binary matrix such that $X_{ij} = 1$ if i is assigned to column j; otherwise, $X_{ij} = 0$. The most common way to solve the LAP is by using the so-called Hungarian Algorithm, which has a computational complexity of $O(n^3)$ [9].

In the case of HDE's selection process, we have a set of individuals (the parents and their offspring) and a set of weight vectors that represent the Pareto Front. The cost of assigning an individual to a weight vector measures how suited this individual is to the part of the Pareto Front that the vector represents. Hence, we can identify which individuals better characterize the Pareto Front (i.e., we can identify the best individuals in the population) by finding the assignment with minimal cost.

The assignment cost can be computed using a scalarizing function (though other methods can also be adopted [7]). Therefore, the cost of assigning the weight vector $\boldsymbol{w}_i$ to individual $\boldsymbol{x}_j$ is defined as follows:

$$c_{ij} = u(\tilde{\boldsymbol{f}}(\boldsymbol{x}_j), \boldsymbol{w}_i) \quad i = 1, \ldots, n, \quad j = 1, \ldots, 2n \tag{11}$$

where u is the scalarizing function, $2n$ is the size of the population considering parents and offspring, n the number of weight vectors, and $\tilde{\boldsymbol{f}}(\boldsymbol{x}_j)$ is the normalized objective vector. This vector is defined as:

$$\tilde{\boldsymbol{f}}(\boldsymbol{x}_j) = [\tilde{f}_1(\boldsymbol{x}_j), \ldots, \tilde{f}_k(\boldsymbol{x}_j)] \tag{12}$$

$$s.t. \quad \tilde{f}_i(\boldsymbol{x}_j) = \frac{f_i(\boldsymbol{x}_j) - z_i^{\min}}{z_i^{\max} - z_i^{\min}}, \quad i = 1, \ldots, k \tag{13}$$

$$z_i^{\min} = \min_{l=1,\ldots,2n} f_i(\boldsymbol{x}_l), \quad i = 1, \ldots, k \tag{14}$$

$$z_i^{\max} = \max_{l=1,\ldots,2n} f_i(\boldsymbol{x}_l), \quad i = 1, \ldots, k \tag{15}$$

where $f_i(\boldsymbol{x}_j)$ is the ith function value of the jth solution. We used in this work the Hungarian algorithm to solve the LAP in which $2n$ must be equal to n. Hence, we added dummy costs (where $c_{ij} = 0$ for $i = n + 1, \ldots, 2n$ and $j = 1, \ldots, 2n$) to match the values, as recommended in [1].

HDE works as follows. First, the algorithm generates the initial population and evaluates it. After that, it generates n weight vectors using the Uniform Design with Hammersley's method (UDH) [9]. Then, during a predefined number of generations, it generates the offspring from the current population using Differential Evolution (DE) and evaluates them. After that, the old and the new population are normalized using Eqs. (12) to (15). Then, the assignment costs are computed using the Tchebycheff function [11]. With these costs, the LAP is constructed, and the best assignment is obtained employing the Hungarian algorithm. Finally, the assigned individuals are selected to become the next generation.

3.1   Drawbacks of HDE's Selection Process

In spite of the excellent performance of HDE in comparison with other algorithms, we identified two drawbacks in its selection process. The first one is that the process sometimes can select duplicated individuals. This problem arises because the assignment costs are the same for repeated solutions since the same value is evaluated in the scalarizing function. Therefore, if these solutions are the best in different weight vectors, the Hungarian Algorithm will prefer them.

For example, we want to select two elements from a set of four individuals such that $x_1 = x_2$, $f(x_1) = [1,2]^T$, $f(x_2) = [1,2]^T$, $f(x_3) = [4,3]^T$, and $f(x_4) = [2,5]^T$. Moreover, suppose we have the following two weight vectors: $w_1 = [1,0]^T$ and $w_2 = [0,1]^T$. Accordingly, their assignment costs using the Tchebycheff function [11] are displayed in Table 1. We can observe that individuals 1 and 2 have the best assignment costs for all the weight vectors. Hence, the Hungarian algorithm will assign either w_1 to x_1 and w_2 to x_2 or vice versa.

Table 1. Example of a case where the Hungarian algorithm selects duplicated solutions. The individuals x_1 and x_2 have the same value and the best assignment cost (highlighted in gray). Therefore, the solutions will be assigned and selected by HDE.

Weight	Individual			
	1	2	3	4
1	0	0	1	$\frac{1}{3}$
2	0	0	$\frac{1}{3}$	1

The second drawback is that the HDE occasionally prefers weakly-dominated solutions over non-dominated ones. To illustrate this fact, we executed HDE over 100 generations with a population size of 120 individuals, using the WFG7 problem [5]. Figure 1 displays, for each generation, the number of non-dominated solutions available in the population of parents and their offspring. Moreover, it shows the number of weakly dominated solutions selected by the HDE. We can see in Fig. 1 that even though, on some occasions, there are more than 120 non-dominated solutions available, HDE still selects 20 or more weakly dominated solutions. Therefore, HDE, in some cases, prefers weakly-dominated solutions over non-dominated ones. We believe that this phenomenon is produced because the scalarizing functions generate at least weakly Pareto solutions [10].

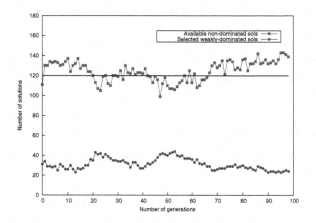

Fig. 1. Execution of the HDE during 100 generations using the WFG7 problem. Squares represent the number of available non-dominated solutions, and circles represent the number of selected weakly-dominated solutions.

4   Our Proposed Approach

This section presents an algorithm that tries to overcome the disadvantages presented above: the Multi-Objective Evolutionary Algorithm Based on the Linear Assignment Problem and the Hypervolume Approximation using Polar Coordinates (MOEA-LAPCO). In the following sections, we will explain the main modules of the MOEA-LAPCO.

4.1   Selection Process

The selection process of MOEA-LAPCO is summarized in Algorithm 1. Its core idea is to split the selection process into two phases. Let $p \in \mathbb{R}$, such that $0 \leq p \leq 50$. The first phase is to discard the p percentage of the parents and their offspring population using the LAP transformation. Then, the second phase is to discard the other $50 - p$ percentage of the population employing the approximation of the hypervolume using polar coordinates. The second phase aims to reduce the number of selected weakly-dominated solutions.

We have to take some considerations before applying the mechanism above. Let $A \subset \mathbb{R}^k$ and $\lambda \in \mathbb{R}^+$, then the reference point for the hypervolume contribution is defined as $z^u = \lambda z$, where all the solutions in A dominate z^u and z is computed in the following way:

$$z = [z_1, \ldots, z_k] \quad \text{s.t.} \quad z_j = \max_{i=1,\ldots,|A|} f_j(x_i), \quad j = 1, \ldots, k \tag{16}$$

Algorithm 1. Select_individuals

Require: Set of weight vectors (w_1), set of weight vectors (w_2), population (Q), uniformly distributed points (θ), population size (n_{pop}), reference point factor (λ), number of objectives (k)

1: $ND \leftarrow$ Obtain the non-dominated solutions from Q

2: $\boldsymbol{z}^{max} = [z_1^{max}, \dots, z_k^{max}]$ s.t. $z_j^{max} = \max_{x \in ND} f_j(\boldsymbol{x})$, $j = \{1, \dots, k\}$

3: $\boldsymbol{z}^{min} = [z_1^{min}, \dots, z_k^{min}]$ s.t. $z_j^{min} = \min_{x \in ND} f_j(\boldsymbol{x})$, $j = \{1, \dots, k\}$

4: $Q' \leftarrow$ Normalize the objective functions of Q employing (12) and (13) using the above $\boldsymbol{z}^{max}$ and $\boldsymbol{z}^{min}$ limits

5: **if** $|Q'| \leq n_{pop}$ **then**

6:      $C \leftarrow$ Compute the assignment cost using Q' and w_1

7:      $I \leftarrow$ Obtain the best assignment in C using the Hungarian algorithm

8: **else**

9:      $C \leftarrow$ Compute the assignment cost using Q' and w_2

10:      $I_H \leftarrow$ Obtain the best assignment in C using the Hungarian algorithm

11:      $I_{ND} \leftarrow$ Obtain the indices of the non-dominated solutions from $A := \{x_i \mid I_H[i] = 1, x_i \in Q'\}$

12:      $nd\_size \leftarrow |\{x_i \mid I_{ND}[i] = 1, x_i \in Q'\}|$

13:      **if** $nd\_size < n_{pop}$ **then**

14:          $I \leftarrow$ Prune_population_with_polar_coordinates(Q', I_H, θ, λ, k)

15:      **else**

16:          $I \leftarrow$ Prune_population_with_polar_coordinates(Q', I_{ND}, θ, λ, k)

17:      **end if**

18: **end if**

19: $P \leftarrow \{x_i \mid I[i] = 1, x_i \in Q\}$

20: **return** P

We observe that the approximation of the hypervolume contribution using polar coordinates is extremely sensitive to the reference point adopted (as pointed out by Deng and Zhang [4]). In particular, we observe that the algorithm's distribution was poor when dominated solutions were considered for computing the point $\boldsymbol{z}$. Therefore, in the selection process, we first compute the non-dominated solutions on the parents and their offspring population (see line 1). If the number of non-dominated solutions is less than n_{pop}, we only employ the LAP transformation to select the individuals for the next generation (see lines 5 to 7). Otherwise, we discard the p percentage of the population using the LAP transformation and discard the remaining individuals using the polar-coordinate approximation of the hypervolume contribution (see lines 8 to 18).

As we mentioned before, the LAP transformation occasionally prefers weakly-dominated solutions over non-dominated ones. Hence, it could happen that even if we have enough non-dominated solutions available, the algorithm returns weakly-dominated solutions. Therefore, we compute the number of non-dominated solutions returned by the LAP before initiating the polar-coordinate discarding process (see line 11). If the number of non-dominated solutions is less than n_{pop}, the pruning process will be carried out over the solutions provided by the LAP. Otherwise, the pruning process will be performed over the selected non-dominated solutions, and the remaining solutions will be discarded. We assume that the duplicated solutions are dominated (except for the first solution to appear). Consequently, the algorithm also removes the duplicated solutions in this phase, decreasing the appearance of these solutions. The last steps are performed in lines 12 to 17.

On the other hand, we changed the normalization limits adopted in HDE (see lines 2 to 4). Instead of obtaining the maximum and minimum values from the whole population, we only consider the non-dominated solutions of the population. Moreover, the normalization is performed using these limits in the original Eqs. (12) and (13).

4.2   Population to Be Pruned

The computational cost of approximating the hypervolume contribution using polar coordinates can be expensive if we do not make some considerations. First, the set θ does not depend on the population to prune and, therefore, it can be the same for all generations. Hence, we compute this set at the beginning of the MOEA-LAPCO algorithm. Second, the values $l_y(\theta^{(i)})$ do not change when a solution is removed. The only information that changes is the Matrix ranking. Therefore, at the beginning of the pruning procedure, we compute the $l_y(\theta^{(i)})^k$ values and store them in a matrix M. We raise the $l_y(\theta^{(i)})$ values to the power k because it does not affect the contribution order and avoids further computational cost. Furthermore, we obtain the indices that sort in descending order of each row of M. The above procedure is displayed in Algorithm 2.

Employing these considerations, the hypervolume contribution can be easily computed (this procedure is displayed in Algorithm 3). First, we initialize the array C of contributions with zeros. Then, we find the best and second-best solution indices for each ith-point in θ. Since we have the matrix I_M, we only have to go through the list $I_M[i]$ to find the first two still selected individuals (see lines 3 and 4). After that, we can obtain the best and second-best elements from M and compute their difference (see lines 5 and 6). Then, we go through the $I_M[i]$ list starting from the index of the best individual, add the difference to the currently selected individuals, and stop when the value $M[i][I_M[j]]$ is different from the best individual (see lines 7 to 18). At the end of the iterations, we multiply the contributions by $\frac{\Phi}{2^k}\frac{1}{kn}$ as in Eq. (6).

Finally, the pruning procedure is displayed in Algorithm 4. The first step is to obtain the reference point from the currently selected individuals. Then, we compute and sort the matrix M using Algorithm 2. Then, we compute the contribution of each individual in the population using Algorithm 3 and remove the one with less contribution. We make the above procedure until the population size is equal to n_{pop}.

4.3   The Final Algorithm: MOEA-LAPCO

Algorithm 5 summarizes the behavior of MOEA-LAPCO. First, the two sets of weight vectors (w_1 and w_2) are generated using the UDH method. The set w_1 is

---

Algorithm 2. Compute_and_sort_M

Require: Normalized objective vectors (Q'), number of elements in Q' (m), list that handles currently selected solutions (I_S), uniformly distributed points (θ), number of points in θ (n), reference point (z^u), number of objectives (k)

1: Initialize matrix M of size $n \times m$ with zeros
2: Initialize matrix I_S of size $n \times m$ with zeros
3: **for** $i = 1$ to n **do**
4:     $c_{im} \leftarrow 0$
5:     **for** $j = 1$ to m **do**
6:         **if** $I_S[j] = 1$ **then**
7:             $\bar{y} \leftarrow Q'[j]$
8:             $l_{\bar{y}}(\theta^{(i)}) \leftarrow \min_{1 \leq l \leq k}(1/\theta_l^{(i)})(z_l^u - \bar{y}_l)$
9:             $M[i][j] = (l_{\bar{y}}(\theta^{(i)}))^k$
10:            $c_{im} \leftarrow c_{im} + 1$
11:            $I_M[i][c_{im}] = j$
12:        **end if**
13:    **end for**
14:    Sort the first c_{im} elements of $I_M[i]$ in descending order such that $I_M[i][x]$ is bigger than $I_M[i][y]$ when $M[i][I_M[i][x]] > M[i][I_M[i][y]]$.
15: **end for**
16: **return** M, I_M, c_{im}

---

used when only the LAP transformation is applied, and the set w_2 is used when the two-phase selection is performed. Therefore, the size of w_2 depends on the parameter p and is calculated as $|w_2| = \frac{p*n_{pop}*2}{100}$. Then, the set θ of uniformly distributed points are generated using Eq. (4). Afterwards, the algorithm continues as the usual MOEAs. The population is generated and evaluated. Then, while a maximum number of evaluations is not reached, a new population is generated from the old one using variation operators. We decided to use the Simulated Binary Crossover (SBX) and the Polynomial-based Mutation (PM) [2] operators. Finally, we select the best individuals with our method adopting the population of parents and their offspring.

5   Experimental Analysis

We evaluated the performance of MOEA-LAPCO with respect to state-of-the-art algorithms. For this purpose, we performed 30 independent runs of HDE [9] comparing DE, HDE with SBX and PM, MOEA/DD [6], NSGA-III [3], and our proposed algorithm. We adopted the WFG1-WFG9 problems from the WFG test suite using 3, 5, 8, and 10 objectives. The position-related parameters were set to $m = 2 \times (k - 1)$ where k is the number of objectives, the distance-related parameters were set to $l = 20$, and the number of variables to $n = m + l$. Finally, we used the hypervolume indicator for the performance assessment.

Regarding the variation operators, SBX and PM were set to $pc = 0.9$, $pm = 1/(\text{number of variables})$, $nc = 20$ and $nm = 20$. Furthermore, we set

---

Algorithm 3. Compute_contribution

Require: matrix (M), indices that sort the solutions in M (I_M), cols of I_M (c_{IM}), list that handles currently selected solutions (I_S), number of objectives (k), number of rows in M (r), number of cols in M (c)

1: $C \leftarrow$ Initialize array of size c with zeros
2: **for** $i = 1$ to r **do**
3:     $jbest \leftarrow \min_{j=1,\ldots,c_{IM}} j$ $s.t.$ $I_S[I_M[i][j]] = 1$
4:     $jsbest \leftarrow \min_{j=1,\ldots,c_{IM}} j$ $s.t.$ $I_S[I_M[i][j]] = 1 \wedge j! = jbest$
5:     $best \leftarrow M[i][I_M[i][jbest]]$
6:     $diff \leftarrow best - M[i][I_M[i][jsbest]]$
7:     $j \leftarrow jbest$
8:     **while** $j < c$ **do**
9:         $idx \leftarrow I_M[j]$
10:         **if** $best = M[i][idx]$ **then**
11:             **if** $I_S[idx] = 1$ **then**
12:                 $C[idx] = C[idx] + diff$
13:             **end if**
14:         **else**
15:             break
16:         **end if**
17:         $j \leftarrow j + 1$
18:     **end while**
19: **end for**
20: **for** $j = 1$ to c **do**
21:     $C[j] = \frac{\Phi}{2^k} \frac{1}{kn} * C[j]$
22: **end for**
23: **return** C

Algorithm 4. Prune_population_with_polar_coordinates

Require: Normalized objective vectors (Q'), list that handles currently selected solutions (I_S), uniformly distributed points (θ), reference point factor(λ), number of objectives (k)

1: $z \leftarrow$ Find the maximum value of each objective from $A := \{x_i \mid I_S[i] = 1, x_i \in Q'\}$
2: $z^u \leftarrow \lambda * z$
3: $r_M \leftarrow |\theta|$
4: $c_M \leftarrow |Q'|$
5: $M, I_M, c_{IM} \leftarrow$ Compute_and_sort_M($Q', c_M, I_S, \theta, r_M, z^u, k$)
6: $num\_sel = c_{IM}$
7: **while** $num\_sel > n_{pop}$ **do**
8:     $C \leftarrow$ Compute_contribution($M, I_M, c_{IM}, I_S, k, r_M, c_M$)
9:     $idx \leftarrow \arg\min_{i=1,\cdots,c_M} C[i]$ $s.t.$ $I_S[i] = 1$
10:     $I_S[idx] = 0$
11:     $num\_sel = num\_sel - 1$
12: **end while**
13: **return** I_S

Algorithm 5. MOEA-LAPCO

Require: Multi-objective problem, population size (n_{pop}), number of uniformly distributed points (n_{hv}), maximum number of evaluations, variation operators' parameters, reference point factor (λ), percentage of solutions to be discarded using the LAP transformation (p)

Ensure: P

1: $w1 \leftarrow$ Generate n_{pop} weights vectors using UDH
2: $w2 \leftarrow$ Generate $\frac{p*n_{pop}*2}{100}$ weights vectors using UDH
3: $\theta \leftarrow$ Generate n_{hv} uniformly distributed points using Eq. (4)
4: Generate initial population P
5: Evaluate population P
6: **while** the maximum number of evaluations is not reached **do**
7:     $P' \leftarrow$ Generate from P the new population using variation operators
8:     Evaluate population P'
9:     $Q \leftarrow P \cup P'$
10:     $P \leftarrow$ Select_individuals(w_1, w_2, Q, θ, n_{pop}, λ, p)
11: **end while**
12: return P

---

the parameters of DE to $F = 1.0$ and $Cr = 0.4$. In the case of the weight vectors of the MOEA/DD and NSGA-III, we used Das and Dennis' approach with the two-layer technique adopted in the NSGA-III for more than five objectives [3]. Concerning the MOEA/DD's parameters, we set $T = 20$, $\delta = 0.9$, and we used the PBI scalarizing function with $\theta = 5$. In the case of MOEA-LAPCO, we set $n_{hv} = 10000$, $\delta = 1.5$, and $p = 50$. Furthermore, we used the Augmented Achievement Scalarizing Function (AASF) [10] with $\alpha = 0.0001$ in both versions of HDE and in the MOEA-LAPCO. Considering the parameters for all the algorithms, we used a population size of 120 for three objectives, 210 for five, 156 for eight, and 276 for ten. In the case of the maximum number of evaluations, we used the population size times 1000 in all the objectives.

Table 2 shows the average and the standard deviation of the hypervolume's values over 30 generations of each algorithm. The best averages are highlighted in dark gray, and the second-bests are highlighted in light gray. Moreover, the symbol "*" indicates that the algorithm is statistically better than the others employing the Wilcoxon rank-sum test with a significance level of 5%. We can observe that MOEA-LAPCO is better than HDE with DE and SBX+PM in almost all the problems, indicating that the new mechanism improves the original versions. On the other hand, the MOEA-LAPCO is better than all the algorithms in 20 out of 36 problems. Remarkably, it is the best in problems WFG3, WFG5, WFG6, WFG8, and WFG9 using 3, 5, and 8 objectives. Moreover, it is the best in the WFG4 and WFG7 problems using 3 and 5 objectives. However, we can notice that it is not the best in any of the problems with ten objectives, suggesting that the algorithm's performance degrades with more than eight objectives. We believe that this happens because the reference point selection mechanism is not good enough for many-objective problems.

Table 2. Average and standard deviation of the hypervolume indicator over 30 generations of MOEA-LAPCO and state-of-the-art algorithms. The best values are highlighted in dark gray, and the second-best values in light gray. The symbol "*" indicates that the algorithm is statistically better than the others.

	k	HDE_DE	HDE_SBX+PM	MOEA-LAPCO	MOEA/DD	NSGA-III
WFG1	3	8.031e−1 (9.0e−3)	8.963e−1 (2.2e−2)	9.214e−1 (2.5e−2)	*1.218e+0 (3.2e−2)	7.749e−1 (3.7e−2)
	5	9.095e−1 (9.5e−3)	1.103e+0 (1.6e−2)	1.157e+0 (1.5e−2)	*1.454e+0 (5.4e−2)	8.978e−1 (3.5e−2)
	8	1.245e+0 (1.7e−2)	1.691e+0 (2.7e−2)	1.746e+0 (1.8e−2)	*1.915e+0 (1.1e−1)	1.384e+0 (1.1e−1)
	10	1.446e+0 (1.8e−2)	2.029e+0 (2.2e−2)	2.077e+0 (2.7e−2)	*2.297e+0 (1.1e−1)	1.958e+0 (1.5e−1)
WFG2	3	1.234e+0 (2.9e−3)	1.189e+0 (8.7e−2)	*1.194e+0 (8.8e−2)	1.168e+0 (8.9e−2)	1.153e+0 (8.9e−2)
	5	1.548e+0 (4.2e−3)	1.572e+0 (6.9e−2)	*1.586e+0 (7.1e−2)	1.537e+0 (6.6e−2)	1.523e+0 (8.2e−2)
	8	*2.134e+0 (1.8e−2)	1.977e+0 (1.7e−1)	2.065e+0 (1.2e−1)	1.957e+0 (9.9e−2)	1.939e+0 (1.5e−1)
	10	*2.591e+0 (1.2e−3)	2.525e+0 (1.3e−1)	2.546e+0 (2.5e−2)	2.292e+0 (3.2e−2)	2.428e+0 (1.1e−1)
WFG3	3	9.105e−1 (3.9e−3)	9.387e−1 (2.1e−3)	*9.466e−1 (1.5e−3)	8.932e−1 (6.0e−3)	9.014e−1 (5.4e−3)
	5	1.109e+0 (5.9e−3)	1.163e+0 (6.5e−3)	*1.186e+0 (6.0e−3)	1.039e+0 (9.5e−3)	1.059e+0 (1.1e−2)
	8	1.426e+0 (7.7e−3)	1.457e+0 (1.2e−2)	*1.494e+0 (2.3e−2)	1.16e+0 (2.2e−2)	1.263e+0 (2.6e−2)
	10	1.694e+0 (8.5e−3)	*1.724e+0 (1.1e−2)	1.703e+0 (3.1e−2)	1.279e+0 (2.0e−2)	1.576e+0 (2.8e−2)
WFG4	3	7.303e−1 (3.9e−3)	7.79e−1 (2.1e−3)	*7.983e−1 (9.6e−4)	7.784e−1 (1.6e−3)	7.565e−1 (3.0e−3)
	5	1.191e+0 (6.3e−3)	1.267e+0 (3.7e−3)	*1.334e+0 (2.4e−3)	1.286e+0 (4.3e−3)	1.210e+0 (8.7e−3)
	8	1.706e+0 (1.5e−2)	1.468e+0 (1.1e−1)	1.54e+0 (4.2e−2)	*1.736e+0 (2.5e−2)	1.595e+0 (4.1e−2)
	10	*2.180e+0 (1.6e−2)	1.972e+0 (9.3e−2)	1.893e+0 (3.6e−2)	2.097e+0 (4.1e−2)	1.959e+0 (3.8e−2)
WFG5	3	7.368e−1 (2.3e−3)	7.419e−1 (4.7e−3)	*7.667e−1 (3.2e−3)	7.434e−1 (3.8e−3)	7.321e−1 (5.1e−3)
	5	1.251e+0 (3.2e−3)	1.237e+0 (3.4e−3)	*1.319e+0 (2.8e−3)	1.258e+0 (3.8e−3)	1.228e+0 (5.3e−3)
	8	1.443e+0 (2.4e−1)	1.640e+0 (1.1e−1)	*1.853e+0 (9.4e−2)	1.681e+0 (2.8e−2)	1.726e+0 (2.5e−2)
	10	1.821e+0 (2.1e−2)	1.964e+0 (3.3e−2)	1.845e+0 (5.7e−2)	2.046e+0 (4.3e−2)	*2.137e+0 (2.6e−2)
WFG6	3	7.056e−1 (6.9e−4)	7.59e−1 (6.4e−3)	*7.783e−1 (5.7e−3)	7.541e−1 (6.2e−3)	7.377e−1 (8.4e−3)
	5	1.206e+0 (1.4e−3)	1.25e+0 (9.5e−3)	*1.314e+0 (8.7e−3)	1.254e+0 (1.1e−2)	1.217e+0 (1.2e−2)
	8	1.802e+0 (2.6e−3)	1.790e+0 (2.5e−2)	*1.934e+0 (3.1e−2)	1.765e+0 (2.8e−2)	1.737e+0 (3.5e−2)
	10	*2.285e+0 (1.4e−3)	2.068e+0 (5.5e−2)	2.143e+0 (9.4e−2)	2.140e+0 (3.9e−2)	2.172e+0 (3.4e−2)
WFG7	3	7.507e−1 (2.2e−3)	7.711e−1 (1.1e−3)	*7.886e−1 (5.7e−4)	7.727e−1 (1.3e−3)	7.599e−1 (2.7e−3)
	5	1.207e+0 (6.1e−3)	1.266e+0 (3.3e−3)	*1.336e+0 (1.3e−3)	1.299e+0 (3.6e−3)	1.245e+0 (1.1e−2)
	8	1.747e+0 (1.9e−2)	1.542e+0 (9.3e−2)	1.764e+0 (1.6e−1)	*1.867e+0 (1.3e−2)	1.709e+0 (3.5e−2)
	10	2.242e+0 (1.6e−2)	2.08e+0 (5.4e−2)	1.95e+0 (8.e−2)	*2.263e+0 (1.7e−1)	2.169e+0 (3.4e−2)
WFG8	3	8.525e−1 (4.8e−3)	8.979e−1 (2.1e−3)	*9.195e−1 (1.3e−3)	9.007e−1 (2.1e−3)	8.717e−1 (5.3e−3)
	5	1.140e+0 (7.1e−3)	1.33e+0 (1.2e−2)	*1.375e+0 (2.e−2)	1.268e+0 (1.2e−2)	1.175e+0 (9.2e−3)
	8	1.598e+0 (1.8e−2)	1.596e+0 (8.1e−2)	*1.874e+0 (9.5e−2)	1.755e+0 (6.4e−2)	1.528e+0 (3.3e−2)
	10	2.116e+0 (1.3e−2)	1.988e+0 (4.3e−2)	1.990e+0 (7.4e−2)	*2.216e+0 (6.1e−2)	1.951e+0 (4.6e−2)
WFG9	3	8.626e−1 (1.5e−3)	8.847e−1 (3.2e−2)	*9.303e−1 (3.e−2)	8.965e−1 (3.0e−2)	8.768e−1 (2.e−2)
	5	1.184e+0 (2.7e−3)	1.169e+0 (4.3e−3)	1.211e+0 (2.3e−3)	1.203e+0 (2.9e−2)	1.167e+0 (1.4e−2)
	8	1.804e+0 (9.9e−3)	1.782e+0 (4.4e−2)	*1.844e+0 (7.0e−2)	1.634e+0 (8.9e−2)	1.68e+0 (5.9e−2)
	10	*2.207e+0 (1.1e−2)	2.108e+0 (5.7e−2)	2.010e+0 (7.0e−2)	1.935e+0 (8.6e−2)	2.075e+0 (4.6e−2)

6   Conclusions and Future Work

In this work, we proposed a new algorithm called "Multi-Objective Evolutionary Algorithm Based on the Linear Assignment Problem and the Hypervolume Approximation using Polar Coordinates" (MOEA-LAPCO). The core idea of the MOEA-LAPCO is to overcome the disadvantages of the HDE selection process by employing an approximation of the hypervolume contribution using Polar Coordinates. Our experimental analysis showed that the algorithm improves the performance of HDE and is competitive with respect to state-of-the-art algorithms. However, the MOEA-LAPCO's performance deteriorates when more than eight objectives are used. As part of our future work, we would like to analyze the reason for the deterioration by analyzing the impact of the reference point in the performance of the MOEA-LAPCO when the number of objectives increases.

Acknowledgements. The first author acknowledges support from CINVESTAV-IPN and CONACyT to pursue graduate studies in computer science. The second author acknowledges support from CONACyT grant no. 1920.

References

1. Burkard, R.E., Dell'Amico, M., Martello, S.: Assignment Problems, Revised Reprint. Other Titles in Applied Mathematics, Society for Industrial and Applied Mathematics (SIAM) (2012)
2. Deb, K., Agrawal, R.B.: Simulated binary crossover for continuous search space. Complex Syst. **9**(2), 115–148 (1995)
3. Deb, K., Jain, H.: An evolutionary many-objective optimization algorithm using reference-point-based nondominated sorting approach, Part I: solving problems with box constraints. IEEE Trans. Evol. Comput. **18**(4), 577–601 (2014). https://doi.org/10.1109/TEVC.2013.2281535
4. Deng, J., Zhang, Q.: Approximating hypervolume and hypervolume contributions using polar coordinate. IEEE Trans. Evol. Comput. **23**(5), 913–918 (2019). https://doi.org/10.1109/TEVC.2019.2895108
5. Huband, S., Barone, L., While, L., Hingston, P.: A scalable multi-objective test problem toolkit. In: Coello Coello, C.A., Hernández Aguirre, A., Zitzler, E. (eds.) EMO 2005. LNCS, vol. 3410, pp. 280–295. Springer, Heidelberg (2005). https://doi.org/10.1007/978-3-540-31880-4_20
6. Li, K., Deb, K., Zhang, Q., Kwong, S.: An evolutionary many-objective optimization algorithm based on dominance and decomposition. IEEE Trans. Evol. Comput. **19**(5), 694–716 (2015). https://doi.org/10.1109/TEVC.2014.2373386
7. Manoatl Lopez, E., Coello Coello, C.A.: IGD$^+$-EMOA: a multi-objective evolutionary algorithm based on IGD$^+$. In: 2016 IEEE Congress on Evolutionary Computation (CEC'2016), pp. 999–1006. IEEE Press, Vancouver, Canada, 24–29 July 2016. https://doi.org/10.1109/CEC.2016.7743898, ISBN 978-1-5090-0623-9
8. Miguel Antonio, L., Molinet Berenguer, J.A., Coello Coello, C.A.: Evolutionary many-objective optimization based on linear assignment problem transformations. Soft. Comput. **22**(16), 5491–5512 (2018)
9. Molinet Berenguer, J.A., Coello Coello, C.A.: Evolutionary many-objective optimization based on Kuhn-Munkres' algorithm. In: Gaspar-Cunha, A., Henggeler Antunes, C., Coello, C.C. (eds.) EMO 2015. LNCS, vol. 9019, pp. 3–17. Springer, Cham (2015). https://doi.org/10.1007/978-3-319-15892-1_1
10. Pescador-Rojas, M., Hernández Gómez, R., Montero, E., Rojas-Morales, N., Riff, M.C., Coello Coello, C.A.: An overview of weighted and unconstrained scalarizing functions. In: Trautmann, H., et al. (eds.) Evolutionary Multi-criterion Optimization, 9th International Conference, Münster, Germany, 19–22 March 2017, EMO 2017. LNCS, vol. 10173, pp. 499–513. Springer, Cham (2017). https://doi.org/10.1007/978-3-319-54157-0_34, ISBN 978-3-319-54156-3
11. Zhang, Q., Li, H.: MOEA/D: a multiobjective evolutionary algorithm based on decomposition. IEEE Trans. Evol. Comput. **11**(6), 712–731 (2007). https://doi.org/10.1109/TEVC.2007.892759

New Solution Creation Operator in MOEA/D for Faster Convergence

Longcan Chen, Lie Meng Pang, and Hisao Ishibuchi[✉]

Guangdong Provincial Key Laboratory of Brain-Inspired Intelligent Computation,
Department of Computer Science and Engineering, Southern University of Science
and Technology, Shenzhen 518055, China
11813009@mail.sustech.edu.cn, {panglm,hisao}@sustech.edu.cn

Abstract. This paper introduces a novel solution generation strategy for MOEA/D. MOEA/D decomposes a multi/many-objective optimization problem into several single-objective sub-problems using a set of weight vectors and a scalarizing function. When a better solution is generated for one sub-problem, it is likely that a further better solution will appear in the improving direction. Examination of such a promising solution may improve the convergence speed of MOEA/D. Our idea is to use the improved directions in the current and previous populations to generate new solutions in addition to the standard genetic operators. To assess the usefulness of the proposed idea, we integrate it into MOEA/D-PBI and use a distance minimization problem to visually examine its behavior. Furthermore, the proposed idea is evaluated on some large-scale multi-objective optimization problems. It is demonstrated that the proposed idea drastically improves the convergence ability of MOEA/D.

Keywords: Evolutionary multi-objective optimization · Large-scale multi-objective optimization · MOEA/D · Solution generation strategy

1 Introduction

Many real-world applications involve multi-objective optimization problems that have conflicting objectives [1]. Without loss of generality, multi-objective optimization problems can be represented as follows:

$$\text{Minimize } \boldsymbol{f}(\boldsymbol{x}) = (f_1(\boldsymbol{x}), f_2(\boldsymbol{x}), ..., f_m(\boldsymbol{x}))^T, \\ \text{subject to } \boldsymbol{x} \in \Omega \tag{1}$$

where $\boldsymbol{x} = (\boldsymbol{x}_1, \boldsymbol{x}_2, ..., \boldsymbol{x}_d)^T$ is a d-dimensional vector of decision variables, Ω is the feasible region, and $f_i(\boldsymbol{x})$ is the i-th objective to be minimized $(i = 1, 2, \ldots, m)$. Since the objectives are conflicting with each other, there is no solution that can optimize all objectives simultaneously. In multi-objective optimization, the final goal is to find a set of Pareto optimal (PO) solutions. Population-based approaches are useful for discovering a set of well-distributed

G. Rudolph et al. (Eds.): PPSN 2022, LNCS 13399, pp. 234–246, 2022.
https://doi.org/10.1007/978-3-031-14721-0_17

and well-converged solutions [1, 2], and evolutionary multi-objective optimization (EMO) is one of the effective approaches.

MOEA/D [3] is one of the most popular decomposition-based EMO algorithms. MOEA/D uses a set of weight vectors $\boldsymbol{W} = (\boldsymbol{w}_1, \boldsymbol{w}_2, \ldots, \boldsymbol{w}_{|P|})^T$ (where $|P|$ is the population size) and a scalarizing function to decompose a multi-objective optimization problem into a set of single-objective sub-problems. Each weight vector $\boldsymbol{w}_i = (\boldsymbol{w}_{i_1}, \boldsymbol{w}_{i_2}, \ldots, \boldsymbol{w}_{i_m})^T$ corresponds to a sub-problem. For a given sub-problem, the scalarizing function is used to calculate the fitness value of a solution. Each weight vector $\boldsymbol{w}_i$ (i-th weight vector, $i = 1, 2, \ldots, |P|$) has a current solution $\boldsymbol{x}_i^{Current}$. When an offspring solution $\boldsymbol{x}_i^{New}$ is better than the current solution $\boldsymbol{x}_i^{Current}$, $\boldsymbol{x}_i^{Current}$ is replaced with $\boldsymbol{x}_i^{New}$. Let us denote the neighborhood of $\boldsymbol{w}_i$ by S_i and its size by $|S_i|$. In each generation, the current solution $\boldsymbol{x}_i^{Current}$ is compared with $|S_i|$ offspring solutions (one by one) generated in the neighborhood S_i. Thus, the current solution can be updated $|S_i|$ times in each generation. The current solution is not updated if there are no better solutions than the current solution.

In MOEA/D, when a better solution is generated for one sub-problem, it is likely that a further better solution will appear in the improving direction. Examination of such a promising solution may improve the convergence speed of MOEA/D. Based on this idea, we use the solutions in the current and previous generations to generate improving directions for sub-problem. By using the improving directions, promising solutions can be generated and used for accelerating the convergence speed of MOEA/D.

The idea of using the information obtained from the current and previous generations to generate new solutions is not entirely new. In the literature, many studies focus on online innovization approaches [4–10]. Online innovization approaches attempt to learn from the current and previous generations. By extracting the patterns or relationships among variables in the decision space, online innovization approaches can accelerate the convergence speed of EMO algorithms. Mittal et al. [4] proposed a learning-based innovized progress operator for EMO algorithms. It uses a machine learning (ML) model to capture the patterns of the variables in the decision space and uses the learned ML model to improve offspring solutions. Ghosh et al. [5] proposed a method that combines user-supplied and machine-learnable patterns and rules to accelerate the convergence speed of EMO algorithms. Mittal et al. [6] proposed an innovized repair operator which uses an ML model to repair the offspring solutions.

In this paper, we propose a solution creation method for MOEA/D by using the improving move of the current and previous solutions corresponding to each sub-problem. This paper is organized as follows. In Sect. 2, we explain the proposed strategy and its implementation. Next, we use computational experiments to demonstrate the usefulness of the proposed strategy in Sect. 3. Finally, we conclude this paper and give some future research directions in Sect. 4.

2   Proposed Strategy and Implementations

In this section, we first explain our idea using a distance minimization problem. In this problem, as Fig. 1 shows, we need to minimize four objectives, which are f_1: Distance to P_1, f_2: Distance to P_2, f_3: Distance to P_3, and f_4: Distance to P_4. When a better solution (i.e., Offspring 4 in Fig. 1) is generated for a weight vector w with the current solution x (i.e., the red point in the figure), it is likely that we will be able to find a further better solution in the improving direction (i.e., a candidate solution as shown by the yellow circle along the red line). Examination of such a promising solution may improve the convergence speed of MOEA/D.

We assume that the current solution is replaced with the candidate solution in Fig. 1. Then as Fig. 2 shows, we also assume that a better solution (i.e., Offspring 5) is found. In this case, we can examine a candidate solution along the improving direction (e.g., Candidate Solution A on the red line). We can also generate another Candidate Solution B by considering both the current improving direction and the previous improving direction.

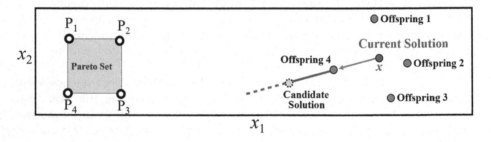

Fig. 1. Illustration of the proposed idea (use of the moves in the current generation). (Color figure online)

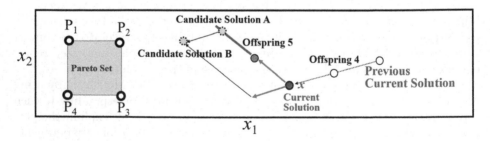

Fig. 2. Illustration of the proposed idea (use of the moves in the current and previous generations).

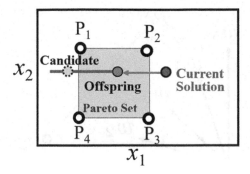

Fig. 3. Illustration of the proposed idea where the candidate is not better than the current solution. (Color figure online)

However, the candidate solutions (e.g., the yellow circle) are not always better than the offspring solution (e.g., the blue circle) as shown in Fig. 3. In this case, the current solution (e.g., the red circle) is replaced with the offspring solution (e.g., the blue circle), not with the candidate solution (the yellow circle).

In this paper, we implement this idea for MOEA/D with the penalty-based boundary intersection (PBI) function ($\theta = 5$) [3]. We propose three different implementations (i.e., Type1, Type2, and Type3) to generate candidate solutions. Type1 implementation uses the moves of the solution for the current sub-problem. Type2 implementation uses the moves of the solution for the current sub-problem and the moves of its neighboring solutions. Type3 implementation uses the moves of all solutions in the population. Additionally, each implementation can be further subdivided into two sub-types. The first sub-type considers only the current improving direction, and the second sub-type considers both the current and previous improving directions.

It should be noted that in the standard MOEA/D implementation, the initial population is randomly generated and assigned to each weight vector, which may affect the performance of the proposed strategy. An example is shown in Fig. 4. Figure 4 illustrates the improving moves of the current solution x_4 for the weight vector w_4. In the figure, the pink curve is the Pareto front, $x_4^{Initial}$ is a randomly generated initial solution, $x_4^{(2)}$ is the current solution after the 2nd generation (which is the best solution among the generated $|S_i|$ offspring solutions in the neighborhood during the 2nd generation), and $x_4^{(3)}$ is the current solution after the 3rd generation. In many cases, the move from $x_4^{Initial}$ to $x_4^{(2)}$ is not a good direction (the red arrow) while the move from $x_4^{(2)}$ to $x_4^{(3)}$ is usually a good direction (the blue arrow).

The implementation of our strategy in each type is explained in the following.

Type1: Independent Formulation for Each Sub-problem. We denote the current solution for the weight vector w_i at the end of the tth generation by $x_i(t)$. In Type1 implementation, when $x_i(t)$ is better than $x_i(t-1)$, a new candidate solution is generated by the proposed strategy with a probability of 0.5. The total

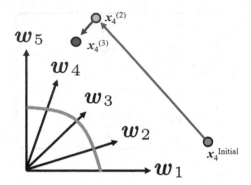

Fig. 4. Illustration of the improvement of the current solution x_4 for the weight vector w_4.

move during the tth generation is defined as $\Delta x_i(t) = x_i(t) - x_i(t-1)$. Since the improving direction in the second generation is not reliable, $\Delta x_i(t)$ is defined as $\Delta x_i(t) = 0$ for $t = 2$. A new candidate solution (i.e., $x_i^{Candidate}$) is generated from $x_i(t)$ and $\Delta x_i(t)$ as explained latter in detail. If $x_i^{Candidate}$ is better than the current solution $x_i(t)$, $x_i(t)$ is replaced with $x_i^{Candidate}$. $x_i^{Candidate}$ is also compared with neighboring solutions. If $x_i^{Candidate}$ is better than a neighboring solution, the neighboring solution is replaced with $x_i^{Candidate}$.

When no solution is improved during the tth generation (i.e., $x_i(t) = x_i(t-1)$), no candidate solution is generated. Even in this case, the current solution can be updated by a candidate solution generated for a neighboring weight vector.

Type1-1: Use of the Current Move. In Type1-1 MOEA/D, we only use the current move to generate candidates. The candidate solution is generated as $x_i^{Candidate} = x_i(t) + \eta \Delta x_i(t)$ where η is a non-negative constant parameter.

Type1-2: Use of the Current and Previous Moves. In Type1-2 MOEA/D, we consider both the current and previous improving directions. The problem is how to define the previous improving direction since the current solution was not always improved during the $(t-1)^{th}$ generation. Thus, we define the previous improving direction using the latest improved generation k before the tth generation as $\Delta x_i(k) = x_i(k) - x_i(k-1)$, where generation k is the latest improved generation ($1 < k < t$) before the tth generation. If there is no improved generation before the tth generation, we define $\Delta x_i(k)$ as $\Delta x_i(k) = 0$. When $k = 2$, we define $\Delta x_i(k)$ as $\Delta x_i(k) = 0$ since the initial solution is randomly assigned. Then, a candidate solution can be generated as $x_i^{Candidate} = x_i(t) + \eta \Delta x_i(t) + \alpha \Delta x_i(k)$ where η and α are non-negative constant parameters.

Type1-2*: One Variant of Type1-2. In Type1-2* MOEA/D, a simplified version of the definition of the candidate solution is to use the current and previous moves as $x_i^{Candidate} = x_i(t) + \eta \Delta x_i(t) + \alpha \Delta x_i(t-1)$. In this variant, the move in the $(t-1)^{th}$ generation is used even if $\Delta x_i(t-1) = 0$. In early generations, it is likely that the current solution is frequently improved. Thus, this variant is similar to Type1-2 MOEA/D. However, in late generations, the current solution is not frequently improved. As a result, this variant is similar to Type1-1 MOEA/D.

Type2: Use of the Moves of Neighboring Solutions. In Type2, when at least one solution in the neighborhood S_i is improved during the tth generation, the candidate solution is generated with a small probability (in our experiment, the probability is set as $5/|S_i|$). The total move during the tth generation is defined as $\Delta x_i(t) = x_i(t) - x_i(t-1)$, and the total move during the tth generation in the neighborhood S_i is defined as $\Delta_{S_i} x_i(t) = \sum_{j \in S_i} (x_j(t) - x_j(t-1))$ where x_i is included in S_i. Since the improving direction in the second generation is not reliable, $\Delta x_i(t)$ and $\Delta_{S_i} x_i(t)$ are defined as $\Delta x_i(t) = 0$ and $\Delta_{S_i} x_i(t) = 0$ for $t = 2$. Then, $x_i^{Candidate}$ is generated as explained below. If $x_i^{Candidate}$ is better than the current solution $x_i(t)$, $x_i(t)$ is replaced with $x_i^{Candidate}$. $x_i^{Candidate}$ is also compared with neighboring solutions. If $x_i^{Candidate}$ is better than a neighboring solution, the neighboring solution is replaced with $x_i^{Candidate}$.

When no solution in the neighborhood S_i is improved during the tth generation (i.e., $x_j(t) = x_j(t-1)$), no candidate solution is generated. Even in this case, the current solution can be updated by a candidate solution generated for a neighboring weight vector.

Type2-1: Use of the Current Move. In Type2-1 MOEA/D, we use the total move in the neighborhood S_i to generate candidates. The candidate solution is generated as $x_i^{Candidate} = x_i(t) + \eta \Delta x_i(t) + \eta_{S_i} \Delta_{S_i} x_i(t)$ where η and η_{S_i} are non-negative constant parameters. In this formulation, $\Delta x_i(t)$ equals to 0 in many cases. However, $\Delta_{S_i} x_i(t)$ is not zero in many cases since all the moves in the neighborhood are summed up.

Type2-2: Use of the Current and Previous Moves. In Type2-2 MOEA/D, we use the current and previous moves in the neighborhood S_i to generate candidates. We define the previous improving direction using the latest improved generation k before the tth generation as $\Delta x_i(k)$ and $\Delta_{S_i} x_i(k)$, where k is the latest improved generation $(1 < k < t)$ where at least one solution in the neighborhood S_i is improved before the tth generation. If there is no improved generation before the tth generation, we define $\Delta x_i(k)$ and $\Delta_{S_i} x_i(k)$ as 0. We also define $\Delta x_i(k)$ and $\Delta_{S_i} x_i(k)$ as 0 when $k = 2$ since the initial solution is randomly assigned. Then, a candidate solution can be generated as $x_i^{Candidate} = x_i(t) + \eta \Delta x_i(t) + \alpha \Delta x_i(k) + \eta_{S_i} \Delta_{S_i} x_i(t) + \alpha_{S_i} \Delta_{S_i} x_i(k)$, where η, η_{S_i}, α and α_{S_i} are non-negative constant parameters.

Type3: Use of the Moves of All Solutions in the Population. In Type3, when at least one solution in the population is improved during the tth generation, the candidate solution is generated with a small probability (in our experiment, the probability is set as $5/|S|$). The total move during the tth generation is defined as $\Delta x_i(t) = x_i(t) - x_i(t-1)$, and the total move during the tth

generation in the population S is defined as $\Delta_S \boldsymbol{x}_i(t) = \sum_{j \in S}(\boldsymbol{x}_j(t) - \boldsymbol{x}_j(t-1))$. For $t = 2$, $\Delta \boldsymbol{x}_i(t)$ and $\Delta_S \boldsymbol{x}_i(t)$ are defined as $\Delta \boldsymbol{x}_i(t) = 0$ and $\Delta_S \boldsymbol{x}_i(t) = 0$. Then, $\boldsymbol{x}_i^{Candidate}$ is generated. When no solution in the population S is improved during the tth generation, no candidate solution is generated.

Type3-1: Use of the Current Move. In Type3-1 MOEA/D, we use the total move in the population S to generate candidates. The candidate solution is generated as $\boldsymbol{x}_i^{Candidate} = \boldsymbol{x}_i(t) + \eta \Delta \boldsymbol{x}_i(t) + \eta_S \Delta_S \boldsymbol{x}_i(t)$, where η and η_S are non-negative constant parameters. It should be noted that all solutions in the population have the same value of $\Delta_S \boldsymbol{x}_i(t)$.

Type3-2: Use of the Current and Previous Moves. In Type3-2 MOEA/D, we use the current and previous moves in the population S to generate candidates. We define the previous improving direction using the latest improved generation k before the tth generation as $\Delta \boldsymbol{x}_i(k)$ and $\Delta_S \boldsymbol{x}_i(k)$, where k is the latest improved generation ($1<k<t$) where at least one solution in the population is improved before the tth generation. The candidate solution is defined as $\boldsymbol{x}_i^{Candidate} = \boldsymbol{x}_i(t) + \eta \Delta \boldsymbol{x}_i(t) + \alpha \Delta \boldsymbol{x}_i(k) + \eta_S \Delta_S \boldsymbol{x}_i(t) + \alpha_S \Delta_S \boldsymbol{x}_i(k)$, where η, η_S, α and α_S are non-negative constant parameters.

To speed up the convergence speed, we try to find a candidate solution in the improving direction. Parameter values decide the position of a candidate solution in the improving direction. For simplicity, in this paper, we set η and α as 1 since we consider that the total move of the solution during each generation has the same weight. When using the total move in the neighborhood or population, we add all moves in the neighborhood or population together. Since the neighborhood and population size may affect the position in the improving direction, we set η_{S_i} and α_{S_i} as $1/|S_i|$, and set η_S and α_S as $1/|S|$.

3   Experimental Study

To examine the usefulness of the proposed strategy (its seven implementations), we use a multi-objective distance minimization problem (MDMP) in the 2-dimensional space [11,12]. The effect can be visually examined by drawing the trajectory of the current solutions. The distance minimization problem is generated by using the following four points in the 2-dimensional space $[1, 1001] \times [1, 1001]$: (2, 6), (6, 2), (2, 2), (6, 6). The four points are intentionally placed in a small region around the corner (1, 1) of the 2-dimensional space in order to examine the effect of the proposed strategy in comparison with the standard implementation of MOEA/D. In this problem, we need to minimize four objectives, which are f_1: Distance to P_1, f_2: Distance to P_2, f_3: Distance to P_3, and f_4: Distance to P_4.

Experimental settings for the 2-dimensional MDMP problem are as follows:

Software Platform. We use PlatEMO [13] as the experimental platform. PlatEMO is an open-source platform based on MATLAB for evolutionary multi-objective optimization.

Parameter Settings. Population size N is set to 56. This setting is based on the number of weight vectors generated by the Das and Dennis method [14]. The termination condition is set to 560 solution evaluations. Each algorithm is applied to each test problem for 31 independent runs.

Performance Metrics. The IGD [15] and IGD$^+$ [16] indicators are used to evaluate the performance of each algorithm.

The experimental results are shown in Table 1. The average values of IGD and IGD$^+$ over 31 runs are summarized in the table. Each algorithm is compared with the standard MOEA/D using the Wilcoxon rank sum test with the significance level of 0.05, in which the symbol "+" means that the compared algorithm is significantly better than the standard MOEA/D, the symbol "−" means that the compared algorithm is significantly worse than the standard MOEA/D, and the symbol "=" means that there is no statistically significant difference between the compared algorithm and the standard MOEA/D. The statistical test results are summarized at the bottom of each table. The best result is highlighted by blue font, and the worst result is highlighted by red font.

As Table 1 shows, almost all algorithms perform well on MDMP. Although Type2-2 MOEA/D performs the worst among all algorithms, there is no statistically significant difference between it and the standard MOEA/D.

To clearly show the convergence ability of MOEA/D with the proposed strategy and the standard MOEA/D, we use Fig. 5 to show the relation between the average IGD$^+$ value (y-axis) and the number of examined solutions (x-axis) for each algorithm.

Table 1. Average IGD$^+$ and IGD Values on MDMP ($d = 2$) obtained by MOEA/D with the proposed strategy and the standard MOEA/D.

Indicator	Type1-1	Type1-2	Type1-2*	Type2-1	Type2-2	Type3-1	Type3-2	MOEA/D
IGD$^+$	0.6962+	0.7265=	0.6416+	0.8352=	0.9375=	0.7526=	0.7925=	0.8004
IGD	1.0345+	1.0950=	0.9932+	1.2195=	1.3138=	1.1216=	1.1941=	1.2015
$+/-/=$	2/0/0	0/0/2	2/0/0	0/0/2	0/0/2	0/0/2	0/0/2	

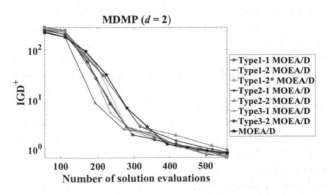

Fig. 5. Average IGD$^+$ value of the current population at each generation over 31 runs on MDMP ($d = 2$).

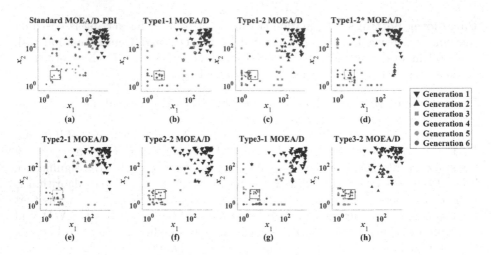

Fig. 6. The current population in the decision space of each algorithm at the first six generations on MDMP ($d = 2$). (Color figure online)

As shown in Fig. 5, MOEA/D with the proposed strategy clearly converges faster than the standard MOEA/D before 300 solution evaluations.

To show the convergence trajectory of each algorithm, we choose a single run with the median IGD^+ value among the 31 runs and plot the population in the decision space at each of the first 6 generations in Fig. 6 (a)–(h). The blue square frame represents the Pareto set. The triangles, squares and dots represent the solutions in the decision space. In Fig. 6, more red dots in the blue square means faster convergence of the algorithm. In Fig. 6 (a), only a small number of red dots are in the blue square frame. However, in Fig. 6 (b)–(h), more red dots are in the blue square frame, which indicates that the proposed strategy can clearly speed up the convergence of MOEA/D in MDMP ($d = 2$).

To further examine the usefulness of the proposed strategy, we use four large-scale MDMPs [17, 18] to test the performance of the MOEA/D with the proposed strategy (its seven implementations). Their decision spaces are 10-, 100-, 500-, and 1000-dimensional, respectively. The decision space of each problem is [0, 100]×[0, 100]× ... ×[0, 100]. Each problem uses the following four points P_1 (1, 1, 0, ..., 0), P_2 (5, 1, 0, ..., 0), P_3 (1, 5, 0, ..., 0), P_4 (5, 5, 0, ..., 0). The four points are intentionally placed in a small region around the corner (0, 0, ..., 0) in order to examine the effect of the proposed strategy in comparison with the standard implementation of MOEA/D. Furthermore, the usefulness of the proposed strategy is also examined on the large-scale three-objective DTLZ1-4 test problems with $d = 500$ and 1000 where d is the number of decision variables.

Our experimental settings are as follow. Population size N is set to 120 on MDMP ($d = 10, 100, 500, 1000$) and 91 on DTLZ1-4 ($d = 500, 1000$). This setting is based on the number of weight vectors generated by the Das and Dennis method [14]. The termination condition is set to 6000, 12000, 60000, and 120000 solution evaluations for MDMP with $d = 10, 100, 500$ and 1000,

Table 2. Average IGD$^+$ and IGD Values on MDMP ($d = 10, 100, 500, 1000$) obtained by the MOEA/D with the proposed strategy and the standard MOEA/D.

Problem	Indicator	Type1-1	Type1-2	Type1-2*	Type2-1	Type2-2	Type3-1	Type3-2	MOEA/D
MDMP	IGD$^+$	0.1965+	0.1932+	0.1935+	0.1997+	0.1974+	0.2032+	0.1989+	0.2199
$d = 10$	IGD	0.3250+	0.3174+	0.3175+	0.3350+	0.3255+	0.3407+	0.3325+	0.3710
MDMP	IGD$^+$	80.770+	6.1398+	16.572+	61.579+	3.4040+	13.828+	2.8008+	262.49
$d = 100$	IGD	80.770+	6.2429+	16.639+	61.579+	3.6002+	13.840+	2.9816+	262.49
MDMP	IGD$^+$	245.75+	23.197+	70.410+	194.20+	24.354+	85.562+	14.220+	782.03
$d = 500$	IGD	245.75+	23.247+	70.419+	194.20+	24.397+	85.562+	14.221+	782.03
MDMP	IGD$^+$	371.12+	59.208+	139.32+	295.12+	43.213+	136.03+	24.485+	1159.9
$d = 1000$	IGD	371.12+	59.219+	139.32+	295.12+	43.230+	136.03+	24.485+	1159.9
+/−/=		8/0/0	8/0/0	8/0/0	8/0/0	8/0/0	8/0/0	8/0/0	

respectively, and 10000 solution evaluations for DTLZ1-4 with $d = 500$ and $d = 1000$. Each algorithm is applied to each test problem for 31 independent runs.

Experimental results on MDMP ($d = 10, 100, 500, 1000$) and DTLZ1-4 ($d = 500, 1000$) are summarized in Tables 2 and 3, respectively.

In Table 2, MOEA/D with any implementation of the proposed strategy performs clearly better than the standard MOEA/D on the large-scale MDMP. Especially, Type3-2 MOEA/D performs clearly the best among all algorithms. In Table 3, the proposed strategy performs clearly better than the standard MOEA/D on the large-scale DTLZ1 and DTLZ3. However, Type1 and Type2 MOEA/D are slightly worse than the standard MOEA/D on DTLZ2 and DTLZ4.

Table 3. Average IGD$^+$ and IGD Values on DTLZ1-4 ($d = 500, 1000$) obtained by the MOEA/D with proposed strategy and the standard MOEA/D.

Problem	Indicator	Type1-1	Type1-2	Type1-2*	Type2-1	Type2-2	Type3-1	Type3-2	MOEA/D
DTLZ1	IGD$^+$	3993.6+	4116.7+	4018.8+	4125.4+	4100.9+	6099.7+	5383.6+	8888.8
$d = 500$	IGD	3993.6+	4116.7+	4018.8+	4125.4+	4100.9+	6099.7+	5383.6+	8888.8
DTLZ1	IGD$^+$	8398.8+	8528.4+	8341.2+	8562.4+	8458.0+	13750+	12297+	22017
$d = 1000$	IGD	8398.8+	8528.4+	8341.2+	8562.4+	8458.0+	13750+	12297+	22017
DTLZ2	IGD$^+$	17.238−	17.315−	17.948−	19.439−	20.212−	15.447=	15.937−	14.910
$d = 500$	IGD	17.238−	17.316−	17.949−	19.440−	20.213−	15.448=	15.937−	14.911
DTLZ2	IGD$^+$	48.929−	49.394−	48.678−	51.909−	52.703−	45.575−	45.847−	44.273
$d = 1000$	IGD	48.929−	49.395−	48.678−	51.910−	52.703−	45.575−	45.848−	44.274
DTLZ3	IGD$^+$	12993+	13134+	12957+	13304+	13051+	20050+	18446+	29554
$d = 500$	IGD	12993+	13134+	12957+	13304+	13051+	20050+	18446+	29554
DTLZ3	IGD$^+$	27139+	27667+	26864+	27382+	27132+	45599+	39454+	74475
$d = 1000$	IGD	27139+	27667+	26864+	27382+	27132+	45599+	39454+	74475
DTLZ4	IGD$^+$	21.359−	19.885−	20.306−	20.505−	22.361−	16.608=	16.949=	17.311
$d = 500$	IGD	21.366−	19.894−	20.314−	20.513−	22.367−	16.622=	16.960=	17.324
DTLZ4	IGD$^+$	54.717−	56.761−	57.274−	58.057−	57.015−	52.648−	53.349−	50.683
$d = 1000$	IGD	54.720−	56.764−	57.276−	58.060−	57.018−	52.651−	53.352−	50.687
+/−/=		8/8/0	8/8/0	8/8/0	8/8/0	8/8/0	8/4/4	8/6/2	

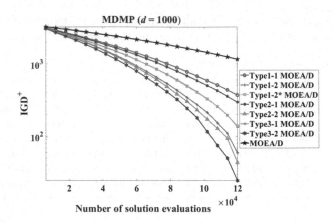

Fig. 7. Average IGD$^+$ value of the current population at each generation over 31 runs on MDMP ($d = 1000$).

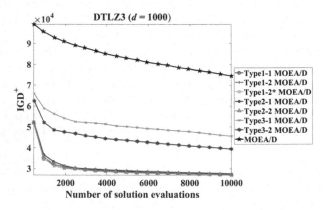

Fig. 8. Average IGD$^+$ value of the current population at each generation over 31 runs on DTLZ3 ($d = 1000$).

Figures 7 and 8 show the relation between the average IGD$^+$ value (y-axis) and the number of examined solutions (x-axis) obtained by the standard MOEA/D and MOEA/D with the proposed strategy on MDMP ($d = 1000$) and DTLZ3 ($d = 1000$).

As shown in Figs. 7 and 8, MOEA/D with any implementation of the proposed strategy converges much faster than the standard MOEA/D. Type3-2 MOEA/D clearly converges the fastest on MDMP ($d = 1000$). However, on DTLZ3 ($d = 1000$), Type3 MOEA/D performs not as well as the MOEA/D with the other implementations (whereas Type3 MOEA/D is much faster than the standard MOEA/D). By comparing between Type1-1 and Type1-2 (and comparing between Type2-1 and Type2-2, and between Type3-1 and Type3-2), we can conclude that the use of the current and previous moves can help MOEA/D converge faster than the use of only the current move. By comparing Type1-2 with Type1-2*, we can conclude that using the move in the $(t\text{-}1)^{th}$ generation even if $\Delta x_i(t-1) = 0$ is not as efficient as using the move in the latest improved generation before the tth generation.

4 Conclusion and Future Work

In this paper, we proposed a novel solution generation operator for MOEA/D. By using the moves of solutions in the current and previous generations, we can generate promising candidate solutions. The experimental studies showed that the proposed strategy significantly speeds up the convergence speed of MOEA/D. In the future, we will compare our proposed algorithms with some state-of-the-art large-scale multi-objective evolutionary algorithms.

One future research topic is to investigate the sensitivity of the performance of the proposed strategy to parameter settings. It is also possible to use a random parameter value instead of a fixed parameter value in the proposed strategy. Another future research topic is to examine the use of information from unsuccessful move attempts where the current solution is not updated.

Acknowledgements. This work was supported by National Natural Science Foundation of China (Grant No. 61876075), Guangdong Provincial Key Laboratory (Grant No. 2020B121201001), the Program for Guangdong Introducing Innovative and Enterpreneurial Teams (Grant No. 2017ZT07X386), The Stable Support Plan Program of Shenzhen Natural Science Fund (Grant No. 20200925174447003), Shenzhen Science and Technology Program (Grant No. KQTD2016112514355531).

References

1. Deb, K.: Multi-objective Optimization Using Evolutionary Algorithms. Wiley, Chichester (2001)
2. Coello, C.A.C., Lamont, G.B., Veldhuizen, D.A.V.: Evolutionary Algorithms for Solving Multi-objective Problems. Springer, New York (2007). https://doi.org/10.1007/978-0-387-36797-2
3. Zhang, Q., Li, H.: MOEA/D: a multiobjective evolutionary algorithm based on decomposition. IEEE Trans. Evol. Comput. **11**(6), 712–731 (2007)
4. Mittal, S., Saxena, D.K., Deb, K., Goodman, E.D.: A learning-based innovized progress operator for faster convergence in evolutionary multi-objective optimization. ACM Trans. Evol. Learn. Optim. **2**(1), 1–29 (2021)
5. Ghosh, A., Deb, K., Averill, R., Goodman, E.: Combining user knowledge and online *innovization* for faster solution to multi-objective design optimization problems. In: Ishibuchi, H., et al. (eds.) EMO 2021. LNCS, vol. 12654, pp. 102–114. Springer, Cham (2021). https://doi.org/10.1007/978-3-030-72062-9_9
6. Mittal, S., Saxena, D.K., Deb, K., Goodman, E.D.: Enhanced innovized repair operator for evolutionary multi- and many-objective optimization. arXiv preprint arXiv:2011.10760 (2020)
7. Ghosh, A., Goodman, E.D., Deb, K., Averill, R., Diaz, A.: A large-scale bi-objective optimization of solid rocket motors using innovization. In: 2020 IEEE Congress on Evolutionary Computation (CEC 2020), pp. 1–8 (2020)

8. Mittal, S., Saxena, D.K., Deb, K.: A unified automated innovization framework using threshold-based clustering. In: 2020 IEEE Congress on Evolutionary Computation (CEC 2020), pp. 1–8 (2020)
9. Mittal, S., Saxena, D.K., Deb, K.: Learning-based multi-objective optimization through ANN-assisted online innovization. In: Proceedings of the 2020 Genetic and Evolutionary Computation Conference Companion (GECCO 2020), pp. 171–172 (2020)
10. Garg, K., Mukherjee, A., Mittal, S., Saxena, D.K., Deb, K.: A generic and computationally efficient automated innovization method for power-law design rules. In: Proceedings of the 2020 Genetic and Evolutionary Computation Conference Companion (GECCO 2020), pp. 161–162 (2020)
11. Ishibuchi, H., Hitotsuyanagi, Y., Tsukamoto, N., Nojima, Y.: Many-objective test problems to visually examine the behavior of multiobjective evolution in a decision space. In: Schaefer, R., Cotta, C., Kołodziej, J., Rudolph, G. (eds.) PPSN 2010. LNCS, vol. 6239, pp. 91–100. Springer, Heidelberg (2010). https://doi.org/10.1007/978-3-642-15871-1_10
12. Ishibuchi, H., Akedo, N., Nojima, Y.: A many-objective test problem for visually examining diversity maintenance behavior in a decision space. In: Proceedings of the 2011 Genetic and Evolutionary Computation Conference Companion (GECCO 2011), pp. 649–656 (2011)
13. Tian, Y., Cheng, R., Zhang, X., Jin, Y.: PlatEMO: a MATLAB platform for evolutionary multi-objective optimization [Educational Forum]. IEEE Comput. Intell. Mag. 12(4), 73–87 (2017)
14. Das, I., Dennis, J.E.: Normal-boundary intersection: a new method for generating the pareto surface in nonlinear multicriteria optimization problems. SIAM J. Optim. 8(3), 631–657 (1998)
15. Coello, C.A.C., Cortés, N.C.: Solving multiobjective optimization problems using an artificial immune system. Genet. Program Evolvable Mach. 6(2), 163–190 (2015)
16. Ishibuchi, H., Masuda, H., Tanigaki, Y., Nojima, Y.: Modified distance calculation in generational distance and inverted generational distance. In: Gaspar-Cunha, A., Henggeler Antunes, C., Coello, C.C. (eds.) EMO 2015. LNCS, vol. 9019, pp. 110–125. Springer, Cham (2015). https://doi.org/10.1007/978-3-319-15892-1_8
17. Ishibuchi, H., Yamane, M., Akedo, N., Nojima, Y.: Many-objective and many-variable test problems for visual examination of multiobjective search. In: 2013 IEEE Congress on Evolutionary Computation (CEC 2013), pp. 1491–1498 (2013)
18. Masuda, H., Nojima, Y., Ishibuchi, H.: Visual examination of the behavior of EMO algorithms for many-objective optimization with many decision variables. In: 2014 IEEE Congress on Evolutionary Computation (CEC 2014), pp. 2633–2640 (2014)

Obtaining Smoothly Navigable Approximation Sets in Bi-objective Multi-modal Optimization

Renzo J. Scholman[1,3]($\boxtimes$)(iD), Anton Bouter[1](iD), Leah R. M. Dickhoff[2](iD),
Tanja Alderliesten[2](iD), and Peter A. N. Bosman[1,3](iD)

[1] Centrum Wiskunde and Informatica, Amsterdam, The Netherlands
{Renzo.Scholman,Anton.Bouter,Peter.Bosman}@cwi.nl
[2] Leiden University Medical Center, Leiden, The Netherlands
{L.R.M.Dickhoff,T.Alderliesten}@lumc.nl
[3] Delft University of Technology, Delft, The Netherlands

Abstract. Even if a Multi-modal Multi-Objective Evolutionary Algorithm (MMOEA) is designed to find solutions well spread over all locally optimal approximation sets of a Multi-modal Multi-objective Optimization Problem (MMOP), there is a risk that the found set of solutions is not smoothly navigable because the solutions belong to various niches, reducing the insight for decision makers. To tackle this issue, a new MMOEAs is proposed: the Multi-Modal Bézier Evolutionary Algorithm (MM-BezEA), which produces approximation sets that cover individual niches and exhibit inherent decision-space smoothness as they are parameterized by Bézier curves. MM-BezEA combines the concepts behind the recently introduced BezEA and MO-HillVallEA to find all locally optimal approximation sets. When benchmarked against the MMOEAs MO_Ring_PSO_SCD and MO-HillVallEA on MMOPs with linear Pareto sets, MM-BezEA was found to perform best in terms of best hypervolume.

Keywords: Evolutionary algorithms · Multi-modal multi-objective optimization · Niching · Bézier curve estimation

1 Introduction

Many real-world optimization problems have multiple conflicting objectives, whereby improvement in one objective often results in the deterioration of another. Multi-Objective Evolutionary Algorithms (MOEAs), like NSGA-II [9], MOEA/D [36], and MO-CMA-ES [17], are widely accepted to be well-suited to solve such Multi-objective Optimization Problems (MOPs) [11]. The aim is to obtain a set of solutions, called the approximation set, such that all solutions are non-dominated and the set itself is close to the set of Pareto-optimal solutions.

Leah R.M. Dickhoff was supported by the Dutch Cancer Society (KWF Kankerbestrijding, Project N.12183) and Elekta.

G. Rudolph et al. (Eds.): PPSN 2022, LNCS 13399, pp. 247–262, 2022.
https://doi.org/10.1007/978-3-031-14721-0_18

Here, a solution x_0 dominates x_1 ($x_0 \succ x_1$) in an MOP with m objectives if $\forall i \in \{0, 1, ..., m-1\} : f_i(x_0) \leq f_i(x_1)$ and $\exists i \in \{0, 1, ..., m-1\} : f_i(x_0) < f_i(x_1)$. The Pareto Set (PS) is $\mathcal{P}_S = \{x_i | \neg \exists x_j : x_j \succ x_i\}$ and the Pareto Front (PF) is $\mathcal{P}_F = \{(f_0(x), \cdots, f_{m-1}(x)) | x \in \mathcal{P}_S\}$.

A more complex type of MOPs is that of Multi-modal MOPs (MMOPs), where the goal is not to find one, but multiple, if not all, (local) PSs. In MMOPs, each of the PSs pertains to a *niche*, a subset of the search space, where a single mode resides, i.e., with one local PS. The PSs may, however, map to the same PF in objective space, similar to having multiple (locally) optimal solutions of the same quality in a single-objective problem, e.g., the sine function. Here we consider MMOPs in the case of real-valued parameters, or continuous optimization. This field has recently gotten more traction, with reviews [31], proposed formal definitions [14] and new visualization techniques [30].

In order to have MOEAs solve MMOPs, they need additional tools that prevent their convergence to a single niche in the landscape [22]. Niching [19] is one of such diversity-preserving tools used by Multi-modal MOEAs (MMOEAs) to effectively and simultaneously search for solutions near the (local) PS in each niche. Niching has been successfully applied to established MMOEAs in the form of the multi-objective particle swarm optimization using ring topology and special crowding distance (MO_Ring_PSO_SCD) algorithm [35] and Omni-optimizer [12] among others.

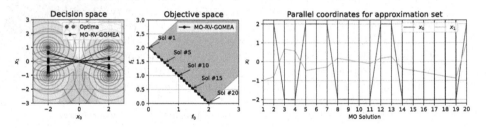

Fig. 1. Approximation set and front with parallel coordinates plot as produced by MO-RV-GOMEA on the MinDist problem: $f_0(x) = min(||x - [1, -1]||, ||x - [-1, 1]||)$ and $f_1(x) = min(||x - [1, 1]||, ||x - [-1, -1]||)$. Shaded blue and red regions correspond to niches with global PSs. (Color figure online)

Most MMOEAs do not explicitly model multiple approximation sets, but include diversity preserving techniques to ensure that solutions from multiple niches are maintained. The result of these MMOEAs is usually given in the form of a single approximation front, often derived from (a subset of) the elitist archive. A decision maker can then investigate this front by traversing the solutions for desired trade-offs. However, the underlying solutions are taken from several distinct niches, which could result in observing a counterintuitive change in decision variable values when navigating the approximation front. Decision makers then might have to investigate all solutions before a correct choice can be made [21,24]. Figure 1 shows such an approximation front and set that contains solutions from both modes on the MinDist problem [23] and demonstrates the counterintuitive changes in decision variable values in the parallel coordinates plot (i.e., if one were to navigate the front by traversing and inspecting the solutions from one extreme to the other). It shows that in objective space a front is found that looks to have approximated the PF to (near) optimality, but the solutions jump around throughout decision space as seen in the parallel coordinates plot.

The issue of counterintuitive navigation along the approximation front has also been explored in recent work, which introduced a new indicator-based MOEA for bi-objective optimization called BezEA [24]. A new problem formulation for population-based MOEAs was introduced whereby they parameterized approximation sets as Bézier curves. This formulation ensures the navigational smoothness of an approximation set, whilst still being able to find good approximation sets when using the HyperVolume indicator (HV) [37]. By design, BezEA disallowes curves to dominate parts of themselves to ensure that the approximation set constitutes a single niche in the landscape.

Recent work that included the concept of niching showed promising results in maintaining multiple approximation sets in a population-based MMOEA called the MO-HillVallEA [23]. The authors extended the concept of Hill-Valley Clustering (HVC) [26] for MOPs to Multi-Objective HVC (MO-HVC) for MMOPs. MO-HillVallEA was found to be capable of finding and preserving approximation sets, one for each niche, in parallel over time by considering Pareto domination per niche. However, MO-HillVallEA produces approximation sets that are not inherently smooth due to slight oscillations around the PS.

In this work, the notions of niching through HVC and Bézier curve parameterizations are combined. The use of niching allows to effectively search the multimodal landscape. The use of Bézier curve parameterizations not only enforces the smooth and intuitive navigability that is desired by decision makers, but also enforces each approximation set to be within a single niche. Furthermore, the use of the HV indicator allows closer convergence to the PS as compared to the Pareto dominance-based algorithms [4]. The new algorithm that we propose is called Multi-Modal Bézier Evolutionary Algorithm (MM-BezEA). The purpose of MM-BezEA is to find all approximation sets for a given MMOP, where each approximation set consists of solutions from a single mode.

In order to combine the techniques of Bézier curve parameterizations and HVC into the proposed algorithm, several contributions are made. First, the problem of how to niche approximation sets in the form of Bézier curve parameterizations is resolved. Second, initialization of approximation sets within a single niche is enabled, as otherwise, clustering becomes ambiguous if these approximation sets span multiple niches.

2   Bézier parameterizations

One of the key features of the newly proposed algorithm is that Bézier parameterizations are used as approximation sets for bi-objective optimization [24]. This allows the algorithm to model the approximation set as a smooth curve in decision space.

2.1   Definition of Solution Set

An ℓ-dimensional Bézier curve $\mathbf{B}(t; C_q)$ can be defined using $q \geq 2$ control points c_i in an ordered set $C_q = \{c_1, ..., c_q\}$, where ℓ is the problem dimensionality and $c_i \in \mathbb{R}^\ell$. The full notation is:

$$\mathbf{B}(t; C_q) = \sum_{i=1}^{q} \binom{q-1}{i-1} (1-t)^{q-i-1} t^{i-1} c_i \text{ for } 0 \leq t \leq 1 \qquad (1)$$

The endpoints of the Bézier curve are always defined by the first and last control points, whilst the other control points are normally not located on the curve. A solution set of given size p, $S_{p,q}(C_q) = \{x_1, ..., x_p\}$ with $x_i \in \mathbb{R}^\ell$, can now be parameterized by a Bézier curve by selecting an evenly spread set of p points x_i. Figure 2 visualizes two solution sets $S_{p,q}(C_q)$ parameterized by Bézier curves. The solution set $S_{p,q}(C_q)$ is formally defined as:

$$S_{p,q}(C_q) = \left\{ \mathbf{B}\left(\frac{0}{p-1}; C_q\right), \mathbf{B}\left(\frac{1}{p-1}; C_q\right), ..., \mathbf{B}\left(\frac{p-1}{p-1}; C_q\right) \right\} \qquad (2)$$

$S_{p,q}(C_q)$ is parameterized for (M)MOEAs by taking the concatenation of the decision variables in the set of control points C_q as a solution [5,24]. This results in a solution being of the form $[c_1, ..., c_q] \in \mathbb{R}^{q \times \ell}$.

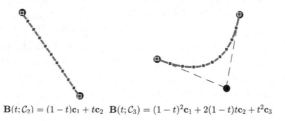

$$\mathbf{B}(t; C_2) = (1-t)c_1 + tc_2 \qquad \mathbf{B}(t; C_3) = (1-t)^2 c_1 + 2(1-t)tc_2 + t^2 c_3$$

Fig. 2. Bézier curves with $q \in \{2, 3\}$ control points in black. Interpolated curve in red with the $p = 11$ points in blue evenly spread in the domain of t along the curve [24]. (Color figure online)

2.2   Evaluation

To evaluate a solution set $S_{p,q}(C_q)$, a number of new functions were previously introduced [24]. These functions are briefly explained in the following paragraphs. Figure 3 illustrates these functions to give the reader a more graphical indication.

A new function $A^{nb}(S_{p,q})$ has been introduced that calculates a navigational Bézier (nb) order o_{nb}. This order is defined as starting from the best solution for objective f_0 to the best solution in f_1. All solutions that are dominated by other solutions on the curve, are omitted from the subset that defines the navigational order. An approximation set $\mathcal{A}_{p,q,o_{nb}}$ is the resulting subset of $S_{p,q}(C_q)$, with only the solution indices as specified in o_{nb}. The quality of the approximation set $\mathcal{A}_{p,q,o_{nb}}$ can now be evaluated, e.g., with the HV indicator [37].

A new constraint func-
tion $C\left(\mathcal{S}_{p,q}, o_{nb}\right)$ was also
introduced. It is employed
in order to not only push
all dominated solutions
on the curve towards
the undominated region
of the search space, but
also to prevent the curve
from intersecting itself in
objective space. This may
for instance happen if a

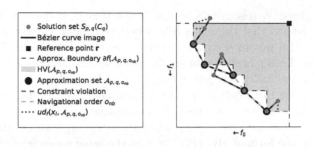

Fig. 3. Evaluation of Bézier parameterizations [24] (Color figure online)

curve stretches across two local PSs, which is not preferential. The constraint function uses the uncrowded distance metric $ud_f(\boldsymbol{x}_i, \mathcal{A})$ [33], which measures the Euclidean distance from a dominated point $\boldsymbol{x}_i$ to the approximation boundary $\partial f(\mathcal{A}_{p,q,o_{nb}})$ in objective space. Furthermore, to further increase pressure towards the unfolding of Bézier curves in objective space, all dominated solutions and those not in $\mathcal{A}_{p,q,o_{nb}}$ are pulled towards their neighbouring solutions on the Bézier curve by taking the Euclidean distance in objective space between these solution and their neighbours as an additional constraint value. All dominated solutions from $\mathcal{S}_{p,q}(C_q)$ now have their uncrowded distance values and the Euclidean distances in objective space to neighbours of those not in $\mathcal{A}_{p,q,o_{nb}}$ summed up as a constraint for the total solution set. In combination with constraint domination [10], this constraint pushes all solutions along the Bézier curve towards the undominated region.

3   Niching Methods

To enable the algorithm proposed in this paper, i.e., MM-BezEA, to effectively search the multi-modal landscape, several previously introduced niching methods are used and combined. These are employed in order to extend the uni-modal search that is originally performed by BezEA. As the number of modes is usually unidentified beforehand, the algorithm needs to be able to adapt to the number of modes present in an MMOP.

3.1   HVC and MO HVC

HVC is a so-called two-stage niching approach that clusters and evolves the population for multi-modal single-objective optimization problems. In each generation, the first stage is used to locate each of the distinct niches, for each of which a core search algorithm is initialized in the second stage.

At the heart of the HVC approach is the Hill-Valley Test (HVT) [34], which can be utilized to determine whether two solutions reside in the same niche. It first determines an edge between two solutions $\boldsymbol{x}_i$ and $\boldsymbol{x}_j$ in the search space. Along this edge, N_t evenly spread points are evaluated, determined by the distance between the two solutions divided by the expected edge length. If any of these N_t test points have a fitness that is worse than that of x_i and x_j, the test detects that there is a *hill* in between them. Consequently, the two solutions are to be put in separate clusters. On the other hand, if all N_t points have equal or better fitness values than both x_i and x_j, these two solutions belong to the same *valley* and are to be clustered together. In order to determine in which order the solutions are to be clustered (i.e., undergo the HVT), the concept of the nearest better tree [27] is employed.

The MO-HillVallEA algorithm [23] expands on the previous HVC approach in the form of MO-HVC. It uses the same concept of the HVT, but now performs clustering for each of the m objective functions separately, which results in m cluster sets. To obtain a single cluster set, the intersection of each pair of clusters from all m clustering sets is taken, similar to the colored regions of Fig. 1.

3.2 Restart Scheme with Elitist Archive

Various algorithms implemented a form of a restart scheme whereby the population size is increased over time. Examples of such schemes are the interleaved multistart scheme [8,16] and the restart-Covariance Matrix Adaptation Evolution Strategy with Increased Population (IPOP-CMA-ES) algorithm [3]. In HillVallEA [26], an elitist archive is combined with a restart scheme, where the population size is doubled after each restart as in IPOP-CMA-ES [3]. By employing the HVT to check if a solution resides in another niche, the elitist archive of HillVallEA is capable of holding the elites for each of the modes, despite it being developed for single objective problems.

To prevent HillVallEA from revisiting already searched modes, it makes use of the elitist archive, which is inspired by the repelling subpopulations (RS-CMSA) algorithm [1] that defines taboo regions close to elites. The steps taken to discard the regions of the search space, for which an elite was already found in one of the earlier populations, start with adding the elites to the population of the current restart. Then, all solutions are clustered using HVC, followed by discarding all clusters that have one of the elites as their best solution. As a result of discarding these regions of the search space, more attention is given to undiscovered parts of the search space after each restart.

4   Multi Modal-Bézier Evolutionary Algorithm

In this section MM-BezEA is described. MM-BezEA is comprised of a combination and modification of techniques described in the previous sections. The most notable of the modifications are the adjustments implemented in HVC in order to apply it to Bézier curve parameterizations, as well as the initialization of approximation sets within niches.

4.1   Clustering Approximation Sets

The Bézier curves are evaluated using the uncrowded HV measure [25]. Since this is a scalar, the HVC approach seems to intuitively allow the clustering of single-objective problems. However, the approximation set $\mathcal{A}_{p,q,o_{nb}}$ that is used in the HV calculation only considers the undominated indices of the Bézier solution set $S_{p,q}(C_q)$ as defined in o_{nb}. Hence, the objective value of a solution set $S_{p,q}(C_q)$ seems highly dependant on how many dominated solutions there are on the Bézier curve due to its orientation and length in decision space.

To enable the clustering of Bézier solution sets $S_{p,q}(C_q)$, the idea behind MO-HVC can be used on the set of control points C_q, as each of these is a single solution as normally defined in MO optimization. Also, since a solution set is defined to be deteriorating in f_0 and improving in f_1 according to o_{nb}, the order of the control points is inverted if $f_0(c_1) < f_0(c_q)$ does not hold [24]. Accordingly, the i-th Bézier solution can be designated to be in the same niche as the j-th Bézier solution if their control points $c_l^i \in C_q^i$ and $c_l^j \in C_q^j$ for $l = \{1, ..., q\}$ are in the same niche. In a general sense, the same HVC approach as used in HillVallEA is used, but inspiration has been taken from the MO-HVC approach to produce a new test for Bézier solution sets, which is shown in Algorithm 1.

Algorithm 1: $[B]$ = Bezier-HillValleyTest($\mathbf{S_i}$, $\mathbf{S_j}$, N_t, f)

Input: Solutions sets $\mathbf{S_i}$, $\mathbf{S_j}$, int N_t, objective functions $f_0, ..., f_{m-1}$
Output: Whether $\mathbf{S_i}$ and $\mathbf{S_j}$ belong to the same niche
for $l = 1, ..., q$ **do**
    $c_{i,l}, c_{j,l} \leftarrow$ control point l of $\mathbf{S_i}$, control point l of $\mathbf{S_j}$
    // Check if $c_{i,l}$ and $c_{j,l}$ are in same niche for all m objectives
    for $k = 0, ..., m - 1$ **do**
        if *HillValleyTest($c_{i,l}$, $c_{j,l}$, N_t, f_k)* **then return** false
return true

4.2   Initialization Within Niches

The original BezEA algorithm initializes all solution sets by sampling from a uniform distribution over the search space. As it is an MOEA that was not designed for multi-modal optimization, the uniform initialization allows solution sets to be initialized within or in between any niche(s). Clustering these solutions with the newly introduced Bézier HVT will result in finding a large number of separate niches as each control point has to be in the same mode. To prevent this, a new initialization method for Bézier solution sets is proposed to enforce their initialization within a niche. First, an iteration of MO-HVC is run on a set of $q \times N$ solutions, N being the population size, that is sampled from a uniform distribution over the search space, where the resulting clusters include all of the x_{test} solutions resulting from applying the Hill-Valley Test. Second, selection

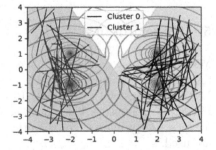

Fig. 4. Initialization of Bézier solutions ($q = 2$) for MinDist

is performed for each cluster proportional to their size in order to reduce their combined size, with test solutions, down to $q \times N$. Lastly, Bézier solution sets $S_{p,q}(C_q)$ are initialized by randomly choosing q solutions as the control points from one single cluster C as produced by MO-HVC if $|C| \geq q$. The result can be seen in Fig. 4 in the case of the example problem MinDist.

4.3 Algorithm Overview

MM-BezEA has a similar structure as the restart scheme in HillVallEA [26] that is described in Sect. 3.2. Every iteration, the combination of initialization of Bézier curves and dismissal of previously optimized clusters with an elite as their best solution takes place. For each of the resulting niches, a core search algorithm is run for one generation, which in the case of MM-BezEA is the RV-GOMEA algorithm [7] that is also used in BezEA. At the end of each generation, the Bézier HVT of Algorithm 1 is used in the HVC

Algorithm 2: $[\mathbb{E}]$ = MM-BezEA(...)

Input: MO function f, popsize N, test points p, control points q, budget

Output : Elitist archive $\mathbb{E} = [\mathcal{E}_0, \mathcal{E}_1, ...]$

$\mathbb{E} = \{\}$

while *budget remaining* **do**

  $\mathcal{P}_{mo}$ = UniformSampling($q \times N$, f)

  $\mathbb{C}_{mo}$ = MO-HillValleyClustering($\mathcal{P}_{mo}$, f)

  $\mathbb{C}$ = InitializeBezierSolutions($\mathbb{C}_{mo}$, q, p, f)

  $\mathbb{C}$ = RemoveElitesFrom($\mathbb{C}$)

  while *budget remaining* **do**

    $\mathcal{P} = \mathbb{E}$

    for $C_i \in \mathbb{C}$ **do**

      $\mathcal{O}_i$ = core_search_algorithm(C_i)

      $\mathcal{P} = \mathcal{P} \cup \mathcal{O}_i$

    $\mathbb{C}_{prev} = \mathbb{C}$

    $\mathbb{C}$ = BezierHillValleyClustering($\mathcal{P}$, f)

    $\mathbb{E}$ = ConstructElitistArchive($\mathbb{C}$, $\mathbb{E}$)

    $\mathbb{C}$ = RemoveElitesFrom($\mathbb{C}$)

    $\mathbb{C}$ = ClusterRegistration($\mathbb{C}$, $\mathbb{C}_{prev}$)

---

step. This step takes all solutions originating from all clusters and clusters them again for the next generation. In between generations, the notion of cluster registration [6] is used on the cluster mean closest in decision space to transfer the parameters for RV-GOMEA between the clusters of each generation. An overview of the algorithm in the form of pseudocode is given in Algorithm 2.

5   Experiments

MM-BezEA is empirically benchmarked on several test problems. The results are compared to MO-HillVallEA [23], MO-RV-GOMEA [8], and MO_Ring_PSO_SCD [35]. MO_Ring_PSO_SCD is implemented through the PlatEMO framework [32], together with a manual implementation of the used metrics and problems. For the other algorithms, original C++ implementations are used.

5.1 Test Problems

Several test problems are employed. First of these is the MinDist problem [23] that was described in the introduction, where linear PSs are to be found. The other employed functions are frequently used in literature, namely OmniTest [12], Two on One [28], and Sympart {1,2,3} [29]. Lastly, several problems are taken from the Multi-modal Multi-objective test Function (MMF) benchmark suite [20] in the form of MMF {1, 2, 12, 14, 15}. A mix of PS and PF shapes have been chosen to determine the capabilities of MM-BezEA on different problem types. Table 1 shows some of the important characteristics for each of the problems.

For all problems with a configurable number of PSs n, it is set to 2, likewise the problem dimension ℓ is fixed to 2. In order to determine the values of the performance indicators, the reference PSs will be made using 5000 points that adhere to the analytical formulas describing the PSs. In the case of Two on One, a very close approximation is used [28].

Table 1. Bi-objective problem characteristics.

Problem	ℓ	PS	PS Shape	PF Shape
MinDist	$[2, \infty) \in \mathbb{Z}$	n	Linear	Convex
Omni Test	$[2, \infty) \in \mathbb{Z}$	3^ℓ	Linear	Convex
Two on One	2	2	Linear	Convex
Sympart 1, 2	2	9	Linear	Convex
Sympart 3	2	9	Non-linear	Convex
MMF 1, 2	2	2	Non-linear	Convex
MMF 12	$[2, \infty) \in \mathbb{Z}$	n	Linear	Disconnected
MMF 14, 15	$[2, \infty) \in \mathbb{Z}$	n	Linear	Concave

5.2 Benchmark Setup

In order to get a fair comparison, each of the algorithms will be given an equally sized budget of 200, 000 function evaluations for each of the problems. This removes the influence of the used programming languages, as the computation time is not limited. The parameters of MO-RV-GOMEA, MO_Ring_PSO_SCD, and MO-HillVallEA are set to the values reported in relevant literature. Furthermore, for each problem and metric, the average over 31 runs will be taken.

The elitist archives sizes $N_\mathbb{E}$ are set to be 1250 for MO-RV-GOMEA and MO-HillVallEA. The population size N is set to 96 for MO-RV-GOMEA and 250 for MO-HillVallEA [23]. MO-RV-GOMEA uses a linkage tree as its linkage model, with a total of 5 clusters. For MO_Ring_PSO_SCD the population size is 800 [35]. For the MM-BezEA algorithm, the number of control points q for each approximation set is set to 2. Just like for the original BezEA algorithm, MM-BezEA is given population sizes of 76 [24]. The number of test points p is set to 7.

5.3 Performance Indicators

The HV indicator [37] is used to see how well the algorithms perform in getting close the PF. As a result of the use of test points in MM-BezEA, the Bézier solutions sets will have a limited amount of points in the approximation set that can be used to calculate the HV values. Therefore, a subset of the approximation set will be taken for the other algorithms to allow a fair comparison based on the HV indicator. Specifically, the same number of test points is selected for a fair comparison by means of greedy Hypervolume Subset Selection (gHSS) [15].

We further use a relatively new performance indicator for multi-modal multi objective optimization, named Pareto Set Proximity (PSP) [35]. It is an indicator that determines how well all PSs are approximated by taking the Cover Rate (CR), that shows how well the extremes of all PSs are captured, divided by the Inverted Generational Distance in decision space (IGDX) [36], which can be used to determine how close the approximation sets are to the PSs. For the IGDX measure, the approximation sets as produced by MM-BezEA are interpolated by taking 1000 intermediate points before determining the IGDX value. This can be performed relatively easily as interpolating these parameterizations does not require any extra fitness evaluations.

Finally, we use a performance indicator regarding smoothness, for which we follow the definition as introduced in the work on BezEA [24]. It captures how smooth an approximation set can be navigated in terms of decision variables by measuring the detour length in decision space when traversing the approximation set from one solution to the next via an intermediate solution, as compared to going to the next solution directly. The smoothness approaches its maximum value of 1 if all solutions would be colinear in decision space, where the lowest possible value is 0. In cases where multiple approximation sets are explicitly determined, like in MO-HillVallEA and MM-BezEA, the average smoothness over all clusters will be taken. In the other cases the smoothness over the entire approximation set is taken.

5.4   Results

Table 2 shows the results for all problems and algorithms per indicator.

The HV results clearly show that all algorithms are capable of performing nearly equally in obtaining a good approximation front. However, MM-BezEA with $q = 2$ does deteriorate in performance on the MMF1 and 2 problems that have non-linear PSs. The deterioration is inherently caused by the chosen parameterizations that create approximation sets which are linear in shape. Another problem instance where a smaller HV for the new algorithm is obtained, is that of MMF12. Here, despite MM-BezEA obtaining the best PSP values, the approximation sets did not fully approximate the actual PSs and did not cover the endpoints.

The PSP indicator shows similar results, except that MO-RV-GOMEA performs worse as it is not an MMOEA and therefore does not explicitly search for multiple niches. Again promising results for MM-BezEA are shown in cases where linear PSs can be found, as seen in Fig. 5a where MM-BezEA approximates all 9 Pareto sets very well. In the problems with non-linear PSs, MO-HillVallEA and MO_Ring_PSO_SCD manage to find better approximations.

The smoothness results show, as intended, that the chosen parameterizations inherently cause smooth approximation sets with a perfect smoothness of 1.0 for MM-BezEA. Other algorithms do not obtain this, except for MO-RV-GOMEA on 2 of the 11 problems. A visualization of the results of MM-BezEA on the

Table 2. Results (avg. ($\pm$ st.dev.)) per problem and algorithm over 31 runs, bold identifies best result with statistical significance (Wilcoxon rank-sum test with $\alpha = 0.05$ and Holm-Bonferroni correction).

	Problem	MM-BezEA	MO-HillVallEA	MO_Ring_PSO_SCD	MO-RV-GOMEA
HV	MinDist	**1.17e+2 ($\pm$9.43e-5)**	1.17e+2 ($\pm$1.62e-2)	1.17e+2 ($\pm$5.95e-3)	1.17e+2 ($\pm$1.36e-3)
	OmniTest	**8.48e+0 ($\pm$2.18e-6)**	8.47e+0 ($\pm$1.96e-3)	8.47e+0 ($\pm$7.34e-4)	8.47e+0 ($\pm$4.82e-4)
	Sympart 1	**1.17e+2 ($\pm$2.16e-5)**	1.17e+2 ($\pm$1.42e-2)	1.17e+2 ($\pm$7.64e-3)	1.17e+2 ($\pm$1.30e-3)
	Sympart 2	**1.17e+2 ($\pm$2.91e-5)**	1.17e+2 ($\pm$7.77e-3)	1.17e+2 ($\pm$8.75e-3)	1.17e+2 ($\pm$4.47e-3)
	Sympart 3	**1.17e+2 ($\pm$9.61e-5)**	1.17e+2 ($\pm$1.61e-2)	1.17e+2 ($\pm$9.21e-3)	1.17e+2 ($\pm$4.91e-3)
	TwoOnOne	1.13e+2 ($\pm$1.33e-4)	1.13e+2 ($\pm$2.39e-4)	**1.13e+2 ($\pm$1.82e-4)**	1.13e+2 ($\pm$1.10e-4)
	MMF 1	6.04e-1 ($\pm$3.86e-2)	**8.05e-1 ($\pm$2.37e-4)**	**8.05e-1 ($\pm$8.69e-5)**	**8.05e-1 ($\pm$6.60e-5)**
	MMF 2	6.34e-1 ($\pm$2.01e-4)	8.04e-1 ($\pm$6.90e-4)	8.04e-1 ($\pm$9.59e-4)	**8.05e-1 ($\pm$1.75e-4)**
	MMF 12	1.78e+0 ($\pm$2.02e-6)	2.06e+0 ($\pm$2.57e-3)	2.06e+0 ($\pm$2.05e-3)	**2.06e+0 ($\pm$1.49e-4)**
	MMF 14	5.63e+0 ($\pm$1.33e-5)	**5.63e+0 ($\pm$7.26e-4)**	5.63e+0 ($\pm$1.92e-3)	5.63e+0 ($\pm$2.23e-4)
	MMF 15	5.56e+0 ($\pm$2.03e-2)	**5.57e+0 ($\pm$6.52e-4)**	5.56e+0 ($\pm$1.54e-3)	**5.57e+0 ($\pm$1.79e-4)**
PSP	MinDist	**3.26e+2 ($\pm$7.31e+1)**	5.02e+1 ($\pm$2.74e+0)	6.97e+1 ($\pm$6.73e+0)	1.21e+0 ($\pm$1.65e+0)
	OmniTest	**2.36e+2 ($\pm$8.81e+1)**	7.13e+1 ($\pm$2.76e+0)	6.90e+1 ($\pm$9.56e+0)	1.36e-1 ($\pm$2.43e-1)
	Sympart 1	**2.67e+2 ($\pm$1.21e+2)**	3.56e+1 ($\pm$1.57e+0)	2.79e+1 ($\pm$3.69e+0)	1.16e-2 ($\pm$2.70e-2)
	Sympart 2	**3.09e+2 ($\pm$8.46e+1)**	3.60e+1 ($\pm$7.42e-1)	2.38e+1 ($\pm$2.46e+0)	1.02e-2 ($\pm$1.71e-2)
	Sympart 3	6.41e+1 ($\pm$7.77e+1)	**4.33e+1 ($\pm$2.10e+0)**	2.64e+1 ($\pm$5.62e+0)	8.09e-3 ($\pm$1.33e-2)
	TwoOnOne	**3.04e+2 ($\pm$2.18e+2)**	4.50e+1 ($\pm$7.21e-1)	2.45e+1 ($\pm$1.03e+1)	2.68e+0 ($\pm$1.11e+0)
	MMF 1	7.22e+0 ($\pm$2.41e+0)	3.17e+1 ($\pm$6.84e-1)	**3.80e+1 ($\pm$6.79e+0)**	1.02e+0 ($\pm$2.89e-1)
	MMF 2	4.00e+0 ($\pm$2.49e+0)	**1.17e+2 ($\pm$1.21e+1)**	5.04e+1 ($\pm$1.51e+1)	2.18e+0 ($\pm$1.41e+0)
	MMF 12	**2.42e+1 ($\pm$7.72e+0)**	1.94e+1 ($\pm$6.46e+0)	1.53e+1 ($\pm$1.46e-1)	8.67e+0 ($\pm$1.36e+0)
	MMF 14	**2.68e+3 ($\pm$4.82e+2)**	3.70e+2 ($\pm$1.03e+1)	2.31e+2 ($\pm$2.16e+1)	1.08e+0 ($\pm$2.84e+0)
	MMF 15	**2.73e+2 ($\pm$4.17e+1)**	2.65e+2 ($\pm$4.58e+0)	2.44e+2 ($\pm$7.25e+0)	2.24e+1 ($\pm$1.40e-2)
Smoothness	MinDist	**1.00e+0 ($\pm$0.00e+0)**	8.09e-1 ($\pm$3.70e-2)	7.63e-1 ($\pm$5.91e-2)	8.94e-1 ($\pm$1.81e-1)
	OmniTest	**1.00e+0 ($\pm$0.00e+0)**	9.23e-1 ($\pm$5.65e-2)	7.28e-1 ($\pm$9.76e-2)	7.16e-1 ($\pm$1.96e-1)
	Sympart 1	**1.00e+0 ($\pm$0.00e+0)**	8.76e-1 ($\pm$2.52e-2)	6.84e-1 ($\pm$9.76e-2)	7.01e-1 ($\pm$2.09e-1)
	Sympart 2	**1.00e+0 ($\pm$0.00e+0)**	8.78e-1 ($\pm$2.14e-2)	5.73e-1 ($\pm$9.30e-3)	7.87e-1 ($\pm$1.68e-1)
	Sympart 3	**1.00e+0 ($\pm$0.00e+0)**	8.70e-1 ($\pm$3.10e-2)	5.29e-1 ($\pm$1.18e-2)	8.30e-1 ($\pm$1.92e-1)
	TwoOnOne	**1.00e+0 ($\pm$0.00e+0)**	7.77e-1 ($\pm$1.21e-2)	7.47e-1 ($\pm$9.36e-3)	7.51e-1 ($\pm$1.59e-1)
	MMF 1	**1.00e+0 ($\pm$0.00e+0)**	9.01e-1 ($\pm$3.98e-2)	7.82e-1 ($\pm$8.49e-3)	5.86e-1 ($\pm$9.13e-2)
	MMF 2	**1.00e+0 ($\pm$0.00e+0)**	9.38e-1 ($\pm$1.26e-2)	5.01e-1 ($\pm$8.92e-3)	8.46e-1 ($\pm$1.60e-1)
	MMF 12	**1.00e+0 ($\pm$0.00e+0)**	8.35e-1 ($\pm$2.73e-2)	6.35e-1 ($\pm$9.86e-3)	**1.00e+0 ($\pm$0.00e+0)**
	MMF 14	**1.00e+0 ($\pm$0.00e+0)**	9.31e-1 ($\pm$1.17e-2)	8.71e-1 ($\pm$1.11e-2)	9.62e-1 ($\pm$1.05e-1)
	MMF 15	**1.00e+0 ($\pm$0.00e+0)**	9.19e-1 ($\pm$1.43e-2)	8.63e-1 ($\pm$8.54e-3)	**1.00e+0 ($\pm$0.00e+0)**

MinDist problem is given in Fig. 5b. This figure depicts the smooth progression of the decision variables values in the parallel coordinates plot for the rightmost approximation set in decision space. It contrasts sharply to the parallel coordinates plot of Fig. 1 when navigating the approximation set as it now shows a smooth course of the decision variable values.

6   Discussion

MM-BezEA did not cover the endpoints of the PSs in the case of MMF12. This can be caused by the fact that the Bézier fitness function will constrain a solution when one of its control points is dominated in objective space by one of the test points. As the endpoints of each part of the discontinuous PF are close to being dominated, i.e., close to the constraint space, it can lead to not entirely capturing the discontinuous pieces of the PF and thus resulting in a lower HV.

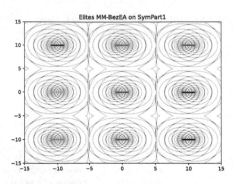
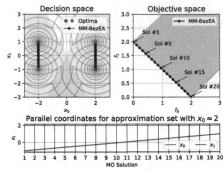

(a) All approximation sets produced by MM-BezEA on SymPart 1 [29]

(b) Approximation sets and front with parallel coordinates plot for one of the approximation sets produced by MM-BezEA on MinDist [23]

Fig. 5. Visualization of results

The HV indicator is a Pareto compliant indicator [13,38], but it does suffer from a downside. In some situations the endpoints of the approximation sets cannot reach the endpoints of the Pareto set because the distribution of points that maximizes the hypervolume does not include the extreme solutions. Even when the number of test points will be set to infinity, the reference point can never be set so that the extremes are captured [2].

Even though the smoothness indicator tries to determine whether an approximation set is smooth by measuring the detour length, it comes down to determining the angle between neighboring solutions. This implies that it only considers linear curves to be perfectly smooth, where there is a straight angle between solutions. When the number of solutions in an approximation set increases, the average distance between the solutions in objective space decreases. As a result of the lower distances and ever so slight oscillations around the PF, the angle between solutions decreases due to which the smoothness indicator will report low smoothness values. In cases where the niches can be separated in a good manner, another definition of smoothness that considers the oscillation around the PS might be more useful.

Future work could investigate the further use of the Bézier parameterizations with more control points to approximate non-linear Pareto sets. Note that the definitions given in this paper already allow for this. Furthermore, no limit on the number of approximation sets can currently be set, which degrades the quality of the approximation sets in highly multi-modal problems as the population is then divided over all niches through HVC [23]. Finally, Bézier simplexes [18] might be usable for problems with more than two objectives.

7   Conclusion

We proposed the algorithm MM-BezEA to search for multiple parameterized approximation sets that define smooth curves in the decision space for bi-objective multi-modal optimization problems. The results show that MM-BezEA is competently capable of locating all modes in a multi-modal landscape as exemplified in various benchmark problems and that the smoothness is indeed enforced by the Bézier parameterizations. Furthermore, MM-BezEA significantly outperformed other algorithms in problems with linear Pareto sets, but was outperformed in problems with non-linear Pareto sets. However, only low-order Bézier curves were used in our experiments, and these results may well be different if higher order curves were used, which the definitions in this paper readily allow.

References

1. Ahrari, A., Deb, K., Preuss, M.: Multimodal optimization by covariance matrix self-adaptation evolution strategy with repelling subpopulations. Evol. Comput. **25**(3), 439–471 (2017). https://doi.org/10.1162/evco_a_00182
2. Auger, A., Bader, J., Brockhoff, D., Zitzler, E.: Theory of the hypervolume indicator: Optimal μ-distributions and the choice of the reference point. In: Proceedings of the Tenth ACM SIGEVO Workshop on Foundations of Genertic Algorithms (FOGA 2009), pp. 87–102. Association for Computing Machinery, New York, NY, USA (2009). https://doi.org/10.1145/1527125.1527138
3. Auger, A., Hansen, N.: A restart CMA evolution strategy with increasing population size. In: 2005 IEEE Congress on Evolutionary Computation. vol. 2, pp. 1769–1776. IEEE, New York, NY, USA (2005). https://doi.org/10.1109/CEC.2005.1554902
4. Berghammer, R., Friedrich, T., Neumann, F.: Convergence of set-based multi-objective optimization, indicators and deteriorative cycles. Theor. Comput. Sci. **456**, 2–17 (2012). https://doi.org/10.1016/J.TCS.2012.05.036
5. Beume, N., Naujoks, B., Emmerich, M.: SMS-EMOA: multiobjective selection based on dominated hypervolume. Eur. J. Oper. Res. **181**(3), 1653–1669 (2007). https://doi.org/10.1016/j.ejor.2006.08.008
6. Bosman, P.A.N.: The anticipated mean shift and cluster registration in mixture-based EDAs for multi-objective optimization. In: Proceedings of the 12th Annual Conference on Genetic and Evolutionary Computation (GECCO 2010), pp. 351–358. Association for Computing Machinery, New York, NY, USA (2010). https://doi.org/10.1145/1830483.1830549
7. Bouter, A., Alderliesten, T., Witteveen, C., Bosman, P.A.N.: Exploiting linkage information in real-valued optimization with the real-valued gene-pool optimal mixing evolutionary algorithm. In: Proceedings of the Genetic and Evolutionary Computation Conference (GECCO 2017), pp. 705–712. Association for Computing Machinery, New York, NY, USA (2017). https://doi.org/10.1145/3071178.3071272
8. Bouter, A., Luong, N.H., Witteveen, C., Alderliesten, T., Bosman, P.A.N.: The multi-objective real-valued gene-pool optimal mixing evolutionary algorithm. In: Proceedings of the Genetic and Evolutionary Computation Conference (GECCO 2017). pp. 537–544. Association for Computing Machinery, New York, NY, USA (2017). https://doi.org/10.1145/3071178.3071274

9. Deb, K., Pratap, A., Agarwal, S., Meyarivan, T.: A fast and elitist multiobjective genetic algorithm: NSGA-II. IEEE Trans. Evol. Comput. **6**, 182–197 (2002). https://doi.org/10.1109/4235.996017

10. Deb, K.: An efficient constraint handling method for genetic algorithms. Comput. Methods Appl. Mech. Eng. **186**(2), 311–338 (2000). https://doi.org/10.1016/S0045-7825(99)00389-8

11. Deb, K.: Multi-Objective Optimization Using Evolutionary Algorithms. John Wiley & Sons Inc, USA (2001)

12. Deb, K., Tiwari, S.: Omni-optimizer: a generic evolutionary algorithm for single and multi-objective optimization. Eur. J. Oper. Res. **185**(3), 1062–1087 (2008). https://doi.org/10.1016/j.ejor.2006.06.042

13. Fleischer, M.: The measure of pareto optima applications to multi-objective metaheuristics. In: Fonseca, C.M., Fleming, P.J., Zitzler, E., Thiele, L., Deb, K. (eds.) EMO 2003. LNCS, vol. 2632, pp. 519–533. Springer, Heidelberg (2003). https://doi.org/10.1007/3-540-36970-8_37

14. Grimme, C., et al.: Peeking beyond peaks: challenges and research potentials of continuous multimodal multi-objective optimization. Comput. Oper. Res. **136**, 105489 (2021). https://doi.org/10.1016/j.cor.2021.105489

15. Guerreiro, A.P., Fonseca, C.M., Paquete, L.: Greedy hypervolume subset selection in low dimensions. Evol. Comput. **24**, 521–544 (2016). https://doi.org/10.1162/EVCO_a_00188

16. Harik, G.R., Lobo, F.G.: A parameter-less genetic algorithm. In: Proceedings of the 1st Annual Conference on Genetic and Evolutionary Computation (GECCO 1999), vol. 1, pp. 258–265. Morgan Kaufmann Publishers Inc., San Francisco, CA, USA (1999)

17. Igel, C., Hansen, N., Roth, S.: Covariance matrix adaptation for multi-objective optimization. Evol. Comput. **15**, 1–28 (2007). https://doi.org/10.1162/evco.2007.15.1.1

18. Kobayashi, K., Hamada, N., Sannai, A., Tanaka, A., Bannai, K., Sugiyama, M.: Bézier simplex fitting: describing Pareto fronts of simplicial problems with small samples in multi-objective optimization. In: Proceedings of the 33rd AAAI Conference on Artificial Intelligence, AAAI 2019, the 31st Innovative Applications of Artificial Intelligence Conference, IAAI 2019 and the 9th AAAI Symposium on Educational Advances in Artificial Intelligence, EAAI 2019, pp. 2304–2313. AAAI press, Palo Alto, CA, USA (Jan 2019)

19. Li, X., Epitropakis, M.G., Deb, K., Engelbrecht, A.: Seeking multiple solutions: an updated survey on niching methods and their applications. IEEE Trans. Evol. Comput. **21**(4), 518–538 (2017). https://doi.org/10.1109/TEVC.2016.2638437

20. Liang, J., Qu, B., Gong, D., Yue, C.: Problem definitions and evaluation criteria for the CEC 2019 special session on multimodal multiobjective optimization. Technical Report (Nov 2018)

21. Luong, N.H., Alderliesten, T., Bel, A., Niatsetski, Y., Bosman, P.A.N.: Application and benchmarking of multi-objective evolutionary algorithms on high-dose-rate brachytherapy planning for prostate cancer treatment. Swarm Evol. Comput. **40**, 37–52 (2018). https://doi.org/10.1016/j.swevo.2017.12.003

22. Mahfoud, S.W.: Niching Methods for Genetic Algorithms. Ph.D. Thesis, University of Illinois at Urbana-Champaign, USA (1996), uMI Order No. GAX95-43663

23. Maree, S.C., Alderliesten, T., Bosman, P.A.N.: Real-valued evolutionary multi-modal multi-objective optimization by hill-valley clustering. In: Proceedings of the Genetic and Evolutionary Computation Conference (GECCO 2019), pp. 568–576. Association for Computing Machinery, New York, NY, USA (2019). https://doi.org/10.1145/3321707.3321759

24. Maree, S.C., Alderliesten, T., Bosman, P.A.N.: Ensuring smoothly navigable approximation sets by Bézier curve parameterizations in evolutionary bi-objective optimization. In: Parallel Problem Solving from Nature - PPSN XVI. pp. 215–228. Springer, Cham (2020)

25. Maree, S.C., Alderliesten, T., Bosman, P.A.N.: Uncrowded hypervolume-based multi-objective optimization with gene-pool optimal mixing. Evol. Comput. 1–24 (2021). https://doi.org/10.1162/evco_a_00303

26. Maree, S.C., Alderliesten, T., Thierens, D., Bosman, P.A.N.: Real-valued evolutionary multi-modal optimization driven by hill-valley clustering. In: Proceedings of the Genetic and Evolutionary Computation Conference (GECCO 2018), pp. 857–864. Association for Computing Machinery, New York, NY, USA (Jul 2018). https://doi.org/10.1145/3205455.3205477

27. Preuss, M.: Niching the CMA-ES via nearest-better clustering. In: Proceedings of the 12th Annual Conference Companion on Genetic and Evolutionary Computation (GECCO 2010), pp. 1711–1718. Association for Computing Machinery, New York, NY, USA (2010). https://doi.org/10.1145/1830761.1830793

28. Preuss, M., Naujoks, B., Rudolph, G.: Pareto set and EMOA behavior for simple multimodal multiobjective functions. In: Runarsson, T.P., Beyer, H.-G., Burke, E., Merelo-Guervós, J.J., Whitley, L.D., Yao, X. (eds.) PPSN 2006. LNCS, vol. 4193, pp. 513–522. Springer, Heidelberg (2006). https://doi.org/10.1007/11844297_52

29. Rudolph, G., Naujoks, B., Preuss, M.: Capabilities of EMOA to detect and preserve equivalent pareto subsets. In: Obayashi, S., Deb, K., Poloni, C., Hiroyasu, T., Murata, T. (eds.) EMO 2007. LNCS, vol. 4403, pp. 36–50. Springer, Heidelberg (2007). https://doi.org/10.1007/978-3-540-70928-2_7

30. Schäpermeier, L., Grimme, C., Kerschke, P.: To boldly show what no one has seen before: a dashboard for visualizing multi-objective landscapes. In: Ishibuchi, H., Zhang, Q., Ishibuchi, H. (eds.) Evolutionary Multi-Criterion Optimization, pp. 632–644. Springer, Cham (2021)

31. Tanabe, R., Ishibuchi, H.: A review of evolutionary multimodal multiobjective optimization. IEEE Trans. Evol. Comput. 24(1), 193–200 (2020). https://doi.org/10.1109/TEVC.2019.2909744

32. Tian, Y., Cheng, R., Zhang, X., Jin, Y.: PlatEMO: a MATLAB platform for evolutionary multi-objective optimization. IEEE Comput. Intell. Mag. 12(4), 73–87 (2017)

33. Touré, C., Hansen, N., Auger, A., Brockhoff, D.: Uncrowded hypervolume improvement: COMO-CMA-ES and the sofomore framework. In: Proceedings of the Genetic and Evolutionary Computation Conference (GECCO 2019), pp. 638–646. Association for Computing Machinery, New York, NY, USA (2019). https://doi.org/10.1145/3321707.3321852

34. Ursem, R.: Multinational evolutionary algorithms. In: Proceedings of the 1999 Congress on Evolutionary Computation-CEC99. vol. 3, pp. 1633–1640. IEEE, New York, NY, USA (1999). https://doi.org/10.1109/CEC.1999.785470

35. Yue, C., Qu, B., Liang, J.: A multi-objective particle swarm optimizer using ring topology for solving multimodal multi-objective problems. IEEE Trans. Evol. Comput. 22, 805–817 (2017). https://doi.org/10.1109/TEVC.2017.2754271

36. Zhang, Q., Li, H.: MOEA/D: A multiobjective evolutionary algorithm based on decomposition. IEEE Trans. Evol. Comput. **11**, 712–731 (2007). https://doi.org/10.1109/TEVC.2007.892759
37. Zitzler, E., Laumanns, M., Thiele, L.: SPEA2: Improving the strength pareto evolutionary algorithm for multiobjective optimization. In: Evolutionary Methods for Design Optimization and Control with Applications to Industrial Problems, pp. 95–100. International Center for Numerical Methods in Engineering, Athens, Greece (Sep 2001)
38. Zitzler, E., Thiele, L., Laumanns, M., Fonseca, C.M., da Fonseca, V.G.: Performance assessment of multiobjective optimizers: an analysis and review. IEEE Trans. Evol. Comput. **7**(2), 117–132 (2003). https://doi.org/10.1109/TEVC.2003.810758

T-DominO

Exploring Multiple Criteria with Quality-Diversity and the Tournament Dominance Objective

Adam Gaier[1]([✉]), James Stoddart[2], Lorenzo Villaggi[2], and Peter J. Bentley[1,3]

[1] Autodesk Research, Bonn, Germany
adam.gaier@autodesk.com
[2] Autodesk Research, New York, USA
[3] University College London, London, UK

Abstract. Real-world design problems are a messy combination of constraints, objectives, and features. Exploring these problem spaces can be defined as a Multi-Criteria Exploration (MCX) problem, whose goals are to produce a set of diverse solutions with high performance across many objectives, while avoiding low performance across any objectives. Quality-Diversity algorithms produce the needed design variation, but typically consider only a single objective. We present a new ranking, T-DominO, specifically designed to handle multiple objectives in MCX problems. T-DominO ranks individuals relative to other solutions in the archive, favoring individuals with balanced performance over those which excel at a few objectives at the cost of the others. Keeping only a single balanced solution in each MAP-Elites bin maintains the visual accessibility of the archive – a strong asset for design exploration. We illustrate our approach on a set of easily understood benchmarks, and showcase its potential in a many-objective real-world architecture case study.

Keywords: Quality-diversity · Generative design · Multi-objective

1 Introduction

Architecture projects must balance a dizzying array of objectives: daylight, views, noise, wind, cost, open spaces, carbon footprint, and ease of construction, to name a few – along with less easily optimized subjective considerations like aesthetics and comfort. In generative design (GD), where algorithms aid design exploration by producing candidate designs, the desired result is not a single solution, but a variety of high performing options [4]. A variety of options is required because the problem has more than one objective, which means there may be many possible solutions. Perhaps more importantly, a varied choice highlights design concepts to stakeholders and decision makers who then select and modify them according to messy human compromises.

© The Author(s) 2022
G. Rudolph et al. (Eds.): PPSN 2022, LNCS 13399, pp. 263–277, 2022.
https://doi.org/10.1007/978-3-031-14721-0_19

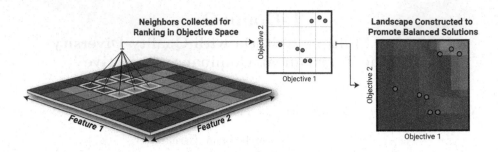

Fig. 1. Calculating the Tournament Dominance Objective (T-DominO)

Though it may resemble multi-objective optimization (MOO) [10], the problem this work focuses on for GD and similar domains is different. We define the problem as a Multi-Criteria Exploration (MCX) problem, whose goals are to:

1. Produce a catalog of diverse solutions
2. with high performance across many objectives
3. while avoiding low performance across any objectives

MCX can be considered an exploratory form of MOO, just as Quality-Diversity (QD) [8,39], is an exploratory form of single-objective optimization. In contrast to MOO, in MCX we do not strive for either uniform coverage of the Pareto front, nor precise proximity to it. Uniform coverage of the front implies coverage of the extremes of the objective space – where solutions earn their place in the Pareto front by dominating on only a subset of objectives. These solutions are uninteresting for MCX, as solutions which disregard user preferences by ignoring some objectives are not useful in practice. Proximity to the front is also less important for exploration – the goal is to generate starting points, not end points. Generated solutions are rarely used without modification, reducing the effort of finding the precise front to an expensive distraction.

The QD approach seems, at first, ideal for solving MCX problems. QD algorithms provide a way of explicitly searching for diversity as defined at a high level by users. Whereas MOO strives for a maximum spread in the objective space, QD searches for spread in a user-defined 'feature'[1] space. As opposed to objectives, these features correspond to different *ways* of solving the problem, not quantities to be minimized. In architecture the number of buildings in a building complex or the distance between them can be explicitly explored with QD in a way that is not possible with MOO.

But QD is designed to explore several features, not to optimize multiple objectives. MAP-Elites [7,33], the most widely used QD algorithm, divides the feature space and searches for the best solution in each partition. The result is a set of optimized designs organized by high level features; the performance of this collection can then be viewed as heat map projected on to the feature space, illuminating the relationship between features and performance.

---

[1] Also referred to in the QD literature as a behavior, descriptor, outcome, or measure.

Recent work has proposed combining MOO and QD by computing a Pareto front in every partition [38]. In the MCX case this 'all of the above' approach is not satisfying – MAP-Elites' intuitive way of organizing, summarizing, and presenting the results depends on finding a single best solution for each partition. An alternate method of reconciling the approaches must be found.

In this work we leverage the insight that the diversity in objective space produced by MOO mechanisms such as crowding distance [12], or reference vectors [11] are unneeded when diversity is enforced by QD. In QD, users can choose the type of variety to explore, and trade-offs in objectives will naturally arise from those choices. When exploration of the objective space is no longer prioritized, non-domination – which favors the extremes of the front to the same degree as the center – ceases to be the most desirable attribute of a solution. In MCX, balanced solutions which perform well on all objectives are preferred. Our second insight is that the 'balance' of a solution can be defined in relation to a population, and that the solutions contained in the MAP-Elites archive can act as that population.

Our approach extends MAP-Elites to the exploration of problems with multiple objectives by introducing the Tournament Dominance Objective (T-DominO), which ranks individuals in a population according to an approximation of their distance to the center of the Pareto front. T-DominO awards poor scores to non-dominated solutions at the extremes of the objective space – those which excel at one objective while doing very poorly at the other – while those at the center of the front receive the highest scores (Fig. 1).

Optimizing MAP-Elites according to T-Domino provides an elegant approach to tackle MCX problems. A simple alternate ranking causes minimal disruption to the core algorithmic machinery while allowing MAP-Elites to discover varied design concepts which balance multiple objectives. Approaches which assume a single objective and single solution in each bin, such as CMA-ME [14], can still be used. Crucially, T-DominO tackles multiple objectives without sacrificing MAP-Elites' intuitive visualization and analysis of solutions, features, and their interaction with objectives – the true goal of the algorithm when used for design.

2   Background

2.1   Generative Design

In design and architecture, experiments with human-machine collaboration are common [2,4,18,26,35,41]. Recent work has demonstrated the viability of GD in real-world applications, from office retrofits [35], large scale trade-shows [36], to neighborhood scale planning [34]. The number of conflicting constraints, preferences, and objectives in these projects makes 'solving' them an ill-defined and impossible task. Optimization tools are typically used at the *beginning* of the design process rather than the end. Optimization algorithms are not used solve problems, but to *explore* them [4,31].

The purpose of GD is less optimization and more communication. Search algorithms are used to understand the possibilities and potential of a problem space. Objectives serve as proxies for preferences, goals, and features of interest that are often difficult or impossible to define mathematically. These objectives signify criteria of a good design or ways of counter-balancing those criteria to prevent extreme solutions which are not aligned with designers intent.

Results are then filtered and categorized in an effort to find qualitatively different design concepts. Typically designs are judged visually first, and only once a set of interesting and varied designs identified is their performance examined. This can be a clumsy process, and for GD to have real success accessibility must be a consideration at every step of the process, including optimization. An intuitive GD approach would not only find a set of solutions which balance performance over several objectives, but explicitly search for the high level diversity that sets solutions apart from each other. This is the goal of MCX.

2.2   Exploration and Optimization with Non-objective Criteria

MOO approaches strive to produce a set of non-dominated solutions that is diverse in the objective space, and as near to the Pareto optimal front as possible [37], but the diversity that interests designers is often not in objective space. Other qualities can be induced with 'helper' objectives in a process known as multiobjectivization [25,27,30]. Helper objectives can optimize quantities unrelated to performance, such as the type of cross sections in a structural frame [21], or the similarity to previous solutions [32], but are still performing minimization. Maximizing or minimizing the number of buildings on a site makes little sense, but understanding the effect of the number of buildings is a valuable insight.

QD approaches such as MAP-Elites [7,33] search for solutions along a continuum of user-defined features, making them ideal for exploration. MAP-Elites has been used for design exploration in domains such as aerodynamics [15–17,23,24], and game design [1,5,19,20], but has been restricted to consideration of a single objective. MAP-Elites operates by first discretizing the feature space into bins, collectively known as a map or archive. Each bin contains a single solution and its corresponding fitness value. New solutions are created by selecting and varying solutions from the map. These new solutions are evaluated and two values produced: a performance measure and a set of coordinates in the feature space. These coordinates indicate the bin to which the solution belongs. The solution is placed in the bin if it is empty, or if the candidate solution has higher performance than the current occupant of the bin, it replaces it.

The elitist nature of MAP-Elites, with only one solution per bin, puts it at odds with the idea of the Pareto front. A concurrent work [38] bridges this gap by introducing a Pareto front in each bin, and replacing fitness tournaments with non-domination. Though this technique is able to find a large set of Pareto fronts, it sacrifices the elegant method of communicating the results. Rather than viewing individual designs and correlations between features and objectives, we are left with a mass of summary statistics – useful for MOO, but not for MCX. In our work we maintain the the elitist nature of MAP-Elites, and instead replace the Pareto front with an alternate formulation of multi-objective performance.

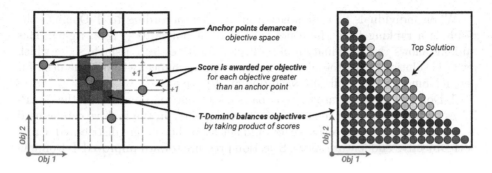

Fig. 2. Using anchor points to calculate T-DominO.

3   Method

When tackling MCX problems, our interest lies in finding solutions which perform well on all objectives in each region of a QD feature space. The Tournament Dominance Objective (T-DominO), introduced here, ranks solutions according to an approximation of their distance to the center of the front, with the most balanced solutions ranking highest. This approximation is calculated through a series of tournaments between a solution and a set of existing points in objective space, or *anchor points* (Fig. 2). An individual is compared to each anchor point on a single objective, and for every anchor point with a lesser or equal objective value one point is awarded. This count is made for every objective, and these counts multiplied.

The T-DominO score of solution with objective values x, compared to a set of anchor points with objective values A is more precisely defined as:

$$
\text{T-DominO}(x, A) = \prod_{n=1}^{objs} \sum_{m=1}^{anchors} f(x_n, A_{mn}), f(x, a) = \begin{cases} 1, & x \geq a \\ 0, & \text{else} \end{cases} \quad (1)
$$

where *objs* and *anchors* is the number of objectives and anchor points.[2]

The integration of T-DominO into MAP-Elites can be summarized as follows:

1. a new individual, based on its feature coordinates, is assigned a bin.
2. a set of anchor points are collected from the history of elites in that bin, the k neighboring bins, and the new individual itself.
3. the T-DominO of the current elite and the challenger are computed based on these anchor points.
4. if the challenger has a greater T-DominO score it replaces the current elite, and the objective values of the replaced elite are stored in the bin to serve as a future anchor point.

---

[2] Or in python: `numpy.prod(numpy.sum(objs >= anchors,axis=0))`.

When individuals in a population are ranked according to T-DominO the result is a ranking from the center of the front outwards (Fig. 2, right). This ranking allows the combination of multiple objectives into a single score which rewards solutions with the highest balanced performance, without the need for penalty functions or arbitrary weighting of objectives.

T-DominO is based around comparisons to anchor points, and the MAP-Elites archive provides a ready source. Existing elites in the archive can be used as a sampling of the objective space, and act as anchor points distributed across both objective and feature space. Selection pressure toward improved T-DominO scores creates high performing solutions which in turn act as anchor points, in a virtuous cycle that leads to ever higher performance.

However, when selection pressure is organized around *only* the current population, cycling can occur. In some circumstances a challenger solution which is better on one objective can replace the current elite, which can in turn be replaced by the original. To prevent this behavior and ensure progress towards better solutions we track the objective values of the previous elites. While each bin continues to contain only a single elite, the objective values of previous elites are maintained to act as anchor points, preventing cycling and creating further refinement of the T-DominO landscape. A simple FIFO buffer of the objective values of a handful of past elites is sufficient.

Neighboring partitions typically have similar performance potential, and so are necessary for creating more fine-grained landscapes, but bins with solutions that dominate all or none of the solutions in the a bin provide no signal to inform selection pressure, and so we can safely limit the anchor points to those contained in the k nearest neighboring bins.

T-Domino allows the simultaneous optimization of several objectives with a single measure, relying on QD diversity mechanisms to prevent convergence on a single point. The output of T-Domino MAP-Elites is ideal for MCX – an archive with a single best balanced solution in each bin. Having a single solution in each bin allows the effects of solution features on that balance to be easily understood and visualized and so contribute to the understanding of the underlying problem, such as the correlation of features and objectives. The creation of a library of designs organized by high level features of the users choosing provides an ideal set of starting points for further refinement.

4   Benchmarks

4.1   Setup

We validate the expected behavior of MAP-Elites with T-DominO on a series of established multi-objective benchmark problems. The purpose of these tests is to validate our claims that T-Domino will:

1. Discover high performing, if not optimal, solutions
2. Produce balanced solutions whose performance does not come at the cost of large trade-offs in a subset of objectives

Benchmark Functions

RastriginMOO. To judge the performance of T-DominO on Multi-Objective QD problems, we test on a version of RastriginMOO as introduced in [38]. The Rastrigin function is a classic optimization benchmark, often used to test QD algorithms because it contains many local minima [6,14]. Here it is converted into a multiobjective benchmark by optimizing a pair of Rastrigin functions with shifted centers. We use a 10-D version with constants added so that every discovered bin has a positive effect on the aggregate QD Score. These objectives can be explicitly defined as:

$$
\begin{cases}
f_1(\mathbf{x}) = 200 - (\sum_{i=1}^{n}[(x_i - \lambda_1)^2 - 10\cos(2\pi(x_i - \lambda_1))]) \\
f_2(\mathbf{x}) = 200 - (\sum_{i=1}^{n}[(x_i - \lambda_2)^2 - 10\cos(2\pi(x_i - \lambda_2))])
\end{cases}
\tag{2}
$$

where $\lambda_1 = 0.0$ and $\lambda_2 = 2.2$ for f_1 and f_2. All parameters are limited to the range $[-2, 2]$, with the feature space defined by the first two parameters.

ZDT3. When spread across the objective space is desired, objectives themselves could be used as features. This use case is demonstrated with the ZDT3 benchmark, a 30 variable problem from the ZDT MOO benchmark problem suite [42] whose hallmark is a set of disconnected Pareto-optimal fronts, and whose first parameter is value of the first objective. Parameter ranges span 0–1 with the first two parameters used as features, enforcing a spread of solutions across the range of the first objective.

DTLZ3. To illustrate T-DominO's bias toward balanced solutions we analyze its performance on DTLZ3, a many-objective benchmark with a tunable number of objectives and variables[13]. We test with 10 parameters and 5 objectives, with the 6th and 7th parameters use as features.[3].

Baseline Approaches

ME Single. MAP-Elites [33] optimizing only a single objective is used to establish an upper and lower bound of performance we can expect from MAP-Elites. Blind to the second objective we can expect it to find the top performing solutions for the first. Equally important, the exploration of all bins without regard to the performance on the second objective establishes a floor for performance – the performance we could expect for having any solution in the bin.

ME Sum. We compare using the T-Domino objective with MAP-Elites [33] using the most naive way of combining multiple objective – simply adding them. Our benchmarks all have well-scaled objectives, but this is typically not the case. To simulate this difficulty we use a weighted sum, with each additional objective values increased by an order of magnitude (e.g. ×1, ×10, ×100...).

---

[3] The first n parameters are explicitly linked to the first n objectives as in ZDT3 – later parameters are used to avoid explicitly exploring the objective space.

NSGA-II. NSGA-II [12] is used as a benchmark for conventional multi-objective optimization without feature space exploration, reaching near the Pareto front on these simple benchmarks. Though it is not our goal to compete with MOO algorithms, they provide a useful metric to contextualize the difference between exploratory approaches and pure optimizers.

Settings. In all MAP-Elites approaches the feature space is partitioned a 20×20 grid, with 2 CMA-ME improvement emitters [14] performing optimization. T-Domino was computed using the neighbors from 4 bins away, using a history of the 10 most recent elites in each bin. Hyperparameters for NSGA-II were kept comparable, a population of 400 matched the 400 bins of the MAP-Elites grids, with the same number of new solutions generated per generation for the same number of generations. A standard implementation of NSGA-II from the PyMoo library [3] is used, as well as the library's formulations for the ZDT3 and DTLZ benchmarks whose the exact formulation is included in the online supplemental. The PyRibs [40] library was used as a basis for all MAP-Elites experiments, with T-DominO implemented as a specialized archive type. All experiments were replicated 30 times, additional plots are provided in the Supplemental.

4.2   Result

Figure 3 illustrates the explored regions of objective and feature space in a single run. Using the NSGA-II solutions to outline the true Pareto front we can see where each MAP-Elites approach concentrates. In the RastriginMOO case, though each version of MAP-Elites explores identical areas of the feature space, the range of possible values in objective space is large. T-Domino produces solutions in the middle of the front, with solutions that strike a balance between the two objectives. In the ZDT3 case we see that by explicitly exploring one of the objectives we can force spread over the objective space, and provide high performing solutions in the other objective.

With more than two objectives the balance seeking property of T-DominO becomes even more pronounced. Parallel coordinate plots (Fig. 3, top right), which plot each solution as a line with one vertex per objective, make clear the differing selection pressure of T-DominO and non-dominated sorting. In contrast to the spiky lines denoting high performance on some objectives and low performance on others, T-DominO's solutions form a flat band of even performance.

The difference of balance is critical when approaching MCX problems. To spread across the five dimensional front, solutions found by NSGA-II must span many areas with solutions that perform poorly on some objectives. If we divide the range of objective values found by NSGA-II into quartiles, only 25% of the solutions found by NSGA-II perform over the bottom quartile on all objectives. If all of these objectives are valued by the user, that means that three quarters of solutions may be discarded immediately – and this will only worsen as the number of objectives grows. In contrast, when T-DominO MAP-Elites' results

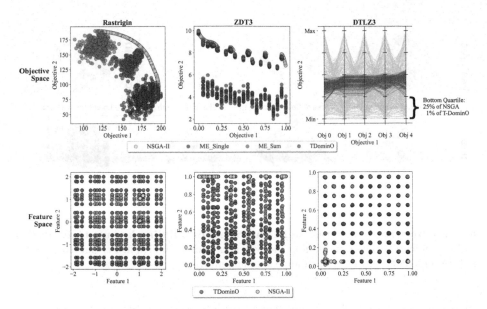

Fig. 3. Benchmark results. *Top:* Objective space as explored by each approach. *Bottom:* Feature space as explored by T-DominO MAP-Elites and NSGA-II. T-DominO points are colored by T-DominO score compute with the entire archive as anchor points.

are judged on the same scale, 99% of the solutions found by T-DominO MAP-Elites perform over the bottom quartile on all objectives.

Visualizing the distribution of found solutions in feature space (Fig. 3, bottom) gives a stark illustration of the main motivation for using a QD approach. The solutions produced by NSGA-II cluster in a tiny portion of the feature space. This region may be Pareto optimal, but QD gives us the ability to explore areas of our choosing.

5   Case Study

5.1   Setup

As a study of the applicability of T-DominO MAP-Elites to MCX problems we explore its use in optimizing building layouts for real-world residential complex. Solutions are produced using wave function collapse (WFC) [22], a popular tool for tile-based procedural content generation in games. WFC is a constraint satisfaction approach which extracts local patterns from a small set of samples, and transforms them into a set of local constraints. The constraints drive generation,

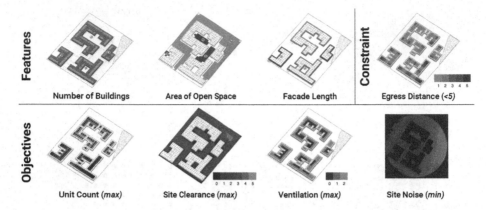

Fig. 4. Objectives, Features, and Constraints of building layout study. Shaded regions indicate portion of building site which cannot be built on. For details on computation of each metric see the online supplemental.

ensuring that every local patch of the output also exists in the set of input examples. We adapt the implementation here [28,29]. Constrained generation systems like WFC are particularly appropriate for the semi-constrained design systems often used in residential building, such as modular or prefabricated units, and do not require the extensive curated datasets of valid designs.

WFC, though constrained, is a stochastic process. At every iteration a tile is 'collapsed', or fixed, with the type chosen stochastically from a list of valid tiles, and new constraints applied to its neighbors. To make this encoding more amenable to optimization we have introduced an evolvable genotype of tiles which are fixed at the beginning of this collapsing process. Children inherit these fixed tiles from parents, in addition to fixing an additional tile from the design produced by the parent or removing one of the tiles that were fixed by the parent. Fixing tiles freezes key portions of the parent design and saves progress toward interesting designs - while still allowing substantial deviation from the parent, as the remainder of the tiles are generated stochastically with WFC. See the supplemental material for set of used tiles and example designs.

The resulting designs are evaluated according to 4 objectives, 1 constraint, and explored along 3 features (illustrated in Fig. 4, more details in the online supplemental). The constraint was handled in a tournament fashion as in [9] – in any tournament where one solution follows the constraint and the other does not, the solution which follows the constraint wins regardless.

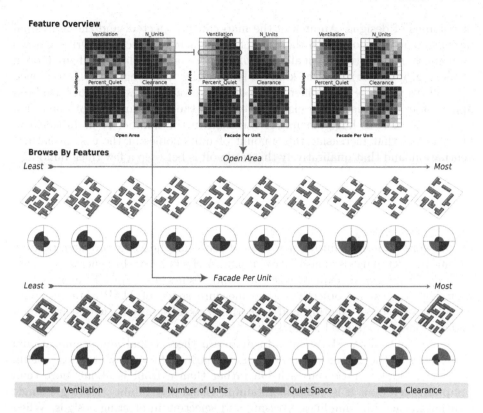

Fig. 5. Exploring building layouts generated with T-DominO MAP-Elites.
Top: 2D views of 3D feature space, solutions in groups of four are identical, colored in
each view by each objective (darker is better). Bottom: A walk through designs that
vary along one feature dimension, with accompanying objective values. Petal plots are
scaled to the final min/max objective value found in the archive. (Color figure online)

5.2   Result

Once a set of designs has been produced, we can extend on MAP-Elites' intuitive
way of optimizing, organizing, and displaying solutions to multiple objectives.
To better understand our 3D feature space we 'flatten' it into a set of 2D views
by creating a set of new 2D archives with the desired feature axes and inserting
all of the solutions from the 3D archive, forcing competition based on only two
features. The result is one map for each pair of the three features (Fig. 5, top).

Each of these views can in turn be split into one map for each objective,
collectively allowing correlations along feature axes and objectives to all be seen
at a glance. Obvious relationships such as an increase in open area resulting
in more units are clear, along with less appreciated connections: with fewer
buildings ventilation is worse – unless buildings have longer facades.

Clear organized grids of solutions open up many avenues for intuitive navigation of the produced solutions. Here we show one possibility, browsing rows

or columns of designs. An area of the map can be selected, and the individual designs displayed along with their performance across objectives. Drilling down on a subset we can see the qualitative differences between large and small open areas (Fig. 5, Center), and the kinds of layouts they each represent. A combined view allows us to see large difference in objective values that may not have been apparent from a qualitative glance: even with the same amount of open area we can see there is huge amount of variation in the number of units that can fit on a site, that increasing this amount of units comes at the cost of natural ventilation, and that qualitatively this trade-off is between a few thick buildings, or several small ones (Fig. 5, Bottom).

6   Discussion

In this work we have defined a new class of problem, the MCX problem, tailored specifically to the needs of generative design. A chief aim of the generative design is to spark ideas and explore concepts, and results are typically explored by browsing designs not objectives. The measure space of MAP-Elites provides an intuitive way of creating and exploring sets of solutions with varied and understandable high level features.

T-DominO allows MAP-Elites to maintain these visual and organizational capabilities in the complex multi-criteria scenarios where they are most useful. Keeping a single solution in each bin rather than a front is about more than computational cost, it is about maintaining visual accessibility. Having a single solution in each bin simplifies browsing and selecting interesting designs. When objectives and features are correlated, the possible objectives values for each feature combination is constrained to a range – so though balanced solutions are found, the larger pattern of objective/feature relations are still clear.

T-DominO allows us to optimize for multiple objectives in a QD setting without making any other fundamental changes to the algorithm. Simple to implement, without adding any appreciable computational burden, T-DominO can be easily integrated into existing approaches. By leaving its elitist character untouched, T-Domino allows MAP-Elites to handle multiple objectives while maintaining its core visualization and presentation strengths. Equipping MAP-Elites with T-DominO allows us to generate diverse sets of well-rounded high performing solutions, creating a powerful tool for tackling MCX problems.

Acknowledgements. The authors would like to thank Renaud Danhaive, Jeffrey Landes, and the entire Spacemaker team for their invaluable site analysis tool and expertise as well as Mark Davis and David Benjamin for their guidance and support.

Supplemental Material and Code. Supplemental material and code available at: https://github.com/agaier/tdomino_ppsn.

References

1. Alvarez, A., Dahlskog, S., Font, J., Togelius, J.: Empowering quality diversity in dungeon design with interactive constrained map-elites. In: 2019 IEEE Conference on Games (CoG), pp. 1–8. IEEE (2019)
2. Arieff, A.: New forms that function better. Technol. Rev. **116**(5), 94–98 (2013). TECHNOL REV 1 MAIN ST, 13 FLR, CAMBRIDGE, MA 02142 USA
3. Blank, J., Deb, K.: pymoo: multi-objective optimization in python. IEEE Access **8**, 89497–89509 (2020)
4. Bradner, E., Iorio, F., Davis, M.: Parameters tell the design story: ideation and abstraction in design optimization. In: Proceedings of the Symposium on Simulation for Architecture & Urban Design, vol. 26. Society for Computer Simulation International (2014)
5. Charity, M., Green, M.C., Khalifa, A., Togelius, J.: Mech-elites: illuminating the mechanic space of gvg-ai. In: International Conference on the Foundations of Digital Games, pp. 1–10 (2020)
6. Cully, A.: Multi-emitter map-elites: improving quality, diversity and data efficiency with heterogeneous sets of emitters. In: Proceedings of the Genetic and Evolutionary Computation Conference, pp. 84–92 (2021)
7. Cully, A., Clune, J., Tarapore, D., Mouret, J.B.: Robots that can adapt like animals. Nature **521**(7553), 503–507 (2015)
8. Cully, A., Demiris, Y.: Quality and diversity optimization: a unifying modular framework. IEEE Trans. Evol. Comput. **22**(2), 245–259 (2017)
9. Deb, K.: An efficient constraint handling method for genetic algorithms. Comput. Methods Appl. Mech. Eng. **186**(2–4), 311–338 (2000)
10. Deb, K.: Multi-objective optimization. In: Search methodologies, pp. 403–449. Springer, Boston (2014). https://doi.org/10.1007/978-1-4614-6940-7_15
11. Deb, K., Jain, H.: An evolutionary many-objective optimization algorithm using reference-point-based nondominated sorting approach, part i: solving problems with box constraints. IEEE Trans. Evol. Comput. **18**(4), 577–601 (2013)
12. Deb, K., Pratap, A., Agarwal, S., Meyarivan, T.: A fast and elitist multiobjective genetic algorithm: Nsga-ii. IEEE Trans. Evol. Comput. **6**(2), 182–197 (2002)
13. Deb, K., Thiele, L., Laumanns, M., Zitzler, E.: Scalable multi-objective optimization test problems. In: Proceedings of the 2002 Congress on Evolutionary Computation, CEC 2002 (Cat. No. 02TH8600), vol. 1, pp. 825–830. IEEE (2002)
14. Fontaine, M.C., Togelius, J., Nikolaidis, S., Hoover, A.K.: Covariance matrix adaptation for the rapid illumination of behavior space. In: Proceedings of the 2020 Genetic And Evolutionary Computation Conference, pp. 94–102 (2020)
15. Gaier, A., Asteroth, A., Mouret, J.B.: Aerodynamic design exploration through surrogate-assisted illumination. In: 18th AIAA/ISSMO Multidisciplinary Analysis And Optimization Conference, p. 3330 (2017)
16. Gaier, A., Asteroth, A., Mouret, J.B.: Data-efficient exploration, optimization, and modeling of diverse designs through surrogate-assisted illumination. In: Proceedings of the Genetic and Evolutionary Computation Conference, pp. 99–106 (2017)
17. Gaier, A., Asteroth, A., Mouret, J.B.: Data-efficient design exploration through surrogate-assisted illumination. Evol. Comput. **26**(3), 381–410 (2018)
18. Gerber, D.J., Lin, S.H., Pan, B., Solmaz, A.S.: Design optioneering: multidisciplinary design optimization through parameterization, domain integration and automation of a genetic algorithm. In: Proceedings of the 2012 Symposium on Simulation for Architecture and Urban Design, pp. 1–8 (2012)

19. González-Duque, M., Palm, R.B., Ha, D., Risi, S.: Finding game levels with the right difficulty in a few trials through intelligent trial-and-error. In: 2020 IEEE Conference on Games (CoG), pp. 503–510. IEEE (2020)
20. Gravina, D., Khalifa, A., Liapis, A., Togelius, J., Yannakakis, G.N.: Procedural content generation through quality diversity. In: 2019 IEEE Conference on Games (CoG), pp. 1–8. IEEE (2019)
21. Greiner, D., Emperador, J.M., Winter, G., Galván, B.: Improving computational mechanics optimum design using helper objectives: an application in frame bar structures. In: Obayashi, S., Deb, K., Poloni, C., Hiroyasu, T., Murata, T. (eds.) EMO 2007. LNCS, vol. 4403, pp. 575–589. Springer, Heidelberg (2007). https://doi.org/10.1007/978-3-540-70928-2_44
22. Gumin, M.: Bitmap and tilemap generation from a single example by collapsing a wave function. GitHub (2016)
23. Hagg, A., Asteroth, A., Bäck, T.: Prototype discovery using quality-diversity. In: Auger, A., Fonseca, C.M., Lourenço, N., Machado, P., Paquete, L., Whitley, D. (eds.) PPSN 2018. LNCS, vol. 11101, pp. 500–511. Springer, Cham (2018). https://doi.org/10.1007/978-3-319-99253-2_40
24. Hagg, A., Wilde, D., Asteroth, A., Bäck, T.: Designing air flow with surrogate-assisted phenotypic niching. In: Bäck, T., et al. (eds.) PPSN 2020. LNCS, vol. 12269, pp. 140–153. Springer, Cham (2020). https://doi.org/10.1007/978-3-030-58112-1_10
25. Handl, J., Lovell, S.C., Knowles, J.: Multiobjectivization by decomposition of scalar cost functions. In: Rudolph, G., Jansen, T., Beume, N., Lucas, S., Poloni, C. (eds.) PPSN 2008. LNCS, vol. 5199, pp. 31–40. Springer, Heidelberg (2008). https://doi.org/10.1007/978-3-540-87700-4_4
26. Holzer, D., Hough, R., Burry, M.: Parametric design and structural optimisation for early design exploration. Int. J. Archit. Comput. 5(4), 625–643 (2007)
27. Jensen, M.T.: Helper-objectives: Using multi-objective evolutionary algorithms for single-objective optimisation. J. Math. Modell. Algo. 3(4), 323–347 (2004)
28. Karth, I.: wfc2019f (2021). https://github.com/ikarth/wfc-2019f
29. Karth, I., Smith, A.M.: Addressing the fundamental tension of pcgml with discriminative learning. In: Proceedings of the 14th International Conference on the Foundations of Digital Games, pp. 1–9 (2019)
30. Knowles, J.D., Watson, R.A., Corne, D.W.: Reducing local optima in single-objective problems by multi-objectivization. In: Zitzler, E., Thiele, L., Deb, K., Coello Coello, C.A., Corne, D. (eds.) EMO 2001. LNCS, vol. 1993, pp. 269–283. Springer, Heidelberg (2001). https://doi.org/10.1007/3-540-44719-9_19
31. Matejka, J., Glueck, M., Bradner, E., Hashemi, A., Grossman, T., Fitzmaurice, G.: Dream lens: exploration and visualization of large-scale generative design datasets. In: Proceedings of the 2018 CHI Conference on Human Factors in Computing Systems, pp. 1–12 (2018)
32. Mouret, J.B.: Novelty-based multiobjectivization. In: New horizons in evolutionary robotics, pp. 139–154. Springer, Berlin (2011). https://doi.org/10.1007/978-3-642-18272-3_10
33. Mouret, J.B., Clune, J.: Illuminating search spaces by mapping elites. arXiv preprint arXiv:1504.04909 (2015)
34. Nagy, D., Villaggi, L., Benjamin, D.: Generative urban design: integration of financial and energy design goals in a generative design workflow for residential neighborhood layout. In: Symposium on Simulation for Architecture and Urban Design (2018)

35. Nagy, D., et al.: Project discover: an application of generative design for architectural space planning. In: Proceedings of the Symposium on Simulation for Architecture and Urban Design, p. 7. Society for Computer Simulation International (2017)
36. Nagy, D., Villaggi, L., Zhao, D., Benjamin, D.: Beyond heuristics: a novel design space model for generative space planning in architecture (2017)
37. Panichella, A.: An adaptive evolutionary algorithm based on non-euclidean geometry for many-objective optimization. In: Proceedings of the Genetic and Evolutionary Computation Conference, pp. 595–603 (2019)
38. Pierrot, T., Richard, G., Beguir, K., Cully, A.: Multi-objective quality diversity optimization. arXiv preprint arXiv:2202.03057 (2022)
39. Pugh, J.K., Soros, L.B., Stanley, K.O.: Quality diversity: a new frontier for evolutionary computation. Frontiers Robot. AI **3**, 40 (2016)
40. Tjanaka, B., Fontaine, M.C., Zhang, Y., Sommerer, S., Dennler, N., Nikolaidis, S.: pyribs: a bare-bones python library for quality diversity optimization (2021). https://github.com/icaros-usc/pyribs
41. Turrin, M., Von Buelow, P., Stouffs, R.: Design explorations of performance driven geometry in architectural design using parametric modeling and genetic algorithms. Adv. Eng. Inform. **25**(4), 656–675 (2011)
42. Zitzler, E., Deb, K., Thiele, L.: Comparison of multiobjective evolutionary algorithms: empirical results. Evol. Comput. **8**(2), 173–195 (2000)

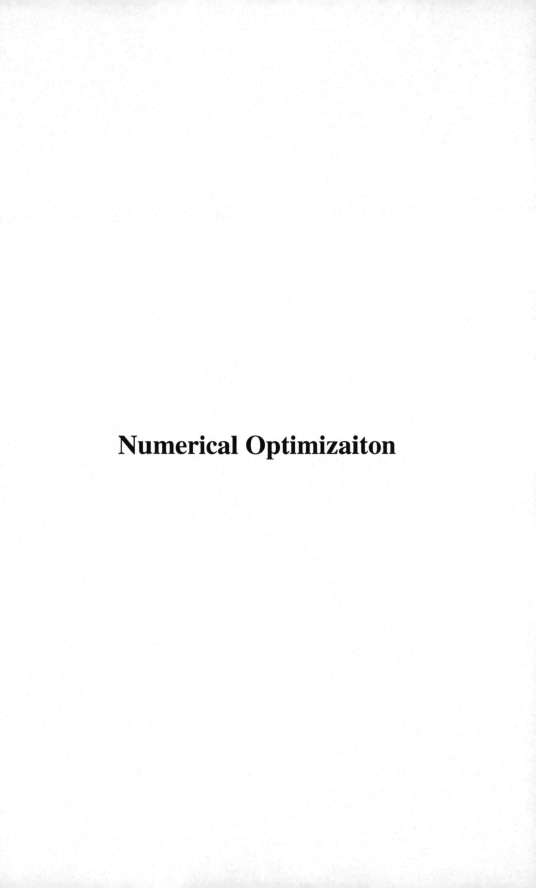

Numerical Optimizaiton

Collective Learning of Low-Memory Matrix Adaptation for Large-Scale Black-Box Optimization

Qiqi Duan[1,3]([⊠]), Guochen Zhou[3], Chang Shao[2,3], Yijun Yang[2,3], and Yuhui Shi[3]([⊠])

[1] Harbin Institute of Technology, Harbin, China
11749325@mail.sustech.edu.cn
[2] University of Technology Sydney, Sydney, Australia
[3] Southern University of Science and Technology, Shenzhen, China
shiyh@sustech.edu.cn

Abstract. The increase of computing power can be continuously driven by parallelism, despite of the end of Moore's law. To cater to this trend, we propose to parallelize the low-memory matrix adaptation evolution strategy (LM-MA-ES) recently proposed for large-scale black-box optimization, aiming at further improving its scalability (w.r.t. CPU cores) in the modern distributed computing platform. To achieve this aim, three key design choices are carefully made and naturally combined within the multilevel learning framework. First, to fit into the memory hierarchy and reduce communication cost, which is critical for parallel performance on modern multi-core computer architectures, the well-known island model with a star interaction network is employed to run multiple concurrent LM-MA-ES instances, each of which can be effectively and serially executed in each separate island owing to its low computational complexity. Second, to support fast convergence under the multilevel learning framework, we adopt Meta-ES to hierarchically exploit the spatial-nonlocal information for global step-size adaptation at the outer-ES level, combined with cumulative step-size adaptation, which exploits the temporal-nonlocal information in the inner-ES (i.e., serial LM-MA-ES) level. Third, a set of fitter individuals at the outer-ES level, represented as (distribution mean, evolution path, transformation matrix)-tuples, are collectively recombined to utilize the desirable genetic repair effect for statistically more stable online learning. Experiments in a clustering computing environment empirically validate the parallel performance of our approach on high-dimensional memory-costly test functions. Its Python code is available at https://github.com/Evolution ary-Intelligence/D-LM-MA.

Keywords: Collective learning · Distributed computing · Evolution strategy

1 Introduction

In the contemporary era, the growth in computing performance can be further driven by software engineering, algorithm advance, and hardware streamlining, though the end of

G. Rudolph et al. (Eds.): PPSN 2022, LNCS 13399, pp. 281–294, 2022.
https://doi.org/10.1007/978-3-031-14721-0_20

Moore's law [1]. Refer to the latest *Science* review [1] for a comprehensive introduction to the improving room of computer power in the post-Moore era. To cater to this computing trend, in this paper we concentrate on application-level parallelism of covariance matrix adaptation evolution strategies (CMA-ES) [5, 8], one of the state-of-the-art randomized search algorithms for black-box optimization (BBO) (see the latest *Nature* review [3]). The increasing popularity of CMA-ES may be attributed to its practical generalizability from invariance against affine transformation and its theoretical foundation recently built on information geometry [9, 24].

More specifically, we focus on parallelism of its latest *low-memory* version called LM-MA-ES [10] for *non-separable* large-scale optimization (LSO), according to the following three reasons. First, due to its $O(nlog(n))$ computational complexity[1] (n is the problem dimension), it provides a very effective alternative to the computationally expensive CMA-ES (with $O(n^2)$ complexity) on LSO problems. Second, it is better suitable for modern multi-core computer architectures than CMA-ES, since it can well exploit the *memory hierarchy*, which is typically critical for parallel performance. Finally, parallel/distributed algorithms [11] are playing an increasingly significant role in modern large-scale machine learning [1].

The main goal of this paper is to increase the *scalability* (w.r.t. CPU cores) and *efficiency* of LM-MA-ES for LSO **in the modern distributed/clustering computing environment**. To reach such a goal, three contributions are made herein:

(1) We analyze the possible opportunities and challenges regarding parallelism of MA-ES, a simplified version of CMA-ES with little performance loss [12], and its LSO version called LM-MA-ES, both quantitatively and qualitatively. And we further point out the drawbacks of two existing solutions in the LSO context. (See Sect. 2 and Fig. 1 for details.)

(2) For LSBBO, we propose a distributed version of LM-MA-ES [10] (D-LM-MA for short) targeted to the modern cloud/clustering computing environment and also provide its open-source Python implementation[2] as a baseline for further parallelism benchmarking. (See Sect. 3 for details.)

First, to fit into *memory hierarchy* and control *communication cost*, we use the well-known island model [13, 14] (with a star interaction topology) to concurrently execute multiple serial instances of LM-MA-ES for each generation.

Second, to support *multilevel learning* [15] and fast convergence, the less-known Meta-ES (also called Nested-ES) [2] is utilized to hierarchically exploit the space-nonlocal information for global step-size adaptation in the outer-ES, which works together with the popular cumulative step-size adaptation (CSA) [8] which exploits the temporal-nonlocal information in the inner-ESs.

Finally, to employ the desirable *genetic repair* effect [16] at the outer-ES level, a set of fitter individuals from different islands (from the inner-ESs), represented as (distribution mean, evolution path, transformation matrix)-tuples, are collectively recombined after each isolation period to generate one statistically more stable (often better) parent for the next generation of the outer-ES.

---

[1] In this paper, the computational complexity is defined w.r.t. one sample per generation.

[2] The code is freely available at https://github.com/Evolutionary-Intelligence/D-LM-MA.

In principle, these simple yet often effective design choices should be naturally extended to parallelize other LSO variants such as rank-one ES [17].

(3) Simulation experiments on a large set of 2000-d test functions with rotated and shifted transformations empirically investigate the parallel advantages (and also cost) of the proposed D-LM-MA when compared with its serial implementation. To the best of our knowledge, it is the first time *in the distributed computing platform* to scale LM-MA-ES to such memory-costly (from rotation operators) benchmarking cases. (See Sect. 4 for details.)

2 Related Work on Distributed ES

Since there have been some well-written introductions to ES (e.g. [19–23, 29]) up to now, in this section we will review only the work related to distributed ES[3] due to the page limit.

Parallelism is an often-claimed advantage of ES [5], since its invention (see e.g. [4, 18]). Recently, its most successful application is direct policy search of deep neural networks (with millions of weights) for episodic reinforcement learning (see [25–27] for examples). Among them, the simplest master-slave model was used to parallelize computationally expensive function evaluations (run on complex simulators). When the task of function evaluations is light-weighted as shown in many real-world applications, however, the master-slave model will suffer from the excessive communication overhead (see Fig. 1 for quantitative analysis on ten test functions). Furthermore, they considered only a highly-customized ES version based on natural ES (NES) without general-purpose strategy parameter adaptation mechanisms, which may result in very slow convergence on other challenging LSO problems (e.g., ill-conditioned) [12]. To alleviate these two problems, another popular island model [13, 14] will be considered in the following, which has a long (more than 20 years) history in the evolutionary computation field [28].

During the past two decades, there have been several open-source software (e.g., DEAP [30], pagmo [13], pCMALib [31]), which provide distributed implementations based on the island model for CMA-ES. However, to our knowledge, all of them only considered the low-dimensional (most < 100) cases and there is still no report about their applications for LSO. It is because the parallelism of CMA-ES for LSO easily suffers from slow convergence rate (typically $O(n^2)$ adaptation time), excessive time occupancy from eigen-decomposition, high communication overhead, and poor usage of memory hierarchy. Instead, its LSO variants (e.g., R1-ES, LM-MA-ES) should be chosen as a baseline for parallelism, as they can better exploit memory hierarchy and control communication cost under their much lower computational complexity.

It is worthwhile noting that there are still two other interesting models (i.e., cellular and hierarchical models) for distributed evolutionary algorithms. However, they have been rarely used for distributed ES till now, and therefore we will not consider them in the following sections.

---

[3] For the latest survey regarding parallel and distributed evolutionary algorithms, please refer to https://github.com/Evolutionary-Intelligence/DistributedEvolutionaryComputation, which is continuously updated in the near future (at least to 2024).

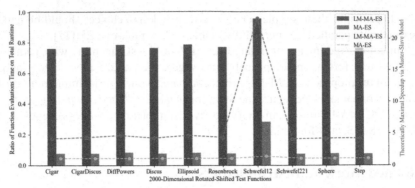

Fig. 1. Ratio of function evaluation time on total runtime (bar, left y-axis) and theoretically maximal speedup of master-slave model (line, right y-axis) obtained by MA-ES and LM-MA-ES on 10 rotated-shifted functions with 2000 dimension. For both MA-ES and LM-MA-ES, even given an infinite parallel resource, a very limited speedup (<5) is obtained on all functions except LM-MA-ES on Schwefel12 (with a much higher time complexity than other functions). The calculation of theoretically maximal speedup is based upon the Amdahl's law [32] under the limit condition.

3   Distributed LM-MA-ES Within Multilevel Learning

In this section, we present the proposed distributed ES framework based on LM-MA-ES (D-LM-MA for brevity), inspired by the latest multilevel learning framework for evolution [15]. Owing to the complexity of distributed algorithms, we focus on three design choices critical for parallel performance and the rationales behind them, while omitting other tedious details. To ensure repeatability and benchmarking, its software implementation using Python [33] is openly available at GitHub.

3.1   Combining Island Model with Meta-eS for Multilevel Learning

In their latest *PNAS* paper, Vanchurin et al. [15] mathematically formulated biological evolution as multilevel learning. According to their ambitious theory, seven principles of evolution underlies **evolvability**, arguably a desirable property for any evolving system. Roughly speaking, "*evolving systems encompass multiple dynamic variables that change on different temporal scales*" [15] and "*slower-changing levels absorb information from faster-changing levels during learning and pass information down to the faster levels for prediction of the state of the environment and the system itself*" [15]. In our opinion, such a *hierarchy of scales* observed in evolution can provide both the philosophical and algorithmic viewpoints regarding the design of distributed evolutionary algorithms, as presented below.

Some evolutionary algorithms (such as Meta-ES, island model, and coevolutionary algorithms) could be seen as concrete algorithmic instances of multilevel learning, as they explicitly exploit the evolution hierarchy of time and/or space scales. Therefore, we can naturally combine the island (aka coarse-grained) model with Meta-ES, resulting in a general-purpose distributed ES framework with online hierarchical learning of strategy parameters, as shown in Fig. 2.

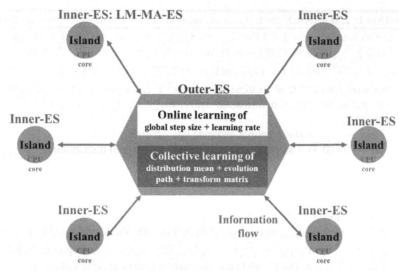

Fig. 2. A simplified diagram of combining the island model with Meta-ES for multilevel learning, where the outer-ES controls online learning of (global step size and learning rate) and collective learning of (distribution mean, evolution path, and transformation matrix) *at a slower-changing timescale* while a set of independent inner-ESs (each of which runs the serial LM-MA-ES for every isolation period) concurrently learn these above strategy parameters *at a faster-changing timescale.*

In this paper, we use the simplest *star* communication topology (with central control) for the island model, which can be regarded as one efficient implementation of the fully-connected topology somewhat. In principle, other more complex topologies can be also employed. Here, an obvious benefit of the star topology against others is the relatively easy understanding and analysis of the resulting distributed algorithm (if properly designed), owing to its simplicity.

For each separate island (often corresponding to one CPU core), a serial LM-MA-ES instance is run in every isolation interval. We prefer to use LM-MA-ES, as presented in Algorithm 1, based on its two most important advantages for LSO: that is, its $O(n \log(n))$ time and space complexity makes it easily exploit the *memory hierarchy* critical for parallel performance and it can efficiently approximate the powerful *invariance* against affine transformation, an essential feature of CMA-ES. Refer to [10] for detailed mathematical derivations and empirical evidences regarding invariance, owing to the limit of space.

After deciding the interaction topology (star) and the computing unit (LM-MA-ES), we need to solve one key design issue for the island model: **what, when**, and **how** to communicate among these islands (i.e., inner-ESs), in order to accelerate convergence and maintain diversity. Here we answer the first two questions (what and when) right now and postpone the last (how) in the following subsection.

Algorithm 1: LM-MA-ES for LSO (which will be inserted into Algorithm 2 as the inner-ES).

1: **Given** $n \in \mathbb{N}^+$, $\lambda = 4 + \lfloor 3\ln n \rfloor$, $\mu = \lfloor \lambda/2 \rfloor$, $m = \lambda$, $\omega_i = (\ln((\lambda + 1)/2) - \ln i)/\sum_{j=1}^{\mu}(\ln((\lambda + 1)/2) - \ln j)$ for $i = (1, ..., \mu)$, $c_{d,i} = 1/(1.5^{i-1}n)$, $c_{c,i} = \lambda/(4^{i-1}n)$ for $i = (1, ..., m)$, $\mu_\omega = 1/\sum_{j=1}^{\mu}\omega_j^2$, and $t \leftarrow 0$

2: **Initialize** mean $x^t \in \mathbb{R}^n$, global step-size $\sigma^t \in \mathbb{R}^+$, learning rate c_σ, evolution path $p_\sigma{}^t \in \mathbb{R}^n$, and transformation matrix $m_i{}^t \in \mathbb{R}^n$ for $i = 1, ..., m$

3: **Repeat**

4:      **For** $i \leftarrow 1, ..., \lambda$ **Do** // sample and evaluate offspring

5:          $z_i{}^t \leftarrow \mathcal{N}(0, I)$ // sample (i.i.d.) from standard normal distribution

6:          $d_i{}^t \leftarrow z_i{}^t$ // temporary direction vector

7:          **For** $j \leftarrow 1, ..., \min(t, m)$ **Do** // sample based on transformation matrix

8:              $d_i{}^t \leftarrow (1 - c_{d,j})d_i{}^t + c_{d,j}m_j{}^t((m_j{}^t)^T d_i{}^t)$

9:          $y_i{}^t \leftarrow f(x^t + \sigma^t d_i{}^t)$ // evaluate offspring's fitness (f: cost function)

10:     $x^{t+1} \leftarrow x^t + \sigma^t \sum_{i=1}^{\mu} \omega_i d_{i:\lambda}{}^t$ // $i{:}\lambda$ means the ith fittest individual based on f

11:     $p_\sigma{}^{t+1} \leftarrow (1 - c_\sigma)p_\sigma{}^t + \sqrt{c_\sigma(2 - c_\sigma)\mu_\omega} \sum_{i=1}^{\mu} \omega_i z_{i:\lambda}{}^t$ // fading cumulation

12:     **For** $i \leftarrow 1, ..., m$ **Do** // update transformation matrix at different scales

13:         $m_i{}^{t+1} \leftarrow (1 - c_{c,i})m_i{}^t + \sqrt{c_{c,i}(2 - c_{c,i})\mu_\omega} \sum_{j=1}^{\mu} \omega_j z_{j:\lambda}{}^t$

14:     $\sigma^{t+1} \leftarrow \sigma^t \exp(\frac{c_\sigma}{2}(\frac{\|p_\sigma{}^{t+1}\|^2}{n} - 1))$ // cumulative step-size adaptation (CSA)

15:     $t \leftarrow t + 1$

16: **Until** one of stopping criteria (e.g., maximal runtime/generations) is met

---

* Refer to [10] for detailed explanations of the above algorithmic flow.

For LM-MA-ES, the settings of the following parameters have a significant impact on its convergence rate: such as, mean of search distribution, global step-size, evolution-path learning rate, evolution path, and transformation matrix. In this paper, we choose these parameters as the basic information source for communication. Note that these parameters[4] are adapted online in the inner-ES at a faster-changing timescale, which can support the construction of multilevel learning in a hierarchical manner.

For the island model, too frequent communications can lead to excessive network overhead and significantly reduce the parallelism level particularly in the distributed computing platform based on commodity servers. On the contrary, too few communications may delay the effect of online learning and result in slow convergence. Therefore, we need to carefully balance the isolation time between a reasonable interval, in order to maximize the parallelism level and obtain fast convergence. We need to tune it for the best performance when solving real-world applications.

3.2 Online and Hierarchical Learning of Strategy Parameters via Meta-eS

It is widely accepted that typically the optimal settings of strategy parameters differ at different evolution stages [2]. As pointed out in [5], the only solution of global step-size control for large populations might be to use a hierarchical method (also referred to as Meta-ES). For optimality, Meta-ES [34–36] maintains and evolves multiple parallel

---

[4] Except the learning rate of evolution path (c_σ), which is fixed during run and often seen as one hyper-parameter for offline tuning.

subpopulations (each corresponding to one inner-ES) with different strategy parameter settings at each isolation period.

In the ES history, two different *nonlocal* self-adaptation methods have been suggested till now, in order to enhance the efficiency of online learning[5] [2]. The first is cumulative step-size adaptation (CSA [8]), which utilizes (possible) correlation between successive promising directions in an exponential smoothing way. The second is the so-called Meta-ES (aka Nested-ES) which employs multiple subpopulations in parallel to hierarchically explore the strategy parameter space.

In this subsection, we try to enjoy the best of both worlds: learning by both *hierarchical use of parallel subpopulations* (Meta-ES) and *online adaptation of strategy parameters in inner-ESs* (CSA). Note that previous Meta-ESs ([34–36]) keep strategy parameters fixed for inner-ESs.

3.3 Collective Learning of Fitness Topology via Multi-recombination

The most essential feature of CMA-ES is the invariance against affine transformation, resembling the topology learning ability of second-order optimizers. When its parallelism is considered for LSO, such a highly desirable feature is expected to be kept. Generally, a large population size may be preferred for LSO owing to its three advantages: 1) better global search ability, 2) less random effect, and 3) massive parallelization [5]. However, in their seminal paper, Hansen et al. [5] has shown that increasing λ alone [6] cannot further promote the efficiency of CMA-ES with only the rank-one update. Their solution is to use a large population size with the extra rank-μ update [7] to increase the parallelism level and exploit more information with fewer generations. However, LM-MA-ES models only the rank-one update (i.e., a small population) and does not consider the rank-μ update (i.e., a large population). Instead, here we use the *structured populations*, well-suitable for the island model.

After each isolation period, the outer-ES collects all necessary information (i.e., distribution mean, evolution path, and transformation matrix[6], all of which are learned in each inner-ES at faster-frequency scales). For updating the distribution mean of the outer-ES yet at lower-frequency scales, we use the *weighted multi-recombination* of its counterparts from all fitter inner-ESs to utilize the well-understood *genetic repair* effect for accelerating convergence. The same operator is also applied to the update of both evolution path and transformation matrix of the outer-ES, in order to obtain statistically more stable estimates even at low-frequency scales. Furthermore, to enhance diversity, the learnt transformation matrix can be **abandoned** (e.g., reinitialized to zero) for some islands (inner-ESs). We find that such a very simple strategy is beneficial for searching in complex landscapes with multiple promising search directions.

3.4 A Distributed ES Framework for Multilevel Learning

In this subsection, we combine the above key design choices (involving island model, Meta-ES, and LM-MA-ES) together to generate a distributed ES framework on multilevel learning for LSBBO, as outlined in Algorithm 2. For validating its scalability, we

---

[5] In this paper, informally, *adaptation* and *learning* are used interchangeably.

[6] It is a low-memory approximation to the full covariance matrix.

choose the latest distributed computing framework called Ray[7] as the execution engine. Perhaps its biggest advantage is that it provides an industry-level unified distributed computing framework for emerging AI applications (e.g., deep learning, reinforcement learning, and hyperparameter tuning [26]) with the excellent fault-tolerance and scheduling abilities. Our preliminary studies [37] as well as some other highly- influenced studies [26, 27] have showed its attractive engineering value for implementing scalable evolutionary algorithms.

Algorithm 2: D-LM-MA for large-scale black-box optimization (LSBBO).

1: **Given** number of parallel islands (inner-ESs) $p \in \mathbb{N}^+$, isolation time $\tau \in \mathbb{R}^{>0}$, number of evolution paths m, recombination weights of outer-ES $\omega_i{}^o$ for $i = (1, \dots, b)$ with $b < p$, generation of outer-ES $g = 0$, a candidate vector of global step-sizes s, a candidate vector of evolution-path learning rates r, a $|s| *$ $|r|$ strategy parameter space matrix (i.e., Cartesian product of s and r), lower and upper bound of initial search space $(l, \mathbf{u})$ with $l < \mathbf{u} \in \mathbb{R}^n$.

2: **Initialize** $x^g \in \mathbb{R}^n$, $\sigma^g \in \mathbb{R}^+$, $c_\sigma{}^g$, $p_\sigma{}^g = 0$, $m_j{}^g = 0$ for $j = 1, \dots, m$.

3: **Repeat**

4:    **For** $i \leftarrow 1, \dots, p$ **Do** // run all islands in parallel

5:      **If** $g == 0$ **Do** // initialize randomly for the first generation

6:        $x^g \leftarrow U(l, \mathbf{u})$ // sample randomly from a uniform distribution U

7:        **If** $i < |s|$ **Do** // run Meta-ES for global step-size (at slow timescales)

8:          $\sigma^{g,i} \leftarrow \sigma^g * sel(s)$ // $sel()$ is a function to return an element

9:        **If** $i < |s| + |r|$ **Do** // run Meta-ES for evolution-path learning rate

10:         $c_\sigma{}^{g,i} \leftarrow sel(r)$

11:       **If** $i < 2 * |s| + |r|$ **Do** // increase diversity of transformation matrix learning

12:         $m_j{}^{g,i} \leftarrow 0$ for $j = 1, \dots, m$ // forgetting

13:         $\sigma^{g,i} \leftarrow \sigma^g * sel(s)$

14:       **If** $i < 2 * |s| + |r| + |s| * |r|$ **Do** // run Meta-ES for online learning

15:         $\sigma^{g,i} \leftarrow \sigma^g * sel(s)$

16:         $c_\sigma{}^{g,i} \leftarrow sel(r)$

17:     **Else:** // inherit with or without mutation, depending on problem at hand

18:       $\sigma^{g,i} \leftarrow inherit(\sigma^g)$

19:       $c_\sigma{}^{g,i} \leftarrow inherit(c_\sigma{}^g)$

20:       $m^{g,i} \leftarrow inherit(m^g)$

21:     // run an inner-ES ($LMMAES$) serially for τ isolation period

22:     $\left(x^{g+1,i}, \sigma^{g+1,i}, c_\sigma{}^{g+1,i}, p_\sigma{}^{g+1,i}, m^{g+1,i}\right) \leftarrow LMMAES(x^g, \sigma^{g,i}, c_\sigma{}^{g,i}, p_\sigma{}^g, m^{g,i}, \tau)$

23:   $x^{g+1} \leftarrow \sum_{j=1}^b \omega_j{}^o x^{g+1,j:b}$ // collective learning of distribution mean

24:   $\sigma^{g+1} \leftarrow \sum_{j=1}^b \omega_j{}^o \sigma^{g+1,j:b}$

25:   $c_\sigma{}^{g+1} \leftarrow \sum_{j=1}^b \omega_j{}^o c_\sigma{}^{g+1,j:b}$

26:   $p_\sigma{}^{g+1} \leftarrow \sum_{j=1}^b \omega_j{}^o p_\sigma{}^{g+1,j:b}$ // collective learning of evolution path

27:   $m_k{}^{g+1} \leftarrow \sum_{j=1}^b \omega_j{}^o m_k{}^{g+1,j:b}$ for $k = 1, \dots, m$ // for transformation matrix

28:   $g \leftarrow g + 1$

29: **Until** one of stopping criteria is met.

---

[7] https://docs.ray.io/en/latest/.

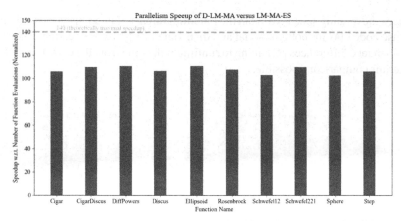

Fig. 3. Parallelism speedup of D-LM-MA versus LM-MA-ES on a set of 2000-d test functions.

4 Numerical Experiments on Clustering Computing Platforms

In this section, we will conduct a series of comparison experiments to study its advantages (and also possible cost) of the proposed D-LM-MA against its serial version (LM-MA-ES) on the contemporary clustering computing platforms. To ensure repeatability and also promote benchmarking, all involved data and source code are freely available at https://github.com/Evolutionary-Intelligence/D-LM-MA.

4.1 Experimental Settings for Large-Scale Black-Box Optimization

Benchmarking Functions. We choose 10 scalable test functions, commonly used in the literature. To validate the invariance against translation and rotation and avoid a biased search (e.g., separability [39]), all of them are shifted and rotated according to [8]. Since here LSO is of our interest, their dimensionality is set to 2000. However, the involved [2000*2000] rotation matrix will result in the memory-costly matrix-vector operator when evaluating, arguably representing many modern black-box LSO applications more or less. To accelerate the fitness evaluation, we utilize the shared-memory trick (see [37] for details). Note that recently Varelas et al. [40] proposed a test suite named *bbob-largescale*, but they did not consider the distributed computing scenario. To our knowledge, there are few literatures (except e.g. [37]) to run distributed ESs on such high-dimensional, memory-costly benchmarking cases.

Hardware and Software Configurations. A private clustering computing platform with a total of 140 physical cores is used in our simulation experiments, consisting of 7 *slightly heterogeneous* commodity servers: the first five equip with Intel Xeon CPU E5-2650v3 (2.30 GHz) while the rest two equip with Intel Xeon CPU E5-2640v4 (2.40 GHz). Each of them runs an Ubuntu 16.04 OS with a 62 GB RAM and 20 physical cores. We select the Python language and its NumPy library highly optimized for numerical computing to program all the involved code. Clearly, the programming quality heavily influences the running speed. To avoid the potential threat to fair comparisons, we use the same base code [44] for both the serial and distributed LM-MA-ES versions. Therefore, the performance differences pertaining to runtime will come from the extra parallel code, to make fair comparisons possible.

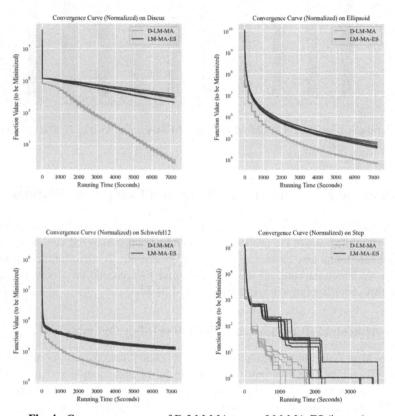

Fig. 4. Convergence curve of D-LM-MA versus LM-MA-ES (in part).

Benchmarking Algorithms. Three ES-based optimizers are compared: MA-ES [12], LM-MA-ES [10], and our proposed D-LM-MA. For hyperparameter settings, we adopt the default values according to their corresponding papers, if not stated otherwise. Since MA-ES shows the worst results on nearly all functions, we will not report them in the following subsections. The reason behind such inferior performance lies at that generally MA-ES needs *quadratic* adaptation time before convergence and the runtime of CMA severely dominates the total runtime.

For D-LM-MA, we set the isolation time to 3 min based on our first intuition, though the ablation study showed that better values exist. To avoid possible scheduling congestion while maximizing the parallelism level, we set the number of islands (inner-ESs) to 250 (larger than 140 physical cores but smaller than 280 logical cores).

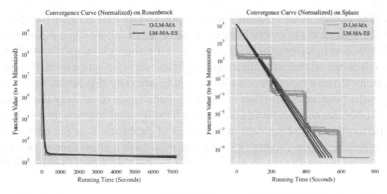

Fig. 5. Convergence curve of D-LM-MA versus LM-MA-ES (continued).

Performance Criteria. Though now there has been a relatively mature methodology for benchmarking serial evolutionary algorithms (such as COCO [41] and NeverGrad [42]), it is still a challenging task to benchmark distributed evolutionary algorithms. For serial optimizers, the number of used function evaluations to reach a preassigned threshold is a good approximator of the needed runtime if it dominates the latter. However, such an indicator does not work well for distributed optimizers, as it can significantly increase with the rapid growth of used computing units under the same runtime. We set the upper bound of runtime (=2 h) as one stopping criterion. As a result, we can easily compare *the quality of results* given the same runtime. We also set the fitness threshold (=1e-10) as another stopping criterion. We execute each optimizer on every function for 7 independent runs.

4.2 Parallel Speedup w.r.t. Total Number of Function Evaluations

For distributed evolutionary algorithms, we always expect to run more function evaluations given the same runtime, in order to maximize parallelism. Here we first calculate the parallel speedup (w.r.t. total number of function evaluations) obtained by D-LM-MA, as compared with its serial version. As seen in Fig. 3, the overall (normalized) parallel efficiency is near 75%, that is, it can concurrently run 105 independent function evaluations on 140 CPU cores. However, a high parallelism level does not necessarily indicate the good rate of convergence, since the poor algorithm design can result in even a worse result than its serial version. In the next subsection we will focus on the more critical performance indicator.

4.3 Performance Comparisons W.R.T. Final Convergence Quality

When compared with its serial version, D-LM-MA obtained much faster convergence rates on 6 of 10 functions (*DiffPowers, Discus, Ellipsoid, Schwefel12, Schwefel221,* and *Step*), as in part shown in Fig. 4. See the online material for complete data, owing to the page limit. For the first four test functions, the main challenge lies that the fact there are multiple promising search directions in their fitness topology. Luckily, D-LM-MA can maintain much more diversity for transform matrices during the multilevel learning process. Take *Discus* as an example. Surprisingly, D-LM-MA achieved a convergence speedup with four orders of magnitude on average when the allowable runtime exhausted. For the last two test functions, the main challenge is that there exists a series of flat plateau, which needs more advanced techniques of global step-size adaptation to pass them. Our hierarchical learning approach can gracefully alleviate this issue, since it always encourages the positive exploration in the much large space of strategy parameters. For *Step*, D-LM-MA spent about half of time to reach the optimal solution than LM-MA-ES.

The performance improvement on one class of problems is at the cost of the performance degradation on another [43]. We observed such a phenomenon on other four functions (*Cigar, CigarDiscus, Rosenbrock*, and *Sphere*), as shown in part in Fig. 5. For the last function, the extra cost from parallelism slightly delays convergence rate, which is very like the early finding [18]. On the first three functions, there is one dominated direction vector. At the outer-ES level, *regress* results in degradation [6].

5 Conclusions

In this paper, we present a novel approach based on multilevel learning [15] to parallelize LM-MA-ES, one of the latest CMA-ES variants for large-scale black-box optimization. Within the multilevel learning framework, three critical design choices (i.e., island model for memory-hierarchy usage, Meta-ES for parameter control, and collective learning via multi-recombination) can be naturally combined to generate a scalable distributed/parallel ES framework in modern industry-level clustering computing platforms. Experiments in a private clustering computing platform demonstrate the effectiveness (and cost) of the proposed approach for memory-costly LSO.

Acknowledgments. This work is partially supported by the Shenzhen Fundamental Research Program under Grant No. JCYJ20200109141235597, the Shenzhen Peacock Plan under Grant No. KQTD2016112514355531, the Program for Guangdong Introducing Innovative and Entrepreneurial Teams under Grant No. 2017ZT07X386, the National Science Foundation of China under Grant No. 61761136008, and the Special Funds for the Cultivation of Guangdong College Students Scientific and Technological Innovation (Climbing Program Special Funds, pdjh2022c0061). Yuhui Shi is the Corresponding Author.

References

1. Leiserson, C.E., et al.: There's plenty of room at the top: what will drive computer performance after Moore's law? Science **368**(6495), p.eaam9744 (2020)
2. Beyer, H.G., Schwefel, H.P.: Evolution strategies–a comprehensive introduction. Nat. Comput. **1**(1), 3–52 (2002)
3. Eiben, A.E., Smith, J.: From evolutionary computation to the evolution of things. Nature **521**(7553), 476–482 (2015)
4. Schwefel, H.P.: Evolutionary learning optimum-seeking on parallel computer architectures. Sydow, A., Tzafestas, S.G., Vichnevetsky, R. (eds.) Systems Analysis and Simulation I. Advances in Simulation, vol. 1, pp. 217–225. Springer, New York (1988). https://doi.org/10.1007/978-1-4684-6389-7_46

5. Hansen, N., Müller, S.D., Koumoutsakos, P.: Reducing the time complexity of the derandomized evolution strategy with covariance matrix adaptation (CMA-ES). Evol. Comput. **11**(1), 1–18 (2003)
6. Schwefel, H.P.: Collective intelligence in evolving systems. In: Wolff, W., Soeder, C.J., Drepper, F.R. (eds.) Ecodynamics. Research Reports in Physics. Springer, Heidelberg, pp. 95–100 (1988). https://doi.org/10.1007/978-3-642-73953-8_8
7. Müller, S.D., Hansen, N., Koumoutsakos, P.: Increasing the serial and the parallel performance of the CMA-evolution strategy with large populations. In: Guervós, J.J.M., Adamidis, P., Beyer, H.G., Schwefel, H.P., Fernández-Villacañas, J.L. (eds.) Parallel Problem Solving from Nature — PPSN VII. PPSN 2002. Lecture Notes in Computer Science, vol. 2439, pp. 422–431. Springer, Berlin, Heidelberg (2002). https://doi.org/10.1007/3-540-45712-7_41
8. Hansen, N., Ostermeier, A.: Completely derandomized self-adaptation in evolution strategies. Evol. Comput. **9**(2), 159–195 (2001)
9. Wierstra, D., Schaul, T., Glasmachers, T., Sun, Y., Peters, J., Schmidhuber, J.: Natural evolution strategies. J. Mach. Learn. Res. **15**(1), 949–980 (2014)
10. Loshchilov, I., Glasmachers, T., Beyer, H.G.: Large scale black-box optimization by limited-memory matrix adaptation. IEEE Trans. Evol. Comput. **23**(2), 353–358 (2019)
11. Bertsekas, D., Tsitsiklis, J.: Parallel and distributed computation: Numerical methods. Athena Scientific (1997)
12. Beyer, H.G., Sendhoff, B.: Simplify your covariance matrix adaptation evolution strategy. IEEE Trans. Evol. Comput. **21**(5), 746–759 (2017)
13. Biscani, F., Izzo, D.: A parallel global multiobjective framework for optimization: pagmo. J. Open Source Softw. **5**(53), 2338 (2020)
14. Ruciński, M., Izzo, D., Biscani, F.: On the impact of the migration topology on the island model. Parallel Comput. **36**(10–11), 555–571 (2010)
15. Vanchurin, V., Wolf, Y.I., Katsnelson, M.I., Koonin, E.V.: Toward a theory of evolution as multilevel learning. Proc. Natl. Acad. Sci. **119**(6), e2120037119 (2022)
16. Beyer, H.G.: An alternative explanation for the manner in which genetic algorithms operate. BioSystems **41**(1), 1–15 (1997)
17. Li, Z., Zhang, Q.: A simple yet efficient evolution strategy for large-scale black box optimization. IEEE Trans. Evol. Comput. **22**(5), 637–646 (2018)
18. Rudolph, G.: Global optimization by means of distributed evolution strategies. In: Schwefel, H.P., Männer, R. (eds.) Parallel Problem Solving from Nature. PPSN 1990. Lecture Notes in Computer Science, vol. 496, pp. 209–213. Springer, Heidelberg (1990). https://doi.org/10.1007/BFb0029754
19. Bäck, T., Hoffmeister, F. and Schwefel, H.P.: A survey of evolution strategies. In Proceedings of International Conference on Genetic Algorithms, pp. 2–9 (1991)
20. Schwefel, H.P., de Brito Mendes, M.A.: 45 years of evolution strategies: Hans-Paul Schwefel interviewed for the genetic argonaut blog. ACM SIGEVOlution **4**(2), 2–8 (2010)
21. Rudolph, G.: Evolutionary strategies. In: Rozenberg, G., Bäck, T., Kok, J.N. (eds.) Handbook of Natural Computing, pp. 673–698. Springer, Heidelberg (2012). https://doi.org/10.1007/978-3-540-92910-9_22
22. Bäck, T., Foussette, C., Krause, P.: Contemporary Evolution Strategies, vol. 86. Springer, Berlin (2013)
23. Hansen, N., Arnold, D.V., Auger, A.: Evolution strategies. In: Kacprzyk, J., Pedrycz, W. (eds.) Springer Handbook of Computational Intelligence. Springer Handbooks, pp. 871–898. Springer, Heidelberg (2015). https://doi.org/10.1007/978-3-662-43505-2_44
24. Yi, S., Wierstra, D., Schaul, T., Schmidhuber, J.: Stochastic search using the natural gradient. In Proceedings of International Conference on Machine Learning, pp. 1161–1168 (2009)
25. Salimans, T., Ho, J., Chen, X., Sidor, S., Sutskever, I.: Evolution strategies as a scalable alternative to reinforcement learning. arXiv preprint arXiv:1703.03864 (2017)

26. Moritz, P., et al.: Ray: A distributed framework for emerging AI applications. In USENIX Symposium on Operating Systems Design and Implementation, pp. 561–577 (2018)
27. Mania, H., Guy, A., Recht, B.: Simple random search of static linear policies is competitive for reinforcement learning. In: Proceedings of Neural Information Processing Systems, pp. 1805–1814 (2018)
28. Alba, E., Tomassini, M.: Parallelism and evolutionary algorithms. IEEE Trans. Evol. Comput. **6**(5), 443–462 (2002)
29. Auger, A., Hansen, N., López-Ibáñez, M., Rudolph, G.: Tributes to Ingo Rechenberg (1934–2021). ACM SIGEVOlution **14**(4), 1–4 (2022)
30. Fortin, F.A., De Rainville, F.M., Gardner, M.A.G., Parizeau, M., Gagné, C.: DEAP: evolutionary algorithms made easy. J. Mach. Learn. Res. **13**(1), 2171–2175 (2012)
31. Müller, C.L., Baumgartner, B., Ofenbeck, G., Schrader, B., Sbalzarini, I.F.: pCMALib: a parallel fortran 90 library for the evolution strategy with covariance matrix adaptation. In: Proceedings of Genetic and Evolutionary Computation Conference, pp. 1411–1418 (2009)
32. Gustafson, J.L.: Reevaluating Amdahl's law. Commun. ACM **31**(5), 532–533 (1988)
33. Harris, C.R., Millman, K.J., Van Der Walt, S.J., et al.: Array programming with NumPy. Nature **585**(7825), 357–362 (2020)
34. Arnold, D.V., MacLeod, A.: Hierarchically organised evolution strategies on the parabolic ridge. In: Proceedings of Annual Conference on Genetic and Evolutionary Computation, pp. 437–444 (2006)
35. Beyer, H.G., Dobler, M., Hämmerle, C., Masser, P.: On strategy parameter control by Meta-ES. In: Proceedings of Annual Conference on Genetic and Evolutionary Computation, pp. 499–506 (2009)
36. Beyer, H.G., Hellwig, M.: Mutation strength control by Meta-ES on the sharp ridge. In: Proceedings of Annual Conference on Genetic and Evolutionary Computation, pp. 305–312 (2012)
37. Duan, Q.Q., Zhou, G.C., Shao, C., Yang, Y.J., Shi, Y.H.: Distributed evolution strategies for large scale optimization. In: Proceedings of Genetic and Evolutionary Computation Conference Companion (2022, Accepted)
38. Moritz, P.C.: Ray: a distributed execution engine for the machine learning ecosystem Doctoral dissertation, UC Berkeley (2019)
39. Whitley, D., Rana, S., Dzubera, J., Mathias, K.E.: Evaluating evolutionary algorithms. Artif. Intell. **85**(1–2), 245–276 (1996)
40. Varelas, K., et al.: Benchmarking large-scale continuous optimizers: the bbob-largescale testbed, a COCO software guide and beyond. Appl. Soft Comput. **97**, 106737 (2020)
41. Auger, A., Hansen, N.: Benchmarking: State-of-the-art and beyond. In: Proceedings of Genetic and Evolutionary Computation Conference Companion, pp. 339–340 (2021)
42. Meunier, L., et al.: Black-box optimization revisited: improving algorithm selection wizards through massive benchmarking. IEEE Trans. Evol. Comput. Early Access (2021)
43. Wolpert, D.H., Macready, W.G.: No free lunch theorems for optimization. IEEE Trans. Evol. Comput. **1**(1), 67–82 (1997)
44. https://github.com/Evolutionary-Intelligence/pypop

Recombination Weight Based Selection in the DTS-CMA-ES

Oswin Krause[✉] [ID]

University of Copenhagen, 2300 Copenhagen, Denmark
oswin.krause@di.ku.dk

Abstract. Surrogate model based Evolution Strategies (like the doubly trained surrogate model CMA-ES, DTS-CMA-ES) use a model of the objective function to reduce the number of function evaluations during optimization. This work investigates to use the expected selection weights averaged over the GP posterior distribution as replacement of the fitness and to guide point-selection for evaluation via the variance of the weights. Results obtained on BBOB show that the proposed technique performs on par with current strategies and allows the usage of surrogate models that are invariant to strictly increasing transformations of the function values. However, initial experiments showed that simple modeling of ranks in the GP does lead to worse results than current GP models of the function values.

Keywords: CMA-ES · DTS-CMA-ES · Gaussian process · Surrogate models · Recombination

1   Introduction

In this paper I consider minimizing a black-box function $f_{\mathrm{opt}} : \mathbb{R}^d \to \mathbb{R}$. I assume that f_{opt} is expensive to evaluate, but noise-free and unconstrained. For this setting, surrogate models are an efficient way to speed up evolution strategies. Recently, there have been many important contributions that combine surrogate models with evolution strategies (ES), especially the CMA-ES [10].

While there are a number of different approaches for surrogate models, for example approaches based on ranking [16,17] or quadratic models [1,4,5,7,12], many algorithms [2,3,13,20,22,23] tend to use Gaussian Processes (GPs) for directly predicting the function values. This is because GPs allow a perfect fit to the observed function values in the noiseless regime, while also providing an estimate of model uncertainty for points that have not been observed. Moreover, since GPs are likelihood based, hyper parameters of the model can be optimized to obtain a better fit to the data. Finally, the use of GPs or other well-understood probabilistic models makes it easy to adopt the successful prior work from Bayesian Optimization (BO [11,21]), especially point-selection criteria for evaluation. The downside is, that GPs are not invariant or equivariant to monotonous transformations of the fitness function.

© The Author(s), under exclusive license to Springer Nature Switzerland AG 2022
G. Rudolph et al. (Eds.): PPSN 2022, LNCS 13399, pp. 295–308, 2022.
https://doi.org/10.1007/978-3-031-14721-0_21

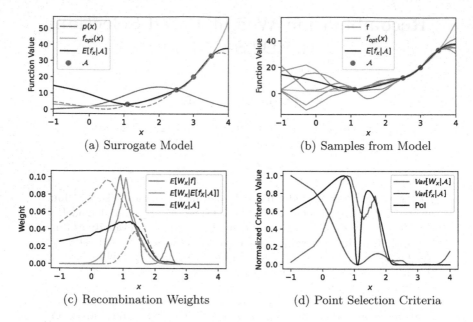

(a) Surrogate Model   (b) Samples from Model

(c) Recombination Weights   (d) Point Selection Criteria

Fig. 1. Visualisation of the idea presented in this work. a) Posterior Mean $\mathbb{E}[f_x|\mathcal{A}]$ (black) and confidence intervals (25%–75%, grey dashed lines) of a Gaussian process fitted to the points in $\mathcal{A}$ (red points) evaluated on the target function f_{opt} (orange line), as well as current search distribution (red line). b) Samples from the Gaussian Process (grey). c) Expected recombination weights $\mathbb{E}[W_x|\mathbb{E}[f_x|\mathcal{A}]]$ (orange), when imputing the mean function, compared to weights from a single sampled function $\mathbb{E}[W_x|f]$. The expectation over all functions from the posterior $\mathbb{E}[W_x|\mathcal{A}]$ (black) with confidence intervals (grey) is very different from $\mathbb{E}[W_x|\mathbb{E}[f_x|\mathcal{A}]]$. d) Comparison of selection criteria. Variance of prediction (blue) and Probability of improvement (black) compared to the proposed weight variance (red). All criteria are scaled to lie between zero and one. (Color figure online)

One of the most successful surrogate-based algorithms for small budgets, the DTS-CMA-ES [19] combines GPs with the CMA-ES by first using BO-based point selection criteria to select which points to evaluate on the objective function, and then use the GP to impute the remaining function-values with the mean-value predicted by the GP.

In this work, I take the DTS-CMA-ES as a starting point to investigate a so-far underexplored area in combining surrogate-based models with evolution strategies: the interplay between surrogate model prediction and weighted selection. Currently, in the DTS-CMA-ES, neither the point selection criteria for evaluation of points, nor prediction of fitness values of unevaluated points takes selection and recombination weights of the underlying ES into account. I hypothesize that this under-utilizes the information captured by the GP, as functions drawn from a GP (see Figs. 1a and 1b) will locate the optimum at different locations (see Fig. 1c), reflecting the uncertainty of the shape of the function after observing the evaluated samples. This uncertainty is missing in the modeled mean.

I propose a novel point-selection criterion, by computing the variance of the selection weights of the CMA-ES (Fig. 1d) and propose to impute the expected recombination weights for all points, instead of the fitness values. For this, I investigate two approaches: 1) a complete replacement of fitness values by the expected recombination weights, and 2) replacement of the recombination weights in the update of the strategy parameters of the CMA-ES. Both approaches effectively decouple the underlying ES from the modeled fitness values. A result of this is that instead of modeling the fitness function, a GP can now be used to model the ranks of evaluated points, which re-introduces invariance to monotonic transformations of the fitness function.

This article is structured as follows. First, the background on GPs, BO, CMA-ES, DTS-CMA-ES and fitting of GP-surrogate models is introduced in Sect. 2. The reader unfamiliar with GPs should read Sect. 2.1 as it also introduces some notation used in the remainder of the article. The main contribution of this article is presented in Sect. 3. Experiments are described in Sect. 4 and results are discussed in Sect. 5. Finally, the article concludes in Sect. 6.

2   Background

2.1   Gaussian Processes

From a birds-eye view, a Gaussian Process (GP) can be seen as a distribution on a Hilbert-space of functions $f : \mathcal{X}^d \to \mathbb{R}$, $\mathcal{X} \subseteq \mathbb{R}^d$. A GP can be fully described by the choice of a mean function $m_{\mathcal{GP}} : \mathcal{X} \to \mathbb{R}$ and a positive definite kernel $k : \mathcal{X} \times \mathcal{X} \to \mathbb{R}$. If we take f as a random function with distribution following that of a GP, it is common to write

$$f \sim \mathcal{GP}(m_{\mathcal{GP}}(\cdot), k(\cdot, \cdot)) \ . \tag{1}$$

In general, f cannot be sampled and instead it is only possible to compute the probabilities of function values of f at a finite set of points $X = \{x_1, \ldots, x_\ell\}$, $x_i \in \mathcal{X}$. We write for the observation of f at points in X

$$f_X = (f_{x_1}, \ldots, f_{x_\ell})^T = (f(x_1), \ldots, f(x_\ell))^T \ .$$

The probability distribution of f_X, $p(f_X) = p(f_{x_1}, \ldots, f_{x_\ell})$ in a GP follows a multivariate normal distribution

$$f_X \sim \mathcal{N}(m_{\mathrm{GP}}(X), K(X)) \ ,$$

where $K(X)$ is a symmetric positive definite $\ell \times \ell$ matrix with element i, j given by $k(x_i, x_j)$. In the following, we denote for sets X and X' the matrix $K(X, X')$ as the matrix with elements $k(x_i, x'_j)$ and identify $K(X) = K(X, X)$.

The fact that all marginal distributions are normal allows efficient inference in GPs. For example, if we are given two sets of points X and X', where we have observed the function values of $f_{X'}$, we can compute the conditional distribution of $f_X|f_{X'}$ easily, via

$$f_X|f_{X'} \sim \mathcal{N}(m(X|X'), K(X|X')) \tag{2}$$

with

$$m(X|X') = m_{GP}(X) + K(X, X')K(X')^{-1}(f_{X'} - m_{GP}(X'))$$
$$K(X|X') = K(X) - K(X, X')K(X')^{-1}K(X, X')^T .$$

An example of a GP is given in Fig. 1a, which shows mean and confidence intervals of the marginal distribution $f_x \in \mathbb{R}$, conditioned on the archive $\mathcal{A}$. In this Figure, we can see that close to a point in the archive the variance shrinks to zero, while points that are far away from points in the archive have a fairly large variance. However, this depiction alone is misleading, as it only shows the variances but ignores the complex covariance structure between points. A way to visualize this structure is by sampling from the conditional distribution for points sampled on a grid, as is visualized in Fig. 1b. Now it becomes clear that the samples drawn from this distribution are not very similar to the mean function. Indeed, while the mean function only implies a single local optimum, the samples drawn have multiple local optima with global optima at both sides of the current best points in the archive.

2.2 Bayesian Optimization

Bayesian optimization (BO) is a branch of optimization that is devoted to optimizing expensive black-box functions and is quite successful on low-dimensional problems [14,15,21]. Most of the basic BO methodology has not changed since the introduction of the EGO algorithm [11]. In EGO, the function f_{opt} is replaced by a probabilistic surrogate model $p(f_x|\mathcal{A})$, derived using Bayes' theorem. Once an archive of known function values $\mathcal{A} = \{(x_1, y_1), \dots\}$, $y_i = f_{opt}(x_i)$ is available, Bayes theorem can be used to derive the posterior distribution based on a chosen prior $p(f)$. The GP in Sect. 2.1 is an example for this with posterior given by equation (2). A BO algorithm can then use this posterior distribution to select the next point to evaluate. This is done by maximizing an acquisition function, or selection criterion $\mathcal{C}_{\mathcal{A}}(x)$. Then, after the algorithm selected the next point, it is evaluated and added to the model. This process repeats until a solution is found that is good enough, or the evaluation budget is exceeded.

There is a large number of potential selection functions that balance the trade-off between selecting a point that likely improves the function value and selecting a point that improves the model. Three classical functions are

- **Upper Confidence Bound (UCB)**: $\mathcal{C}_{\mathcal{A}}(x) = -\mathbb{E}[f_x|\mathcal{A}] + \alpha\sqrt{\mathrm{Var}[f_x|\mathcal{A}]}$,
- **Probability of Improvement (PoI)**: $\mathcal{C}_{\mathcal{A}}(x) = P(f_x \leq \min_i y_i - \epsilon|\mathcal{A})$,
- **Expected Improvement (EI)**: $\mathcal{C}_{\mathcal{A}}(x) = -\int_{-\infty}^{\min_i y_i} P(f_x|\mathcal{A})f_x df_x$,
- **Model variance** $\mathcal{C}_{\mathcal{A}}(x) = \mathrm{Var}[f_x|\mathcal{A}]$.

Both PoI and UCB have a hyper parameter α or ϵ, that each govern the exploration/exploitation trade-off. In both cases, the larger the parameter, the more the model explores. In UCB this is driven by giving more weight to the model uncertainty, while in PoI ϵ governs the minimum amount of improvement over the best function value in the archive. As an extreme example on the other end, the model variance just minimizes the model uncertainty. See Fig. 1d for an example of PoI and model variance.

2.3 CMA-ES

The CMA-ES [10] is an evolution strategy that in each iteration t represents its search distribution by a normal distribution $\mathcal{N}(m_t, \sigma_t^2 C_t)$, where $\sigma_t > 0$ is called the step-size and C_t is a covariance matrix. Further, the algorithm keeps track of two evolution paths $p_{C,t}$ and $p_{\sigma,t}$ that are meant to learn long term trends in step direction and length. In each iteration, the steps are:

1. **Sample Population**. The algorithm samples a population

$$x_1, \ldots, x_\lambda \sim \mathcal{N}(m_t, \sigma_t^2 C_t)$$

2. **Ranked weights.** The sampled population is evaluated to obtain $y_i = f(x_i)$. Then the points/function-value pairs are reordered such that $y_1 \leq y_2 \leq \cdots \leq y_\lambda$. Finally, each point is assigned a weight according to its rank, w_i. The weights are chosen such, that the best $\mu = \lfloor \frac{\lambda}{2} \rfloor$ points are assigned a non-zero weight. To be more precise, $w_i = \frac{w_i'}{\sum_{i=1}^{\lambda} w'}$ and

$$w_i' = \begin{cases} \log\left(\frac{\lambda+1}{2}\right) - \log(i), & \text{if } i \leq \mu \\ 0, \text{otherwise} \end{cases}.$$

3. **Update of Strategy parameters** The algorithm now updates the strategy parameters.[1] The update of the mean is simply the weighted mean of sampled points $m_{t+1} = \sum_{i=1}^{\lambda} w_i x_i$. Then, two evolution paths are updated. For this, first a normalized step is computed

$$y_t = \sqrt{\mu_{\text{eff}}} \frac{m_{t+1} - m_t}{\sigma_t} \tag{3}$$

This normalization ensures that if samples are ranked randomly, then $y_t \sim \mathcal{N}(0, C_t)$. Especially, the normalization with $\mu_{\text{eff}} = 1/\sum_{i=1}^{\lambda} w_i^2$ corrects for the loss in variance due to the weighted mean. With this normalized value, the update of the paths then reads

$$p_{\sigma,t+1} = (1 - c_\sigma)p_{\sigma,t} + \sqrt{c_\sigma(2 - c_\sigma)} C_t^{-1/2} y_t$$

$$p_{C,t+1} = (1 - c_c)p_{C,t} + \sqrt{c_c(2 - c_c)} y_t .$$

---

[1] The description of the $h_\sigma \in \{0, 1\}$ mechanism is missing for brevity.

Next the step-size is updated via the Cumulative Step-Size adaptation [18]

$$\sigma_{t+1} = \sigma_t \exp\left(\frac{c_\sigma}{d_\sigma}\left(\frac{\|p_{\sigma,t+1}\|}{\mathbb{E}[\chi(d)]} - 1\right)\right) ,$$

where $\mathbb{E}[\chi(d)]$ is the expectation of a χ variable with d degrees of freedom. Finally, C_t is updated via:

$$C_{t+1} = (1 - c_1 - c_\mu)C_t + \frac{c_\mu}{\sigma_t^2}\sum_{i=1}^{\mu} w_i(x_i - m_t)(x_i - m_t)^T + c_1 p_{C,t+1} p_{C,t+1}^T$$

The parameters $c_\sigma, d_\sigma, c_c, c_1, c_\mu > 0$ are learning rates. Further, the default population size λ is chosen as $\lambda = 4 + \lfloor 3\log(d) \rfloor$.

2.4   DTS-CMA-ES

The doubly trained surrogate CMA-ES, DTS-CMA-ES is an extension of the CMA-ES in the area of expensive optimization [3,19]. It changes the evaluation of sampled points in the CMA-ES, while keeping the sampling and update mechanism unchanged. It stores all evaluated points and function-values in an archive $\mathcal{A} = \{(x_1, y_1), \dots\}$ and uses them to fit a GP to model of f_{opt} around the current mean m_t of the search distribution. The DTS-CMA-ES performs the following steps:

1. Fit a Gaussian Process using points in $\mathcal{A}$
2. Query the current population $x_1, \dots, x_\lambda$ from the CMA-ES
3. Select n_{orig} points with indices $I = \{I_1, \dots, I_{n_{\mathrm{orig}}}\} \subset \{1, \dots, \lambda\}$ from the population that maximize the selection criterion $\mathcal{C}_\mathcal{A}(x_i)$
4. Evaluate $f_{I_k} = f(x_{I_k})$ for the selected points and add pairs (x_{I_k}, f_{I_k}), $k = 1, \dots, n_{\mathrm{orig}}$ to $\mathcal{A}$
5. Fit the surrogate model with the updated $\mathcal{A}$, and set $f_i = \mathbb{E}[f_{x_i}]$ for $i \notin I$.
6. Update the CMA-ES using the computed values of f_i

In a larger study [3] the authors compared many selection criteria $\mathcal{C}_\mathcal{A}$ commonly used in BO (including EI) and concluded that both model variance and PoI performed best, where PoI performed better on uni-modal functions and model variance performed better on multi-modal functions. For PoI ϵ was chosen as $\epsilon = 0.05(\max_i y_i - \min_i y_i)$. Another selection criterion that investigated the change of ranking uncertainty of the model was deemed to not be superior to these two approaches.

Further, the authors investigated the hyper parameters of the algorithm As selection threshold $r_{\max}$ the authors chose a multiple of the 99th percentile of a $\chi(d)$ distributed variable, $r_{\max} = 4\chi(0.99, d)$. For the population size λ, the authors concluded that doubling the population to $\lambda = 8 + \lfloor 6\log(d) \rfloor$ greatly improved performance. The number of evaluated points should scale with the population size and here it was found that $n_{\mathrm{orig}} = \lceil 0.05\lambda \rceil$ performed best, which for dimensions $d < 20$ is one.

2.5 Model Fitting

In this section, I expand on Sect. 2.1 and explain how the GP is fit as part of the DTS-CMA-ES. The fitting procedure closely follows the strategy described in [19] with all selected hyper parameters. As kernel, the Matern-5/2 kernel is used

$$k(x,y) = \alpha_k^2 \frac{2^{1-\nu}}{\Gamma(\nu)} z^\nu K_\nu(z), \; z = \sqrt{2\nu} \frac{\|x - y\|}{\sigma_k} \; ,$$

where lengthscale parameter σ_k and variance scale α_k are parameters, $\nu = 5/2$ is fixed and K_ν is the modified Bessel function. This kernel leads to a prior in the Hilbert-space of functions that are twice continuously differentiable. Further, the mean function is constant zero, $m_{GP}(x) = 0$.

Fitting and evaluating a population X on the GP in the DTS-CMA-ES consists of the following steps:

1. Select and normalize points

$$\mathcal{A}_{\text{fit}} = \left\{ (T(x), \frac{y - m_y}{\sigma_y}) \mid \|T(x)\| \leq r_{\max}, \; (x,y) \in \mathcal{A} \right\} \; ,$$

where $T(x) = \frac{C_t^{-1/2}}{\sigma_t}(x - m_t)$ and m_y, σ_y are selected such, that the function values in $\mathcal{A}_{\text{fit}}$ have mean zero and unit variance.
2. Find the optimal parameters σ_k, α_k for the Matern kernel by maximizing the marginal likelihood of the GP using $\mathcal{A}_{\text{fit}}$.
3. For a new population of sampled points X, first apply T to all points in X, then compute the conditional distribution of f_X according to Eq. (2) and correct by adding m_y to each entry of the mean of the conditional distribution.

The main idea of the first step is that fitting a Gaussian process on a function with high conditioning is difficult, since this would require learning a kernel with full covariance matrix. Instead, we use the fact that the CMA-ES tends to learn a covariance $\sigma_t^2 C_t$ which is some approximation of the inverse hessian matrix. The transformation $T(x)$ normalizes the coordinate system such, that the search distribution becomes $\mathcal{N}(0, I)$, which then simplifies the local shape of the function in the transformed coordinate system.

Further, as in this coordinate system the current population is standard normally distributed, we can use the norm of a point in the archive $\|T(x)\|$ to assess whether it is relevant for fitting. This turns the global GP model into a local model, as when maximizing the likelihood, points that are further away than $r_{\max}$ do not affect the fit. As a consequence, the fitting procedure in Step 2 will focus on points in the archive that are likely candidates for sampling, instead of being affected of areas far away that might require very different kernel parameters to be fit correctly.

3 Fully Weight-Based DTS-CMA-ES

The DTS-CMA-ES as described in Sect. 2.4 has two disadvantages. First, it is not invariant to monotonous transformations of the objective function. Even though the underlying CMA-ES uses ranking, the GP used to model the function-values is neither invariant, nor equivariant to monotonous increasing transformations of the function values. Thus, the choice of points of the selection criteria and even the relative ranking of the selected mean values will differ.

Second, the way the GP is used to select and impute fitness values does not take the selection mechanism of the CMA-ES into account. To see this, let $\mathcal{W}$: $\mathbb{R}^\lambda \to \mathbb{R}^\lambda$, be the function that for the vector of function values of the population $f_X = (f_{x_1}, \ldots, f_{x_\lambda})$ computes the rank-based recombination weights, $W_X = \mathcal{W}(f_X)$ of the CMA-ES. Further, let $p(x)$ be the current search distribution and let X be the current population.

The GP conditioned on the archive of previously observed function evaluations $\mathcal{A}$ gives rise to a random variable $f_X|\mathcal{A}$, the distribution of function values at X when drawing a function from the posterior distribution of the GP. Then, $W_X|\mathcal{A} = \mathcal{W}(f_X|\mathcal{A})$ also becomes a random variable: each function drawn from the GP can result in different ranks, and thus weights, even when the population is kept fixed. The differences can be large, as can be seen from the samples drawn in Fig. 1c and the expected weight assigned to a point for a single function f drawn from the GP shown in Fig. 1c (blue line). As a result, the expectation of function values $\mathbb{E}[W_X|\mathcal{A}]$ can differ a lot from $\mathcal{W}(\mathbb{E}[f_x|\mathcal{A}])$, the rank-weights using the mean of the posterior distribution of the GP. Again, this is visualized in Fig. 1c (orange and black lines), where also the average is taken over all populations to obtain the average weight of a sampled point x.

As a result, imputing the mean as fitness value for the CMA-ES in the DTS-CMA-ES loses variance information learned by the GP and using $\mathbb{Var}[f_X|\mathcal{A}]$ or PoI does only incompletely reflect the actual uncertainty in the weighting process—for example it will over-estimate the variance of points that are likely assigned weight zero and under-estimate the variance of points that are assigned large weights.

I therefore propose the following changes: instead of using PoI or GP posterior variance as selection criterion in step 3 of the algorithm (Sect. 2.4), use $\mathbb{Var}[W_X|\mathcal{A}]$ (see Fig. 1d), and instead of imputing the mean in step 5 for all unevaluated points, either impute the expected rank weight $\mathbb{E}[W_X|\mathcal{A}]$ for all points, including the evaluated points, or directly replace the weights used by the CMA-ES by $\mathbb{E}[W_X|\mathcal{A}]$. Since these weights can not be computed analytically, Monte-Carlo sampling can be used by directly sampling function values from the posterior distribution of the GP and computing the rank weights for each sample.

If the weights of the CMA-ES are replaced, one must also change Eq. (3) to compute μ_{eff} from the imputed weights, as otherwise the expected lengths of the evolution paths will change. The result is a close approximation to taking the expected step, when sampling over all possible samples $f|\mathcal{A}$ from the GP.

Replacing the predicted and evaluated fitness by the weight also allow us to re-introduce invariance to monotonous transformations: when replacing the GP by a model that is invariant to monotonous transformations, then no further changes are needed if the algorithm uses rank-weights instead of fitness values.

4   Experiments

I compare seven variants of the DTS-CMA-ES (see Sect. 2.4) with each other and the CMA-ES on the BBOB single-objective benchmark suite of 24 functions [9] on dimensions $d \in \{2, 3, 5, 10, 20\}$. To compare the effect of invariance, I further compare the algorithms on functions $f_1^\alpha(x) = (x^2)^\alpha$, $\alpha \in \{1/4, 1/2, 1, 3/2, 2\}$ for $d = 5$.

The variants of the DTS-CMA-ES include the previously evaluated versions [3], using either PoI (DTS-PoI-CMA-ES) or GP variance (DTS-Var-CMA-ES). The new variants are: DTSV-CMA-ES that just changes the selection criterion to the newly proposed criterion, DTSVE-CMA-ES, that additionally imputes the expected rank-weight for all points and DTSW-CMA-ES that uses the new criterion and replaces the rank weights in the CMA-ES by the computed mean. For the last two algorithms, I also add two invariant versions, I-DTSW-CMA-ES and I-DTSVE-CMA-ES. Invariance is introduced by fitting the GP to the numerical ranks of the points in the archive, instead of using the function values. All variants use the same model and fitting process as described in Sect. 2.5 and I use GPy for implementing the GP [6]. For the CMA-ES, I adapted the pycma package [8] version 3.2.2 without active-learning or mirrored sampling, thus following Sect. 2.3.

For number of evaluations, I picked $n_{\text{orig}} = 1$ for $d < 10$ and $n_{\text{orig}} = 2$ for $d \in \{10, 20\}$. Estimates for $\mathbb{E}[W_X|\mathcal{A}]$ and $\mathbb{Var}[W_X|\mathcal{A}]$ are based on 100 samples from the GP posterior. In all variants, the full population is evaluated in the first evaluation. This empirically led to more stability of all variants of the DTS-CMA-ES.

The initial mean m_0 is taken from the proposed starting point of the BBOB functions and the initial variance is taken as $\sigma_0 = 8/3$, following [3]. Restarts are performed whenever the largest eigenvalue of $\sigma_t C_t$ is smaller than 10^{-7} or the largest eigenvalue is larger than 10^{10}. In that case, the search distribution is reset to the initial values and the population size is doubled as is n_{orig}. Further, the archive is sub-sampled to ensure that no points are closer than $\sigma_0/4$ to each other. This prevents numerical instabilities due to very close points, while informing the model of the previous evaluations. As budget, $50d$ function evaluations were used. Restarts therefore only happened rarely.

5   Results and Discussion

The results of the algorithm on BBOB are given in Fig. 2. For space-reasons I only show results for $d =\in \{3, 5, 20\}$, but results for $d = 2$ and $d = 10$ are qualitatively similar to $d = 3$ and $d = 20$, respectively. Comparing the average results on all functions (Figs. 2a and 2c and 2e), there are three groups of algorithms: the non-invariant DTS-CMA-ES variants (DTS-PoI-CMA-ES, DTS-Var-CMA-ES, DTSV-CMA-ES, DTSV-CMA-ES, DTSVE-CMA-ES and DTSW-CMA-ES) all showed comparable performance, with no clear winner. The invariant algorithms (I-DTSVE-CMA-ES and I-DTSW-CMA-ES) performed clearly worse, but still better than the CMA-ES without a surrogate model.

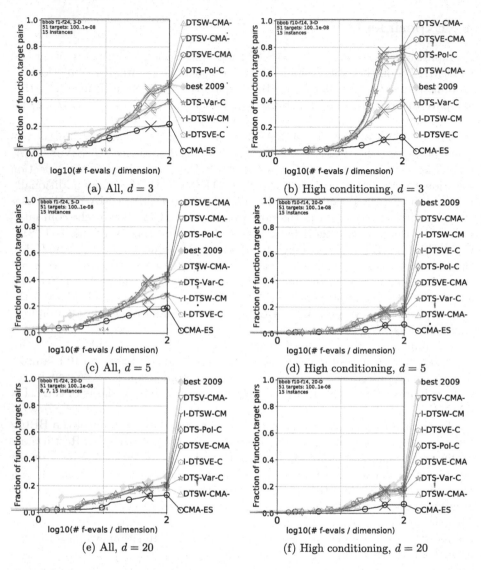

Fig. 2. Bootstrapped empirical cumulative distribution of the number of objective function evaluations divided by dimension #fevals/d for 51 targets with target precision in $10^{[-8..2]}$ for $d \in \{3, 5, 20\}$. As reference algorithm, the best algorithm from BBOB 2009 is shown as light thick line with diamond markers. Left, a)c)e) results averaged over all 24 functions, Right, b)d)e) results of the functions with high condition number ($f_{10}-f_{14}$). Not shown are results for $d = 2$ and $d = 10$, which are visually similar to $d = 3$ and $d = 20$, respectively.

The function group with the largest difference over BBOB was the group of functions with high conditioning, $f_{10} - f_{14}$, shown in Figs. 2b and 2d and 2f. Here the invariant algorithms performed markedly worse, especially for $d = 5$.

This is probably because optimizing problems with high conditioning requires the CMA-ES to adapt the covariance matrix to the right shape before being able to make any progress. Thus, at this small budget even small differences in the adaptation speed can lead to large differences in the evaluation. This is supported by the results shown in Fig. 3. Figure 3a shows that all algorithms obtained similar convergence speeds on Sphere, however the invariant algorithms and DTS-Var-CMA-ES were slightly slower. Thus the relative slow progress on f_{10}, Fig. 3b, was likely a result of slower learning of the covariance matrix. But in general, learning the covariance matrix worked as is illustrated by f_{14}, Fig. 3c, where the covariance matrix becomes more ill-conditioned the closer the points are to the optimum. Here, the algorithms performed more similar.

For the other function groups (not shown), especially multi-modal functions, there were no significant differences, as none of the algorithms showed a better capability to discover better local optima at this budget. However, for multi-modal functions, all DTS-CMA-ES algorithms performed on par or better than the best 2009 baseline.

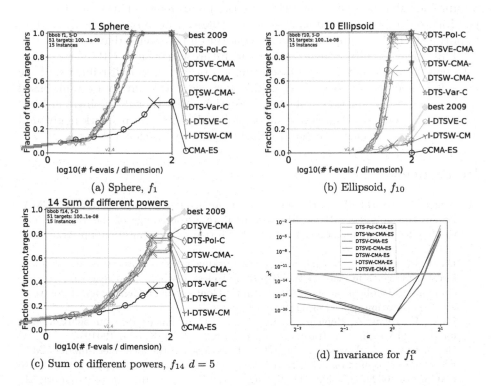

(a) Sphere, f_1

(b) Ellipsoid, f_{10}

(c) Sum of different powers, f_{14} $d = 5$

(d) Invariance for f_1^α

Fig. 3. Selected results. a)b)c) Bootstrapped empirical cumulative distribution of the number of objective function evaluations divided by dimension #fevals/d for 51 targets with target precision in $10^{[-8..2]}$ for $d = 5$ and functions f_1, f_{10} and f_{14}. As reference algorithm, the best algorithm from BBOB 2009 is shown as light thick line with diamond markers. d) Test for invariance to monotonic transformations in $d = 5$. Shown are the smallest function value achieved for the different algorithms on $f_1^\alpha = (x^2)^\alpha$ for $\alpha \in \{1/4, 1/2, 1, 1.5, 2\}$. For $\alpha = 1$ the ECDF graph is shown in a).

Finally, the results for invariance to monotonous transformations are shown in Fig. 3d. Here, it can be seen that when moving away from the simple quadratic function, performance of the non-invariant algorithm varied, and for $\alpha = 2$ the algorithms failed to make any progress. For the invariant algorithms, performance is expected to be the same as for $\alpha = 1$.

6   Conclusion

In this work, I investigated the combination of the recombination weights of the CMA-ES with the variance estimates provided by the GP surrogate model. I proposed to use the weight variance as criterion for deciding which point to evaluate, and proposed to use their mean as replacement for the function values (or direct replacement of the recombination weights) in the underlying CMA-ES. I finally used this to develop a variant of the DTS-CMA-ES that is fully invariant to monotonous transformations of the function values.

The results showed that using statistics on the recombination weights instead of function values has negligible effects on the performance. This shows that it is possible to completely avoid using function values from surrogate models. More importantly, the new approach has no exploration parameter that needs tuning.

This opens up interesting directions in future work: since GP-inference can take noise into account and gives an estimate of the posterior distribution of the noise-free function, the techniques in this work open the way to use rank-based statistics on noisy function evaluations. Further, even though the invariant GP used in this work is rather simple and performed poorly, future work might find better models that close the performance gap in the noiseless case.

References

1. Auger, A., Brockhoff, D., Hansen, N.: Benchmarking the local metamodel CMA-ES on the noiseless BBOB'2013 test bed. In: Proceedings of the 15th Annual Conference Companion on Genetic and Evolutionary Computation, pp. 1225–1232 (2013)
2. Bajer, L., Pitra, Z., Holeňa, M.: Benchmarking gaussian processes and random forests surrogate models on the BBOB noiseless testbed. In: Proceedings of the Companion Publication of the 2015 Annual Conference on Genetic and Evolutionary Computation, pp. 1143–1150 (2015)
3. Bajer, L., Pitra, Z., Repický, J., Holeňa, M.: Gaussian process surrogate models for the CMA evolution strategy. Evol. Comput. **27**(4), 665–697 (2019)
4. Bouzarkouna, Z., Auger, A., Ding, D.Y.: Investigating the local-meta-model CMA-ES for large population sizes. In: Di Chio, C., et al. (eds.) EvoApplications 2010. LNCS, vol. 6024, pp. 402–411. Springer, Heidelberg (2010). https://doi.org/10.1007/978-3-642-12239-2_42
5. Bouzarkouna, Z., Auger, A., Ding, D.Y.: Local-meta-model CMA-ES for partially separable functions. In: Proceedings of the 13th Annual Conference on Genetic and Evolutionary Computation, pp. 869–876 (2011)
6. GPy: A Gaussian process framework in Python. http://github.com/SheffieldML/GPy (Since 2012)

7. Hansen, N.: A global surrogate assisted CMA-ES. In: Proceedings of the Genetic and Evolutionary Computation Conference, pp. 664–672 (2019)
8. Hansen, N., Akimoto, Y., Baudis, P.: CMA-ES/pycma on Github, February 2019. https://doi.org/10.5281/zenodo.2559634
9. Hansen, N., Auger, A., Ros, R., Mersmann, O., Tušar, T., Brockhoff, D.: COCO: a platform for comparing continuous optimizers in a black-box setting. Optim. Methods Softw. **36**(1), 114–144 (2021)
10. Hansen, N., Ostermeier, A.: Completely derandomized self-adaptation in evolution strategies. Evol. Comput. **9**(2), 159–195 (2001)
11. Jones, D.R., Schonlau, M., Welch, W.J.: Efficient global optimization of expensive black-box functions. J. Global Optim. **13**(4), 455–492 (1998)
12. Kern, S., Hansen, N., Koumoutsakos, P.: Local meta-models for optimization using evolution strategies. In: Runarsson, T.P., Beyer, H.-G., Burke, E., Merelo-Guervós, J.J., Whitley, L.D., Yao, X. (eds.) PPSN 2006. LNCS, vol. 4193, pp. 939–948. Springer, Heidelberg (2006). https://doi.org/10.1007/11844297_95
13. Koza, J., Tumpach, J., Pitra, Z., Holeňa, M.: Using past experience for configuration of Gaussian processes in Black-Box Optimization. In: Simos, D.E., Pardalos, P.M., Kotsireas, I.S. (eds.) LION 2021. LNCS, vol. 12931, pp. 167–182. Springer, Cham (2021). https://doi.org/10.1007/978-3-030-92121-7_15
14. Le Riche, R., Picheny, V.: Revisiting Bayesian optimization in the light of the COCO benchmark. Struct. Multidiscip. Optim. **64**(5), 3063–3087 (2021)
15. Liu, Z., et al.: Towards automated deep learning: analysis of the AutoDL challenge series 2019. In: NeurIPS 2019 Competition and Demonstration Track, pp. 242–252. PMLR (2020)
16. Loshchilov, I., Schoenauer, M., Sebag, M.: Comparison-based optimizers need comparison-based surrogates. In: Schaefer, R., Cotta, C., Kołodziej, J., Rudolph, G. (eds.) PPSN 2010. LNCS, vol. 6238, pp. 364–373. Springer, Heidelberg (2010). https://doi.org/10.1007/978-3-642-15844-5_37
17. Loshchilov, I., Schoenauer, M., Sebag, M.: Self-adaptive surrogate-assisted covariance matrix adaptation evolution strategy. In: Proceedings of the 14th Annual Conference on Genetic and Evolutionary Computation, pp. 321–328 (2012)
18. Ostermeier, A., Gawelczyk, A., Hansen, N.: Step-size adaptation based on non-local use of selection information. In: Davidor, Y., Schwefel, H.-P., Männer, R. (eds.) PPSN 1994. LNCS, vol. 866, pp. 189–198. Springer, Heidelberg (1994). https://doi.org/10.1007/3-540-58484-6_263
19. Pitra, Z., Bajer, L., Holeňa, M.: Doubly trained evolution control for the surrogate CMA-ES. In: Handl, J., Hart, E., Lewis, P.R., López-Ibáñez, M., Ochoa, G., Paechter, B. (eds.) PPSN 2016. LNCS, vol. 9921, pp. 59–68. Springer, Cham (2016). https://doi.org/10.1007/978-3-319-45823-6_6
20. Pitra, Z., Hanuš, M., Koza, J., Tumpach, J., Holeňa, M.: Interaction between model and its evolution control in surrogate-assisted CMA evolution strategy. In: Proceedings of the Genetic and Evolutionary Computation Conference, pp. 528–536 (2021)
21. Turner, R., et al.: Bayesian optimization is superior to random search for machine learning hyperparameter tuning: analysis of the black-box optimization challenge 2020. In: NeurIPS 2020 Competition and Demonstration Track, pp. 3–26. PMLR (2021)

22. Ulmer, H., Streichert, F., Zell, A.: Evolution strategies assisted by Gaussian processes with improved preselection criterion. In: The 2003 Congress on Evolutionary Computation, 2003, CEC 2003, vol. 1, pp. 692–699. IEEE (2003)
23. Yang, J., Arnold, D.V.: A surrogate model assisted (1+1)-es with increased exploitation of the model. In: Proceedings of the Genetic and Evolutionary Computation Conference, pp. 727–735 (2019)

The (1+1)-ES Reliably Overcomes Saddle Points

Tobias Glasmachers[✉] [iD]

Department for Computer Science, Institute for Neural Computation,
Ruhr-University Bochum, Bochum, Germany
`tobias.glasmachers@ini.rub.de`

Abstract. It is known that step size adaptive evolution strategies (ES) do not converge (prematurely) to regular points of continuously differentiable objective functions. Among critical points, convergence to minima is desired, and convergence to maxima is easy to exclude. However, surprisingly little is known on whether ES can get stuck at a saddle point. In this work we establish that even the simple (1+1)-ES reliably overcomes most saddle points under quite mild regularity conditions. Our analysis is based on drift with tail bounds. It is non-standard in that we do not even aim to estimate hitting times based on drift. Rather, in our case it suffices to show that the relevant time is finite with full probability.

1 Introduction

The question how optimization algorithms handle saddle points is a classic subject. In the standard analysis of gradient-based optimization, it is easy to rule out premature convergence to a regular point. In contrast, excluding convergence to saddle points requires considerable effort [4].

In evolutionary computation, the situation is no different. Akimoto et al. [3] established that many optimizers cannot converge to a regular point of the objective function under the rather basic assumption that they successfully diverge on a linear slope.

Prior work on the behavior of evolution strategies in the presence of a saddle

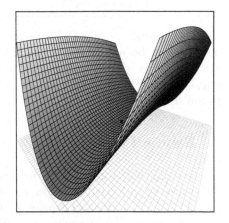

Fig. 1. Graph of a difficult saddle point.

point seems to be sparse. We need to highlight that usually in optimization the goal is *not* to get stuck at a saddle point, but rather to proceed to a (local) optimum. This is different from the goal of locating saddle points by means of optimization techniques, in cases where these saddles are of interest by themselves [1]. That line of work on "saddle point optimization", also called min-max-problems, is unrelated to our research question.

G. Rudolph et al. (Eds.): PPSN 2022, LNCS 13399, pp. 309–319, 2022.
https://doi.org/10.1007/978-3-031-14721-0_22

In our own prior work [5], we conducted a detailed analysis of conditions under which convergence of the (1+1)-ES to the global optimum can be guaranteed, on an extremely wide class of functions. In that work, premature convergence to saddle points can only be excluded if the success probability in the saddle point exceeds the target success rate of $1/5$ in the limit of small step sizes. On the other hand, for some extremely deceptive saddle points of sharp ridges, a positive probability for premature convergence is proven.

There is a considerable gap between the two cases. While existing guarantees do not apply to these cases, empirical evidence indicates—maybe surprisingly— that already the simple (1+1)-ES reliably overcomes even extremely ill-conditioned saddle points. In the present paper we close this gap by cementing the empirical evidence with a proof.

Algorithm 1: (1+1)-ES with 1/5-success rule

1: **input** $m_0 \in \mathbb{R}^d$, $\sigma_0 > 0$, $f : \mathbb{R}^d \to \mathbb{R}$, **parameter** $\alpha > 1$
2: **for** $t = 1, 2, \ldots$, *until stopping criterion is met* **do**
3:     sample $x_t \sim \mathcal{N}(m_t, \sigma_t^2 I)$
4:     **if** $f(x_t) \leq f(m_t)$ **then**
5:         $m_{t+1} \leftarrow x_t$ ▷ move to the better solution
6:         $\sigma_{t+1} \leftarrow \sigma_t \cdot \alpha$ ▷ increase the step size
7:     **else**
8:         $m_{t+1} \leftarrow m_t$ ▷ stay where we are
9:         $\sigma_{t+1} \leftarrow \sigma_t \cdot \alpha^{-1/4}$ ▷ decrease the step size

---

We consider the (1+1)-ES as specified in Algorithm 1. This version of the method can be attributed to Kern et al. [8]. It was used in the recent analysis [2]. For a given algorithm state (m, σ), we define the success probability $p_{\mathrm{succ}}(m, \sigma) = \Pr\left(f(x) \leq f(m)\right)$. It plays a key role for analyzing step size adaptation in the (1+1)-ES.

2   Saddle Points

In the following, we define various types of critical points of a continuously differentiable objective function $f : \mathbb{R}^d \to \mathbb{R}$. A point $x^* \in \mathbb{R}^d$ is called *critical* if $\nabla f(x^*) = 0$, and *regular* otherwise. A critical point is a *local minimum/maximum* if there exists $r > 0$ such that it is minimal/maximal within an open ball $B(x^*, r)$. If x^* is critical but neither (locally) minimal nor maximal, then it is a *saddle point*.

If f is twice continuously differentiable then most critical points are well characterized by their second order Taylor expansion

$$f(x) = f(x^*) + (x - x^*)^T H(x - x^*) + o(\|x - x^*\|^2).$$

The eigenvalues of the Hessian H determine its type: if all eigenvalues are positive/negative then it is a minimum/maximum. If both positive and negative eigenvalues exist then it is a saddle point. Zero eigenvalues are not informative, since the behavior of the function in the corresponding eigenspaces is governed by higher order terms.[1]

Therefore, a prototypical problem exhibiting a saddle point is the family of objective functions

$$f_a(x) = \sum_{i=1}^{d} a_i x_i^2$$

with parameter $a \in \mathbb{R}^d$. We assume that there exists $b \in \{1, \ldots, d-1\}$ such that $a_i < 0$ for all $i \leq b$ and $a_i > 0$ for all $i > b$. In all cases, the origin $x^* = 0$ is a saddle point. The eigenvalues of the Hessian are the parameters a_i. Therefore, every saddle point of a twice continuously differentiable function with non-zero eigen values of the Hessian is well approximated by an instance of f_a after applying translation and rotation operations, to which the (1+1)-ES is invariant. This is why analyzing the (1+1)-ES on f_a covers an extremely general case.

We observe that f_a is scale invariant, see also Fig. 2: $f_a(c \cdot x) = c^2 \cdot f_a(x)$ holds, and hence $f_a(x) < f_a(x') \Leftrightarrow f_a(c \cdot x) < f(c \cdot x')$ for all $x, x' \in \mathbb{R}^d$ and $c > 0$. This means that level sets look the same on all scales, i.e., they are scaled versions of each other. Also, the f-ranking of two points $x, x' \in \mathbb{R}^d$ agrees with the ranking of the $c \cdot x$ versus $c \cdot x'$.

Related to the structure of f_a we define the following notation. For $x \in \mathbb{R}^d$ we define $x_-, x_+ \in \mathbb{R}^d$ as the projections of x onto the first b components and onto the last $d - b$ components, respectively. To be precise, we have $(x_-)_i = x_i$ for $i \in \{1, \ldots, b\}$ and $(x_+)_i = x_i$ for $i \in \{b+1, \ldots, d\}$, while the remaining components of both vectors are zero. We obtain $x = x_- + x_+$.

For the two-dimensional case, three instances are plotted in Fig. 2. The parameter a controls the difficulty of the problem. The success probability of the (1+1)-ES at the saddle point $m = 0$ equals $p_{\text{succ}}(0, \sigma) = \cot^{-1}(\sqrt{|a_2/a_1|})$, which decays to zero for $|a_2| \gg |a_1|$. This is a potentially fatal problem for the (1+1)-ES, since it may keep shrinking its step size and converge prematurely [5].

The contribution of this paper is to prove that we do not need to worry about this problem. More technically precise, we aim to establish the following theorem:

Theorem 1. *Consider the sequence of states $(m_t, \sigma_t)_{t \in \mathbb{N}}$ of the (1+1)-ES on the function f_a. Then, with full probability, there exists $T \in \mathbb{N}$ such that for all $t \geq T$ it holds $f_a(m_t) < 0$.*

It ensures that the (1+1)-ES surpasses the saddle point with full probability in finite time (iteration T). This implies in particular that the saddle point is not a limit point of the sequence $(m_t)_{t \in \mathbb{N}}$ (see also Lemma 1 below).

---

[1] It should be noted that a few interesting cases exist for zero eigenvalues (which should be improbable in practice), like the "Monkey saddle" $f(x) = x_1^3 - 3x_1 x_2^2$. We believe that this case can be analyzed with the same techniques as developed below, but it is outside the scope of this paper.

Fig. 2. Level sets of different instances of f_a for $a = (-4, 1)$ (left), $a = (-1, 1)$ (middle), and $a = (-1, 20)$ (right), centered onto the saddle point. The scale of the axes is irrelevant since the problem is scale-invariant. The shaded areas correspond to positive function values. Problem difficulty increases from left to right, since the probability of sampling a "white" point (negative function value) in the vicinity of the saddle point shrinks.

3   Preliminaries

In this section, we prepare definitions and establish auxiliary results. We start by defining the following sets: $D_a^- = f_a^{-1}(\mathbb{R}_{<0})$, $D_a^0 = f_a^{-1}(\{0\})$, and $D_a^+ = f_a^{-1}(\mathbb{R}_{>0})$. They form a partition of the search space $\mathbb{R}^d$.

For a vector $x \in \mathbb{R}^d$ we define the semi-norms

$$\|x\|_- = \sqrt{-\sum_{i=1}^{b} a_i x_i^2} \quad \text{and} \quad \|x\|_+ = \sqrt{\sum_{i=b+1}^{d} a_i x_i^2}.$$

The two semi-norms are Mahalanobis norms in the subspaces spanned by eigenvectors with negative and positive eigenvalues of the Hessian of f_a, respectively, when interpreting the Hessian with negative eigenvalues flipped to positive as an inverse covariance matrix. In other words, $f_a(x) = \|x\|_+^2 - \|x\|_-^2$ holds. Furthermore, we have $\|x_+\|_+ = \|x\|_+$, $\|x_-\|_- = \|x\|_-$, $\|x_-\|_+ = 0$, and $\|x_+\|_- = 0$.

In the following, we exploit scale invariance of f_a by analyzing the stochastic process (m_t, σ_t) in a normalized state space. We map a state to the corresponding normalized state by

$$(m, \sigma) \mapsto \left(\frac{m}{\|m\|_+}, \frac{\sigma}{\|m\|_+} \right) = (\tilde{m}, \tilde{\sigma}).$$

This normalization is different from the normalizations m/σ and $m/(d\sigma)$, which give rise to a scale-invariant process when minimizing the Sphere function [2]. The different normalization reflects the quite different dynamics of the (1+1)-ES on f_a.

We are particularly interested in the case $m \in D_a^+$, since we need to exclude the case that the (1+1)-ES stays in that set indefinitely. Due to scale invariance, this condition is equivalent to $\tilde{m} \in D_a^+$. We define the set

$$M = \left\{ x \in \mathbb{R}^d \,\middle|\, \|x\|_+ = 1 \right\}.$$

The state space for the normalized states $(\tilde{m}, \tilde{\sigma})$ takes the form $M \times \mathbb{R}_{>0}$. We also define the subset $M_0^+ = M \cap (D_a^+ \cup D_a^0)$. The reason to include the zero level set is that closing the set makes it compact. Its boundedness can be seen from the reformulation $M_0^+ = \left\{ m \in \mathbb{R}^d \,\middle|\, \|m\|_+ = 1 \text{ and } \|m\|_- \leq 1 \right\}$. In the following, compactness will turn out to be very useful, exploiting the fact that on a compact set, every lower semi-continuous function attains its infimum.

The success probability $p_{\text{succ}}(m, \sigma)$ is scale invariant, and hence it is well-defined as a function of the normalized state $(\tilde{m}, \tilde{\sigma})$. It is everywhere positive. Indeed, it is uniformly lower bounded by $p_{\min} = \min(p^*, \frac{1}{2}) > 0$, where $p^* = p_{\text{succ}}(0, 1)$ denotes the success probability in the saddle point (which is independent of the step size, and depends only on a). The following two lemmas deal with the success rate in more detail.

Lemma 1. *If there exists $T \in \mathbb{N}$ such that $m_T \in D_a^0 \cup D_a^-$ then with full probability, the saddle point $0 \in \mathbb{R}^d$ of f_a is not a limit point of the sequence $(m_t)_{t \in \mathbb{N}}$.*

Proof. Due to elitism, the sequence m_t can jump from D_a^+ to D_a^0 and then to D_a^-, but not the other way round. In case of $m_T \in D_a^-$ all function values for $t > T$ are uniformly bounded away from zero by $f(m_t) \leq f(m_T) < 0$. Therefore $f(m_t)$ cannot converge to zero, and m_t cannot converge to the saddle point.

Now consider the case $m_T \in D_a^0$. For all $m \in D_a^0$ and all $\sigma > 0$, the probability of sampling an offspring in D_a^- is positive, and it is lower bounded by $p_{\min}$, which is positive and independent of σ. Not sampling an offspring $m_t \in D_a^-$ for n iterations in a row has a probability of at most $(1 - p_{\min})^n$, which decays to zero exponentially quickly. Therefore, with full probability, we obtain $m_t \in D_a^-$ eventually.   □

However, $p_{\min}$ being positive is not necessarily enough for the (1+1)-ES to escape the saddle point, since for $p_{\min} < 1/5$ it may stay inside of D_a^+, keep shrinking its step size, and converge prematurely [5]. In fact, based on the choice of the parameter a of f_a, $p_{\min}$ can be arbitrarily small. In the following lemma, we therefore prepare a drift argument, ensuring that the normalized step size remains in or at least always returns to a not too small value.

Lemma 2. *There exists a constant $0 < \tilde{\sigma}_{40\%} \leq \infty$ such that $p_{succ}(\tilde{m}, \tilde{\sigma}) \geq 2/5$ holds for all states fulfilling $\tilde{m} \in M_0^+$ and $\tilde{\sigma} \leq \tilde{\sigma}_{40\%}$.*

Proof. It follows immediately from the geometry of the level sets (see also Fig. 2) that for each fixed $\tilde{m} \in M_0^+$ (actually for $m \neq 0$), it holds

$$\lim_{\tilde{\sigma} \to 0} p_{\text{succ}}(\tilde{m}, \tilde{\sigma}) = \frac{1}{2} \quad \text{and} \quad \lim_{\tilde{\sigma} \to \infty} p_{\text{succ}}(\tilde{m}, \tilde{\sigma}) = p^*.$$

Noting that $p_{\text{succ}}(\tilde{m}, \tilde{\sigma})$ is continuous between these extremes, we define a pointwise critical step size as

$$\tilde{\sigma}_{40\%}(\tilde{m}) = \arg\min_{\tilde{\sigma} > 0} \left\{ p_{\text{succ}}(\tilde{m}, \tilde{\sigma}) \leq 2/5 \right\}.$$

With the convention that arg min over an empty set is ∞, this definition makes $\tilde{\sigma}_{40\%} : M_0^+ \to \mathbb{R} \cup \{\infty\}$ a lower semi-continuous function. Due to compactness of M_0^+ it attains its minimum $\tilde{\sigma}_{40\%} > 0$. $\square$

4   Drift of the Normalized State

In this section we establish two drift arguments. They apply to the following drift potential functions:

$$V(\tilde{m}, \tilde{\sigma}) = \log(\tilde{\sigma})$$
$$W(\tilde{m}, \tilde{\sigma}) = \|\tilde{m}\|_-$$
$$\Phi(\tilde{m}, \tilde{\sigma}) = \beta \cdot V(\tilde{m}, \tilde{\sigma}) + W(\tilde{m}, \tilde{\sigma})$$

The potentials govern the dynamics of the step size $\tilde{\sigma}$, of the mean $\tilde{m}$, and of the combined process, namely the (1+1)-ES. The trade-off parameter $\beta > 0$ will be determined later. Where necessary we extend the definitions to the original state by plugging in the normalization, e.g., resulting in $W(m, \sigma) = \frac{\|m\|_-}{\|m\|_+}$.

For a normalized state $(\tilde{m}, \tilde{\sigma})$ let $(\tilde{m}', \tilde{\sigma}')$ denote the normalized successor state. We measure the drift of all three potentials as follows:

$$\Delta^V(\tilde{m}, \tilde{\sigma}) = \mathbb{E}[V(\tilde{\sigma}') - V(\tilde{\sigma})]$$
$$\Delta^W(\tilde{m}, \tilde{\sigma}) = \mathbb{E}[\min\{W(\tilde{m}') - W(\tilde{m}), 1\}]$$
$$\Delta^\Phi(\tilde{m}, \tilde{\sigma}) = \beta \cdot \Delta^V(\tilde{m}, \tilde{\sigma}) + \Delta^W(\tilde{m}, \tilde{\sigma})$$

As soon as $W(\tilde{m}) > 1$, $\tilde{m} \in D_a^-$ holds and the (1+1)-ES has successfully passed the saddle point according to Lemma 1. Therefore we aim to show that the sequence $W(\tilde{m}_t)$ keeps growing, and that is passes the threshold of one. To this end, we will lower bound the progress Δ^W of the truncated process.

Truncation of particularly large progress in the definition of Δ^W, i.e., W-progress larger than one, serves the purely technical purpose of making drift theorems applicable. This sounds somewhat ironic, since a progress of more than one on W immediately jumps into the set D_a^- and hence passes the saddle. On the technical side, an upper bound on single steps is a convenient prerequisite. Its role is to avoid that the expected progress is achieved by very few large steps while most steps make no or very litte progress, which would make it impossible to bound the runtime based on expected progress. Less strict conditions allowing for rare large steps are possible [6,9]. The technique of bounding the single-step progress instead of the domain of the stochastic process was introduced in [2].

The speed of the growth of W turns out to depend on $\tilde{\sigma}$. In order to guarantee growth at a sufficient pace, we need to keep the normalized step size from decaying to zero too quickly. Indeed, we will show that the normalized step size drifts away from zero by analyzing the step-size progress Δ^V.

The following two lemmas establish the drift of mean $\tilde{m}$ and step size $\tilde{\sigma}$.

Lemma 3. *Assume $\tilde{m} \in M_0^+$. There exists a constant B_1 such that $\Delta^V(\tilde{m}, \tilde{\sigma}) \geq B_1$ holds. Furthermore, there exist constants $B_2 > 0$ and $\tilde{\sigma}^* \in (0, \tilde{\sigma}_{40\%}]$ such that for all $\tilde{\sigma} \leq \tilde{\sigma}^*$ it holds $\Delta^V(\tilde{m}, \tilde{\sigma}) \geq B_2$.*

Lemma 4. *Assume $\tilde{m} \in M_0^+$. The W-progress $\Delta^W(\tilde{m}, \tilde{\sigma})$ is everywhere positive. Furthermore, for each $\tilde{\sigma}^* \in (0, \tilde{\sigma}_{40\%}]$ there exists a constant $C > 0$ depending on $\tilde{\sigma}^*$ such that it holds $\Delta^W(\tilde{m}, \tilde{\sigma}) \geq C$ if $\tilde{\sigma} \geq \tilde{\sigma}^*$.*

The proofs of these lemmas contain the main technical work.

Proof (of Lemma 3). From the definition of $\tilde{\sigma}_{40\%}$, for $\tilde{\sigma} \leq \tilde{\sigma}_{40\%}$, we conclude that the probability of sampling a successful offspring is at least $2/5$. In case of an unsuccessful offspring, $\tilde{\sigma}$ shrinks by the factor $\alpha^{-1/4}$. For a successful offspring it is multiplied by $\alpha \cdot \frac{\|m\|_+}{\|m'\|_+}$, where the factor $\alpha > 1$ comes from step size adaptation, and the fraction is due to the definition of the normalized state.

The dependency on m and m' is inconvenient. However, for small step size $\tilde{\sigma}$ we have $\|m'\| \approx \|m\|$, simply because modifying m with a small step results in a similar offspring, which is then accepted as the new mean m'. In the limit we have

$$\lim_{\tilde{\sigma} \to 0} \mathbb{E}\left[\log\left(\frac{\|m\|_+}{\|m'\|_+}\right)\right] = 0.$$

This allows us to apply the same technique as in the proof of Lemma 2. The function $(\tilde{m}, \tilde{\sigma}) \mapsto \mathbb{E}\left[\log\left(\frac{\|m\|_+}{\|m'\|_+}\right)\right]$ is continuous. We define a pointwise lower bound through the lower semi-continuous function

$$\tilde{m} \mapsto \arg\min_{0 < \tilde{\sigma} \leq \tilde{\sigma}_{40\%}}\left\{\mathbb{E}\left[\log\left(\frac{\|m\|_+}{\|m'\|_+}\right)\right] \leq \frac{1}{\sqrt{\alpha}}\right\},$$

where the arg min over the empty set shall take the value $\sigma_{40\%}$. We define $\tilde{\sigma}^*$ as its infimum. It is attained, since M_0^+ is compact, and hence positive.

For $\tilde{\sigma} \leq \tilde{\sigma}^*$ we obtain the following drift condition:

$$\Delta^V(\tilde{m}, \tilde{\sigma}) \geq \frac{2}{5} \cdot \left[\log(\alpha^{-\frac{1}{2}}) + \log(\alpha)\right] - \left(1 - \frac{2}{5}\right) \cdot \frac{1}{4} \cdot \log(\alpha)$$

$$= \frac{1}{5} \cdot \log(\alpha) - \frac{3}{20} \cdot \log(\alpha) = \frac{1}{20} \cdot \log(\alpha) > 0$$

For $\tilde{\sigma} > \tilde{\sigma}^*$ we consider the worst case of a success rate of zero. Then we obtain

$$\Delta^V(\tilde{m}, \tilde{\sigma}) \geq -\frac{1}{4} \cdot \log(\alpha).$$

Hence, the statement holds with $B_1 = -\frac{1}{4} \cdot \log(\alpha)$ and $B_2 = \frac{1}{20} \cdot \log(\alpha)$. $\square$

Proof (of Lemma 4). We start by showing that Δ^W is always positive. We decompose the domain of the sampling distribution (which is all of $\mathbb{R}^d$) into spheres of fixed radius $r = \|\tilde{m}' - \tilde{m}\|$ and show that the property holds, conditioned to the success region within each sphere. Within each sphere, the distribution is uniform.

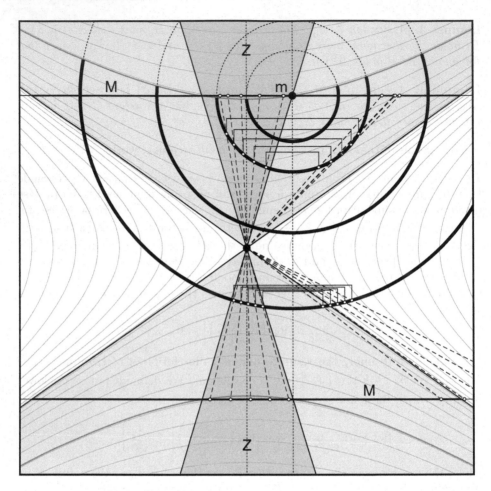

Fig. 3. Geometric illustration of the proof of Lemma 4. The figure shows the saddle point (center), level sets of f_a (thin lines), the point m, and the set M (two horizontal lines), the thick part of which is M_0^+. The area D_a^- has a white background, while D_a^+ is the gray area. The dark gray area is the set Z. The figure displays spheres of different radii into which the sampling distribution is decomposed. The spheres are drawn as dotted circles, and as bold solid arcs in the region of successful offspring, outperforming m. The thickened arcs indicate sets of corresponding points. Ten pairs of corresponding points are shown, five each for two different spheres.

Each sphere makes positive and negative contributions to $W(x) - W(\tilde{m}) = \frac{\|x\|_-}{\|x\|_+} - \|\tilde{m}\|_-$. Within the set

$$Z = \left\{ z \in \mathbb{R}^d \ \middle| \ \frac{\|z\|_-}{\|z\|_+} < \|\tilde{m}\|_- \right\}$$

the contributions are negative. The set is illustrated in Fig. 3. Outside of Z, contributions are positive. We aim to show that overall, for each sphere, the expectation is positive. To this end, we define pairs of corresponding points such that the negative contribution of one point is (more than) compensated by the positive contribution of the other. For our argument, it is important that the Lebesgue measure of each subset $S \subset Z$ is at most as large the Lebesgue measure of the set of corresponding points outside of Z. This property will be fulfilled by construction, and with equality.

For each successful offspring in Z we define a corresponding point outside of Z on the same sphere. Corresponding points are mirrored at the symmetry axis through m. More precisely, for $z \in Z$ we define $z'_- = 2m_- - z_-$ and $z'_+ = z_+$. It holds $\tilde{m}_- - z_- = z'_- - \tilde{m}_-$, and we call this difference $\delta = \tilde{m}_- - z_-$.

By projecting both points onto the normalized state space M we obtain their contributions to the expectation. This amounts to following the dashed lines in Fig. 3. Adding the contributions of z and z' yields

$$W(z) + W(z') - 2\,W(\tilde{m}) = \frac{\|z\|_-}{\|z\|_+} + \frac{\|z'\|_-}{\|z'\|_+} - 2\|\tilde{m}\|_-$$
$$= \frac{\|\tilde{m} - \delta\|_- + \|\tilde{m} + \delta\|_-}{\|z\|_+} - 2\|\tilde{m}\|_-$$
$$\geq \|\tilde{m} - \delta\|_- + \|\tilde{m} + \delta\|_- - 2\|\tilde{m}\|_- \geq 0\,.$$

The first inequality holds because of $\|z\|_+ = \|z'\|_+ \leq 1$ (note that the set M corresponds to $\|\cdot\|_+ = 1$, see also Fig. 3). The second step is the triangle inequality of the semi-norm $\|\cdot\|_-$. Both inequalities are strict outside of a set of measure zero.

Truncating progress $W(z') - W(\tilde{m})$ larger than one does not pose a problem. This is because $W(\tilde{m}) - W(z) < 1$ is obtained from the fact that $\tilde{m}$ and z are both contained in D_a^+, and this is where W takes values in the range $[0, 1)$. We obtain $W(z) - W(\tilde{m}) + 1 > 0$ in the truncated case.

Integrating the sum over all corresponding pairs on the sphere, and noting that there are successful points outside of Z which do not have a successful corresponding point inside but not the other way round, we see that the expectation of $W(\tilde{m}') - W(\tilde{m})$ over the success region of each sphere is positive.

Integration over all radii $r > 0$ completes the construction. In the integration, the weights of different values of r depend on $\tilde{\sigma}$ (by means of the pdf of a χ-distribution scaled by $\tilde{\sigma}$). Since the integrand is non-negative, we conclude that $\Delta^W(\tilde{m}, \tilde{\sigma}) > 0$ holds for all $\tilde{\sigma} > 0$.

In the limit $\tilde{\sigma} \to \infty$, the expected progress in case of success converges to one (due to truncation), and hence the expected progress converges to p^*. This allows us to exploit compactness once more. The expectation of the truncated progress $\Delta^W(\tilde{m}, \tilde{\sigma})$ is continuous as a function of the normalized state. We define a pointwise lower bound as

$$C(\tilde{m}) = \min_{\tilde{\sigma} \geq \tilde{\sigma}_{40\%}} \left\{ \Delta^W(\tilde{m}, \tilde{\sigma}) \right\}.$$

$C(\tilde{m})$ is a continuous function, and (under slight misuse of notation) we define C as its infimum over the compact set M_0^+. Since the infimum is attained, it is positive. $\qquad\square$

Now we are in the position to prove the theorem.

Proof (of Theorem 1). Combining the statements of Lemma 3 and 4 we obtain

$$\Delta^{\Phi}(\tilde{m}, \tilde{\sigma}) \geq \theta := \min\{\beta B_2, C + \beta B_1\}$$

for all $\tilde{m} \in M_0^+$ and $\tilde{\sigma} > 0$. The choice $\beta = \frac{-C}{2B_1}$ results in $\theta = \min\{B_2, C/2\} > 0$. The constant θ is a bound on the additive drift of Φ, hence we can apply additive drift with tail bound (e.g., Theorem 2 in [9] with additive drift as a special case, or alternatively inequality (2.9) in Theorem 2.3 in [6]) to obtain the following: Let

$$T = \min\left\{t \in \mathbb{N} \,\middle|\, \Phi(\tilde{m}_t, \tilde{\sigma}_t) > 1\right\}$$

denote the waiting time for the event that Φ reaches or exceeds one (called the first hitting time). Then the probability of T exceeding $T_0 \in \mathbb{N}$ decays exponentially in T_0. Therefore, with full probability, the hitting time T is finite. $\Phi(\tilde{m}_T, \tilde{\sigma}_T) > 1$ is equivalent to $f(m_T) < 0$. For all $t > T$, the function value stays negative, due to elitism.   □

5   Discussion and Conclusion

We have established that the (1+1)-ES does not get stuck at a (quadratic) saddle point, irrespective of its conditioning (spectrum of its Hessian), with full probability. This is all but a trivial result since the algorithm is suspectable to premature convergence if the success rate is smaller than 1/5. For badly conditioned problems, close to the saddle point, the success rate can indeed be arbitrarily low. Yet, the algorithm passes the saddle point by avoiding it "sideways": While approaching the level set containing the saddle point, there is a systematic sidewards drift away from the saddle. This keeps the step size from decaying to zero, and the saddle is circumvented.

In this work we are only concerned with quadratic functions. Realistic objective functions to be tackled by evolution strategies are hardly ever so simple. Yet, we believe that our analysis is of quite general value. The reason is that the negative case, namely premature convergence to a saddle point, is an inherently local process, which is dominated by a local approximation like the second order Taylor polynomial around the saddle point. Our analysis makes clear that as long as the saddle is well described by a second order Taylor approximation with a full-rank Hessian matrix, then the (1+1)-ES will not converge prematurely to the saddle point. We believe that our result covers the most common types of saddle points. Notable exceptions are sharp ridges, plateaus, and Monkey saddles.

The main limitation of this work is not the covered class of functions, but the covered algorithms. The analysis sticks closely to the (1+1)-ES with its success-bases step size adaptation mechanism. There is no reason to believe that a fully fledged algorithm like the covariance matrix adaptation evolution strategy (CMA-ES) [7] would face more problems with a saddle than the simple (1+1)-ES, and to the best of our knowledge, there is no empirical indication thereof.

In fact, our intuition is that most algorithms should profit from the sidewards drift, as long as they manage to break the symmetry of the problem, e.g., through randomized sampling. Yet, it should be noted that our analysis does not easily extend to non-elitist algorithms and step size adaptation methods other than success-based rules.

References

1. Akimoto, Y.: Saddle point optimization with approximate minimization oracle. Tech. Rep. 2103.15985 (2021). arXiv.org
2. Akimoto, Y., Auger, A., Glasmachers, T.: Drift theory in continuous search spaces: expected hitting time of the (1+1)-es with 1/5 success rule. In: Proceedings of the Genetic and Evolutionary Computation Conference (GECCO). ACM (2018)
3. Akimoto, Y., Nagata, Y., Ono, I., Kobayashi, S.: Theoretical analysis of evolutionary computation on continuously differentiable functions. In: Genetic and Evolutionary Computation Conference, pp. 1401–1408. ACM (2010)
4. Dauphin, Y.N., Pascanu, R., Gulcehre, C., Cho, K., Ganguli, S., Bengio, Y.: Identifying and attacking the saddle point problem in high-dimensional non-convex optimization. In: Ghahramani, Z., Welling, M., Cortes, C., Lawrence, N.D., Weinberger, K.Q. (eds.) Advances in Neural Information Processing Systems, vol. 27, pp. 2933–2941 (2014)
5. Glasmachers, T.: Global convergence of the (1+1) evolution strategy. Evol. Comput. J. (ECJ) **28**(1), 27–53 (2020)
6. Hajek, B.: Hitting-time and occupation-time bounds implied by drift analysis with applications. Adv. Appli. Prob. **14**(3), 502–525 (1982)
7. Hansen, N., Ostermeier, A.: Completely derandomized self-adaptation in evolution strategies. Evol. Comput. **9**(2), 159–195 (2001)
8. Kern, S., Müller, S.D., Hansen, N., Büche, D., Ocenasek, J., Koumoutsakos, P.: Learning probability distributions in continuous evolutionary algorithms-a comparative review. Nat. Comput. **3**(1), 77–112 (2004)
9. Lehre, P.K., Witt, C.: General drift analysis with tail bounds. Tech. Rep. 1307.2559 (2013). arXiv.org

Real-World Applications

Evolutionary Time-Use Optimization for Improving Children's Health Outcomes

Yue Xie[1]([✉]), Aneta Neumann[1], Ty Stanford[2], Charlotte Lund Rasmussen[3,4], Dorothea Dumuid[2], and Frank Neumann[1]

[1] The University of Adelaide, Adelaide, SA, Australia
yue.xie@adelaide.edu.au
[2] Alliance for Research in Exercise, Nutrition and Activity, Allied Health and Human Performance, University of South Australia, Adelaide, SA, Australia
[3] Norwegian University of Science and Technology, Trondheim, Norway
[4] Department of Physical Education and Sport Sciences, University of Limerick, Limerick, Ireland

Abstract. How someone allocates their time is important to their health and well-being. In this paper, we show how evolutionary algorithms can be used to promote health and well-being by optimizing time usage. Based on data from a large population-based child cohort, we design fitness functions to explain health outcomes and introduce constraints for viable time plans. We then investigate the performance of evolutionary algorithms to optimize time use for four individual health outcomes with hypothetical children with different day structures. As the four health outcomes are competing for time allocations, we study how to optimize multiple health outcomes simultaneously in the form of a multi-objective optimization problem. We optimize one-week time-use plans using evolutionary multi-objective algorithms and point out the trade-offs achievable with respect to different health outcomes.

Keywords: Real-world application · Time-use optimization · Single-objective optimization · Multi-objective optimization

1 Introduction

Evolutionary algorithms (EAs) are bio-inspired randomized optimization techniques and have been very successfully applied to various real-world combinatorial optimization problems [21,25,28]. Evolutionary algorithms use a population of search points in the decision space of a given optimization problem to solve the problem. Moreover, many real-world optimization problems consist of several conflicting objectives that must be optimized simultaneously. No single solution can optimize multiple objectives, instead a set of trade-off optimal solutions is obtained. EAs can approximate multiple optimal solutions in a single run, which make EAs popular in solving multi-objective optimization problems [12,15].

G. Rudolph et al. (Eds.): PPSN 2022, LNCS 13399, pp. 323–337, 2022.
https://doi.org/10.1007/978-3-031-14721-0_23

A real-world multi-objective optimization problem is "How should children spend their time (i.e. sleeping, sedentary behaviour and physical activity) to optimize their health, well-being, and cognitive development?" [9,10]. The importance of this problem has led governing bodies and health authorities such as the World Health Organization (WHO) to provide guidelines for daily durations of sleep, screen time, and physical activity [31]. Such guidelines for school-aged children (5–12 years) currently recommend 9–11 h of sleep, no more than 2 h of sedentary screen time, and at least 1 h of moderate-to-vigorous physical activity (MVPA) per day [31]. However, these guidelines are primarily underpinned by systematic reviews collating evidence of how the duration of a single behaviour, such as MVPA, is associated with a single measure of health or well-being [31]. These studies show whether more or less of behaviour is beneficially associated with the outcome [9,10,31], rather than identifying optimal durations, which would be required to support recommendations for daily durations of the behaviour. Almost no studies have attempted to define optimal durations for these activity behaviours for a single health outcome, let alone for multiple health and well-being outcomes.

To address the lack of evidence for optimal time-use allocations, a recent study [17] used compositional linear regression [18] to model the relationship between how children allocated their daily time to four activities (sleep, sedentary behaviour, light physical activity (LPA) and MVPA) and twelve outcomes spanning physical, mental and cognitive health domains. Compositional data analysis enabled all four activities to be included in a single model whilst ensuring their constant-sum constraint to 24 h was respected [1]. Using published compositional data methods, the raw activity data of minutes per day were expressed as a set of isometric log-ratios [29]. With these compositional regression models, [19] estimated values of the outcomes for every possible and feasible combination of sleep, sedentary behaviour, LPA and MVPA duration were calculated. Optimal daily duration of the activities were derived for each of the twelve health outcomes from the average "time-use composition" associated with the best 5% of estimated values for the respective health outcomes.

It remains unknown how to perform the best multi-objective optimisation of time use for overall health and well-being. The method developed by [19] is computationally intensive for four activities requiring almost 4 million iterations of different possible time-use scenarios. This method becomes unfeasible with a large number of daily activities (e.g., activities such as chores, sport, transport, school, sleep, quiet time, social time, screen time, etc.) routinely collected by time-use recalls [34]. Additionally, varying constraints to daily time use, which may limit application to the real world, were not considered.

The research described in this paper extends previous work proposed in [17] by considering four decision variables: daily time allocation to sleep, sedentary behaviour, LPA and MVPA, and four health objectives for children: body mass index (BMI), cognition, life satisfaction and fitness. Firstly, we formulate the one-day time-use optimization problem as a single-objective problem in continuous space by optimizing one of the four presented health outcomes. Then, we extend the one-day time-use schedule to one week and present multi-objective optimization models for the time-use optimization problem.

EAs are introduced to develop time-use optimization approaches that incorporate daily and weekly time constraint schedules and provide decision-making tools for trading off multiple health outcomes against each other. For single-objective time-use optimization, we evaluate the performance of the differential evolution (DE) algorithm [37] with different operators, particle swarm optimization (PSO) [26] and covariance matrix adaptation evolutionary strategy (CMA-ES) [22,23] to optimize health outcomes in different day structures. For multi-objective time-use optimization, we investigate the performance of the multi-objective evolutionary algorithm based on decomposition (MOEA/D) [40], Non-dominated sorting genetic algorithm (NSGA-II) [16] and Strong Pareto evolutionary algorithm 2 (SPEA2) [42].

The paper is organized as follows. We introduce the data set used in Sect. 1.1. Section 2 describes application of our time-use optimization models for different health outcomes, and to different day constraints. The proposed optimization methods are described in Sect. 3. The results of the optimization experiments are described in Sect. 4. Conclusions and avenues for future work are presented in Sect. 5.

1.1 Data Description

This study uses data from a large population-based child cohort to illustrate the real-world application of a novel time-use optimisation procedure. Data were from the Child Health CheckPoint study [11], a cross-sectional module nested between waves 6 and 7 of the Longitudinal Study of Australian Children (LSAC) [20]. Child participants of the LSAC birth cohort (commenced in 2004 with n = 5107) that were retained to Wave 6 ($n = 3764$) were invited to take part in Child Health CheckPoint (2015–16) when they were 11–12 years old. Of these, $n = 1874$ (50%) consented to participate via written informed consent from their parent/guardian. Ethical approval for CheckPoint was granted by The Royal Children's Hospital (Melbourne) Human Research Ethics Committee (HREC33225D) and the Australian Institute of Family Studies Ethics Committee (AIFS14-26).

Participants were fitted with a wrist-worn accelerometer (GENEActive, Activinsights Ltd, UK) by a trained researcher, with instructions to wear the device 24 h a day for eight days. Following the return of the device, activity data were downloaded and processed following published procedures [17,20] to determine the average daily minutes spent in sleep, sedentary time, LPA and MVPA.

BMI was derived from the child participant's measured height (Invicta 10955 stadiometer) and weight (2-limb Tanita BC-351 or 4-limb InBody 230). BMI was calculated as weight (kg)/height (m)^2 and expressed as age- and sex-specific z-scores [32]. The cognition score was derived from the NIH Picture Vocab test, which asks the child to select on an iPad a picture that best represents the meaning of words they hear through headphones [39]. A higher score indicates better receptive vocabulary, which represents cognition. Life satisfaction was obtained from the 5-item Brief Multi-Dimensional Students' Life Satisfaction Scale, with a higher score indicating higher satisfaction with their family life, friendships, school experience and themselves, where they live, and their overall life [36]. Fitness was obtained from a cycle ergometer test which was used to determine the estimated maximal work rate from which VO2max (predicted maximal aerobic power) was estimated. A higher VO2max indicates better aerobic fitness [7].

2   The Time-Use Optimization Models

In this section, we first list the notations and descriptions of health outcomes and decision variables in Table 1(a). Column *Optimal* lists the definition of optimal value of each objective. Then, we introduce a general model for the one-day time-use optimization problem without considering any specific day structure or health outcome.

$$
\text{obj:}\quad f(x) = \hat{\beta}_0 + \hat{\beta}_1 z_1 + \hat{\beta}_2 z_2 + \hat{\beta}_3 z_3 + \hat{\beta}_4 z_1 z_1 + \hat{\beta}_5 z_1 z_2
$$
$$
+ \hat{\beta}_6 z_1 z_3 + \hat{\beta}_7 z_2 z_2 + \hat{\beta}_8 z_2 z_3 + \hat{\beta}_9 z_3 z_3 \tag{1}
$$

$$
\text{s.t.}\quad z_1 = \sqrt{\frac{3}{4}} \ln\left(\frac{x_1}{\sqrt[3]{x_2 x_3 x_4}}\right), z_2 = \sqrt{\frac{2}{3}} \ln\left(\frac{x_2}{\sqrt{x_3 x_4}}\right), z_3 = \sqrt{\frac{1}{2}} \ln\left(\frac{x_3}{x_4}\right)
$$

$$
\sum_{i=1}^{4} x_i = 1440 \tag{2}
$$

$$
x_i^l \le x_i \le x_i^u \quad \forall i = \{1, \dots, 4\} \tag{3}
$$

The decision vector of this model can be expressed as $x = \{x_1, x_2, x_3, x_4\}$ which consists of four activity variables (sleep, sedentary time, LPA, MVPA). The objective function (1) shows how to calculate health outcomes based on values of the decision variables and parameters. Where $\beta_0, \beta_1, \dots, \beta_9$ are unknown regression coefficients to be estimated, they are different in the objective function of each health outcome. Here, those regression coefficients are estimated using the data described in Sect. 1.1. We list the estimated values $\hat{\beta}_i, i = \{1, \dots, 9\}$ for different health outcomes in Table 1 (b) and introduce how to obtain those values in Sect. 2.1. Constraint (2) forces the sum of decision variables of the problem equal to the total minutes (1440 min min) per day. We introduce a closure operation (see Algorithm 1) to tackle this constraint and make the working progress of any search algorithm fast to achieve a feasible solution. Upper and lower bounds on each decision variables are enforced by constraint (3), where x_i^l denotes the lower bound of x_i and x_i^u denotes the upper bound of x_i. The upper and lower bounds are different according to the day structure considered.

Without loss of generality, we study six different hypothetical day structures. We label these day structures to reflect real-world scenarios: *Studious day (STD)*, *Sporty day (SPD)*, *After-School Job day (ASJD)*, *Sporty Weekend day (SPWD)*, *Studious/screen weekend day (STWD)* and *Working weekend day (WWD)*. The lower and upper bounds of the decision variables are set to suit the day-above-day structures, as advised by an external child behavioural epidemiologist, and by considering the empirical activity durations found in the underlying data (please refer to Table 2). These replace the 24-h constraint (3) which is present in a general model.

Table 1. Notation and values of parameters

(a) Description of notation			(b) Estimated regression coefficients				
Notation	Description	Optimal	Notation	f_1	f_2	f_3	f_4
f_1	Body mass index (BMI)	min $|f_1|$	β_0	0.23307	2.3508268	12395.053	68.85903
f_2	Cognition (vocab) objective	max f_2	β_1	−0.59691	−0.032037	2255.008	−17.84326
f_3	Life satisfaction objective	max f_3	β_2	0.05029	0.0670568	−885.351	−1.77607
f_4	Fitness (VO2max) objective	max f_4	β_3	0.68497	−0.003155	1264.635	−11.25996
			β_4	0	0	0	3.15694
x_1	Minutes of sleeping		β_5	0	0	0	13.88458
x_2	Minutes of sedentary behaviour		β_6	0	0	0	−5.12788
x_3	Minutes of LPA		β_7	0	0	0	−6.85649
x_4	Minutes of MVPA		β_8	0	0	0	2.69689
			β_9	0	0	0	2.52276

Table 2. Values of lower bounds and upper bounds

		Studious day	Sporty day	After-school job day	Sporty weekend day	Studious/screen weekend day	Working weekend day
Sleep	LB	360	360	360	420	420	360
	UB	720	720	720	720	720	720
Sedentary	LB	690	480	480	210	270	210
	UB	900	900	900	900	900	900
LPA	LB	150	210	220	210	150	390
	UB	480	480	480	480	480	480
MVPA	LB	1	61	1	61	1	1
	UB	210	210	210	210	210	210

2.1 Model Parameter Estimation

Estimates of the model parameters ($\hat{\beta}_i, i = 1, \ldots, 9$) in Eq. (1) are calculated using least-squares multiple linear regression on the CheckPoint data. It is not possible to use all the untransformed time-use predictors simultaneously in the linear model as they are *linearly dependent* which in turn prohibits the matrix inverse calculation in estimating the parameter estimates. The *isometric log ratio* (*ilr*) transformation is a widely used transformation of the predictors to remove the linear dependence in the predictors [18].

For each outcome variable, f_1, f_2, f_3, f_4, the Box-Cox transformation is applied after removing predictor effects for variance stabilisation, and improvement in the normality of the residuals [8]. Quadratic terms of the time-use *ilr* predictors are considered for each outcome model which correspond to the model terms associated with the parameters $\beta_4, \ldots, \beta_9$ in Eq. (1). If the quadratic terms do not significantly improve the model fit statistically at the $\alpha = 0.05$ level (ANOVA F-test), the model parameters $\beta_4, \ldots, \beta_9$ are set to 0 (i.e., only linear *ilr* terms remain). For more information about fitting quadratic compositional terms in linear regression, we refer to Chapter 5 of [6].

Algorithm 1: Closure Operation

Input: Decision vector $\{x_1, x_2, x_3, x_4\}$

$a = \sum_{i=1}^{4} x_i$;

for $i = 1$ *to* 4 **do**

$\quad | \quad x_i = \frac{1440 x_i}{a}$;

return the decision variables.

Table 3. Different mixture of one-week plan

Index	Studious day	Sporty day	After-school job day	Sporty weekend day	Studious/screen weekend day	Working weekend day
1	3	1	0	1	1	1
2	3	0	2	0	1	1
3	3	2	0	0	1	1
4	2	2	1	0	2	0
5	2	2	0	1	0	2
6	2	2	1	1	1	0

The full fit of the linear model also includes covariates of age, sex and puberty status and their associated coefficients. The sample average covariates are then used (age $= 12$, female/male $= 1{:}1$ and puberty status $=$ "Midpubertal"). The estimated effects of these covariates, and the intercept term of the model, are included as the β_0 term in Eq. (1). The objective functions therefore become the prediction for the theoretical average child in the sample. A sample with missing values in either the outcome or the predictors is removed in each model fit as data are reasonably assumed to be missing at random [35]. Diagnostic plots of each model are observed to ensure the model assumptions are reasonable. All analysis is performed in R version 4.0.3 [33].

2.2 One Week Plan

We extend the one-day problem to a one-week problem by mixing different day structures, given seven days where each day has four decision variables $x_d = \{x_{d1}, x_{d2}, x_{d3}, x_{d4}\}$. Different mixtures shown in Table 3 were used to make the one-week plans more realistic. The number listed in each column shows how many of each day type are planned for the week. The objective function for a one-week plan is $F(x) = \sum_{d=1}^{7} f(x_d)$ which is subject to the constraints of each included day.

2.3 Multi-objectives Problem

Now, we introduce a multi-objective model for time-use optimization. A multi-objective model involves finding solutions to optimize the problem defined by at least two conflicting objectives. The multi-objective model of time-use optimization can be defined as follows.

$$Objs: \quad M(x) = [f_1(x), f_2(x), f_3(x), f_4(x)] \qquad (4)$$

$$s.t. \quad \sum_{i=1}^{4} x_i = 1440 \qquad (5)$$

$$x_i^l \le x_i \le x_i^u \quad \forall i = \{1, \ldots, 4\} \qquad (6)$$

where x denotes a solution, $f_i(x) \to \mathbb{R}$ denotes the ith objective function to be optimized. Since there are four single objectives studied in this paper, we investigate all combinatorial objectives as multi-objective problems.

2.4 Fitness Function

We investigate the performance of different evolutionary algorithms for single-objective and multi-objective time-use optimization problems. The fitness of a solution x considers all constraints of one-day time-use optimization and one-week time-use optimization $h(x)$ and $H(x)$ separately.

$$h(x) = (u(x), f(x)) \qquad (7)$$

$$H(x) = (U(x), F(x)), \qquad (8)$$

where $u(x) = \sum_{i=1}^{4} \max\{0, x_i - x_i^u, x_i^l - x_i\}$ and $U(x) = \sum_{d=1}^{7} \sum_{i=1}^{4} \max\{0, x_{di} - x_{di}^u, x_{di}^l - x_{di}\}$. We optimize h and H with respect to lexicographic order, i.e. $h(x) \ge h(y)$ holds *iff* $u(x) < u(y) \lor (u(x) = u(y) \land f(x) \ge f(y))$ for objective f_2, f_3 and f_4, $u(x) < u(y) \lor (u(x) = u(y) \land |f(x)| \le |f(y)|)$ for objective f_1. Therefore, for the time-use optimization problem, any infeasible solution that violates the boundary constraints is worse than any feasible solution. Among solutions that meet all constraints, we aim to optimize the objective function.

3 Evolutionary Algorithms for the Time-Use Optimisation Problem

The algorithms that follow are classified into two classes. The first one contains single-objective evolutionary algorithms (Sect. 3.1), and the second has multi-objective evolutionary algorithms (Sect. 3.2). In this section, we only list the algorithms implemented in this study without detailed descriptions. Moreover, when implementing the presented algorithm for solving time-use optimization, Algorithm 1 is conducted before evaluating a generated solution.

3.1 Single-objective Evolutionary Algorithms

For the single-objective time-use optimization, we compare three evolutionary algorithms to optimize all health outcomes in different day structures.

Differential Evolution (DE) [14,37] is a well known global search heuristic using a binomial crossover and a mutation operator. We evaluate two mutation operators *DE/rand/1* and *DE/current-to-rand/1* for the single-objective time -use optimization problem. The population size is set to 50, and other control parameters are $F = 0.5$, $C_r = 0.5$.

Particle Swarm Optimization (PSO) [2,27], is a type of swarm intelligence evolutionary algorithm, with population size 50, $c_1 = 1$, $c_2 = 1$. For more understanding the working processes of PSO, we refer to [4,5,24,26,38,41].

Covariance matrix adaptation evolutionary strategy (CMA-ES) [22,23] is a self-adaptive evolution strategy that solves non-linear non-convex optimization problems in continuous domains. We implement the CMA-ES using $\lambda = 10$ and $\sigma = 0.3$.

3.2   Multi-objective Evolutionary Algorithms

For multi-objective time-use optimization, three multi-objective evolutionary algorithms are considered here.

Multi-objective evolutionary algorithm based on decomposition (MOEA/D) is a decomposition based algorithm commonly used to solve multi-objective optimisation problems [40]. We use the standard version of MOEA/D with the Tchebycheff approach, and population size is set to 100.

Non-dominated sorting genetic algorithm (NSGA-II) [16] is a fast non-dominated sorting procedure for ranking solutions in its selection step. It has been shown to be efficient when dealing with two objective optimization problems. We apply the NSGA-II with SBX operator and set the population size to 100.

Strong pareto evolutionary algorithm 2 (SPEA2) [42] is one of the most popular evolutionary multiple objective algorithms for dealing with optimization problems. We apply the SPEA2 with binary tournament selection and population size 100.

4   Experiments

This section shows detailed optimization results comparing the different evolutionary algorithms. Firstly, to evaluate the performance of the single-objective algorithms we investigate one-day instances of six different day structures with boundary constraints (Table 2) against four single objectives. Secondly, we evaluate the performance of the multi-objective algorithms on six different mixtures of one-week instances (Table 3), taking *Sporty day* as an exemplar with all the combinations of objectives for bio-objective optimization.

For each optimization algorithm with the configurations above, we execute 30 runs and report the statistic results using the Kruskal-Wallis test with 95% confidence intervals and follow-up with Bonferroni adjustments to account for multiple comparisons [13]. All experiments are performed using Jmetal of version 5.11, which is based on the description included in [30], and carried out on a MacBook Pro with an M1 chip.

Table 4. Mean (mean) and standard deviation (std) of 30 runs (print four decimal places). Best mean values are highlighted in **Best mean** by comparing results one-day single-objective time-use optimization problem

Day struct	Health outcomes	DE/rand/1 (1)			DE/current-to-rand/1 (2)			PSO (3)			CMA-ES (4)			Best results			
		Mean	Std	Stat	Mean	Std	Stat	Mean	Std	Stat	Mean	Std	Stat	x_1	x_2	x_3	x_4
Studious day	BMI	1.8343E−09	4.04E−09	2,4	4.6657E−05	5.21E−05	4	**1.1039E−13**	5.82E−13	2,4	0.0012	4.41E−19		392	713	150	185
	Congnition	2.5187	4.52E−16	2,4	2.5187	1.15E−05	4	**2.5187**	4.52E−16	2,4	2.5155	2.26E−15		389	900	150	1
	Life satisfaction	12445.2233	1.85E−12	2,4	12445.1461	0.21	4	**12445.2233**	1.85E−12	2,4	12331.6566	1.85E−12		465	690	150	136
	Fitness	60.4817	4.34E−14	4	60.4817	6.34E−14	4	**60.4817**	4.34E−14	4	60.1741	2.17E−14		390	690	150	210
Sporty day	BMI	9.4090E−08	2.97E−08	2,4	4.6941E−05	7.06E−05	4	**3.8719E−16**	1.40E−15	1,2,4	0.0191	3.53E−18		597	489	210	144
	Congnition	2.4423	3.62E−04	2,4	2.4418	1.36E−15		**2.4426**	8.72E−05	1,2,4	2.4419	9.03E−16	2	360	819	210	61
	Life satisfaction	13116.2140	9.25E−12	2,4	13116.1700	0.09	4	**13118.2140**	9.25E−12	1,2,4	13044.5263	0		573	480	210	178
	Fitness	62.2440	2.17E−14	2,4	62.2343	8.68E−03	4	**62.2440**	2.17E−14	2,4	61.0977	2.17E−14		360	480	390	210
After-school job day	BMI	**0.3387**	0	4	0.3388	4.84E−04	4	**0.3387**	0	4	0.4325	2.82E−16		420	480	330	210
	Congnition	2.4942	4.14E−04		2.4932	4.39E−04		2.4964	8.56E−05	1,2	**2.4964**	9.03E−16	1,2	360	790	330	1
	Life satisfaction	**12135.2479**	3.70E−12	2,4	12135.1980	0.07	4	**12135.2479**	3.70E−12	2,4	12026.5782	1.85E−12		481	480	330	149
	Fitness	**62.2440**	2.17E−14	2,4	62.2392	3.58E−03	4	**62.2440**	2.17E−14	2,4	61.6931	4.34E−14		360	480	390	210
Sporty weekend day	BMI	3.4636E−07	3.32E−07	2,4	1.0618E−04	7.15E−05	4	**9.2353E−14**	3.28E−13	2,4	0.0202	3.53E−18		718	279	297	146
	Life satisfaction	14453.8050	6.57		**14459.7788**	3.70E−12	1,3	14442.1353	1.82E+01		**14459.7788**	3.70E−12	1,3	720	240	240	210
	Congnition	2.4338	6.71E−04	2	2.4327	4.21E−04		**2.4356**	0	1,2,	**2.4356**	0	1,2	420	784	210	61
	Fitness	60.8883	3.49E−03	3,4	**60.8928**	1.07E−05	1,3,4	60.8739	5.79E−04	4	55.8222	7.23E−15		441	558	221	210
Studious weekend day	BMI	1.5481E−09	1.98E−09	2,4	4.2805E−05	5.09E−05	4	**5.9270E−13**	2.24E−13	2,4	0.0012	4.41E−19		458	690	150	142
	Congnition	**2.5187**	4.52E−16	2,4	2.5187	1.54E−05	4	**2.5187**	4.52E−16	2,4	2.5155	2.26E−15		389	900	150	1
	Life satisfaction	**12445.2233**	1.85E−12	2,4	12445.1818	0.08	4	**12445.2233**	1.85E−12	2,4	12331.6566	1.85E−12		458	690	150	142
	Fitness	**60.4817**	4.34E−14	4	60.4817	3.80E−14	4	**60.4817**	4.34E−14	4	60.1741	2.17E−14		390	690	150	210
Working weekend day	BMI	0.0589	4.14E−17	4	0.0589	2.12E−17	4	**0.0589**	4.53E−17	4	0.1068	7.06E−17		630	210	390	210
	Congnition	2.4876	8.97E−04		2.4858	1.40E−03		2.4900	5.45E−05	1,2	**2.4900**	9.03E−16	1,2	360	753	390	1
	Life satisfaction	**13809.0070**	5.55E−12	2,4	13808.9369	0.10	4	**13809.0070**	5.55E−12	2,4	13794.5028	1.85E−12		641	210	390	199
	Fitness	**62.2804**	2.17E−14	2,4	62.2804	0	4	**62.2804**	2.17E−14	2,4	55.7913	2.17E−14		360	454	416	210

4.1 Results of Single-objective Time-Use Optimization

Table 4 and Table 5 list the results obtained of one-day instances and one-week instances separately. We provide the results from 30 independent runs with 25,000 generation for all instances. The *mean* denotes the average objective value of the 30 runs and *std* denotes standard deviation. Since we aim to minimize the absolute value of BMI, the results listed in the *BMI* rows are absolute values. The best solutions are bold and shadowed in each row. We also report the decision variables (rounded to minutes) of the optimal solution for each health outcome of each day structure in Table 4.

Column *stat* lists the results of statistical comparisons between the algorithms. If two algorithms can be compared significantly, then the index of algorithms that list in each column is significantly worse than the current algorithm. For example, the first row in Table 4 shows that PSO and DE/rand/1 are significantly better than DE/current-to-rand/1 and CMA-ES when optimizing the BMI of Studious day, and DE/current-to-rand/1 is significantly better than CMA-ES. However, there is no significant difference between the performance of DE/rand/1 and PSO. As can be seen from the table of one-day instances, the results obtained by the PSO are better than other algorithms in nearly all cases. DE algorithm with *DE/rand/1* operator is the second best algorithm, outperforming the DE algorithm with *DE/current-to-rand/1* operator and CMA-ES in many instances, while CMA-ES shows an advantage when aiming to optimize Cognition for many day structures. Moreover, as observed in the *std* columns, the standard deviation of 30 runs of all the evaluated algorithms in most instances is close to zero. Therefore, we can argue that for the single-objective optimization, the results obtained by the investigated algorithms, especially the DE/rand/1 and PSO, are close to optimal.

Table 5. Mean (mean) and standard deviation (std) of 30 runs (print four decimal places). Best mean values are highlighted in **Best mean** by comparing results one-week single-objective time-use optimization problem

Day struct	Health outcomes	DE/rand/1 (1)			DE/current-to-rand/1 (2)			PSO (3)			CMA-ES (4)		
		Mean	Std	Stat	Mean	Std	Stat	Mean	Std	Stat	Mean	Std	Stat
1	BMI	0.0780	3.86E−03	2,4	1.3504	0.1327		0.0657	0.0171	1,2,4	0.8056	0.1326	2
	Cognition	17.4336	7.12E−06	2,3,4	17.4133	0.0045	3,4	17.3504	0.0184	4	17.2625	0.0268	
	Life satisfaction	91084.2750	0.3571	2,4	86871.4104	565.3453		91065.6882	55.3775	2,4	86950.4715	475.0648	
	Fitness	427.1767	0.0260	2,4	383.2675	3.7084		426.1141	3.1869	2,4	406.9240	3.5495	2
2	BMI	0.7480	0.0022	2,4	2.3733	0.2167		0.7368	0.0016	1,2,4	1.3006	0.0251	2
	Cognition	17.5453	6.34E−06	2,3,4	17.5230	0.0043	3,4	17.4626	0.0196	4	17.3739	0.0239	
	Life satisfaction	87860.0591	0.3618	2,4	83699.4248	609.0356		87857.4099	13.1646	2,4	84854.7392	48.1036	2
	Fitness	428.5592	0.0260	2,3,4	378.0875	4.7320		423.9175	7.3229	2,4	419.0847	0.0000	2
3	BMI	0.0742	0.0036	2,4	1.5243	0.1372		0.0802	0.0530	2,4	1.0385	0.1753	2
	Cognition	17.4428	5.70E−06	2,3,4	17.4222	0.0038	3,4	17.3614	0.0244	4	17.2726	0.0220	
	Life satisfaction	89821.7778	0.5145	2,3,4	85591.7374	509.1520		89780.8014	97.5659	2,4	85694.0353	540.8402	2
	Fitness	428.5469	0.0356	2,3,4	385.2941	4.2711		425.9137	4.6540	2,4	410.7629	4.9936	2
4	BMI	0.3538	0.0032	2,4	1.7570	0.1912		0.3525	0.0393	2,4	0.9861	0.1783	2
	Cognition	17.4516	5.87E−06	2,3,4	17.4326	0.0042	3,4	17.3689	0.0199	4	17.3010	0.0118	
	Life satisfaction	88148.1600	0.2810	3,4	88148.1600	0.2810	3,4	88144.4983	12.2982	4	84944.1618	533.3457	
	Fitness	428.5187	0.0274	2,3,4	387.9432	3.4146		426.5169	3.4843	2,4	419.7527	5.2723	2
5	BMI	0.1364	0.0041	2,4	1.4189	0.1520	4	0.1190	0.0035	1,2,4	1.5034	0.1565	
	Cognition	17.3224	2.92E−06	2,3,4	17.3016	0.0052	3,4	17.2863	0.0169	4	17.1519	0.0205	
	Life satisfaction	93118.9359	0.5341	2,3,4	88787.2339	309.6765	4	93081.6400	86.8573	2,4	87702.1107	661.4579	
	Fitness	430.6356	0.0405	2,3,4	392.2353	4.3508		429.6756	2.2065	2,4	402.4828	4.8568	2
6	BMI	0.3587	0.0036	2,4	1.5370	0.1611		0.3422	0.0097	1,2,4	1.0645	0.1762	2
	Cognition	17.3655	4.15E−06	2,3,4	17.3466	0.0041	3,4	17.3085	0.0220	4	17.1923	0.0238	
	Life satisfaction	90081.1467	0.6989	2,3,4	86438.6857	477.3889	4	90075.1480	20.6442	2,4	86225.2387	628.1738	
	Fitness	428.8659	0.0321	2,3,4	390.9420	4.6767		426.5522	4.5375	2,4	413.3007	4.2733	2

Table 5 presents the summary statistic for the results of one-week single-objective instances. A closer inspection of the table shows that DE/rand/1 outperforms the other algorithms in most instances, and PSO outperforms the last two algorithms. Therefore, these results suggest that for solving the single-objective optimization problem, DE algorithm with *DE/rand/1* operator and PSO both perform well. PSO is preferred for one-day instances, and the DE algorithm with *DE/rand/1* operator is preferred for solving one-week instances.

4.2   Results of Multi-objective Time-Use Optimization

To compare the difference between evolutionary multi-objective optimization algorithms, we analyze the experimental results of two, three and four objectives, respectively. For performance evaluation, we use hypervolume [3,43] as the metric. The hypervolume statistics are provided in Table 6. The best hypervolume is highlighted and bold for each combination of objectives in each row. It can be seen from the *stat* results in the table that SPEA significantly outperforms the other algorithms for two-objective optimization instances, and NSGA-II outperforms the other two algorithms for three- and four-objective optimization instances.

Table 6. Multi-objective optimization hypervolume statistics

Combine of health outcomes	MOEA/D (1)					NSGA-II (2)					SPEA2 (3)				
	Best	Worst	Median	Std	Stat	Best	Worst	Median	Std	Stat	Best	Worst	Median	Std	Stat
BMI & Cognition	0.9895	0.9894	0.9895	7.87E−06		0.9898	0.9897	0.9898	8.67E−06	1	0.9898	0.9898	0.9898	9.13E−06	1,2
BMI & Life satisfaction	0.9747	0.9382	0.9451	7.34E−03		0.9985	0.9984	0.9985	1.98E−05	1	0.9988	0.9986	0.9987	2.58E−05	1,2
BMI & Fitness	0.9841	0.9837	0.9839	1.04E−04		0.9841	0.9780	0.9840	1.46E−03	1	0.9841	0.9841	0.9841	6.00E−06	1,2
Cognition & Life satisfaction	0.9794	0.9780	0.9788	4.10E−04		0.9975	0.9967	0.9969	1.59E−04	1	0.9978	0.9970	0.9972	2.08E−04	1,2
Cognition & Fitness	0.9959	0.9956	0.9958	6.48E−05		0.9976	0.9891	0.9952	2.48E−03		0.9977	0.9976	0.9976	1.85E−05	1,2
Life satisfaction & Fitness	0.9961	0.9770	0.9926	7.98E−03		0.9974	0.9774	0.9970	7.07E−03	1	0.9976	0.9971	0.9973	1.53E−04	1,2
BMI & Cognition & Life satisfaction	0.9745	0.9712	0.9714	7.58E−04		0.9874	0.9866	0.9870	2.27E−04	1,3	0.9869	0.9671	0.9816	5.16E−03	1
BMI & Cognition & Fitness	0.9708	0.9690	0.9701	4.81E−04		0.9751	0.9724	0.9738	7.03E−04	1,3	0.9750	0.9396	0.9725	9.90E−03	
Cognition & Life satisfaction & Fitness	0.9759	0.9572	0.9726	4.43E−03		0.9925	0.9780	0.9864	4.12E−03	1,3	0.9769	0.9607	0.9735	3.31E−03	
BMI & Cognition & Life satisfaction & Fitness	0.9613	0.9556	0.9569	1.33E−03		0.9706	0.9653	0.9683	1.27E−03	1,3	0.9677	0.9463	0.9601	5.07E−03	

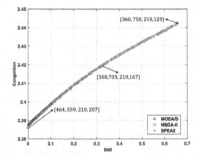

(a) Median HV Run of optimizing *BMI* and *Cognition*

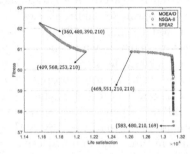

(b) Median HV Run of optimizing *Life satisfaction* and *Fitness*

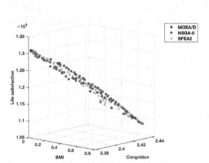

(c) Median HV Run of optimizing *BMI*, *Cognition* and *Life satisfaction*

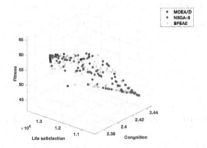

(d) Median HV Run of optimizing *Cognition*, *Life satisfaction* and *Fitness*

Fig. 1. Results obtained for multi-objective model of sporty day

The bio-objective results obtained in a median hypervolume run for each algorithm are plotted in Fig. 1. Figure 1 (a) shows that the trade-off fronts of optimizing the first two objectives achieved by SPEA2 are more generally distributed in the Pareto front than MOEA/D and NSGA-II. Similarly, Fig. 1 (b) indicates that the trade-off solutions obtained by MOEA/D and NSGA-II are clustered in a small area of the solution space. Moreover, for three-objective optimization (Fig. 1 (c) and (d)), NSGA-II and SPEA2 generate better Pareto solutions in comparison with MOEA/D. On Fig. 1 (a) and (b), selected optimized solutions are shown to reflect optimal daily activity durations if one individual outcome is preferred above another (near to the respective axes) or if the outcomes are equally preferred (near the mid-point of the Pareto front).

5   Conclusion

The way children spend their time on sleep, sedentary behaviour and physical activity (LPA and MVPA) affects their health and well-being. The main goal of the current study is to implement evolutionary algorithms on daily allocations to optimize children's health outcomes. Based on a real-world data set, we introduce single- and multi-objective optimization models and design fitness functions of one-day and one-week problems. Our experimental results show that when tackling the single-objective problem, DE algorithm with $DE/rand/1$ and PSO outperforms other proposed algorithms on both one-day instances and one-week instances. Moreover, the SPEA2 has a higher hypervolume than NSGA-II and MOEA/D in two-objective optimization instances for the multi-objective problem. In comparison, NSGA-II has a higher hypervolume than the other algorithms in three and four objectives instances. Overall, this study strengthens the idea that evolutionary algorithms can be used to enhance our understanding of how children can allocate their daily time to optimize their health and well-being. Parents are concerned about their children's sleep, screen time and physical activity, and they want evidence-based guidance on how much time should be spent in these behaviours. However, it is unlikely to be feasible to expect families to follow strict daily time allocation schedules. The evidence generated from the application of optimization algorithms may be better understood as general advice, and primarily serve to inform public health guidelines for children's time-use behaviours. Population-level surveillance of guideline compliance can help inform public health policy, track secular trends overtime and to evaluate the effectiveness of public health interventions.

Acknowledgements. This work has been supported by NHMRC Ideas grant 1186123, by ARC grant FT200100536, and by the South Australian Government through the Research Consortium "Unlocking Complex Resources through Lean Processing". Dorothea Dumuid is supported by NHMRC Fellowship 1162166 and by the Centre of Research Excellence in Driving Global Investment in Adolescent Health funded by NHMRC 1171981. The CheckPoint study was supported by the NHMRC [1041352; 1109355]; the National Heart Foundation of Australia [100660]; The Royal Children's Hospital Foundation [2014-241]; the Murdoch Children's Research Institute (MCRI); The University of Melbourne; the Financial Markets Foundation for Children [2014-055, 2016-310]; and the Australian Department of Social Services (DSS). Research at the MCRI is supported by the Victorian Government's Operational Infrastructure

Support Program. The funders played no role in the study design, data collection and analysis, decision to publish, or preparation of the manuscript.

References

1. Aitchison, J.: The statistical analysis of compositional data. J. Roy. Stat. Soc.: Ser. B (Methodol.) **44**(2), 139–160 (1982)
2. AlRashidi, M.R., El-Hawary, M.E.: A survey of particle swarm optimization applications in electric power systems. IEEE Trans. Evol. Comput. **13**(4), 913–918 (2009)
3. Auger, A., Bader, J., Brockhoff, D., Zitzler, E.: Hypervolume-based multiobjective optimization: theoretical foundations and practical implications. Theor. Comput. Sci. **425**, 75–103 (2012)
4. Banks, A., Vincent, J., Anyakoha, C.: A review of particle swarm optimization. Part I: background and development. Nat. Comput. **6**(4), 467–484 (2007)
5. Banks, A., Vincent, J., Anyakoha, C.: A review of particle swarm optimization. Part II: hybridisation, combinatorial, multicriteria and constrained optimization, and indicative applications. Nat. Comput. **7**(1), 109–124 (2008)
6. Van den Boogaart, K.G., Tolosana-Delgado, R.: Analyzing Compositional Data with R, vol. 122. Springer, Cham (2013). https://doi.org/10.1007/978-3-642-36809-7
7. Boreham, C., Paliczka, V., Nichols, A.: A comparison of the PWC170 and 20-MST tests of aerobic fitness in adolescent schoolchildren. J. Sports Med. Phys. Fitness **30**(1), 19–23 (1990)
8. Box, G.E.P., Cox, D.R.: An analysis of transformations. J. R. Stat. Soc. Ser. B (Methodol.) **26**(2), 211–252 (1964)
9. Carson, V., et al.: Systematic review of sedentary behaviour and health indicators in school-aged children and youth: an update. Appl. Physiol. Nutr. Metab. **41**(6), S240–S265 (2016)
10. Chaput, J.P., et al.: Systematic review of the relationships between sleep duration and health indicators in school-aged children and youth. Appl. Physiol. Nutr. Metab. **41**(6), S266–S282 (2016)
11. Clifford, S.A., Davies, S., Wake, M.: Child health checkpoint: cohort summary and methodology of a physical health and biospecimen module for the longitudinal study of Australian children. BMJ Open **9**(Suppl. 3) (2019)
12. Coello, C.A.C., van Veldhuizen, D.A., Lamont, G.B.: Evolutionary Algorithms for Solving Multi-objective Problems, Genetic Algorithms and Evolutionary Computation, vol. 5. Kluwer (2002)
13. Corder, G.W., Foreman, D.I.: Nonparametric Statistics: A Step-by-Step Approach. Wiley, Hoboken (2014)
14. Das, S., Suganthan, P.N.: Differential evolution: a survey of the state-of-the-art. IEEE Trans. Evol. Comput. **15**(1), 4–31 (2010)
15. Deb, K.: Multi-objective Optimization Using Evolutionary Algorithms. Wiley-Interscience Series in Systems and Optimization. Wiley (2001)
16. Deb, K., Pratap, A., Agarwal, S., Meyarivan, T.: A fast and elitist multiobjective genetic algorithm: NSGA-II. IEEE Trans. Evol. Comput. **6**(2), 182–197 (2002)
17. Dumuid, D., et al.: Goldilocks days: optimising children's time use for health and well-being. J. Epidemiol. Community Health **76**, 301–308 (2021)
18. Dumuid, D., et al.: Compositional data analysis for physical activity, sedentary time and sleep research. Stat. Methods Med. Res. **27**(12), 3726–3738 (2018)

19. Dumuid, D., et al.: Balancing time use for children's fitness and adiposity: evidence to inform 24-hour guidelines for sleep, sedentary time and physical activity. PLoS ONE **16**(1), e0245501 (2021)

20. Gray, M., Smart, D.: Growing up in Australia: the longitudinal study of Australian children is now walking and talking. Fam. Matters **79**, 5–13 (2008)

21. Han, L., Wang, H.: A random forest assisted evolutionary algorithm using competitive neighborhood search for expensive constrained combinatorial optimization. Memet. Comput. **13**(1), 19–30 (2021). https://doi.org/10.1007/s12293-021-00326-9

22. Hansen, N.: The CMA evolution strategy: a comparing review. In: Towards a New Evolutionary Computation, Studies in Fuzziness and Soft Computing, vol. 192, pp. 75–102. Springer, Cham (2006). https://doi.org/10.1007/3-540-32494-1_4

23. Hansen, N., Ostermeier, A.: Completely derandomized self-adaptation in evolution strategies. Evol. Comput. **9**(2), 159–195 (2001)

24. Houssein, E.H., Gad, A.G., Hussain, K., Suganthan, P.N.: Major advances in particle swarm optimization: theory, analysis, and application. Swarm Evol. Comput. **63**, 100868 (2021)

25. Jakob, W.: Applying evolutionary algorithms successfully: a guide gained from real-world applications. CoRR arXiv:2107.11300 (2021)

26. Kennedy, J., Eberhart, R.: Particle swarm optimization. In: Proceedings of ICNN'95-International Conference on Neural Networks, vol. 4, pp. 1942–1948. IEEE (1995)

27. Lee, K., Kim, J.: Multiobjective particle swarm optimization with preference-based sort and its application to path following footstep optimization for humanoid robots. IEEE Trans. Evol. Comput. **17**(6), 755–766 (2013)

28. Li, X., Bonyadi, M.R., Michalewicz, Z., Barone, L.: Solving a real-world wheat blending problem using a hybrid evolutionary algorithm. In: IEEE Congress on Evolutionary Computation, pp. 2665–2671. IEEE (2013)

29. Mateu-Figueras, G.: The principle of working on coordinates. In: Pawlowsky-Glahn, V., Buccianti, A. (eds.) compositional Data Analysis: Theory and Applications (2011)

30. Nebro, A.J., Durillo, J.J., Vergne, M.: Redesigning the jMetal multi-objective optimization framework. In: GECCO (Companion), pp. 1093–1100. ACM (2015)

31. Okely, A.D., et al.: A collaborative approach to adopting/adapting guidelines. The Australian 24-hour movement guidelines for children (5–12 years) and young people (13–17 years): an integration of physical activity, sedentary behaviour, and sleep. Int. J. Behav. Nutr. Phys. Act. **19**(1), 1–21 (2022)

32. Onis, M.D., Onyango, A.W., Borghi, E., Siyam, A., Nishida, C., Siekmann, J.: Development of a WHO growth reference for school-aged children and adolescents. Bull. World Health Organ. **85**, 660–667 (2007)

33. R Core Team: R: A Language and Environment for Statistical Computing. R Foundation for Statistical Computing, Vienna, Austria (2020). https://www.R-project.org/

34. Ridley, K., Olds, T.S., Hill, A.: The multimedia activity recall for children and adolescents (MARCA): development and evaluation. Int. J. Behav. Nutr. Phys. Act. **3**(1), 1–11 (2006)

35. Saha, C., Jones, M.P.: Asymptotic bias in the linear mixed effects model under non-ignorable missing data mechanisms. J. R. Stat. Soc. Ser. B (Stat. Methodol.) **67**(1), 167–182 (2005)

36. Seligson, J.L., Huebner, E.S., Valois, R.F.: Preliminary validation of the brief multidimensional students' life satisfaction scale (BMSLSS). Soc. Indic. Res. **61**(2), 121–145 (2003)
37. Storn, R., Price, K.: Differential evolution-a simple and efficient heuristic for global optimization over continuous spaces. J. Global Optim. **11**(4), 341–359 (1997)
38. Wang, D., Tan, D., Liu, L.: Particle swarm optimization algorithm: an overview. Soft. Comput. **22**(2), 387–408 (2017). https://doi.org/10.1007/s00500-016-2474-6
39. Weintraub, S., et al.: Cognition assessment using the NIH Toolbox. Neurology **80**(11 Supplement 3), S54–S64 (2013)
40. Zhang, Q., Li, H.: MOEA/D: a multiobjective evolutionary algorithm based on decomposition. IEEE Trans. Evol. Comput. **11**(6), 712–731 (2007)
41. Zhang, Y., Wang, S., Ji, G.: A comprehensive survey on particle swarm optimization algorithm and its applications. Math. Probl. Eng. **2015** (2015)
42. Zitzler, E., Laumanns, M., Thiele, L.: SPEA 2: improving the strength pareto evolutionary algorithm. TIK-report 103 (2001)
43. Zitzler, E., Thiele, L.: Multiobjective evolutionary algorithms: a comparative case study and the strength pareto approach. IEEE Trans. Evol. Comput. **3**(4), 257–271 (1999)

Iterated Local Search for the eBuses Charging Location Problem

César Loaiza Quintana[1][✉][iD], Laura Climent[2][iD], and Alejandro Arbelaez[2][iD]

[1] University College Cork, Cork, Ireland
c.loaizaquintana@cs.ucc.ie
[2] Universidad Autónoma de Madrid, Madrid, Spain
{laura.climent,alejandro.arbelaez}@uam.es

Abstract. Electric buses (eBuses) will be the mainstream in mass urban transportation in the near future. Thus, installing the charging infrastructure in convenient locations will play a critical role in the transition to eBuses. Taking this into account, in this paper we propose an iterated local search algorithm to optimize the location of charging stations while satisfying certain properties of the transportation system, e.g., satisfying the demand and ensuring that the limited driving range of the buses will not impact the service. The effectiveness of our approach is demonstrated by experimenting with a set of problem instances with real data from 3 Irish cities, i.e., Limerick, Cork, and Dublin. We compare our approach against a MIP-based solution. Results show that our approach is superior in terms of scalability and its anytime behavior.

Keywords: Local search · Charging location problem · Electric buses

1 Introduction

The transition to a fleet of eBuses will require a considerable up-front investment in the charging infrastructure. The actual cost of a charging station varies depending on the desired charging infrastructure. The starting cost is now estimated at €10K with additional fees proportional to the required power, e.g., €10K for a fast charging point with approximately 600 kW capacity as well as substantial upgrades and cabling costs that can certainly increase the total cost [9]. The daily power consumption of an eBus depends on multiple factors, such as bus type, weight, weather, and road conditions. A standard double-decker working 12 h daily consumes approx. 1.5 kWh per kilometer. However, the power consumption might increase or decrease due to multiple factors.

Range anxiety is one of the main concerns for transitioning to a clean and sustainable bus transportation network. Nowadays, eBuses can travel approx. up to 200–300 KM on a full charge and the charging time varies depending on the technology from a couple of minutes (e.g., with fast-charging stations) to hours (e.g., with slow overnight charging). Therefore, the charging infrastructure must be implemented in a way that the charging time and limited driving range will not impact the quality of service of the transportation system. In this paper,

© The Author(s) 2022
G. Rudolph et al. (Eds.): PPSN 2022, LNCS 13399, pp. 338–351, 2022.
https://doi.org/10.1007/978-3-031-14721-0_24

we propose an iterated local search to efficiently tackle the charging location problem, i.e., finding the optimal location of charging stations while satisfying certain properties of the transportation system.

This paper is organized as follows: Sect. 2 describes previous work on the charging location problem for electric vehicles. Section 3 formally describes the charging location problem. Section 4 presents our new meta-heuristic solution. Section 5 presents our empirical evaluation. Finally, Sect. 6 presents some concluding remarks.

2   Related Work

A considerable amount of work has been devoted to the charging location problem for light-duty vehicles (e.g., personal use cars) rather than heavy-duty vehicles (e.g., buses). We remark that public buses operate daily schedules for a given set of routes and buses consume more energy than cars. In this line, in [4, 5, 14] the authors propose a set of heuristic solutions to identify suitable charging stations for electric vehicles, so that the vehicles can reach their destinations without running out of power. [15] takes into account a cost model for the charging location problem by considering different technologies, bio-diesel, bio-gas, and electric to assign a type of bus to each route.

In another line of work, the Green Vehicle Routing Problem (Green VRP) is an extension of the traditional VRP [3]. This extension aims at finding the optimal routes traveled by the fleet of vehicles to visit a set of customers while taking into account certain refueling constraints. We recall that we aim at transitioning to eBuses while maintaining the same service. [2] outlines numerous variations of the VRP for electric vehicles.

Closer to our research, [12] considered a fleet of vehicles, each one with well-defined routes within a network. This work proposes a MIP model whose objective was to minimize the cost of the charging stations, which had to be located into a subset of the stops. As the author suggests this work could be adapted to a bus transportation system by considering timetables, however, he neglects them in his validation carried out over random-generated networks.

[13] proposes a MIP model for the eBus charging location problem with fixed routes but without taking into consideration certain operational constraints, e.g., fixed timetables or overlapping constraints to prevent multiple buses using the same charging unit at the same time. In [10] the authors take into account the impact of timetables of the original routes by allowing charging when the buses are not moving (dwelling times), e.g., during the rest time of the drivers, therefore, the authors keep the original timetables. Instead, in this paper, we propose a more flexible approach to allow small enough deviations in the original timetable leading to a more general version of the problem. [1] provides more details of the described behavior of the charging location problem.

3   The Charging Location Problem

Conceived in the context of a transition to eBuses, the charging location problem aims to identify, within the set of stations, the locations for installing the minimum number of electrical chargers. These electrical chargers must allow a fleet of eBuses to keep the operation of the system with minimal disruptions (deviations) from the original timetables. In this context, we assume that the charging stations are able to recharge the buses during their normal operation, i.e., in the time between the arrival and departure of the buses in the stations.

Figure 1 outlines an instance of the system for a single bus. In this example, we assume that the bus visits 12 stations. For each station we have three timestamps, i.e., actual arrival time (left), expected arrival time (middle), and departure time (right). In order to reduce the disruptions in the transition from regular diesel buses to eBuses, we fix a maximum deviation time μ from the original timetables (2 min in this example). Therefore, eBuses are not allowed to arrive more than μ minutes after (or before) the timetabled stop. We also associate each trip between two stations with the required time and power to complete the trip. The number of points sitting below each charging station denotes the amount of power (in kWh) gained at the charging station (yellow stations in this example), e.g., in station 5 the bus gains 2 kWh units of power.

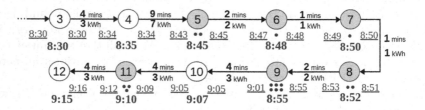

Fig. 1. Example of a solution to the problem for a single bus

Hereafter, we formally describe the charging location problem. Table 1 outlines terms and variables commonly used through the paper.

A public bus transportation system is composed of a set of buses $\mathcal{B}$ traveling across a set of stations ST, where each bus follows a sequence of ordered stops S_b visited in a workday. Thus, each stop $s_{bi} \in S_b$ has associated a station and a timetable that the bus should comply with (st_{bi} and T_{bi}).

We build our formulation of the charging location problem on top of a previous MIP model [1]. The objective function (1) is to minimize the number of installed chargers. Constraints 2–5 ensure buses are not running out of energy during their trips, by maintaining the battery within safe levels. Furthermore, Constraints 6–9 are in charge of the scheduling of the buses, which must stay within a reasonable deviation time Δt of μ with respect to the original timetables. Constraint 10 enforces a minimum reasonable charging time.

Table 1. Description of variables and constants.

Constants	
$\mathcal{B}, ST, SM$	set of buses, set of stations, and security margin between chargers
$\mathcal{S}_b$	sequence of ordered stops visited by bus b in a workday, $\mathcal{S}_b = \{s_{b1}, \ldots, s_{bn}\}$, where n is the number of stops
st_{bi}	i-th station in the path of a bus b in a workday
τ_{bj}	timetabled arrival time of the bus b at its j-th stop
μ	max. time disruption allowed (considering the scheduled times)
T_{ji}, D_{ji}	time and energy needed to complete the trip between stations j and i
C_{min}, C_{max}	min. and max. capacity battery levels of the buses
ψ, μ, β, α	min. charging time, max. deviation time from original timetable, max. charging time per cycle, and charging rate in kWh per minute.

Variables	
t_{bj}	actual arrival time of bus b at the arrival to its j-th stop
Δt_{bj}	time difference between the arrival and scheduled times
c_{bj}	current battery level of bus b at the arrival to its j-th stop
e_{bj}	energy re-charged of bus b at its j-th stop
ct_{bj}	recharging time of bus b at its j-th stop
x_{bj}	Boolean var. denoting whether we recharge bus b at its j-th stop
x_i	Boolean var. denoting whether we install a charging unit at station i
Z_{bdij}	Boolean var. denoting whether buses b and d are using the same charging station ($i = j$) or not
z_{bdi}	Boolean var. indicating if bus b charges after the bus d at station i

$$min \sum_{c \in ST} x_c \tag{1}$$

$$C_{min} \leq c_{bi} \qquad\qquad \forall_{b \in B} \forall_{s_i \in \mathcal{S}_b \setminus \{s_0\}} \tag{2}$$

$$c_{bi} + e_{bi} \leq C_{max} \qquad\qquad \forall_{b \in B} \forall_{s_i \in \mathcal{S}_b \setminus \{s_0\}} \tag{3}$$

$$c_{bi} \leq c_{bj} + e_{bj} - D_{ji} \qquad\qquad \forall_{b \in B} \forall_{s_i \in \mathcal{S}_b \setminus \{s_0\}, j = i-1} \tag{4}$$

$$\alpha \cdot ct_{bi} \geq e_{bi} \qquad\qquad \forall_{b \in B} \forall_{s_i \in \mathcal{S}_b \setminus \{s_0\}, j = i-1} \tag{5}$$

$$t_{bi} \geq t_{bj} + ct_{bj} - T_{ji} \qquad\qquad \forall_{b \in B} \forall_{s_i \in \mathcal{S}_b \setminus \{s_0\}, j = i-1} \tag{6}$$

$$\Delta t_{bi} \geq t_{bi} - \tau_{bi} \qquad\qquad \forall_{b \in B} \forall_{s_i \in \mathcal{S}_b \setminus \{s_0\}} \tag{7}$$

$$\Delta t_{bi} \geq \tau_{bi} - t_{bi} \qquad\qquad \forall_{b \in B} \forall_{s_i \in \mathcal{S}_b \setminus \{s_0\}} \tag{8}$$

$$\Delta t_{bi} \leq \mu \qquad\qquad \forall_{b \in B} \forall_{s_i \in \mathcal{S}_b \setminus \{s_0\}} \tag{9}$$

$$ct_{bi} \geq \psi \cdot x_{bi} \qquad\qquad \forall_{b \in B} \forall_{s_i \in \mathcal{S}_b \setminus \{s_0\}} \tag{10}$$

$$x_{st_{bi}} \geq x_{bi} \qquad\qquad \forall_{b \in B} \forall_{s_i \in \mathcal{S}_b \setminus \{s_0\}} \tag{11}$$

$$\beta \cdot x_{bi} \geq ct_{bi} \qquad\qquad \forall_{b \in B} \forall_{s_i \in \mathcal{S}_b \setminus \{s_0\}} \tag{12}$$

$$Z_{bdij} \leq x_{bi} \qquad\qquad \forall_{b,d \in B | b \neq d} \forall_{s_i \in \mathcal{S}_b}, \forall_{s_j \in \mathcal{S}_d | st_{bi} = st} \tag{13}$$

$$Z_{bdij} \leq x_{dj} \qquad\qquad \forall_{b,d \in B | b \neq d} \forall_{s_i \in \mathcal{S}_b}, \forall_{s_j \in \mathcal{S}_d | st_{bi} = st} \tag{14}$$

$$x_{bi} + x_{dj} \leq Z_{bdij} + 1 \qquad\qquad \forall_{b,d \in B | b \neq d} \forall_{s_i \in \mathcal{S}_b}, \forall_{s_j \in \mathcal{S}_d | st_{bi} = st} \tag{15}$$

$$t_{bi} \geq t_{dj} + ct_{dj} - M z_{bdi} \qquad \forall_{b,d \in B|b \neq d} \forall_{s_i \in \mathcal{S}_b}, \forall_{s_j \in \mathcal{S}_d|st_{bi}=st} \qquad (16)$$

$$t_{dj} \geq t_{bi} + ct_{bi} - M z_{dbj} \qquad \forall_{b,d \in B|b \neq d} \forall_{s_i \in \mathcal{S}_b}, \forall_{s_j \in \mathcal{S}_d|st_{bi}=st} \qquad (17)$$

$$z_{bdi} + z_{dbj} - (1 - Z_{bdij}) \leq 1 \qquad \forall_{b,d \in B|b \neq d} \forall_{s_i \in \mathcal{S}_b}, \forall_{s_j \in \mathcal{S}_d|st_{bi}=st} \qquad (18)$$

Constraint 11 determines the selected charging stations. Constraint 12 prevents overheating by limiting charging time. Finally, Constraints 13–18 prevent multiple buses from using the same charger at the same time. We refer the reader to [1] for more details of the MIP formulation.

4   The Iterated Local Search

Iterated Local Search (ILS) is a popular meta-heursitic technique to solve complex combinatorial optimization problems [7,11]. Generally speaking, the algorithm produces a single-transformation chain of solutions that iteratively improves the objective function. First, in a local search phase, the algorithm tries to improve an initial solution by performing small changes, until a local minimum ($s*$) is reached. Then, in a perturbation phase, the algorithm performs random changes to the current incumbent solution ($s*$) to escape difficult regions of the search (e.g., a plateau), and to produce a perturbed solution (s'). Next, the perturbed solution (s') is given as an input for a new local search phase that generates a new local minimum (s''). Finally, an acceptance criterion decides which will be the new incumbent solution ($s*$) to continually repeat the process from the perturbation phase. The idea is that the algorithm will primarily prefer the best solution between $s*$ and s''. However, in order to diversify the search, the algorithm selects the most recent local minimum (s'') with a certain probability. In this paper, unless stated otherwise, we use a 5% probability to update the incumbent solution.

In the context of the charging location problem, we define an open station as a location (i.e., bus station) with an installed charging unit iff at least one bus is relying on the station to recharge the battery. Therefore, during the local search phase, we aim at closing as many open stations as possible. Alternatively, during the perturbation phase, we aim at diversifying the search by randomly opening stations.

Initial Solution

We propose a simple greedy algorithm to compute an initial solution. We assume that the fleet of eBuses start operations with full capacity. The algorithm commands a given bus b to recharge the battery at a given stop $s_{bj} \in S_b$ if there is not enough power to reach the next stop s_{bj+1}. Furthermore, b will recharge the battery of the bus with, at least, the minimum required power to reach s_{bj+1}. We note that considering that b starts with full capacity, all the stations on the path of the bus might be open once the bus depletes its initial charge.

Furthermore, if b is expected to arrive at s_{bj+1} before the allowed scheduled time (see Constraints 7–9), then b uses this additional time (dwell time) to recharge the battery and get additional power. The algorithm repeats this process for each bus $b \in B$ in lexicographical order. Additionally, whenever we get an overlapping conflict (i.e., two buses are attempting to use the charging station at the same time, see Constraints 13–18), we delay the arrival time of the current bus until the charging station is available.

Let us illustrate the process to compute the initial solution with Fig. 1. We assume that the initial capacity of the bus is enough to reach station 5, at this point the bus needs to recharge the battery with at least 2 kWh; otherwise, the eBus runs out of battery before reaching station 6. Therefore, the bus arrives at station 5 at 8:43 and recharges the battery for 2 min to leave the station at 8:45. Then, the bus arrives at station 6 at 8:47 to recharge the battery again for 1 min. The algorithm repeats the same process until arriving at station 9 at 8:55. Here, our greedy algorithm initially recharges the bus for 3 min (enough to reach station 10), however, by doing so, the bus would arrive at the next station at 9:02, 3 min ahead of the earliest allowed time. We recall the bus is constrained to arrive at station 10 at 9:05. This additional time is known as dwell. Therefore, our greedy construction of the initial solution uses this time to provide extra power (6 kWh). Consequently, station 10 remains closed as the bus does not need to use the station to reach station 11.

Local Search

The main purpose of the local search (Algorithm 1), given a solution s with a set of opened stations (*stations* - line 1), is to close as many as it can. Thus, the algorithm repeatedly attempts close open stations in a certain order. The selection of an open station (*SelectStation*), which is a function itself, is passed as a parameter. In particular, we explore two different selection strategies: *Random* selects a station uniformly at random and *MinActivity* selects the station with the fewest number of scheduled recharges.

Lines 4–15 form the core of the local search algorithm. Lines 5–8 select and attempt to close the station. Line 9 removes the selected station from the candidate set. After exhausting the list of candidate open stations we verify whether the algorithm reached a local minimum (Lines 10–12). Otherwise, we repeat the process with the remaining open stations in the solution. Let us recall that stations that could not be closed at a certain point of the execution may become closable after modifying the schedule of nearby stations.

The amount of charged energy remains constant during the whole process, and it is set to the minimum required amount. Therefore, when the algorithm closes or opens a charging station, it redistributes or borrows energy to/from the neighboring stations. The proportion of energy going/coming to/from previous and following stations is random.

The *CloseStation* operator (Algorithm 2) attempts to stop all charging activity of the buses in a given installed charging station. Therefore, in order to ensure the operation of the system, we redistribute the energy provided by the closing station towards adjacent open stations. We refer to these adjacent stations as *previous* and *next* open charging stations.

Algorithm 1. LocalSearch(s, *SelectStation*)

1: *stations* $\leftarrow \{c | c \in ST \land x_c = 1\}$
2: *bestObj, obj* $\leftarrow |stations|$
3: *isLocalMin* $\leftarrow false$
4: **while not** isLocalMin **do**
5:    $c \leftarrow SelectStation(stations)$
6:    **if** $CloseStation(c)$ **then**
7:       $obj \leftarrow obj - 1$
8:       $x_c \leftarrow 0$
9:       $stations \leftarrow stations - \{c\}$
10:    **if** $|stations| = 0$ **then**
11:       **if** $obj \geq bestObj$ **then**
12:          $isLocalMin \leftarrow true$
13:       **else**
14:          $bestObj \leftarrow obj$
15:          $stations \leftarrow \{c | c \in ST \land x_c = 1\}$

For instance, Fig. 2 represents two different bus routes (the blue and the green one). Both routes are sharing two locations but only one charging station in location 6. Thus, the charging stations are highlighted in yellow. Furthermore, the small points near the charging stations depict units of energy delivered to each bus. Figure 2a shows the green bus recharging three units of energy at charging station 6 while the blue bus gets one unit of energy at charging station 2. In Fig. 2b, when closing the charging station 6, the three units of energy provided by the charging station 6 to the green bus are redistributed as follows: two-thirds to the next station and one-third to the previous station where the bus was recharging in the original solution. This proportion is decided uniformly at randomly for each stop. In this particular example, the energy is evenly reallocated for the blue bus. Figure 2c outlines the resulting state of the solution after completing the close operation.

Algorithm 2. *CloseStation*(selectedStation)

1: $openStopsPerStation \leftarrow \{s_{bi} | \forall_{b \in B} \forall_{s_{bi} \in S_b}, st_{bi} = selectedStation \land x_{bi} = 1\}$
2: **for** s_{bi} in openStopsPerStation **do**
3:    **if** $FeasibleCloseStop(s_{bi})$ **then** $CloseStop(s_{bi})$ **else return** false
4: **return** true

Algorithm 2 (*CloseStation*) checks whether a charging station can be closed or not under certain circumstances and apply the operation. To this end, *FeasibleCloseStop* verifies if a given bus b can keep operating once the power gotten from recharging at s_{bi} is moved to the *Previous* and the *Next* open stations in the path of b. *CloseStop* completes the close operations and updates the arrival times, charging times, and the capacity of the batteries of the buses as prescribed

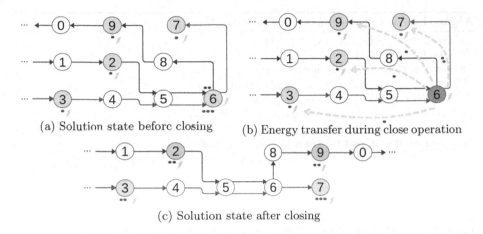

(a) Solution state before closing   (b) Energy transfer during close operation

(c) Solution state after closing

Fig. 2. Example of the close operation applied to station 6 (Color figure online)

by the *FeasibleCloseStop* algorithm. Taking this into account, *CloseStation* indicates whether the operation is successful for the entire fleet of buses relying on the input charging station, i.e., the system is capable of redistributing the power provided by the station without violating the operational constraints. We remark that the close operation might be performed partially as once the algorithm finds a non-closable stop, it might have already closed multiple stops for certain buses. We decided not to undo partial closes as it increases the complexity of the operation.

FeasibileCloseStop (Algorithm 3) verifies if the power provided by the m-th station in the path of b can be redistributed to adjacent stations. To this end, Line 1 calculates an ordered set with the open stations in the path of b. Line 2 calculates ρ the proportion of energy that goes to the next open stop (1-ρ goes to the previous one). Lines 3 and 4 define some variables that verify the feasibility of the constraints. Lines 5 and 6 calculate the index of the next (n) and previous (k) charging stations in the path of the bus. We remark that the next and previous charging stations are not always the same as we constantly open and close stations.

Lines 7–15 verify the feasibility of re-distributing energy to n. ne denotes the actual amount of power that b gains at n. nct denotes the additional time that the bus needs to gain the extra power at n. Lines 10–11 check whether the current capacity of the battery is enough to reach n and the new charging time is within the limits. Lines 12–15 check the overlapping constraints and prevent multiple buses of using the same charging station at the same time. We mark that in this scenario there is no need to check for potential time delays as the bus arrives at n earlier than expected.

Lines 16–26 verify the feasibility of re-distributing energy backwards to k. Whereas the risk of running out of power is inexistent the bus could experiment delays in the timetable. The new extra charging time at k could potential delay

Algorithm 3. $FeasibleCloseStop(selectedStop_{bm})$

1: $openStops_b \leftarrow \{s_{bi}|\forall_{s_{bi}\in S_b}, x_{bi} = 1\}$
2: $\rho \leftarrow Random(0, 1)$
3: $notOverflow, notDelay, notDeplete \leftarrow true$
4: $notExtraPCt, notExtraNCt, notOverlap \leftarrow true$
5: $n \leftarrow NextIndex(openStops_b, selectedStop_{bm})$
6: $k \leftarrow PreviousIndex(openStops_b, selectedStop_{bm})$
7: **if** n **is not** nil **then**
8:    $ne \leftarrow$ **if** k **is not** nil **then** $(1 - \rho) \cdot e_{bm}$ **else** e_{bm}
9:    $nct \leftarrow \frac{ne}{\alpha}$
10:   $notDeplete \leftarrow c_{bn} - ne \geq C_{min}$
11:   $notExtraNCt \leftarrow ct_{bn} + nct \leq \beta$
12:   **for** d in $\mathcal{B}$ $|d \neq b$ **do**
13:       **for** s_{di} in S_d $|st_{bi} = st_{bn} \land x_{di} = 1$ **do**
14:           $notOverlapTmp \leftarrow t_{bn} + ct_{bn} + nct \leq t_{di} \lor t_{bn} + ct_{bn} + nct \geq t_{di} + ct_{di}$
15:           $notOverlap \leftarrow notOverlap \land notOverlapTmp$
16: **if** k **is not** nil **then**
17:    $pe \leftarrow$ **if** n **is not** nil **then** $\rho \cdot e_{bm}$ **else** e_{bm}
18:    $pct \leftarrow \frac{pe}{\alpha}$
19:    $max\Delta t_{bm} \leftarrow max(\{\Delta t_{bk+1}...\Delta t_{bm}\})$
20:    $notDelay \leftarrow \mu \geq max\Delta t_{bm} + pct$
21:    $notOverflow \leftarrow c_{bk} + e_{bk} + pe \leq C_{max}$
22:    $notExtraPCt \leftarrow ct_{bk} + pct \leq \beta$
23:    **for** d in $\mathcal{B}$ $|d \neq b$ **do**
24:        **for** s_{di} in S_d $|st_{bi} = st_{bk} \land x_{di} = 1$ **do**
25:            $notOverlapTmp \leftarrow t_{bk} + ct_{bk} + pct \leq t_{di} \lor t_{bk} + ct_{bk} + pct \geq t_{di} + ct_{di}$
26:            $notOverlap \leftarrow notOverlap \land notOverlapTmp$
27: $notExtraCt \leftarrow notExtraPCt \land notExtraNCt$
28: $isFeasible \leftarrow notExtraCt \land notOverflow \land notDelay \land notDeplete \land notOverlap$
29: **return** isFeasible

all the arrival times of the bus in the stations between the closing station and the previous charger. Let us define pct and $max\Delta_{bi}$ (Lines 18 and 19). pct denotes the additional time that the bus needs to gain the power at k and $max\Delta_{bi}$ denotes the maximum delay a bus is currently experiencing during the trip between the previous station and the closing one. Line 20 ensures that the extra time spent in the previous charger is not disrupting the timetables beyond the threshold μ. Besides checking a battery overflow in Line 21, there is no need to further inspect that the energy is within the limits. Line 22 checks that the charging time is within the limit. Lines 23–26 check for overlapping charging events.

Let us illustrate the verification process with Fig. 2, and let us assume that the bus recharges one unit of power per minute. Firstly, we recall that the close operation acts on a single bus stop at a time. Thus, the sequence of open stops of the green bus (Algorithm 3 - line 1) includes the stops of the bus at stations 3, 6, and 7. Furthermore, since the selected stop corresponds to the stop at station

6 in that new sequence, the previous and next functions return the stops at stations 3 and 7 respectively.

Now, let us consider the first case to transfer energy towards the previous open stop at stop 3. In the current solution (before the operation) the bus departs from station 3 with at least 1 unit of power. After the close operation, the bus departs with an additional unit of power (we depict this behavior in Fig. 2c). Thus, it is impossible to run out of power during the trip between stations 3 and 6. On the other hand, in the new solution, the bus is spending more time recharging at station 3. This extra time introduces a delay when arriving at station 4, 5, and 6. Therefore, it is paramount to check that the delay of a minute is not violating the maximum allowed disruption time (in the timetable) in any of those stops (checking the most delayed timetable is enough).

Similarly, it can be seen graphically that the amount of power available for traveling between stations 6 and 7 is reduced after applying the operation. Therefore, we must ensure that the trip can be completed. Lastly, since the bus is saving two recharging minutes before departing from station 6, there is no way the bus is delayed when arriving at 7.

Perturbation

The perturbation phase aims at escaping from difficult regions (i.e., local minimum) while maintaining an appropriate balance between diversification and intensification. The operator iteratively tries to open a given station st by attempting to recharge the battery of certain buses while stopping at the station. To this end, we randomly redistribute the gained power at adjacent charging stations of st (i.e., *Next* and *Previous*). Thus, we label st as open iff at least one bus relies on the station to recharge the bus. We attempt to open an additional percentage of the opened stations of the incumbent solution, which determines the level of perturbation.

Let us illustrate the process of opening a station with Fig. 2 by looking at the example the other way around. Then Fig. 2c denotes the incumbent solution and attempting to open station 6. Furthermore, let us assume that the yellow arrows in Fig. 2b are pointing in the opposite direction. In this context, the blue bus will attempt to transfer a random portion of the gained power at stations 9 and 2 into station 6. Therefore, in order to open station 6 (blue bus), it is important to verify that the redistribution of the power lost at station 2 leaves the bus with enough power to complete the trip between stations 2 and 6 and that any potential delay (derived from charging at station 6) is not impacting the quality of service.

5 Evaluation

In this paper we use a real dataset with the operations of three Irish cities, i.e., Limerick, Cork, and Dublin.[1] We implemented the MIP model with CPLEX

---

[1] The GPS location of the bus stations and timetables are available at https://buseireann.ie/ and http://www.dublinbus.ie/.

12.10 and conducted our experiments on a 2.5 GHz Intel Xeon W-2175 processor with 64 GB of memory running Ubuntu 18.04.5. We used CPLEX with its default parameters, including the parallel optimizer with 12 threads and two time limits (10 and 120 min). Furthermore, we executed our ILS algorithm on each instance 10 times (each time with a different random seed) and reported the median number of open stations (out of the 10 executions) with a 10-minute timeout. All series of experiments had a standard deviation of less than 1.

Table 2 outlines the performance of our ILS algorithm with two selection strategies (i.e., *Random* and *MinActivity*). We recall that *MinActivity* attempts to close the open station with the fewest number of charging events. The first column displays the number of buses and the total number of bus stops per city. We simulate multiple scenarios varying the C_{max} value and timetable disruptions (i.e., max μ). Moreover, in all our experiments, we establish that the charging rate is 1 kWh per minute (i.e., $\alpha = 1$), and the buses consume 1 kWh per km. Additionally, we limit the charging time per cycle of the buses (β) to the time needed to recharge the batteries to up to 80% of the max. capacity. Furthermore, we set the level of perturbation to 20% after trying other values (10% and 40%) that did not suggest a significant difference. Bold numbers indicate the method was able to find the optimal solution, whereas the highlighted cells point out the method with the best performance. Lastly, it is worth noticing that we generate initial solutions for Cork and Limerick with up to 455 charging stations, whereas for Dublin we need 1666 stations. Also, the number of local minima found during an execution varies in the order of tens of thousands for Dublin, hundreds of thousands for Cork, and millions for Limerick.

In these experiments we observe that our ILS algorithm finds the optimal solution for small-size (i.e., Limerick) and mid-size instances (i.e., Cork) with $C_{max} = 180$ kWh and 200 kWh. Notably, ILS dominates the performance for the remaining instances with at least the same performance as CPLEX with 12 times more computational resources.

Furthermore, it is remarkable that CPLEX is unable to find reasonable solutions, within 10 min, for 6 instances (out of 10) for our Cork and Dublin datasets. Interestingly, CPLEX with 120 min recommends the installation of more than two hundred of charging units for one of the most constrained instances (i.e., Cork with 120 kWh and max $\mu = 4$), whereas our ILS algorithm quickly computes a solution with only 6 charging units, Similarly, our ILS algorithm finds a solution with approx. 40% fewer stations (15 vs. 28 charging stations) for the most constrained Dublin dataset, i.e., 200 kWh with max $\mu = 6$. Finally, in general our *MinActivity* selection strategy performs slightly better than *Random*. We attribute this to the fact that this heuristic favors stations where the close operator is more likely to succeed.

Table 2. Empirical evaluation for Limerick, Cork, and Dublin. Bold numbers denote the optimal solution and highlighted cells outline the best performance.

City	C_{max}	μ (mins)	CPLEX		ILS (10 mins)	
			10 mins	120 mins	Random	MinActivity
	120	4	5	**4**	4	4
	150		**4**	**4**	4	4
Limerick	180		**2**	**2**	2	2
23 buses	200		**2**	**2**	2	2
8,417 stops	120	6	**4**	**4**	4	4
	150		**4**	**4**	4	4
	180		**2**	**2**	2	**2**
	200		**2**	**2**	2	**2**
	120	4	565	220	7	6
	150		-	5	5	4
Cork	180		5	**3**	3	3
65 buses	200		**3**	**3**	3	3
29,724 stops	120	6	-	8	7	6
	150		567	6	5	4
	180		4	**3**	3	3
	200		**3**	**3**	3	3
Dublin 163 buses	200	6	2636	28	17	15
36,609 stops	200	10	-	17	17	15

6 Conclusions

In this paper we have proposed an efficient ILS algorithm to tackle the Charging Location Problem for eBuses. Our proposed algorithm relies on two simple operations, i.e., opening and closing charging stations in a given solution. Furthermore, our approach assumes that the eBuses recharge a constant amount of energy during a workday, therefore, the energy is locally distributed (e.g., when closing a station), among the adjacent stations of the ones affected by the operators. The effectiveness is demonstrated by experimenting with a set of real instances from three Irish cities (i.e., Limerick, Cork, and Dublin). We compared our algorithm against a MIP-based solution and our ILS solution is notably better than CPLEX in terms of the number of installed charging stations, generating results with up to 40% fewer stations in less than 10% of the time for some of the most difficult instances.

In the future, we plan to extend our ILS with robustness to tackle the Charging Location Problem, so that the system is resilient to failures in the transportation network. Also, we plan to perform an extensive parameter tuning using tools like ParamILS [8] or Calibra [6].

Acknowledgements. This work received funding from the Sustainable Energy Authority of Ireland (SEAI) under the RDD 2019 programme - Grant No 19/RDD/519. The authors would like to thank the anonymous reviewers for their comments and suggestions which helped to improve the paper.

References

1. Arbelaez, A., Climent, L.: Transition to eBuses with minimal timetable disruptions. In: International Symposium on Combinatorial Search (SoCS). Association for the Advancement of Artificial Intelligence (2020)
2. Erdelić, T., Carić, T.: A survey on the electric vehicle routing problem: variants and solution approaches. J. Adv. Transp. **2019**(54), 1–48 (2019)
3. Erdoğan, S., Miller-Hooks, E.: A green vehicle routing problem. Transp. Res. Part E Logistics Transp. Rev. **48**(1), 100–114 (2012)
4. Funke, S., Nusser, A., Storandt, S.: Placement of loading stations for electric vehicles: no detours necessary! In: AAAI, pp. 417–423. AAAI Press (2014)
5. Funke, S., Nusser, A., Storandt, S.: Placement of loading stations for electric vehicles: allowing small detours. In: ICAPS 2016, pp. 131–139. AAAI Press (2016)
6. Hoos, H.H.: Automated algorithm configuration and parameter tuning. In: Hamadi, Y., Monfroy, E., Saubion, F. (eds.) Autonomous Search, pp. 37–71. Springer, Heidelberg (2011). https://doi.org/10.1007/978-3-642-21434-9_3
7. Hoos, H.H., Stützle, T.: Stochastic Local Search: Foundations and Applications. Elsevier, Amsterdam (2004)
8. Hutter, F., Hoos, H.H., Leyton-Brown, K., Stützle, T.: ParamILS: an automatic algorithm configuration framework. J. Artif. Intell. Res. **36**, 267–306 (2009)
9. Karlsson, E.: Charging infrastructure for electric city buses: an analysis of grid impact and costs (2016)
10. Kunith, A., Mendelevitch, R., Goehlich, D.: Electrification of a city bus network-an optimization model for cost-effective placing of charging infrastructure and battery sizing of fast-charging electric bus systems. Int. J. Sustain. Transp. **11**(10), 707–720 (2017)
11. Stützle, T., Ruiz, R.: Iterated local search. In: Martí, R., Pardalos, P.M., Resende, M.G.C. (eds.) Handbook of Heuristics, pp. 579–605. Springer, Cham (2018). https://doi.org/10.1007/978-3-319-07124-4_8
12. Wang, I.L., Wang, Y., Lin, P.C.: Optimal recharging strategies for electric vehicle fleets with duration constraints. Transp. Res. Part C Emerg. Technol. **69**, 242–254 (2016)
13. Wang, X., Yuen, C., Hassan, N.U., An, N., Wu, W.: Electric vehicle charging station placement for urban public bus systems. IEEE Trans. Intell. Transp. Syst. **18**(1), 128–139 (2016)
14. Wang, Y.W., Lin, C.C.: Locating road-vehicle refueling stations. Transp. Res. Part E Logistics Transp. Rev. **45**(5), 821–829 (2009)
15. Xylia, M., Leduc, S., Patrizio, P., Kraxner, F., Silveira, S.: Locating charging infrastructure for electric buses in Stockholm. Transp. Res. Part C Emerg. Technol. **78**, 183–200 (2017)

Multi-view Clustering of Heterogeneous Health Data: Application to Systemic Sclerosis

Adán José-García[1]([✉]), Julie Jacques[1,2], Alexandre Filiot[3], Julia Handl[6], David Launay[4], Vincent Sobanski[3,5], and Clarisse Dhaenens[1]

[1] Univ. Lille, CNRS, Centrale Lille, UMR 9189 CRIStAL, 59000 Lille, France
adan.josegarcia@univ.lille.fr
[2] FGES, Université Catholique de Lille, 59000 Lille, France
[3] Univ. Lille, Inserm, CHU Lille, U1286, INFINITE, 59000 Lille, France
[4] Univ. Lille, Inserm, CHU Lille, Service de Médecine Interne et Immunologie Clinique, CeRAINO, U1286, INFINITE, 59000 Lille, France
[5] Institut Universitaire de France (IUF), Paris, France
[6] Alliance Manchester Business School, University of Manchester, Manchester, UK

Abstract. Electronic health records (EHRs) involve heterogeneous data types such as binary, numeric and categorical attributes. As traditional clustering approaches require the definition of a single proximity measure, different data types are typically transformed into a common format or amalgamated through a single distance function. Unfortunately, this early transformation step largely pre-determines the cluster analysis results and can cause information loss, as the relative importance of different attributes is not considered. This exploratory work aims to avoid this premature integration of attribute types prior to cluster analysis through a multi-objective evolutionary algorithm called MVMC. This approach allows multiple data types to be integrated into the clustering process, explore trade-offs between them, and determine consensus clusters that are supported across these data views. We evaluate our approach in a case study focusing on systemic sclerosis (SSc), a highly heterogeneous auto-immune disease that can be considered a representative example of an EHRs data problem. Our results highlight the potential benefits of multi-view learning in an EHR context. Furthermore, this comprehensive classification integrating multiple and various data sources will help to understand better disease complications and treatment goals.

Keywords: Clustering · Multi-view clustering · Systemic sclerosis · Multi-objective optimization

1 Introduction

Many real-world applications consist of heterogeneous datasets comprising multiple attribute types, including binary, numerical, and categorical features. For example, electronic health records (EHRs) in medicine consist of heterogeneous

G. Rudolph et al. (Eds.): PPSN 2022, LNCS 13399, pp. 352–367, 2022.
https://doi.org/10.1007/978-3-031-14721-0_25

structured and unstructured data elements, including demographic information, diagnoses, laboratory results, medication prescriptions, and free-text clinical notes [28,37]. In this regard, unsupervised machine learning methods are often used to discover homogeneous groups from unlabeled data because limited information is known about the classes' distribution in these heterogeneous datasets. However, most clustering algorithms are limited to working on a single specific attribute type (i.e. numerical or nominal).

Two approaches are mainly used to address this heterogeneous data clustering problem: (i) methods based on features transformation such as discretization and (ii) methods that directly use a proximity measure designed to handle mixed-attribute types such as the Gower distance. Despite their popularity, those approaches either yield substantial information loss (i) or require the selection of the "best" proximity measure beforehand (ii).

This work explores multi-view clustering to integrate multiple attribute types (data views) into the clustering process. First, specialized dissimilarity measures are used to create views, each characterized by a specific attribute type in the heterogeneous dataset. Then, the multi-view clustering algorithm explores tradeoffs between the views to discover consensus clusters supported across all views. This approach was applied and evaluated in a case study of systemic sclerosis (SSc), a highly heterogeneous disease that can be considered a representative example of an EHRs data problem.

2   Background and Related Work

With the advent of so-called *big-data*, most real-world problems now involve multiple, heterogeneous data sources. Dealing with mixed types of attributes remains challenging for the clustering and clinical communities as conventional clustering algorithms require a single common data format (e.g. numerical or categorical). In the present section, we look at this heterogeneous data clustering problem through the lens of distance-based methods. A more complete, exhaustive review of other research fields, e.g., hierarchical [13,19], model-based [6,16,22] and neural network-based clusterings [5], will be addressed in future work. With this in mind, we recall that no single *best* clustering method exists in a general sense [15,17,36], but rather a wide variety of clustering techniques that must be carefully selected depending on the data at hand, especially in a clinical setting.

2.1   Distance-Based Clustering on Heterogeneous Data

Most conventional, e.g., distance-based clustering algorithms work with numerical-only or categorical-only data. Two main approaches are usually followed to deal with mixed-type data [3,15,38,40]: (i) methods based on features transformation [7,8,41] and (ii) methods that cluster the heterogeneous data types directly [2,9,10,18,21,29,39].

Data transformation-based methods aim to first unify the data format and then apply a distance-based clustering method, such as K-means [38]. It consists in either discretizing numerical variables into nominal ones (needed for K-modes clustering) or reciprocally encoding nominal attributes into continuous ones (needed for K-means clustering). Although those transformations are commonly used for clustering, it involves a potentially substantial information loss, as the clustering results strongly rely on either the cut-points (which may be inappropriate) or the coding mechanism and its underlying assumptions. Alternative approaches have been proposed to address this limitation. Wei et al. [41] proposed a mutual information-based unsupervised feature transformation (UFT) for non-numerical variables, avoiding the need for manual coding. Another popular approach is to use dimensionality reduction techniques, such as Factor Analysis of Mixed Data [8], in complement to some clustering techniques.

On the other hand, most distance-based clustering methods use a single proximity measure designed to handle mixed-data types [2,9,10,18,21,29]. The Gower distance is a widespread example of such a measure, which may be best suited depending on the data clustering structure [9]. Ahmad et al. [2] proposed a K-means algorithm based on a weighted combination of the Euclidean distance and the co-occurrence of discrete values, addressing some limitations of previous K-prototypes algorithm from Huang et al. [21]. Further work has been published by Ahmad et al. [4] on a novel K-means initialization technique for mixed data, called *initKmix*, which may outperform random initialization methods on several heterogeneous datasets. Recently, Budiaji et al. [10] proposed a simple and fast K-medoids algorithm (SFKM) combined with a generalized distance function (GDF), allowing more flexible trade-offs between numerical, binary, and categorical variables. Similarly, Harikumar and Surya [18] proposed a K-medoids approach based on a similarity measure in the form of a triplet. Among the wide range of mixed-types-based proximity measures, one can also cite the work of Li et al. [29] focusing on similarity-based agglomerative clustering (SBAC), an algorithm based on the Goodall dissimilarity.

For a given dataset, most of the above methods require the selection of the "best" proximity measure (or "best" weighting of distinct proximity measures) in advance. Therefore, finding more generic, adaptive trade-offs between the contributions of the different data types remains challenging. Multi-view clustering [1,27] potentially addresses these limitations by dividing the dataset into subsets, called *views*, each characterized by a given data type, and then treats them simultaneously. In this work, we explore the use of multi-view clustering to integrate multiple data views during the clustering process.

2.2   From Single to Multi-objective Clustering

In view of the complementarity between different distance functions, the optimal cluster structures could be better identified using multiple proximity measures *simultaneously* [11,14,25–27,30,31]. As said, traditional clustering algorithms require the choice of a single proximity measure such as the Euclidean, Hamming or Cosine distance. One approach is to assign weights to the different proximity measures [6,12,20,21]. However, the appropriate weighting is hard to determine without any prior knowledge of the data itself, and the reliability of the information provided by the distance measures.

Multi-view clustering algorithms can integrate multiple dissimilarity matrices simultaneously in order to find consensus clusters that are consistent across the different data views [14,27], and yield high-quality clustering results that optimally balance the contribution of each data source [26]. Recent research has reported some first steps to exploit the intrinsic multi-criterion nature of the multi-view problems [25–27,30,31].

Liu et al. [31] presented a multi-objective evolutionary algorithm (based on NSGA-II [30]) that simultaneously considers two different distance measures (Euclidean and Path distances). Each individual is represented using a label-based encoding of size N (number of data points) which is then evaluated using the intra-cluster variance with respect to both distance measures. Afterward, Liu et al. [30] extended this work by proposing a fuzzy clustering approach based on a multi-objective differential evolution algorithm. In this approach, a centroid-based codification is used to represent the candidate clustering solutions. However, these methods are currently limited to two views due to the lack of generality of the Pareto dominance-based approaches. In this regard, Jose-Garcia et al. [25,27] proposed a many-objective approach to multi-view data clustering that exploits the benefits of complementary information sources taken from multiple dissimilarity matrices. Additionally, this multi-view clustering algorithm allows scaling with respect to the number of data views.

3   Multi-view Clustering Approach

The proposed methodology aims to provide a solution in the context of cluster analysis to deal with heterogeneous data characterized by multiple attribute types. First, the data is decomposed into several subsets according to the attribute types. Subsequently, a suitable proximity measure is chosen for each data subset generating a dissimilarity matrix. Finally, a multi-objective evolutionary clustering algorithm uses all dissimilarity matrices as data views to find consensus clusters across the data views. This approach is illustrated in a general way in Fig. 1 and described in detail in the following sections.

3.1   Construction of the Data Views

Multi-view clustering algorithms use multiple feature spaces (data views) simultaneously. The construction and selection of data views is an important step for the accurate functioning of the algorithm. In this setting, each view represents a given data source that describes a specific perspective of a phenomenon. In this regard, in the presence of a heterogeneous dataset, we propose to create different views for different types of attributes, e.g. binary, numerical and categorical. Therefore, the database is decomposed into subsets of attributes according to

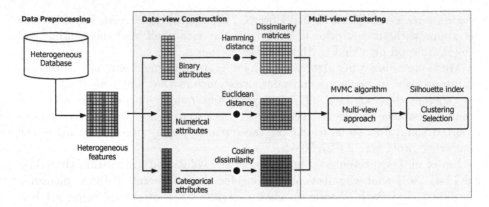

Fig. 1. Main stages and components of the proposed multi-view clustering methodology for a heterogeneous dataset (Color figure online).

their data types, resulting in many feature spaces. Then, for each data-type feature space, an appropriate proximity measure is used to generate a dissimilarity matrix representing a particular data view of the overall heterogeneous problem. To the best of our knowledge, this is the first time an unsupervised multi-view approach for clustering a heterogeneous database has been proposed and evaluated. This is because such approaches usually work on homogeneous data spliced across several datasets.

3.2   Multi-view Clustering Algorithm: MVMC

The MVMC algorithm is a multi-objective evolutionary approach to multi-view clustering that was developed to identify all optimal trade-offs between available data views [27]. It allows scalability to a significant number of views through the use of a many-objective optimizer. Specifically, MVMC uses a decomposition-based optimizer, MOEA/D [34], as the underlying search engine for its clustering approach. Furthermore, it employs a medoid-based representation, a representation that is more general than centroids, as it can be used both for problems defined in terms of feature spaces or dissimilarity matrices. In its current implementation, MVMC uses a fixed number of medoids, so requires the desired number of clusters as input.

MVMC focuses on a single cluster-quality criterion, but aims to optimize it concerning each view, resulting in a multi-objective optimization problem with as many clustering criteria as data views. Let $\mathbf{C}^r$ and $\mathbf{w}^r$ be the partition and weight vector, respectively, corresponding to the r-th subproblem. Also, let $\{D_1, \ldots, D_M\}$ denote M dissimilarity matrices, which represent M different data views and are each considered by a separate objective. MVMC then uses the within-cluster scatter as the optimization criterion, which, for the m-th objective of the r-th subproblem, is computed as:

$$f_m(\mathbf{C}^\tau) = \sum_{c_k \in \mathbf{C}^\tau} \sum_{i,j \in c_k} d_m(i,j) \ , \tag{1}$$

where $d_m(i,j)$ is the dissimilarity between the points i and j as defined in D_m.

MVMC overcomes one major dilemma of previous attempts at designing representations for multi-view clustering: how to ensure that these are scalable without biasing the representation or decoding step toward one particular dissimilarity space. Specifically, the limitations of other representations are:

- For representations that are dissimilarity space agnostic, with each gene directly encoding cluster membership for each data point, the search space increases exponentially with the dataset size, affecting their scalability to large data.
- Representations that employ cluster prototypes in the form of centroids require the centroid to be represented in one or a concatenation of the feature spaces, which implies a single fixed weighting between views.
- Representations employing cluster prototypes (whether centroids or medoids) require a decoding step involving the assignment of data points to clusters. This step relies on using one or a sum of several dissimilarity functions, implying a single fixed weighting between views.

MVMC overcomes this issue by exploiting the availability of an explicit weight vector for each sub-problem in decomposition-based optimizers. Furthermore, employing a medoid-based encoding and accessing subproblem-specific weights in the decoding step avoids any prior bias towards one particular dissimilarity space whilst benefiting from a compact representation.

3.3 Selection of Clustering Solutions

The Silhouette index is often considered to be a more effective measure of cluster validity, as it combines both within and between-cluster variation of a partition. Unlike within-cluster scatter, maximizing the Silhouette index is potentially suitable for solution selection across a range of different numbers of clusters. For a given clustering solution $\mathbf{C}$ with N data points, the Silhouette index Sil($\mathbf{C}$) can be defined as the sum of individual Silhouette indexes $\{SW(i) \,|\, i = 1, ..., N\}$ [35]:

$$\mathrm{Sil}(\mathbf{C}) = \frac{1}{N} \sum_{i=1}^{N} SW(i) = \frac{1}{N} \sum_{i=1}^{N} \frac{b_i - a_i}{\max\{a_i, b_i\}} \tag{2}$$

where a_i represents the average distance from i to all other data points in its cluster. b_i represents the minimum distance of i to another cluster, where the distance between i and another cluster is calculated as the average distance from i to all data points in that cluster.

MVMC generates a set of non-dominated clustering solutions, but a single solution is usually required in practice. For this purpose, a model selection approach based on the Silhouette index is used [27]. This approach computes the index from a weighted dissimilarity matrix obtained from the weights assigned to the different data views during the clustering task.

4   Experimental Study

4.1   CHUL Database and Data-View Configurations

In this work, the different clustering methods were assessed and compared using the SSc patient database of the *Centre Hospitalier Universitaire de Lille* (herein referred to as CHUL[1] database). The CHUL database was created in 2014 and held clinical information of 550 SSc patients with regular, detailed follow-up visits recorded on a standardized case-report form. Currently, the database contains more than 1500 patterns (patient visits) and nearly 400 attributes (e.g. demographic information, physical examination, laboratory exams, medical analyses). Two experienced clinicians (VS and DL authors) selected 39 relevant attributes, of which 22 are binary, 16 are numerical, and three are categorical (or nominal). In addition, data from the most recent visit of each patient were considered, limiting the analysis to 530 patterns. As a result, the clustering task was performed on 530 patterns described by 39 attributes with heterogeneous types. Three data views were generated from the CHUL database and used in the multi-view clustering algorithm:

- *Binary view*, {Bin}. This view is based on the binary dissimilarity data matrix computed with the Hamming distance on the 22 binary attributes.
- *Numerical view*, {Num}. This view is based on the numeric dissimilarity data matrix computed with the Euclidean distance on the 16 numerical attributes (integer and double data types) of the CHUL database.
- *Categorical view*, {Str}. This view is based on the categorical dissimilarity data matrix computed with the Cosine similarity measure on the 3 categorical attributes of the CHUL database.

For the MVMC algorithm, different view combinations of those data views were considered: {Bin,Num}, {Bin,Str}, {Num,Str}, and {Bin,Num,Str}. In addition, the {Num,Gower} configuration was considered, where the {Gower} view is a dissimilarity matrix created using the Gower distance from the union of the binary and categorical attributes in the CHUL dataset.

4.2   Reference Methods

To indicate baseline performance for the studied SSc data problem, we compare MVMC against two well-known and conceptually different clustering algorithms: K-medoids [33] and WARD hierarchical clustering method [38]. Our experiments apply K-medoids and WARD methods on four dissimilarity matrices, {HAM}, {EUC}, {COS}, and {GOWER}, using Hamming, Euclidean, Cosine, and Gower distances, respectively. These matrices were obtained from the entire CHUL dataset by transforming all attributes into numerical values.

---

[1] SSc patients in the Internal Medicine Department of University Hospital of Lille, France, between October 2014 and December 2021 as part of the FHU PRECISE project (PREcision health in Complex Immune-mediated inflammatory diseaSEs); sample collection and usage authorization, CPP 2019-A01083-54.

The Silhouette scores obtained by WARD and K-medoids methods on each dissimilarity matrix were also computed, giving rise to possible comparisons between single-view and multi-view algorithms[2].

4.3 Parameter Settings

The settings for MVMC adopted in our experiments are as follows [27]. The population size is NP $= 100$, the number of generations is $G_{max} = 100$, the recombination probability is Pr $= 0.5$, the mutation probability is Pm $= 0.03$, and the neighborhood size is T $= 10$.

For the stochastic clustering methods analyzed and compared in this study, MVMC and K-medoids, a total of 31 independent executions were performed. In all cases, statistical significance is evaluated using the Kruskal–Wallis test, considering a significance level of $\alpha = 0.05$ and Bonferroni correction.

5 Results and Discussions

This section presents a series of experiments conducted on the CHUL dataset (530 patterns, 39 attributes) where different views and corresponding dissimilarity

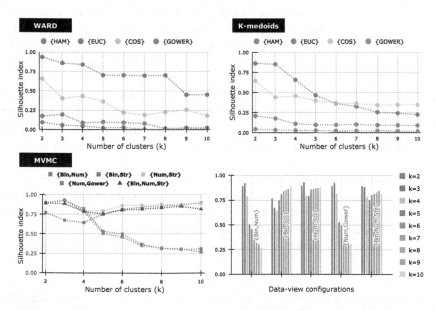

Fig. 2. Illustration of the clustering performance obtained by the different algorithm configurations when varying the number of clusters, $K = \{k \mid 2 \leqslant k \leqslant 10\}$. (Color figure online)

---

[2] Note that the Silhouette score is intended to compare different partitions produced by a single method. Usually, the Rand index is preferred to the Silhouette score to compare two solutions when a ground-truth partition is available [35].

measures are considered according to attribute types. As described in Sect. 4.1, four dissimilarity matrices and five data-view configurations are used by two single-view, WARD and K-medoids algorithms, and the multi-view approach MVMC.

5.1 Clustering Performance

This first experiment aims to analyze the clustering performance of the clustering algorithms with the number of clusters. Thus, the results obtained by WARD and K-medoids will serve as a reference (baseline) when compared with those obtained by the multi-view approach, MVMC. The WARD and K-medoids algorithms were used to separately cluster the four dissimilarity matrices {HAM}, {EUC}, {COS}, and {GOWER}, whereas MVMC used five different data-views combinations.

Table 1. Clustering performance in terms of the Silhouette index obtained by the different algorithm configurations when varying k, K = {k | 2 ≤ k ≤ 10}. The best Silhouette value scored for each algorithm configuration has been shaded and highlighted in bold and, additionally, the statistically best ($\alpha = 0.05$) results are highlighted in boldface.

Alg.	Data views	Number of clusters (k)								
		k = 2	k = 3	k = 4	k = 5	k = 6	k = 7	k = 8	k = 9	k = 10
WARD	{HAM}	**0.095**	0.060	0.048	0.025	0.031	0.006	0.006	0.010	0.013
	{EUC}	**0.937**	**0.861**	0.840	0.703	0.703	0.698	0.701	0.452	0.451
	{COS}	**0.657**	0.405	0.433	0.364	0.219	0.190	0.227	0.256	0.182
	{GOWER}	0.175	**0.196**	0.090	0.100	0.094	0.079	0.017	0.026	0.026
K-medoids	{HAM}	**0.043**	0.036	0.030	0.029	0.025	0.025	0.025	0.021	0.019
	{EUC}	**0.861**	**0.851**	0.656	0.465	0.362	0.327	0.255	0.242	0.227
	{COS}	**0.644**	0.440	0.457	0.397	0.369	0.362	0.342	0.347	0.349
	{GOWER}	**0.208**	0.178	0.112	0.098	0.098	0.103	0.094	0.092	0.091
MVMC	{Bin,Num}	0.894	**0.922**	0.787	0.506	0.461	0.352	0.321	0.308	0.271
	{Bin,Str}	0.770	0.674	0.643	0.751	**0.811**	**0.842**	**0.857**	**0.867**	**0.895**
	{Num,Str}	0.898	**0.933**	0.797	0.793	**0.860**	**0.863**	**0.876**	**0.876**	**0.886**
	{Num,Gower}	0.895	**0.925**	0.819	0.533	0.499	0.369	0.316	0.302	0.310
	{Bin,Num,Str}	0.892	0.891	0.785	0.753	**0.803**	**0.815**	**0.833**	**0.849**	**0.815**

The experiment was conducted as follows. First, for each clustering algorithm and each data view, a collection of ℂ partitions were generated by varying the number of clusters k in the range K = {k | 2 ≤ k ≤ 10}. Then, in a second step, each clustering solution in collection ℂ was evaluated using the Silhouette index. Usually, the partition(s) with the best index values are considered the final solutions that best fit the data problem. This procedure is commonly used when the number of clusters is unknown and needs to be determined using a cluster validity index. For this purpose, the Silhouette index is well known and has performed satisfactorily in practice [24]. The results of this analysis are summarized in Fig. 2, with more detailed results, and their statistical significance, presented in Table 1.

The average Silhouette index values tend to decrease as the number of clusters increases from two to ten for traditional single-view algorithms, i.e., the Silhouette index suggests that the most appropriate number of clusters is at the beginning of the range of explored clusters. Then, it is observed that the index quickly loses its discriminative ability to find other suitable underlying structures in this highly heterogeneous dataset. Moreover, this monotonous decreasing convergence behavior is observed in both single-view algorithms and is independent of the type of proximity measure used in the experiments.

On the other hand, regarding the clustering performance obtained by the multi-view clustering algorithm MVMC using the five data-view configurations, it is observed that in general, (i) the algorithm obtained higher average Silhouette values than traditional clustering approaches and (ii) that the Silhouette values are changing as the number of clusters increases (i.e. the values increase and decrease). In addition, two types of Silhouette convergences are observed concerning the performance of the different data configurations. First, configurations {Bin, Num}

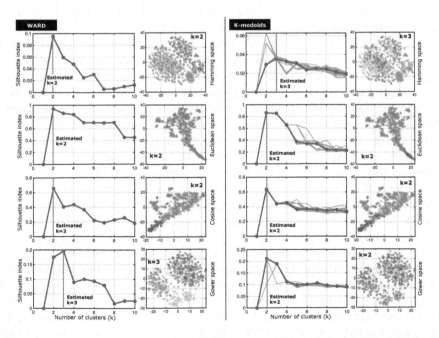

Fig. 3. Best clustering solutions obtained by WARD (left) and K-medoids (right) algorithms using the Silhouette index. For each subfigure, the median convergence plot is shown in blue. The best solution is marked in red. The corresponding clustering solution is visualized in the embedding space associated with a proximity measure. (Color figure online)

and {Num,Gower} obtained very similar convergence results: they start by slightly increasing, up to a certain k, and then start to decrease as the number of clusters increases further. Second, for data configurations {Bin,Str}, {Num,Str}, and {Bin,Num,Str} the Silhouette values increase, decrease, increase again to a certain threshold, and then remain constant. These Silhouette index fluctuations indicate that multiple suitable cluster structures are encountered across the range of explored clusters. Thus, in the following subsection, we investigate the selection of the most appropriate clustering solutions.

5.2   Selection of Clustering Solutions

An important problem in cluster analysis is to determine the number of clusters from the inherent information in a clustering structure [23]. Thus, the following experiment aims to find both the most appropriate number of clusters and its corresponding clustering solution from a collection of solutions using the Silhouette index. This experiment was conducted as follows. First, the solutions(s) with the highest Silhouette value(s) are selected among the collection of solutions generated by a clustering algorithm. Subsequently, the chosen solution(s) is visualized in an embedded two-dimensional feature space, obtained from a dissimilarity matrix using the t-SNE [32] projection technique (parameters: $n\_components = 2$, $n\_iter = 100$, $perplexity = 30$). The resulting clustering solutions of this analysis are presented in Figs. 3–4.

Figure 3 presents the selected solutions for the two single-view algorithms. In general, we can observe that the choice of the distance function over the original heterogeneous dataset considerably influences the two-dimensional distribution of t-SNE projections. Furthermore, there is a clear tendency for the Silhouette index to discover two clusters in most scenarios, except for configurations WARD$_{\{GOWER\}}$ and K-medoids$_{\{EUC\}}$, where the number of groups is three.

Regarding the clustering solutions generated by the multi-view approach MVMC, from Fig. 5 (Appendix), it is clear that the determined number of clusters is three as the Silhouette index obtained its highest point value at this point, k = 3. Figure 4 illustrates the generated clustering solutions for the data-view configurations, {Bin,Str}, {Num,Str}, and {Bin,Num,Str}. Two solutions with the best Silhouette values were chosen for each configuration in this scenario. Firstly, we observe that the best clustering solutions tend to be found at the knee of the Pareto front approximations (PFAs), red box in the PFA, representing trade-offs between the views involved. These compromise points suggest that the consensus clustering solution exploits pieces of information from all the multiple data views in a complementary manner. As a result, the multi-view clustering setting reveals three and six clusters (inflection points in convergence plots). Interestingly, the combination of the (mixed) data-view contributions produces embedded feature spaces with observable groups, particularly for the six-cluster solutions, as illustrated in Fig. 4.

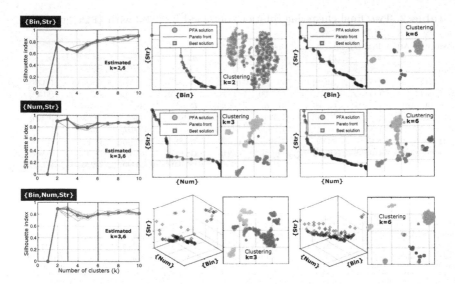

Fig. 4. MVMC clustering solutions for the configurations, {Bin,Str}, {Num,Str}, and {Bin,Num,Str}. Each configuration includes the convergence plots shown in blue and gray, with the two best solutions marked red. Then, for each selected solution, (i) the Pareto front approximation (PFAs) and (ii) the clustering solution, which is visualized in a weighted embedding space associated with the data views in the configuration. (Color figure online)

Finally, Table 2 presents two clustering solutions (**P** and **G**) obtained by the MVMC algorithm with the data-view configuration {Bin,Num,Str}. The first clustering solution contains two clusters and is shown in the first two columns in gray. In contrast, the second solution involves six groups and is described in the last six columns in light blue. Regarding clinical relevance, solution **P** exhibits two groups of patients separated on the basis of the presence of ILD, and interestingly not regarding the cutaneous involvement (historical subclassification [37]). The six-cluster solution provided a better delineation of six homogeneous groups, which best captured the patients' variability in terms of the disease severity as expressed by the EUSTAR and Medsger scores. G1 included the majority of patients with mild disease. G4 and G6 were mostly patients with diffuse cutaneous involvement. G2, G3, and G4 were patients with ILD and different degrees of severity as shown by the FVC and DLCO values. PH was found with a high prevalence in G2, G3, G4 and G6, but DLCO values unveiled that G2 and G4 were the most severe regarding gas exchange capacity.

Table 2. Two final clustering solutions obtained by MVMC with {Bin,Num,Str}.

Descriptive Atts.[a]	P(k = 2)		G(k = 6)					
	P1	P2	G1	G2	G3	G4	G5	G6
Cluster Size	177	353	255	70	68	50	50	37
Sex (m,f)	(25,75)	(12,88)	(10,90)	(29,71)	(13,87)	(38,62)	(18,82)	(14,86)
SSc Type (dc,lc,sc)	(40,59,1)	(10,72,18)	(0,82,18)	(29,69,3)	(13,87,0)	(92,8,0)	(0,66,34)	(81,19,0)
Active DU (y,n)	(60,40)	(42,58)	(41,59)	(54,46)	(53,47)	(68,32)	(36,64)	(65,35)
Active SRC (y,n)	(3,97)	(0,100)	(0,100)	(4,96)	(0,100)	(4,96)	(0,100)	(0,100)
ILD (y,n)	(98,2)	(6,94)	(1,99)	(100,0)	(85,15)	(100,0)	(24,76)	(3,97)
PH (y,n)	(12,88)	(8,92)	(7,93)	(11,89)	(15,85)	(16,84)	(6,94)	(11,89)
Calcinosis (y,n)	(10,90)	(13,87)	(14,86)	(6,94)	(18,82)	(6,94)	(4,96)	(19,81)
Joint Sx (y,n)	(34,66)	(41,59)	(40,60)	(31,69)	(37,63)	(34,66)	(42,58)	(43,57)
Intestinal Sx (y,n)	(27,73)	(30,70)	(31,69)	(23,77)	(32,68)	(28,72)	(16,84)	(43,57)
mRSS	8.78±7.6	5.73±4.8	4.30±3.3	8.58±8.2	7.05±6.0	11.28±7.4	5.63±5.7	10.47±6.2
LVEF	63.44±28.6	64.74±23.3	63.91±5.5	60.85±4.4	65.05±4.5	62.84±6.9	61.20±6.5	65.06±5.4
FVC	87.41±27.1	102.13±29.4	107.83±19.1	83.49±23.6	101.95±16.3	85.54±24.1	106.07±20.4	103.57±21.9
DLCO	55.54±16.5	69.38±21.9	74.21±22.0	54.78±18.8	68.08±18.0	56.07±19.4	73.84±19.1	70.92±17.1
Score EUSTAR	1.70±1.5	1.55±1.3	1.42±1.1	1.59±1.3	1.77±1.5	2.38±1.8	1.61±1.3	2.32±1.6
Score Medsger	1.41±0.8	1.25±0.7	1.46±0.8	1.67±0.9	1.77±0.9	1.67±0.8	1.71±1.0	2.17±1.2

[a] Sex: m (male), f (female); SSC Type: dc / lc (diffuse / limited cutaneous), sc (sine scleroderma); DU: digital ulceration; SRC: scleroderma renal crisis; ILD: interstitial lung disease; PH: pulmonary hypertension; Sx: symptoms; mRSS: mean Rodnan skin score; LVEF: left ventricular injection fraction; FVC: forced vital capacity; DLCO: diffusion lung capacity for carbon monoxide; EUSTAR: european scleroderma trials and research.

6   Conclusion

This work explores the benefits of multi-view clustering to identify groups of systemic sclerosis (SSc) patients, a highly heterogeneous auto-immune disease, within electronic health records (EHRs) capturing several types of attributes. Our approach avoids the premature integration of attribute types before cluster analysis through a multi-objective evolutionary algorithm called MVMC. MVMC integrates multiple data types into the clustering process in the form of data views, explores trade-offs between them, and determines consensus clusters supported across these views. This comprehensive classification integration of multiple and various data sources helped to discover meaningful clustering solutions ($P_{k=2}$ and $G_{k=6}$) that will help to better understand disease complications and treatment goals.

Acknowledgments. The authors are grateful to the University of Lille, CHU Lille, and INSERM, founded by the MEL through the I-Site cluster humAIn@Lille.

Appendix

This Appendix includes figures complementing the results of the experiments presented in Sect. 5. From Fig. 5 (Appendix), it is clear that the determined number of clusters is three as the Silhouette index obtained its highest point value at this point, $k = 3$. Also, from the Pareto front approximations obtained by these configurations, a substantial inference of the {Num} view is observed over the {Bin} and {Gower} views, respectively. Accordingly, the clustering solutions and the weighted embedding space are remarkably similar between these two data-view configurations.

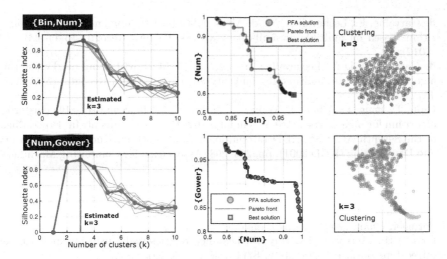

Fig. 5. MVMC clustering solutions for two data-view configurations, {Bin,Num} and {Num,Gower}. Each configuration includes (i) the convergence plots shown in blue and gray, with the best solution marked red; (ii) the Pareto front approximation corresponding to the estimated k value; (iii) the clustering solution, which is visualized in a weighted embedding space associated with the data views in the configuration. (Color figure online)

References

1. Abdullin, A., Nasraoui, O.: Clustering heterogeneous data sets. In: American Web Congress, pp. 1–8. IEEE (2012)
2. Ahmad, A., Dey, L.: A k-mean clustering algorithm for mixed numeric and categorical data. Data Knowl. Eng. **63**(2), 503–527 (2007)
3. Ahmad, A., Khan, S.S.: Survey of state-of-the-art mixed data clustering algorithms. IEEE Access **7**, 31883–31902 (2019)
4. Ahmad, A., Khan, S.S.: initKmix-a novel initial partition generation algorithm for clustering mixed data using k-means-based clustering. Expert Syst. Appl. **167**, 114149 (2021)
5. Aljalbout, E., Golkov, V., Siddiqui, Y., Strobel, M., Cremers, D.: Clustering with deep learning: taxonomy and new methods (2018). arXiv:1801.07648
6. Banfield, J.D., Raftery, A.E.: Model-based gaussian and non-gaussian clustering. Biometrics **49**(3), 803–821 (1993)
7. Basel, A.J., Rui, F., Nandi, K.A.: Integrative cluster analysis in bioinformatics. John Wiley & Sons, USA (2015)
8. Bécue-Bertaut, M., Pagés, J.: Multiple factor analysis and clustering of a mixture of quantitative, categorical and frequency data. Comput. Stat. Data Anal. **52**(6), 3255–3268 (2008)
9. Ben Ali, B., Massmoudi, Y.: K-means clustering based on gower similarity coefficient: a comparative study. In: International Conference on Modeling, Simulation and Applied Optimization (ICMSAO), pp. 1–5. IEEE (2013)
10. Budiaji, W., Leisch, F.: Simple k-medoids partitioning algorithm for mixed variable data. Algorithms **12**(9), 177 (2019)

11. de Carvalho, F., Lechevallier, Y., de Melo, F.M.: Partitioning hard clustering algorithms based on multiple dissimilarity matrices. Pattern Recogn. **45**(1), 447–464 (2012)
12. de Carvalho, F.D.A., Lechevallier, Y., de Melo, F.M.: Partitioning hard clustering algorithms based on multiple dissimilarity matrices. Pattern Recogn. **45**(1), 447–464 (2012)
13. Chiu, T., Fang, D., Chen, J., Wang, Y., Jeris, C.: A robust and scalable clustering algorithm for mixed type attributes in large database environment. In: Proceedings of the Seventh ACM SIGKDD International Conference on Knowledge Discovery and Data Mining (KDD 2001), pp. 263–268. Association for Computing Machinery, New York, NY, USA (2001)
14. de Carvalho, F., Lechevallier, Y., Despeyroux, T., de Melo, F.M.: Advances in knowledge discovery and management. In: Zighed, F., Abdelkader, G., Gilles, P., Venturini, B.D. (eds.) Multi-view Clustering on Relational Data, pp. 37–51. Springer, Cham (2014). https://doi.org/10.1007/978-3-319-02999-3_3
15. Foss, A.H., Markatou, M., Ray, B.: Distance metrics and clustering methods for mixed-type data. Int. Stat. Rev. **87**(1), 80–109 (2019)
16. Fraley, C., Raftery, A.E.: How many clusters? which clustering method? answers via model-based cluster analysis. Comput. J. **41**(8), 578–588 (1998)
17. Green, P.E., Rao, V.R.: A note on proximity measures and cluster analysis. J. Mark. Res. **3**(6), 359–364 (1969)
18. Harikumar, S., Surya, P.V.: K-medoid clustering for heterogeneous datasets. Procedia Comput. Sci. **70**, 226–237 (2015)
19. Hsu, C.C., Chen, C.L., Su, Y.W.: Hierarchical clustering of mixed data based on distance hierarchy. Inf. Sci. **177**(20), 4474–4492 (2007)
20. Huang, J., Ng, M., Rong, H., Li, Z.: Automated variable weighting in k-means type clustering. IEEE Trans. Pattern Anal. Mach. Intell. **27**(5), 657–668 (2005)
21. Huang, Z.: Clustering large data sets with mixed numeric and categorical values. In: The Pacific-Asia Conference on Knowledge Discovery and Data Mining, pp. 21–34 (1997)
22. Hunt, L., Jorgensen, M.: Clustering mixed data. WIREs Data Min. Knowl. Disc. **1**(4), 352–361 (2011)
23. José-García, A., Gómez-Flores, W.: Automatic clustering using nature-inspired metaheuristics: a survey. Appl. Soft Comput. **41**, 192–213 (2016)
24. José-García, A., Gómez-Flores, W.: A survey of cluster validity indices for automatic data clustering using differential evolution. In: Proceedings of the Genetic and Evolutionary Computation Conference, pp. 314–322. ACM Press (2021). https://doi.org/10.1145/3449639.3459341
25. José-García, A., Handl, J.: On the interaction between distance functions and clustering criteria in multi-objective clustering. In: Ishibuchi, H., Zhang, Q., Cheng, R., Li, K., Li, H., Wang, H., Zhou, A. (eds.) EMO 2021. LNCS, vol. 12654, pp. 504–515. Springer, Cham (2021). https://doi.org/10.1007/978-3-030-72062-9_40
26. José-García, A., Handl, J., Gómez-Flores, W., Garza-Fabre, M.: Many-view clustering: an illustration using multiple dissimilarity measures. In: Genetic and Evolutionary Computation Conference - GECCO 2019, pp. 213–214. ACM Press, Prague, Czech Republic (2019)
27. José-García, A., Handl, J., Gómez-Flores, W., Garza-Fabre, M.: An evolutionary many-objective approach to multiview clustering using feature and relational data. Appl. Soft Comput. **108**, 107425 (2021)
28. Landi, I., et al.: Deep representation learning of electronic health records to unlock patient stratification at scale. NPJ Digital Med. **3**(1), 96 (2020)

29. Li, C., Biswas, G.: Unsupervised learning with mixed numeric and nominal data. IEEE Trans. Knowl. Data Eng. **14**(4), 673–690 (2002)
30. Liu, C., Chen, Q., Chen, Y., Liu, J.: A fast multiobjective fuzzy clustering with multimeasures combination. Math. Prob. Eng. **2019**, 1–21 (2019)
31. Liu, C., Liu, J., Peng, D., Wu, C.: A general multiobjective clustering approach based on multiple distance measures. IEEE Access **6**, 41706–41719 (2018)
32. Van der Maaten, L., Hinton, G.: Visualizing data using t-SNE. J. Mach. Learn. Res. **9**(11), 2579–2605 (2008)
33. MacQueen, J.: Some methods for classification and analysis of multivariate observations. In: Proceedings of the Fifth Berkeley Symposium on Mathematical Statistics and Probability, pp. 281–297. University of California Press (1967)
34. Zhang, Q., Li, H.: MOEA/D: a multiobjective evolutionary algorithm based on decomposition. IEEE Trans. Evol. Comput. **11**(6), 712–731 (2007)
35. Rousseeuw, P.J.: Silhouettes: a graphical aid to the interpretation and validation of cluster analysis. J. Comput. Appl. Math. **20**, 53–65 (1987)
36. Shirkhorshidi, A.S., Aghabozorgi, S., Wah, T.Y.: A comparison study on similarity and dissimilarity measures in clustering continuous data. PLOS ONE **10**(12), e0144059 (2015)
37. Sobanski, V., Giovannelli, J., Allanore, Y., et al.: Phenotypes determined by cluster analysis and their survival in the prospective european scleroderma trials and research cohort of patients with systemic sclerosis. Arthritis Rheumatol. **71**(9), 1553–1570 (2019)
38. Theodoridis, S., Koutrumbas, K.: Pattern Recognition. Elsevier Inc., Amsterdam (2009)
39. Vandromme, M., Jacques, J., Taillard, J., Jourdan, L., Dhaenens, C.: A biclustering method for heterogeneous and temporal medical data. IEEE Trans. Knowl. Data Eng. **34**(2), 506–518 (2022)
40. van de Velden, M., Iodice D'Enza, A., Markos, A.: Distance-based clustering of mixed data. WIREs Comput. Stat. **11**(3), e1456 (2019)
41. Wei, M., Chow, T., Chan, R.: Clustering heterogeneous data with k-means by mutual information-based unsupervised feature transformation. Entropy **17**(3), 1535–1548 (2015)

Specification-Driven Evolution of Floor Plan Design

Katarzyna Grzesiak-Kopeć⬛, Barbara Strug(✉)⬛, and Grażyna Ślusarczyk⬛

Institute of Applied Computer Science, Jagiellonian University,
ul. Łojasiewicza 11, Kraków, Poland
{katarzyna.grzesiak-kopec,barbara.strug,grazyna.slusarczyk}@uj.edu.pl

Abstract. Generating floor plan designs is a challenging task that requires from an architect both engineering knowledge and creativity. Various computer-aided design tools are used to improve the efficiency of the design process, the most promising of which are intelligent computational models. In this paper a human-computer interaction based framework for multi-storey houses floor plan design is proposed, where the generation of possible solutions is driven by the evolutionary search directed by the user-defined criteria. The constraints and requirements specified by the user provide the basis for the definition of the requirement-weighted fitness function and can be modified during the evolution process. In the first stage of evolution the layouts for one floor are generated. Floor plans for other floors are generated in the next stage, which allows for introducing additional constraints regarding the position of structural elements (such as load-bearing walls or stairs) that cannot be mutated, and thus adjust these plans to the ones generated earlier. The genotypes of individuals are represented by the vectors of numerical values of points representing endpoints of room walls. This structure allows for representing any rectilinear rooms. A case study of the floor plan design for a two-storey house is presented.

Keywords: Evolutionary design · Floor plan optimization · Design constraints and requirements

1 Introduction

Creating house floor plan designs is a challenging task as the architect has to take into account many constraints and requirements, and determine the location of rooms, their sizes, accessibility and adjacency relations among them. House design is a time-consuming iterative process, requiring multiple rounds of refinements. The ability to automatically generate feasible floor plans could significantly reduce design costs in the real estate industry. Therefore, there is a growing interest in advanced generative and optimization models by architects and building engineers. Computer-aided tools for floor plan design should support the designer in decision making on the basis of the initial visualization of constraints and users requirements and provide design knowledge required for reasoning about design solutions.

G. Rudolph et al. (Eds.): PPSN 2022, LNCS 13399, pp. 368–381, 2022.
https://doi.org/10.1007/978-3-031-14721-0_26

Over the past 50 years, a great deal of research has been done in the field of computer-aided architectural design with the main goal of automatically generating floor plans, which would be treated by designers as preliminary layouts to be further modified and adapted by them [31,33]. In most existing approaches the graph-based representation of floor layouts is used [8,29,33]. In [30] an evolutionary technique based on a graph representation of genotypes was proposed. Generating floor plans for given adjacency graphs is computationally demanding as it requires specifying a geometric interpretation. Moreover, the proper arrangement of rooms may not exist, or the generation can result in layouts which may not be architecturally and aesthetically meaningful.

Early methods formulated the problem as iterative optimization [20,21]. In [5,9,23] the generation of floor plans is based on shape grammars which define iterative generation processes. In [7], an agent system combined with shape grammars was used to support floor layout designs. However, shape grammar interpreters are difficult to implement, as matching parametric shapes is still a challenging problem [3,16].

In recent years, many approaches to the automation of floor plan design based on artificial intelligence and machine learning methods have been developed. In [21], stochastic optimization and a supervised learning algorithm based on Bayesian networks trained to learn distributions of architectural components is presented. In [14] a graph neural network generating floor plans is presented, in [26,32] convolutional neural networks (CNNs) are used for indoor scene generation while in [4,25] the ability of Generative Adversarial Networks (GANs) to floor plan generation is described. In [34] deep learning approach to generate floor plans without specifying any constraints is presented. Deep learning frameworks for interior design are also used in [18,27,35]. The construction of floor plans using simulated annealing is discussed in [2]. Most of the above-mentioned methods are not suitable for complex layouts design, as their time complexity is very high due to the stochastic nature of the algorithms used. In methods, where graphs describing layouts are represented within deep networks, a large amount of training data is needed to attain good accuracy.

Other methods applied for floor plan generation are based on evolutionary algorithms. In [28] a hybrid evolutionary algorithm is developed to generate a set of floor plans in the early design stages, while in [1] an evolutionary approach to design an interactive layout solver is used. A non-data-driven approach is based on neuroevolution of augmenting network topologies (NEAT) [15]. NEAT involves the use of genetic algorithms to find the most fitting topology of a floor plan in a given initial configuration. However, its performance significantly depends on the initially chosen topology [13] and the control of complexity within inner networks [17]. In [22] simulated annealing and genetic algorithms are used to optimize mathematically defined design objectives and constraints. The ability of the model is dependent on the predefined mathematical functions.

In order to create floor plans using small and more controlled data sets, which allow for generating an incremental level of spatial quality, in this paper, an evolutionary approach is proposed. The evolutionary technique is an efficient method for creating a variety of topologically distinct, but still valid, design solutions for a given floor layout problem. It allows for visual exploration of the designer preliminary ideas which is necessary in the conceptual phase of design. In our approach the crossover operation is not used, while the mutation mechanism is constructed in such a way to control the scope of the introduced changes. More radical changes may be expected at the beginning of the evolution, while with time weak changes are preferred. In the evaluation module the domain knowledge is required to decide which generated potential solutions are to be kept in future evolution cycles. In this manner, a gradual improvement in the overall quality of the proposed solutions is obtained.

It should be noted that while the most of the above mentioned papers deal with single floor layouts, the design of plans for a multi-storey building involves more than just the repetition of single floor generation. Design of many floors is a more complex problem as it requires additional knowledge and constraints in order to make the whole construction structurally matched.

In the proposed framework for multi-story house floor plan generation, the evolutionary search for new solutions consists of two stages. In the first stage of evolution, the best floor plans for one floor are generated according to the specified constraints and requirements. Then the constraints regarding the position of structural elements (like load-bearing walls or columns, walls with sewer lines) and the placement of certain room types (like stairs or lifts) are determined by the user for the selected solutions. In the second stage of evolution, in which room layouts for remaining floors are generated, these constraints are taken into account. They determine walls that cannot be mutated, which means that the generation in this stage of evolution is additionally driven by the required placement of the specified elements.

The house floor plans are generated so as to comply with specifications based on building code constraints, the designers knowledge, and the requirements determined in collaboration with the customer. In these specifications, the required number of rooms, their areas and functions as well as the rules of their arrangement can be determined. Design constraints and many requirements are to be fully satisfied. There are also requirements which are to be fulfilled to some degree, and the ones the fulfillment of which would be desirable but is not absolutely necessary [27]. Therefore, the weights determining the importance of requirements are defined. Then, the quality of the generated solutions and the degree to which they meet the design specifications can be properly assessed.

In our approach, the initial population of floor plans is created by the designer based on the given specification of the problem. The genotypes corresponding to these solutions are represented in the form of numerical vectors consisting of numbers of nodes representing points where the walls of the rooms meet. Such a representation makes the description of evolutionary operators very simple and allows for fast computing of all mutation types. It is also efficient as the visualization of the obtained floor plans is straightforward. This type of a structure allows for representing not only rectangular spaces but any rectilinear rooms as well and implements the actual dimensions of the designed floor layout. During the process of evolution, the genotype vectors are modified by mutation. After each evolutionary step, a new generation of phenotypes being floor plans corresponding to the obtained population is rendered. The fitness function, which evaluates generated floor plans, takes into account the specified design constraints and requirements. By using optimization techniques, the proposed framework generates floor plans together with an assessment that determines the degree of their compliance with the design task specifications.

The paper is organized as follows. In Sect. 2 the proposed framework for visual floor plan generation is described. The details concerning the floor plan representation and the elements of the used evolution algorithm are presented. The case study illustrating the proposed approach is considered in Sect. 3. The conclusions are drawn in Sect. 4.

2   Visual Floor Plan Generation Framework

Usually during the conceptual stage of design an architect expresses ideas graphically, evaluates them, and adjust through an iterative revision process cooperating with a customer. During this time- and cost-consuming process the design diagram itself is being explored as an essential stage of the design process and the design thinking [19]. Considering that it is usually difficult to pinpoint clear objective metrics to optimize solutions, many trials are the key to find a satisfactory layout. Computer-aided design reducing the amount of human time involvement would significantly speed up the whole process and drastically cut the costs. In [11], the Visual Floor Plan Generation Framework (VFPGF) for quick, efficient and effective one-story floor layouts generation was proposed. It benefits from the dynamic character of the design context modified by a designer and the evolutionary programming generation engine. The human-machine cooperation process not only allows to find the personalized final solution faster but also facilitates the definition of design rules thanks to the use of the evolution-driven machine learning approach.

In this paper, we present the adaptation of the previous solution to the design of multi-storey buildings (see Fig. 1). The designer, together with the customer, starts the design process with the initial definition of the requirements (and their importance coefficients) and the constraints. Then, the designer, based on his/her own professional experience, indicates a set of initial floor plan solutions constituting the starting population for the evolutionary programming generation engine. The domain knowledge base is an important part of the system and influences the direction of evolution. It can contain both the information derived from design standards (for example from the building code) and any additional information from the design domain, like availability of infrastructure specific information for the area where the building is going to be located. During the ongoing generation of possible solutions, the designer explores various layouts and can modify the machine learning parameters to change the course of evolution and achieve better results. At the same time, the evolutionary process

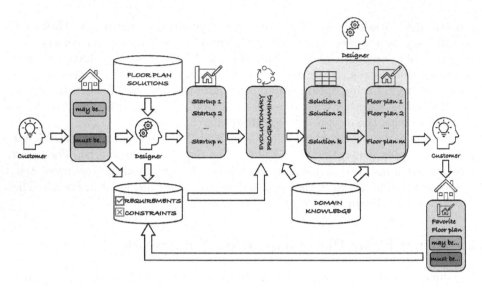

Fig. 1. Visual Floor Plan Generation Framework (VFPGF).

stimulates the creativity of the designer who can experiment with many potential design ideas simultaneously. The bidirectional human-machine collaboration is crucial for the quality of the finally generated solutions.

Floor layout planning does not have measure and optimize evaluation metrics, which would be easy to define. Nevertheless, it is always accompanied by in-depth study of the customer utility zones. The specific customer environmental conditions are basic guidelines to architectural design inspection. Therefore, floor layout planning can be defined as the search process of the best floor plan behavior that adapts to the given requirements and constraints context. Given such an optimization problem, evolutionary programming (EP) can be applied to solve it. Instead of focusing on the genotypic evolution, EP emphasizes the phenotypic one and explores a space of possible behaviors. It finds the best solution by adapting a set of semi-optimal behaviors where the search process is driven solely by the mutation and the selection. Recombination operators are not used and the individual representation is strictly problem dependent. The fitness function is relative and evaluates individual behavior with respect to some chosen subgroup of the current population and calculates their behavioral error. Also, selection is based on competition and those individuals that exhibit the best behavior compared to the competing group have the highest probability of entering the next generation. This competition for survival applies to both the parents and the offspring [6].

The proposed approach has been pretested using Python in the Jupyter Notebook, which was selected for the possibility of visual and interactive verification of the ongoing evolutionary design process.

2.1 Floor Plan Representation

The aim of the authors of the proposed approach was to indicate the simplest possible representation for the room layouts, which could be easily mutated and assessed by the fitness function. Taking into account the assumption that all rooms are polygons, a real-valued sequence of points was adopted. A room is uniquely defined by a set of points $P = \{p_1, \ldots, p_n\}$ where $p_i = (x, y) \in R^2, 1 \leq i \leq N$. Each point represents the place where the walls of the rooms meet. Rooms are identified by a flood fill algorithm [12] application. The individual floor plan $I = \{'R1' : [p_{11}, p_{12}, \ldots, p_{1j_1}], 'R2' : [p_{21}, \ldots, p_{2j_2}], \ldots, 'Rk' : [p_{k1}, \ldots, p_{kj_k}]\}$ where $R_1, \ldots, R_k$ denote recognized rooms and $[p_{m1}, \ldots, p_{mj_m}]$ is a sequence of points of all walls' connections in a room R_m.

Both, selected points, and whole rooms may be labelled as immobile. When a room is immobile then all points in this room are immobile too. This special tag indicates those elements of the layout that cannot be changed during evolution. In this way, the designer can mark not only structural elements, such as load-bearing walls, but also fragments of solutions preferred by the customer (e.g. the location of the living room or the mezzanine).

2.2 Population Initialization

Generating the initial population for a floor plan design task is not trivial. It would be very desirable to uniformly cover the given domain of the layout optimization problem, but it is unlikely to ever happen in the case of design tasks which, in the course of a creative process, allow for obtaining surprising and at the same time functional solutions. Therefore, the designer domain knowledge is irreplaceable, and she/he arbitrary points out a set of startup solutions from a floor plan solutions database.

2.3 Fitness Function

The evolutionary process is directed by the fitness function which determines the quality of the obtained solution. The fitness function $F(I)$ for an individual I provides an absolute measure of how well the individual fulfills all the constraints and the degree to which it achieves the objectives of the design task. In other words, how well it behaves in a given environment and solves the design problem. The constraints can be handled either by penalty functions or by problem reformulation to unconstrained one [10,24]. In our case, they are defined as binary functions, with value of 0 meaning that the design is to be rejected and the value of 1 meaning that a given design may be further evaluated. The constraints may have different origin i.e. they may refer to the building code, local area regulations, fire protection code or other rules that must be followed without any exceptions. They may also originate from technological requirements, for example the design must be made of predefined types of material or has a predefined external area. Finally, constraints may be provided by the customer in the form of unchangeable requirements, for example a house must have two storeys or it must have a garage.

The requirements, on the other hand, are assigned an importance coefficient. They may originate from the designer body of knowledge accumulated over the time or be provided by the customer. They define elements that should/can be present in the floor layout. Such requirements include customer expectations such as: the floor should have at least four bedrooms, it should have a living room on the first floor, all bedrooms should be on the same floor etc. They define measurable plan properties like a number of spaces, minimal and maximal dimensions, areas orientations and their relative positions etc.

Each requirement is defined as $Rq_i = (name_i, value_i, storey_i)$, where $name_i$ is the requirement name, $value_i$ is its required value and $storey_i$ specifies the floor that this requirement concerns. To make this formalism flexible additional symbols are used, i.e. $>, <$ denote more than and less than, respectively. For example $Rq_1 = (spaces, <> 10, 0)$ represents the requirement that there are about 10 spaces on the ground floor. Moreover to each requirement a weight w_i ranging from 0 to 1, and an evaluation function Req_i are assigned (see Fig. 2).

If all the constraint $Con_j(I) \in \{0, 1\}(j = 1, \ldots, m)$ are met by the individual, its final fitness score is computed as the weighted sum of the degrees to which the requirements are fulfilled, which can be summarized as follows:

$$F(I) = \begin{cases} -\infty & , \exists j \leq m : Con_j = 0 \\ \sum_{i=1}^{n} w_i Req_i(I) & , otherwise \end{cases} \tag{1}$$

2.4   Mutation

The main objective of a mutation operator is to introduce variation (noise) in the population and produce new candidate solutions. It may be applied one or more times to each parent and produce one or more offspring. In the early stages of evolution, it should introduce great diversity to the population and dynamically explore the search space. Unlike in the last phase of evolution, where major changes are not desired, and the aim of the mutation is to exploit the obtained results to fine tune them. The exploration exploitation trade-off in EP is modified by strategy parameters.

In the proposed approach three mutation operators are defined: a new point can be added, a point can be deleted, and a point can be moved. The operators may be applied individually or in a randomly selected sequence. In order to achieve only feasible solutions, some remedial steps must be taken. A new point may be added on the existing wall or inside a room. In both cases some extra points have to be added to produce valid floor plan solution. On the other hand, deleting a point requires indication of another point connected to it by a wall

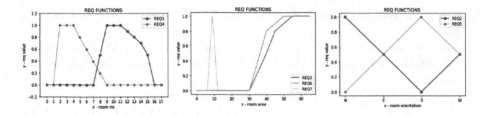

Fig. 2. Requirements evaluation functions.

and these two points are deleted together. Applying a mutation operator, which moves a point can require the greatest number of additional changes (for detail see [11]).

Since the proposed mutation operators introduce great diversity to the population, the following scheme was adopted:

1. The Gaussian distribution, where mutation operators are weaker in exploration (due to narrow tail) but facilitate the fine-tuning of the final solutions, is selected.
2. The mutation operators are applied individually.
3. The dynamic strategy that over time reduces the probability of a mutation occurring is implemented.

2.5 Selection

The survival in EP is usually based on a relative fitness measure. Both parents and offspring take part in this competition where the goal is to enter the next generation. It expresses how well an individual performs in a population or in its subset. Such a subset may be selected with a use of various selection operators: random, proportional, tournament, rank-based, elitism, hall of fame etc. To preserve the high diversity in the population a low selective pressure operator, in our approach a roulette-wheel proportional selection is applied. The relative fitness is calculated as the number of competitors that are worse adapted and have a lower fitness result.

3 Case Study

Let us consider an example of designing a two storey house. The customer gives the following requirements: a double garage, a kitchen with a pantry, a dinning room, a living room, a boiler, an extra room on the ground floor, and at least three bedrooms on the first floor. She/he also gives the upper boundary of the building area which is $24 \times 24 \, \text{m}^2$ and emphasizes that a garage is essential. Based on these demands, the building code and further designer inquiry, the following constraints and requirements have been distinguished:

$Con_1 = (storeys, 2, *)$: two floors in the house,

Con_2: the ordinate and abscissa are within the range $<0, 17>$ (the building area),

$Con_3 = (border\_spaces > 35, 1, 0)$: one space larger than $35\,m^2$ adjacent to the external wall of the building on the ground floor with one wall $6\,m$ long (a garage),

$Con_4 = (wall < 1, 0, *)$: no wall shorter than $1.0\,m$.

$Rq_1 = (spaces, <> 10, 0)$: there should be about 10 spaces on the ground floor, $w_1 = 0.5$,

$Rq_2 = (border\_spaces > 35, N, 0)$: the space from Con_3 should be oriented to the north, $w_2 = 0.6$,

$Rq_3 = (adj\_spaces, \geq 40, 0)$: there should be two adjacent rooms (dining, living) together at least $40\,m^2$ on the ground floor (the bigger the better); it is not applied to space from Con_3, $w_3 = 0.95$,

$Rq_4 = (spaces\_area, \geq 10, 0)$: there should be additional two spaces at least $10\,m^2$ on the ground floor (kitchen, room), $w_4 = 0.8$,

$Rq_5 = (bigger\_space, S, 0)$: the bigger room from Rq_3 should be oriented to the south, $w_5 = 0.7$,

$Rq_6 = (spaces \geq 15, 3, 1)$: there should be three spaces on the first floor each at least $15\,m^2$ (the bigger the better), $w_6 = 0.8$,

$Rq_7 = (spaces <> 9, 1, 1)$: there should be additional space having around $9\,m^2$ on the first floor (bathroom), $w_7 = 0.7$.

It should be noted that we assume that constraints and requirements are checked in a specified order, so later defined requirements can relate to the entities which fulfil earlier defined ones. In order to evaluate requirements fulfillment appropriate rating functions must be provided (see Fig. 2). Taking into account the importance coefficients agreed with the customer, the highest possible fitness value for an individual I would be calculated as follows: $F(I_1) = 0.5 \cdot 1 + 0.6 \cdot 1 + 0.95 \cdot 1 + 0.8 \cdot 1 + 0.7 \cdot 1 = 3.55$ for the ground floor and $F(I_2) = 0.8 \cdot 1 + 0.7 \cdot 1 = 1.5$ for the first floor. The lowest fitness value for individuals in our example is 0 (for layouts with many spaces but relatively small ones) when none of the requirements is fulfilled.

Since the case of a two-storey house is considered, the solution generation is carried out in two steps, two evolutionary searches, one for each floor. First, considering all the above-mentioned criteria, designer selects a subset of his/her floor plan solutions for the ground floor which is the basis for the startup population. In our example this core subset consists of eight elements. Then the full initial population of 100 individuals is generated by applying all three possible mutations to core elements.

When the evolutionary generation stops, the designer presents to the customer a bunch of the most promising individuals, i.e., the ones with the highest evaluation values. The customer indicates the floor plans that suit him best and become the basis for creating the initial population for the next stage of evolution.

The evolution process runs only once to get different solutions for a particular floor layout. The termination condition is defined as the number of epochs which is specified by the designer (in this study - 50). If there is no satisfying floor layout among the best evaluated solutions, the user can continue the process of generating next populations or change some design requirements. Different requirements for our case study as well as different case studies need other runs starting from the beginning.

The fitness function evaluates individuals behavior with respect to the subgroup of 30 elements selected by a roulette-wheel from the current population. The "behavioral error" decreases in subsequent generations, so the average fitness increases. However, in this case, fitness is a relative matter, and it should be noted that in designing floor layouts there is no optimal solution, as different customers can have different preferences and tastes, and thus may choose the individuals with lower evaluation values as the best fit for them. Therefore the performance of different runs is not comparable and the statistic analysis of them can be misleading.

In our example, the floor plans depicted in Fig. 3 denoted by A.0 and B.0 were selected as the best ones for the customer. The fitness value of these individuals equals $F(A.0) = 3.3$ and $F(B.0) = 3.36$, respectively. The lowest rated requirement for the individual A.0 is $Rq_1 = 0.5$ since the number of generated spaces is 16 while $<> 10$ is expected. For the individual B.0 the lowest value has $Rq_3 = 0.8$ as the total area of possible dining and living rooms is not as large as desired.

For the selected ground floor layouts additional constraints regarding the position of structural elements that cannot be mutated must be provided. There are three types of elements to be marked by the designer as immutable. The first one concerns all the load bearing walls that have to remain unchanged on the subsequent floors. Such a wall is marked with a wide yellow line in Fig. 3. The second type of the restricted elements are all walls containing water pipes - while the customer may choose any available space as a kitchen or bathroom it is expected that all spaces that need access to water would be located in such a way that the number of vertical water pipes is limited. Such elements are depicted in Fig. 3 as blue lines. Moreover, in the case of a multi-storey building, a vertical communication must be provided and it must occupy the same space on adjacent levels. When a space for vertical communication is decided, (e.g. stairs) it must be carried over to adjacent levels without changes. In Fig. 3 the walls surrounding such a space are denoted by red lines.

Having decided on the placement of all the required immutable elements the evolutionary process for the first floor can be started in the same way as for the ground floor. In order to facilitate making decisions on the additional constraints, the labels determining the functions of spaces have been assigned. The example labeling for this case study is presented in Fig. 3, where the shaded areas represent balconies or terraces. Figures 3.A.1 and 3.A.2, and Figs. 3.B.1 and 3.B.2 present two different first floor layouts generated for the earlier selected ground floors A.0 and B.0, respectively. Even though the first floor solutions for each ground floor are significantly different, they all have the highest possible fitness value equal to 1.5.

Fig. 3. Two example floor plans A.0 and B.0 for the ground floor and corresponding plans for the first floor: A1 and A2, B1 and B2 respectively. (Color figure online)

4   Conclusions

This paper presents a new model of human-computer interaction in specification-driven evolution of floor plan designs. The main goal of the research is to propose a computer-aided design framework that not only generates a single floor plan but floor layouts for multi-storey buildings as well. The proposed approach has been successfully applied to two-storey house layout generation. In order to address the search for a design solution better, the evolutionary process is driven by the requirements and constraints agreed by the designer and the customer. It is performed semi-automatically in incremental stages: one stage for one floor. After generating each floor, the elements which are to be immutable in next generation steps are marked by a human. In this way it is possible to avoid generating numerous unfeasible solutions i.e. such that have load carrying walls moved or inconsistent vertical communication and water connections. Moreover, the sequential evaluation of the requirements results in high quality solutions which are more promising for the customer.

In future we plan several improvements. In the first step the formalism used to define requirements and constraints is to be extended by adding wildcards to denote conditions for all floors to allow expressing conditions like "there should be about 20 spaces in the building".

References

1. Bahrehmand, A., Batard, T., Marques, R., Evans, A., Blat, J.: Optimizing layout using spatial quality metrics and user preferences. Graph. Models **93**(C), 2538 (2017). https://doi.org/10.1016/j.gmod.2017.08.003
2. Bao, F., Yan, D.M., Mitra, N.J., Wonka, P.: Generating and exploring good building layouts. ACM Trans. Graph. **32**(4), 1–10 (2013). https://doi.org/10.1145/2461912.2461977
3. Beirao, J.N.: CityMaker: designing grammars for urban design, Doctoral thesis (2012). https://doi.org/10.4233/uuid:16322ba7-6c37-4c31-836b-bc42037ea14c
4. Chaillou, S.: ArchiGAN: a generative stack for apartment building design (2019). https://developer.nvidia.com/blog/archigan-generative-stack-apartment-building-design/
5. Duarte, J.: A discursive grammar for customizing mass housing: the case of Siza's houses at Malagueira. Autom. Constr. **14**(2 SPEC.ISS.), 265–275 (2005). https://doi.org/10.1016/j.autcon.2004.07.013
6. Engelbrecht, A.P.: Computational Intelligence: An Introduction, 2nd edn. Wiley, New York (2007)
7. Grabska, E., Grzesiak-Kopeć, K., Ślusarczyk, G.: Designing floor-layouts with the assistance of curious agents. In: Alexandrov, V.N., van Albada, G.D., Sloot, P.M.A., Dongarra, J. (eds.) ICCS 2006. LNCS, vol. 3993, pp. 883–886. Springer, Heidelberg (2006). https://doi.org/10.1007/11758532_115
8. Grabska, E., Achwa, A., Lusarczyk, G.: New visual languages supporting design of multi-storey buildings. Adv. Eng. Informatics **26**, 681–690 (2012)
9. Grzesiak-Kope, K., Ogorzaek, M.: Intelligent 3D layout design with shape grammars. In: Proceedings of the 2013 6th International Conference on Human System Interactions (HSI), pp. 265–270, Sopot, Poland, June 2013
10. Grzesiak-Kope, K., Oramus, P., Ogorzaek, M.: Hypergraphs and extremal optimization in 3D integrated circuit design automation. Adv. Eng. Inform. **33**(C), 491501 (2017). https://doi.org/10.1016/j.aei.2017.06.004
11. Grzesiak-Kopeć, K., Strug, B., Ślusarczyk, G.: Evolutionary methods in house floor plan design. Appl. Sci. **11**(17), 8229 (2021). https://doi.org/10.3390/app11178229, https://www.mdpi.com/2076-3417/11/17/8229
12. Henrich, D.: Space-efficient region filling in raster graphics. Vis. Comput. **10**, 205–215 (2005)

13. Hohenheim, J., Fischler, M., Zarubica, S., Stucki, J.: Combining neuro-evolution of augmenting topologies with convolutional neural networks, January 2017
14. Hu, R., Huang, Z., Tang, Y., Van Kaick, O., Zhang, H., Huang, H.: Graph2Plan: learning floorplan generation from layout graphs. ACM Trans. Graph. **39**(4), 118:1–118:14 (2020). https://doi.org/10.1145/3386569.3392391
15. Ibrahim, M.Y., Sridhar, R., Geetha, T.V., Deepika, S.S.: Advances in neuroevolution through augmenting topologies a case study. In: 2019 11th International Conference on Advanced Computing (ICoAC), pp. 111–116 (2019)
16. Krishnamurti, R.: Explicit design space? AI EDAM **20**, 95–103 (2006). https://doi.org/10.1017/S0890060406060082
17. Le Goff, L.K., Hart, E., Coninx, A., Doncieux, S.: On Pros and Cons of evolving topologies with novelty search. In: ALIFE 2021: The 2021 Conference on Artificial Life, ALIFE 2020: The 2020 Conference on Artificial Life, pp. 423–431, July 2020. https://doi.org/10.1162/isal_a_00291
18. Li, J., Yang, J., Hertzmann, A., Zhang, J., Xu, T.: LayoutGAN: generating graphic layouts with wireframe discriminators. CoRR arXiv: abs/1901.06767 (2019)
19. Liu, H., Tang, M.: Evolutionary design in a multi-agent design environment. Appl. Soft Comput. **6**, 207–220 (2006). https://doi.org/10.1016/j.asoc.2005.01.003
20. Martin, J.: Procedural house generation: a method for dynamically generating floor plans. In: Symposium on Interactive 3D Graphics and Games (2006)
21. Merrell, P., Schkufza, E., Koltun, V.: Computer-generated residential building layouts. ACM Trans. Graph. **29**, 1–12 (2010). https://doi.org/10.1145/1866158.1866203
22. Michalek, J., Choudhary, R., Papalambros, P.: Architectural layout design optimization. Eng. Optim. **34**, 461–484 (2002). https://doi.org/10.1080/03052150214016
23. Müller, P., Wonka, P., Haegler, S., Ulmer, A., Van Gool, L.: Procedural modeling of buildings. In: ACM SIGGRAPH 2006 Papers, SIGGRAPH 2006, p. 614623. Association for Computing Machinery, New York, NY, USA (2006). https://doi.org/10.1145/1179352.1141931
24. Myung, H., Kim, J.-H.: Lagrangian-based evolutionary programming for constrained optimization. In: Yao, X., Kim, J.-H., Furuhashi, T. (eds.) SEAL 1996. LNCS, vol. 1285, pp. 35–44. Springer, Heidelberg (1997). https://doi.org/10.1007/BFb0028519
25. Nauata, N., Chang, K., Cheng, C., Mori, G., Furukawa, Y.: House-GAN: relational generative adversarial networks for graph-constrained house layout generation. CoRR arXiv: abs/2003.06988 (2020)
26. Ritchie, D., Wang, K., Lin, Y.: Fast and flexible indoor scene synthesis via deep convolutional generative models. CoRR arXiv: abs/1811.12463 (2018)
27. Ritchie, D., Wang, K., Lin, Y.A.: Fast and flexible indoor scene synthesis via deep convolutional generative models. In: Proceedings of the IEEE/CVF Conference on Computer Vision and Pattern Recognition (CVPR), June 2019
28. Rodrigues, E., Gaspar, A.R., Gomes, Á.: An approach to the multi-level space allocation problem in architecture using a hybrid evolutionary technique. Autom. Constr. **35**, 482–498 (2013). https://doi.org/10.1016/j.autcon.2013.06.005, https://www.sciencedirect.com/science/article/pii/S0926580513001027
29. Ślusarczyk, G.: Graph-based representation of design properties in creating building floorplans. Comput. Aided Des. **95**(C), 2439 (2018). https://doi.org/10.1016/j.cad.2017.09.004

30. Strug, B., Grabska, E., Ślusarczyk, G.: Supporting the design process with hypergraph genetic operators. Adv. Eng. Inform. **28**(1), 1127 (2014). https://doi.org/10.1016/j.aei.2013.10.002
31. Upasani, N., Shekhawat, K., Sachdeva, G.: Automated generation of dimensioned rectangular floorplans. CoRR arXiv: abs/1910.00081 (2019)
32. Wang, K., Savva, M., Chang, A.X., Ritchie, D.: Deep convolutional priors for indoor scene synthesis. ACM Trans. Graph. **37**(4), 1–14 (2018). https://doi.org/10.1145/3197517.3201362
33. Wang, X.Y., Yang, Y., Zhang, K.: Customization and generation of floor plans based on graph transformations. Autom. Constr **94**(C), 405–416 (2018). https://doi.org/10.1016/j.autcon.2018.07.017
34. Wu, W., Fu, X.M., Tang, R., Wang, Y., Qi, Y.H., Liu, L.: Data-driven interior plan generation for residential buildings. ACM Trans. Graph. (SIGGRAPH Asia) **38**(6), 1–12 (2019)
35. Zou, C., Colburn, A., Shan, Q., Hoiem, D.: LayoutNet: reconstructing the 3D room layout from a single RGB image. In: Proceedings of the IEEE Conference on Computer Vision and Pattern Recognition (CVPR), June 2018

Surrogate-Assisted Multi-objective Optimization for Compiler Optimization Sequence Selection

Guojun Gao[1,2], Lei Qiao[3(✉)], Dong Liu[1,2], Shifei Chen[1,2], and He Jiang[1,2(✉)]

[1] School of Software, Dalian University of Technology, Dalian, China
{ggj_gao,chenshifei}@mail.dlut.edu.cn, {dongliu,jianghe}@dlut.edu.cn
[2] Key Laboratory for Ubiquitous Network and Service Software of Liaoning Province, Dalian, China
[3] Beijing Institute of Control Engineering, Beijing, China
fly2moon@aliyun.com

Abstract. Compiler developers typically design various optimization options to produce optimized programs. Generally, it is a challenging task to identify a reasonable set of optimization options (i.e., compiler optimization sequence) in modern compilers. Optimization objectives, in addition to the target architecture and source code of the program, influence the selection of optimization sequences. Current applications are often required to optimize two or more conflicting objectives simultaneously, such as execution time and code size. Existing approaches employ evolutionary algorithms to find appropriate optimization sequences to trade off the above two objectives. However, since program compilation and execution are time-consuming, and the two objectives are inherently conflicting, applying evolutionary algorithms faces the diverse objectives influence and computationally expensive problem. In this study, we present a surrogate-assisted multi-objective optimization approach. To speed up the convergence, it employs a fast global search based on non-dominated sorting. The approach then uses two surrogate models for each objective to generate approximate fitness evaluations rather than using actual expensive evaluations. Extensive experiments on the benchmark suite cBench show that our approach outperforms the baseline NSGA-II on hypervolume by an average of 11.7%. Furthermore, experiments verify that the surrogate model contributes to solving the computationally expensive problem and taking fewer actual fitness evaluations.

Keywords: Multi-objective · Compiler optimization sequence selection · Surrogate model

1 Introduction

Today, the compiler is one of the most important foundations of the complex software infrastructure, and it has been used to generate optimized executable binaries for several decades [4,14]. Modern compilers provide numerous compiler

G. Rudolph et al. (Eds.): PPSN 2022, LNCS 13399, pp. 382–395, 2022.
https://doi.org/10.1007/978-3-031-14721-0_27

optimization options to satisfy a wide range of complex optimization requirements (e.g., code size and execution time). GCC, for example, offers hundreds of optimization options. As a result, it is impractical to select the best compiler optimization sequence from massive optimization options to optimize programs by hand. Despite the fact that compilers provide some predefined standard optimization levels (-O1,-O2,-O3,-Os, etc.) with a fixed optimization sequence, they fail to achieve the best performance on every program [3,4,9].

Besides, when selecting an appropriate compiler optimization sequence, the selection is indeed influenced not only by the program source code and the target architecture but also by the optimization objectives [8]. For the compiled object code, however, the code size and execution time are two conflicting objectives. The program will be expanded if you pursue the program's execution time. On the contrary, pursuing smaller executable code frequently results in slower program execution speed [21].

In previous studies, multi-objective optimization algorithms were used to select optimization sequences to make a trade-off between code size and execution time. Lokuciejewski et al. [16,17] applied SPEA2, NSGA-II, and IBEA to select optimization sequences that reduced execution time and code size. Experiments show that the performance of these algorithms is significantly improved compared with standard optimization levels. Chebolu and Wankar [8] used a novel genetic algorithm based on weighted value functions to obtain a faster execution time than the binary code generated by "-Ofast", while the code size does not exceed "-OS". Ansel et al. [2] introduced "OpenTuner", an open-source framework for constructing multi-objective optimized compiler optimization sequences. The framework supports the user-defined setting of multiple search methods to search for the appropriate optimization sequence.

However, the key issue in applying the multi-objective optimization algorithm in selecting compiler optimization sequences is its efficiency. On the one hand, multi-objective optimization needs to search the whole objective space, and the conflicting objectives may influence the convergence rate to some extent [15]. On the other hand, because compiling and executing a program is time-consuming, the computational cost of fitness function evaluations is high, and the number of evaluations required to obtain Pareto-optimal solutions is restricted. This brings up the computationally expensive problem.

To address the above two challenges, we propose a novel surrogate-assisted multi-objective optimization approach that efficiently selects promising optimization sequences. The approach contains two key components: a fast global search and surrogate models. Firstly, a fast global search based on non-dominated sorting is adopted to deliberate the overall performance on different objectives. Secondly, surrogate models are employed to approximate the expensive fitness functions to avoid a lot of actual compilation and execution. In the iterative process, two surrogate models for execution time and code size are used to determine which individuals can be incorporated into the non-dominated solution set. These individuals are then compiled and executed to obtain fitness values. Our goal is to develop an efficient approach for the computationally expensive multi-objective problem with over one hundred variables. Here, we use random forest surrogate models to approximate the expensive compilation.

Furthermore, we study our approach experimentally on the compiler GCC and the benchmark suite cBench [11]. Experimental results demonstrate that our approach performs averagely better than NSGA-II by 11.7% on hypervolume. Besides, the impact of the parameters: crossover rate and mutation rate, is investigated. It shows that configuring the two parameters has no noticeable effect on our approach. Additionally, a comparison between the convergence rate of our approach and a variant without surrogate models indicates that the surrogate models contribute to the efficiency of compiler optimization sequence selection. The contributions of this paper are summarized as follows:

- A novel approach is proposed for efficiently selecting optimization sequences to reduce program code size and speed up execution.
- Surrogate models and a fast global based on non-dominated sorting are developed to overcome the challenges of the computationally expensive problem and the influence of diverse objectives.
- Experimental studies conducted on the compiler GCC and the benchmark suite cBench show that our approach not only identifies better compiler optimization sequences, but also involves fewer expensive fitness evaluations.

The remainder of this paper is organized as follows. Section 2 describes the background of the compiler optimization sequence selection. Section 3 provides a detailed description of the proposed approach. Studies comparing the proposed approach with NSGA-II and the experimental results are presented in Sect. 4. Finally, Sect. 5 concludes the paper with a summary and some ideas for future work.

2   Background

In this section, we introduce the background knowledge of compiler optimization sequence selection as well as the multi-objective optimization for the compiler optimization sequence selection.

2.1   Compiler Optimization Sequence Selection

The compiler provides plentiful optimization options to satisfy different performance requirements. A set of compiler optimization options forms an optimization sequence. We define the problem as **compiler optimization sequence selection** if we disregard the order of the compiler optimization options and instead focus on whether an optimization option is applied. Many previous studies [1,5] have shown that the interactions among optimization options are so complicated that they have a significant impact on program performance.

The problem of compiler optimization sequence selection can be formalized as follows:

Let a Boolean vector $seq = \{o_1, o_2, o_3, ..., o_n\}$ be a compiler optimization sequence, and n is the number of optimization options under analysis. The ith element o_i in seq represents the ith optimization option used. Besides, the value of an optimization option o_i is $o_i = 1$ or $o_i = 0$, indicating whether the optimization option is turned on or off.

Given a program P that is being optimized, one or more of the effective optimization sequences $\overline{seq}$ will be found and provided to the compiler to generate smaller or faster machine object code.

Furthermore, the search space $S = \{0, 1\}^n$ of the problem has an exponential space. For instance, if we only analyze $n = 10$ optimization options and select the right optimization sequence, we need to explore a total state space of $2^n = 1024$. In practice, there are far more than ten, even hundreds of optimization options available when using the compiler to optimize the program. Thus, developers face a significant challenge in manually selecting an appropriate optimization sequence. Automatic methods are necessary to be introduced.

2.2 Multi-objective Optimization for Compiler Optimization Sequence Selection

This study focuses on the optimization of execution time and code size of the machine object code. The program's execution time should be as short as possible during the process of compiler optimization sequence selection using multi-objective optimization. Additionally, to save storage space, the code size should be as small as possible. For easy understanding and unified representation, we design two fitness functions to calculate execution speedup and code size reduction, respectively (as shown in Eqs. 2 and 3 in Sect. 3). In summary, we set the maximum execution speedup and the maximum code size reduction as two optimization objectives in our study.

The multi-objective optimization problem of compiler optimization sequence selection is formulated in Eq. 1 as follows:

$$\max_{seq \in S} (Fitness_s(seq), Fitness_t(seq)) \tag{1}$$

where $Fitness_s(seq)$ is the fitness functions for code size and $Fitness_t(seq)$ is the fitness functions for execution time. The compiler optimization sequence is denoted by seq, and the search space is denoted by the set S. Finding a feasible compiler optimization sequence that maximizes code size reduction and execution speedup at the same time is unthinkable. As a result, we seek to investigate and identify Pareto optimal solutions that cannot be improved in any objective without degrading others.

3 The Proposed Approach

This paper proposes a surrogate-assisted multi-objective optimization algorithm for optimization sequence selection to enhance the performance in terms of execution time and code size. In this section, we will first present the solution representation and the fitness function in our approach. Following that, our approach is described in detail. Finally, we introduce the surrogate model that we used in our approach.

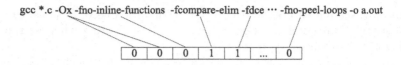

Fig. 1. The solution representation

3.1 Representation

For the current study of selecting the best optimization sequences, a candidate solution represents an optimization sequence that is composed of optimization options from the four standard optimization levels (-O1,-O2,-O3, and -Os). The solution representation in this paper is shown in Fig. 1.

Here, we use a vector to represent a solution, with the first two dimensions representing the encoding of four standard optimization levels and the remaining dimensions representing compiler options. For example, we set the first two dimensions to '00', '01', '10', and '11' to represent -O1, -O2, -O3, and -Os, respectively. The rest dimensions' values are Boolean values that are represented as genes on a chromosome. We set the value to 1 or 0, where 1 indicates that we select the optimization option to optimize the program and 0 indicates that the optimization option is disabled. As illustrated in Fig. 1, the compiler is then informed of the optimization selection information via -f⟨optimization option⟩ and -fno-⟨optimization option⟩, which indicate the selection or non-selection of the optimization option.

Besides, a solution has a fixed size, which includes two label bits (the first two dimensions) and all optimization options used in this study. Because the compiler GCC has its own logic when invoking the optimization option, the order of the optimization options that we provided has no effect on the results. We keep the same order of the optimization options that the compiler is given.

3.2 Fitness Function

The fitness function is intended to assess the quality of a candidate solution. In our study, we look at two objectives: execution time and code size. To measure the code size reduction and execution time speedup of a candidate optimization sequence seq, two fitness functions $Fitness_s(seq)$ and $Fitness_t(seq)$ are defined. As shown in Eqs. 2 and 3, the two fitness functions use the default -O0 and -O3 as bases, respectively. The optimization level -O0 does not optimize the program at all, while -O3 provides the most aggressive optimization on execution time.

$$Fitness_s(seq) = code\_size(\text{-O0}) - code\_size(seq) \qquad (2)$$

where $code\_size(seq)$ represents the code size of the executable file when the optimization sequence seq is applied to the program. We apply the 'size' command to the generated executable file, which yields the code size corresponding to the optimization sequence.

$$Fitness_t(seq) = execution\_time(\text{-O3})/execution\_time(seq) \qquad (3)$$

where $execution\_time(seq)$ represents the execution time of the executable file. Similarly, we use the 'time' command to get the execution time while the executable file is running.

3.3 Surrogate-Assisted Multi-objective Optimization Algorithm

We develop a novel surrogate-assisted multi-objective optimization algorithm to address the challenges of diverse objectives influence and the computationally expensive problem mentioned in Sect. 1. The pseudo-code of the proposed approach is presented in Algorithm 1.

First of all, the approach generates some parameters and functions, such as population size p_{num}, crossover rate p_α, mutation rate p_β, optimization sequence seq, program P, and the fitness functions for execution time $Fitness_t(seq)$ and code size $Fitness_s(seq)$. A set of solutions in $population$ is randomly initialized before the iteration begins (line 2). These solutions are then evaluated using the actual expensive fitness functions (i.e., $Fitness_t(seq)$ and $Fitness_s(seq)$). The non-dominated solutions are archived in $Archive$ (lines 3–5). Meanwhile, two surrogate models for execution time and code size are constructed in line 6 to predict the fitness values of optimization sequences.

In the main loop (lines 7–22), we apply crossover, mutation, and selection to generate a new population in the evolutionary process. Concretely, the crossover operator and mutation operator are utilized to produce offspring (lines 8–9). Here, we use the traditional single-point crossover [10] to generate child solutions and bit flip mutation to mutate the solution. Then, using two surrogate models, $Surrogate_t()$ and $Surrogate_s()$, we evaluate all solutions in the offspring and obtain their approximate execution time and code size (line 10).

Afterward, the new population is reproduced by combining the parent and offspring populations using $nondominated\_sorting(population)$ and $selection$ $(population, p_{num})$ (lines 11–13). Before applying the selection mechanism, a fast non-dominated sorting procedure is used in $population$ to divide the population into non-dominated fronts $F = \{F1, F2, ..., Fn\}$. The non-dominated solutions in $population$ belong to the front $F1$. The solutions in $F1$ are then discounted temporarily, and the non-dominated solutions in the remaining population form the next front $F2$. Repeating the above procedure until all solutions in $population$ are assigned to a front. During the selection process, solutions are chosen from the front $F1$ to Fn, which can help maintain the elitism while also generating good solutions. Furthermore, for solutions along the same front, crowding distance is used to select suitable solutions until the number of $population$ reaches p_{num}.

Following that, all solutions in the new population that are offspring are reevaluated using $Fitness_t(seq)$ and $Fitness_s(seq)$ to get their actual fitness values (lines 14–18). In lines 19–20, the new population is sorted again, and the non-dominated solutions are saved in $Archive$. Next, two surrogate models, $Surrogate_t()$ and $Surrogate_s()$ are updated until they achieve a high level of accuracy. When the program reaches the stopping criterion, the evolutionary process is terminated.

Finally, the achieved solutions in *Archive* are regarded as Pareto optimal solutions.

Algorithm 1: Surrogate-assisted Multi-objective optimization algorithm

---

Input: population size p_{num}, crossover rate p_α, mutation rate p_β, optimization sequence seq, program P, the fitness function of the objective execution time $Fitness\_t(seq)$, the fitness function of the objective code size $Fitness\_s(seq)$
Output: pareto optimal solutions *Archive*

1 **begin**
2       $population \longleftarrow Initialize(p_{num})$
3   Evaluate *population* using $Fitness_t(seq)$ and $Fitness_s(seq)$
4       $population \longleftarrow nondominated\_sorting(population)$
5       $Archive \longleftarrow$ the_non-dominated_solutions in *population*
6   Construct two surrogate models for each objective, $Surrogate_t()$ and $Surrogate_s()$
     // Generating the next population in the evolutionary process
7   **while** *Stopping criterion is not met* **do**
8           $offspring \longleftarrow crossover(population, p_\alpha)$
9           $offspring \longleftarrow mutation(offspring, p_\beta)$
10   Evaluate *offspring* using two surrogate models $Surrogate_t()$ and $Surrogate_s()$
11          $population \longleftarrow population \cup offspring$
12          $population \longleftarrow no - ndominated\_sorting(population)$
13          $population \longleftarrow selection(population, p_{num})$
14   **for** $seq_i \in population$ **do**
15   **if** seq_i *in offspring* **then**
16   Evaluate seq_i using $Fitness_t(seq)$ and $Fitness_s(seq)$
17   **end**
18   **end**
19          $population \longleftarrow nondominated\_sorting(population)$
20          $Archive \longleftarrow$ the non-dominated solutions in *population*
21   Update two surrogate models $Surrogate_t()$ and $Surrogate_s()$
22   **end**
23   **return** *pareto optimal solutions Archive*
24 **end**

3.4 Surrogate Model

Random Forest. The random forest [6] is an ensemble learning algorithm based on the bootstrap sampling technique that consists of several decision trees to solve classification or regression tasks. It is one of the most commonly used methods in numerous studies as the surrogate model [7,13,19]. For classification or regression tasks, the random forest constructs a multitude of decision trees and uses the result of most trees selected or the mean value of all trees returned to make the prediction. Therefore, the random forest outperforms any individual tree in terms of performance.

In this study, we also utilize the random forest as a surrogate model. Because we are concentrating on the compiler optimization options, and their values are Boolean values. Besides, the current study has demonstrated that the random forest as a surrogate model is very suitable for approximating such problems with discrete decision variables [13]. Additionally, the random forest has advantages in dealing with high-dimensional problems and can avoid over-fitting problems [12, 20]. Since over one hundred optimization options are considered in optimization sequence selection, we use the random forest in our work to approximate two objectives: the execution time and code size.

Surrogate Model Managing. As shown in Algorithm 1, two random forest models, $Surrogate_t()$ and $Surrogate_s()$, are constructed to approximate the execution time and code size rather than using expensive fitness evaluations. In the iterative procedure, we evaluate the solutions in offspring using two random forest models to obtain the predicted values of the two objectives. It should be noted that because the fitness values of all offspring solutions are approximated using surrogate models, the approximation error and prediction accuracy of surrogate models should be considered. We estimate two random forest models using the root mean square error (RMSE) and record $RMSE = \{R_1, R_2\}$. It is the square root of the difference between the predicted and actual value. The formula for calculation is shown below:

$$R_i = \sqrt{\frac{1}{N} \sum_{j=1}^{N} (\widehat{y_i^j} - y_i^j)^2} \tag{4}$$

where R_i is the root mean square error of the ith objective ($i = 1$ or 2), y_i^j represents the actual value of the ith objective of the jth solution, $\widehat{y_i^j}$ represents the predicted value of the ith objective of the jth solution obtained by the surrogate model, and N is the number of solutions in the current population. The lower the RSME, the better the predictive accuracy of the surrogate model.

Based on the mechanism of new population production, some solutions are introduced into the newly reproduced population by merging the parent population and the offspring, and then their actual fitness values are obtained using the actual compilation. These new reevaluated solutions with actual fitness values are employed to update two surrogate models until the RMSE threshold is satisfied. Hench, we train the surrogate models using the solutions in the initial population and these reevaluated solutions using fitness functions during iteration.

4 Experimental Results

4.1 Experimental Setup

In this section, we empirically evaluate the performance of our proposed approach and its ability to select promising compiler optimization sequences. Here, we investigate and answer the following Research Questions (RQs):

> **RQ1: How does our approach stack up against the baseline?**
> **RQ2: How do parameters impact the outcome of our approach?**
> **RQ3: How do surrogate models help our approach be more efficient?**

For the above three RQs, RQ1 investigates the effectiveness of our approach and the quality of the selected compiler optimization sequences. RQ2 intends to analyze the impact of the choice of the crossover and mutation rate to set competitive parameters. RQ3 seeks to determine whether the surrogate models improve the efficiency of our approach.

Table 1. The programs in cBench benchmark suite

No.	Program	No.	Program	No.	Program
1	automotive_bitcount	12	consumer_tiff2bw	23	security_blowfish_e
2	automotive_qsort1	13	consumer_tiff2rgba	24	security_pgp_d
3	automotive_susan_c	14	consumer_tiffdither	25	security_pgp_e
4	automotive_susan_e	15	consumer_tiffmedian	26	security_rijndael_d
5	automotive_susan_s	16	network_dijkstra	27	security_rijndael_e
6	bzip2d	17	network_patricia	28	security_sha
7	bzip2e	18	office_ispell	29	telecom_adpcm_c
8	consumer_jpeg_c	19	office_ghostscript	30	telecom_adpcm_d
9	consumer_jpeg_d	20	office_rsynth	31	telecom_CRC32
10	consumer_lame	21	office_stringsearch1	32	telecom_gsm
11	consumer_mad	22	security_blowfish_d		

In this study, we conduct experiments on the benchmark suite cBench with the compiler GCC 9.4.0. cBench, which is commonly used in compiler auto-tuning involving various programs such as embedding functions and desktop programs. The 32 programs in cBench are listed in Table 1. Similarly, the employment of GCC is due to its popularity and provision of plenty of optimization options. The complete list of 107 optimization options is available at the GCC official website[1]. The experiments are then run 15 times with different random seeds to reduce the impact of their stochastic nature and produce reasonable results. In addition, all experiments are carried out on a machine equipped with an Intel Core i9 2.8 GHz CPU and 32 GB of memory running Ubuntu 20.04.

4.2   Experimental Results

Performance Metric. We use the popular hypervolume (HV) [22] to assess the quality of the Pareto front in the comparisons between our approach and the baseline. This metric measures the volume of the objective space between the Pareto front and the reference point. Before computing HV, we use the min-max normalization to normalize the objective values and choose the point [1,1] as the reference point to maintain the accuracy of HV. Due to the aim of maximizing both objectives, a lower HV value indicates higher quality.

Investigation of RQ1. To investigate whether our approach can effectively generate suitable optimization sequences in RQ1. We adopt the HV values to compare the Pareto set explored by our approach with the baseline NSGA-II. Besides, the Wilcoxon rank sum test with a significance level of 0.05 is used to assess the statistical significance of the difference between the two algorithms.

---

[1] https://gcc.gnu.org/onlinedocs/gcc-9.4.0/gcc/Optimize-Options.html#Optimize-Options.

Table 2. The statistical results of our approach and NSGA-II on cBench

No.	HV(NSGA-II)	HV(Our)	p-value	No.	HV(NSGA-II)	HV(Our)	p-value
1	0.5817	0.5148	1.16E−02	17	0.9092	0.8214	1.69E−05
2	0.4578	0.3994	4.89E−01	18	0.6261	0.5385	1.12E−04
3	0.1246	0.1075	3.48E−03	19	0.5737	0.4764	3.97E−04
4	0.2266	0.2042	1.58E−01	20	0.7986	0.7384	5.12E−05
5	0.2129	0.1898	2.18E−03	21	0.0341	0.0300	3.55E−04
6	0.5088	0.4699	7.48E−05	22	0.4466	0.3877	6.28E−03
7	0.4928	0.4413	4.57E−02	23	0.4358	0.3868	7.91E−06
8	0.5613	0.4736	1.91E−04	24	0.6511	0.5897	4.63E−06
9	0.3729	0.3384	7.36E−03	25	0.5632	0.5224	1.14E−04
10	0.6700	0.5732	7.16E−05	26	0.8474	0.7556	1.77E−01
11	0.5790	0.5301	3.21E−01	27	0.9125	0.8421	8.65E−06
12	0.8182	0.7040	6.66E−05	28	0.2798	0.2408	3.61E−05
13	0.7371	0.6694	7.85E−05	29	0.4991	0.4476	8.91E−06
14	0.6443	0.5462	2.01E−05	30	0.5113	0.4335	3.68E−05
15	0.5396	0.4827	2.50E−03	31	0.6537	0.5449	1.34E−05
16	0.0654	0.0554	3.17E−03	32	0.7065	0.5919	2.19E−05
Average					0.5325	0.4702	

In our experiments, the population size p_{num} is set to 100. Then, the crossover rate and mutation rate are set to 0.9 and 0.01, respectively, as the recommended parameter settings in RQ2 (the impact of crossover and mutation rate will be investigated later). Table 2 shows the HV values and p-values obtained using our approach and NSGA-II. The first and fifth columns represent the numerical order of the programs. The second and sixth columns are the HV values of NSGA-II. The HV values of our approach are then listed in columns three and seven. The p-values are shown in columns four and eight. Finally, the average HV values of the two algorithms are provided at the bottom of the table.

It can be seen from the table, our approach can explore a better Pareto set and achieve better performance in the majority of programs. On average, our approach achieves 0.4702, while NSGA-II achieves 0.5325. When compared to NSGA-II, our approach improves by 11.7% and performs reasonably well on HV values. There are two major reasons for this result. One is the utilization of random forest, as a surrogate model, which is appropriate for high-dimensional discrete problems and reduces the actual solution evaluations. Another is that the surrogate model management mechanism adds potential solutions to the training set to update the surrogate model, thus improving the search procedure and quality. Besides, as shown in Table 2, the p-values on 28 programs are less than 0.05, implying that there is a significant difference between our approach and NSGA-II for the majority of programs (28/32). In summary, these results reveal that our approach outperforms NSGA-II in terms of finding optimization sequences.

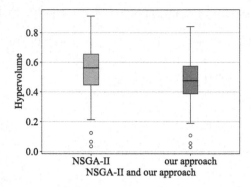

Fig. 2. The boxplot of HV values for two algorithms

In addition, to visually compare the performance of the two algorithms, Fig. 2 plots the HV values of the two algorithms in a box plot. A similar conclusion can be drawn from Fig. 2 that our approach performs better than NSGA-II (the box of our approach is lower, which indicates better results).

Answer to RQ1. By comparing our approach to NSGA-II, we demonstrate that our approach outperforms NSGA-II, which can achieve a 11.7% improvement in HV values. As a result, our approach can effectively explore the Pareto set and select better optimization sequences.

Investigation of RQ2. In this RQ, we attempt to investigate the impact of two parameters: crossover rate and mutation rate. A feasible set of parameters may result in preferable results with improved performance. To examine the impact of the crossover rate, we change its values while keeping the mutation rate fixed values, and vice-versa. Besides, the two parameters with varying probabilities are as follows: crossover rate = {0.5, 0.7, 0.9}, mutation rate = {0.01, 0.05, 0.1}. Similar to RQ1, the experiments are carried out on cBench and GCC.

Figure 3 depicts the impact results of these two parameters. We make the following three observations based on Fig. 3(a) and Fig. 3(b). First, our approach converges faster as the crossover rate increases. Second, we find that the convergence of our approach becomes slower when the mutation rate ranges from 0.01 to 0.1. As expected, the higher the crossover rate and the lower the mutation rate, the faster convergence occurs. Third, in terms of convergence rate, the experimental results of different parameter settings show no noticeable difference in crossover and mutation rates.

Answer to RQ2. The findings of the parameters impact analysis show that our approach is not very sensitive to the crossover and mutation rates that are set. In our experiment, we set the crossover rate to 0.9, and the mutation rate to 0.01.

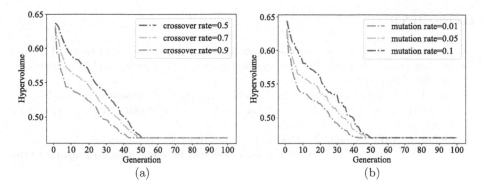

Fig. 3. The impact of crossover rate and mutation rate

Investigation of RQ3. To gain a better understanding of our approach, we investigate whether the two surrogate models can effectively improve the efficiency of our approach in RQ3. The number of actual fitness evaluations is applied as in [18] to investigate the value of surrogate models. By comparing the number of actual fitness evaluations required using our approach and a variant without surrogate models to reach convergence and stop the iteration, the results show that our approach converges when the number of actual fitness evaluations is 4400, while the variant requires 5600, i.e., more fitness evaluations. Besides, based on the conclusion of RQ1, we can conclude that our approach performs better and can find more suitable compiler optimization sequences while requiring fewer actual fitness evaluations.

Answer to RQ3. Taking into account the actual fitness evaluation of our approach and a variant without surrogate models, we conclude that the two surrogate models aid in improving the efficiency of our approach and make a contribution to solving the computationally expensive problem.

5   Conclusion and Future Work

In this paper, we present a novel surrogate-assisted multi-objective optimization algorithm to improve the efficiency of evolutionary algorithms for compiler optimization sequence selection. To address the diverse objectives influence and the computationally expensive problem, the proposed approach combines a fast global search based on non-dominated sorting with two random forests as surrogate models. The experimental results on cBench show that our proposed approach achieves better performance than NSGA-II and requires fewer actual fitness evaluations.

Despite the promising results, this work is still preliminary. Because of the complexity and variety of programs, different appropriate surrogate models will be designed for different programs in the future to improve efficiency. In addition, the dimensionality reduction method will be considered on account of the high-dimensional problem with a large number of compiler optimization options.

References

1. Agakov, F., et al.: Using machine learning to focus iterative optimization. In: Proceedings of the International Symposium on Code Generation and Optimization, pp. 295–305 (2006)
2. Ansel, J., et al.: OpenTuner: an extensible framework for program autotuning. In: Proceedings of the 23rd International Conference on Parallel Architectures and Compilation, pp. 303–316 (2014)
3. Ashouri, A.H., Bignoli, A., Palermo, G., Silvano, C., Kulkarni, S., Cavazos, J.: MICOMP: mitigating the compiler phase-ordering problem using optimization subsequences and machine learning. ACM Trans. Archit. Code Optim. **14**(3), 29 (2017)
4. Ashouri, A.H., Killian, W., Cavazos, J., Palermo, G., Silvano, C.: A survey on compiler autotuning using machine learning. ACM Comput. Surv. **51**(5), 1–42 (2018)
5. Ashouri, A.H., Mariani, G., Palermo, G., Park, E., Cavazos, J., Silvano, C.: COBAYN: compiler autotuning framework using Bayesian networks. ACM Trans. Archit. Code Optim. (TACO) **13**(2), 1–25 (2016)
6. Breiman, L.: Random forests. Mach. Learn. **45**(1), 5–32 (2001)
7. Cáceres, L.P., Bischl, B., Stützle, T.: Evaluating random forest models for irace. In: Proceedings of the Genetic and Evolutionary Computation Conference Companion, pp. 1146–1153 (2017)
8. Chebolu, N.A.B.S., Wankar, R.: Multi-objective exploration for compiler optimizations and parameters. In: Murty, M.N., He, X., Chillarige, R.R., Weng, P. (eds.) MIWAI 2014. LNCS (LNAI), vol. 8875, pp. 23–34. Springer, Cham (2014). https://doi.org/10.1007/978-3-319-13365-2_3
9. Chen, J., Xu, N., Chen, P., Zhang, H.: Efficient compiler autotuning via Bayesian optimization. In: 2021 IEEE/ACM 43rd International Conference on Software Engineering (ICSE), pp. 1198–1209. IEEE (2021)
10. Deb, K., Agrawal, R.B.: Simulated binary crossover for continuous search space. Complex Syst. **9**(2), 115–148 (1995)
11. Fursin, G.: Collective benchmark (cBench), a collection of open-source programs with multiple datasets assembled by the community to enable realistic benchmarking and research on program and architecture optimization (2010). http://cTuning.org/cbench
12. Gu, Q., Wang, D., Jiang, S., Xiong, N., Jin, Y.: An improved assisted evolutionary algorithm for data-driven mixed integer optimization based on Two_Arch. Comput. Ind. Eng. **159**, 107463 (2021)
13. Gu, Q., Wang, Q., Li, X., Li, X.: A surrogate-assisted multi-objective particle swarm optimization of expensive constrained combinatorial optimization problems. Knowl.-Based Syst. **223**, 107049 (2021)
14. Hall, M., Padua, D., Pingali, K.: Compiler research: the next 50 years. Commun. ACM **52**(2), 60–67 (2009)
15. Hong, W., Yang, P., Wang, Y., Tang, K.: Multi-objective magnitude-based pruning for latency-aware deep neural network compression. In: Bäck, T., et al. (eds.) PPSN 2020. LNCS, vol. 12269, pp. 470–483. Springer, Cham (2020). https://doi.org/10.1007/978-3-030-58112-1_32
16. Lokuciejewski, P., Plazar, S., Falk, H., Marwedel, P., Thiele, L.: Multi-objective exploration of compiler optimizations for real-time systems. In: 2010 13th IEEE International Symposium on Object/Component/Service-Oriented Real-Time Distributed Computing, pp. 115–122. IEEE (2010)

17. Lokuciejewski, P., Plazar, S., Falk, H., Marwedel, P., Thiele, L.: Approximating pareto optimal compiler optimization sequences-a trade-off between WCET, ACET and code size. Softw. Pract. Experience **41**(12), 1437–1458 (2011)
18. Sun, C., Ding, J., Zeng, J., Jin, Y.: A fitness approximation assisted competitive swarm optimizer for large scale expensive optimization problems. Memetic Comput. **10**(2), 123–134 (2018)
19. Sun, Y., Wang, H., Xue, B., Jin, Y., Yen, G.G., Zhang, M.: Surrogate-assisted evolutionary deep learning using an end-to-end random forest-based performance predictor. IEEE Trans. Evol. Comput. **24**(2), 350–364 (2019)
20. Valdiviezo, H.C., Van Aelst, S.: Tree-based prediction on incomplete data using imputation or surrogate decisions. Inf. Sci. **311**, 163–181 (2015)
21. Zhou, Y.Q., Lin, N.W.: A study on optimizing execution time and code size in iterative compilation. In: 2012 Third International Conference on Innovations in Bio-Inspired Computing and Applications, pp. 104–109. IEEE (2012)
22. Zitzler, E., Thiele, L.: Multiobjective optimization using evolutionary algorithms—a comparative case study. In: Eiben, A.E., Bäck, T., Schoenauer, M., Schwefel, H.-P. (eds.) PPSN 1998. LNCS, vol. 1498, pp. 292–301. Springer, Heidelberg (1998). https://doi.org/10.1007/BFb0056872

Theoretical Aspects of Nature-Inspired Optimization

A First Runtime Analysis of the NSGA-II on a Multimodal Problem

Benjamin Doerr and Zhongdi Qu[✉]

Laboratoire d'Informatique (LIX), Ecole Polytechnique, CNRS, Institut
Polytechnique de Paris, Palaiseau, France
doerr@lix.polytechnique.fr, d.q.1124@gmail.com

Abstract. Very recently, the first mathematical runtime analyses of the multi-objective evolutionary optimizer NSGA-II have been conducted. We continue this line of research with a first runtime analysis of this algorithm on a benchmark problem consisting of two multimodal objectives. We prove that if the population size N is at least four times the size of the Pareto front, then the NSGA-II with four different ways to select parents and bit-wise mutation optimizes the OneJumpZeroJump benchmark with jump size $2 \leq k \leq n/4$ in time $O(Nn^k)$. When using fast mutation, a recently proposed heavy-tailed mutation operator, this guarantee improves by a factor of $k^{\Omega(k)}$. Overall, this work shows that the NSGA-II copes with the local optima of the OneJumpZeroJump problem at least as well as the global SEMO algorithm.

Keywords: NSGA-II · Multimodal problem · Runtime analysis

1 Introduction

The mathematical runtime analysis of evolutionary algorithms (EAs) has contributed significantly to our understanding of these algorithms, given advice on how to set their parameters, and even proposed new algorithms [6,14,20,23]. Most of the insights, however, have been obtained by regarding artificially simple algorithms such as the $(1 + 1)$ EA, the fruit fly of EA research.

In contrast, the recent work [31] succeeded in analyzing the *non-dominated sorting genetic algorithm II (NSGA-II)* [12], the multi-objective EA (MOEA) most used in practice [32]. This line of research was almost immediately followed up in [7] and [30]. These three works, just like the majority of the theoretical works on MOEAs, only regard multi-objective problems composed of unimodal objectives (see Sect. 2 for more details).

In this work, we continue the runtime analysis of the NSGA-II with a first analysis on a problem composed of two multi-modal objectives, namely the ONEJUMPZEROJUMP problem proposed in [17]. This problem, defined on bit strings of length n, is a natural multi-objective analogue of the single-objective JUMP problem, which might be the multimodal problem most studied

G. Rudolph et al. (Eds.): PPSN 2022, LNCS 13399, pp. 399–412, 2022.
https://doi.org/10.1007/978-3-031-14721-0_28

in single-objective runtime analysis. The JUMP problem (and the two objectives of the ONEJUMPZEROJUMP problem) come with a difficulty parameter $k \in [1..n] := \{1, \ldots, n\}$, which is the width of the valley of low fitness around the global optimum. Consequently, typical hillclimbers at some point need to flip the right k bits, which is difficult already for moderate sizes of k. For the multi-objective ONEJUMPZEROJUMP problem the situation is similar. Here the Pareto front is not a connected set in the search space $\{0,1\}^n$, but there are solutions which can only be reached from other points on the Pareto front by flipping k bits, which creates a challenge similar to the single-objective case.

Our Results: We conduct a mathematical runtime analysis of the NSGA-II algorithm on the ONEJUMPZEROJUMP problem with jump sizes $k \in [2..\frac{1}{4}n]$. We allow that k is functionally dependent on n and let all asymptotic notation be with respect to n. Since the runtimes we observe are at least exponential in k, the restriction of $k \le \frac{1}{4}n$, done mostly to avoid some not very interesting technicalities, is not a harsh restriction. As *runtime*, we consider the number of fitness evaluations until the full Pareto front (that is, at least one individual for each Pareto-optimal objective value) is contained in the parent population of the NSGA-II. As in [31], we assume that the population size N of the NSGA-II is sufficiently large, here at least four times the size of the Pareto front (since a population size equal to the Pareto front size does not suffice to find the Pareto front even of the simple ONEMINMAX problem [31], this assumption appears justified). We regard the NSGA-II with four different ways to select the parents (each individual once ("fair selection"), uniform, N independent binary tournaments, and N binary tournaments from two random permutations of the population ("two-permutation tournament scheme")), with bit-wise mutation with mutation rate $\frac{1}{n}$, and, for the theoretical analyses, without crossover. We prove that this algorithm on the ONEJUMPZEROJUMP problem with jump size k has an expected runtime of at most $(1 + o(1))KNn^k$, where K is a small constant depending on the selection method. Hence for $N = \Theta(n)$, the NSGA-II satisfies the same asymptotic runtime guarantee of $O(n^{k+1})$ as the (mostly relevant in theory) algorithm global SEMO (GSEMO), for which a runtime guarantee of $(1 + o(1))1.5e(n - 2k + 3)n^k$ was shown in [17].

Since it has been observed many times that a heavy-tailed mutation operator called *fast mutation* can significantly speed up leaving local optima [1–3,5,9,13, 16–19,25,29], we also regard the NSGA-II with this mutation operator. Similar to previous works, we manage to show a runtime guarantee which is lower by a factor of $k^{\Omega(k)}$ (see Theorem 3 for a precise statement of this result). This result suggests that the NSGA-II, similar to many other algorithms, profits from fast mutation when local optima need to be left.

2   Previous Works

For reasons of space, we only briefly mention the most relevant previous works. For a more detailed account of the literature, we refer to these works or the survey [8].

The first mathematical runtime analysis of the NSGA-II [31] showed that this algorithm can efficiently find the Pareto front of the ONEMINMAX and LOTZ bi-objective problems when the population size N is at least some constant factor larger than the size of the Pareto front (which is $n+1$ for these problems). In this case, once an objective value of the Pareto front is covered by the population, it remains so for the remaining run of the algorithm. This is different when the population size is only equal to the size of the Pareto front. Then such values can be lost, and this effect is strong enough that for an exponential number of iterations a constant fraction of the Pareto front is not covered [31]. Nevertheless, also in this case the NSGA-II computes good approximations of the Pareto front as the first experiments in [31] and a deeper analysis in [30] show.

The most recent work [7] extends [31] in several directions. (i) For the NSGA-II using crossover, runtime guarantees for the ONEMINMAX, COCZ, and LOTZ problems are shown which agree with those in [31]. (ii) By assuming that individuals with identical objective value appear in the same or inverse order in the sortings used to compute the crowding distance, the minimum required population size is lowered to $2(n+1)$. (iii) A stochastic tournament selection is proposed that reduces the runtimes by a factor of $\Theta(n)$ on LOTZ and $\Theta(\log n)$ on the other two benchmarks.

The ONEMINMAX, COCZ, and LOTZ benchmarks are all composed of two unimodal objectives, namely functions isomorphic to the benchmarks ONEMAX and LEADINGONES from single-objective EA theory. The theory of MOEA has strongly focused on such benchmarks, a benchmark composed of multimodal objectives was only proposed and analyzed in [17].

Besides the definition of the ONEJUMPZEROJUMP problem, the main results in that work are that the SEMO algorithm cannot optimize this benchmark, that the GSEMO takes time $O((n - 2k + 3)n^k)$ (where the implicit constants can be chosen independent of n and k), and that the GSEMO with fast mutation with power-law exponent $\beta > 1$ succeeds in time $O((n-2k+3)k^{-k+\beta-0.5}n^k(\frac{n}{n-k})^{n-k})$ (where the implicit constant can be chosen depending on β only). A slightly weaker, but still much better bound than for bit-wise mutation was shown for the GSEMO with the stagnation detection mechanism of Rajabi and Witt [26] (we omit the details for reasons of space).

3 Preliminaries

3.1 The NSGA-II Algorithm

We only give a brief overview of the algorithm here due to space constraints, and refer to [12] for a more detailed description of the general algorithm and to [31] for more details on the particular version of the NSGA-II we regard.

The algorithm starts with a random initialization of a parent population of size N. At each iteration, N children are generated from the parent population via a mutation method, and N individuals among the combined parent and children population survive to the next generation based on their ranks in the non-dominated sorting and, as tie-breaker, the crowding distance.

Ranks are determined recursively. All individuals that are not strictly dominated by any other individual have rank 1. Given that individuals of rank $1, \ldots, i$ are defined, individuals of rank $i + 1$ are those only strictly dominated by individuals of rank i or smaller. Clearly, individuals of lower ranks are preferred.

The crowding distance, denoted by $\text{cDis}(x)$ for an individual x, is used to compare individuals of the same rank. To compute the crowding distances of individuals of rank i with respect to a given objective function f_j, we first sort the individuals in ascending order according to their f_j objective values. The first and last individuals in the sorted list have infinite crowding distance. For the other individuals, their crowding distance is the difference between the objective values of its left and right neighbors in the sorted list, normalized by the difference of the minimum and maximum values. The final crowding distance of an individual is the sum of its crowding distances with respect to each objective function.

At each iteration, the critical rank i^* is the rank such that if we take all individuals of ranks smaller than i^*, the total number of individuals will be less than or equal to N, but if we also take all individuals of rank i^*, the total number of individuals will be over N. Thus, all individuals of rank smaller than i^* survive to the next generation, and for individuals of rank i^*, we take the individuals with the highest crowding distance, breaking ties randomly, so that in total exactly N individuals are kept.

3.2   The ONEJUMPZEROJUMP Benchmark

Let $n \in \mathbb{N}$ and $k = [2..n/4]$. The bi-objective function $\text{ONEJUMPZEROJUMP}_{n,k} = (f_1, f_2) : \{0,1\}^n \to \mathbb{R}^2$ is defined by

$$f_1(x) = \begin{cases} k + |x|_1, & \text{if } |x|_1 \leq n - k \text{ or } x = 1^n, \\ n - |x|_1, & \text{else;} \end{cases}$$

$$f_2(x) = \begin{cases} k + |x|_0, & \text{if } |x|_0 \leq n - k \text{ or } x = 0^n, \\ n - |x|_0, & \text{else.} \end{cases}$$

The aim is to maximize both f_1 and f_2. The first objective is the classical $\text{JUMP}_{n,k}$ function. It has a valley of low fitness around its optimum, which can be crossed only by flipping the k correct bits, if no solutions of lower fitness are accepted. The second objective is isomorphic to the first, with the roles of zeroes and ones exchanged. According to Theorem 2 of [17], the Pareto set of the $\text{ONEJUMPZEROJUMP}_{n,k}$ function is $S^* = \{x \in \{0,1\}^n \mid |x|_1 = [k..n-k] \cup \{0, n\}\}$, and the Pareto front F^* is $\{(a, 2k + n - a) \mid a \in [2k..n] \cup \{k, n+k\}\}$, making the size of the front is $n - 2k + 3$. We define the inner part of the Pareto set by $S_I^* = \{x \mid |x|_1 \in [k..n-k]\}$, the outer part by $S_O^* = \{x \mid |x|_1 \in \{0, n\}\}$, the inner part of the Pareto front by $F_I^* = f(S_I^*) = \{(a, 2k + n - a) \mid a \in [2k..n]\}$, and the outer part by $F_O^* = f(S_O^*) = \{(a, 2k + n - a) \mid a \in \{k, n+k\}\}$.

4 Runtime Analysis for the NSGA-II

In this section, we prove our runtime guarantees for the NSGA-II, first with bit-wise mutation with mutation rate $\frac{1}{n}$ (Subsect. 4.1), then with fast mutation (Subsect. 4.2). For reasons of space, most mathematical proofs had to be omitted from this extended abstract. They can be found in the preprint [15].

The obvious difference to the analysis for ONEMINMAX in [31] is that with ONEJUMPZEROJUMP, individuals with between one and $k - 1$ zeroes or ones are not optimal. Moreover, all these individuals have a very low fitness in both objectives. Consequently, such individuals usually will not survive into the next generation, which means that the NSGA-II at some point will have to generate the all-ones string from a solution with at least k zeroes (unless we are extremely lucky in the initialization of the population). This difference is the reason for the larger runtimes and the advantage of the fast mutation operator.

A second, smaller difference which however cannot be ignored in the mathematical proofs is that the very early populations of a run of the algorithm may contain zero individual on the Pareto front. This problem had to be solved also in the analysis of LOTZ in [31], but the solution developed there required that the population size is at least 5 times the size of the Pareto front (when tournament selection was used). For ONEJUMPZEROJUMP, we found a different argument to cope with this situation that, as all the rest of the proof, only requires a population size of at least 4 times the Pareto front size.

We start with a few general observations that apply to both cases. A crucial observation, analogous to a similar statement in [31], is that with sufficient population size, objective values of rank-1 individuals always survive to the next generation.

Lemma 1. *Consider one iteration of the NSGA-II algorithm optimizing the* ONEJUMPZEROJUMP$_{n,k}$ *benchmark, with population size $N \geq 4(n - 2k + 3)$. If in some iteration t the combined parent and offspring population R_t contains an individual x of rank 1, then the next parent population P_{t+1} contains an individual y such that $f(y) = f(x)$. Moreover, if an objective value on the Pareto front appears in R_t, it will be kept in all future iterations.*

Proof. Let F_1 be the set of rank-1 individuals in R_t. To prove the first claim, we need to show that for each $x \in F_1$, there is a $y \in P_{t+1}$ such that $f(x) = f(y)$. Let $S_{1.1}, \ldots, S_{1.|F_1|}$ be the list of individuals in F_1 sorted by ascending f_1 values and $S_{2.1}, \ldots, S_{2.|F_1|}$ be the list of individuals sorted by ascending f_2 values, which were used to compute the crowding distances. Then there exist $a \leq b$ and $a' \leq b'$ such that $[a..b] = \{i \mid f_1(S_{1.i}) = f_1(x)\}$ and $[a'..b'] = \{i \mid f_2(S_{2.i}) = f_2(x)\}$. If any one of $a = 1$, $a' = 1$, $b = |F_1|$, or $b' = |F_1|$ is true, then there is an individual $y \in F_1$ satisfying $f(y) = f(x)$ of infinite crowding distance. Since there are at most $4 < N$ individuals of infinite crowding distance, y is kept in P_{t+1}. So consider the case that $a, a' > 1$ and $b, b' < |F_1|$. By the definition of the crowding distance, we have that $\text{cDis}(S_{1.a}) \geq \frac{f_1(S_{1.a+1}) - f_1(S_{1.a-1})}{f_1(S_{1.|F_1|}) - f_1(S_{1.1})} \geq \frac{f_1(S_{1.a}) - f_1(S_{1.a-1})}{f_1(S_{1.|F_1|}) - f_1(S_{1.1})}$. Since $f_1(S_{1.a}) - f_1(S_{1.a-1}) > 0$ by the definition of a, we have $\text{cDis}(S_{1.a}) > 0$.

Similarly, we have $\text{cDis}(S_{1.a'}), \text{cDis}(S_{1.b}), \text{cDis}(S_{1.b'}) > 0$. For $i \in [a + 1..b - 1]$ and $S_{1.i} = S_{2.j}$ for some $j \in [a' + 1..b' - 1]$, we have that $f_1(S_{1.i-1}) = f_1(x) = f_1(S_{1.i+1})$ and $f_2(S_{2.j-1}) = f_2(x) = f_2(S_{2.j+1})$. So $\text{cDis}(S_{1.i}) = 0$. Therefore, for each $f(x)$ value, there are at most 4 individuals with the same objective value and positive crowding distances. By Corollary 6 in [17], $|F_1| \leq n - 2k + 3$. So the number of rank-1 individuals with positive crowding distances is at most $4(n - 2k + 3) \leq N$ and therefore they will all be kept in P_{t+1}.

The second claim then follows since if $x \in R_t$ and $f(x)$ is on the Pareto front, we have $x \in F_1$. By the first claim, $x \in P_{t+1}$ and therefore $x \in R_{t+1}$. The same reasoning applies for all future iterations. $\square$

For our analysis, we divide a run of the NSGA-II algorithm optimizing the ONEJUMPZEROJUMP$_{n,k}$ benchmark into the following stages.

- Stage 1: $P_t \cap S_I^* = \emptyset$. In this stage, the algorithm tries to find the first individual with objective value in F_I^*.
- Stage 2: There exists a $v \in F_I^*$ such that $v \notin f(P_t)$. In this stage, the algorithm tries to cover the entire set F_I^*.
- Stage 3: $F_I^* \subseteq f(P_t)$, but $F_O^* \nsubseteq f(P_t)$. In this stage, the algorithm tries to find the extremal values of the Pareto front.

By Lemma 1, once the algorithm has entered a later stage, it will not go back to an earlier stage. Thus, we can estimate the expected number of iterations needed by the NSGA-II algorithm by separately analyzing each stage.

A mutation method studied in [31] is to flip one bit selected uniformly at random. For reasons of completeness, we prove in the following lemma the natural result that the NSGA-II with this mutation operator with high probability is not able to cover the full Pareto front of the ONEJUMPZEROJUMP$_{n,k}$ benchmark.

Lemma 2. *With probability $1 - N \exp(-\Omega(n))$, the NSGA-II algorithm using one-bit flips as mutation operator does not find the full Pareto front of the* ONEJUMPZEROJUMP$_{n,k}$ *benchmark, regardless of the runtime.*

Proof. Since $k \leq n/4$, a simple Chernoff bound argument shows that a random initial individual is in S_I^* with probability $1 - \exp(-\Omega(n))$. By a union bound, we have $P_0 \subseteq S_I^*$ with probability $1 - N \exp(-\Omega(n))$. We argue that in this case, the algorithm can never find an individual in S_O^*.

We observe that any individual in S_I^* strictly dominates any individual in the gap regions of the two objectives, that is, with between 1 and $k - 1$ zeroes or ones. Consequently, in any population containing at least one individual from S_I^*, such a gap individual can never have rank 1, and the only rank 1 individuals are those on the Pareto front. Hence if P_t for some iteration t contains only individuals on the Pareto front, P_{t+1} will do so as well.

By induction and our assumption $P_0 \subseteq S_I^*$, we see that the parent population will never contain an individual with exactly one one-bit. Since only from such a parent the all-zeroes string can be generated (via one-bit mutation), we will never have the all-zeroes string in the population. $\square$

In the light of Lemma 2, the one-bit flip mutation operator is not suitable for the optimization of ONEJUMPZEROJUMP. We therefore do not consider this operator in the following runtime analyses.

4.1 Runtime Analysis for the NSGA-II Using Bit-Wise Mutation

In this section, we analyze the complexity of the NSGA-II algorithm when mutating each bit of each selected parent with probability $\frac{1}{n}$. We consider four different ways of selecting the parents for mutation: (i) fair selection (selecting each parent once), (ii) uniform selection (selecting one parent uniformly at random for N times), (iii) via N independent tournaments (for N times, uniformly at random sample 2 different parents and conduct a binary tournament between the two, i.e., select the one with the lower rank and, in case of tie, select the one with the larger crowding distance, and, in case of tie, select one randomly), and (iv) via a two-permutation tournament scheme (generate two random permutations π_1 and π_2 of P_t and conduct a binary tournament between $\pi_j(2i-1)$ and $\pi_j(2i)$ for all $i \in [1..N/2]$ and $j \in \{1,2\}$; this is the selection method used in Deb's implementation of the NSGA-II when ignoring crossover [12]).

Lemma 3. *Using population size $N \geq 4(n - 2k + 3)$, bit-wise mutation for variation, and any parent selection method, stage 1 needs in expectation at most $e(\frac{4k}{3})^k$ iterations.*

Proof. Suppose x is selected for mutation during one iteration of stage 1 and $|x|_1 = i$. Then $i < k$ or $i > n - k$. If $i < k$, then the probability of obtaining an individual with k 1-bits is at least $\binom{n-i}{k-i}(\frac{1}{n})^{k-i}(1 - \frac{1}{n})^{n-(k-i)} \geq (\frac{n-i}{n(k-i)})^{k-i}(1 - \frac{1}{n})^{n-1} > \frac{1}{e}(\frac{3}{4(k-i)})^{k-i} \geq \frac{1}{e}(\frac{3}{4k})^k$ (where the second to last inequality uses the assumption that $i < k \leq \frac{n}{4}$). If $i > n - k$, then the probability of obtaining an individual with $n - k$ 1-bits is at least $\binom{i}{i-(n-k)}(\frac{1}{n})^{i-(n-k)}(1 - \frac{1}{n})^{2n-i-k} \geq (\frac{i}{n(i-n+k)})^{i-n+k}(1 - \frac{1}{n})^{n-1} > \frac{1}{e}(\frac{3}{4(i-n+k)})^{i-n+k} \geq \frac{1}{e}(\frac{3}{4k})^k$ (where the second to last inequality uses the assumption that $i > n - k \geq \frac{3}{4n}$). Hence each iteration with probability at least $\frac{1}{e}(\frac{3}{4k})^k$ marks the end of stage 1. Consequently, stage 1 ends after in expectation at most $(\frac{1}{e}(\frac{3}{4k})^k)^{-1} = e(\frac{4k}{3})^k$ iterations. $\square$

For the remaining two stages, we first regard the technically easier fair and uniform selection methods.

Lemma 4. *Using population size $N \geq 4(n - 2k + 3)$, selecting parents using fair or uniform selection, and using bit-wise mutation for variation, stage 2 needs in expectation $O(n \log n)$ iterations.*

For reasons of space, we omit the proof, which is very similar to the corresponding part of the analysis on ONEMINMAX [31]. Different arguments, naturally, are needed in the following analysis of stage 3.

Lemma 5. *Using population size $N \geq 4(n - 2k + 3)$ and bit-wise mutation for variation, stage 3 needs in expectation at most $2en^k$ iterations if selecting parents using fair selection, and $2\frac{e^2}{e-1}n^k$ iterations if using uniform selection.*

Proof. Consider one iteration t of stage 3. We know that there is an $x \in P_t$ such that $|x|_1 = k$. Denote the probability that x is selected at least once to be mutated in this iteration by p_1. Conditioning on x being selected, denote the probability that all k 1-bits of x are flipped in this iteration by p_2. Then the probability of generating 0^n in this iteration is at least $p_1 p_2$. Since by Lemma 1, x is kept for all future generations, we need at most $\frac{1}{p_1 p_2}$ iterations to obtain 0^n. With fair selection, we have $p_1 = 1$ and with uniform selection, $p_1 = 1 - (1 - \frac{1}{N})^N \geq 1 - \frac{1}{e}$. On the other hand, $p_2 = (\frac{1}{n})^k (1 - \frac{1}{n})^{n-k} \geq \frac{1}{en^k}$. So the expected number $\frac{1}{p_1 p_2}$ of iterations to obtain 0^n is bounded by en^k if using fair selection, and by $\frac{e^2}{e-1} n^k$ if using uniform selection. The case for obtaining 1^n is symmetrical. Therefore, the expected total number of iterations needed to cover the extremal values of the Pareto front is at most $2en^k$ if using fair selection, and $2\frac{e^2}{e-1} n^k$ if using uniform selection. □

Combining the lemmas, we immediately obtain the runtime guarantee.

Theorem 1. *Using population size $N \geq 4(n - 2k + 3)$, selecting parents using fair or uniform selection, and mutating using bit-wise mutation, the NSGA-II needs in expectation at most $(1 + o(1))KNn^k$ fitness evaluations to cover the entire Pareto front of the $\textsc{OneJumpZeroJump}_{n,k}$ benchmark, where $K = 2e$ for fair selection and $K = 2\frac{e^2}{e-1}$ for uniform selection.*

In the above result, we have given explicit values for the leading constant K to show that it is not excessively large, but we have not tried to optimize this constant. In fact, it is easy to see that the 2 could be replaced by 1.5 by taking into account that the expected time to find the first extremal point is only half the time to find a particular extremal point. Since we have no non-trivial lower bounds at the moment, we find it too early to optimize the constants.

We now turn to the case where the mutating parents are chosen using one of two ways of *binary tournaments*, namely, via N independent tournaments and the two-permutation tournament scheme.

Theorem 2. *Using population size $N \geq 4(n - 2k + 3)$, selecting parents using N independent tournaments or the two-permutation tournament scheme, and mutating using bit-wise mutation, the NSGA-II takes in expectation at most $(1 + o(1))KNn^k$ fitness evaluations to cover the entire Pareto front of the $\textsc{OneJumpZeroJump}_{n,k}$ benchmark, where $K = 2\frac{e^2}{e-1}$ if using N independent tournaments, and $K = \frac{8}{3}e$ if using the two-permutation tournament scheme.*

The proof of this result follows the outline of the proof of Theorem 1, but needs some technical arguments from [31] on the probability that an individual next to an uncovered spot on the Pareto front is chosen to be mutated.

4.2   Runtime Analysis for the NSGA-II Using Fast Mutation

We now consider the NSGA-II with heavy-tailed mutation, i.e., the mutation operator proposed in [13] and denoted by $\text{MUT}^\beta(\cdot)$ in [17], a work from which we shall heavily profit in the following.

Let $\beta > 1$ be a constant (typically below 3). Let $D^\beta_{n/2}$ be the distribution such that if a random variable X follows the distribution, then $\Pr[X = \alpha] = (C^\beta_{n/2})^{-1}\alpha^{-\beta}$ for all $\alpha \in [1..n/2]$, where n is the size of the problem and $C^\beta_{n/2} := \sum_{i=1}^{n/2} i^{-\beta}$. In an application of the mutation operator $\mathrm{MUT}^\beta(\cdot)$, first an α is chosen according to the distribution $D^\beta_{n/2}$ (independent from all other random choices of the algorithm) and then each bit of the parent is flipped independently with probability α/n. Let $x \in \{0,1\}^n$, $y \sim \mathrm{MUT}^\beta(x)$, and $H(x,y)$ denote the Hamming distance between x and y. Then, by Lemma 13 of [17], we have

$$P^\beta_j := \Pr[H(x,y) = j] = \begin{cases} (C^\beta_{n/2})^{-1}\Theta(1) & \text{for } j = 1; \\ (C^\beta_{n/2})^{-1}\Omega(j^{-\beta}) & \text{for } j \in [2..n/2]. \end{cases}$$

Theorem 3. *Using population size $N \geq 4(n - 2k + 3)$, selecting parents using fair or uniform selection, and mutating with the $\mathrm{MUT}^\beta(\cdot)$ operator, the NSGA-II takes at most $(1 + o(1))\frac{1}{P^\beta_k}NK\binom{n}{k}$ fitness evaluations in expectation to cover the entire Pareto front of the* $\mathrm{ONEJUMPZEROJUMP}_{n,k}$ *benchmark, where $K = 2$ for fair selection, $K = \frac{2e}{e-1}$ for uniform selection and selection via N independent binary tournaments, and $K = \frac{8}{3}$ for the two-permutation binary tournament scheme.*

Noting that $\binom{n}{k}$ is by a factor of $k^{\Omega(k)}$ smaller than n^k, whereas $1/P^\beta_k$ is only $O(k^\beta)$, we see that the runtime guarantee for the heavy-tailed operator is by a factor of $k^{\Omega(k)}$ stronger than our guarantee for bit-wise mutation. Without a lower bound on the runtime in the bit-wise setting, we cannot claim that the heavy-tailed algorithm is truly better, but we strongly believe so (we do not see a reason why the NSGA-II with bit-wise mutation should be much faster than what our upper bound guarantees).

We note that it is easy to prove a lower bound of $\Omega(n^k)$ for the runtime of the NSGA-II with bit-wise mutation (this is a factor of N below our upper bound, which stems from pessimistically assuming that in each iteration, N times a parent is selected that has a $\Theta(n^{-k})$ chance of generating an extremal point of the Pareto front). For k larger than, say, $\log(N)$, this weak lower bound would suffice to show that the heavy-tailed NSGA-II is asymptotically faster. We spare the details and hope that at some time, we will be able to prove tight lower bounds for the NSGA-II.

We omit the formal proof of Theorem 3, which is not too different from the proofs of Theorems 1 and 2, for reasons of space.

When $k \leq \sqrt{n}$, the runtime estimates above can be estimated further as follows. In [17], it was shown that

$$P^\beta_i \geq \begin{cases} \frac{\beta-1}{e\beta} & \text{for } i = 1; \\ \frac{\beta-1}{4\sqrt{2\pi}e^{8\sqrt{2}+13}\beta}i^{-\beta} & \text{for } i \in [2..\lfloor\sqrt{n}\rfloor]. \end{cases}$$

Also, for $k \leq \sqrt{n}$, a good estimate for the binomial coefficient is $\binom{n}{k} \leq \frac{n^k}{k!}$ (losing at most a constant factor of e, and at most a $(1 + o(1))$-factor when $k = o(\sqrt{n})$.

Hence the runtime estimate from Theorem 3 for $k \leq \sqrt{n}$ becomes

$$(1 + o(1))K \frac{4\sqrt{2\pi}e^{8\sqrt{2}+13}\beta}{\beta - 1} N k^\beta \frac{n^k}{k!},$$

which is a tight estimate of the runtime guarantee of Theorem 3 apart from constants independent of n and k. In any case, this estimate shows that for moderate values of k, our runtime guarantee for the heavy-tailed NSGA-II is better by a factor of $\Theta(k! k^{-\beta})$, which is substantial already for small values of k.

5   Experiments

To complement our theoretical results, we also experimentally evaluate the runtime of the NSGA-II algorithm on the ONEJUMPZEROJUMP benchmark.

Settings: We implemented the algorithm as described in Sect. 3 in Python (the code can be found at https://github.com/deliaqu/NSGA-II). We use the following settings.

– Problem size n: 20 and 30.
– Jump size k: 3.
– Population size N: In our theoretical analysis, we have shown that with $N = 4(n - 2k + 3)$, the algorithm is able to recover the entire Pareto front. To further explore the effect of the population size, we conduct experiments with this population size, with half this size, and with twice this size, that is, for $N \in \{2(n - 2k + 3), 4(n - 2k + 3), 8(n - 2k + 3)\}$.
– Parent selection: For simplicity, we only experiment with using N independent binary tournaments.
– Mutation operator: Following our theoretical analysis, we consider two mutation operators, namely bit-wise mutation (flipping each bit with probability $\frac{1}{n}$) and fast mutation, that is, the heavy-tailed mutation operator $\text{MUT}^\beta(\cdot)$.
– Number of independent repetitions per setting: 50. This number is a compromise between the longer runtimes observed on a benchmark like ONEJUMPZEROJUMP and the not very concentrated runtimes (for most of our experiments, we observed a corrected sample standard deviation between 50% and 80% of the mean, which fits to our intuition that the runtimes are dominated by the time to find the two extremal points of the Pareto front).

Experimental Results: Table 1 contains the average runtime (number of fitness evaluations done until the full Pareto front is covered) of the NSGA-II algorithm when using bit-wise mutation and the heavy-tailed mutation operator. The most obvious finding is that the heavy-tailed mutation operator already for these small problem and jump sizes gives significant speed-ups.

While our theoretical results are valid only for $N \geq 4(n - 2k + 3)$, our experimental data suggests that also with the smaller population size $N = 2(n - 2k + 3)$

Table 1. Average runtime of the NSGA-II with bit-wise mutation and heavy-tailed mutation operator on the ONEJUMPZEROJUMP benchmark with $k = 3$.

	$n = 20$		$n = 30$	
	Bit-wise	Heavy-tailed	Bit-wise	Heavy-tailed
$N = 2(n - 2k + 3)$	264932	178682	1602552	785564
$N = 4(n - 2k + 3)$	366224	188213	1777546	1080458
$N = 8(n - 2k + 3)$	529894	285823	2836974	1804394

Table 2. Average runtime of the NSGA-II with bit-wise mutation and crossover on the ONEJUMPZEROJUMP benchmark with $k = 3$

	$n = 20$	$n = 30$	$n = 40$
$N = 2(n - 2k + 3)$	68598	265993	773605
$N = 4(n - 2k + 3)$	45538	205684	510650
$N = 8(n - 2k + 3)$	68356	316500	635701

the algorithm is able to cover the entire Pareto front of the ONEJUMPZERO-JUMP benchmark. We suspect that this is because even though theoretically there could be 4 individuals in each generation with the same objective value and positive crowding distances, empirically this happens relatively rarely and the expected number of individuals with the same objective value and positive crowding distances is closer to 2. We also note that with a larger population, e.g., $N = 8(n - 2k + 3)$, naturally, the runtime increases, but usually by significantly less than a factor of two. This shows that the algorithm is able to profit somewhat from the larger population size.

Crossover: Besides fast mutation, two further mechanisms were found that can speed up the runtime of evolutionary algorithms on (single-objective) jump functions, namely the stagnation-detection mechanism of Rajabi and Witt [16, 26–28] and crossover [4,10,11,21]. We are relatively optimistic that stagnation detection, as together with the global SEMO algorithm [17], can provably lead to runtime improvements, but we recall from [17] that the implementation of stagnation detection is less obvious for MOEAs. For that reason, we ignore this approach here and immediately turn to crossover, given that no results proving a performance gain from crossover for the NSGA-II exist, and there are clear suggestions on how to use it in [11].

Inspired by [11], we propose and experimentally analyze the following variant of the NSGA-II. The basic algorithm is as above. In particular, we also select N parents via independent tournaments. We partition these into pairs. For each pair, with probability 90%, we generate two intermediate offspring via a 2-offspring uniform crossover (that is, for each position independently, with probability 0.5, the first child inherits the bit from the first parent, and otherwise from the second parent; the bits from the two parents that are not inherited by the first child make up the second child). We then perform bit-wise mutation on these two intermediate offspring. With the remaining 10% probability, mutation is performed directly on the two parents.

Table 2 contains the average runtimes for this algorithm. We observe that crossover leads to massive speed-ups (which allows us to also conduct experiments for problem size $n = 40$). More detailedly, comparing the runtimes for $n = 30$ and bit-wise mutation (which is fair since the crossover version also uses this mutation operator), the crossover-based algorithm only uses between 8% and 15% percent of the runtime of the mutation-only algorithm.

We note that different from the case without crossover, with $N = 2(n-2k+3)$, the algorithm consistently takes more time than with $N = 4(n - 2k + 3)$. We suspect that the smaller population size makes it less likely that the population contains two parents from which crossover can create a profitable offspring.

6   Conclusions and Future Works

In this first mathematical runtime analysis of the NSGA-II on a bi-objective multimodal problem, we have shown that the NSGA-II with a sufficient population size performs well on the ONEJUMPZEROJUMP benchmark and profits from heavy-tailed mutation, all comparable to what was shown before for the GSEMO algorithm.

Due to the more complicated population dynamics of the NSGA-II, we could not prove an interesting lower bound. For this, it would be necessary to understand how many individuals with a particular objective value are in the population – note that this number is trivially one for the GSEMO, which explains why for this algorithm lower bounds could be proven [17]. Understanding better the population dynamics of the NSGA-II and then possibly proving good lower bounds is an interesting and challenging direction for future research.

A second interesting direction is to analyze how the NSGA-II with crossover optimizes the ONEJUMPZEROJUMP benchmark. Our experiments show clearly that crossover can lead to significant speed-ups here. Again, we currently do not have the methods to analyze this algorithm, and we speculate that a very good understanding of the population dynamics is necessary to solve this problem. We note that the only previous work [7] regarding the NSGA-II with crossover does not obtain faster runtimes from crossover. Besides that work, we are only aware of two other runtime analyses for crossover-based MOEAs, one for the multi-criteria all-pairs shortest path problem [22], the other also for classic benchmarks, but with an initialization that immediately puts the extremal points of the Pareto front into the population [24]. So it is unlikely that previous works can be used to analyze the runtime of the crossover-based NSGA-II on ONEJUMPZEROJUMP.

Acknowledgment. This work was supported by a public grant as part of the Investissements d'avenir project, reference ANR-11-LABX-0056-LMH, LabEx LMH.

References

1. Antipov, D., Buzdalov, M., Doerr, B.: Fast mutation in crossover-based algorithms. In: Genetic and Evolutionary Computation Conference, GECCO 2020, pp. 1268–1276. ACM (2020)
2. Antipov, D., Buzdalov, M., Doerr, B.: First steps towards a runtime analysis when starting with a good solution. In: Bäck, T., et al. (eds.) PPSN 2020, Part II. LNCS, vol. 12270, pp. 560–573. Springer, Cham (2020). https://doi.org/10.1007/978-3-030-58115-2_39
3. Antipov, D., Buzdalov, M., Doerr, B.: Lazy parameter tuning and control: choosing all parameters randomly from a power-law distribution. In: Genetic and Evolutionary Computation Conference, GECCO 2021, pp. 1115–1123. ACM (2021)
4. Antipov, D., Buzdalov, M., Doerr, B.: Fast mutation in crossover-based algorithms. Algorithmica 84, 1724–1761 (2022)
5. Antipov, D., Doerr, B.: Runtime analysis of a heavy-tailed $(1 + (\lambda, \lambda))$ genetic algorithm on jump functions. In: Bäck, T., et al. (eds.) PPSN 2020, Part II. LNCS, vol. 12270, pp. 545–559. Springer, Cham (2020). https://doi.org/10.1007/978-3-030-58115-2_38
6. Auger, A., Doerr, B. (eds.): Theory of Randomized Search Heuristics. World Scientific Publishing, Hackensack (2011)
7. Bian, C., Qian, C.: Running time analysis of the non-dominated sorting genetic algorithm II (NSGA-II) using binary or stochastic tournament selection. In: Rudolph, G., et al. (eds.) PPSN 2022. LNCS, vol. 13399, pp. xx-yy. Springer, Cham (2022)
8. Brockhoff, D.: Theoretical aspects of evolutionary multiobjective optimization. In: Auger, A., Doerr, B. (eds.) Theory of Randomized Search Heuristics, pp. 101–140. World Scientific Publishing (2011)
9. Corus, D., Oliveto, P.S., Yazdani, D.: Automatic adaptation of hypermutation rates for multimodal optimisation. In: Foundations of Genetic Algorithms, FOGA 2021, pp. 4:1–4:12. ACM (2021)
10. Dang, D., et al.: Escaping local optima with diversity mechanisms and crossover. In: Genetic and Evolutionary Computation Conference, GECCO 2016, pp. 645–652. ACM (2016)
11. Dang, D., et al.: Escaping local optima using crossover with emergent diversity. IEEE Trans. Evol. Comput. 22, 484–497 (2018)
12. Deb, K., Pratap, A., Agarwal, S., Meyarivan, T.: A fast and elitist multiobjective genetic algorithm: NSGA-II. IEEE Trans. Evol. Comput. 6, 182–197 (2002)
13. Doerr, B., Le, H.P., Makhmara, R., Nguyen, T.D.: Fast genetic algorithms. In: Genetic and Evolutionary Computation Conference, GECCO 2017, pp. 777–784. ACM (2017)
14. Doerr, B., Neumann, F. (eds.): Theory of Evolutionary Computation-Recent Developments in Discrete Optimization. Springer, Cham (2020). cs.adelaide.edu.au/~frank/papers/TheoryBook2019-selfarchived.pdf
15. Doerr, B., Qu, Z.: A first runtime analysis of the NSGA-II on a multimodal problem. CoRR abs/2204.07637 (2022)
16. Doerr, B., Rajabi, A.: Stagnation detection meets fast mutation. In: Pérez Cáceres, L., Verel, S. (eds.) Evolutionary Computation in Combinatorial Optimization, EvoCOP 2022. LNCS, vol. 13222, pp. 191–207. Springer, Cham (2022). https://doi.org/10.1007/978-3-031-04148-8_13

17. Doerr, B., Zheng, W.: Theoretical analyses of multi-objective evolutionary algorithms on multi-modal objectives. In: Conference on Artificial Intelligence, AAAI 2021, pp. 12293–12301. AAAI Press (2021)
18. Friedrich, T., Göbel, A., Quinzan, F., Wagner, M.: Heavy-tailed mutation operators in single-objective combinatorial optimization. In: Auger, A., Fonseca, C.M., Lourenço, N., Machado, P., Paquete, L., Whitley, D. (eds.) PPSN 2018, Part I. LNCS, vol. 11101, pp. 134–145. Springer, Cham (2018). https://doi.org/10.1007/978-3-319-99253-2_11
19. Friedrich, T., Quinzan, F., Wagner, M.: Escaping large deceptive basins of attraction with heavy-tailed mutation operators. In: Genetic and Evolutionary Computation Conference, GECCO 2018, pp. 293–300. ACM (2018)
20. Jansen, T.: Analyzing Evolutionary Algorithms - The Computer Science Perspective. Springer, Heidelberg (2013). https://doi.org/10.1007/978-3-642-17339-4
21. Jansen, T., Wegener, I.: The analysis of evolutionary algorithms - a proof that crossover really can help. Algorithmica 34, 47–66 (2002)
22. Neumann, F., Theile, M.: How crossover speeds up evolutionary algorithms for the multi-criteria all-pairs-shortest-path problem. In: Schaefer, R., Cotta, C., Kołodziej, J., Rudolph, G. (eds.) PPSN 2010, Part I. LNCS, vol. 6238, pp. 667–676. Springer, Heidelberg (2010). https://doi.org/10.1007/978-3-642-15844-5_67
23. Neumann, F., Witt, C.: Bioinspired Computation in Combinatorial Optimization - Algorithms and Their Computational Complexity. Springer, Heidelberg (2010). https://doi.org/10.1007/978-3-642-16544-3
24. Qian, C., Yu, Y., Zhou, Z.: An analysis on recombination in multi-objective evolutionary optimization. Artif. Intell. 204, 99–119 (2013)
25. Quinzan, F., Göbel, A., Wagner, M., Friedrich, T.: Evolutionary algorithms and submodular functions: benefits of heavy-tailed mutations. Nat. Comput. 20(3), 561–575 (2021). https://doi.org/10.1007/s11047-021-09841-7
26. Rajabi, A., Witt, C.: Self-adjusting evolutionary algorithms for multimodal optimization. In: Genetic and Evolutionary Computation Conference, GECCO 2020, pp. 1314–1322. ACM (2020)
27. Rajabi, A., Witt, C.: Stagnation detection in highly multimodal fitness landscapes. In: Genetic and Evolutionary Computation Conference, GECCO 2021, pp. 1178–1186. ACM (2021)
28. Rajabi, A., Witt, C.: Stagnation detection with randomized local search. In: Zarges, C., Verel, S. (eds.) EvoCOP 2021. LNCS, vol. 12692, pp. 152–168. Springer, Cham (2021). https://doi.org/10.1007/978-3-030-72904-2_10
29. Wu, M., Qian, C., Tang, K.: Dynamic mutation based pareto optimization for subset selection. In: Huang, D.-S., Gromiha, M.M., Han, K., Hussain, A. (eds.) ICIC 2018, Part III. LNCS (LNAI), vol. 10956, pp. 25–35. Springer, Cham (2018). https://doi.org/10.1007/978-3-319-95957-3_4
30. Zheng, W., Doerr, B.: Better approximation guarantees for the NSGA-II by using the current crowding distance. In: Genetic and Evolutionary Computation Conference, GECCO 2022. ACM (2022). arxiv.org/abs/2203.02693
31. Zheng, W., Liu, Y., Doerr, B.: A first mathematical runtime analysis of the Non-Dominated Sorting Genetic Algorithm II (NSGA-II). In: Conference on Artificial Intelligence, AAAI 2022. AAAI Press (2022). arxiv.org/abs/2112.08581
32. Zhou, A., Qu, B.Y., Li, H., Zhao, S.Z., Suganthan, P.N., Zhang, Q.: Multiobjective evolutionary algorithms: a survey of the state of the art. Swarm Evol. Comput. 1, 32–49 (2011)

Analysis of Quality Diversity Algorithms for the Knapsack Problem

Adel Nikfarjam[(✉)], Anh Viet Do, and Frank Neumann

Optimisation and Logistics, School of Computer Science,
The University of Adelaide, Adelaide, Australia
adel.nikfarjam@adelaide.edu.au

Abstract. Quality diversity (QD) algorithms have been shown to be very successful when dealing with problems in areas such as robotics, games and combinatorial optimization. They aim to maximize the quality of solutions for different regions of the so-called behavioural space of the underlying problem. In this paper, we apply the QD paradigm to simulate dynamic programming behaviours on knapsack problem, and provide a first runtime analysis of QD algorithms. We show that they are able to compute an optimal solution within expected pseudo-polynomial time, and reveal parameter settings that lead to a fully polynomial randomised approximation scheme (FPRAS). Our experimental investigations evaluate the different approaches on classical benchmark sets in terms of solutions constructed in the behavioural space as well as the runtime needed to obtain an optimal solution.

Keywords: Quality diversity · Runtime analysis · Dynamic programming

1 Introduction

Computing diverse sets of high quality solutions has recently gained significant interest in the evolutionary computation community under the terms Evolutionary Diversity Optimisation (EDO) and Quality Diversity (QD). With this paper, we contribute to the theoretical understanding of such approaches algorithms by providing a first runtime analysis of QD algorithms. We provide rigorous results for the classical knapsack problem and carry out additional experimental investigations on the search process in the behavioral space.

Diversity is traditionally seen as a mechanism to explore niches in a fitness landscape and prevent premature convergence during evolutionary searches. On the other hand, the aim of EDO is to explicitly maximise the structural diversity of a set of solutions, which usually have to fulfill some quality criteria. The concept was first introduced in [29] in a continuous domain. Later, EDO has been adopted to evolve a set of images [1] and benchmark instances for traveling salesperson problem (TSP) [12]. The star-discrepancy and the indicators from multi-objective evolutionary algorithms have been incorporated in EDO for the same purpose as the

G. Rudolph et al. (Eds.): PPSN 2022, LNCS 13399, pp. 413–427, 2022.
https://doi.org/10.1007/978-3-031-14721-0_29

previous studies in [16] and [17], respectively. More recently, EDO has been investigated in context of computing a diverse set of solutions for several combinatorial problems such as TSP in [7, 18, 19], the quadratic assignment problem [8], the minimum spanning tree problem [3], the knapsack problem [2], the optimisation of monotone sub-modular functions [15], and traveling thief problem [21].

On the other hand, QD explores a predefined behavioural space to find niches. It recently has gained increasing attention among the researchers in evolutionary computation. The optimisation paradigm first emerged in the form of novelty search, in which the goal is to find solutions with unique behaviours aside from the quality of solutions [14]. Later, a mechanism is introduced in [6] to solely retain best-performing solutions while exploring new behaviours. An algorithm, named MAP-Elite is introduced in [5] to plot the distribution of high-performing solutions in a behavioural space. MAP-Elite is shown efficient in developing behavioural repertoire. QD was coined as a term, and defined as a concept in [23, 24]. The paradigm has been widely applied in the context of robotic and gaming [10, 11, 25, 26, 31]. More recently, QD has been adopted for a multi-component combinatorial problem, namely traveling thief problem [20]. Bossek and Neumann [4] generated diverse sets of TSP instances by the use of QD. To the best of our knowledge, the use of QD in solving a combinatorial optimisation problem is limited to an empirical study [20]. Although the QD-based algorithm has been shown to yield very decent results, theoretical understandings of its performance have not yet been established.

In this work, we contribute to this line of research by theoretically and empirically studying QD for the knapsack problem (KP), with a focus on connections between populating behavioural spaces and constructing solutions in dynamic programming (DP) manner. The use of evolutionary algorithms building populations of specific structure to carry out dynamic programming has been studied in [9, 13, 27]. We consider a more natural way of enabling dynamic programming behavior by using QD algorithms with appropriately defined behavioural spaces. To this end, we define two behavioural spaces based on weights, profits and the subset of the first i items, as inspired by dynamic programming (DP) [28] and the classic fully polynomial-time approximation scheme (FPTAS) [30]. Here, the scaling factor used in the FPTAS adjusts the niche size along the weight/profit dimension. We formulate two simple mutation-only algorithms based on MAP-Elite to populate these spaces. We show that both algorithms mimic DP and find an optimum within pseudo-polynomial expected runtime. Moreover, we show that in the profit-based space, the algorithm can be made into a fully polynomial-time randomised approximation scheme (FPRAS) with an appropriate choice of the scaling value. Our experimental investigation on various instances suggests that these algorithms significantly outperforms $(1+1)$EA and $(\mu+1)$EA, especially in hard cases. With this, we demonstrate the ability of QD-based mechanisms to imitate DP-like behaviours in KP, and thus its potential value in black-box optimisers for problem with recursive subproblem structures.

The remainder of the paper is structured as follows. We formally define the knapsack problem, the behavioural spaces, and the algorithms in Sect. 2. Next,

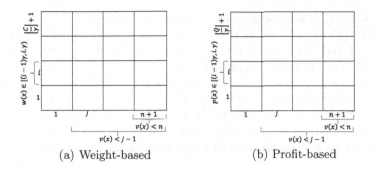

(a) Weight-based   (b) Profit-based

Fig. 1. The representation of the empty maps in the behavioral spaces.

we provide a runtime analysis for the algorithms in Sect. 3. In Sect. 4, we examine the distribution of high-quality knapsack solutions in the behavioural spaces and compare QD-based algorithms to other EAs. Finally, we finish with some concluding remarks.

2   Quality-Diversity for the Knapsack Problem

The knapsack problem is defined on a set of items I, where $|I| = n$ and each item i corresponds to a weight w_i and a profit p_i. Here, the goal is to find a selection of item $x = (x_1, x_2, \ldots, x_n)$ that maximise the profit while the weight of selected items is constrained to a capacity C. Here, x is the characteristic vector of the selection of items. Technically, KP is a binary linear programming problem: let $w = (w_1, \ldots, w_n)$ and $p = (p_1, \ldots, p_n)$, find arg $\max_{x \in \{0,1\}^n} \{p^T x \mid w^T x \leq C\}$. We assume that all items have weights in $(0, C]$, since any item violating this can be removed from the problem instance.

In this section, we introduce two MAP-Elite based algorithms exploring two different behavioral spaces. To determine behaviour of a solution in a particular space, a behaviour descriptor (BD) is required. MAP-Elite is an EA, where a solution competes with other solutions with a similar BD. MAP-Elites discretizes a behavioural space into a grid to define the similarity and acceptable tolerance of difference in two descriptors. Each cell in the grid corresponds with a BD type, and only best solution with that particular BD is kept in the cell.

For KP, we formulate the behavioral spaces based on the two ways in which the classic dynamic programming approach is implemented [28], i.e. profit-based and weight-based sub-problem divisions. Let $v(x)$ be the function returning the index of the last item in solution x: $v(x) = \max_i \{i \mid x_i = 1\}$.

2.1   Weight-Based Space

For the weight-based approach, $w(x)$ and $v(x)$ serve as the BD, where $w(x) = w^T x$. Figure 1a outlines an empty map in the weight-based behavioural space. To exclude infeasible solutions, the weight dimension is restricted to $[0, C]$.

As depicted, the behavioural space consists of $(\lfloor C/\gamma \rfloor + 1) \times (n+1)$ cells, in which cell (i,j) includes the best solution x (i.e. maximizing $p(x) = p^T x$) with $v(x) = j - 1$ and $w(x) \in [(i-1)\gamma, i\gamma)$. Here, γ is a factor to determine the size of each cell. The algorithm is initiated with a zero string 0^n. Having a parent is selected uniformly at random from the population, we generate a single offspring by standard flip mutation. If $w(x) \leq C$, we find the cell corresponding with the solution BD. We check the cells one by one. If the cells are empty, x is store in the cell; otherwise, the solution with highest profit remains in the corresponding cell. These steps are continued until a termination criterion is met.

2.2   Profit-Based Space

For the profit-based approach, $p(x)$ and $v(x)$ serve as the BD. Figure 1b depicts the profit-based behavioural space with $(\lfloor Q/\gamma \rfloor + 1) \times (n+1)$ cells where $Q = \sum_{i \in I} p_i$. Here, the selection in each cell minimizes the weight, and cell (i,j) includes a solution x with $v(x) = j - 1$ and $p(x) \in [(i-1)\gamma, i\gamma)$. Otherwise, the parent selection and the operator are the same as in weight-based MAP-Elite. After generating the offspring, we determine the cell associating with the BD $((v(x), p(x)))$. If the cell are empty the solution is stored in the cells; otherwise, the solution with the lower weight $w(x)$ is kept in the cell. The steps are continued until a termination criterion is met.

2.3   DP-Based Filtering Scheme

In classical MAP-Elites, the competition between solutions is confined within each cell. However, in this context, the mapping from solution space to behaviour space is transparent enough in both cases that a dominance relation between solutions in different cells can be determined; a property exploited by the DP approach. Therefore, in order to reduce the population size and speed up the search for the optimum, we incorporate a filtering scheme that forms the core of the DP approach. Given solutions x_1 and x_2 with $v(x_1) \geq v(x_2)$ and $w(x_1) = w(x_2)$; then, x_1 dominates x_2 in the weight-based space if $p(x_1) > p(x_2)$. To filter out the dominated solutions, we relax the restriction that each BD corresponds to only one cell and redefine acceptable solutions for Cell (i,j) in the weight-based space: $v(x) \leq j - 1$ and $w(x) \in [(i-1)\gamma, i\gamma)$. This means a particular BD is acceptable for multiple cells, and MAP-Elite algorithms must check all the cells accepting the offspring. Algorithm 1 outlines the MAP-Elite algorithm exploring this space; this is referred to as weight-based MAP-Elites.

The same scheme can be applied to the profit-based space, where cell (i,j) accepts solution x with $v(x) \leq j - 1$ and $p(x) \in [(i-1)\gamma, i\gamma)$. In this case, the dominance relation is formulated to minimise weight. Algorithm 2 sketches the profit-based MAP-Elites.

3   Theoretical Analysis

In this section, we give some runtime results for Algorithm 1 and 2 based on expected time, as typically done in runtime analyses. Here, we use "time" as

---

Algorithm 1. weight-based MAP-Elites

Input: weights $\{w_i\}_{i=1}^n$, C, profits $\{p_i\}_{i=1}^n$, γ

1: $P \leftarrow \{0^n\}$ // P is indexed from 1, 0^n is an all-zero string
2: $A \leftarrow 0_{n+1 \times \lfloor C/\gamma \rfloor + 1}$ // $0_{n+1 \times \lfloor C/\gamma \rfloor + 1}$ is an all-zero matrix
3: $B \leftarrow 0$,
4: **while** Termination criteria are not met **do**
5:    $i \leftarrow Uniform(\{1, \ldots, |P|\})$
6:    Get x from flipping each bit in $P(i)$ independently with probability $1/n$
7:    **if** $w(x) \leq C$ **then**
8:      $W' \leftarrow \lfloor w(x)/\gamma \rfloor + 1$
9:      **if** $A_{v(x)+1,W'} = 0$ **then**
10:        $P \leftarrow P \cup \{x\}$ // x is indexed last in P
11:        $A_{v(x)+1,W'} \leftarrow |P|$
12:      **else if** $p(x) > p(P(A_{v(x)+1,W'}))$ **then**
13:        $P(A_{v(x)+1,W'}) \leftarrow x$
14:      **for** j from $v(x) + 2$ to $n + 1$ **do**   // DP-based filtering scheme
15:        **if** $A_{j,W'} = 0$ Or $p(x) > p(P(A_{j,W'}))$ **then**
16:          $A_{j,W'} \leftarrow A_{v(x)+1,W'}$
17:      **if** $p(x) > B$ **then**
18:        $B \leftarrow p(x)$
19: **return** B

Algorithm 2. profit-based MAP-Elites

Input: Weights $\{w_i\}_{i=1}^n$, C, profits $\{p_i\}_{i=1}^n$, γ

1: $P \leftarrow \{0^n\}$ // P is indexed from 1, 0^n is an all-zero string
2: $A \leftarrow 0_{n+1 \times \sum_{i=1}^n p_i + 1}$ // $0_{n+1 \times \lfloor C/\gamma \rfloor + 1}$ is an all-zero matrix
3: $B \leftarrow 0$,
4: **while** Termination criteria are not met **do**
5:    $i \leftarrow Uniform(\{1, \ldots, |P|\})$
6:    Get x from flipping each bit in $P(i)$ independently with probability $1/n$
7:    $G \leftarrow \lfloor p(x)/\gamma \rfloor + 1$
8:    **if** $A_{v(x)+1,G} = 0$ **then**
9:      $P \leftarrow P \cup \{x\}$ // x is indexed last in P
10:      $A_{v(x)+1,G} \leftarrow |P|$
11:    **else if** $w(x) < w(P(A_{v(x)+1,G}))$ **then**
12:      $P(A_{v(x)+1,G}) \leftarrow x$
13:    **for** j from $v(x) + 2$ to $n + 1$ **do**   // DP-based filtering scheme
14:      **if** $A_{j,G} = 0$ Or $w(x) < w(P(A_{j,G}))$ **then**
15:        $A_{j,G} \leftarrow A_{v(x)+1,G}$
16:    **if** $w(x) \leq C$ **then**
17:      **if** $p(x) > B$ **then**
18:        $B \leftarrow p(x)$
19: **return** B

---

a shorthand for "number of fitness evaluations", which in this case equals the number of generated solutions during a run of the algorithm. We define $a \wedge b$ and $a \vee b$ to be the bit-wise AND and bit-wise OR, respectively, between two equal length bit-strings a and b. Also, we denote k-length all-zero and all-one bit-strings by 0^k and 1^k, respectively. For convenience, we denote the k-size prefix of $a \in \{0,1\}^n$ with $a^{(k)} = a \wedge 1^k 0^{n-k}$, and the k-size suffix with $a_{(k)} = a \wedge 0^{n-k} 1^k$.

It is important to note that in all our proofs, we consider solution y replacing solution x during a run to imply $v(y) \leq v(x)$. Since this holds regardless of whether filtering scheme outlined in Sect. 2.3 is used, our results should apply to both cases, as we use the largest possible upper bound of population size. Note that this filtering scheme may not reduce the population size in some cases.

We first show that with $\gamma = 1$ (no scaling), Algorithm 1 ensures that prefixes of optimal solutions remain in the population throughout the run, and that these increase in sizes within a pseudo-polynomial expected time. For this result, we assume all weights are integers.

Theorem 1. *Given $\gamma = 1$ and $k \in [0,n]$, within expected time $e(C+1)n^2 k$, Algorithm 1 achieves a population P such that for any $j \in [0,k]$, there is an optimal solution x^* where $x^{*(j)} \in P$.*

Proof. Let P_t be the population at iteration $t \geq 0$, S be the set of optimal solutions, $S_j = \{s^{(j)} \mid s \in S\}$, $X_t = \max\{h \mid \forall j \in [0,h], S_j \cap P_t \neq \emptyset\}$, and $H(x,y)$ be the Hamming distance between x and y, we have $S_n = S$. We see that for any $j \in [0, X_t]$, any $x \in S_j \cap P_t$ must be in $P_{>t}$, since otherwise, let y be the solution replacing it, and $y^* = y \vee x^*_{(n-j)}$ for any $x^* \in S$ where $x = x^{*(j)}$, we would have $p(y^*) - p(x^*) = p(y) - p(x) > 0$ and $w(y^*) = w(x^*) \leq B$, a contradiction. Additionally, if $x \in S_i \cap S_j$ for any $0 \leq i < j \leq n$, then $x \in \bigcap_{h=i}^{j} S_h$. Thus, if $X_t < n$, then $S_{X_t} \cap S \cap P_t = \emptyset$, so for all $x \in S_{X_t} \cap P_t$, there is $y \in S_{>X_t}$ such that $H(x,y) = 1$. We can then imply from the algorithm's behaviour that for any $j \in [0, n-1]$, $Pr[X_{t+1} < j \mid X_t = j] = 0$ and

$$Pr[X_{t+1} > j \mid X_t = j] \geq \frac{1}{n}\left(1 - \frac{1}{n}\right)^{n-1} \frac{|S_{X_t} \cap P_t|}{|P_t|} \geq \frac{1}{en \max_h |P_h|}.$$

Let T be the minimum integer such that $X_{t+T} > X_t$, then the expected waiting time in a binomial process gives $E[T \mid X_t < j] \leq en \max_h |P_h|$ for any $j \in [1,n]$. Let T_k be the minimum integer such that $X_{T_k} \geq k$, we have for any $k \in [0,n]$, $E[T_k] \leq \sum_{i=1}^{k} E[T \mid X_t < i] \leq en \max_h |P_h| k$, given that $0^n \in S_0 \cap P_0$. Applying the bound $\max_h |P_h| \leq (C+1)n$ yields the claim.   □

We remark that with $\gamma > 1$, Algorithm 1 may fail to maintain prefixes of optimal solutions during a run, due to rounding error. That is, assuming there is $x = x^{*(j)} \in P_t$ at step t and for some $j \in [0,n]$ and optimal solution x^*, a solution y may replace x if $p(y) > p(x)$ and $w(y) < w(x) + \gamma$. It is possible that $y^* = y \vee x^*_{(n-j)}$ is infeasible (i.e. when $C < w(x^*) + \gamma$), in which case the algorithm may need to "fix" y with multiple bit-flips in one step. The expected runtime till optimality can be derived directly from Theorem 1 by setting $k = n$.

Corollary 1. *Algorithm 1, run with $\gamma = 1$, finds an optimum within expected time $e(C + 1)n^3$.*

Using the notation $Q = \sum_{i=1}^{n} p_i$, we have the following result for Algorithm 2, which is analogous to Theorem 1 for Algorithm 1.

Theorem 2. *Given $k \in [0, n]$, and let z be an optimal solution, within expected time $e\left(\lfloor Q/\gamma \rfloor + 1\right) n^2 k$, Algorithm 2 achieves a population P such that, if $\gamma > 0$ is such that p_i/γ is integer for every item i in z, then for any $j \in [0, k]$, there is a feasible solution x where*

- *there is an integer m such that $p(x^{(j)}), p(z^{(j)}) \in [m\gamma, (m + 1)\gamma)$,*
- *$x_{(n-j)} = z_{(n-j)}$,*
- *$x^{(j)} \in P$.*

Moreover, for other γ values, the first property becomes $p(x^{(j)}), p(z^{(j)}) \in [m\gamma, (m + j + 1)\gamma)$.

Proof. The proof proceeds similarly as that of Theorem 1. We have the claim holds for $k = 0$ since the empty set satisfies the properties for $j = 0$ (i.e. x and z would be the same). For other k values, it suffices to show that if there is such a solution x for some $j \in [0, k]$: 1) any solution y replacing $x^{(j)}$ in a future step must be the j-size prefix of another solution with the same properties, and 2) at most one bit-flip is necessary to have it also hold for $j + 1$.

1) Let y be the solution replacing $x^{(j)}$, we have $p(y), p(x^{(j)}) \in [m\gamma, (m + 1)\gamma)$ for some integer m, and $w(y) < w(x^{(j)})$. Let $y^* = y \vee z_{(n-j)}$, we have $p(y), p(z^{(j)}) \in [m\gamma, (m + 1)\gamma)$, and $w(y^*) - w(x) = w(y) - w(x^{(j)}) < 0$, implying y^* is feasible. Therefore, y^* possess the same properties as x. Note that this also holds for the case where $p(x^{(j)}), p(z^{(j)}) \in [m\gamma, (m + j + 1)\gamma)$. In this case, $p(y) \in [m\gamma, (m + j + 1)\gamma)$.

2) If this also holds for $j+1$, no further step is necessary. Assuming otherwise, then z contains item $j + 1$, the algorithm only needs to flip the position $j + 1$ in $x^{(j)}$, since x and z shares $(n - j - 1)$-size suffix, and the p_{j+1} is a multiple of γ. Since this occurs with probability at least $1/en \max_h |P_h|$, the rest follows identically, save for $\max_h |P_h| \leq (\lfloor Q/\gamma \rfloor + 1) n$. If p_{j+1} is a not multiple of γ, then $p(x^{(j+1)})$ may be mapped to a different profit range from $p(z^{(j+1)})$. The difference is increased by at most 1 since $p(x^{(j+1)}) - p(x^{(j)}) = p(z^{(j+1)}) - p(z^{(j)})$, i.e. if $p(x^{(j)}), p(z^{(j)}) \in [m\gamma, (m + j + 1)\gamma)$ for some integer m, then $p(x^{(j+1)}), p(z^{(j+1)}) \in [m'\gamma, (m' + j + 2)\gamma)$ for some integer $m' \geq m$. Since x can be replaced in a future step by another solution with a smaller profit due to rounding error, the difference can still increase, so the claim holds non-trivially. $\square$

Theorem 2 gives us the following profit guarantees of Algorithm 2 when $k = n$. Here OPT denotes the optimal profit.

Corollary 2. *Algorithm 2, run with $\gamma > 0$, within expected time $e\left(\lfloor Q/\gamma \rfloor + 1\right) n^3$ obtains a feasible solution x where $p(x) = OPT$ if p_i/γ is integer for all $i = 1, \ldots, n$, and $p(x) > OPT - \gamma n$ otherwise.*

Proof. If p_i/γ is integer for all $i = 1, \ldots, n$, then $|p(a) - p(b)|$ is a multiple of γ for any solutions a and b. Since by Theorem 1, x is feasible and $p(x) > OPT - \gamma$, it must be that $p(x) = OPT$. For the other case, Theorem 1 implies that $p(x), OPT \in [m\gamma, (m + n + 1)\gamma)$ for some integer m. This means $p(x) > OPT - \gamma n$. □

Using this property, we can set up a FPRAS with an appropriate choice of γ, which is reminiscent of the classic FPTAS for KP based on DP. As a reminder, x is a $(1 - \epsilon)$-approximation for some $\epsilon \in (0, 1)$ if $p(x) \geq (1-\epsilon)OPT$. The following corollary is obtained from the fact that $Q \leq n \max_i\{p_i\}$, and $\max_i\{p_i\} \leq OPT$.

Corollary 3. *For some $\epsilon \in (0, 1)$, Algorithm 2, run with $\gamma = \epsilon \max_i\{p_i\}/n$, obtains a $(1 - \epsilon)$-approximation within expected time $e\left(\lfloor n^2/\epsilon \rfloor + 1\right) n^3$.*

For comparison, the asymptotic runtime of the classic FPTAS achieving the same approximation guarantee is $O(n^2 \lfloor n/\epsilon \rfloor)$ [30].

4 Experimental Investigations

In this section, we experimentally examine the two MAP-Elite based algorithms. The experiments can be categorised in three sections. First, we illustrate the distribution of high-performing solutions in the two behavioural spaces. Second, we compare Algorithm 1 and 2 in terms of population size and ratio in achieving the optimums over 30 independent runs. Finally, we compare between the best MAP-Elite algorithm and two baseline EAs, namely $(1 + 1)$EA and $(\mu + 1)$EA. These baselines are selected due the same size of offspring in each iteration. For the first round of experiments, three instances from [22] are considered. There is a strong correlation between the weight and profit of each items in the first instance. The second and third instances are not correlated, while the items have similar weights in third instance. The termination criterion is set to the maximum fitness evaluations of Cn^2. We also set $\gamma \in \{1, 5, 25\}$. For the second and third rounds of experiments, we run algorithms on 18 test instances from [22], and change the termination criterion to either achieving the optimal value or the maximum CPU-time of 7200 s.

Figure 2 illustrates the high-performing solutions obtained by Algorithm 1 in the weight-based space. As shown on the figure, the best solutions can be found the right top of the space. One can expected it since in that area of the space, solutions get to involve most items and the most of the knapsack's capacity, whereby on the left bottom of the space a few items and a small proportion of C can be used. Algorithm 1 can successfully populates the most of the space in instance 1, 2, while we can see most of the space is empty in instance 3. This is because the weights are uniformly distributed within $[1000, 1010]$, while C is set to 4567. As shown on the figure, the feasible solutions can only pick 4 items. Figure 2 also shows that the DP-based filtering removes many dominated solutions that contribute to convergence rate and pace of the algorithm.

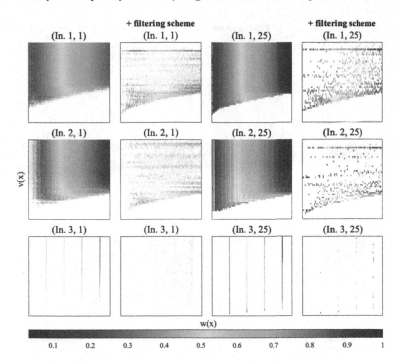

Fig. 2. The distribution of high-performing solutions in the weight-based behavioral space. The title of sub-figures show (Ints. No, γ). Colors are scaled to OPT.

Figure 3 shows the best-performing solutions obtained by Algorithm 2 in the profit-based space. It can be observed that we can only populate the half of space by Algorithm 2 or any other algorithm. To have a solution with profit of Q, the solutions needs to pick all items. This means that it is impossible to populate any other cells except cell $(n+1), (Q+1)$. On the contrary of the weight-based space, we can have both feasible and infeasible solutions in the profit-based space. For example, the map is well populated in instance 3, but mostly contains infeasible solution. Figure 4 depicts the trajectories of population size of Algorithm 1 and 2. The figure shows Algorithm 1 results in significantly smaller $|P|$ than Algorithm 2. For example, the final population size of Algorithm 1 is equal to 37 in instance 3, where $\gamma = 25$, while it is around 9000 for Algorithm 2. This is because we can limit the first space to the promising part of it $(w(x) \leq W)$, but we do not have the similar advantage for the profit-based space; the space accept the full range of possible profits $(p(x) \leq Q)$. We believe this issue can cause an adverse effect on the efficiency of MAP-Elites in reaching optimality, based on theoretical observations. This is explored further in our second experiment, where we look at the actual run-time to achieve the optimum.

Table 1 and 2 show the ratio of Algorithm 1 and Algorithm 2 in achieving the optimum for each instances in 30 independent runs, respectively. The tables also presents the mean of fitness evaluations for the algorithms to hit the optimal

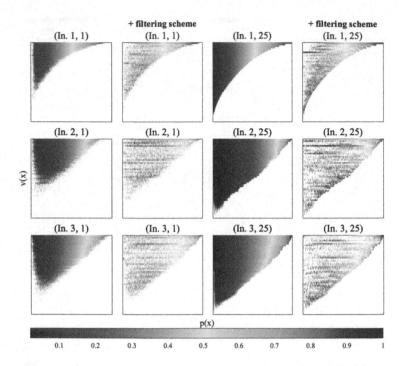

Fig. 3. The distribution of high-performing solutions in the profit-based behavioral space. Analogous to Fig. 2.

Table 1. Number of fitness evaluations needed by Algorithm 1 to obtain the optimal solutions.

Inst.	n	C	U	$\gamma = 1$		$\gamma = 5$		$\gamma = 25$	
				Mean	Time	Mean	Time	Mean	Time
1	50	4029	1.37e+09	1.53e+06	2.74e+01	3.76e+05	4.35e+00	1.61e+05	1.96e+00
2	50	2226	7.57e+08	5.32e+05	8.14e+00	1.74e+05	2.32e+00	5.86e+04	9.75e−01
3	50	4567	1.55e+09	2.43e+04	3.48e−01	1.12e+04	1.81e−01	5.82e+03	1.08e−01
4	75	5780	6.63e+09	5.30e+06	8.28e+01	1.45e+06	2.07e+01	4.12e+05	5.60e+00
5	75	3520	4.04e+09	3.63e+06	7.11e+01	1.15e+06	2.44e+01	3.49e+05	5.96e+00
6	75	6850	7.86e+09	1.17e+05	2.21e+00	4.09e+04	5.91e−01	1.44e+04	2.32e−01
7	100	8375	2.28e+10	2.42e+07	4.75e+02	6.57e+06	1.33e+02	6.60e+07	1.34e+03
8	100	4815	1.31e+10	9.56e+06	2.02e+02	2.73e+06	5.07e+01	8.66e+05	1.30e+01
9	100	9133	2.48e+10	6.18e+05	1.06e+01	1.78e+05	3.26e+00	5.79e+04	1.34e+00
10	123	10074	5.10e+10	3.56e+07	6.65e+02	9.90e+06	1.77e+02	2.55e+07	4.71e+02
11	123	5737	2.90e+10	2.05e+07	5.40e+02	5.12e+06	9.38e+01	1.47e+06	3.54e+01
12	123	11235	5.68e+10	1.45e+06	3.74e+01	3.38e+05	5.27e+00	1.21e+05	1.90e+00
13	151	12422	1.16e+11	5.04e+07	9.15e+02	1.48e+07	2.73e+02	7.51e+06	1.86e+02
14	151	6924	6.48e+10	4.27e+07	9.75e+02	1.24e+07	2.73e+02	3.35e+06	5.71e+01
15	151	13790	1.29e+11	3.18e+06	9.87e+01	6.73e+05	1.70e+01	2.35e+05	3.94e+00

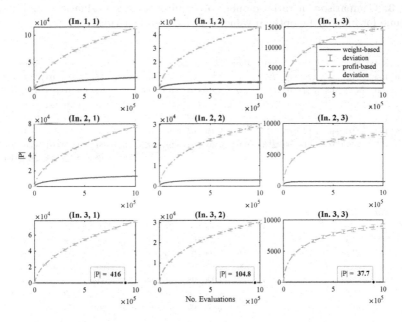

Fig. 4. Means and standard deviations of population sizes over fitness evaluations (the filtering scheme is used).

Table 2. Number of fitness evaluations needed by Algorithm 2 to obtain the optimal solutions

Inst.	n	Q	$\gamma = 1$			$\gamma = 5$			$\gamma = 25$		
			Mean	Ratio	U	Mean	Ratio	U	Mean	Ratio	U
1	50	53928	1.15e+07	100	1.83e+10	3.68e+06	100	e.66e+09	1.21e+06	100	7.33e+08
2	50	23825	5.36e+06	100	8.10e+09	1.34e+06	100	1.62e+09	4.00e+05	100	3.24e+08
3	50	24491	3.86e+06	100	8.32e+09	1.27e+06	100	1.66e+09	1.27e+06	100	3.33e+08
4	75	78483	5.07e+07	100	9.00e+10	1.50e+07	100	1.8e+10	4.03e+06	100	3.6e+09
5	75	37237	2.74e+07	100	4.27e+10	7.36e+06	100	8.54e+09	2.32e+06	100	1.71e+09
6	75	38724	1.93e+07	100	4.44e+10	5.63e+06	100	8.88e+09	2.95e+06	100	1.78e+09
7	100	112635	2.24e+08	97	3.06e+11	6.67e+07	100	6.12e+10	1.72e+07	100	1.22e+10
8	100	48042	6.76e+07	100	1.31e+11	1.82e+07	100	2.61e+10	5.03e+06	100	5.22e+09
9	100	52967	7.99e+07	100	1.44e+11	2.86e+07	100	2.88e+10	1.18e+08	87	5.76e+09
10	123	135522	3.35e+08	87	6.86e+11	1.05e+08	100	1.37e+11	7.34e+07	100	2.74e+10
11	123	57554	1.47e+08	100	2.91e+11	3.58e+07	100	5.82e+10	8.87e+06	100	1.16e+10
12	123	63116	1.71e+08	97	3.19e+11	8.45e+07	100	5.38e+10	2.66e+07	100	1.28e+10
13	151	166842	3.81e+08	13	1.56e+12	1.39e+08	100	3.12e+11	5.69e+07	100	6.25e+10
14	151	70276	3.15e+08	77	6.58e+11	8.86e+07	100	1.32e+11	1.96e+07	100	2.63e+10
15	151	76171	2.64e+08	90	7.13e+11	9.58e+07	100	1.42e+11	3.16e+07	100	2.85e+10
16	194	227046	2.94e+08	0	4.51e+12	3.33e+08	23	9.01e+11	1.30e+08	100	1.8e+11
17	194	92610	3.43e+08	0	1.84e+12	2.22e+08	97	3.68e+11	5.80e+07	100	7.35e+10
18	194	97037	3.55e+08	0	1.93e+12	2.13e+08	87	3.85e+11	9.98e+07	100	7.7e+10

Table 3. Comparison in ratio, number of required fitness evaluations and required CPU time for hitting the optimal value in 30 independent runs.

Inst.	n	QD				$(1+1)$EA				$(\mu+1)$EA			
		Ratio	Mean	Time	Stat	Ratio	Mean	Time	Stat	Ratio	Mean	Time	Stat
1	50	100	1.53e+06	2.74e+01	2^-3^-	40	1.32e+09	4.49e+03	1^+3^*	40	4.59e+08	4.92e+03	1^+2^*
2	50	100	5.32e+05	8.14e+00	2^*3^*	100	5.30e+05	2.02e+00	1^*3^*	100	6.10e+05	6.21e+00	1^*2^*
3	50	100	2.43e+04	3.48e−01	2^+3^+	100	1.01e+04	4.38e−02	1^-3^*	100	1.21e+04	1.50e−01	1^-2^*
4	75	100	5.30e+06	8.28e+01	2^*3^*	97	1.20e+08	5.70e+02	1^*3^*	100	3.46e+07	6.90e+02	1^*2^*
5	75	100	3.63e+06	7.11e+01	2^*3^*	100	6.34e+07	3.44e+02	1^*3^*	100	3.16e+07	4.44e+02	1^*2^*
6	75	100	1.17e+05	2.21e+00	2^+3^+	100	1.30e+04	8.98e−02	1^-3^-	100	2.14e+04	3.20e−01	1^-2^+
7	100	100	2.42e+07	4.75e+02	2^-3^-	63	5.72e+08	3.26e+03	1^+3^*	43	2.51e+08	4.92e+03	1^+2^*
8	100	100	9.56e+06	2.02e+02	2^+3^+	100	2.33e+06	1.46e+01	1^-3^*	100	3.56e+06	5.81e+01	1^-2^*
9	100	100	6.18e+05	1.06e+01	2^+3^+	100	3.72e+04	2.24e−01	1^-3^*	100	5.03e+04	8.84e−01	1^-2^*
10	123	100	3.56e+07	6.65e+02	2^-3^-	77	3.90e+08	2.44e+03	1^+3^*	47	2.34e+08	4.70e+03	1^+2^*
11	123	100	2.05e+07	5.40e+02	2^*3^+	97	1.38e+08	1.13e+03	1^*3^*	87	6.10e+07	1.41e+03	1^-2^*
12	123	100	1.45e+06	3.74e+01	2^+3^+	100	6.55e+04	5.24e−01	1^-3^*	100	6.71e+04	1.34e+00	1^-2^*
13	151	100	5.04e+07	9.15e+02	2^*3^*	97	1.25e+08	1.07e+03	1^*3^*	87	8.65e+07	2.14e+03	1^*2^*
14	151	100	4.27e+07	9.75e+02	2^+3^+	100	1.20e+07	1.06e+02	1^-3^*	100	1.10e+07	3.33e+02	1^-2^*
15	151	100	3.18e+06	9.87e+01	2^+3^+	100	1.17e+05	1.09e+00	1^-3^*	100	1.09e+05	2.59e+00	1^-2^*
16	194	100	1.58e+08	4.22e+03	2^-3^-	57	4.99e+08	4.21e+03	1^+3^*	47	2.34e+08	5.47e+03	1^+2^*
17	194	100	1.18e+08	2.25e+03	2^-3^*	57	4.29e+08	3.91e+03	1^+3^*	40	2.07e+08	4.87e+03	1^*2^*
18	194	100	7.76e+06	1.50e+02	2^+3^+	100	1.17e+05	1.42e+00	1^-3^*	100	1.37e+05	4.71e+00	1^-2^*

value or reach the limitation of CPU time. Table 1 shows that the ratio is 100% for Algorithm 1 on all instances and all $\gamma \in \{1, 5, 25\}$. On the other hand, Algorithm 2 cannot achieve the optimums in all 30 runs, especially in large instances when $\gamma = 1$. However, increasing γ to 25 enables the algorithm to obtain the optimum in the most instances with exception of instance 9. Moreover, the number of fitness evaluations required for Algorithm 2 is considerably higher than that of Algorithm 1. We can conclude that Algorithm 1 is more time-efficient than Algorithm 2, confirming our theoretical findings. This also suggests that the rounding errors are not detrimental to these algorithms' performances.

For the last round of the experiments, we compare Algorithm 1 to two well-known EAs in the literature, $(1 + 1)$EA and $(\mu + 1)$EA. Table 3 presents the ratio of the three algorithms in achieving the optimum and the mean of fitness evaluations required for them to reach the optimum. As shown on the table, the performances of $(1 + 1)$EA and $(\mu + 1)$EA deteriorate on the strongly correlated instances. It seems that $(1 + 1)$EA and $(\mu + 1)$EA are prone to get stuck in local optima, especially in instances with a strong weights-profits correlation. On the other hand, the MAP-Elite algorithm performs equally good in all instances through the diversity of solutions. Moreover, the mean of its runtime is significantly less in the half of instances although the population size of Algorithm 1 can be significantly higher that the other two EAs.

5   Conclusions

In this study, we examined the capability of QD approaches and in particular, MAP-Elite in solving knapsack problem. We defined two behavioural spaces inspired by the classic DP approaches, and two corresponding MAP-Elite-based algorithms operating on these spaces. We established that they imitate the exact DP approach, and one of them behaves similarly to the classic FPTAS for KP under a specific parameter setting, making it a FPRAS. We then compared the runtime of the algorithms empirically on instances of various properties related to their hardness, and found that the MAP-Elite selection mechanism significantly boosts efficiency of EAs in solving KP in terms of convergence ratio, especially in hard instances. Inspecting the behavioural spaces and population sizes reveals that smaller populations correlate to faster optimisation, demonstrating a well-known trade-off between optimisation and exploring behavioural spaces.

It is an open question to which extent MAP-Elites can simulate DP-like behaviours in other problems with recursive subproblem structures. Moreover, it might be possible to make such approaches outperform DP via better controls of behavioural space exploration, combined with more powerful variation operators.

Acknowledgements. This work was supported by the Australian Research Council through grants DP190103894 and FT200100536.

References

1. Alexander, B., Kortman, J., Neumann, A.: Evolution of artistic image variants through feature based diversity optimisation. In: GECCO, pp. 171–178. ACM (2017)
2. Bossek, J., Neumann, A., Neumann, F.: Breeding diverse packings for the knapsack problem by means of diversity-tailored evolutionary algorithms. In: GECCO, pp. 556–564. ACM (2021)
3. Bossek, J., Neumann, F.: Evolutionary diversity optimization and the minimum spanning tree problem. In: GECCO, pp. 198–206. ACM (2021)
4. Bossek, J., Neumann, F.: Exploring the feature space of TSP instances using quality diversity. CoRR abs/2202.02077 (2022)
5. Clune, J., Mouret, J., Lipson, H.: Summary of "the evolutionary origins of modularity". In: GECCO (Companion), pp. 23–24. ACM (2013)
6. Cully, A., Mouret, J.: Behavioral repertoire learning in robotics. In: GECCO, pp. 175–182. ACM (2013)
7. Do, A.V., Bossek, J., Neumann, A., Neumann, F.: Evolving diverse sets of tours for the travelling salesperson problem. In: GECCO, pp. 681–689. ACM (2020)
8. Do, A.V., Guo, M., Neumann, A., Neumann, F.: Analysis of evolutionary diversity optimisation for permutation problems. In: GECCO, pp. 574–582. ACM (2021)
9. Doerr, B., Eremeev, A.V., Neumann, F., Theile, M., Thyssen, C.: Evolutionary algorithms and dynamic programming. Theor. Comput. Sci. **412**(43), 6020–6035 (2011)
10. Fontaine, M.C., et al.: Illuminating Mario scenes in the latent space of a generative adversarial network. In: AAAI, pp. 5922–5930. AAAI Press (2021)

11. Fontaine, M.C., Togelius, J., Nikolaidis, S., Hoover, A.K.: Covariance matrix adaptation for the rapid illumination of behavior space. In: GECCO, pp. 94–102. ACM (2020)

12. Gao, W., Nallaperuma, S., Neumann, F.: Feature-based diversity optimization for problem instance classification. Evol. Comput. **29**(1), 107–128 (2021)

13. Horoba, C.: Analysis of a simple evolutionary algorithm for the multiobjective shortest path problem. In: FOGA, pp. 113–120. ACM (2009)

14. Lehman, J., Stanley, K.O.: Abandoning objectives: evolution through the search for novelty alone. Evol. Comput. **19**(2), 189–223 (2011)

15. Neumann, A., Bossek, J., Neumann, F.: Diversifying greedy sampling and evolutionary diversity optimisation for constrained monotone submodular functions. In: GECCO, pp. 261–269. ACM (2021)

16. Neumann, A., Gao, W., Doerr, C., Neumann, F., Wagner, M.: Discrepancy-based evolutionary diversity optimization. In: GECCO, pp. 991–998. ACM (2018)

17. Neumann, A., Gao, W., Wagner, M., Neumann, F.: Evolutionary diversity optimization using multi-objective indicators. In: GECCO, pp. 837–845. ACM (2019)

18. Nikfarjam, A., Bossek, J., Neumann, A., Neumann, F.: Computing diverse sets of high quality TSP tours by EAX-based evolutionary diversity optimisation. In: FOGA, pp. 9:1–9:11. ACM (2021)

19. Nikfarjam, A., Bossek, J., Neumann, A., Neumann, F.: Entropy-based evolutionary diversity optimisation for the traveling salesperson problem. In: GECCO, pp. 600–608. ACM (2021)

20. Nikfarjam, A., Neumann, A., Neumann, F.: On the use of quality diversity algorithms for the traveling thief problem. CoRR abs/2112.08627 (2021)

21. Nikfarjam, A., Neumann, A., Neumann, F.: Evolutionary diversity optimisation for the traveling thief problem. CoRR abs/2204.02709 (2022)

22. Polyakovskiy, S., Bonyadi, M.R., Wagner, M., Michalewicz, Z., Neumann, F.: A comprehensive benchmark set and heuristics for the traveling thief problem. In: GECCO, pp. 477–484. ACM (2014)

23. Pugh, J.K., Soros, L.B., Stanley, K.O.: Quality diversity: a new frontier for evolutionary computation. Front. Robot. AI **3**, 40 (2016)

24. Pugh, J.K., Soros, L.B., Szerlip, P.A., Stanley, K.O.: Confronting the challenge of quality diversity. In: GECCO, pp. 967–974. ACM (2015)

25. Rakicevic, N., Cully, A., Kormushev, P.: Policy manifold search: exploring the manifold hypothesis for diversity-based neuroevolution. In: GECCO, pp. 901–909. ACM (2021)

26. Steckel, K., Schrum, J.: Illuminating the space of beatable lode runner levels produced by various generative adversarial networks. In: GECCO Companion, pp. 111–112. ACM (2021)

27. Theile, M.: Exact solutions to the traveling salesperson problem by a population-based evolutionary algorithm. In: Cotta, C., Cowling, P. (eds.) EvoCOP 2009. LNCS, vol. 5482, pp. 145–155. Springer, Heidelberg (2009). https://doi.org/10.1007/978-3-642-01009-5_13

28. Toth, P.: Dynamic programming algorithms for the zero-one knapsack problem. Computing **25**(1), 29–45 (1980)

29. Ulrich, T., Thiele, L.: Maximizing population diversity in single-objective optimization. In: GECCO, pp. 641–648. ACM (2011)

30. Vazirani, V.V.: Approximation Algorithms. Springer, Heidelberg (2001). https://doi.org/10.1007/978-3-662-04565-7
31. Zardini, E., Zappetti, D., Zambrano, D., Iacca, G., Floreano, D.: Seeking quality diversity in evolutionary co-design of morphology and control of soft tensegrity modular robots. In: GECCO, pp. 189–197. ACM (2021)

Better Running Time of the Non-dominated Sorting Genetic Algorithm II (NSGA-II) by Using Stochastic Tournament Selection

Chao Bian and Chao Qian[✉]

State Key Laboratory for Novel Software Technology, Nanjing University,
Nanjing 210023, China
{bianc,qianc}@lamda.nju.edu.cn

Abstract. Evolutionary algorithms (EAs) have been widely used to solve multi-objective optimization problems, and have become the most popular tool. However, the theoretical foundation of multi-objective EAs (MOEAs), especially the essential theoretical aspect, i.e., running time analysis, is still largely underdeveloped. The few existing theoretical works mainly considered simple MOEAs, while the non-dominated sorting genetic algorithm II (NSGA-II), probably the most influential MOEA, has not been analyzed except for a very recent work considering a simplified variant without crossover. In this paper, we present a running time analysis of the standard NSGA-II for solving LOTZ, the commonly used bi-objective optimization problem. Specifically, we prove that the expected running time (i.e., number of fitness evaluations) is $O(n^3)$ for LOTZ, which is the same as that of the previously analyzed simple MOEAs, GSEMO and SEMO, as well as the NSGA-II without crossover. Next, we introduce a new parent selection strategy, stochastic tournament selection (i.e., k tournament selection where k is uniformly sampled at random), to replace the binary tournament selection strategy of NSGA-II, decreasing the upper bound on the required expected running time to $O(n^2)$. Experiments are also conducted, suggesting that the derived running time upper bounds are tight. We also empirically compare the performance of the NSGA-II using the two selection strategies on the widely used benchmark problem ZDT1, and the results show that stochastic tournament selection can help the NSGA-II converge faster.

1 Introduction

Multi-objective optimization, which requires optimizing several objective functions simultaneously, arises in many areas. Since the objectives are usually conflicting, there does not exist a single solution that can perform well on all these objectives. Thus, the goal of multi-objective optimization is to find a set of

This work was supported by the NSFC (62022039) and the Jiangsu NSF (BK20201247). Chao Qian is the corresponding author. Due to space limitation, proof details are available at https://arxiv.org/abs/2203.11550.

Pareto optimal solutions (or the Pareto front, i.e., the set of objective vectors of the Pareto optimal solutions), representing different optimal trade-offs between these objectives. Evolutionary algorithms (EAs) [2] are a kind of randomized heuristic optimization algorithms, inspired by natural evolution. They maintain a set of solutions, i.e., a population, and iteratively improve the population by reproducing new solutions and selecting better ones. Due to the population-based nature, EAs are very popular for solving multi-objective optimization problems, and have been widely used in many real-world applications [4].

Compared with practical applications, the theoretical foundation of EAs is still underdeveloped, which is mainly because the sophisticated behaviors of EAs make theoretical analysis quite difficult. Though much effort has been devoted to the essential theoretical aspect, i.e., running time analysis, leading to a lot of progress [1,10,28,34] in the past 25 years, most of them focused on single-objective optimization, while only a few considered the more complicated scenario of multi-objective optimization. In the following, we briefly review the results of running time analyses on multi-objective EAs (MOEAs).

The running time analysis of MOEAs started from GSEMO, a simple MOEA which employs the bit-wise mutation operator to generate an offspring solution in each iteration and keeps the non-dominated solutions generated-so-far in the population. For GSEMO solving the bi-objective optimization problems LOTZ and COCZ, the expected running time has been proved to be $O(n^3)$ [16] and $O(n^2 \log n)$ [3,30], respectively, where n is the problem size. SEMO is a counterpart of GSEMO, which employs the local mutation operator, one-bit mutation, instead of the global bit-wise mutation operator. Laumanns et al. [21] proved that the expected running time of SEMO solving LOTZ and COCZ are $\Theta(n^3)$ and $O(n^2 \log n)$, respectively. Giel and Lehre [17] considered another bi-objective problem OneMinMax, and proved that both GSEMO and SEMO can solve it in $O(n^2 \log n)$ expected running time. Doerr et al. [9] also proved a lower bound $\Omega(n^2/p)$ for GSEMO solving LOTZ, where $p < n^{-7/4}$ is the mutation rate, i.e., the probability of flipping each bit when performing bit-wise mutation.

Later, the analyses of GSEMO were conducted on multi-objective combinatorial optimization problems. For bi-objective minimum spanning trees (MST), GSEMO was proved to be able to find a 2-approximation of the Pareto front in expected pseudo-polynomial time [24]. For multi-objective shortest paths, a variant of GSEMO can achieve an $(1 + \epsilon)$-approximation in expected pseudo-polynomial time [18,26], where $\epsilon > 0$. Laumanns et al. [20] considered GSEMO and its variant for solving a special case of the multi-objective knapsack problem, and proved that the expected running time of the two algorithms for finding all the Pareto optimal solutions are $O(n^6)$ and $O(n^5)$, respectively.

There are also studies that analyze GSEMO for solving single-objective constrained optimization problems. By optimizing a reformulated bi-objective optimization problem that optimizes the original objective and a constraint-related objective simultaneously, GSEMO can reduce the expected running time significantly for achieving a desired approximation ratio. For example, by reformulating the set cover problem into a bi-objective problem, Friedrich et al. [12] proved that GSEMO and SEMO can solve a class of set cover instances in

$O(mn(\log c_{\max} + \log n))$ expected running time, which is better than the exponential expected running time of (1+1)-EA, i.e., the single-objective counterpart to GSEMO, where m, n and $c_{\max}$ denote the size of the ground set, the size of the collection of subsets, and the maximum cost of a subset, respectively. More evidence has been proved on the problems of minimum cuts [25], minimum cost coverage [31], MST [27] and submodular optimization [15]. Note that we concern inherently multi-objective optimization problems in this paper.

Based on GSEMO and SEMO, the effectiveness of some strategies for multi-objective evolutionary optimization has been analyzed. For example, Laumanns et al. [21] showed the effectiveness of greedy selection by proving that using this strategy can reduce the expected running time of SEMO from $O(n^2 \log n)$ to $\Theta(n^2)$ for solving the COCZ problem. Qian et al. [30] showed that crossover can accelerate filling the Pareto front by comparing the expected running time of GSEMO with and without crossover for solving the artificial problems COCZ and weighted LPTNO (a generalization of LOTZ), as well as the combinatorial problem multi-objective MST. The effectiveness of some other mechanisms, e.g., heuristic selection [29], diversity [13], fairness [14,21], and diversity-based parent selection [6] have also been examined.

Though GSEMO and SEMO share the general structure of MOEAs, they have been much simplified. To characterize the behavior of practical MOEAs, some efforts have been devoted to analyzing MOEA/D, which is a popular MOEA based on decomposition [32]. Li et al. [23] analyzed a simplified variant of MOEA/D without crossover for solving COCZ and weighted LPTNO, and proved that the expected running time is $\Theta(n \log n)$ and $\Theta(n^2)$, respectively. Huang et al. [19] also considered a simplified MOEA/D, and examined the effectiveness of different decomposition approaches by comparing the running time for solving two many-objective problems mLOTZ and mCOCZ, where m denotes the number of objectives.

Surprisingly, the running time analysis of the non-dominated sorting genetic Algorithm II (NSGA-II) [8], the probably most influential MOEA, has been rarely touched. The NSGA-II enables to find well-spread Pareto-optimal solutions by incorporating two substantial features, i.e., non-dominated sorting and crowding distance, and has become the most popular MOEA for solving multi-objective optimization problems [7]. To the best of our knowledge, the only attempt is a very recent work, which, however, considered a simplified version of NSGA-II without crossover, and proved that the expected running time is $O(n^2 \log n)$ for OneMinMax and $O(n^3)$ for LOTZ [33].

In this paper, we present a running time analysis for the standard NSGA-II. We prove that the expected running time of NSGA-II is $O(n^3)$ for solving LOTZ. Note that the running time upper bound is the same as that of GSEMO and SEMO [16,17,21,30], implying that the NSGA-II does not have an advantage over simplified MOEAs on LOTZ if the derived upper bound is tight.

Next, we introduce a new parent selection strategy, i.e., stochastic tournament selection, which samples a number k uniformly at random and then performs k tournament selection. By replacing the original binary tournament selection of NSGA-II with stochastic tournament selection, we prove that the expected running time of NSGA-II can be improved to $O(n^2)$ for LOTZ. We also conduct experiments, suggesting that the derived upper bounds are tight. Furthermore, we empirically examine the performance of the NSGA-II using the two selection strategies on the widely used benchmark problem ZDT1 [35]. The results show that stochastic tournament selection can help the NSGA-II converge faster, disclosing its potential in practical applications.

2 Preliminaries

In this section, we first introduce multi-objective optimization, and then introduce the procedure of NSGA-II.

2.1 Multi-objective Optimization

Multi-objective optimization requires to simultaneously optimize two or more objective functions, as shown in Definition 1. We consider maximization here, while minimization can be defined similarly. The objectives are usually conflicting, and thus there is no canonical complete order in the solution space $\mathcal{X}$. The comparison between solutions relies on the *domination* relationship, as presented in Definition 2. A solution is *Pareto optimal* if there is no other solution in $\mathcal{X}$ that dominates it. The set of objective vectors of all the Pareto optimal solutions constitutes the *Pareto front*. The goal of multi-objective optimization is to find the Pareto front, that is, to find at least one corresponding solution for each objective vector in the Pareto front.

Definition 1 (Multi-objective Optimization). *Given a feasible solution space $\mathcal{X}$ and objective functions $f_1, f_2, \ldots, f_m$, multi-objective optimization can be formulated as $\max_{x \in \mathcal{X}} \left(f_1(x), f_2(x), \ldots, f_m(x) \right)$.*

Definition 2 (Domination). *Let $f = (f_1, f_2, \ldots, f_m) : \mathcal{X} \to \mathbb{R}^m$ be the objective vector. For two solutions x and $y \in \mathcal{X}$:*

- *x weakly dominates y (denoted as $x \succeq y$) if $\forall 1 \leq i \leq m, f_i(x) \geq f_i(y)$;*
- *x dominates y (denoted as $x \succ y$) if $x \succeq y$ and $f_i(x) > f_i(y)$ for some i;*
- *x and y are incomparable if neither $x \succeq y$ nor $y \succeq x$.*

2.2 NSGA-II

The NSGA-II Algorithm [8] as presented in Algorithm 1 is a popular MOEA, which incorporates two substantial features, i.e., non-dominated sorting and crowding distance. NSGA-II starts from an initial population of N random solutions (line 1). In each generation, it employs binary tournament selection N times to generate a parent population P' (line 4), and then applies one-point crossover and bit-wise mutation on the $N/2$ pairs of parent solutions to generate N offspring solutions (lines 5–9). Note that the two adjacent selected solutions form a

Algorithm 1. NSGA-II Algorithm [8]

Input: objective functions $f_1, f_2 \ldots, f_m$, population size N
Output: N solutions from $\{0, 1\}^n$
 1: $P \leftarrow N$ solutions uniformly and randomly selected from $\{0,1\}^n$;
 2: **while** criterion is not met **do**
 3:    $Q = \emptyset$;
 4:    apply binary tournament selection N times to generate a parent population P'
      of size N;
 5:    **for** each consecutive pair of the parent solutions x and y in P' **do**
 6:       apply one-point crossover on x and y to generate two solutions x' and y',
         with probability 0.9;
 7:       apply bit-wise mutation on x' and y' to generate x'' and y'', respectively;
 8:       add x'' and y'' into Q
 9:    **end for**
10:    partition $P \cup Q$ into non-dominated sets $F_1, F_2, \ldots$;
11:    let $P = \emptyset$, $i = 1$;
12:    **while** $|P \cup F_i| < N$ **do**
13:       $P = P \cup F_i$, $i = i + 1$
14:    **end while**
15:    assign each solution in F_i with a crowding distance;
16:    sort the solutions in F_i by crowding distance in descending order, and add the
      first $N - |P|$ solutions into P
17: **end while**
18: **return** P

---

pair, and thus the N selected solutions form $N/2$ pairs. The one-point crossover operator first selects a crossover point $i \in \{1, 2, \ldots, n\}$ uniformly at random, where n is the problem size, and then exchanges the first i bits of two solutions. The bit-wise mutation operator flips each bit of a solution independently with probability $1/n$. Note that for real-coded solutions (which, however, are not considered in this paper), the one-point crossover operator and bit-wise mutation operator can be replaced by other operators, e.g., the simulated binary crossover (SBX) operator and polynomial mutation operator [8]. The binary tournament selection presented in Definition 3 picks two solutions randomly from the population P with or without replacement, and then selects a better one (ties broken uniformly). Note that we consider the strategy with replacement in this paper.

Definition 3 (Binary Tournament Selection). *The binary tournament selection strategy first picks two solutions from the population P uniformly at random, and then selects a better one with ties broken uniformly.*

After generating N offspring solutions, the best N solutions in the current population P and the offspring population Q are selected as the population in the next generation (lines 10–16). In particular, the solutions in the current and offspring populations are partitioned into non-dominated sets $F_1, F_2, \ldots$ (line 10), where F_1 contains all the non-dominated solutions in $P \cup Q$, and F_i ($i \geq 2$) contains all the non-dominated solutions in $(P \cup Q) \setminus \cup_{j=1}^{i-1} F_j$. Note that we use the notion $\text{rank}(\boldsymbol{x}) = i$ to denote that $\boldsymbol{x}$ belongs to F_i. Then, the solutions in $F_1, F_2, \ldots$ are added into the next population (lines 12–14), until the population size exceeds N. For the critical set F_i whose inclusion makes the population size larger than N, the crowding distance is computed for each of the contained solutions (line 15). Finally, the solutions in F_i with large crowding distance are selected to fill the remaining population slots (line 16).

When using binary tournament selection (line 4), the selection criterion is based on the crowded-comparison, i.e., $\boldsymbol{x}$ is superior to $\boldsymbol{y}$ (denoted as $\boldsymbol{x} \succ_c \boldsymbol{y}$) if

$$\text{rank}(\boldsymbol{x}) < \text{rank}(\boldsymbol{y}) \ \ \text{OR} \ \ \text{rank}(\boldsymbol{x}) = \text{rank}(\boldsymbol{y}) \wedge \text{dist}(\boldsymbol{x}) > \text{dist}(\boldsymbol{y}). \quad (1)$$

Intuitively, the crowding distance of a solution means the distance between its closest neighbour solutions, and a solution with larger crowding distance is preferred so that the diversity of the population can be preserved as much as possible. Note that when computing the crowding distance, we assume that the relative positions of the solutions with the same objective vector are unchanged or totally reversed when the solutions are sorted w.r.t. some objective function. Such requirement can be met by any *stable* sorting algorithm, e.g., the bubble sort or merge sort, which maintains the relative order of items with equal keys (i.e., values). What's more, the built-in sorting functions in MATLAB, e.g., sortrows() and sort(), can also satisfy the requirement. Under such assumption, the population size needed to find the Pareto front can be reduced (a detailed discussion is provided after Theorem 1).

In line 6 of Algorithm 1, the probability of using crossover has been set to 0.9, which is the same as the original setting and also commonly used [8]. However, the theoretical results derived in this paper can be directly generalized to the scenario where the probability of using crossover belongs to $[\Omega(1), 1 - \Omega(1)]$.

3   Running Time Analysis of NSGA-II

In this section, we analyze the expected running time of the standard NSGA-II in Algorithm 1 solving the bi-objective pseudo-Boolean problem LOTZ, which is widely used in MOEAs' theoretical analyses [9,21,30].

The LOTZ problem presented in Definition 4 aims to maximize the number of leading 1-bits and the number of trailing 0-bits of a binary bit string. The Pareto front of LOTZ is $\mathcal{F} = \{(0, n), (1, n-1), \ldots, (n, 0)\}$, and the corresponding Pareto optimal solutions are $0^n, 10^{n-1}, \ldots, 1^n$.

Definition 4 (LOTZ [21]). *The LOTZ problem of size n is to find n bits binary strings which maximize $\boldsymbol{f}(\boldsymbol{x}) = \left(\sum_{i=1}^{n} \prod_{j=1}^{i} x_j, \sum_{i=1}^{n} \prod_{j=i}^{n} (1 - x_j) \right)$, where x_j denotes the j-th bit of $\boldsymbol{x} \in \{0, 1\}^n$.*

We prove in Theorem 1 that the NSGA-II can find the Pareto front in $O(n^2)$ expected number of generations, i.e., $O(n^3)$ expected number of fitness evaluations, because the generated N offspring solutions need to be evaluated in each iteration. Note that the running time of an EA is usually measured by the number of fitness evaluations, because evaluating the fitness of a solution is often the most time-consuming step in practice. The main proof idea can be summarized as follows. The NSGA-II first employs the mutation operator to find the two solutions with the largest number of leading 1-bits and the largest number of trailing 0-bits, i.e., 1^n and 0^n, respectively; then employs the crossover operator to find the whole Pareto front.

Theorem 1. *For the NSGA-II solving LOTZ, if using binary tournament selection and a population size N such that $2n + 2 \leq N = O(n)$, then the expected number of generations for finding the Pareto front is $O(n^2)$.*

Note that Zheng et al. [33] proved that for NSGA-II using bit-wise mutation (without crossover), the expected running time is $O(Nn^2)$ if the population size N is at least $5n + 5$. Thus, the requirement for the population size N is relaxed from $5n + 5$ to $2n + 2$. The main reason for the relaxation is that under the assumption in Sect. 2.2 (i.e., the order of the solutions with the same objective vector is unchanged or totally reversed when the solutions are sorted according to some f_j), there exist at most two solutions with i leading 1-bits such that their ranks are equal to 1 and crowding distances are larger than 0, for each $i \in \{0, 1, \ldots, n\}$. Meanwhile, for any objective vector in the Pareto front that has been obtained by the algorithm, there is at least one corresponding solution in the population such that its rank is equal to 1 and crowding distance is larger than 0, implying that the solution will occupy one of the $2n + 2$ slots, and thus be maintained in the population. Without such assumption, there may exist more solutions with crowding distance larger than 0 for each objective vector in the Pareto front, thus requiring a larger population size.

4   NSGA-II Using Stochastic Tournament Selection

In the previous section, we have proved that the expected running time of the standard NSGA-II is $O(n^3)$ for LOTZ, which is the same as that of the previously analyzed simple MOEAs, GSEMO and SEMO [21, 30]. Next, we introduce a new parent selection strategy, i.e., the stochastic tournament selection, into the NSGA-II, and show that the expected running time needed to find the whole Pareto front can be reduced to $O(n^2)$.

4.1   Stochastic Tournament Selection

As the crowded-comparison $\succ_c$ in Eq. (1) actually gives a total order of the solutions in the population P, binary tournament selection can be naturally extended to k tournament selection [11], as presented in Definition 5, where k is a parameter such that $1 \leq k \leq N$. That is, k solutions are first picked from P uniformly at random, and then the solution with the smallest rank is selected. If several solutions have the same smallest rank, the one with the largest crowding distance is selected, with ties broken uniformly.

Definition 5 (k Tournament Selection). *The k tournament selection strategy first picks k solutions from the population P uniformly at random, and then selects the best one with ties broken uniformly.*

Note that a larger k implies a larger selection pressure, i.e., a larger probability of selecting a good solution, and thus the value of k can be used to control the selection pressure of EAs [11]. However, this also brings about a new issue, i.e., how to set k properly. In order to reduce the risk of setting improper values of k as well as the overhead of tuning k, we introduce a natural strategy, i.e., stochastic tournament selection in Definition 6, which first selects a number k randomly, and then performs the k tournament selection. In this paper, we consider that the tournament candidates are picked with replacement.

Definition 6 (Stochastic Tournament Selection). *The stochastic tournament selection strategy first selects a number k from $\{1, 2, \dots, N\}$ uniformly at random, where N is the size of the population P, and then employs the k tournament selection to select a solution from the population P.*

In each generation of NSGA-II, we need to select N parent solutions independently, and each selection may involve the comparison of several solutions, which may lead to a large number of comparisons. To improve the efficiency of stochastic tournament selection, we can first sort the solutions in the population P, and then perform the parent selection procedure. Specifically, each solution $\boldsymbol{x}_i$ ($1 \le i \le N$) in P is assigned a number $\pi(i)$, where $\pi : \{1, 2, \dots, N\} \to \{1, 2, \dots, N\}$ is a bijection such that

$$\forall 1 \le i, j \le N, i \ne j : \boldsymbol{x}_i \succ_c \boldsymbol{x}_j \Rightarrow \pi(i) < \pi(j). \tag{2}$$

That is, a solution with a smaller number is better. Note that the number $\pi(\cdot)$ is assigned randomly if several solutions have the same rank and crowding distance. Then, we sample a number k randomly from $\{1, 2, \dots, N\}$ and pick k solutions from P at random, where the solution with the lowest $\pi(\cdot)$ value is finally selected.

Lemma 1 presents the property of stochastic tournament selection, which will be used in the following theoretical analysis. It shows that any solution (even the worst solution) in P can be selected with probability at least $1/N^2$, and any solution belonging to the best $O(1)$ solutions in P (with respect to $\succ_c$) can be selected with probability at least $\Omega(1)$. Note that for binary tournament selection, the probability of selecting the worst solution (denoted as $\boldsymbol{x}^w$) is $1/N^2$, because $\boldsymbol{x}^w$ is selected if and only if the two solutions picked for competition are both $\boldsymbol{x}^w$; the probability of selecting the best solution (denoted as $\boldsymbol{x}^b$) is $1 - (1 - 1/N)^2 = 2/N - 1/N^2$, because $\boldsymbol{x}^b$ is selected if and only if $\boldsymbol{x}^b$ is picked at least once. Thus, compared with binary tournament selection, stochastic tournament selection can increase the probability of selecting the top solutions, and meanwhile maintain the probability of selecting the bottom solutions. Note that such probability is similar to the power law distribution in [6].

Lemma 1. *If using stochastic tournament selection, any solution in P can be selected with prob. at least $1/N^2$. Furthermore, a solution $\boldsymbol{x}_i \in P$ with $\pi(i) = O(1)$ can be selected with prob. $\Omega(1)$, where $\pi : \{1, 2, \ldots, N\} \to \{1, 2, \ldots, N\}$ is a bijection satisfying Eq. (2).*

4.2   Running Time Analysis

We prove that the expected number of generations of the NSGA-II using stochastic tournament selection is $O(n)$ (implying $O(n^2)$ expected running time) for solving LOTZ, in Theorem 2. The proof idea of Theorem 2 is similar to that of Theorem 1. That is, the NSGA-II first employs the mutation operator to find the solutions that maximize each objective function, and then employs the crossover operator to quickly find the remaining objective vectors in the Pareto front. However, the utilization of stochastic tournament selection can make the NSGA-II select prominent solutions, i.e., solutions maximizing each objective function, with larger probability, making the crossover operator easier fill in the remaining Pareto front and thus reducing the total running time.

Theorem 2. *For the NSGA-II solving LOTZ, if using stochastic tournament selection and a population size N such that $2n + 2 \leq N = O(n)$, then the expected number of generations for finding the Pareto front is $O(n)$.*

5   Experiments

In the previous sections, we have proved that when binary tournament selection is used in the NSGA-II, the expected number of generations is $O(n^2)$ for LOTZ; when stochastic tournament selection is used, the expected number of generations can be improved to $O(n)$. But as the lower bounds on the running time have not been derived, the comparison may be not strict. Thus, we conduct experiments to examine the tightness of these upper bounds. We also conduct experiments on the widely used benchmark problem ZDT1 [35], to examine the performance of the stochastic tournament selection in more realistic scenarios.

5.1   LOTZ Problem

For the LOTZ problem, we examine the performance of NSGA-II when the problem size n changes from 10 to 100, with a step of 10. On each problem size n, we run the NSGA-II 1000 times independently, and record the number of generations until the Pareto front is found. Then, the average number of generations and the standard deviation of the 1000 runs are reported in Fig. 1(a). To show the relationship between the average number of generations and the problem size n clearly, we also plot the estimated ratio, i.e., the average number of generations divided by the problem size n, in Fig. 1(b).

From the left subfigures of Fig. 1(a) and 1(b), we can observe that the average number of generations is approximately $\Theta(n^2)$, suggesting that the upper bound $O(n^2)$ derived in Theorem 1 is tight. By the right subfigures of Fig. 1(a) and 1(b), the average number of generations is clearly a linear function of n, which suggests that the upper bound $O(n)$ derived in Theorem 2 is also tight.

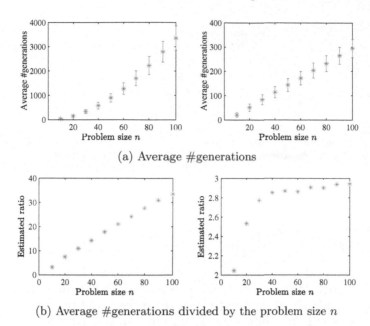

(a) Average #generations

(b) Average #generations divided by the problem size n

Fig. 1. Average #generations and the estimated ratio of the NSGA-II using binary tournament selection or stochastic tournament selection for solving the LOTZ problem. Left subfigure: the NSGA-II using binary tournament selection; right subfigure: the NSGA-II using stochastic tournament selection.

5.2   ZDT1 Problem

The ZDT1 problem presented in Definition 7 is a widely used benchmark to test the practical performance of MOEAs [35]. It has 30 continuous decision variables with each variable taking value from $[0, 1]$. As suggested in [8], we use 30 bits (i.e., a binary string of length 30) to code each decision variable. The Pareto front of ZDT1 is $\mathcal{F} = \{(f_1, 1 - \sqrt{f_1}) \mid f_1 \in [0, 1]\}$. Note that ZDT1 is a minimization problem, and thus we need to change the domination relationship in Definition 2 from "$f_i(\boldsymbol{x}) \geq (\text{or} >) f_i(\boldsymbol{y})$" to "$f_i(\boldsymbol{x}) \leq (\text{or} <) f_i(\boldsymbol{y})$" accordingly.

Definition 7 (ZDT1 [35]). *The ZDT1 problem is to find a 30-dimensional decision vector* $\boldsymbol{x} = (x_1, x_2, \ldots, x_{30})$ *which minimizes* $\boldsymbol{f}(\boldsymbol{x}) = \left(x_1, g(\boldsymbol{x})\left(1 - \sqrt{x_1/g(\boldsymbol{x})}\right)\right)$, *where* $\forall 1 \leq i \leq 30 : x_i \in [0, 1]$, *and* $g(\boldsymbol{x}) = 1 + 9 \cdot \sum_{i=2}^{30} x_i/29$.

Different from the LOTZ problem, the Pareto front of the ZDT1 problem is an uncountable set. Thus, instead of examining the running time for finding the whole Pareto front, we run NSGA-II for a fixed number of generations, i.e., 300, and examine the quality of the obtained population. To measure the quality of a set of solutions, we use the inverted generational distance (IGD) indicator, which has been widely used in multi-objective optimization [5,22]. As presented in Definition 8, $\text{IGD}(R, A)$ intuitively means the average distance of

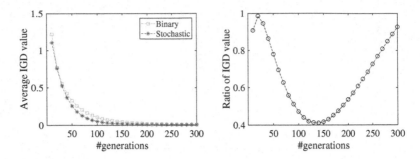

Fig. 2. Average IGD value of the NSGA-II using binary or stochastic tournament selection for solving the ZDT1 problem. Left subfigure: average IGD value of the NSGA-II vs. #generations; right subfigure: average IGD value of the NSGA-II using stochastic tournament selection divided by that of the NSGA-II using binary tournament selection vs. #generations.

the reference points in R to the objective vectors in A, where the reference points are usually sampled from the Pareto front in advance, and the set A consists of objective vectors of the solutions in the population. It is straightforward to see that a smaller IGD value implies a better approximation of the population to the Pareto front, in terms of both convergence and diversity.

Definition 8 (IGD [5]). *Given a set $R = \{r^1, r^2, \ldots, r^l\}$ of reference points and a set A of objective vectors, the IGD value of the set A with respect to R is defined as $IGD(R, A) = \frac{1}{l} \sum_{i=1}^{l} \min_{a \in A} d_2(r^i, a)$, where $d_2(r^i, a)$ denotes the Euclidean distance between r^i and a.*

In our experiments, we sample 200 points uniformly from the Pareto front as the reference set R, and set the population size N to 100. We run the NSGA-II 1000 times independently, and report the average IGD value of the 1000 runs every 10 generations. From Fig. 2(a), we can observe that (i) initially, the two selection strategies achieve similar performance; (ii) in the intermediate stage, the NSGA-II using stochastic tournament selection converges to the Pareto front faster than the NSGA-II using binary tournament selection; (iii) finally, the two selection strategies both achieve IGD value very close to 0, implying a good approximation ability of the two strategies. We also plot the ratio of the average IGD value obtained by the NSGA-II using stochastic tournament selection and binary tournament selection in Fig. 2(b). We can observe that the ratio is always at most 1, and decreases rapidly in the initial optimization procedure, implying that stochastic tournament selection is always better, and can help the NSGA-II converge faster. As time goes by, the advantage of stochastic tournament selection diminishes, because the NSGA-II using the two strategies have both found objective vectors which approximate the Pareto front well.

To better visualize the performance of the NSGA-II using the two selection strategies, we also plot the objective vectors obtained by the NSGA-II every 50 generations in one of the runs. Figure 3 shows that the objective vectors

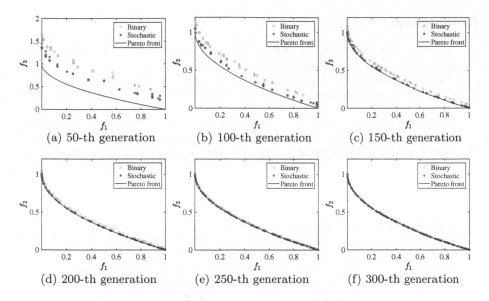

Fig. 3. Objective vectors obtained by the NSGA-II using binary or stochastic tournament selection for solving the ZDT1 problem.

obtained by the NSGA-II using stochastic tournament selection are always evenly distributed along the Pareto front, and gradually converge, suggesting the good spread ability of stochastic tournament selection.

In summary, the two selection strategies can achieve similar performance when given long enough time, but stochastic tournament selection can help the NSGA-II converge faster. The reason may be that the second objective value of the ZDT1 problem can be consecutively decreased by decreasing the value of $x_2, x_3, \ldots, x_{30}$, i.e., a currently good solution is helpful in the subsequent optimization process; and stochastic tournament selection can take advantage of these good solutions more efficiently.

6   Conclusion

In this paper, we theoretically analyze the running time of the NSGA-II solving the bi-objective problem LOTZ, and derive the upper bound that is the same as that of the previously analyzed simple MOEAs, GSEMO and SEMO. Then, we propose a new parent selection strategy, stochastic tournament selection, to replace the binary tournament selection strategy of the NSGA-II, and prove that the NSGA-II using the new strategy can find the Pareto front of LOTZ with a much smaller running time upper bound. Experimental results suggest that the derived upper bounds on LOTZ are tight, and also show the superior performance of stochastic tournament selection on the widely used benchmark problem ZDT1. In the future, we will analyze the lower bounds on the running time to make the comparison strict, and it is also interesting to examine the effectiveness of stochastic tournament selection on more problems.

References

1. Auger, A., Doerr, B.: Theory of Randomized Search Heuristics: Foundations and Recent Developments. World Scientific, Singapore (2011)
2. Bäck, T.: Evolutionary Algorithms in Theory and Practice: Evolution Strategies, Evolutionary Programming, Genetic Algorithms. Oxford University Press, Oxford (1996)
3. Bian, C., Qian, C., Tang, K.: A general approach to running time analysis of multi-objective evolutionary algorithms. In: Proceedings of the 27th International Joint Conference on Artificial Intelligence (IJCAI 2018), Stockholm, Sweden, pp. 1405–1411 (2018)
4. Coello Coello, C.A., Lamont, G.B.: Applications of Multi-Objective Evolutionary Algorithms. World Scientific, Singapore (2004)
5. Coello Coello, C.A., Sierra, M.R.: A general approach to running time analysis of multi-objective evolutionary algorithms. In: Proceedings of the Mexican International Conference on Artificial Intelligence (MICAI 2004), Mexico City, Mexico, pp. 688–697 (2004)
6. Covantes Osuna, E., Gao, W., Neumann, F., Sudholt, D.: Design and analysis of diversity-based parent selection schemes for speeding up evolutionary multi-objective optimisation. Theoret. Comput. Sci. **832**, 123–142 (2020)
7. Deb, K.: Multi-objective optimisation using evolutionary algorithms: an introduction. In: Multi-objective Evolutionary Optimisation for Product Design and Manufacturing, pp. 3–34. Springer, London (2011). https://doi.org/10.1007/978-0-85729-652-8_1
8. Deb, K., Pratap, A., Agarwal, S., Meyarivan, T.: A fast and elitist multiobjective genetic algorithm: NSGA-II. IEEE Trans. Evol. Comput. **6**(2), 182–197 (2002)
9. Doerr, B., Kodric, B., Voigt, M.: Lower bounds for the runtime of a global multi-objective evolutionary algorithm. In: Proceedings of the 2013 IEEE Congress on Evolutionary Computation (CEC 2013), Cancun, Mexico, pp. 432–439 (2013)
10. Doerr, B., Neumann, F. (eds.): Theory of Evolutionary Computation: Recent Developments in Discrete Optimization, Natural Computing Series, Springer, Cham (2020). https://doi.org/10.1007/978-3-030-29414-4
11. E. Eiben, A., E. Smith, J.: Introduction to Evolutionary Computing. Springer-Verlag, Berlin (2015). https://doi.org/10.1007/978-3-662-05094-1
12. Friedrich, T., He, J., Hebbinghaus, N., Neumann, F., Witt, C.: Approximating covering problems by randomized search heuristics using multi-objective models. Evol. Comput. **18**(4), 617–633 (2010)
13. Friedrich, T., Hebbinghaus, N., Neumann, F.: Plateaus can be harder in multi-objective optimization. Theoret. Comput. Sci. **411**(6), 854–864 (2010)
14. Friedrich, T., Horoba, C., Neumann, F.: Illustration of fairness in evolutionary multi-objective optimization. Theoret. Comput. Sci. **412**(17), 1546–1556 (2011)
15. Friedrich, T., Neumann, F.: Maximizing submodular functions under matroid constraints by evolutionary algorithms. Evol. Comput. **23**(4), 543–558 (2015)
16. Giel, O.: Expected runtimes of a simple multi-objective evolutionary algorithm. In: Proceedings of the 2003 IEEE Congress on Evolutionary Computation (CEC 2003), Canberra, Australia, pp. 1918–1925 (2003)
17. Giel, O., Lehre, P.K.: On the effect of populations in evolutionary multi-objective optimisation. Evol. Comput. **18**(3), 335–356 (2010)
18. Horoba, C.: Analysis of a simple evolutionary algorithm for the multiobjective shortest path problem. In: Proceedings of the 10th International Workshop on Foundations of Genetic Algorithms (FOGA 2009), Orlando, FL, pp. 113–120 (2009)

19. Huang, Z., Zhou, Y., Luo, C., Lin, Q.: A runtime analysis of typical decomposition approaches in MOEA/D framework for many-objective optimization problems. In: Proceedings of the 30th International Joint Conference on Artificial Intelligence (IJCAI 2021), Virtual, pp. 1682–1688 (2021)
20. Laumanns, M., Thiele, L., Zitzler, E.: Running time analysis of evolutionary algorithms on a simplified multiobjective knapsack problem. Nat. Comput. **3**, 37–51 (2004)
21. Laumanns, M., Thiele, L., Zitzler, E.: Running time analysis of multiobjective evolutionary algorithms on pseudo-Boolean functions. IEEE Trans. Evol. Comput. **8**(2), 170–182 (2004)
22. Li, M., Yao, X.: Quality evaluation of solution sets in multiobjective optimisation: a survey. ACM Comput. Surv. **52**(2), 26:1–38 (2020)
23. Li, Y., Zhou, Y., Zhan, Z., Zhang, J.: A primary theoretical study on decomposition-based multiobjective evolutionary algorithms. IEEE Trans. Evol. Comput. **20**(4), 563–576 (2016)
24. Neumann, F.: Expected runtimes of a simple evolutionary algorithm for the multi-objective minimum spanning tree problem. Eur. J. Oper. Res. **181**(3), 1620–1629 (2007)
25. Neumann, F., Reichel, J., Skutella, M.: Computing minimum cuts by randomized search heuristics. Algorithmica **59**, 323–342 (2011)
26. Neumann, F., Theile, M.: How crossover speeds up evolutionary algorithms for the multi-criteria all-pairs-shortest-path problem. In: Proceedings of the 11th International Conference on Parallel Problem Solving from Nature (PPSN 2010), Krakow, Poland, pp. 667–676 (2010)
27. Neumann, F., Wegener, I.: Minimum spanning trees made easier via multi-objective optimization. Nat. Comput. **5**, 305–319 (2006)
28. Neumann, F., Witt, C.: Bioinspired Computation in Combinatorial Optimization: Algorithms and Their Computational Complexity. Springer-Verlag, Berlin (2010). https://doi.org/10.1007/978-3-642-16544-3
29. Qian, C., Tang, K., Zhou, Z.H.: Selection hyper-heuristics can provably be helpful in evolutionary multi-objective optimization. In: Proceedings of the 14th International Conference on Parallel Problem Solving from Nature (PPSN 2016), Edinburgh, Scotland, pp. 835–846 (2016)
30. Qian, C., Yu, Y., Zhou, Z.H.: An analysis on recombination in multi-objective evolutionary optimization. Artif. Intell. **204**, 99–119 (2013)
31. Qian, C., Yu, Y., Zhou, Z.H.: On constrained Boolean Parto optimization. In: Proceedings of the 24th International Joint Conference on Artificial Intelligence (IJCAI 2015), Buenos Aires, Argentina, pp. 389–395 (2015)
32. Zhang, Q., Li, H.: MOEA/D: a multiobjective evolutionary algorithm based on decomposition. IEEE Trans. Evol. Comput. **11**(6), 712–731 (2007)
33. Zheng, W., Liu, Y., Doerr, B.: A first mathematical runtime analysis of the non-dominated sorting genetic algorithm II (NSGA-II). In: Proceedings of the 36th AAAI Conference on Artificial Intelligence (AAAI 2022), Virtual (to appear 2022)
34. Zhou, Z.H., Yu, Y., Qian, C.: Evolutionary Learning: Advances in Theories and Algorithms. Springer, Singapore (2019). https://doi.org/10.1007/978-981-13-5956-9
35. Zitzler, E., Deb, K., Thiele, L.: Comparison of multiobjective evolutionary algorithms: empirical results. Evol. Comput. **8**(2), 173–195 (2000)

Escaping Local Optima with Local Search: A Theory-Driven Discussion

Tobias Friedrich[1], Timo Kötzing[1(✉)], Martin S. Krejca[2],
and Amirhossein Rajabi[3]

[1] Hasso Plattner Institute, University of Potsdam, Potsdam, Germany
{Tobias.Friedrich,Timo.Koetzing}@hpi.de
[2] Sorbonne University, CNRS, LIP6, Paris, France
Martin.Krejca@lip6.fr
[3] Technical University of Denmark, Kgs. Lyngby, Denmark
amraj@dtu.dk

Abstract. Local search is the most basic strategy in optimization settings when no specific problem knowledge is employed. While this strategy finds good solutions for certain optimization problems, it generally suffers from getting stuck in local optima. This stagnation can be avoided if local search is modified. Depending on the optimization landscape, different modifications vary in their success.

We discuss several features of optimization landscapes and give analyses as examples for how they affect the performance of modifications of local search. We consider modifying *random local search* by restarting it and by considering larger search radii. The landscape features we analyze include the *number* of local optima, the *distance* between different optima, as well as the *local landscape* around a local optimum. For each feature, we show which modifications of local search handle them well and which do not.

Keywords: Local search · Theory · Run time analysis

1 Introduction

For optimizing a given objective function, the following strategy is widely used. Start with any, possibly randomly generated, solution. Check *neighboring* solutions, where just a few defining properties of the solution are altered, for having better quality. Whenever you find a better solution, let it replace the previous solution and continue from there. This is the general concept of *local search*.

Basic local search already finds good solutions for a variety of problems [1, 15, 16, 25] by *hillclimbing*, i.e., going up the gradient until a peak in objective value is found. This simple greedy behavior can be very beneficial, e.g., in settings where no additional knowledge about the problem to be optimized is available, so-called *black box optimization*. The main drawback is when local search gets stuck in a local optimum where all nearby solutions do not have better quality

© The Author(s), under exclusive license to Springer Nature Switzerland AG 2022
G. Rudolph et al. (Eds.): PPSN 2022, LNCS 13399, pp. 442–455, 2022.
https://doi.org/10.1007/978-3-031-14721-0_31

than the local optimum, while the quality of solutions in other parts of the search space is significantly better. Overcoming the issue of local optima is a long-standing and frequently addressed problem.

One common way to escape local optima is to introduce randomness into how many local changes are performed when modifying a single solution. Prominent examples of this strategy are evolutionary algorithms (EAs [24]), which typically allow to modify solutions to vast extents, larger modifications commonly having a lower probability of occurring. Although this approach potentially allows to escape local optima, it also has some drawbacks. As an example, if better solutions require larger modifications to the current solution, the probability of making such a change may be very small [5]. Moreover, defining a mechanism that allows to change solutions in a manner such that each solution can be produced (a *global* operator) requires greater knowledge of the search space, e.g., when defining the probabilities for each possible change. In contrast, local changes are usually well understood and easy to implement.

In this article, we study *random local search* (RLS), a very basic local-search variant that maintains a single solution. In an iterative manner, it modifies this solution only slightly, i.e., locally. If the new solution is at least as good as the current, the current solution is updated to the new one, otherwise not. It is clear that RLS ceases improving the maintained solution once a search point is found whose direct neighbors have strictly worse objective-function value.

In order to overcome local optima, we consider two simple, different modifications to RLS: *restarts* and *larger search radii*. Restarts modify the way that the maintained solution is selected by always accepting the new solution when a restart is triggered. In addition, the distribution from which the new solution is drawn may be changed. Larger search radii modify the way that a new solution is created by considering solutions that are not direct neighbors of the current solution. This can be done by considering a local operator (i.e., creating solutions in a certain distance) or a global operator (i.e., creating any solution).

We study RLS and its modifications on various functions (see Fig. 1), containing different types of local optima. Our goal is to understand how the modifications of RLS cope with these local optima. We are particularly interested in an overview of which different characteristics of the optimization landscape favor which modifications and which not.

We aim to raise awareness about the usefulness of modifications to RLS in various settings. To this end, our analyzes do not aim for *depth* (i.e., giving a narrow but sophisticated analysis of a single setting), as is frequent in theory research, but instead for *breadth*. We note that we consider a local optimum to be points in the search space such that all directly neighboring points are worse in objective-function value. Allowing for neighboring points to have equal values results in *plateaus* and in completely different discussions. For recent results on plateaus, we refer the interested reader to the literature [2, 4].

Contributions. Our results concern four landscape characteristics. We give an intuitive description as well as key insights for each characteristic below.

(1) Section 3: A **basin of attraction** of a local optimum x is the part of the landscape from where local search can find the local optimum x.

Key Insight: Restarts are beneficial and better than larger search radii if the basin of attraction of the global optimum is large.

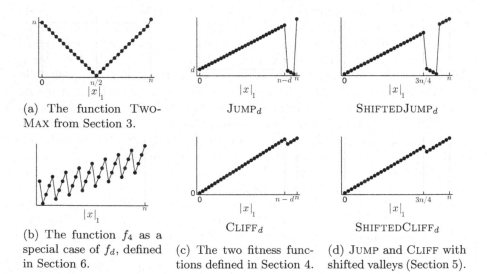

(a) The function TWO-MAX from Section 3.

JUMP$_d$

SHIFTEDJUMP$_d$

(b) The function f_4 as a special case of f_d, defined in Section 6.

(c) The two fitness functions defined in Section 4.

CLIFF$_d$

(d) JUMP and CLIFF with shifted valleys (Section 5).

SHIFTEDCLIFF$_d$

Fig. 1. Most of the fitness functions that we analyze in Sects. 3 to 6.

(2) Section 4: Between a local optimum and the global optimum is a *valley* of worse objective-function values that needs to be crossed. Depending on the values within the valley, this is a **deceptive valley** (leading back to the local optimum) or it provides **guiding information**.

Key Insight: Restarts are very beneficial for exploiting guiding information. However, they fail in the case of deception, where larger search radii prove useful and comparable to global operators.

(3) Section 5: The difficulty in crossing valleys depends on whether on the other side of the valley there is a **single or multiple targets** to transition to.

Key Insight: Both modifications of RLS are unaffected by the number of targets. In contrast, a global operator improves the performance drastically.

(4) Section 6: An algorithm might encounter **iterated local optima**, i.e., it has to cross multiple, consecutive valleys to find the global optimum.

Key Insight: The structure of each local optimum is essential. Warm restarts may help majorly if the local structure has guiding information (i.e., is well suited) but fail in case of deceptive information. When using larger search radii, the performance is unaffected by the shape of the valley. It is far slower in case of guiding information but better in case of deception.

Paper Outline. In Sect. 2, we give the details of all algorithms considered, followed by the technical sections considering the four mentioned landscape characteristics in turn. Last, we provide a discussion and conclusions in Sect. 7, where we go into more detail about the general learnings from the analyzes.

Algorithm 1: The framework for trajectory-based heuristics, requiring the potentially parametrized subroutines MUTATE and SELECT as well as a fitness function f.

---

1 $x^{(0)} \leftarrow$ individual drawn uniformly at random from $\{0,1\}^n$;
2 **for** $t \in \mathbb{N}$ **do**
3 $\quad \mid \quad y \leftarrow \text{MUTATE}(x^{(t)})$;
4 $\quad \mid \quad x^{(t+1)} \leftarrow \text{SELECT}_f(x^{(t)}, y)$;

---

2   Definitions and Algorithms

We let $\mathbb{N}$ denote the set of all natural numbers, including 0, and let $\mathbb{R}$ denote the set of all reals. For all $a, b \in \mathbb{R}$, let $[a..b] := [a, b] \cap \mathbb{N}$ denote the set of natural numbers from at least a to at most b. Further, for all $a \in \mathbb{R}$, let $[a] := [1..a]$.

We consider the maximization of pseudo-Boolean functions of dimension $n \in \mathbb{N}_{\geq 1}$, that is, functions $\{0,1\}^n \to \mathbb{R}$. Throughout this article, let n always denote the dimension of the objective function under consideration. All asymptotics (that is, big-Oh notation) are with respect to this n.

We call a pseudo-Boolean function f a *fitness function*, and we refer to bit strings as *individuals*. For each $x \in \{0,1\}^n$, let $|x|_1$ denote the number of 1s in x, and let $|x|_0$ denote its number of 0s. Further, for each $i \in [n]$, let x_i denote the bit at position i in x. We say that we flip bit i when we refer to the value $1 - x_i$. We call $f(x)$ the *fitness* of x. For $x, y \in \{0,1\}^n$, we call $d_{\mathrm{H}}(x, y) := |\{i \in [n] \mid x_i \neq y_i\}|$ the *Hamming distance* of x and y. Last, for all $i \in [n]$, we call the set of all individuals with distance i to x the *i-neighborhood of x*.

Given an algorithm A and a fitness function f, we call the number of fitness function evaluations (number of calls to f) that A performs until is finds a global maximum of f for the first time the *run time* of A.

2.1   Algorithms

We consider modifications to RLS. All of these algorithms follow the framework of a trajectory-based heuristic for optimizing a fitness function f (Algorithm 1). Each such heuristic evolves iteratively a trajectory $(x^{(t)})_{t \in \mathbb{N}}$ of individuals (the *current* individuals). The initial individual $(x^{(0)})$ is drawn uniformly at random from the search space $\{0,1\}^n$. For all $t \in \mathbb{N}$, the individual $x^{(t+1)}$ is determined via two, potentially parametrized, subroutines: MUTATE and SELECT. The subroutine MUTATE: $\{0,1\}^n \to \{0,1\}^n$ gets $x^{(t)}$ as input (the *parent*) and returns a modified copy of $x^{(t)}$, denoted by y (the *offspring*). We call this process *mutation*, and we say that $x^{(t)}$ is *mutated*. After mutation, utilizing f, the subroutine

SELECT: $(\{0,1\}^n)^2 \to \{0,1\}^n$ selects either $x^{(t)}$ or y as a starting point for the next iteration, and the result is assigned to $x^{(t+1)}$. We refer to this process as *selection*. We allow mutation and selection to take into account additional information, such as the number of iterations since the last improvement was found.

RLS employs *elitist* selection, i.e., if the fitness of the offspring is at least that of the parent, the offspring is selected. During mutation, RLS flips exactly one bit in its parent, which it chooses uniformly at random. Since this approach leads RLS to getting stuck in local optima where the 1-neighborhood is strictly worse, we consider the following modifications of RLS, each of which adjusts selection and/or mutation: restarts and larger search radii.

Restarts. This approach refers to changing selection after a certain amount of non-improving iterations such that it *always* accepts the offspring. In addition, a restart strategy may change how the offspring is generated (i.e., mutation). There are two straightforward ways that we consider: (1) create an individual sampled uniformly at random, that is, start a new run of RLS (*cold* restart), or (2) create offspring normally but always accept it (*warm* restart). We refer to RLS with cold restarts as cr-RLS and to the variant with warm restart as wr-RLS. Both variants have a parameter $R \in \mathbb{R}_{>0}$. If there are more than $n \ln R$ non-improving iterations, the restart is initiated. The parameter R bounds the probability of failing to find the possible improvement. The probability of not finding an improvement in such a situation is at most $1/R$ [22, Lemma 2].

Larger Search Radii. This approach refers to employing mutations that search beyond the 1-neighborhood. One modification following this pattern is *variable neighborhood search* (VNS [10]), for which many different versions exist. We consider the one displayed in Algorithm 2, which creates offspring with increasing distance from the current individual, exploiting each neighborhood fully before going to the next. Each neighborhood is explored randomly, stopping at the first improvement, and each individual in the neighborhood is created at most once. This guarantees to explore all neighborhoods *eventually*. However, as the neighborhood sizes grow exponentially until distance $n/2$ to the parent, it takes a considerable amount of time to get to larger distances.

Adding Global Mutations. Last, we further add a global mutation to RLS in order to see how much the previous algorithms are hampered by relying on local mutations. Since global mutation serves the same purpose as VNS, we remove the VNS modification. The resulting algorithm is effectively an evolutionary algorithm that uses a local search as mutation. This algorithm is called the $(1 + 1)$ *memetic algorithm* ($(1 + 1)$ MA [17]; Algorithm 4). After creating its offspring by flipping each of the n bits independently with probability $1/n$, it then aims at improving it via the *first-improvement local search* (FILS; Algorithm 3). FILS creates a random permutation π over $[n]$ and flips each bit of its input in the order they appear in π, keeping those and only those flips that improve the individual. Note that FILS flips bits in potentially improved individuals.

Algorithm 2: VNS maximizing fitness function f.

---

1 $x^{(0)} \leftarrow$ individual drawn uniformly at random from $\{0,1\}^n$;

2 $s \leftarrow 1$;

3 **for** $t \in \mathbb{N}$ **do**

4   $\quad$ $y \leftarrow x^{(t)}$;

5   $\quad$ $\Gamma \leftarrow$ the ordered s-neighborhood of $x^{(t)}$, where the order is chosen uniformly at random;

6   $\quad$ **for** $i \in [|\Gamma|]$ **do**

7    $\quad\quad$ $y \leftarrow \Gamma(i)$;

8    $\quad\quad$ **if** $f(y) > f(x^{(t)})$ **then** break the loop iterating over i;

9   $\quad$ **if** $f(y) > f(x^{(t)})$ **then**

10   $\quad\quad$ $x^{(t+1)} \leftarrow y$;

11   $\quad\quad$ $s \leftarrow 1$;

12   $\quad$ **else**

13   $\quad\quad$ $x^{(t+1)} \leftarrow x^{(t)}$;

14   $\quad\quad$ $s \leftarrow \min\{s+1, n\}$;

Algorithm 3: First-Improvement Local Search (FILS) of an individual x, maximizing fitness function f.

---

1 $\pi \leftarrow$ permutation over $[n]$ chosen uniformly at random;

2 **for** $i \in [n]$ **do**

3   $\quad$ $y \leftarrow$ copy of x with bit $\pi(i)$ flipped;

4   $\quad$ **if** $f(y) > f(x)$ **then** $x \leftarrow y$;

5 **return** x;

---

3   Basins of Attraction

A basin of attraction [11] is, intuitively, the area of the search space around a local optimum x such that a local-search algorithm ends up in x (in this sense, the local optimum "attracts" the search points in the basin). Note that some search points might lead to different local optima depending on the random choices of the local-search algorithm (in which case they would be counted to all reachable local optima with the probability to reach the local optimum).

A large basin of attraction around a global optimum x^* is good, as it makes it more likely for the local search to find x^*. For the same reason, a large basin of attraction around a local but not global optimum y is bad, as the local search cannot escape y once it gets to its basin of attraction. Thus, the amount and shape of basins of attraction drastically influence how well local search performs.

We briefly discuss this property of search spaces by considering the case of only two local maxima – one being the global maximum. We model this problem via the function TwoMax: $\{0,1\}^n \to \mathbb{R}$ defined in [9], where one local maximum is the all-0s string 0^n, and the other one is the all-1s string 1^n, which is also the

---

Algorithm 4: $(1 + 1)$ MA maximizing fitness function f.

---
1 $x^{(0)} \leftarrow$ individual drawn uniformly at random from $\{0,1\}^n$;
2 **for** $t \in \mathbb{N}$ **do**
3    $\quad y \leftarrow$ flip each bit in a copy of $x^{(t)}$ with probability $\frac{1}{n}$;
4    $\quad z \leftarrow$ apply FILS to y;
5    $\quad$ **if** $f(z) \geq f(x^{(t)})$ **then** $x^{(t+1)} \leftarrow z$;
6    $\quad$ **else** $x^{(t+1)} \leftarrow x^{(t)}$;

Table 1. Results for run times on TwoMax for different algorithms. Highlights show good run times.

Algorithm	Run time	
RLS	∞ with prob. at least 0.5	[14]
cr-RLS, $R = \omega(n)$	**Expected $O(n \log(nR))$**	
wr-RLS, $R = \omega(n)$	$n^{\Omega(n)}$ with prob. at least $(1 - o(1))0.5$	
VNS	$\Omega(2^n)$ with prob. at least 0.5	
$(1 + 1)$ MA	$n^{\Omega(n)}$ with prob. at least 0.5	

unique global maximum.[1] Both maxima have a basin of attraction that consists of an easy slope toward it, and both basins have the same size. More formally, for all $x \in \{0,1\}^n$ we define

$$\text{TwoMax}(x) = \begin{cases} n + 1, & x = 1^n; \\ \max\{|x|_0, |x|_1\}, & \text{otherwise}; \end{cases}$$

which we aim to maximize; see Fig. 1a for a depiction. Note that a slightly different version of TwoMax, containing both 0^n and 1^n as global maxima, was already defined and analyzed in [18,26].

For this setting we get the following theorem about the performance of various local search algorithms.

Theorem 1. *Regarding run times of RLS, cr-RLS, wr-RLS, VNS, and (1 + 1) MA on* TwoMax, *we get Table 1.*

Intuitively, since the basin of 0^n in TwoMax consists of half the search space, RLS gets stuck at a non-global maximum with probability 1/2 (by symmetry);

---

[1] The optimum of all test functions in this paper is given by the all-1 string, which leads to the observation that the optimum can be found in constant time by just conjecturing this string. Still theoretical research analyzes such functions, because (a) we can nonetheless observe the behavior of different algorithms on these functions, giving insights into the algorithms; and (b) these functions are representatives of much wider classes of functions with either isomorphic or at least similar properties, but for a theoretical analysis we restrict ourselves to the clean case where the rule "more 1 s means closer to the optimum" holds.

this was noted by [14]. For wr-RLS, VNS and the $(1 + 1)$ MA, the same reasoning applies, with the potential to leave again once stuck at the non-global optimum, but at a stiff price.

Since both basins of TwoMax are large, a cheap way of escaping 0^n is to restart RLS. Choosing a reasonable restart parameter R, the expected run time is not only finite but also very efficient.

Table 2. Results for run times on JUMP_d and CLIFF_d, where $d = O(1)$, $d \geq 2$, for different algorithms. Highlights show good run times.

Algorithm	Run time on JUMP_d	Run time on CLIFF_d
RLS	∞ with prob. $1 - o(1)$	∞ with prob. $1 - o(1)$
cr-RLS	Expected $\Omega(2^n)$	Expected $\Omega(2^n/n^d)$
wr-RLS, $R = \Omega(n)$	$n^{\omega(n)}$	**Expected $\Theta(n^3 \log R)$**
VNS	**Expected $\Theta(n^d)$**	Expected $\Theta(n^d)$
$(1 + 1)$ MA	Expected $\Theta(n^{d+1})$	**Expected $\Theta(n^3)$**

Note that the constant 0.5 is essentially due to the basin of the non-global optimum being a 0.5 portion of the search space. The observation about RLS getting stuck and cr-RLS being efficient can thus be generalized in dependence of how large the basin of the global optimum is. We omit this generalization.

4   Deceptive Valleys vs. Guiding Information

Given a local optimum, a *valley* is the area of the search space that has lower fitness than the local optimum but that has to be crossed to arrive at the global optimum. We consider crossing two kinds of valleys. Two well-established fitness functions to model this setting are JUMP [8] and CLIFF [12], parametrized by $d \in \mathbb{N}$, determining the width of the valley. The two functions model two extremes regarding the shape of the valley: In JUMP, the valley contains deceptive fitness signals, guiding the search back to the local optimum, while in CLIFF the fitness signal points to the global optimum. Formally, for all $x \in \{0, 1\}^n$, let

$$\text{JUMP}_d(x) = \begin{cases} |x|_1 + d, & \text{if } |x|_1 \leq n - d \vee |x|_1 = n; \\ |x|_0, & \text{otherwise}; \end{cases}$$

$$\text{CLIFF}_d(x) = \begin{cases} |x|_1, & \text{if } |x|_1 \leq n - d; \\ |x|_1 - d + 1/2, & \text{otherwise}. \end{cases}$$

Both functions are functions of unitation, i.e., the fitness only depends on the number of 1 s of the evaluated solution (see Fig. 1c). Note that there are far more search points with about $n/2$ 0 s than with just a few 0 s (where the valley is), so any local search starts, with high probability, somewhere in the middle and encounters the valley on the way to the global optimum. As a result, with high probability, RLS ends up in a local optimum without chance of escaping. Thus, cold restarts do not lead to successful optimization in polynomial time.

One way to overcome the valley is by finding a local optimum (in distance d of the global optimum) and then creating the global optimum with a single mutation. This is what VNS does. Note that, in this case, the exact layout of the valley is of no importance. This is very different for algorithms which can explore valleys. The $(1 + 1)$ MA and wr-RLS both suffer from the presence of deceptive information, while making good use of guiding information.

Theorem 2. *Regarding run times of RLS, cr-RLS, wr-RLS, VNS, and $(1 + 1)$ MA on* JUMP *and* CLIFF, *we get Table 2.*

The idea of the proof for the $(1 + 1)$ MA is as follows. When currently in a local optimum, with probability $\Theta(1/n)$, samples a search point in the valley just one step closer to the optimum and then, with probability $\Theta(1/n)$ runs up the slope to the global optimum (an otherwise returns to the local optimum).

5   Single Target vs. Multiple Targets

In Sect. 4, we discuss crossing a valley to reach one specific point. In this section, we address the question of what changes if there is not just one point on the other side of the valley, but multiple. To this end, we consider again two fitness functions; they are variants of JUMP and CLIFF from Sect. 4 but suitably shifted into an area of the search space with more than one point after the valley. The case of JUMP was first considered in [3,21]. We make the following formal definitions. Let $d \in \mathbb{N}$. For all $x \in \{0,1\}^n$,

$$\text{SHIFTEDJUMP}_d(x) = \begin{cases} |x|_1 + d, & \text{if } |x|_1 \leq 3n/4 \text{ or } |x|_1 \geq 3n/4 + d; \\ |x|_0, & \text{otherwise}; \end{cases}$$

$$\text{SHIFTEDCLIFF}_d(x) = \begin{cases} |x|_1, & \text{if } |x|_1 \leq 3n/4; \\ |x|_1 - d + 1/2, & \text{otherwise}. \end{cases}$$

Note that the depictions of the functions in Fig. 1d are somewhat misleading: It looks like there is still only one solution directly after the valley. However, since the search space is not the integers from 0 to n, but rather all bit strings $\{0,1\}^n$, there are indeed a lot of points on the other side of the valley at a distance of d to any local optimum: A local optimum has exactly $n/4$ many 0s, and flipping any d of those 0s gives a solution on the other side of the valley (i.e., a point with a fitness higher than that of the local optimum). Thus, for constant d, there are indeed $\Theta(n^d)$ search points just on the other side of the valley.

In Sect. 4, we show that the VNS and the $(1 + 1)$ MA behave basically the same for crossing a deceptive valley: they need to make the jump to the other side of the valley in one go. In this section, we show a major difference. For VNS, after finding a local optimum, this algorithm first searches neighborhoods of distance less than d before finally picking a distance of d for the search. This implies that a lot of time is wasted searching through unrewarding parts of the search space. In contrast to this, the global mutation of the $(1 + 1)$ MA enables stepping over the valley in one jump *of constant probability*. This is also the behavior exhibited by the $(1 + 1)$ EA (see [3]).

Theorem 3. *Regarding run times of RLS, cr-RLS, wr-RLS, VNS, and (1 + 1) MA on* SHIFTEDJUMP *and* SHIFTEDCLIFF, *we get Table 3.*

Table 3. Results for run times on SHIFTEDJUMP$_d$ and SHIFTEDCLIFF$_d$, where $d = O(1)$, $d \geq 2$, for different algorithms. Highlights show good run times.

Algorithm	SHIFTEDJUMP$_d$	SHIFTEDCLIFF$_d$
RLS	∞ with prob. $1 - o(1)$	∞ with prob. $1 - o(1)$
cr-RLS	Expected $2^{\Omega(n)}$	Expected $\Omega(2^n \binom{n}{n/4-1}^{-1})$
wr-RLS, $R = \Omega(n)$	$n^{\omega(n)}$	$\Theta(n \log(R))$
VNS	Expected $\Theta(n^{d-1})$	Expected $\Theta(n^{d-1})$
$(1 + 1)$ MA	**Expected $\Theta(n)$**	**Expected $\Theta(n)$**

6 Iterated Local Optima

In Sect. 4, we show that non-elitist algorithms can have a big advantage in crossing fitness valleys. In this section, we point out one drawback of such algorithms, namely that they can fail and essentially have to restart optimization from a bad part of the search space. Let us suppose, e.g., that the valley is crossed successfully with probability p and otherwise a complete reoptimization has to be made. If only a single valley has to be crossed, this success probability gives $1/p$ attempts and reoptimizations in expectation, which might still be acceptable. However, the success probability decreases exponentially with the number of optima to be crossed in approaching the global optimum.

This is modeled by the following fitness functions inspired by combining the HURDLE fitness function [19] and the RIDGE fitness function [20]. HURDLE consists of multiple CLIFF-like structures, leading to a sequence of local optima. RIDGE is a fitness function where the algorithm has a path of "width 1" to climb to go up to the global optimum. In order to make comparisons with only one local optimum on a ridge, we also define a version of CLIFF on a ridge. For any

$d \in \mathbb{N}$ (denoting the length of the valley) and for all $i \in [0..n]$ and $x \in \{0,1\}^n$,

$$f_d(i) = \begin{cases} 2n, & \text{if } i = n; \\ f_d(i+1) + 2d - 3, & \text{if } d \text{ divides } n - i; \\ f_d(i+1) - 2, & \text{otherwise}; \end{cases}$$

$$\text{HurdleRidge}_d(x) = \begin{cases} n + f_d(|x|_1), & \text{if } x = 1^{|x|_1} 0^{n-|x|_1}; \\ |x|_0, & \text{otherwise}. \end{cases}$$

$$\text{CliffRidge}_d(x) = \begin{cases} |x|_0, & \text{if } x \neq 1^{|x|_1} 0^{n-|x|_1}; \\ n + |x|_1, & \text{if } |x|_1 \leq n - d; \\ n + |x|_1 - d + 1/2, & \text{otherwise}. \end{cases}$$

Note that, for $i \in [0..n]$, (see also Fig. 1b for a depiction)

$$f_d(i) = 2i - (2d+1)|\{j \in [i..n-1] \mid d \text{ divides } n - i\}|.$$

Most search points in HurdleRidge point to the solution 0^n; this is the starting point of the path to the global optimum 1^n. Along this path the fitness is steadily increasing, but once every d steps it goes down $d - 1$, leading to $\Theta(n/d)$ valleys of width d to be crossed.

Table 4. Results for run times on HurdleRidge$_d$ and CliffRidge$_d$, where $d = O(1)$, $d \geq 2$, for different algorithms. Highlights show good run times.

Algorithm	HurdleRidge$_d$	CliffRidge$_d$
RLS	∞ with prob. $1 - o(1)$	∞ with prob. $1 - o(1)$
cr-RLS	Expected $2^{\Omega(n)}$	Expected $2^{\Omega(n)}$
wr-RLS, $R = \Omega(n \log n)$	$\boldsymbol{O(n^3 + n^2 \log R)}$	$\forall c : \Omega(n^c)$ with prob. $1 - o(1)$
VNS	Expected $\Theta(n^{d+1})$	Expected $\Theta(n^d)$
(1 + 1) MA	**Expected $\boldsymbol{\Theta(n^3)}$**	**Expected $\boldsymbol{\Theta(n^3)}$**

For elitist algorithms, optimization proceeds by crossing each of the $\Theta(n)$ many local optima one after the other. In contrast to this result, non-elitist algorithms have a chance to *fail* crossing a fitness valley. This is not a big problem if there is only one valley to be crossed, resulting in an acceptable optimization time on CliffRidge. But on HurdleRidge there are linearly many valleys to cross, so even some small failure probability per crossing leads almost surely to failing to optimize. This happens for warm restarts for HurdleRidge$_d$. Note that this result does not generalize to the (1 + 1) MA, since it can recover from a failure when trying to cross a valley by reverting to the best-so-far solution.

Theorem 4. *Regarding run times of RLS, cr-RLS, wr-RLS, VNS and (1 + 1) MA on HurdleRidge and CliffRidge, we get Table 4.*

7   Discussion and Conclusion

We have seen many different strategies for overcoming local optima. The first strategy, applicable to any randomized algorithm, is to just run the algorithm multiple times (cr-RLS). This leads to a very diverse set of starting points for local search and can boost the success probability of any algorithm which starts off with a reasonable success probability. One problem in this area is to decide when to restart. For RLS, this decision is somewhat easily made, since after about $n \log n$ iterations without improvements, all neighbors have been considered at least once with high probability, so no further improvement occurs. In practice, also small improvements might be a sign of stagnation and can be used as a signal to restart the algorithm. An extreme version of searching with restarts is random search, where no local optimization is employed. This strategy is popular when the fitness landscape is extremely rugged (which blocks local optimization) and different parts of the landscape are very different. Simple grid search optimization also falls into this category.

In Sect. 4, we have seen that giving up elitism in favor of being able to make use of guiding information in the valley might be valuable. Some of the first algorithms that made use of this idea were the Metropolis algorithm and Simulated Annealing, which in turn suffer in their ability to climb simple gradients; for a theoretical comparison with elitist search heuristics, see [13]. Both the Metropolis Algorithm and Simulated Annealing behave like RLS, but they accept worse offspring with a certain probability depending on the fitness difference to the parent. This makes the algorithms sensitive to the fitness values, in contrast to the non-elitist (and elitist) algorithms considered in this paper based on restarts (accepting worse moves only rarely). The advantage of rare (warm) restarts is that other moves can be elitist and thus able to find local optima. Since there are typically more potential worsening moves than improving moves, it is vital to reject worsening moves most of the time.

Another strategy for overcoming local optima is to look further than just the direct neighborhood. This is the idea behind VNS. However, sometimes a lot of samples are wasted locally before attempting a larger jump as, for example, $(1 + 1)$ MA does, see Sect. 5. This is the principle domain of global search heuristics, such as the well-studied $(1 + 1)$ EA. Taking this idea one step further gives the so-called fast $(1 + 1)$ EA [6], sampling offspring at far distances significantly more frequently than the $(1 + 1)$ EA, while still sampling search points at a distance of 1 with constant probability. Another idea is to adjust the search distance distribution whenever progress stagnates; this idea, so-called *stagnation detection*, was analyzed in [7, 21–23]. Note that it is typically fruitful to spend a lot of time searching the local neighborhood in order to exploit local structure.

The different test functions considered are abstractions of what real-world optimization problems look like. In particular, they study the different features in isolation. In Sect. 6, we discussed a test function where a complex test function is constructed by iterating the setting of a local optimum. We saw that in this more complex setting, an algorithm that is successful without this iterated setting is now unsuccessful. Iterated obstacles are generally no bigger problem for elitist algorithms than non-iterated obstacles, but non-elitism has to be applied more carefully. The $(1 + 1)$ MA provides a hybrid, where non-elitism is allowed, but the algorithm might revert to the best-so-far search point.

In conclusion, we see that there is no universally best strategy to do so (which is known for a long time), but properties of the fitness landscape can inform about what algorithms could be efficient. In this paper, we studied the connections between the properties of the fitness landscape and the success of various strategies. In general, since most of the variants do not hamper the ability of local search to find local optima, it is advisable to use *some* variant that can escape local optima. However, the choice of *which* variant to choose depends on the fitness landscape of the problem to optimize. Thus, if one has some knowledge about the optimization problem, that is, one faces a *gray*-box and not a *black*-box scenario, incorporating this knowledge into the choice of how to escape local optima is a very useful or even crucial step in order to get best possible results.

Acknowledgments. This work was supported by a grant by the Independent Research Fund Denmark (DFF-FNU 8021-00260B), and by the Paris Île-de-France Region via the European Union's Horizon 2020 research and innovation program under the Marie Skłodowska-Curie grant agreement No. 945298-ParisRegionFP.

References

1. Aarts, E., Aarts, E.H., Lenstra, J.K.: Local Search in Combinatorial Optimization. Princeton University Press, Princeton (2003)
2. Antipov, D., Doerr, B.: Precise runtime analysis for plateau functions. ACM Trans. Evol. Learn. Optim. **1**(4), 13:1–13:28 (2021). https://doi.org/10.1145/3469800
3. Bambury, H., Bultel, A., Doerr, B.: Generalized jump functions. In: Proceedings of GECCO 2021, pp. 1124–1132. ACM (2021). https://doi.org/10.1145/3449639.3459367
4. Bian, C., Qian, C., Tang, K., Yu, Y.: Running time analysis of the (1+1)-EA for robust linear optimization. Theor. Comput. Sci. **843**, 57–72 (2020). https://doi.org/10.1016/j.tcs.2020.07.001
5. Doerr, B., Le, H.P., Makhmara, R., Nguyen, T.D.: Fast genetic algorithms. In: Proceedings of GECCO 2017, pp. 777–784. ACM Press (2017)
6. Doerr, B., Le, H.P., Makhmara, R., Nguyen, T.D.: Fast genetic algorithms. In: Bosman, P.A.N. (ed.) Proceedings of GECCO 2017, pp. 777–784. ACM (2017). https://doi.org/10.1145/3071178.3071301
7. Doerr, B., Rajabi, A.: Stagnation detection meets fast mutation. In: Proceedings of EvoCOP 2022, pp. 191–207. Springer, Cham (2022). https://doi.org/10.1007/978-3-031-04148-8_13
8. Droste, S., Jansen, T., Wegener, I.: On the analysis of the (1+1) evolutionary algorithm. Theor. Comput. Sci. **276**, 51–81 (2002)
9. Friedrich, T., Oliveto, P.S., Sudholt, D., Witt, C.: Analysis of diversity-preserving mechanisms for global exploration. Evol. Comput. **17**(4), 455–476 (2009)
10. Hansen, P., Mladenovic, N.: Variable neighborhood search. In: Martí, R., Pardalos, P.M., Resende, M.G.C. (eds.) Handbook of Heuristics, pp. 759–787. Springer, Cham (2018). https://doi.org/10.1007/978-3-319-07124-4_19

11. Horn, J., Goldberg, D.E.: Genetic algorithm difficulty and the modality of fitness landscapes. In: Proceedings of FOGA 1995, vol. 3, pp. 243–269. Elsevier (1995)
12. Jagerskupper, J., Storch, T.: When the plus strategy outperforms the comma strategy and when not. In: 2007 IEEE Symposium on Foundations of Computational Intelligence, pp. 25–32. IEEE (2007)
13. Jansen, T., Wegener, I.: A comparison of simulated annealing with a simple evolutionary algorithm on pseudo-Boolean functions of unitation. Theor. Comput. Sci. **386**(1), 73–93 (2007). https://doi.org/10.1016/j.tcs.2007.06.003, https://www.sciencedirect.com/science/article/pii/S0304397507004811
14. Jansen, T., Zarges, C.: Example landscapes to support analysis of multimodal optimisation. In: Handl, J., Hart, E., Lewis, P.R., López-Ibáñez, M., Ochoa, G., Paechter, B. (eds.) PPSN 2016. LNCS, vol. 9921, pp. 792–802. Springer, Cham (2016). https://doi.org/10.1007/978-3-319-45823-6_74
15. Johnson, D.S.: Local optimization and the Traveling Salesman Problem. In: Paterson, M.S. (ed.) ICALP 1990. LNCS, vol. 443, pp. 446–461. Springer, Heidelberg (1990). https://doi.org/10.1007/BFb0032050
16. Neumann, F., Witt, C.: Bioinspired Computation in Combinatorial Optimization - Algorithms and Their Computational Complexity. Springer, Cham (2010). https://doi.org/10.1007/978-3-642-16544-3
17. Nguyen, P.T.H., Sudholt, D.: Memetic algorithms outperform evolutionary algorithms in multimodal optimisation. Artif. Intell. **287**, 103345 (2020). https://doi.org/10.1016/j.artint.2020.103345
18. Pelikan, M., Goldberg, D.E.: Genetic algorithms, clustering, and the breaking of symmetry. In: Schoenauer, M., et al. (eds.) PPSN 2000. LNCS, vol. 1917, pp. 385–394. Springer, Heidelberg (2000). https://doi.org/10.1007/3-540-45356-3_38
19. Prügel-Bennett, A.: When a genetic algorithm outperforms hill-climbing. Theoret. Comput. Sci. **320**(1), 135–153 (2004)
20. Quick, R.J., Rayward-Smith, V.J., Smith, G.D.: Fitness distance correlation and Ridge functions. In: Eiben, A.E., Bäck, T., Schoenauer, M., Schwefel, H.-P. (eds.) PPSN 1998. LNCS, vol. 1498, pp. 77–86. Springer, Heidelberg (1998). https://doi.org/10.1007/BFb0056851
21. Rajabi, A., Witt, C.: Stagnation detection in highly multimodal fitness landscapes. In: Proceedings of GECCO 2021. ACM Press (2021)
22. Rajabi, A., Witt, C.: Stagnation detection with randomized local search. In: Zarges, C., Verel, S. (eds.) EvoCOP 2021. LNCS, vol. 12692, pp. 152–168. Springer, Cham (2021). https://doi.org/10.1007/978-3-030-72904-2_10
23. Rajabi, A., Witt, C.: Self-adjusting evolutionary algorithms for multimodal optimization. Algorithmica **84**, 1694–1723 (2022). https://doi.org/10.1007/s00453-022-00933-z. Preliminary version in GECCO 2020
24. Simon, D.: Evolutionary Optimization Algorithms. Wiley, Hoboken (2013)
25. Stützle, T.: Applying iterated local search to the permutation flow shop problem. Technical report, Citeseer (1998)
26. Van Hoyweghen, C., Goldberg, D.E., Naudts, B.: From TwoMax to the Ising model: easy and hard symmetrical problems. Generations **11**(01), 10 (2001)

Evolutionary Algorithms
for Cardinality-Constrained Ising Models

Vijay Dhanjibhai Bhuva[1,2], Duc-Cuong Dang[2]([✉])[iD], Liam Huber[1],
and Dirk Sudholt[2][iD]

[1] Max-Planck-Institut für Eisenforschung, Düsseldorf, Germany
[2] University of Passau, Passau, Germany
duccuong.dang@uni-passau.de

Abstract. The Ising model is a famous model of ferromagnetism, in
which atoms can have one of two spins and atoms that are neighboured
prefer to have the same spin. Ising models have been studied in evolutionary computation due to their inherent symmetry that poses a challenge
for evolutionary algorithms.

Here we study the performance of evolutionary algorithms on a variant
of the Ising model in which the number of atoms with a specific spin is
fixed. These cardinality constraints are motivated by problems in materials science in which the Ising model represents chemical species of the
atom and the frequency of spins is constrained by the chemical composition of the alloy being modelled. Under cardinality constraints, mutating
spins independently becomes infeasible, thus we design and analyse different mutation operators of increasing complexity that swap different
atoms to maintain feasibility. We prove that randomised local search with
a naive swap operator finds an optimal configuration in $\Theta(n^4)$ expected
worst case time. This time is drastically reduced by using more sophisticated operators such as identifying and swapping clusters of atoms with
the same spin. We show that the most effective operator only requires
$O(n)$ iterations to find an optimal configuration.

Keywords: Ising model · Randomised local search · Constrained
optimisation · Runtime analysis · Graph bisection

1 Introduction

Introduced in 1925 [14] by Ernst Ising, the Ising model is a model of ferromagnetism that consists of an undirected weighted graph whose vertices represent
atoms typically arranged in a lattice structure. (We will use the terms vertex
and atom interchangeably.) Each atom has a spin from $\{+1, -1\}$ and, for a ferromagnetic material, atoms prefer to have the same spin as their neighbours. On
finite lattices, edges are often considered "wrapping around" at the boundaries
so that each vertex has the same number of neighbours and this local property
is repeated in all directions.

G. Rudolph et al. (Eds.): PPSN 2022, LNCS 13399, pp. 456–469, 2022.
https://doi.org/10.1007/978-3-031-14721-0_32

At 0 Kelvin, entropy of mixing among the spins has no benefit and the system is driven towards a ground state that minimises the potential energy of the system. We can view this as an optimisation problem, where the goal is to find an optimal configuration of spins and the fitness is the number of edges for which both end points have the same spin. This means that all configurations where all atoms have the same spin are optimal. While these solutions are easy to write down, *finding* one with evolutionary algorithms (EAs) can be difficult.

This is because the problem has the inherent property of spin-flip symmetry: inverting all spins yields a solution with the same fitness. This symmetry also applies to smaller parts of the graph and it gives rise to *synchronisation problems* [12]: if different parts of a graph evolve clusters of atoms with the same spin, these clusters may use different spins. Then it might be necessary to alter large parts of a cluster to escape from local optima. The Ising model has attracted some interest from the EA community. It was shown empirically that adding niching techniques was particular effective for EAs on Ising models [12]. Fischer and Wegener [9] confirmed this in theoretical runtime analyses for the one-dimensional Ising model on n vertices, for which EAs using crossover and fitness sharing can outperform mutation-only EAs by a factor of $\Theta(n)$. Sudholt [24] considered the Ising model on binary trees and showed that crossover and fitness sharing provide an exponential speedup over mutation-only EAs. Fischer [8] further analysed the Metropolis algorithm on the two-dimensional Ising model and proved that, despite the existence of many hard local optima, it can find global optima in expected polynomial time. More recently, the 2-colouring problem, which is equivalent to the Ising model on bipartite graphs, has been studied theoretically in the context of dynamic optimisation, where the graph may change over time [5,6]. Ising models can also be found as benchmarks in modern software packages for EAs such as the IOHprofiler software [7].

We consider a variant of this model where a cardinality constraint is imposed, that is, the only permitted configurations are those that contain the same number of -1 spins and $+1$ spins. The motivation for this cardinality constraint stems from materials science. Instead of considering the Ising spin to represent magnetic spin on the atoms of a crystalline solid, we can instead use the spin to represent chemical species of the atom. Unlike magnetic moments, the frequency of each Ising spin is then constrained in the grand canonical ensemble by the chemical composition of the alloy being modeled. In this work we restrict ourselves to systems with only two spins and a ferromagnetic Ising interaction matrix, and always accept fitness-improving swaps (which is equivalent to a physical temperature of 0 K in Monte Carlo schemes used to capture thermodynamic averages). These restrictions drive the system to perfect phase separation of the two spins, but are not intrinsically necessary for the cardinality-constrained spin swaps discussed here – in principle we are free to use an arbitrary number of spins with arbitrary interaction matrices, or even to evaluate the fitness by some other means, e.g. with a cluster expansion description [18]. Thus, although we treat simple binary systems here, we hope that these tools may be a useful addition to the existing toolbox for modern materials science problems, such as exploring short range ordering in high entropy alloys [13,26].

From a computer science perspective, although here we only consider lattice graphs, the cardinality constraint significantly increases the difficulty of the problem of finding the Ising model's ground state on general graphs. Optimal configurations of the unconstrained Ising model are easy to state (identical spins), however adding the cardinality constraints further requires that: (i) all spins must be present in the configuration and (ii) they must be present in equal quantity. Imposing only (i) while minimising the number of edges with end points of different spins yields the MINCUT problem [17], which is in $\mathcal{P}$ and well-studied, but the solution is non-trivial. Requiring both (i) and (ii) implies a graph bisection problem: partitioning the vertices of a graph in two equal-sized sets such that the number of cut edges is minimised. This problem is $\mathcal{NP}$-hard and hard to approximate unless $\mathcal{P} = \mathcal{NP}$ [1]. Thus we also hope that having a solid understanding of the behaviours of operators on simple instances can help the design of better algorithms for graph bisection and related problems (e. g. [16]) in the future. The problem of finding the ground state for the generalisation known as the Ising spin glass model, in which a subset of the edges may prefer the interactions of opposite spins on its vertices, is already $\mathcal{NP}$-hard on non-planar lattices [2], however such systems are out of the scope of our study.

The cardinality constraint has a major impact on EA design. While previous studies on unconstrained Ising models used operators inverting individual spins (e. g. one-bit flips or standard bit mutation, encoding spins -1 or 1 as bit values 0 and 1), these operators may easily create infeasible solutions. Thus, we use operators that maintain feasibility by swapping individual spins.

We provide a theoretical runtime analysis for the constrained Ising model in one dimension, i. e. on n-vertex cycle graphs and show that the choice of operators plays a key role for the performance of a simple EA, randomised local search. While a naive swap operator swapping the spins of two randomly chosen atoms uses $\Theta(n^4)$ expected steps in the worst case, designing swap operators using a Gray Box Optimisation approach yields much better results. Restricting swaps to atoms that are neighboured to atoms of the opposite spins, so-called boundary swaps, yields an improved runtime bound of $O(n^2 \log n)$. We then introduce a new operator that identifies clusters of atoms with identical spins and tries to swap whole clusters instead of individual spins. This again speeds up optimisation, reflected in an upper bound of $O(n^{4/3})$ expected generations, at the expense of an increased execution time. Combining clustering with boundary swaps even gives a bound of $O(n)$ generations, which is optimal. Experiments show that our new operators also have significant advantages for 2D and 3D Ising models with varying neighbourhoods and for more than two types of spins.

Our work addresses a hot topic: analysing EAs on problems with constraints [3,4,10,11,21,22]. It aims to advance our understanding of EAs for problems with permutation representations, for which rigorous theoretical studies are scarce [19, 20,23,25]. Finally, we showcase using the Ising model how insights from runtime analyses can inspire new operators with improved performance.

In this extended abstract, many proofs are omitted or sketched.

2   Preliminaries

The natural logarithm is denoted $\ln(\cdot)$ and that of base 2 is denoted $\log(\cdot)$. The n-th harmonic number is $H_n := \sum_{i=1}^{n} \frac{1}{i}$ and $H_n \leq 1 + \ln n = O(\log n)$.

We consider an Ising model given as a graph $G = (V, E)$ where the $n := |V|$ vertices represent spins and edges connect two neighbouring spins. In our model, the choices of spin labels, e.g. $\{-1, 1\}$ versus $\{0, 1\}$, do not matter, thus we consider half of the spins are 0 and half the spins are 1; and n is always even. A configuration is an assignment of spins to the vertices, and it is only feasible if the aforementioned condition is respected. The fitness $f(x)$ of a configuration x is the number of monochromatic edges, i.e. edges where both end points have the same spin. The goal is to maximise this fitness or, equivalently, to minimise the number of dichromatic edges.

Randomised Local Search (RLS) is a simple EA that repeatedly applies a mutation operator OP to produce new solutions, and those that do not worsen the current fitness replace their parent.

Algorithm 1. RLS(x) using elementary mutation operator OP

1: **while** optimum not found **do**
2:      Generate y by applying OP on x, denoted $y := \text{OP}(x)$.
3:      If $f(y) \geq f(x)$, let $x := y$.

Concrete instantiations of OP will be defined later. RLS can be generalised towards an EA with a global search operator by executing a random number of operations in sequence. However, we will show that one operation is sufficient for the one-dimensional Ising model and focus on RLS for simplicity.

3   Runtime Analyses for One-Dimensional Ising Model

In the one-dimensional cardinality-constrained Ising model, the graph G consists of a cycle with an even number n of vertices. We use theoretical runtime analysis to provide rigorous bounds on the expected optimisation time, i.e., the expected number of generations until a global optimum is found.

The current configuration can be seen as a sequence of *blocks*, a maximal sequence of atoms with the same spin (i.e. a sequence that cannot be extended by adding adjacent atoms). Blocks may wrap around the boundaries. Since blocks have maximal length, subsequent blocks have alternating spins. This means that the number of blocks i is always an even number and the fitness is $f(x) = n - i$ since every block has a unique dichromatic edge to the following block.

Mutations increasing the number of blocks are always rejected as then the fitness decreases. We call a vertex v a *boundary vertex* if it has at least one neighbour of the opposite spin.

3.1   Results for Single Swaps

We start with a simple swap operator, denoted SWAP, which picks a pair of atoms of opposite spins uniformly at random and then swaps their spins.

Algorithm 2. SWAP (x) mutation operator

1: Choose an atom i of x with spin 0 uniformly at random and choose an atom j of x with spin 1 uniformly at random.
2: Return y as a copy of x but with the spins of i and j are swapped.

The following result shows that an optimal configuration is found in expected time bounded by $O(n^4)$, for every initial configuration. Despite the high degree 4 of this polynomial, this bound is asymptotically tight as there are configurations for which RLS indeed requires $\Theta(n^4)$ iterations in expectation.

Theorem 1. *From any initial configuration, RLS using* SWAP *optimises the cardinality-constrained 1D Ising model in expected time at most*

$$\frac{4\ln(2) - \pi^2/6 - 1}{64} \cdot n^4 + \frac{5(\ln(2) - 1/2)\, n^3}{16} = O(n^4).$$

There exists an initial configuration from which the above upper bound is asymptotically tight and RLS needs $\Omega(n^4)$ *expected time.*

Proof Sketch. We follow and refine the analysis of RLS on the unconstrained 1D Ising model by Fischer and Wegener [9]. Note that a block of length 1, i.e. a single vertex v, can be removed by swapping v with a boundary vertex of the opposite spin that is not adjacent to v. This improves the fitness by at least 2.

If all blocks contain at least two atoms, it is not possible to eliminate a block in one swap. Following [9], we argue that the lengths of blocks can change over time until a block is reduced to a length of 1, enabling improving swaps.

Let ℓ_t denote the length of a shortest block at time t. If the current configuration has i blocks, by the pigeon hole principle, the size of the smallest block is at most $\ell_t \leq \lfloor n/i \rfloor$. We consider the expected time for a smallest block to disappear and model the process as a Markov chain with states $\{0, 1, \ldots, \lfloor n/i \rfloor\}$ that reflect the length of the shortest block.

If $\ell_t \geq 2$, all blocks have length at least 2 and all blocks have two boundary vertices. Hence, there are exactly i boundary vertices for each spin. W.l.o.g. assume that the vertices are labelled in ascending order from 0 to $n-1$ and assume that there is a smallest block involving vertices $1, \ldots, \ell_t$ with spins of 1. Now, if mutation swaps the spins of position 1 with that of a boundary vertex with spin 0, other than the vertex at position 0, the block is shortened by 1 and we have $\ell_{t+1} = \ell_t - 1$. If we swap position 0 (which has spin 0) with that of a boundary vertex of spin 1 other than that at position 1, the block is lengthened by 1 and $\ell_{t+1} = \ell_t + 1$ (unless there is another block of length ℓ_t; in that case $\ell_{t+1} = \ell_t$). The same arguments apply symmetrically for swaps concerning positions ℓ_t and $\ell_t + 1$. Note that ℓ_t can also be decreased by 1 in case another block of length ℓ_t is shortened. Hence, $P(\ell_{t+1} = \ell_t - 1) \geq 2(i-1)/(n/2)^2 =: p$ and $P(\ell_{t+1} = \ell_t + 1) \leq 2(i-1)/(n/2)^2 = p$. The full proof shows that with the remaining probability mass, $\ell_{t+1} = \ell_t$ (i.e. there are no other transitions).

From the largest state $\ell_t = \lfloor n/i \rfloor$ we may shorten the block with probability p. If the block lengthens, another block becomes the shortest block and its length is at most $\lfloor n/i \rfloor$. Roughly speaking, the time until a shortest block disappears is dominated by a fair random walk with transition probabilities $p = 2(i-1)/(n/2)^2$ to neighbouring states for all states in $\{2, \ldots, \lfloor n/i \rfloor - 1\}$, a reflecting state $\lfloor n/i \rfloor$ and an absorbing state 0. (Transition probabilities from state 1 differ slightly.) The expected waiting time for a transition to another state is $O(1/p) = O(n^2/i)$ and the expected number of transitions to absorption in a fair random walk on states $\{0, \ldots, \lfloor n/i \rfloor\}$ is $O(n^2/i^2)$. Together, the expected time for reaching an improvement, starting with any configuration with i blocks is at most $O(n^4/i^3)$. Summing up these times for all values of i yields an upper bound of $O(n^4)$ since $\sum_{i=1}^{n} 1/i^3 = O(1)$. The full proof uses a rigorous and precise analysis of the Markov chain and works out leading constants from the statement.

For the second statement, assume that the initial configuration has $i = 4$ blocks, each of length $n/4$. Then for each block, while no block has decreased its length to 1, the probability of lengthening it is $p = 2(i - 1)/(n/2)^2 = 24/n^2$ and the probability of shortening it is $p = 24/n^2$ as well. By standard Chernoff bounds, the probability that after εn^4 steps a fixed block has decreased its length to at most 1 is at most $1/8$, if $\varepsilon > 0$ is chosen as a sufficiently small constant. By a union bound, the probability that there is a block whose length has reduced to at most 1 within this time is at most $1/2$. Hence, with probability at least $1/2$, εn^4 generations are not sufficient. This establishes a lower bound of $\varepsilon/2 \cdot n^4$. $\square$

3.2   Swapping only Boundary Atoms

The worst-case expected optimisation time of RLS with SWAP of $\Theta(n^4)$ shows that performance scales poorly with the number of vertices, n. A main reason is that swapping spins of vertices that are not boundary vertices locally worsens the fitness by 2. Hence, the fitness can only worsen, or remain neutral, in such a step. In a typical run, blocks will reduce in number and grow in size and then the chances of choosing to swap boundary vertices are slim.

We argue that this can be easily remedied by adapting the mutation operator. If we use a Gray Box approach and exploit knowledge about the current configuration, we can redefine the operator to only swap boundary vertices. That is, in step 1 of Algorithm 2, we only pick a pair of boundary vertices of opposite spins uniformly at random, and we refer to this operator as BOUNDARY.

The following result shows that BOUNDARY eliminates many idle steps and the expected optimisation time significantly improves to $O(n^2 \log n)$. This can be proven similarly to Theorem 1 since all beneficial swaps considered in the proof of Theorem 1 are also possible with BOUNDARY. While SWAP chose a pair i, j of atoms to swap uniformly at random from $(n/2)^2$ possible pairs, BOUNDARY chooses from at most i^2 possible pairs since there are at most i boundary vertices of any spin. Thus, the associated random walk has better transition probabilities.

Theorem 2. *From any initial configuration, RLS using* BOUNDARY *optimises the cardinality-constrained 1D Ising model in expected time at most*

$$\frac{n^2 H_n}{8} + \frac{n^2}{2} + \frac{5n\left(H_n - 1\right)}{8} = O(n^2 \log n).$$

3.3 Swapping Clusters of Atoms

Swapping boundary vertices improves performance; however, progress is still slow since only single atoms are swapped. We now design an operator able to find *clusters* of identical spins, to swap whole clusters instead of individual spins.

We first pick two atoms i and j with different spins uniformly at random. Then a Breadth-First-Search (BFS) starting from vertex i and exploring the vertices that have the same spin as i is performed. We do the same *in parallel* for j, starting a BFS on all vertices that have the same spin as j. We make sure that these two BFSs are *synchronised*, thus they explore the same depth at each time step. As soon as one BFS call is finished (say the one at i) we also stop the other BFS call (then this one is at j). At that point, we have identified a cluster of vertices with the same spin as i that are all connected to i, let this vertex set be denoted as V_i. Similarly, the other BFS gives another cluster V_j of vertices of the opposite spin, and we have $|V_i| = |V_j|$ due to the synchronisation. We swap the spins in these two vertex sets and maintain the cardinality constraint. We refer to this operator as CLUSTER. Its execution time on 1D, 2D and 3D lattices is proportional to the size of the smaller cluster and $\Theta(n)$ in the worst case; while clusters are small, the average execution time may be much smaller. It is no larger than the execution time for a fitness evaluation.

Algorithm 3. CLUSTER(x) operator

1: Choose an atom i of x with spin 0 and an atom j of x with spin 1 uniformly at random.
2: Run two parallel Breadth-First-Searches (BFS), starting in i and j, respectively, and restricted to the subgraph induced by vertices of the same spin. Stop after one of the BFSs has explored its whole subgraph. Let $V_i := \text{BFS}(x, i)$ and $V_j := \text{BFS}(x, j)$ be the vertex sets of vertices explored during these BFSs.
3: Return y by swapping the spins in V_i and V_j of x.

We show that RLS with CLUSTER only needs $O(n^{4/3})$ expected iterations.

Theorem 3. *For any initial configuration, RLS using* CLUSTER *optimises the cardinality/constrained 1D Ising model in an expected number of iterations of at most*

$$\frac{3 \cdot 4^{1/3} n^{4/3}}{8} + \frac{2^{1/3} n^{2/3}}{4} = O(n^{4/3}).$$

Proof. Let B_i be the block of atoms with spin 0 that contains atom i and let B_j be the block of atoms with spin 1 that contains atom j. BFS starting at atom i will return a set $V_i \subseteq B_i$ and BFS starting at atom j will return a set $V_j \subseteq B_j$.

We also know that $|V_i| = |V_j| = \min\{|B_i|, |B_j|\}$ since the two BFS calls run synchronously and stop when one of the BFS calls has explored a whole block.

Now, if $|B_i| = |B_j|$ then $V_i = B_i$ and $V_j = B_j$ and swapping the spins in B_i and B_j will erase both blocks, improving the fitness by 4.

Assume $|B_i| \neq |B_j|$ and w. l. o. g. $|B_i| < |B_j|$. Then $V_i = B_i$ and $V_j \subsetneq B_j$. Swapping the spins in V_i and V_j will erase B_i. If V_j does *not* contain a boundary atom in B_j, a new block of 0-spins will be created inside of B_j, and the fitness will be unchanged. Otherwise, V_j contains exactly one boundary vertex of B_j. B_j will be shortened to a length of $|B_j| - |V_j| > 0$ and the fitness increases by 2.

We give a lower bound on the probability of improving the fitness in a cluster swap, when two blocks have been fixed.

Lemma 4. *Assume the indices i, j used in BFS are chosen uniformly at random within a block of 0-spins of size a and a block of 1-spins of size b. Then the probability that the sets V_i and V_j returned by the BFS calls will both contain a boundary vertex is at most*

$$\frac{\min\{a, b\}}{\max\{a, b\}}.$$

Proof. Let B_i and B_j denote the respective blocks of 0-spins and 1-spins. When $a = b$, the sets V_i, V_j will contain the whole blocks, that is, $V_i = B_i$ and $V_j = B_j$, and both sets will contain boundary vertices with probability 1 as claimed.

If $a < b$ then V_j contains a boundary vertex if and only if a boundary vertex is found during the first a vertices visited by BFS on B_j, starting at index j. Within a iterations of BFS, j and all vertices within a graph distance of $\lfloor (a - 1)/2 \rfloor$ are reached. This implies that, if j is chosen as one of the leftmost $1 + \lfloor (a - 1)/2 \rfloor$ vertices, the left boundary will be reached. Likewise, if j is chosen as one of the rightmost $1 + \lfloor (a - 1)/2 \rfloor$ vertices, the right boundary will be reached. Note that these two sets are disjoint since $|B_j| = b > a + 1 = 2(1 + (a - 1)/2) \geq 2(1 + \lfloor (a - 1)/2 \rfloor)$. Thus, there are $2(1 + \lfloor (a - 1)/2 \rfloor) \geq 2(1 + (a - 2)/2) = a$ possible choices of j that lead to the discovery of a boundary vertex.

Since by assumption j is chosen uniformly at random from b positions, the sought probability is at least $\frac{a}{b} = \frac{\min\{a,b\}}{\max\{a,b\}}$. The case $a > b$ is symmetric and yields a term of $\frac{b}{a} = \frac{\min\{a,b\}}{\max\{a,b\}}$. $\qquad\square$

We now bound the probability of choosing indices i and j such that a boundary atom is included in the bigger block. Recall that the number of 0-blocks is equal to the number of 1-blocks as spins are alternating from block to block. Denote the sizes of all 0-blocks as $a_1, \ldots, a_r$ and the sizes of all 1-blocks as $b_1, \ldots, b_r$ and note that $a_1 + \cdots + a_r = n/2$ and $b_1 + \cdots + b_r = n/2$. The index i is chosen uniformly at random from all 0-spins. Alternatively, we may imagine that in order to choose i, we first choose a 0-block a_k with probability $a_k/(n/2)$ and then choose i uniformly at random within said block. It is easy to see that the latter approach also creates a uniform distribution over all 0-spins.

Imagining the same two-step procedure for choosing i and j, two blocks a_k and b_m are chosen with probability $a_k/(n/2) \cdot b_m/(n/2)$. Applying Lemma 4 for all combinations of blocks yields that the probability of choosing a block with a boundary vertex is at least

$$\sum_{k,m\in[r]} \frac{a_k}{n/2} \cdot \frac{b_m}{n/2} \cdot \frac{\min\{a_k, b_m\}}{\max\{a_k, b_m\}} = \frac{1}{(n/2)^2} \left(\sum_{k,m\in[r], a_k\leq b_m} a_k^2 + \sum_{k,m\in[r], a_k> b_m} b_m^2 \right)$$

where the equality follows since all summands with $a_k \leq b_m$ simplify to $a_k b_m \cdot a_k/b_m = a_k^2$ and all summands with $a_k > b_m$ simplify to $a_k b_m \cdot b_m/a_k = b_m^2$. The following lemma, whose proof is omitted, bounds this sum from below.

Lemma 5. *For any two sequences of natural numbers, $a_1, \ldots, a_r \in \mathbb{N}$ and $b_1, \ldots, b_r \in \mathbb{N}$, with $a_1 + \cdots + a_r = n/2$ and $b_1 + \cdots + b_r = n/2$,*

$$\frac{1}{(n/2)^2} \left(\sum_{k,m\in[r], a_k\leq b_m} a_k^2 + \sum_{k,m\in[r], a_k> b_m} b_m^2 \right) \geq \max\left\{ \frac{1}{r}, \frac{r^2}{(n/2)^2} \right\}.$$

The maximum from the statement of Lemma 5 is $1/r$ for $r \leq (n/2)^{2/3}$ and $\frac{r^2}{(n/2)^2}$ for $r > (n/2)^{2/3}$. The expected waiting time to increase the fitness is thus at most r for $r \leq (n/2)^{2/3}$ and at most $\frac{(n/2)^2}{r^2}$ for $r > (n/2)^{2/3}$.

Note that the fitness is $n - 2r$ since there are $2r$ blocks. Since the algorithm does not accept any fitness decreases, r is non-increasing over time. Summing over all values of r, the expected optimisation time is at most

$$\sum_{r=1}^{(n/2)^{2/3}} r + \sum_{r=(n/2)^{2/3}+1}^{n/2} \frac{(n/2)^2}{r^2}.$$

The first sum is $\frac{(n/2)^{2/3}((n/2)^{2/3}+1)}{2} = \frac{4^{1/3} n^{4/3}}{8} + \frac{2^{1/3} n^{2/3}}{4}$ and the second sum is at most $(n/2)^2 \int_{(n/2)^{2/3}}^{\infty} \frac{1}{r^2}\, dr = (n/2)^2 \cdot \frac{1}{(n/2)^{2/3}} = \frac{4^{1/3} n^{4/3}}{4}$, proving the claim. $\square$

Finally, we combine our previous ideas of focusing on boundary vertices and swapping clusters of atoms. This means in line 1 of Algorithm 3, instead of picking i and j uniformly from all atoms with the respective spin, we pick them uniformly from all *boundary* atoms with the respective spin. We refer to this mutation operator as BCLUSTER. RLS with this operator improves the fitness in every iteration by at least 2, thus at most $n/2$ iterations are required.

Theorem 6. *For any initial configuration, RLS with BCLUSTER optimises the cardinality/constrained 1D Ising model in at most $n/2$ iterations.*

4 Numerical Experiments

To accompany these theoretical results, we have performed numerical experiments. All our code and the workflows necessary to reproduce the figures shown

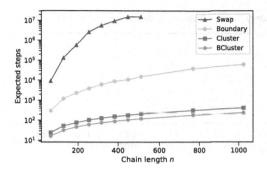

Fig. 1. Median number of iterations using SWAP (blue triangles), BOUNDARY (green circles), CLUSTER (red squares), and BCLUSTER (brown hexagons). (Color figure online)

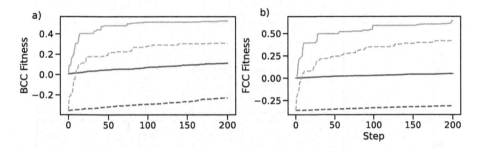

Fig. 2. Evolution of fitness for BOUNDARY (blue) and BCLUSTER (orange) for the (a) body-centered cubic (BCC) and (b) face-centered cubic (FCC) lattices for both two- (solid) and three-spin (dashed) systems. (Color figure online)

here can be found at https://github.com/liamhuber/pyiron_ising and are built on top of the open-source pyiron IDE [15]. For each of the four operators (SWAP, BOUNDARY, CLUSTER, BCLUSTER), we performed 100 independent simulations and measured the median optimisation time for 1D chains of length 64 through 1024. These results are shown on a logarithmic scale in Fig. 1, where it is clear that the more sophisticated swapping routines outperform the simple SWAP operator by many orders of magnitude and have better scaling behaviour, and that cluster-based approaches significantly outperform the approaches which only use single pairs (at the expense of a larger execution time).

We can also take advantage of our numerical infrastructure to explore more complex situations, namely higher dimensions and more spins. In Fig. 3 we look at the first 200 steps of the BOUNDARY and BCLUSTER algorithms applied to two- and three-spins on a 2D square lattice with dimension 32×32, which has four neighbours – one in each Cartesian direction. The fitness is normalised to $[-1, +1]$ and follows a common presentation in physics. (This does not affect the performance of RLS.) The fitness function takes an average over all spin-spin interactions: $f = \frac{1}{n \cdot n_{\text{neigh}}} \sum_{ij} s_i^T M_{ij} s_j$, where s is the one-hot spin vector,

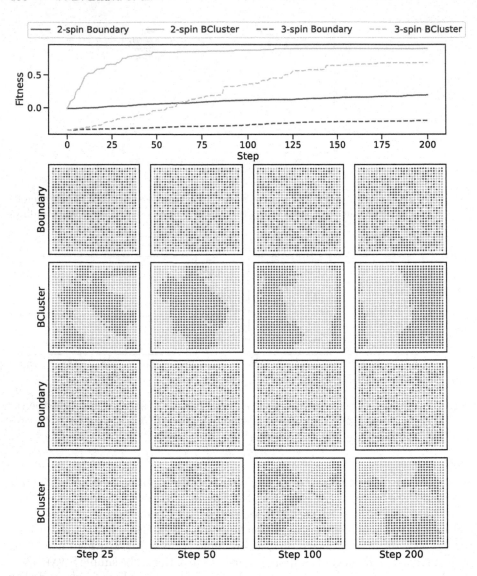

Fig. 3. Fitness as a function of steps for BOUNDARY (blue) and BCLUSTER (orange) operators with two (solid) and three (dashed) spins, along with accompanying snapshots for the square-lattice 2D system. (Color figure online)

n_{neigh} is the number of neighbours (e.g. 4 in the square 2D case), and M is a square matrix with 1's on the diagonal and -1's elsewhere. With this formulation, the initial fitness for the three-spin system is sub-zero since there is a smaller fraction of like-spin vertices available. For both systems the performance of the cluster-based BCLUSTER operator is significantly better. A more intuitive representation of these results is shown below the fitness curve, where snapshots of the system are shown through time with spins distinguished by both colour and shape. Although BOUNDARY does provide some conglomeration of like spins, which is especially obvious to the eye for the two-spin case, the BCLUSTER operator actually achieves reasonable phase separation, and in the two-spin case has already made good progress in minimising the interfacial area.

We also performed these computations for $8 \times 8 \times 8$ 3D lattices (see Fig. 2): a body-centered cubic lattice (BCC – formed from two interpenetrating simple cubic lattices, giving 8 neighbours) and a face-centered cubic lattice (FCC – formed by close packing spheres with a three-layer ABCABC repetition, giving 12 neighbours). These topologies occur frequently in nature, e.g. the structures of iron and aluminium, respectively, at standard temperature and pressure. Here too we see the superiority of the cluster-based algorithm.

To facilitate the use of BCLUSTER in higher dimensions we have added one additional step: the truncation of the clusters. Without this, the increased connectivity of the graph in higher dimensions quickly leads to the formation of two super-clusters in which all like spins are connected. Under these conditions a cluster swap leads to a simple spin inversion and no further progress is possible. Here we have chosen a maximum cluster size uniformly from the interval $[1, n/10]$ before beginning the BFS and stopped BFS prematurely if this size is reached. The details of this constraint are arbitrarily chosen, although from Fig. 2 it is clear that our choices do not destroy the advantage that the BCLUSTER operator holds over BOUNDARY.

5 Conclusions

We have shown that a Gray Box approach and a careful design of swap operators can lead to a significant speed up in the expected optimisation time of RLS on the Ising model on n-vertex cycle graphs. Swapping vertices chosen uniformly at random requires $\Theta(n^4)$ expected time in the worst case. Focusing on boundary vertices in this operator speeds this up by a factor of $\Omega(n^2/\log n)$. Swapping clusters of vertices reduces the expected optimisation time to at most $O(n^{4/3})$, and combining the two ideas gives $O(n)$ expected time. However, cluster swaps have a possibly larger execution time that depends on the size of clusters.

The increased effectiveness of improved operators can be clearly observed in the experiments as well. The data suggests that our upper bound of $O(n^{4/3})$ for RLS with the CLUSTER operator might not be tight as the median runtime appears to be $\Theta(n)$. Further theoretical investigation is required to get tight bounds and to quantify the precise overhead incurred by cluster swaps. We also demonstrated empirically that swapping clusters of vertices performs exceptionally well on high dimension lattices.

References

1. Andreev, K., Räcke, H.: Balanced graph partitioning. Theory Comput. Syst. **39**(6), 929–939 (2006)
2. Barahona, F.: On the computational complexity of Ising spin glass models. J. Phys. A Math. Gen. **15**(10), 3241–3253 (1982)
3. Bian, C., Feng, C., Qian, C., Yu, Y.: An efficient evolutionary algorithm for subset selection with general cost constraints. In: The Thirty-Fourth AAAI Conference on Artificial Intelligence, AAAI 2020, pp. 3267–3274. AAAI Press (2020)
4. Bian, C., Qian, C., Neumann, F., Yu, Y.: Fast pareto optimization for subset selection with dynamic cost constraints. In: Proceedings of the Thirtieth International Joint Conference on Artificial Intelligence, IJCAI 2021, pp. 2191–2197 (2021)
5. Bossek, J., Neumann, F., Peng, P., Sudholt, D.: More effective randomized search heuristics for graph coloring through dynamic optimization. In: Proceedings of the Genetic and Evolutionary Computation Conference (GECCO 2020), pp. 1277–1285. ACM (2020)
6. Bossek, J., Neumann, F., Peng, P., Sudholt, D.: Time complexity analysis of randomized search heuristics for the dynamic graph coloring problem. Algorithmica **83**(10), 3148–3179 (2021)
7. Doerr, C., Ye, F., Horesh, N., Wang, H., Shir, O.M., Bäck, T.: Benchmarking discrete optimization heuristics with IOH profiler. Appl. Soft Comput. **88**, 106027 (2020)
8. Fischer, S.: A polynomial upper bound for a mutation-based algorithm on the two-dimensional ising model. In: Deb, K. (ed.) GECCO 2004. LNCS, vol. 3102, pp. 1100–1112. Springer, Heidelberg (2004). https://doi.org/10.1007/978-3-540-24854-5_108
9. Fischer, S., Wegener, I.: The one-dimensional Ising model: mutation versus recombination. Theoret. Comput. Sci. **344**(2–3), 208–225 (2005)
10. Friedrich, T., Göbel, A., Neumann, F., Quinzan, F., Rothenberger, R.: Greedy maximization of functions with bounded curvature under partition matroid constraints. In: The Thirty-Third AAAI Conference on Artificial Intelligence, AAAI 2019, pp. 2272–2279. AAAI Press (2019)
11. Friedrich, T., Kötzing, T., Lagodzinski, J.A.G., Neumann, F., Schirneck, M.: Analysis of the (1+1) EA on subclasses of linear functions under uniform and linear constraints. Theoret. Comput. Sci. **832**, 3–19 (2020)
12. Goldberg, D.E., Van Hoyweghen, C., Naudts, B.: From TwoMax to the Ising model: easy and hard symmetrical problems. In: Proceedings of the Genetic and Evolutionary Computation Conference (GECCO 2002), pp. 626–633. Morgan Kaufmann (2002)
13. Ikeda, Y., Grabowski, B., Körmann, F.: Ab initio phase stabilities and mechanical properties of multicomponent alloys: a comprehensive review for high entropy alloys and compositionally complex alloys. Mater. Charact. **147**, 464–511 (2019)
14. Ising, E.: Beitrag zur Theorie des Ferromagnetismus. Z. Phys. **31**(1), 253–258 (1925)
15. Janssen, J., et al.: pyiron: an integrated development environment for computational materials science. Comput. Mater. Sci. **163**, 24–36 (2019)
16. Jin, Y., Xiong, B., He, K., Hao, J.-K., Li, C.-M., Fu, Z.-H.: Clustering driven iterated hybrid search for vertex bisection minimization. IEEE Trans. Comput. (2021, Early Access)

17. Karger, D.R., Stein, C.: A new approach to the minimum cut problem. J. ACM **43**(4), 601–640 (1996)
18. Laks, D.B., Ferreira, L., Froyen, S., Zunger, A.: Efficient cluster expansion for substitutional systems. Phys. Rev. B **46**(19), 12587 (1992)
19. Nallaperuma, S., Neumann, F., Sudholt, D.: Expected fitness gains of randomized search heuristics for the traveling salesperson problem. Evol. Comput. **25**, 673–705 (2017)
20. Neumann, F.: Expected runtimes of evolutionary algorithms for the Eulerian cycle problem. Comput. Oper. Res. **35**(9), 2750–2759 (2008). ISSN 0305–0548
21. Qian, C., Zhang, Y., Tang, K., Yao, X.: On multiset selection with size constraints. In: McIlraith, S.A., Weinberger, K.Q. (eds.) Proceedings of the Thirty-Second AAAI Conference on Artificial Intelligence (AAAI 2018), pp. 1395–1402. AAAI Press (2018)
22. Roostapour, V., Neumann, A., Neumann, F., Friedrich, T.: Pareto optimization for subset selection with dynamic cost constraints. Artif. Intell. **302**, 103597 (2022)
23. Scharnow, J., Tinnefeld, K., Wegener, I.: The analysis of evolutionary algorithms on sorting and shortest paths problems. J. Math. Model. Algorithms **3**(4), 349–366 (2004)
24. Sudholt, D.: Crossover is provably essential for the Ising model on trees. In: Proceedings of the Genetic and Evolutionary Computation Conference (GECCO 2005), pp. 1161–1167. ACM Press (2005)
25. Theile, M.: Exact solutions to the traveling salesperson problem by a population-based evolutionary algorithm. In: Cotta, C., Cowling, P. (eds.) EvoCOP 2009. LNCS, vol. 5482, pp. 145–155. Springer, Heidelberg (2009). https://doi.org/10.1007/978-3-642-01009-5_13
26. Wu, Y., et al.: Short-range ordering and its effects on mechanical properties of high-entropy alloys. J. Mater. Sci. Technol. **62**, 214–220 (2021)

General Univariate
Estimation-of-Distribution Algorithms

Benjamin Doerr[1（✉）] and Marc Dufay[2]

[1] LIX, CNRS, École Polytechnique, Institut Polytechnique de Paris,
Palaiseau, France
`doerr@lix.polytechnique.fr`
[2] École Polytechnique, Institut Polytechnique de Paris, Palaiseau, France

Abstract. We propose a general formulation of a univariate estimation-of-distribution algorithm (EDA). It naturally incorporates the three classic univariate EDAs *compact genetic algorithm, univariate marginal distribution algorithm* and *population-based incremental learning* as well as the *max-min ant system* with iteration-best update. Our unified description of the existing algorithms allows a unified analysis of these; we demonstrate this by providing an analysis of genetic drift that immediately gives the existing results proven separately for the four algorithms named above. Our general model also includes EDAs that are more efficient than the existing ones and these may not be difficult to find as we demonstrate for the ONEMAX and LEADINGONES benchmarks.

Keywords: Estimation of distribution algorithms · Genetic drift ·
Running time analysis · Theory

1   Introduction

Estimation-of-distribution algorithms (EDAs) are a class of iterated randomized search heuristics proposed first in the 1990s [21]. Different from genetic algorithms (GAs), which evolve a set P ("population") of good solutions for a given problem, EDAs evolve a probability distribution ("probabilistic model") on the set of possible solutions, hopefully in the way that good solutions have a higher probability assigned to them. Since it is clear that a set P of solutions can be represented by a probability distribution (namely the uniform distribution on P), EDAs (with an appropriate probabilistic model) have a much richer way of transporting information from one iteration to the next than genetic algorithms.

Several results show that this theoretical advantage can be turned into a true advantage when running the EDA in the right way. For example, it was shown that the more cautious way of updating the probabilistic model of EDAs (as opposed to the only alternatives of a GA, which are to accept or discard a solution) can lead to a high robustness to noise [15, 16]. The fact that EDAs can sample with a larger variance was shown to be advantageous for leaving local

G. Rudolph et al. (Eds.): PPSN 2022, LNCS 13399, pp. 470–484, 2022.
https://doi.org/10.1007/978-3-031-14721-0_33

optima [5, 8, 18, 38]. In [7], it was demonstrated that the probabilistic model developed by an EDA allows to obtain much more diverse good solutions than what can be achieved by population-based algorithms.

Due to their higher simplicity, the most studied form of EDAs are *univariate* ones, which sample the variables of each solution independently. When restricting ourselves to pseudo-Boolean optimization, that is, the solutions are bit-strings of length n, then this means that the probabilistic model can be described by a *frequency vector* $p = (p_1, \ldots, p_n) \in [0, 1]^n$ such that a sample $x \in \{0, 1\}^n$ from this model satisfies

$$\Pr[x_i = 1] = p_i \text{ independently for all } i \in [1..n] := \{1, \ldots, n\}. \tag{1}$$

The three classic univariate EDAs are *population-based incremental learning (PBIL)* [2], the *univariate marginal distribution algorithm (UMDA)* [28], and the *compact genetic algorithm (cGA)* [17]. As observed in [22], the *max-min ant system (MMAS)* [35] with iteration-best pheromone update also is a univariate EDA (when used for pseudo-Boolean optimization). We note that the UMDA and this MMAS are special cases of PBIL. Unfortunately, with very few results existing for the PBIL, this connection so far could not be exploited extensively.

So far, these four algorithms have mostly been discussed separately, and for many aspects, only one or two of the four algorithms have been regarded. For example, there are only two mathematical analysis on how EDAs cope with Gaussian noise and these regards only the cGA [16] and the MMAS [15]. For the question how EDAs cope with local optima, the existing runtime analyses only regard the cGA [5, 18, 38] and the MMAS [3]. This leaves many questions unanswered.

We also note that many arguments used in the past were specific to the particular algorithm regarded. For example, the analyses in [5, 18] exploit that the cGA enjoys the property that if the sample with better fitness is closer to the optimum, then the model update will reduce the expected distance of the samples from the optimum. The MMAS does not have this property and consequently, a different proof approach was necessary in [3].

Our Results: In this work, we try to improve this situation by proposing a simple, yet general class of EDAs that includes the four algorithms mentioned above. Our hope is that by thus distilling the common features of these algorithms, it becomes easier to find analyses that apply simultaneously to all four algorithms. We demonstrate that this is indeed possible by proving a quantitative statement on the genetic drift effect in our EDA class. This result contains as special cases the results (separately) proven in [12].

Our second hope is that the large class of EDAs defined by our model also contains algorithms with better performance than the four known algorithms. With elementary non-rigorous arguments, we design such an EDA and show via an experimental analysis that it is at least twice as fast at the cGA and UMDA with optimized parameters on the ONEMAX benchmark. We note that this new algorithm is in no way more complicated than the known special cases of our general model – it just profits from wider ranges of allowed parameters.

2   Previous Work

For reasons of space and since several good surveys and textbooks are available, we describe here only the works that are really close to ours. For a general introduction to EDAs and details on applications, we refer to the surveys [19, 25,31].

Our work, while not purely mathematical, nevertheless is regarding EDAs more from a theoretical perspective. A very recent survey on the state of the art of the theory of EDAs is [22], broader introductions to theoretical approaches in evolutionary computation include [1,10,20,29]. As can easily be deduced from this survey, the theoretical understanding of EDAs is far from complete and for many basic questions, e.g., the runtime on the simple ONEMAX benchmark, a complete answer is still missing. What can also be observed from this survey is that essentially all previous works regard only a single univariate EDA. There are few exceptions, e.g., in [36] both the cGA and the MMAS is analyzed, but also in these cases the results for different algorithms are proven separately.

The only previous work we are aware of that undertakes an attempt towards a unified treatment of univariate EDAs is [14]. There, the framework of an n-Bernoulli-λ-EDA is defined. This framework is very general and includes not only our EDA model, but in fact all univariate EDAs which sample a fixed number λ of offspring according to (1) and then update the probabilistic model p via any deterministic function ϕ that takes as arguments the current model and the offspring together with their fitness. Not surprisingly, in such an extremely general model it is hard to prove meaningful results, and consequently, the particular results in [14] need non-trivial additional assumptions: To show that a stable EDA is not balanced, in particular the additional assumption is made that whenever the EDA optimizes a function with neutral i-th bit, then at all times t the sampling frequency $p_i(t)$ satisfies $\mathrm{Var}[p_i(t+1) \mid p_i(t)] = -ap_i(t)^2 + bp_i(t) + c$ for suitable $a, b, c \in \mathbb{R}$ with $0 < a < 1$, see [14, Theorem 10] (this notion has been relaxed to the requirement that $\inf\{\mathrm{Var}[p_i(t+1) + \mathbf{1}[p_i(t) \notin [d, 1-d]] \mid p_i(t)] \mid t \in \mathbb{N}\} > 0$ for some $d = o(1)$ in [23, Theorem 6.11]). Similarly, the runtime analysis on the LEADINGONES benchmark relies on two specific assumptions how the frequencies behave during the optimization process [14, Theorem 12]. There is no doubt that also with these restrictions, the results in [14] are strong and impressive, but the need for the restrictions suggests that the n-Bernoulli-λ-EDA model is too general to admit strong results covering the whole model (and this is where we hope that our more narrow model is more effective).

There have also been some attempts to encompass EDAs in a model even wider. One of them is by defining these algorithms as *model-based search* algorithms which rely on a parameterized probabilistic model as opposed to *instance-based search* algorithms which rely on a population of solutions [39]. A model-based search algorithm is described by its probabilistic model and the way it updates its model and some parallels can be made between univariate EDAs and gradient-based methods. Another approach described in [30] is by turning existing EDAs into a continuous-time black-box optimization method using the *information-geometric optimization* (IGO) method which can then be turned back into algorithms using time discretization. Existing univariate algorithms like cGA or PBIL can be retrieved using this method. However, these approaches result in a model that is too general to obtain running time results or to obtain ideas how to set the parameter of the algorithms.

3   Univariate EDA: Classic and New

In this section, we first describe briefly the four existing algorithms mentioned in the introduction and then derive from these a general model encompassing all four. We shall write $x \sim \text{Sample}(p)$ to denote that $x \in \{0,1\}^n$ is sampled according to the univariate model described by the frequency vector $p \in [0,1]^n$, that is, that x satisfies (1). We assume that each call of this sampling procedure is stochastically independent from all other samplings and possibly other random decisions of the algorithm. When an algorithm optimizing a function f samples λ individuals, we denote these by $x[1], \ldots, x[\lambda]$ and we denote by $\tilde{x}[1], \ldots, \tilde{x}[\lambda]$ the sorting of these by decreasing (worsening) fitness f, with ties broken randomly. All algorithms initialize the univariate model as $p = (\frac{1}{2}, \ldots, \frac{1}{2})$, which gives the uniform distribution on the search space $\{0,1\}^n$. In their main loop, all sample a certain number of solutions und update the model based on the fitness of the solutions. We first describe all algorithms in the basic version without artificial frequency margins, then propose our general EDA model (also without frequency margins), and finally discuss how to include such margins.

The *compact genetic algorithm (cGA)* [17] samples only two solutions and modifies the frequency vector by a scalar multiple of the difference between the better and the worse solution, that is, $p \leftarrow p + \frac{1}{K}(\tilde{x}[1] - \tilde{x}[2])$. Here K is the only algorithm parameter called *hypothetical population size*. In other words, a frequency p_i does not change if the two samples agree in the i-th bit, and it moves by an additive term of $\frac{1}{K}$ towards the bit value of the better solution otherwise. Usually, K is taken as an even integer since this automatically keeps the frequencies in the range $[0,1]$. For other values of K, one would need to cap the frequencies after the update into the interval $[0,1]$.

The *univariate marginal distribution algorithm (UMDA)* [28] with parameters $\lambda, \mu \in \mathbb{Z}_{\geq 1}$ samples λ solutions and updates the model to the average of the μ best solutions, that is, $p \leftarrow \frac{1}{\mu} \sum_{i=1}^{\mu} \tilde{x}[i]$.

The *max-min ant system (MMAS)* [35] with iteration-best update besides the sample size λ has the *learning rate* $\rho \in]0,1]$ (*pheromone evaporation rate* in the ant colony optimization language) as second parameter. Only the best offspring is used for the model update and it enters the model with weight ρ, that is, the model update is $p \leftarrow (1 - \rho)p + \rho\tilde{x}[1]$.

Population-based incremental learning (PBIL) [2] selects μ out of λ solutions and combines their average weighted by ρ with the current model: $p \leftarrow (1 - \rho)p + \rho\frac{1}{\mu} \sum_{i=1}^{\mu} \tilde{x}[i]$. Consequently, PBIL has as special cases both the UMDA (by taking $\rho = 1$) and the MMAS (by taking $\mu = 1$).

The pseudocodes for these four algorithms are given in Algorithms 1 to 4. As can easily be seen, in all four cases the new model is a linear combination of the samples and the old model. This suggests the following *general univariate EDA model.* Let $\lambda \in \mathbb{Z}_{\geq 1}$ the sample size and $\gamma_0, \gamma_1, \ldots, \gamma_\lambda \in \mathbb{R}$ such that $\sum_{i=0}^\lambda \gamma_i = 1$. The general univariate EDA in its main loop samples λ solutions and updates the frequency vector to $p \leftarrow \gamma_0 p + \sum_{i=1}^\lambda \gamma_i \tilde{x}[i]$, where this is to be understood that frequencies below zero or above one are replaced by zero or one. The complete pseudocode is given in Algorithm 5.

Algorithm 1: The cGA with parameter $K > 0$, maximizing a given function $f : \{0,1\}^n \rightarrow \mathbb{R}$.

---

1   $p(0) = \left(\frac{1}{2}, \ldots, \frac{1}{2}\right) \in [0,1]^n$
2   **for** $t = 1, 2, \ldots$ **do**
3   $\quad$ $x[1] \sim \text{Sample}(p(t-1))$
4   $\quad$ $x[2] \sim \text{Sample}(p(t-1))$
5   $\quad$ **if** $f(x[1]) \geq f(x[2])$ **then**
6   $\quad\quad$ $p(t) = p(t-1) + \frac{1}{K}(x[1] - x[2])$
7   $\quad$ **else**
8   $\quad\quad$ $p(t) = p(t-1) + \frac{1}{K}(x[2] - x[1])$
9   $\quad$ $p(t) = \max(0, \min(1, p(t)))$

Algorithm 2: The UMDA with parameters $\lambda \in \mathbb{Z}_{\geq 1}$ and $\mu \in [1..\lambda]$.

---

1   $p(0) = \left(\frac{1}{2}, \ldots, \frac{1}{2}\right) \in [0,1]^n$
2   **for** $t = 1, 2, \ldots$ **do**
3   $\quad$ **for** $i = 1, 2, \ldots, \lambda$ **do**
4   $\quad\quad$ $x[i] \sim \text{Sample}(p(t-1))$
5   $\quad$ Sort the individuals into $\tilde{x}[1], \ldots, \tilde{x}[\lambda]$ ordered by worsening fitness
6   $\quad$ *%% Update the frequency*
7   $\quad$ $p(t) = \frac{1}{\mu} \sum_{i=1}^\mu \tilde{x}[i]$

We immediately see that the general univariate EDA contains the four algorithms above as special cases. We obtain the cGA by taking $\lambda = 2$, $\gamma_0 = 1$, $\gamma_1 = \frac{1}{K}$, and $\gamma_2 = -\frac{1}{K}$. For the UMDA with parameters λ and μ, we use the same λ and the weights $\gamma_0 = 0$, $\gamma_1 = \cdots = \gamma_\mu = \frac{1}{\mu}$ and $\gamma_{\mu+1} = \cdots = \gamma_\lambda = 0$. The MMAS results from taking $\gamma_0 = 1 - \rho$, $\gamma_1 = \rho$, and $\gamma_2 = \cdots = \gamma_\lambda = 0$. Finally, PBIL is the general EDA with $\gamma_0 = 1 - \rho$, $\gamma_1 = \cdots = \gamma_\mu = \frac{\rho}{\mu}$, and $\gamma_{\mu+1} = \cdots = \gamma_\lambda = 0$.

Algorithm 3: The MMAS with parameters $\lambda \in \mathbb{Z}_{\geq 1}$ and evaporation factor $\rho \in \;]0, 1]$.

---

1   $p(0) = \left(\frac{1}{2}, \ldots, \frac{1}{2}\right) \in [0, 1]^n$
2   **for** $t = 1, 2, \ldots$ **do**
3   **for** $i = 1, 2, \ldots, \lambda$ **do**
4         $x[i] \sim \text{Sample}(p(t-1))$
5   Find an individual with the best fitness $\tilde{x}[1]$
6   %% Update the frequency
7      $p(t) = (1 - \rho)p(t-1) + \rho\tilde{x}[1]$

Algorithm 4: PBIL with parameters $\rho \in \;]0, 1]$, $\lambda \in \mathbb{N}$ and $\mu \in [1..\lambda]$.

---

1   $p(0) = \left(\frac{1}{2}, \ldots, \frac{1}{2}\right) \in [0, 1]^n$
2   **for** $t = 1, 2, \ldots$ **do**
3   **for** $i = 1, 2, \ldots, \lambda$ **do**
4         $x[i] \sim \text{Sample}(p(t-1))$
5   Sort the individuals into $\tilde{x}[1], \ldots, \tilde{x}[\lambda]$ ordered by their fitness
6   %% Update the frequency
7      $p(t) = (1 - \rho)p(t-1) + \frac{\rho}{\mu}\sum_{i=1}^{\mu}\tilde{x}[i]$

---

4   Genetic Drift

Genetic drift is the phenomenon that the sampling frequencies of the probabilistic model move in some direction not because of the feedback from the fitness, but by an unfortunate accumulation of the small random movements that occur when there is no clear signal from the fitness. Genetic drift is problematic in that it can move frequencies close to the boundary values 0 and 1, where they tend to stay longer. This phenomenon and its drawbacks were first discussed in the series of works [32–34]. After a long sequence of fundamental results such as [4, 12–14, 24, 27, 36, 37], mostly runtime analyses which only apply to a regime with low genetic drift, we now understand this phenomenon quite well. For reasons of completeness, we note that EDAs can also be successful in regimes with genetic drift, see, e.g., the runtimes results [4, 37] for the UMDA on ONEMAX and LEADINGONES when the population size is logarithmic, but the general understanding is that genetic drift is dangerous and examples like the analyses of the UMDA on the DLB problem [9, 26] show that genetic drift can lead to drastic performance losses.

The tightest quantitative statements on genetic drift were given in [12]. They were proven via separate analyses for the cGA and PBIL (which imply the corresponding results for the UMDA and MMAS). With our general model for univariate EDAs, we can now provide a unified analysis for these classic algorithms (and all algorithms that will be defined in the future that fit into this model).

Algorithm 5: Our general EDA algorithm defined by $(\gamma_i)_{i=0,\ldots,n}$ such that $\sum_{i=0}^{\lambda} \gamma_i = 1$.

---

1 $p(0) = \left(\frac{1}{2}, \ldots, \frac{1}{2}\right) \in [0,1]^n$
2 **for** $t = 1, 2, \ldots$ **do**
3 $\quad$ %%Sample the individuals
4 $\quad$ **for** $i = 1, 2, \ldots, \lambda$ **do**
5 $\quad\quad$ %%Generate the i-th individual $x[i]$
6 $\quad\quad$ $x_t[i] \sim \text{Sample}(p(t-1))$
7 $\quad$ Sort the individuals into $\tilde{x}_t[1], \ldots, \tilde{x}_t[\lambda]$ by worsening fitness
8 $\quad$ %% Update the frequency
9 $\quad$ $p(t) = \max(0, \min(1, \gamma_0 p(t-1) + \sum_{i=1}^{\lambda} \gamma_i \tilde{x}[i]))$

Genetic drift is usually studied by regarding a neutral bit, that is, a bit that has no influence on the fitness (note that such results imply similar results for bits that are neutral only for a certain time as in the LEADINGONES benchmark or bits that have a preference for one value as in monotonic functions, see [12]). By symmetry, the expected value of the sampling frequency of a neutral bit is always $\frac{1}{2}$ (and in fact, the distribution of this frequency is also symmetric around $\frac{1}{2}$). Nevertheless, as discussed above, the random fluctuations stemming from the updates of the probabilistic model will move this frequency towards the boundary values 0 and 1, and this is the phenomenon of genetic drift. Genetic drift can be quantified, e.g., via statements on the first time that the frequency leaves some middle ground, e.g., the interval $[\frac{1}{3}, \frac{2}{3}]$.

In the remainder of this section, let us assume that the first bit of our objective function f is neutral. Then this bit has no influence on the selection, and consequently for all $i \in [1..\lambda]$, we have $\tilde{x}_1[i] \sim \mathcal{B}(p_1(t-1))$. For simplicity, we write $x_t^i = \tilde{x}_1[i], p_t = p_1(t)$ for all $t \geq 0, i \in [1..\lambda]$. We will also assume that we are not in a totally degenerate case, so there exists $i \in [1..\lambda]$ such that $\gamma_i \neq 0$.

Lemma 1. *The sequence* $\left(\frac{p_t(1-p_t)}{(1-\sum_{i=1}^{\lambda} \gamma_i^2)^t}\right)_{t \geq 0}$ *with respect to the filtration* $(p_t)_{t \geq 0}$ *is a martingale.*

We note that this result is quite beautiful because it gives a good insight on the behavior of a neutral bit and no approximation was needed, allowing us to obtain a martingale and not a supermartingale or a submartingale like what is usually the case. For reasons of space, the formal proof of this and the other results of this paper had to be omitted. They can be found in the appendix of the preprint [6].

Using this result, we can find an upper bound on the expected time for a neutral bit frequency to move away from 1/2.

Lemma 2. *Let* $T_L = \min\{t \geq 0, p_t \leq 1/3 \text{ or } p_t \geq 2/3\}$ *be the first time for a neutral bit to leave* $[1/3, 2/3]$. *Then* $E[T_L] = \mathcal{O}\left(\frac{1}{\sum_{i=1}^{\lambda} \gamma_i^2}\right)$.

To obtain a lower bound and more precise concentration results, we can use a Hoeffding inequality in a way similar, but more general than what was done in [12].

Lemma 3. *For all $T \in \mathbb{N}$ and $\delta > 0$, we have*

$$P\left[\forall t \in [0..T], |p_t - 1/2| < \delta\right] \geq 1 - 2\exp\left(\frac{-\delta^2}{2T\sum_{i=1}^{\lambda}\gamma_i^2}\right).$$

With $T_0 = \frac{\left(\sum_{i=1}^{\lambda}\gamma_i^2\right)^{-1}}{4 \cdot 36 \log n}$ and a union bound, we obtain the following guarantee that neutral frequencies stay away from the boundaries.

Corollary 1. *Assuming that all bits are independent and neutral, with high probability, before iteration T_0, all bits frequencies stay within the range $[1/3, 2/3]$.*

As in [12, part VI], this result can be extended to bits with a preference. For a fitness function f, we say that it is *weakly preferring* 1 in bit i if for all $(x_1, \ldots, x_{i-1}, x_{i+1}, \ldots, x_n) \in \{0,1\}^{n-1}$ we have

$$f(x_1, \ldots, x_{i-1}, 1, x_{i+1}, \ldots, x_n) \geq f(x_1, \ldots, x_{i-1}, 0, x_{i+1}, \ldots, x_n).$$

Many common fitness functions like OneMax or LeadingOnes are *weakly preferring* 1 in any bit.

Corollary 2. *If the fitness function is weakly preferring a 1 on all of its bits, then we have $P\left[\forall i \in [1..n], \forall t \in [0..T_0], p_i^t \geq 1/3\right] = 1 - o(1)$.*

5   Optimizing the $(\gamma_i)_i$

A second advantage of our general formulation of univariate EDAs, besides giving unified proofs, could be that this broad class of algorithms contains EDAs that are superior to the four special cases that have been regarded in the past. To help finding such algorithms, we now discuss the influence on the γ_i on the optimization progress. Since different γ_i might be profitable in different stages of the optimization progress, we analyze their effect in a single iteration, that is, we condition on the current frequency vector. To ease the notation, let us call this frequency vector p (without any time index). Let $\tilde{x}[1], \ldots, \tilde{x}[\lambda]$ denote the λ samples taking in this iteration, sorted already by decreasing fitness. Then, ignoring the influence of frequency boundaries, the next frequency vector p' satisfies $p' = \gamma_0 p + \sum_{i=1}^{\lambda} \gamma_i \tilde{x}[i]$.

We would like to have an idea of what the optimal (γ_i) with respect to minimizing the expected convergence time to reach the optimal solution would look like. To do so, we look during a single iteration for the OneMax function at the best distribution of (γ_i) while keeping the genetic drift minimal. During iteration t, let $X(t)$ be a random variable following distribution $(p_i(t))_i$, we want to maximize $E[f(X(t+1))]$ knowing the previous distribution. ONEMAX being linear, using the linearity of expectation on all the different bits, we have

$$E[f(X(t+1))] = \gamma_0 E[f(X(t))] + \sum_{i=1}^{\lambda} \gamma_i E[f(\tilde{x}[i])]$$

$$= \left(1 - \sum_{i=1}^{\lambda} \gamma_i\right) E[f(X(t))] + \sum_{i=1}^{\lambda} \gamma_i E[f(\tilde{x}[i])]$$

$$= E[f(X(t))] + \sum_{i=1}^{\lambda} \gamma_i \left(E[f(\tilde{x}[i])] - E[f(X(t))]\right).$$

Let us assume that $(\tilde{\gamma}_i)_i$ are optimal for the current iteration and let $\delta = \sum_{i=1}^{\lambda} \tilde{\gamma}_i^2$ be the genetic drift. Because this iteration maximizes the expected outcome of the next distribution while minimizing the genetic drift, it is a solution to

$$\text{Maximize: } E[f(X(t))] + \sum_{i=1}^{\lambda} \gamma_i \left(E[f(\tilde{x}[i])] - E[f(X(t))]\right)$$

$$\text{Subject to: } \sum_{i=1}^{\lambda} \gamma_i^2 \leq \delta$$

Both the function to optimize and the constraint are polynomial so differentiable. Moreover the set solution to the constraint is bounded and closed, so it is compact. Therefore an optimal solution exists and we can use the method of Lagrange multipliers to find it: there exists a Lagrange multiplier $\alpha \leq 0$ such that

$$\begin{bmatrix} E[f(\tilde{x}[1])] - E[f(X(t))] \\ E[f(\tilde{x}[2])] - E[f(X(t))] \\ \ldots \\ E[f(\tilde{x}[\lambda])] - E[f(X(t))] \end{bmatrix} + \alpha \begin{bmatrix} 2\tilde{\gamma}_1 \\ 2\tilde{\gamma}_2 \\ \ldots \\ 2\tilde{\gamma}_\lambda \end{bmatrix} = 0.$$

So $(\tilde{\gamma}_i)_i$ are proportional to $(E[f(\tilde{x}[i])] - E[f(X(t))])_i$. Because $(\tilde{x}[i])$ are sorted according to their fitness, $(E[f(\tilde{x}[i])])_i$ is decreasing so $(\tilde{\gamma}_i)_i$ should also be decreasing.

6   Designing New Univariate EDAs

In this section, we propose two new univariate EDAs (that is, EDAs within our framework with γ_i that do not lead to one of the four classical algorithms) and analyze them via experimental means. Given the momentary state of the art in mathematical runtime analysis of EDAs, it seems out of reach to conduct a mathematical runtime analysis precise enough to make visible the influence of the γ_i on the runtime. The main insight derived from this part of our work is that with not much effort, one can find univariate EDAs which outperform the classic univariate EDAs. We conduct this line of research for the two classic benchmarks ONEMAX and LEADINGONES.

OneMax: Since univariate EDAs sample the bits independently and since in the ONEMAX benchmark each bit contributes the same to the fitness, we expect a somewhat regular behavior in a set of independent samples: Those with best fitness will have many bits set correctly, those with lowest fitness with miss many bit values. This, together with the considerations of the previous section, suggests to give more weights to better samples in the frequency update, and to do this in a somewhat continuous manner. One way of doing so is taking

$$\gamma_0 = 1 - \beta \sum_{i=1}^{\lambda} (1 - \tfrac{i}{\lambda/2}) \approx 1 \text{ and } \gamma_i = \beta(1 - \tfrac{i}{\lambda/2}) \text{ for } i \in [1..\lambda], \qquad (2)$$

where β is a positive number still to be determined. While not perfectly symmetric, essentially here $\tilde{x}[i]$ and $\tilde{x}[\lambda - i]$ have weights of opposite sign, hence γ_0 is essentially one.

We compare this new EDA with the two classic ones UMDA and cGA with optimized parameters. We do not regard the other two classic EDAs since with their learning rate ρ they are structurally quite different and it is less understood what are good parameter settings for these. We note that there is no indication in the literature that the MMAS or PBIL with their slightly cautious learning mechanism could outperform the other two algorithms on a simple unimodal benchmark such as ONEMAX.

For the UMDA and cGA, we determine good parameter values as follows. For the UMDA, we chose to fix λ as $\lfloor \log n\sqrt{n} \rfloor$ since both theoretical and experimental results show that this leads to good performances [37]. We use the same value of λ for our EDA. Still for the UMDA, we set $\mu = \lfloor \lambda/3 \rfloor$ as this gave the best expected runtimes in the experiments we conducted to opitmize the parameters of the UMDA. For cGA, the only parameters that needs to be determined is the hypthetical population size K. From [11, Fig. 1], we know that the expected runtime of the cGA on ONEMAX is roughly a unimodal function in K.[1] Since β in our algorithm plays a similar role as K in the cGA (namely it regulates the strength of the model update), we expect a similar unimodal dependence on β for our algorithms, which we confirm in experiments. For that reason, for each problem size n we determined the optimal values for K and β via ternary search.

Figure 1 displays the average (in 200 runs) runtime of these three algorithms for different problems sizes. These results show that our general algorithm with a gamma distribution that was not used in previous algorithms is about twice as fast as the optimized UMDA and cGA. This suggest that it is not too difficult to find in our broad class of univariate EDAs new algorithms which are significantly faster than the classic algorithms.

---

[1] We know that [27] proved that the runtime of the cGA on ONEMAX is not unimodal in K when n is large enough, but apparently this asymptotic results becomes relevant only for very large population sizes.

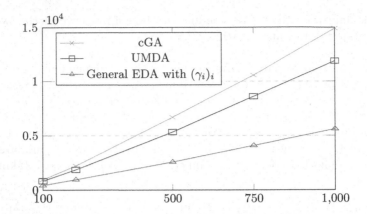

Fig. 1. Average running times (in fitness evaluations) of cGA (with optimized value of K), UMDA (with fixed $\lambda = \lfloor \log n \sqrt{n} \rfloor$ and optimized value $\mu = \lambda/3$), and our general algorithm with fixed gamma as in (2) and β optimized, on the ONEMAX benchmark with problem size between $n = 100$ and $n = 1000$.

LeadingOnes: We undertook a similar work for the LEADINGONES benchmark. In this function, the bits do not contribute independently to the fitness, so our considerations valid in the design of the EDA above are not valid anymore. More detailedly, search points with low fitness reveal very little information how good solutions look like. For this reason, we design our new EDA in the way that such solutions are not taken into account for the model update. Without any optimizing, we set the cutoff for this regime at $\lambda/3$, that is, we have $\tilde{\gamma}_i = 0$ for all $i > \lambda/3$. For the remaining samples, we expect some positive information towards the optimum, and again we expect this to be stronger for better solutions, so we take $\tilde{\gamma}_i$ proportional to $\lfloor \lambda/3 \rfloor - (i - 1)$. With no particular reason, we decided to define an EDA resembling the UMDA, that is, we take $\tilde{\gamma}_0 = 0$ and

$$\tilde{\gamma}_i = \frac{\lfloor \lambda/3 \rfloor - (i - 1)}{\sum_{j=1}^{\lfloor \lambda/3 \rfloor} \lfloor \lambda/3 \rfloor - (j - 1)} \tag{3}$$

for all $i \in [1..\lambda/3]$.

In Fig. 2, we experimentally compare the EDA just designed, the EDA designed in the previous subsection, and the UMDA with parameters optimized (for LEADINGONES) as described in the previous subsection. As expected, the running time of our general algorithm with the $(\gamma_i)_i$ chosen in the previous subsection is not very good (roughly by 25% worse that the UMDA). The EDA just designed, however, beats the UMDA with optimized parameters by roughly 20%. This again shows that with moderately effort, one can find superior EDAs in the class of univariate EDAs defined in this work.

We admit that the ONEMAX and LEADINGONES benchmarks are well-understood, so designing a better univariate EDA for a complicated real-world problem will require more work. Nevertheless, we are optimistic that using intuitive ideas such as the ones above, e.g., a continuous dependence of the γ_i on the rank i, together with some trial-and-error experimentation can lead to good EDAs (better than the classic ones) also for more complex problems.

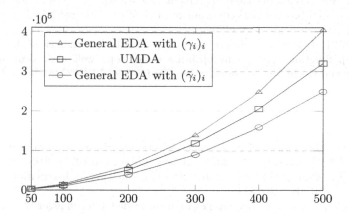

Fig. 2. Average running times (in fitness evaluations) over 200 runs of the classic UMDA (with optimized parameters) and the two EDAs designed in this section, on LEADINGONES with problem size between $n = 50$ and $n = 500$. The $\tilde{\gamma}_i$ chosen with consideration of elementary properties of LEADINGONES clearly outperform the other two algorithms.

7   Conclusion

In this work, we proposed a general formulation of a univariate EDA. It captures the three main univariate EDAs and the MMAS ant colony optimizer with iteration-best update. Our formulation allows to phrase proofs, so far conducted individually for the different algorithms, in a unified manner. We demonstrate this for a recent quantitative analysis of genetic drift. We are optimistic that our formulation also allows to conduct some of the existing runtime analyses in a unified manner. This would be particularly interesting as here many results have been shown only for some of the classic algorithms, e.g., the runtime analyses on the ONEMAX and JUMP benchmarks as well as the results on noisy optimization. However, given the high complexity of the existing analyses for particular algorithms, this might be a challenging task.

Our general formulation also allows to define new univariate EDAs, which might turn out to be superior to the existing ones. With intuitive arguments, we define such EDAs and show experimentally that they beat existing EDAs for the ONEMAX and LEADINGONES benchmarks. We are optimistic that this approach can be profitable also for other optimization problems.

Acknowledgment. This work was supported by a public grant as part of the Investissements d'avenir project, reference ANR-11-LABX-0056-LMH, LabEx LMH.

References

1. Auger, A., Doerr, B. (eds.): Theory of Randomized Search Heuristics. World Scientific Publishing (2011). https://doi.org/10.1142/7438
2. Baluja, S.: Population-based incremental learning: A method for integrating genetic search based function optimization and competitive learning. Tech. rep., Carnegie Mellon University (1994)
3. Benbaki, R., Benomar, Z., Doerr, B.: A rigorous runtime analysis of the 2-MMAS$_{ib}$ on jump functions: ant colony optimizers can cope well with local optima. In: Genetic and Evolutionary Computation Conference, GECCO 2021, pp. 4–13. ACM (2021). https://doi.org/10.1145/3449639.3459350
4. Dang, D.-C., Lehre, P.K., Nguyen, P.T.H.: Level-based analysis of the univariate marginal distribution algorithm. Algorithmica 81(2), 668–702 (2018). https://doi.org/10.1007/s00453-018-0507-5
5. Doerr, B.: The runtime of the compact genetic algorithm on jump functions. Algorithmica 83(10), 3059–3107 (2020). https://doi.org/10.1007/s00453-020-00780-w
6. Doerr, B., Dufay, M.: General univariate estimation-of-distribution algorithms (2022). CoRR abs/2206.11198
7. Doerr, B., Krejca, M.S.: Bivariate estimation-of-distribution algorithms can find an exponential number of optima. In: Genetic and Evolutionary Computation Conference, GECCO 2020, pp. 796–804. ACM (2020). https://doi.org/10.1145/3377930.3390177
8. Doerr, B., Krejca, M.S.: A simplified run time analysis of the univariate marginal distribution algorithm on LeadingOnes. Theoret. Comput. Sci. 851, 121–128 (2021). https://doi.org/10.1016/j.tcs.2020.11.028
9. Doerr, B., Krejca, M.S.: The univariate marginal distribution algorithm copes well with deception and epistasis. Evol. Comput. 29, 543–563 (2021). https://doi.org/10.1162/evco_a_00293
10. Doerr, B., Neumann, F. (eds.): Theory of Evolutionary Computation-Recent Developments in Discrete Optimization. Springer, Cham (2020). https://doi.org/10.1007/978-3-030-29414-4, http://www.lix.polytechnique.fr/Labo/Benjamin.Doerr/doerr_neumann_book.html
11. Doerr, B., Zheng, W.: From understanding genetic drift to a smart-restart parameter-less compact genetic algorithm. In: Genetic and Evolutionary Computation Conference, GECCO 2020, pp. 805–813. ACM (2020). https://doi.org/10.1145/3377930.3390163
12. Doerr, B., Zheng, W.: Sharp bounds for genetic drift in estimation-of-distribution algorithms. IEEE Trans. Evol. Comput. 24, 1140–1149 (2020). https://doi.org/10.1109/TEVC.2020.2987361
13. Droste, S.: A rigorous analysis of the compact genetic algorithm for linear functions. Nat. Comput. 5, 257–283 (2006). https://doi.org/10.1007/s11047-006-9001-0
14. Friedrich, T., Kötzing, T., Krejca, M.S.: EDAs cannot be balanced and stable. In: Genetic and Evolutionary Computation Conference, GECCO 2016, pp. 1139–1146. ACM (2016). https://doi.org/10.1145/2908812.2908895
15. Friedrich, T., Kötzing, T., Krejca, M.S., Sutton, A.M.: Robustness of ant colony optimization to noise. Evol. Comput. 24, 237–254 (2016). https://doi.org/10.1162/EVCO_a_00178
16. Friedrich, T., Kötzing, T., Krejca, M.S., Sutton, A.M.: The compact genetic algorithm is efficient under extreme Gaussian noise. IEEE Trans. Evol. Comput. 21, 477–490 (2017). https://doi.org/10.1109/TEVC.2016.2613739

17. Harik, G.R., Lobo, F.G., Goldberg, D.E.: The compact genetic algorithm. IEEE Trans. Evol. Comput. **3**, 287–297 (1999). https://doi.org/10.1109/4235.797971
18. Hasenöhrl, V., Sutton, A.M.: On the runtime dynamics of the compact genetic algorithm on jump functions. In: Genetic and Evolutionary Computation Conference, GECCO 2018, pp. 967–974. ACM (2018). https://doi.org/10.1145/3205455.3205608
19. Hauschild, M., Pelikan, M.: An introduction and survey of estimation of distribution algorithms. Swarm Evol. Comput. **1**, 111–128 (2011). https://doi.org/10.1016/j.swevo.2011.08.003
20. Jansen, T.: Analyzing Evolutionary Algorithms - The Computer Science Perspective. Springer (2013). https://doi.org/10.1007/978-3-642-17339-4
21. Juels, A., Baluja, S., Sinclair, A.: The equilibrium genetic algorithm and the role of crossover (1993), (Unpublished)
22. Krejca, M.S., Witt, C.: Theory of estimation-of-distribution algorithms. In: Theory of Evolutionary Computation. LNCS, pp. 405–442. Springer, Cham (2020). https://doi.org/10.1007/978-3-030-29414-4_9
23. Krejca, M.S.: Theoretical Analyses of Univariate Estimation-of-Distribution Algorithms. Ph.D. thesis, Universität Potsdam (2019)
24. Krejca, M.S., Witt, C.: Lower bounds on the run time of the univariate marginal distribution algorithm on OneMax. Theoret. Comput. Sci. **832**, 143–165 (2020). https://doi.org/10.1016/j.tcs.2018.06.004
25. Larrañaga, P., Lozano, J.A. (eds.): Estimation of Distribution Algorithms. Genetic Algorithms and Evolutionary Computation. Springer, New York (2002). https://doi.org/10.1007/978-1-4615-1539-5
26. Lehre, P.K., Nguyen, P.T.H.: On the limitations of the univariate marginal distribution algorithm to deception and where bivariate EDAs might help. In: Foundations of Genetic Algorithms, FOGA 2019, pp. 154–168. ACM (2019). https://doi.org/10.1145/3299904.3340316
27. Lengler, J., Sudholt, D., Witt, C.: The complex parameter landscape of the compact genetic algorithm. Algorithmica **83**(4), 1096–1137 (2020). https://doi.org/10.1007/s00453-020-00778-4
28. Mühlenbein, H., Paaß, G.: From recombination of genes to the estimation of distributions I. Binary parameters. In: Voigt, H.-M., Ebeling, W., Rechenberg, I., Schwefel, H.-P. (eds.) PPSN 1996. LNCS, vol. 1141, pp. 178–187. Springer, Heidelberg (1996). https://doi.org/10.1007/3-540-61723-X_982
29. Neumann, F., Witt, C.: Bioinspired Computation in Combinatorial Optimization - Algorithms and Their Computational Complexity. Springer (2010). https://doi.org/10.1007/978-3-642-16544-3
30. Ollivier, Y., Arnold, L., Auger, A., Hansen, N.: Information-geometric optimization algorithms: a unifying picture via invariance principles. J. Mach. Learn. Res. **18**, 1–65 (2017)
31. Pelikan, M., Hauschild, M.W., Lobo, F.G.: Estimation of distribution algorithms. In: Kacprzyk, J., Pedrycz, W. (eds.) Springer Handbook of Computational Intelligence, pp. 899–928. Springer, Heidelberg (2015). https://doi.org/10.1007/978-3-662-43505-2_45
32. Shapiro, J.L.: The sensitivity of PBIL to its learning rate, and how detailed balance can remove it. In: Foundations of Genetic Algorithms, FOGA 2002, pp. 115–132. Morgan Kaufmann (2002)
33. Shapiro, J.L.: Drift and scaling in estimation of distribution algorithms. Evol. Comput. **13**, 99–123 (2005). https://doi.org/10.1162/1063656053583414

34. Shapiro, J.L.: Diversity loss in general estimation of distribution algorithms. In: Runarsson, T.P., Beyer, H.-G., Burke, E., Merelo-Guervós, J.J., Whitley, L.D., Yao, X. (eds.) PPSN 2006. LNCS, vol. 4193, pp. 92–101. Springer, Heidelberg (2006). https://doi.org/10.1007/11844297_10

35. Stützle, T., Hoos, H.H.: MAX-MIN ant system. Futur. Gener. Comput. Syst. **16**, 889–914 (2000). https://doi.org/10.1016/S0167-739X(00)00043-1

36. Sudholt, D., Witt, C.: On the choice of the update strength in estimation-of-distribution algorithms and ant colony optimization. Algorithmica **81**, 1450–1489 (2019). https://doi.org/10.1007/s00453-018-0480-z

37. Witt, C.: Upper bounds on the running time of the univariate marginal distribution algorithm on onemax. Algorithmica **81**(2), 632–667 (2018). https://doi.org/10.1007/s00453-018-0463-0

38. Witt, C.: On crossing fitness valleys with majority-vote crossover and estimation-of-distribution algorithms. In: Foundations of Genetic Algorithms, FOGA 2021, pp. 2:1–2:15. ACM (2021). https://doi.org/10.1145/3450218.3477303

39. Zlochin, M., Birattari, M., Meuleau, N., Dorigo, M.: Model-based search for combinatorial optimization: a critical survey. Ann. Oper. Res. **131**, 373–395 (2004). https://doi.org/10.1023/B:ANOR.0000039526.52305.af

Population Diversity Leads to Short Running Times of Lexicase Selection

Thomas Helmuth[1], Johannes Lengler[2], and William La Cava[3(✉)]

[1] Hamilton College, Clinton, NY, USA
thelmuth@hamilton.edu
[2] ETH Zürich, Zürich, Switzerland
[3] Boston Children's Hospital, Harvard Medical School, Boston, MA, USA
william.lacava@childrens.harvard.edu

Abstract. In this paper we investigate why the running time of lexicase parent selection is empirically much lower than its worst-case bound of $O(N \cdot C)$. We define a measure of population diversity and prove that high diversity leads to low running times $O(N + C)$ of lexicase selection. We then show empirically that genetic programming populations evolved under lexicase selection are diverse for several program synthesis problems, and explore the resulting differences in running time bounds.

Keywords: Lexicase selection · Population diversity · Running time analysis

1 Introduction

Semantic selection methods have been of increased interest as of late in the evolutionary computation community [16,22] due to the observed improvements over more traditional selection methods (such as tournament selection) that only consider the behavior of individuals in aggregate. One such method is lexicase selection [10,19], a parent selection method originally proposed for genetic programming. Since then, the original algorithm and its variants have found success in different domains, including program synthesis [8], symbolic regression [13,18], evolutionary robotics [17], and learning classifier systems [1].

Although an active research community has illuminated many aspects of lexicase selection's behavior via experimental analyses [4,5,7,17], theoretical analyses of lexicase selection have been slower to develop. Previous theoretical work has looked at the probability of selection under lexicase, and also made connections between lexicase selection and Pareto optimization [12]. A study focusing on ecological theory provided insights into the efficacy of lexicase selection [3]. Additionally, the running time of a simple hill climbing algorithm utilizing lexicase selection has been analyzed for the bi-objective leading ones trailing zeroes benchmark problem [11]. However, the recursive nature of lexicase selection, and its step-wise dependence on the behavior of subsets of the population, make it difficult to analyze.

G. Rudolph et al. (Eds.): PPSN 2022, LNCS 13399, pp. 485–498, 2022.
https://doi.org/10.1007/978-3-031-14721-0_34

We focus this paper on a particular gap in the theory of lexicase selection, which is an understanding of its running time. Although the worst-case complexity is known to be $O(N \cdot C)$, where N is the population size and C is the set of training cases, empirical data suggest the worst-case condition is extremely rare in practice [12]. Our goal is to explain this discrepancy through a combination of theory and experiment.

1.1 Our Contributions

We find that the observed running time of lexicase selection can be explained with *population diversity*, by which we mean the phenotypic/behavioral diversity of individuals in a population. Our contributions are threefold:

1. We introduce a new way of measuring population diversity, Definitions 2 and 3, which we call *ε-Cluster Similarity*, or *ε-Similarity* for short. Here, for different values of the parameter ε, we obtain a measure of how similar the population is, where small ε-Cluster Similarity corresponds to high diversity. As we show, this measure is not directly tied to other measures of diversity like the average phenotypical distance (Sect. 2.2) or the mean of the behavioral covariance matrix (Fig. 3).
2. We prove mathematically that lexicase selection is fast when applied to populations which are diverse. More precisely, we show that with low ε-Cluster Similarity, the expected running time of lexicase selection drops from $O(N \cdot C)$ to $O(N + C)$, where the hidden constants depend on the parameter ε and on the quantity k that measures ε-Cluster Similarity.
3. Finally, we show empirically for several program synthesis problems [8] that genetic programming populations are indeed diverse in our sense (have low ε-Similarity). We investigate which parameter ε gives the best running time guarantees for lexicase selection, and we find that the running time guarantees are substantially better than the trivial running time bound of $N \cdot C$.

Our findings apply to both discrete and continuous problems and population behaviors. Although we restrict our analysis to vanilla lexicase selection, we note the results generalize to other variants, including ϵ-lexicase selection [14][1] and down-sampled lexicase selection [9].

2 Preliminaries

2.1 Lexicase Selection

Lexicase selection is used to select a parent for reproduction in a given population. Unlike many common parent selection methods, lexicase selection does not aggregate an individual's performance into a single fitness value. Instead, it considers the loss (errors) on different training *cases* (a.k.a. samples/examples)

---

[1] Our results hold for the original variant, later dubbed "static" ε-lexicase selection [12].

Algorithm 1. Lexicase Selection applied to a population $\mathcal{N}$ without duplicates, with discrete loss/error $L(n, c)$ on training cases $c \in \mathcal{C}$ and individual $n \in \mathcal{N}$. Returns an individual selected to be a parent. $\mathcal{N}_t$ is the remaining candidate pool at step t, $\mathcal{C}'$ is the set of remaining training cases.

LEX($\mathcal{N}$, $\mathcal{C}$, L):
    $\mathcal{C}' \leftarrow \mathcal{C}$; $t \leftarrow 0$; $\mathcal{N}_0 \leftarrow \mathcal{N}$;
    while $|\mathcal{N}_t| > 1$:
        $c \leftarrow$ random choice from $\mathcal{C}'$
        $\ell^* \leftarrow \min\{L(n, c) \mid n \in \mathcal{N}_t\}$
        $\mathcal{N}_{t+1} \leftarrow \{n \in \mathcal{N}_t \mid L(n, c) = \ell^*\}$
        $\mathcal{C}' \leftarrow \mathcal{C}' \setminus \{c\}$
        $t \leftarrow t + 1$
    return unique element from $\mathcal{N}_t$

independently, never comparing (even indirectly) the results on one training case with those on another.

Lexicase selection begins by putting the entire population into a candidate pool. As a preprocessing step, all phenotypical *duplicates* are removed from the pool, i.e., if several individuals give the same loss on all training cases, all but one are removed in the following filtering steps. Then lexicase selection repeatedly selects a new training case t at random, and removes all individuals from the current candidate pool that do not achieve the best loss on case t within the current pool. This process is repeated until the candidate pool contains only a single individual. If the remaining individual has phenotypical duplicates, the selected parent is taken at random from among these behavioral clones. We formalize the algorithm in Algorithm 1.

We remark that the process is guaranteed to end up with a candidate pool of size one: whenever the candidate pool contains at least two individuals, they are not duplicates due to preprocessing. Hence, they differ on at least one training case c, and one of them is filtered out when c is considered. So after all training cases have been processed, it is not possible that the candidate pool contains more than one individual.

Note that the described procedure selects a single individual from the population. In order to gather enough parents for the next generation, it is typically performed $O(N)$ times, where N is the population size. An exception is the preprocessing step that only needs to be performed once each generation. Moreover, finding duplicates can be efficiently implemented via a hash map. Thus preprocessing is usually not the bottleneck of the procedure, and we will focus in this paper on the remaining part: the repeated reduction of the selection pool via random training cases. To exclude the effect of preprocessing, we will assume that the initial population is already free of duplicates.

In case of real-valued (non-discrete) losses, one typically uses a variant known as ϵ-lexicase selection [14]. (Note this use of ϵ is distinct from that used in Sect. 2.2). In the original algorithm, later dubbed "static" ϵ-lexicase selection [12], phenotypic behaviors are binarized prior to lexicase selection, such that individuals within ϵ of the population-wide minimal loss on c have an error of 0, and otherwise an error of 1. Our results extend naturally to this version of ϵ-lexicase selection.

In contrast, in the "dynamic" and "semi-dynamic" variants of ϵ-lexicase selection, lexicase selection removes those individuals whose loss is larger than $\ell^* + \epsilon$, where ℓ^* is the minimal loss in the *current candidate pool* [12]. Our results may extend to these scenarios, but the framework becomes more complicated. Here, it is no longer possible to separate the preprocessing step (i.e., de-duplication) from the actual selection mechanism. Of course, it is still possible to define two individuals $n_1, n_2 \in \mathcal{N}$ as *duplicates* if they differ by at most δ on all training cases. But this is no longer a transitive relation, i.e., it may happen that n_1 and n_2 are duplicates, n_2 and n_3 are duplicates, but n_1 and n_3 are not duplicates. For these reasons, it is necessary to handle duplicates indirectly during the execution of the algorithm. To avoid these complications, we only present Algorithm 1 in the case of discrete losses and without duplicates, but we do include the case of real-valued losses in our analysis.

The worst-case running time of lexicase selection is $O(N \cdot C)$, where $N := |\mathcal{N}|$ is the population size and $C := |\mathcal{C}|$ is the number of training cases. The problem is that in an iteration of the while-loop, it may happen that $\mathcal{N}_{t+1} = \mathcal{N}_t$, i.e., that the candidate pool does not shrink. This is more likely for binary losses. Then, it may happen that $\mathcal{N}_t = \mathcal{N}$ for many iterations of the while-loop, and then computing ℓ^* and $\mathcal{N}_{t+1}$ needs time $O(N)$. Since the while-loop is executed up to C times, this leads to the worst-case runtime $O(N \cdot C)$. Recall that we usually want to run the procedure $O(N)$ times to select all parents for the next generation, which then takes time $O(N^2 \cdot C)$.

One might hope that the expected runtime is much better than the worst-case runtime. This is the case for many classical algorithms like quicksort, but for populations with an unfavorable loss profile, it is not the case here. Consider a population of individuals which have the same losses on all training cases except a single case c, and on case c they all have different losses. Then the candidate pool does not shrink before this case c is found, and finding this case needs $C/2$ iterations in expectation. Thus the expected runtime is still of order $O(N \cdot C)$.[2]

So in order to give better bounds on the running time of lexicase selection, it is required to have some understanding of the involved populations. This is precisely the contribution of this paper: we define a notion of diversity that provably leads to a small running time of lexicase selection, and we empirically show in several genetic programming settings that populations are diverse with respect to this measure.

---

[2] This worst-case example does not hold if the losses are binary, but even that does not help much. It is possible to construct a population of N individuals without duplicates that differ only on $\log_2 N$ binary training cases, and are identical on all other training cases. In this situation, the candidate pool does not shrink before at least one of those training cases is found, and in expectation this takes $C / \log_2 N$ iterations. Thus the expected runtime in this situation is at least $O(N \cdot C / \log N)$, which is not much better than $O(N \cdot C)$.

2.2 ε-Cluster Similarity

We now come to our first main contribution, a new way of measuring diversity. The measure is in phenotype space, so it measures for a training set C how similar the individuals perform on this training set. We first introduce a useful notion, which is the *phenotypical distance* of two individuals.

Definition 1 (Phenotypical Distance). *Consider two individuals m, n that are compared on a set C of training cases. The phenotypical distance between m and n is the number of training cases in which m and n have different losses.*

If the losses are real-valued, then for $\delta > 0$, the phenotypical δ-distance between m and n is the number of training cases in which the losses of m and n differ by at least δ.

We next define the ε-Cluster Similarity. To get an intuition, think of a set of individuals such that *all* individuals inside this set have pairwise small phenotypical distance. Let us call this a *cluster*. Then the ε-Cluster Similarity is the smallest k such that no cluster of size k exists. Or in other words, $k - 1$ is the size of the biggest cluster. Formally, we obtain the following definition.

Definition 2 (ε-Cluster Similarity). *Let $\mathcal{N}$ be a population of individuals, and C be a set of training cases with discrete losses, for example binary losses. Let $\varepsilon \in [0, 1]$. Then the ε-Cluster Similarity is defined to be the minimal $k \geq 2$ such that among every set of k different individuals in $\mathcal{N}$, there are at least two individuals $m, n \in \mathcal{N}$ with phenotypical distance at least $\varepsilon|C|$.*

If instead C is a training set with real-valued losses, then let $\varepsilon \in [0, 1]$ and $\delta > 0$. Then the ε-Cluster Similarity for δ-distance is defined as the minimal $k \geq 2$ such that among every set of k different individuals in $\mathcal{N}$, there are at least two individuals $m, n \in \mathcal{N}$ with phenotypical δ-distance at least $\varepsilon|C|$.

A few remarks are in order to understand the definition better. Firstly, the ε-Cluster Similarity k is a *decreasing* measure of diversity, i.e., less similarity means more diversity and vice versa. Moreover, the value k is increasing in ε: we are only satisfied with two individuals of distance at least $\varepsilon|C|$, which is harder to achieve for larger values of ε. Therefore, we may need a larger set to ensure that it contains a pair of individuals with such a large distance. In other words, a larger value of ε means that we are more restrictive in counting individuals as different, which yields larger value of k: the population is more similar with respect to a more restrictive measure of difference. There is an important tradeoff between ε and k: larger values of ε (which are desirable in terms of diversity, since we search for individuals with larger distances) lead to larger values of k (which is undesirable since we only find such individuals in larger sets).

Second, having small ε-Cluster Similarity is a rather weak notion of diversity: it does not require that *all* pairs of individuals are different from each other. For example, if the population consists of clusters of $k - 1$ individuals which are pairwise very similar, then the ε-Cluster Similarity is k as long as the clusters have distances at least $\varepsilon|\mathcal{C}|$ from each other. We just forbid that there is a cluster of size k such that *every* pair of individuals in the cluster has small distance.

On the other hand, the ε-Cluster Similarity may be a finer measure than, say, the average phenotypical distance in the population. For example, consider a population that consists only of two clusters of almost identical individuals, but the clusters are in opposite corners of the phenotype space, i.e., they differ on almost all training cases. Then the average phenotypical distance is extremely large, $\approx |\mathcal{C}|/2$, which would suggest high diversity. But even for absurdly high $\varepsilon = 0.9$, we would find $k = |\mathcal{N}|/2 + 1$, i.e., a very low diversity according to our definition. It is not hard to see that in this example the expected running time of lexicase selection is $\Omega(|\mathcal{N}| \cdot |\mathcal{C}|)$: in the first step one of the clusters will be removed completely, but afterwards it is very hard to make any further progress. Hence, this example shows that *average phenotypical distance does not predict the running time of lexicase selection well*: even though the example has large average phenotypical distance ("large diversity" in that sense), the running time of lexicase selection is very high. The main theoretical insight of this paper is that this discrepancy can never happen with ε-Cluster Similarity. Whenever ε-Cluster Similarity is low (large diversity), then the expected running time of lexicase selection is small.

To give the reader another angle to grasp the definition of ε-Cluster Similarity, we give a second, equivalent definition in terms of graph theory.

Definition 3 (ε-Cluster Similarity, Equivalent Definition). *Let $\mathcal{N}$ be a population of search points, and $\mathcal{C}$ be a set of training cases with discrete losses. Let $\varepsilon \in [0, 1]$. We define a graph $G = (V, E)$ as follows. The vertex set $V := \mathcal{N}$ is identical with the population. Between any two vertices $m, n \in \mathcal{N}$, we draw an edge if and only if the individuals m and n have the same loss in more than $(1 - \varepsilon)|\mathcal{C}|$ training cases. Then the ε-Cluster Similarity is $k := \alpha + 1$, where α is the clique number of G, i.e., α is the size of the largest clique of the graph G.*

If $\mathcal{C}$ is a training set with real-valued losses and $\delta > 0$ a parameter, then we use the same vertex set for $G = G(\delta)$, but we draw an edge between m and n if and only if the losses of m and n differ by at most δ for more than $(1 - \varepsilon)|\mathcal{C}|$ training cases. Then the ε-Cluster Similarity for δ-distance is again $k := \alpha + 1$, where α is the clique number of $G(\delta)$.

3   Theoretical Result: Low ε-Cluster Similarity Leads to Small Running Times

In this section, we prove mathematically, that a high diversity (i.e., a small ε-Similarity) leads to a small expected running time for lexicase selection.

3.1   Preliminaries

To proof our main theoretical result, we will use the following theorem, known as Multiplicative Drift Theorem [2,15], which is a standard tool in the theory of evolutionary computation.

Theorem 1 (Multiplicative Drift).   *Let* $(X_t)_{t\geq 0}$ *be a sequence of non-negative random variables with a finite state space* $\mathcal{S} \subseteq \mathbb{R}_0^+$ *such that* $0 \in \mathcal{S}$. *Let* $s_{\min} := \min(\mathcal{S} \setminus \{0\})$, *let* $T := \inf\{t \geq 0 \mid X_t = 0\}$, *and for* $t \geq 0$ *and* $s \in \mathcal{S}$ *let* $\Delta_t(s) := E[X_t - X_{t+1} \mid X_t = s]$. *Suppose there exists* $\delta > 0$ *such that for all* $s \in \mathcal{S} \setminus \{0\}$ *and all* $t \geq 0$ *the drift is*

$$\Delta_t(s) \geq \delta s. \tag{1}$$

Then

$$E[T] \leq \frac{1 + E[\ln(X_0/s_{\min})]}{\delta}. \tag{2}$$

Now we can give our theoretical results. Note that the following theorems refer to a single execution of lexicase selection, i.e., M refers to the complexity of finding a single parent via lexicase selection. The following theorem says that the running time is low, $O(|\mathcal{N}| + |\mathcal{C}|)$ if the population has large ε-Cluster Similarity. As common in theoretical running time analysis, we give the running time in terms of *evaluations*, where an evaluation is an execution (or lookup) of $L(n, c)$ for an individual n and a training case c. The running time is proportional to the number of evaluations.

Theorem 2. *Let* $0 < \varepsilon < 1$. *Consider lexicase selection on a population* $\mathcal{N}$ *without duplicates and with* ε-*Cluster Similarity of* $k \in \mathbb{N}$. *Let* M *be the number of evaluations until the population pool is reduced to size 1. Then*

$$E[M] \leq \frac{4|\mathcal{N}|}{\varepsilon} + 2k|\mathcal{C}|.$$

Proof. Consider any two individuals $m, n \in \mathcal{N}$. Assume that both m, n are still in the candidate pool after some selection steps (i.e., after some iterations of the while-loop have been processed). Then m and n can not differ in any of the processed cases $\mathcal{C} \setminus \mathcal{C}'$, because otherwise one of them would have been removed from the population. Therefore, the $\varepsilon|\mathcal{C}|$ cases in which m and n differ are all still contained in $\mathcal{C}'$. In particular, if we choose a new case from $\mathcal{C}'$ at random, then the probability that m and n differ in this case is at least $\varepsilon|\mathcal{C}|/|\mathcal{C}'| \geq \varepsilon$. Note that this holds throughout the algorithm and for any two individuals m, n that are still candidates.

Now we turn to the computation. Let X_t be the number of remaining individuals after t executions of the while-loop. We define $Y_t := X_t$ if $X_t \geq 2k$ and $Y_t := 0$ if $X_t < 2k$. Let T' be the first point in time when $Y_{T'} = 0$ (and thus, $X_{T'} < 2k$).

If $X_t \geq 2k$, then we split the population before the $t+1$-st step into pairs as follows. Since $X_t \geq k$, there are at least two individuals which differ in at least $\varepsilon|\mathcal{C}|$ cases, so we pick two such individuals and pair them up. We can iterate this until there are less than k unpaired individuals left. Therefore, we are able to pair up at least $X_t - (k-1) > X_t - X_t/2 = X_t/2$ individuals, forming at least $X_t/4$ pairs. For each pair, there is a chance of ε that the two differ in the case of the $t+1$-st step, in which case at least one of them is eliminated. Hence, for every $x \geq 2k$,

$$E[X_{t+1} \mid X_t = x] \leq x - \tfrac{\varepsilon x}{4} = x(1 - \tfrac{\varepsilon}{4}).$$

Now let us assume that $Y_t = y > 0$ (and thus $y \geq 2k$). Since $Y_{t+1} \leq X_{t+1}$ by definition, we obtain

$$E[Y_{t+1} \mid Y_t = y] \leq E[X_{t+1} \mid X_t = y] \leq y(1 - \tfrac{\varepsilon}{4}). \tag{3}$$

The advantage of Y_t is that the above bound holds for all $y \geq 0$ (it is trivial for $y = 0$), whereas the corresponding bound for X_t may not hold for $0 < x < 2k$.

Now we bound M by splitting it into the running time M_1 before step T', and the running time M_2 after and including step T'. For M_1, we proceed as follows.

$$M_1 = \sum_{t=0}^{T'-1} X_t = \sum_{t=0}^{T'-1} Y_t = \sum_{t=0}^{\infty} Y_t,$$

because $Y_t = 0$ for $t \geq T'$. Applying (3) iteratively to Y_t, we obtain

$$E[Y_t] \leq (1 - \tfrac{\varepsilon}{4})^t Y_0,$$

where $Y_0 = |\mathcal{N}|$. Plugging this in, we get

$$E[M_1] = \sum_{t=0}^{\infty} E[Y_t] \leq \sum_{t=0}^{\infty} (1 - \tfrac{\varepsilon}{4})^t (|\mathcal{N}| - 1) \leq \tfrac{4|\mathcal{N}|}{\varepsilon},$$

where in the last step we have used the formula $\sum_{t=0}^{\infty} q^t = 1/(1-q)$ for geometric series with $q = 1 - \varepsilon/4$. It remains to bound M_2, and we use a simple bound. Since every case occurs at most once, and since the population size is at most $2k$, we have deterministically $M_2 \leq 2k \cdot |\mathcal{C}|$.

4   Empirical Evaluation in Program Synthesis

We evaluated the theoretical bounds given by Theorem 2 on examples using genetic programming to solve program synthesis benchmark problems. The purpose of this evaluation is to 1) find out how diverse the populations are according to ε-Cluster Similarity; 2) measure the extent to which the new bounds shrink our estimates of the running time of lexicase selection, relative to the known worst-case bounds; 3) evaluate the sensitivity of ε-Cluster Similarity to the parameter, ε, across several problems; and 4) determine how ε-Cluster Similarity compares to a more standard diversity metric in real data.[3]

---

[3] Experiment code: https://github.com/cavalab/lexicase_runtime.

4.1   Experimental Setup

We investigate these aims using 8 program synthesis problems taken from the General Program Synthesis Benchmark Suite [8]. These problems require solution programs to manipulate multiple data types and exhibit different control structures, similar to the types of programs we expect humans to write. Among the 8 problems there are 5 different expected output types (Boolean, integer, float, vector of integers, and string), allowing us to test against multiple data types. In particular, we note that two problems (compare-string-lengths and mirror-image) have Boolean outputs, which we expect to have higher ε-Cluster Similarity values due to having fewer possible output values.

Our experiments were conducted using PushGP, which evolves programs in the Push programming language [20, 21]. PushGP is expressive in the data types and control structures it allows, and has been used previously with these problems [8]. We use Clojush, the Clojure implementation of Push, in our experiments.[4] Each run evolves a population of 1000 individuals for a maximum of 300 generations using lexicase selection and UMAD mutation (without crossover) [6]. We conduct 100 runs of genetic programming per problem.

For each problem, trial, generation, and selection event, we calculated the ε-Cluster Similarity for $\varepsilon \in [0.05, 0.6]$ in increments of 0.05. Using these values, we calculated the bound on the expected running time of lexicase selection according to Theorem 2. For comparison, we also calculated 1) the worst-case complexity of lexicase selection at those operating points and 2) the average pair-wise covariance of the population error vectors. We calculated worst-case running time as $N \cdot C$, neglecting constants, in order to make our comparison to the new running time calculation conservative.

4.2   Results

Figure 1 visualizes the new running time bound as a fraction of the bound given by the worst-case complexity, $N \cdot C$. Across problems, the running time bound given by Theorem 2 ranges from approximately 10–70% of the worst-case complexity bound, indicating much lower expected running times. On average over all problems, the bound given by Theorem 2 is 24.7% of the worst-case bound on running time.

Figure 2 shows the components of Theorem 2 as a function of ε, as well as the total expected running time bound. For small values of ε, the $4|\mathcal{N}|/\varepsilon$ term dominates, whereas for larger values of ε, the $2k|\mathcal{C}|$ term dominates. The observed behavior agrees with our intuition, since larger values of ε lead to larger values of k. The value of ε corresponding to the lowest bound on running time varies by problem, with an average value of 0.29.

Figure 3 compares the new diversity metric (Definition 3) to a more typical definition of behavioral diversity: the mean of the covariance matrix given by

---

[4] https://github.com/lspector/Clojush.

population errors. In general, we observe that ε-Cluster Similarity does not correlate strongly with mean covariance, suggesting that it does indeed measure a different aspect of phenotypic diversity as suggested in Sect. 2.2.

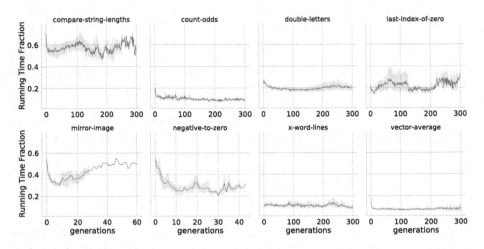

Fig. 1. New running time bound divided by the previously known worst-case bound, as a function of evolutionary generations. The y-axis shows the ratio of both bounds using measurements of relevant parameters. The filled region represents confidence interval of the estimates over all trials.

5   Discussion

We see in Fig. 1 that the new running time bound is below the old bound, and sometimes substantially lower by a factor of 5–10. Thus a substantial part of the discrepancy between the (old) worst-case running time bound and the empirically observed fast running times can be explained by the fact that populations in real-world data are diverse according to our new measure. Since the theoretical analysis is still a worst-case analysis (over all populations), we do not expect that the new bound can explain the whole gap in all situations, but it does explain a substantial factor.

In Fig. 2, we investigate which choice of ε gives the best running time bound. Note that ε is a parameter that can be chosen freely. Every choice of ε gives a value k for the ε-Cluster Similarity, which in turn gives a running time bound. While in Fig. 1 we plotted the best bound that can be achieved with any ε, Fig. 2 shows for each ε the bound that can be obtained with this ε. It is theoretically clear that the term $4N/\varepsilon$ (yellow) is decreasing in ε and $2kC$ (red) is increasing (since k increases with ε). The bound (blue) is the sum of these two terms, and we observe that very small and very large choices of ε often give worse bounds. However, often the blue curve shows some range in which it is rather flat, indicating that there is a large range of ε that gives comparable bounds. In particular, it seems that the range $0.15 \le \varepsilon \le 0.25$ often gives reasonable bounds.

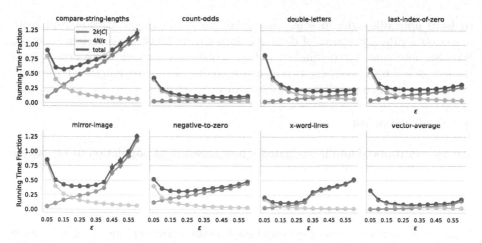

Fig. 2. The ratio of new running time bound and previous known worst-case bound as a function of ε. Optimal values vary by problem but we note flat regions for many problems suggesting a broad range of possible ε values that give similar running time bounds. (Color figure online)

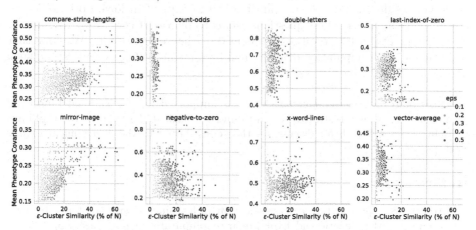

Fig. 3. A comparison of ε-Cluster Similarity (x-axis) and the mean of the covariance matrix of population error (y-axis), colored by ε. ε-Cluster Similarity (k) is plotted as a percent of the population size, N. In most cases (6/8), we observe little relation between the two measures, suggesting ε-Cluster Similarity is indeed measuring a distinct aspect of population diversity, one that is particularly relevant to the running time of lexicase selection.

In Fig. 3 we compare the ε-Cluster Similarity (k normalized as a percent of N) with another diversity measure: the mean of the covariance matrix of population error. If both measures of diversity were highly correlated, we would expect that for any fixed ε (points of the same color), points with larger x-value would consistently have smaller y-value (since k is an inverse measure of diversity). However, in many plots this correlation is spurious at best. It even appears to have opposite signs in some cases. We conclude that ε-Cluster Similarity measures a different aspect of diversity than the mean of the covariance matrix. We also note that, interestingly, the two problems for which there *does* appear to be a relation between the two measures (compare-string-lengths and mirror-image) are the two problems with boolean error vectors.

6   Conclusions

We have introduced and investigated a new measure of population diversity, the ε-Cluster Similarity. We have theoretically proven that large population diversity makes lexicase selection fast, and empirically confirmed that populations are diverse in real-world examples of genetic programming. Thus we have concluded that diverse populations can explain a substantial part of the discrepancy between the general worst-case bound for lexicase selection, and the fast running time in practice.

Naturally, the question arises whether populations in other areas than genetic programming are also diverse with respect to this measure. Moreover, what other consequences does a diverse population have? For example, does it lead to good crossover results? Does it help against getting trapped in local optima? While the intuitive answer to these questions seems to be Yes, it is not easy to pinpoint such statements with rigorous experimental or theoretical results. We hope that our new notion of population diversity can be a means to better understand such questions.

Acknowledgements. William La Cava was supported by the National Library of Medicine and National Institutes of Health under award R00LM012926. We would like to thank Darren Strash for discussions that contributed to the development of this work.

References

1. Aenugu, S., Spector, L.: Lexicase selection in learning classifier systems. In: Proceedings of the Genetic and Evolutionary Computation Conference, pp. 356–364 (2019)
2. Doerr, B., Johannsen, D., Winzen, C.: Multiplicative drift analysis. Algorithmica **64**(4), 673–697 (2012)
3. Dolson, E., Ofria, C.: Ecological theory provides insights about evolutionary computation. In: Proceedings of the Genetic and Evolutionary Computation Conference Companion, GECCO 2018, pp. 105–106. Association for Computing Machinery, New York, NY, USA (2018). https://doi.org/10.1145/3205651.3205780
4. Helmuth, T., McPhee, N.F., Spector, L.: Effects of lexicase and tournament selection on diversity recovery and maintenance. In: Proceedings of the 2016 on Genetic and Evolutionary Computation Conference Companion, pp. 983–990. ACM (2016). http://dl.acm.org/citation.cfm?id=2931657
5. Helmuth, T., McPhee, N.F., Spector, L.: The impact of hyperselection on lexicase selection. In: Proceedings of the 2016 on Genetic and Evolutionary Computation Conference, pp. 717–724. ACM (2016). http://dl.acm.org/citation.cfm?id=2908851

6. Helmuth, T., McPhee, N.F., Spector, L.: Program synthesis using uniform mutation by addition and deletion. In: Proceedings of the Genetic and Evolutionary Computation Conference, GECCO 2018, pp. 1127–1134. ACM, Kyoto, Japan, 15–19 July 2018. https://doi.org/10.1145/3205455.3205603
7. Helmuth, T., Pantridge, E., Spector, L.: On the importance of specialists for lexicase selection. Genet. Program. Evolvable Mach. **21**(3), 349–373 (2020). https://doi.org/10.1007/s10710-020-09377-2
8. Helmuth, T., Spector, L.: General program synthesis benchmark suite. In: GECCO 2015: Proceedings of the 2015 conference on Genetic and Evolutionary Computation Conference, Madrid, Spain, pp. 1039–1046. ACM, 11–15 July 2015. https://doi.org/10.1145/2739480.2754769
9. Helmuth, T., Spector, L.: Explaining and exploiting the advantages of downsampled lexicase selection. In: Artificial Life Conference Proceedings, pp. 341–349. MIT Press, 13–18 July 2020. https://doi.org/10.1162/isal_a_00334, https://www.mitpressjournals.org/doi/abs/10.1162/isal_a_00334
10. Helmuth, T., Spector, L., Matheson, J.: Solving uncompromising problems with lexicase selection. IEEE Trans. Evol. Comput. **19**(5), 630–643 (2015). https://doi.org/10.1109/TEVC.2014.2362729
11. Jansen, T., Zarges, C.: Theoretical analysis of lexicase selection in multi-objective optimization. In: Auger, A., Fonseca, C.M., Lourenço, N., Machado, P., Paquete, L., Whitley, D. (eds.) Parallel Problem Solving from Nature - PPSN XV, pp. 153–164. Springer, Cham (2018). https://doi.org/10.1007/978-3-319-99259-4_13
12. La Cava, W., Helmuth, T., Spector, L., Moore, J.H.: A probabilistic and multi-objective analysis of lexicase selection and epsilon-lexicase selection. Evol. Comput. **27**(3), 377–402 (2019). https://doi.org/10.1162/evco_a_00224, https://arxiv.org/pdf/1709.05394
13. La Cava, W., et al.: Contemporary symbolic regression methods and their relative performance. In: Proceedings of the Neural Information Processing Systems Track on Datasets and Benchmarks, vol. 1, December 2021
14. La Cava, W., Spector, L., Danai, K.: Epsilon-lexicase selection for regression. In: Proceedings of the Genetic and Evolutionary Computation Conference 2016, GECCO 2016, New York, NY, USA, pp. 741–748. ACM (2016). https://doi.org/10.1145/2908812.2908898
15. Lengler, J.: Drift analysis. In: Theory of Evolutionary Computation. NCS, pp. 89–131. Springer, Cham (2020). https://doi.org/10.1007/978-3-030-29414-4_2
16. Liskowski, P., Krawiec, K., Helmuth, T., Spector, L.: Comparison of semantic-aware selection methods in genetic programming. In: Proceedings of the Companion Publication of the 2015 Annual Conference on Genetic and Evolutionary Computation, GECCO Companion 2015, New York, NY, USA, pp. 1301–1307. ACM (2015). https://doi.org/10.1145/2739482.2768505
17. Moore, J.M., Stanton, A.: Tiebreaks and diversity: isolating effects in lexicase selection. In: The 2018 Conference on Artificial Life, pp. 590–597 (2018). https://doi.org/10.1162/isal_a_00109
18. Orzechowski, P., La Cava, W., Moore, J.H.: Where are we now? A large benchmark study of recent symbolic regression methods. In: Proceedings of the 2018 Genetic and Evolutionary Computation Conference, GECCO 2018, April 2018. https://doi.org/10.1145/3205455.3205539, tex.ids: orzechowskiWhereAreWe2018a arXiv: 1804.09331

19. Spector, L.: Assessment of problem modality by differential performance of lex-icase selection in genetic programming: a preliminary report. In: Proceedings of the Fourteenth International Conference on Genetic and Evolutionary Computation Conference Companion, pp. 401–408 (2012). http://dl.acm.org/citation.cfm?id=2330846

20. Spector, L., Klein, J., Keijzer, M.: The Push3 execution stack and the evolution of control. In: GECCO 2005: Proceedings of the 2005 conference on Genetic and Evolutionary Computation, Washington DC, USA, vol. 2, pp. 1689–1696. ACM Press, 25–29 June 2005. https://doi.org/10.1145/1068009.1068292

21. Spector, L., Robinson, A.: Genetic programming and autoconstructive evolution with the push programming language. Genet. Program. Evolvable Mach. 3(1), 7–40 (2002). http://hampshire.edu/lspector/pubs/push-gpem-final.pdf, https://doi.org/10.1023/A:1014538503543

22. Vanneschi, L., Castelli, M., Silva, S.: A survey of semantic methods in genetic programming. Genet. Program. Evolvable Mach. 15(2), 195–214 (2014). https://doi.org/10.1007/s10710-013-9210-0

Progress Rate Analysis of Evolution Strategies on the Rastrigin Function: First Results

Amir Omeradzic[✉][iD] and Hans-Georg Beyer[iD]

Research Center Business Informatics, Vorarlberg University of Applied Sciences,
Hochschulstraße 1, 6850 Dornbirn, Austria
{amir.omeradzic,hans-georg.beyer}@fhv.at
http://homepages.fhv.at/hgb

Abstract. A first order progress rate is derived for the intermediate multi-recombinative Evolution Strategy $(\mu/\mu_I, \lambda)$-ES on the highly multimodal Rastrigin test function. The progress is derived within a linearized model applying the method of so-called noisy order statistics. To this end, the mutation-induced variance of the Rastrigin function is determined. The obtained progress approximation is compared to simulations and yields strengths and limitations depending on mutation strength and distance to the optimizer. Furthermore, the progress is iterated using the dynamical systems approach and compared to averaged optimization runs. The property of global convergence within given approximation is discussed. As an outlook, the need of an improved first order progress rate as well as the extension to higher order progress including positional fluctuations is explained.

Keywords: Evolution Strategies · Rastrigin function · Progress rate analysis · Global optimization

1 Introduction

Evolution Strategies (ES) [12,13] are well-recognized Evolutionary Algorithms suited for real-valued non-linear optimization. State-of-the-art ES such as the CMA-ES [8] or its simplification [5] are also well-suited for locating global optimizers in highly multimodal fitness landscapes. While the CMA-ES was originally mainly intended for non-differentiable optimization problems, but yet regarded as a locally acting strategy, it was already in [7] observed that using a large population size can make the ES a strategy that is able to locate the global optimizer among a huge number of local optima. This is a surprising observation when considering the ES as a strategy that acts mainly local in the search space following some kind of gradient or natural gradient [3,6,11]. As one can easily check using standard (highly) multimodal test functions such as Rastrigin, Ackley, and Griewank to name a few, this ES property is not intimately related to the covariance matrix adaptation (CMA) ES which generates non-isotropic

© The Author(s) 2022
G. Rudolph et al. (Eds.): PPSN 2022, LNCS 13399, pp. 499–511, 2022.
https://doi.org/10.1007/978-3-031-14721-0_35

correlated mutations, but can also be found in $(\mu/\mu_I, \lambda)$-ES with *isotropic* mutations. Therefore, if one wants to understand the underlying working principles how the ES locates the global optimizer, the analysis of the $(\mu/\mu_I, \lambda)$-ES should be the starting point.

The question regarding why and when optimization algorithms – originally designed for local search – are able to locate global optima has gained attention in the last few years. A recurring idea comes from relaxation procedures that transform the original multimodal optimization problem into a convex optimization problem called Gaussian continuation [9]. Gaussian continuation is nothing else but a convolution of the original optimization problem with a Gaussian kernel. As has been shown in [10], using the right Gaussian, Rastrigin-like functions can be transformed into a convex optimization problem, thus making it accessible to gradient following strategies. However, this raises the question how to perform the convolution efficiently. One road followed in [14] uses high-order Gauss-Hermite integration in conjunction with a gradient descent strategy yielding surprisingly good results. The other road coming to mind is approximating the convolution by Gaussian sampling. This resembles the procedure ES do: starting from a parental state, offspring are generated by Gaussian mutations. The problem is, however, that in order to get a reliable gradient, a huge number of samples, i.e. offspring in ES must be generated in order to get reliable convolution results. The number of offspring needed to get reliable estimates seems much larger than the offspring population size needed in ES experiments conducted in [7] showing approximately a linear relation between problem dimension N and population size for the Rastrigin function. Therefore, understanding the ES performance from viewpoint of Gaussian relaxation does not seem to help much.

The approach followed in this paper will incorporate two main concepts, namely a progress rate analysis as well as its application within the so-called evolution equations modeling the transition dynamics of the ES [2]. The progress rate measure yields the expected positional change in search space between two generations depending on location, strategy and test function parameters. Aiming to investigate and understand the dynamics of globally converging ES runs, the progress rate is an essential quantity to model the expected evolution dynamics over many generations.

This paper provides first results of a scientific program that aims at an analysis of the performance of the $(\mu/\mu_I, \lambda)$-ES on Rastrigin's test function based on a first order progress rate. After a short introduction of the $(\mu/\mu_I, \lambda)$-ES, the N-dimensional first order progress will be defined and an approximation will be derived resulting in a closed form expression. The predictive power and its limitations will be checked by one-generation experiments. The progress rate will then be used to simulate the ES dynamics on Rastrigin using difference equations. This simulation will be compared with real runs of the $(\mu/\mu_I, \lambda)$-ES. In a concluding section a summary of the results and outlook of the future research will be given.

2   Rastrigin Function and Local Quality Change

The real-valued minimization problem defined for an N-dimensional search vector $\mathbf{y} = (y_1, ..., y_N)$ is performed on the Rastrigin test function f given by

$$f(\mathbf{y}) = \sum_{i=1}^{N} f_i(y_i) = \sum_{i=1}^{N} y_i^2 + A - A\cos(\alpha y_i), \tag{1}$$

with A denoting the oscillation amplitude and $\alpha = 2\pi$ the corresponding frequency. The quadratic term with superimposed oscillations yields a finite number of local minima M for each dimension i, such that the overall number of minima scales exponentially as M^N posing a highly multimodal minimization problem. The global optimizer is at $\hat{\mathbf{y}} = \mathbf{0}$.

For the progress rate analysis in Sect. 4 the local quality function $Q_{\mathbf{y}}(\mathbf{x})$ at $\mathbf{y}$ due to mutation vector $\mathbf{x} = (x_1, ..., x_N)$ is needed. In order to reuse results from noisy progress rate theory it will be formulated for the *maximization* case of $F(\mathbf{y}) = -f(\mathbf{y})$ with $F_i(y_i) = -f_i(y_i)$, such that local quality change yields

$$Q_{\mathbf{y}}(\mathbf{x}) = F(\mathbf{y} + \mathbf{x}) - F(\mathbf{y}) = f(\mathbf{y}) - f(\mathbf{y} + \mathbf{x}). \tag{2}$$

$Q_{\mathbf{y}}(\mathbf{x})$ can be evaluated for each component i independently giving

$$Q_{\mathbf{y}}(\mathbf{x}) = \sum_{i=1}^{N} Q_i(x_i) = \sum_{i=1}^{N} f_i(y_i) - f_i(y_i + x_i) \tag{3}$$

$$= \sum_{i=1}^{N} -\left(x_i^2 + 2y_i x_i + A\cos(\alpha y_i)(1 - \cos(\alpha x_i)) + A\sin(\alpha y_i)\sin(\alpha x_i)\right). \tag{4}$$

A closed form solution of the progress rate appears to be obtainable only for a linearized expression of $Q_i(x_i)$. A first approach taken in this paper is based on a Taylor expansion for the mutation x_i and discarding higher order terms

$$Q_i(x_i) = F_i(y_i + x_i) - F_i(y_i) = \frac{\partial F_i}{\partial y_i} x_i + O(x_i^2) \tag{5}$$

$$\approx (-2y_i - \alpha A\sin(\alpha y_i)) x_i =: -f_i' x_i, \tag{6}$$

using the following derivative terms

$$k_i = 2y_i \quad \text{and} \quad d_i = \alpha A\sin(\alpha y_i), \quad \text{such that} \quad \frac{\partial f_i}{\partial y_i} = f_i' = k_i + d_i. \tag{7}$$

A second approach is to consider only the linear term of Eq. (4) and neglect all non-linear terms denoted by $\delta(x_i)$ according to

$$Q_i(x_i) = -2y_i x_i - x_i^2 - A\cos(\alpha y_i)(1 - \cos(\alpha x_i)) - A\sin(\alpha y_i)\sin(\alpha x_i) \tag{8}$$

$$= -2y_i x_i + \delta(x_i) \approx -2y_i x_i = -k_i x_i. \tag{9}$$

The linearization using f_i' is a local approximation of the function incorporating oscillation parameters A and α. Using only k_i (setting $d_i = 0$) discards oscillations by approximating the quadratic term via $k_i = \partial(y_i^2)/\partial y_i = 2y_i$ with negative sign due to maximization. Both approximations will be evaluated later.

3   The $(\mu/\mu_I, \lambda)$-ES with Normalized Mutations

The Evolution Strategy under investigation consists of a population of μ parents and λ offspring ($\mu < \lambda$) per generation g. Algorithm 1 is presented below and offspring variables are denoted with overset "$\sim$".

Population variation is achieved by applying an isotropic normally distributed mutation $\mathbf{x} \sim \sigma \mathcal{N}(0, \mathbf{1})$ with strength σ to the parent recombinant in Lines 6 and 7. The recombinant is obtained using intermediate recombination of all μ parents equally weighted in Line 11. Selection of the $m = 1, ..., \mu$ best search vectors $\mathbf{y}_{m;\lambda}$ (out of λ) according to their fitness is performed in Line 10.

Note that the ES in Algorithm 1 operates under constant normalized mutation σ^* in Lines 3 and 12 using the spherical normalization

$$\sigma^* = \frac{\sigma^{(g)} N}{\left\| \mathbf{y}^{(g)} \right\|} = \frac{\sigma^{(g)} N}{R^{(g)}}. \tag{10}$$

This property ensures global convergence of the algorithm as the mutation strength $\sigma^{(g)}$ decreases if and only if the residual distance $\left\| \mathbf{y}^{(g)} \right\| = R^{(g)}$ decreases. While σ^* is not known during black-box optimizations, it is used here to investigate the dynamical behavior of the ES using the first order progress rate approach to be developed in this paper. Incorporating self-adaptation of σ or cumulative step-size adaptation remains for future research.

Algorithm 1. $(\mu/\mu_I, \lambda)$-ES with constant σ^*

1: $g \leftarrow 0$
2: $\mathbf{y}^{(0)} \leftarrow \mathbf{y}^{(\text{init})}$
3: $\sigma^{(0)} \leftarrow \sigma^* \left\| \mathbf{y}^{(0)} \right\| / N$
4: **repeat**
5:     **for** $l = 1, ..., \lambda$ **do**
6:         $\tilde{\mathbf{x}}_l \leftarrow \sigma^{(g)} \mathcal{N}_l(0, \mathbf{1})$
7:         $\tilde{\mathbf{y}}_l \leftarrow \mathbf{y}^{(g)} + \tilde{\mathbf{x}}_l$
8:         $\tilde{f}_l \leftarrow f(\tilde{\mathbf{y}}_l)$
9:     **end for**
10:     $(\tilde{\mathbf{y}}_{1;\lambda}, \ldots, \tilde{\mathbf{y}}_{\mu;\lambda}) \leftarrow \text{sort} \left(\tilde{\mathbf{y}} \text{ w.r.t. ascending } \tilde{f} \right)$
11:     $\mathbf{y}^{(g+1)} \leftarrow \frac{1}{\mu} \sum_{m=1}^{\mu} \tilde{\mathbf{y}}_{m;\lambda}$
12:     $\sigma^{(g+1)} \leftarrow \sigma^* \left\| \mathbf{y}^{(g+1)} \right\| / N$
13:     $g \leftarrow g + 1$
14: **until** termination criterion

---

4   Progress Rate

4.1   Definition

Having introduced the Evolution Strategy, we are interested in the expected one-generation progress of the optimization on the Rastrigin function (1) before investigating the dynamics over multiple generations.

A first order progress rate φ_i for the i-th component between two generations $g \rightarrow g+1$ can be defined as the expectation value over the positional difference of the parental components

$$\varphi_i = \mathrm{E}\left[y_i^{(g)} - y_i^{(g+1)} \,\middle|\, \sigma^{(g)}, \mathbf{y}^{(g)} \right] = y_i^{(g)} - \mathrm{E}\left[y_i^{(g+1)} \,\middle|\, \sigma^{(g)}, \mathbf{y}^{(g)} \right], \quad (11)$$

given mutation strength $\sigma^{(g)}$ and the position $\mathbf{y}^{(g)}$. First, an expression for $\mathbf{y}^{(g+1)}$ is needed, see Algorithm 1, Line 11. It is the result of mutation, selection and recombination of the $m = 1, ..., \mu$ offspring vectors yielding the highest fitness, such that $\mathbf{y}^{(g+1)} = \frac{1}{\mu} \sum_{m=1}^{\mu} \tilde{\mathbf{y}}_{m;\lambda} = \frac{1}{\mu} \sum_{m=1}^{\mu} (\mathbf{y}^{(g)} + \mathbf{x})_{m;\lambda}$. Considering the i-th component, noting that $\mathbf{y}^{(g)}$ is the same for all offspring and setting $(\mathbf{x}_{m;\lambda})_i = x_{m;\lambda}$ one has

$$y_i^{(g+1)} = \frac{1}{\mu} \sum_{m=1}^{\mu} (y_i^{(g)} + x_{m;\lambda}) = y_i^{(g)} + \frac{1}{\mu} \sum_{m=1}^{\mu} x_{m;\lambda}. \quad (12)$$

Taking the expectation $\mathrm{E}\left[y_i^{(g+1)}\right]$, setting $x = \sigma z = \sigma \mathcal{N}(0,1)$ and inserting the expression back into (11) yields

$$\varphi_i = -\frac{1}{\mu} \mathrm{E}\left[\sum_{m=1}^{\mu} x_{m;\lambda} \,\middle|\, \sigma^{(g)}, \mathbf{y}^{(g)} \right] = -\frac{\sigma}{\mu} \mathrm{E}\left[\sum_{m=1}^{\mu} z_{m;\lambda} \,\middle|\, \sigma^{(g)}, \mathbf{y}^{(g)} \right]. \quad (13)$$

Therefore progress can be evaluated by averaging over the expectations of μ selected mutation contributions. In principle this task can be solved by deriving the induced order statistic density $p_{m;\lambda}$ for the m-th best individual and subsequently solving the integration over the i-th component

$$\varphi_i = -\frac{1}{\mu} \sum_{m=1}^{\mu} \int_{-\infty}^{\infty} x_i \, p_{m;\lambda}(x_i | \sigma^{(g)}, \mathbf{y}^{(g)}) \mathrm{d}x_i. \quad (14)$$

However, the task of computing expectations of sums of order statistics under noise disturbance has already been discussed and solved by Arnold in [1]. Therefore the problem of Eq. (13) will be reformulated in order to apply the solutions provided by Arnold.

4.2   Expectations of Sums of Noisy Order Statistics

Let z be a random variate with density $p_z(z)$ and zero mean. The density is expanded into a Gram-Charlier series by means of its cumulants κ_i $(i \geq 1)$ according to [1, p. 138, D.15]

$$p_z(z) = \frac{1}{\sqrt{2\pi\kappa_2}} e^{-\frac{z^2}{2\kappa_2}} \left(1 + \frac{\gamma_1}{6} \operatorname{He}_3 \left(\frac{z}{\sqrt{\kappa_2}} \right) + \frac{\gamma_2}{24} \operatorname{He}_4 \left(\frac{z}{\sqrt{\kappa_2}} \right) + ... \right), \quad (15)$$

with expectation $\kappa_1 = 0$, variance κ_2, skewness $\gamma_1 = \kappa_3/\kappa_2^{3/2}$, excess $\gamma_2 = \kappa_4/\kappa_2^2$ (higher order terms not shown) and He_k denoting the k-th order probabilist's Hermite polynomials. For the problem at hand, see Eq. (13), the mutation variate $z \sim \mathcal{N}(0,1)$ with $\kappa_2 = 1$ and $\kappa_i = 0$ for $i \neq 2$ yielding a standard normal density.

Furthermore, let $\epsilon \sim \mathcal{N}(0,\sigma_\epsilon^2)$ model additive noise disturbance, such that resulting observed values are $v = z + \epsilon$. Selection of the m-th largest out of λ values yields

$$v_{m;\lambda} = (z + \mathcal{N}(0,\sigma_\epsilon^2))_{m;\lambda}, \quad (16)$$

and the distribution of selected source terms $z_{m;\lambda}$ follows a noisy order statistic with density $p_{m;\lambda}$. Given this definition and a linear relation between $z_{m;\lambda}$ and $v_{m;\lambda}$ the method of Arnold is applicable.

In our case the i-th mutation component $x_{m;\lambda}$ of Eq. (13) is related to selection via the quality change defined in Eq. (3). Maximizing the fitness $F_i(y_i + x_i)$ conforms to maximizing quality $Q_i(x_i)$ with $F_i(y_i)$ being a constant offset.

Aiming at an expression of form (16) and starting with (3), we first isolate component Q_i from the remaining $N-1$ components denoted by $\sum_{j \neq i} Q_j$. Then, approximations are applied to both terms yielding

$$Q_\mathbf{y}(\mathbf{x}) = Q_i(x_i) + \sum_{j \neq i} Q_j(x_j) \quad (17)$$

$$\approx -f_i' x_i + \mathcal{N}(E_i, D_i^2), \quad (18)$$

with linearization (6) applied to $Q_i(x_i)$. Additionally, $\sum_{j \neq i} Q_j \simeq \mathcal{N}(E_i, D_i^2)$, as the sum of independent random variables asymptotically approaches a normal distribution in the limit $N \to \infty$ due to the Central Limit Theorem. This is ensured by Lyapunov's condition provided that there are no dominating components within the sum due to largely different values of y_j. The corresponding Rastrigin quality variance $D_i^2 = \operatorname{Var}[\sum_{j \neq i} Q_j(x_j)]$ is calculated in the supplementary material (https://github.com/omam-evo/paper/blob/main/ppsn22/PPSN22_OB22.pdf). As the expectation $E_i = \operatorname{E}[\sum_{j \neq i} Q_j(x_j)]$ is only an offset to $Q_\mathbf{y}(\mathbf{x})$ it has no influence on the selection and its calculation can be dropped.

Using $x_i = \sigma z_i$ and $f_i' = \operatorname{sgn}(f_i')\,|f_i'|$, expression (18) is reformulated as

$$Q_\mathbf{y}(\mathbf{x}) = -\operatorname{sgn}(f_i')\,|f_i'|\sigma z_i + E_i + \mathcal{N}(0,D_i^2) \quad (19)$$

$$\frac{Q_\mathbf{y}(\mathbf{x}) - E_i}{|f_i'|\sigma} = \operatorname{sgn}(-f_i')\,z_i + \mathcal{N}\left(0, \frac{D_i^2}{(f_i'\sigma)^2} \right). \quad (20)$$

The decomposition using sign function and absolute value is needed for correct ordering of selected values w.r.t. z_i in (20).

Given result (20), one can define the linearly transformed quality measure $v_i := (Q_\mathbf{y}(\mathbf{x}) - E_i)/|f_i'|\sigma$ and noise variance $\sigma_\epsilon^2 := (D_i/f_i'\sigma)^2$, such that the selection of mutation component $\text{sgn}\,(-f_i')\,z_i$ is disturbed by a noise term due to the remaining $N - 1$ components. A relation of the form (16) is obtained up to the sign function.

In [1] Arnold calculated the expected value of arbitrary sums S_P of products of noisy ordered variates containing ν factors per summand

$$S_P = \sum_{\{n_1, \ldots, n_\nu\}} z_{n_1;\lambda}^{p_1} \cdots z_{n_\nu;\lambda}^{p_\nu}, \tag{21}$$

with random variate z introduced in Eqs. (15) and (16). The vector $P = (p_1, \ldots, p_\nu)$ denotes the positive exponents and distinct summation indices are denoted by the set $\{n_1, \ldots, n_\nu\}$. The generic result for the expectation of (21) is provided in [1, p. 142, D.28] and was adapted to account for the sign difference between (16) and (20) resulting in possible exchanged ordering. Performing simple substitutions in Arnold's calculations in [1] and recalling that in our case $\gamma_1 = \gamma_2 = 0$, the expected value yields

$$\mathrm{E}\,[S_P] = \text{sgn}\,(-f_i')^{\|P_1\|}\sqrt{\kappa_2}^{\|P_1\|}\frac{\mu!}{(\mu - \nu)!}\sum_{n=0}^{\nu}\sum_{k \geq 0}\zeta_{n,0}^{(P)}(k)h_{\mu,\lambda}^{\nu-n,k}. \tag{22}$$

Note that expression (22) deviates from Arnold's formula only in the sign in front of $\sqrt{\kappa_2}$. The coefficients $\zeta_{n,0}^{(P)}(k)$ are defined in terms of a noise coefficient a according to

$$a = \sqrt{\frac{\kappa_2}{\kappa_2 + \sigma_\epsilon^2}} \quad \text{with } \zeta_{n,0}^{(P)}(k) = \text{Polynomial}(a), \tag{23}$$

for which tabulated results are presented in [1, p. 141]. The coefficients $h_{\mu,\lambda}^{i,k}$ are numerically obtainable solving

$$h_{\mu,\lambda}^{i,k} = \frac{\lambda - \mu}{\sqrt{2\pi}}\binom{\lambda}{\mu}\int_{-\infty}^{\infty}\text{He}_k\,(x)\,e^{-\frac{1}{2}x^2}[\phi(x)]^i[\Phi(x)]^{\lambda-\mu-1}[1 - \Phi(x)]^{\mu-i}dx. \tag{24}$$

Now we are in the position to calculate expectation (13) using (22). Since $z \sim \mathcal{N}(0,1)$, it holds $\kappa_2 = 1$. Identifying $P = (1)$, $\|P\|_1 = 1$ and $\nu = 1$ yields

$$\mathrm{E}\left[\sum_{m=1}^{\mu} z_{m;\lambda}\right] = \text{sgn}\,(-f_i')\frac{\mu!}{(\mu - 1)!}\sum_{n=0}^{1}\sum_{k \geq 0}\zeta_{n,0}^{(1)}(k)h_{\mu,\lambda}^{1-n,k}$$

$$= \text{sgn}\,(-f_i')\,\mu\zeta_{0,0}^{(1)}(0)h_{\mu,\lambda}^{1,0} = -\,\text{sgn}\,(f_i')\,\mu a c_{\mu/\mu,\lambda}, \tag{25}$$

with $\zeta_{1,0}^{(1)}(k) = 0$ for any k, and $\zeta_{0,0}^{(1)}(k) \neq 0$ only for $k = 0$ yielding a. The expression $h_{\mu,\lambda}^{1,0}$ is equivalent to the progress coefficient definition $c_{\mu/\mu,\lambda}$

[2, p. 216]. Inserting (25) back into (13), using $a = \sqrt{1/(1 + (D_i/f_i'\sigma)^2)} = |f_i'|\sigma/\sqrt{(f_i'\sigma)^2 + D_i^2}$ with the requirement $a > 0$, and noting that $f_i' = \text{sgn}\,(f_i')\,|f_i'|$ one finally obtains for the i-th component first order progress rate

$$\varphi_i(\sigma, \mathbf{y}) = c_{\mu/\mu,\lambda}\frac{f_i'(y_i)\sigma^2}{\sqrt{(f_i'(y_i)\sigma)^2 + D_i^2(\sigma, (\mathbf{y})_{j\neq i})}}. \tag{26}$$

The population dependency is given by progress coefficient $c_{\mu/\mu,\lambda}$. The fitness dependent parameters are contained in f_i', see (7), and in D_i^2 calculated in the supplementary material (https://github.com/omam-evo/paper/blob/main/ppsn22/PPSN22_OB22.pdf). For better readability the derivative f_i' and variance D_i^2 are not inserted into (26). An exemplary evaluation of D_i^2 as a function of the residual distance R using normalization (10) is also shown in the supplementary material.

4.3   Comparison of Simulation and Approximation

Figure 1 shows an experimentally obtained progress rate compared to the result of (26). Due to large N one exemplary φ_i-graph is shown on the left, and corresponding $i = 1, ..., N$ errors are shown on the right.

The left plot shows the progress rate over a σ-range of $[0, 1]$. This magnitude was chosen in order to study the oscillation, as the frequency $\alpha = 2\pi$. The initial position was chosen randomly to be on the sphere surface $R = 10$.

The red dashed curve uses f_i' as linearization, while the blue dash-dotted curve assumes $f_i' = k_i$ (with $d_i = 0$), see also (7). As f_i' approximates the quality change locally, agreement for the progress is given only for very small mutations σ. For larger σ very large deviation may occur, depending on the local derivative.

The blue curve $\varphi_i(k_i)$ neglects the oscillation ($d_i = 0$) and therefore follows the progress of the quadratic function $f(\mathbf{y}) = \sum_i y_i^2$ for large σ with very good agreement. Due to a *linearized* form of $Q_i(x_i)$ in (6) neither approximation can reproduce the oscillation for moderately large σ.

To verify the approximation quality, the error between (26) and simulation is displayed on the right side of Fig. 1 for all $i = 1, ..., N$. It was done for small $\sigma = 0.1$ and large $\sigma = 1$. The deviations are very similar in magnitude for all i, given randomly chosen y_i. Note that for $\sigma = 1$ the red points show very large errors compared to blue, which was expected.

Figure 2 shows the progress rate φ_i over σ^*, for $i = 2$ as in Fig. 1, with $\mathbf{y}$ randomly on the surface radii $R = \{100, 10, 1, 0.1\}$. Using σ^* the mutation σ is normalized by the residual distance R with spherical normalization (10). Far from the origin with $R = \{100, 10\}$ the quadratic terms are dominating giving better results using $\varphi_i(k_i)$. Reaching $R = 1$ local minima are more relevant and mixed results are obtained with $\varphi_i(f_i')$ better for smaller σ^* and $\varphi_i(k_i)$ for larger σ^*. Within the global attractor $R = 0.1$ the local structure dominates and $\varphi_i(f_i')$ yields better results. These observations will be relevant analyzing the dynamics in Fig. 3 where both approximations show strengths and weaknesses.

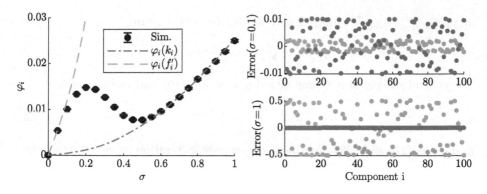

Fig. 1. One-generation experiments with $(150/150, 300)$-ES, $N = 100$, $A = 10$ are performed and quantity (11) is measured averaging over 10^5 runs. Left: φ_i over σ for $i = 2$ at position $y_2 \approx 1.19$, where $\mathbf{y}$ was chosen randomly such that $\|\mathbf{y}\| = R = 10$. Right: error measure $\varphi_i - \varphi_{i,\mathrm{sim}}$ between (26) and simulation for $i = 1, ..., N$ evaluated at $\sigma = \{0.1, 1\}$. The colors are set according to the legend. (Color figure online)

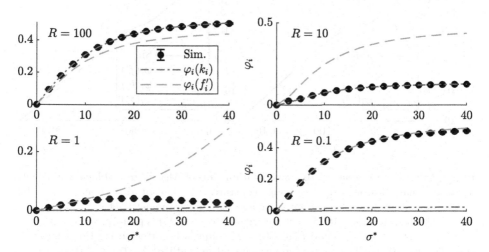

Fig. 2. One-generation progress φ_i $(i = 2)$ over normalized mutation σ^* for $(150/150, 300)$-ES, $N = 100$, $A = 1$ and $R = \{100, 10, 1, 0.1\}$. Simulations are averaged over 10^5 runs. These experiments are preliminary investigations related to the dynamics shown in Fig. 3 with $\sigma^* = 30$. Given a constant σ^* the approximation quality varies over different magnitudes of R.

5 Evolution Dynamics

As we are interested in the dynamical behavior of the ES, averaged real optimization runs from Algorithm 1 will be compared to the iterated dynamics using progress result (26) by applying the dynamical systems approach [2]. Neglecting fluctuations, i.e., $y_i^{(g+1)} = \mathrm{E}\left[y_i^{(g+1)} \big| \sigma^{(g)}, \mathbf{y}^{(g)}\right]$ the mean value dynamics for the mapping $y_i^{(g)} \rightarrow y_i^{(g+1)}$ immediately follows from (11) giving

$$y_i^{(g+1)} = y_i^{(g)} - \varphi_i(\sigma^{(g)}, \mathbf{y}^{(g)}). \tag{27}$$

The control scheme of $\sigma^{(g)}$ was introduced in Eq. (10) and yields simply

$$\sigma^{(g)} = \sigma^* \left\| \mathbf{y}^{(g)} \right\| / N. \tag{28}$$

Equations (27) and (28) describe a deterministic iteration in search space and rescaling of mutations according to the residual distance. For a convergence analysis, we are interested in the dynamics of $R^{(g)} = \left\| \mathbf{y}^{(g)} \right\|$ rather than the actual position values $\mathbf{y}^{(g)}$. Hence in Fig. 3 the $R^{(g)}$-dynamics of the conducted experiments is shown.

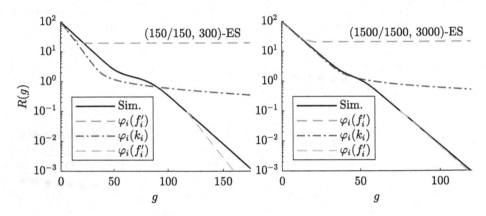

Fig. 3. Comparing average of 100 optimization runs of Algorithm 1 (black, solid) with iterated dynamics from Eq. (27) under constant $\sigma^* = 30$ for $A = 1$ and $N = 100$. Large populations sizes are chosen to ensure global convergence (left: $\mu = 150$; right: $\mu = 1500$; constant $\mu/\lambda = 0.5$). Iteration using progress (26) is performed for both $f_i' = k_i + d_i$ (red/orange dashed) and $f_i'(d_i = 0) = k_i$ (blue dash-dotted) using Equations (27) and (28). The orange dashed iteration was initialized with $R^{(0)} = 0.1$ and translated to the corresponding position of the simulation for easier comparison. The evaluation of quality variance $D_i^2(R)$ is shown in the supplementary material (https://github.com/omam-evo/paper/blob/main/ppsn22/PPSN22_OB22.pdf). (Color figure online)

In Fig. 3, all runs of Algorithm 1 exhibit global convergence with the black line showing the average. The left and right plots differ by population size. Iteration $\varphi_i(k_i)$, blue dash-dotted curve, also converges globally, though very slowly and therefore not shown entirely. The convergence behavior of iteration $\varphi_i(f_i')$, red and orange dashed curves, strongly depends on the initialization and is discussed below.

Three phases can be observed for the simulation. It shows linear convergence at first being followed by a slow-down due to local attractors. Reaching the global attractor the convergence speed increases again. Iteration $\varphi_i(k_i)$ is able to model the first two phases to some degree. Within the global attractor the slope information d_i is missing such that the progress is largely underestimated.

Iteration $\varphi_i(f_i')$ converges first, but yields a stationary state with $R^{st} \approx 20$ when the progress φ_i becomes dominated by derivative term d_i. Starting from $R^{(0)} = 10^2$ the stationary y_i^{st} are either fixed or alternating between coordinates depending on σ, D_i, k_i, and d_i. This effect is due to attraction of local minima and due to the deterministic iteration disregarding fluctuations. It occurs also with varying initial positions. Initialized at $R^{(0)} = 10^{-1}$ orange iteration $\varphi_i(f_i')$ is globally converging.

It turns out that the splitting point of the two approximations in Fig. 3 occurs at a distance R to the global optimizer where the ES approaches the attractor region of the "first" local minima. For the model parameters considered in the experiment this is at about $R \approx 28.2$ – the distance of the farest local minimizer to the global optimizer (obtained by numerical analysis).

Plots in Fig. 3 differ by population size. The convergence speed, i.e. the slopes, show better agreement for large populations, which can be attributed to the fluctuations neglected in (27). Investigations on unimodal funtions Sphere [2] and Ellipsoid [4] have shown that progress is decreased by fluctuations due to a loss-term scaling with $1/\mu$, which agrees with Fig. 3. On the left the iterated progress is faster due to neglected but present fluctuations, while on the right better agreement is observed due to insignificant fluctuations. These observations will be investigated in future research.

6   Summary and Outlook

A first order progress rate φ_i was derived for the $(\mu/\mu_I, \lambda)$-ES by means of noisy order statistics in (26) on the Rastrigin function (1). To this end, the mutation induced variance of the quality change D_i^2 is needed. Starting from (4) a derivation yielding D_i^2 has been presented in the supplementary material. Furthermore, the approximation quality of φ_i was investigated using Rastrigin and quadratic derivatives f_i' and k_i, respectively, by comparing with one-generation experiments.

Linearization f_i' shows good agreement for small-scale mutations, but very large deviations for large mutations. Conversely, linearization k_i yields significantly better results for large mutations as the quadratic fitness term dominates. A progress rate modeling the transition between the regimes is yet to be determined. First numerical investigations of (14) including all terms of (4) indicate that nonlinear terms are needed for a better progress rate model, which is an open challenge and part of future research.

The obtained progress rate was used to investigate the dynamics by iterating (27) using (28) and comparing with ES runs. Iteration via f_i' only converges globally if initialized close to the optimizer, since local attraction is strongly dominating. Dynamics via k_i converges globally independent of initialization, but the observed rate matches only for the initial phase and for very large populations.

This confirms the need for a higher order progress rate modeling the effect of fluctuations, especially when function evaluations are expensive and small populations must be used. Additionally, an advanced progress rate formula is needed combining effects of global and local attraction to model all three phases of the dynamics correctly.

The investigations done so far are a first step towards a full dynamical analysis of the ES on the multimodal Rastrigin function. Future investigations must also include the complete dynamical modeling of the mutation strength control. One aim is the tuning of mutation control parameters such that the global convergence probability is increased while still maintaining search efficiency. Our final goal will be the theoretical analysis of the full evolutionary process yielding also recommendations regarding the choice of the minimal population size needed to converge to the global optimizer with high probability.

Acknowledgments. This work was supported by the Austrian Science Fund (FWF) under grant P33702-N. Special thanks goes to Lisa Schönenberger for providing valuable feedback and helpful discussions.

References

1. Arnold, D.: Noisy Optimization with Evolution Strategies. Kluwer Academic Publishers, Dordrecht (2002)
2. Beyer, H.G.: The Theory of Evolution Strategies. Natural Computing Series. Springer, Heidelberg (2001). https://doi.org/10.1007/978-3-662-04378-3
3. Beyer, H.G.: Convergence analysis of evolutionary algorithms that are based on the paradigm of information geometry. Evol. Comput. **22**(4), 679–709 (2014). https://doi.org/10.1162/EVCO_a_00132
4. Beyer, H.G., Melkozerov, A.: The dynamics of self-adaptive multi-recombinant evolution strategies on the general ellipsoid model. IEEE Trans. Evol. Comput. **18**(5), 764–778 (2014). https://doi.org/10.1109/TEVC.2013.2283968
5. Beyer, H.G., Sendhoff, B.: Simplify your covariance matrix adaptation evolution strategy. IEEE Trans. Evol. Comput. **21**(5), 746–759 (2017). https://doi.org/10.1109/TEVC.2017.2680320
6. Glasmachers, T., Schaul, T., Sun, Y., Wierstra, D., Schmidhuber, J.: Exponential natural evolution strategies. In: Branke, J., et al., (ed.) GECCO 2010: Proceedings of the Genetic and Evolutionary Computation Conference, pp. 393–400. ACM, New York (2010)
7. Hansen, N., Kern, S.: Evaluating the CMA evolution strategy on multimodal test functions. In: Yao, X., et al. (eds.) PPSN 2004. LNCS, vol. 3242, pp. 282–291. Springer, Heidelberg (2004). https://doi.org/10.1007/978-3-540-30217-9_29
8. Hansen, N., Müller, S., Koumoutsakos, P.: Reducing the time complexity of the derandomized evolution strategy with covariance matrix adaptation (CMA-ES). Evol. Comput. **11**(1), 1–18 (2003)
9. Mobahi, H., Fisher, J.: A theoretical analysis of optimization by Gaussian continuation. In: Proceedings of the Twenty-Ninth AAAI Conference on Artificial Intelligence, pp. 1205–1211. AAAI Press (2015)
10. Müller, N., Glasmachers, T.: Non-local optimization: imposing structure on optimization problems by relaxation. In: Foundations of Genetic Algorithms, vol. 16, pp. 1–10. ACM (2021). https://doi.org/10.1145/3450218.3477307

11. Ollivier, Y., Arnold, L., Auger, A., Hansen, N.: Information-geometric optimization algorithms: a unifying picture via invariance principles. J. Mach. Learn. Res. **18**(18), 1–65 (2017)
12. Rechenberg, I.: Evolutionsstrategie: Optimierung technischer Systeme nach Prinzipien der biologischen Evolution. Frommann-Holzboog Verlag, Stuttgart (1973)
13. Schwefel, H.P.: Numerical Optimization of Computer Models. Wiley, Chichester (1981)
14. Zhang, J., Bi, S., Zhang, G.: A directional Gaussian smoothing optimization method for computational inverse design in nanophotonics. Mater. Des. **197**, 109213 (2021). https://doi.org/10.1016/j.matdes.2020.109213

Running Time Analysis of the (1+1)-EA Using Surrogate Models on OneMax and LeadingOnes

Zi-An Zhang, Chao Bian, and Chao Qian$^{(\boxtimes)}$

State Key Laboratory for Novel Software Technology, Nanjing University, Nanjing 210023, China
{zhangza,bianc,qianc}@lamda.nju.edu.cn

Abstract. Evolutionary algorithms (EAs) have been widely applied to solve real-world optimization problems. However, for problems where fitness evaluation is time-consuming, the efficiency of EAs is usually unsatisfactory. One common approach is to utilize surrogate models, which apply machine learning techniques to approximate the real fitness function. Though introducing noise, using surrogate models can reduce the time of fitness evaluation significantly, and has been shown useful in many applications. However, the theoretical analysis (especially the essential theoretical aspect, running time analysis) of surrogate-assisted EAs has not been studied. In this paper, we make a preliminary attempt by analyzing the running time of the (1+1)-EA using two typical kinds of pre-selection surrogate for solving the OneMax and LeadingOnes problems. The results imply that the running time can be significantly reduced when the surrogate model is accurate enough and properly used.

Keywords: Evolutionary algorithm · Surrogate model · Running time analysis

1 Introduction

EAs, a kind of randomized heuristic optimization algorithm [2], have been widely used in real-world applications. However, the fitness (i.e., objective) evaluation for real-world problems is often very time-consuming. For example, in aerodynamic design [15], it is often necessary to carry out computational fluid dynamics simulations to evaluate the performance of a given structure, which is computationally expensive. Other examples include protein design, drug design, and material design [15]. The expensive fitness evaluation has limited the efficiency of EAs largely.

Much effort thus has been devoted to reducing the computational time in both the design and application of EAs. One popular idea is using machine learning models, called *surrogate models*, to approximate the real objective functions [15]. Specifically, it first samples some solutions from the solution space and

This work was supported by the National Science Foundation of China (62022039).

evaluates their true fitness, and then uses them to train a learning model, which will be used to evaluate the newly generated solutions during the evolutionary process. Surrogate-assisted EAs have been widely used to solve real-world problems, e.g., the design of turbine blades, airfoils, forging, and vehicle crash tests [13]. Note that the idea of surrogate model also appears in other optimization methods, e.g., in Bayesian optimization where Gaussian processes are used as surrogate models [16,19].

However, it has been found that if only the surrogate model is used for fitness evaluation, EAs are very likely to converge to a false optimum [14]. Therefore, the surrogate model should be used together with the original fitness function in a proper way, leading to the issue of surrogate management [12,13,15]. Preselection is a widely used surrogate management strategy, which first expands the number of candidate offspring solutions and then uses the surrogate model to filter out some unpromising ones before the real fitness evaluation. Typical preselection mechanisms include the Regression model-based PreSelection (RPS) [10] which predicts the fitness of a solution, the Classification model-based PreSelection (CPS) [24] which predicts the probability of a solution being good, and the binary Relation Classification-based PreSelection (RCPS) [9] which predicts whether a solution is better than another one.

In contrast to the wide application of EAs, the theoretical foundation of EAs is underdeveloped due to their sophisticated behaviors. Much effort has been devoted to analyzing the essential theoretical aspect, i.e., running time complexity, of EAs [1,7,17,25]. The running time analysis starts from simple EAs solving synthetic problems. For example, one classical result is that the expected running time of the (1+1)-EA on OneMax and LeadingOnes is $\Theta(n \log n)$ and $\Theta(n^2)$, respectively [8]. Meanwhile, general running time analysis approaches, e.g., drift analysis [5,6,11,18], fitness-level methods [4,20,21], and switch analysis [3,22,23], have also been proposed. However, to the best of our knowledge, running time analysis of surrogate-assisted EAs has not been touched.

This paper aims at moving the first step towards running time analysis of surrogate-assisted EAs by considering the (1+1)-EA using the RPS and RCPS surrogates. Specifically, we first introduce a concept of (k,δ)-RPS surrogate, which generates k candidate offspring solutions in each iteration and predicts the fitness of a candidate solution wrong with probability δ, and then prove that the (1+1)-EA using the (k,δ)-RPS surrogate with $k = c/\delta$ (where c is a positive constant) and $\delta < 1/2$ can solve OneMax and LeadingOnes in $O(n + \delta n \log n)$ and $O(\max\{n, \delta n^2\})$ expected running time, respectively. The results show that the performance of EAs can be significantly improved, as long as δ is given appropriate values, e.g., $\delta = O(1/n)$. We also prove that the above upper bounds on the expected running time hold for the (1+1)-EA using the (k,δ)-RCPS surrogate, where δ denotes the probability of predicting the relation between any two offspring solutions wrong.

The rest of this paper starts with some preliminaries. Then, the running time analysis of the (1+1)-EA using the RPS and RCPS surrogates is presented in Sects. 3 and 4, respectively. Section 5 concludes the paper.

2   Preliminaries

In this section, we first introduce EAs, surrogate models and problems studied in this paper, respectively, and then present the analysis tools that we use throughout this paper.

2.1   (1+1)-EA

The (1+1)-EA as described in Algorithm 1 is a simple EA for maximizing pseudo-Boolean functions over $\{0,1\}^n$. It reflects the common structure of EAs, and has been widely used in the running time analysis of EAs [1,17]. The (1+1)-EA maintains only one solution during the optimization procedure (i.e., the population size is 1), and repeatedly improves the current solution by using bit-wise mutation (i.e., line 3) and selection (i.e., lines 4–6).

Algorithm 1. (1+1)-EA

Given a function $f : \{0,1\}^n \rightarrow \mathbb{R}$ to be maximized:

1: $\boldsymbol{x} :=$ uniformly randomly selected from $\{0,1\}^n$;
2: **repeat**
3:     $\boldsymbol{x}' :=$ flip each bit of $\boldsymbol{x}$ independently with probability $1/n$;
4:     **if** $f(\boldsymbol{x}') \geq f(\boldsymbol{x})$ **then**
5:         $\boldsymbol{x} := \boldsymbol{x}'$
6:     **end if**
7: **until** the termination condition is met

2.2   Surrogate Models

In this paper, we incorporate the widely used preselection surrogate model [9,10] into the (1+1)-EA. As described in Algorithm 2, the (1+1)-EA using preselection has the same general procedure as the original (1+1)-EA, i.e., it randomly generates an initial solution and improves it repeatedly. However, it inserts two key subprocedures: surrogate training (i.e., lines 1–2) and preselection (i.e., lines 5–8), which aim to train the surrogate model using the sampled data and use the surrogate model to select a promising solution, respectively. In the following, we present two specific preselection surrogates, i.e., RPS and RCPS surrogates, that will be studied in this paper.

RPS Surrogate tries to learn a mapping from the solution space to the objective space, based on a training set $P = \{\langle \boldsymbol{x}_i, f(\boldsymbol{x}_i) \rangle \,|\, i = 1, ..., N\}$, and then employs the mapping to predict the fitness value of newly generated candidate solutions. Note that $f(\boldsymbol{x})$ denotes the true fitness value of a solution. Specifically, we first sample a set of solutions from the solution space, and then employ a regression learning method, e.g., regression tree, to learn the mapping $\mathfrak{M}$. That is, line 2 of Algorithm 2 changes to

$$\mathfrak{M} = \text{RegressorTrain}(\{\langle \boldsymbol{x}_i, f(\boldsymbol{x}_i) \rangle \,|\, i = 1, ..., N\}).$$

Algorithm 2. (1+1)-EA with Preselection

1: Conduct a training data set P;
2: $\mathfrak{M} = \text{SurrogateTrain}(P)$;
3: $\boldsymbol{x} :=$ uniformly randomly selected from $\{0,1\}^n$;
4: **repeat**
5:    **for** $i = 1$ to k **do**
6:       $\boldsymbol{u}_i :=$ flip each bit of $\boldsymbol{x}$ independently with probability $1/n$
7:    **end for**
8:    $\boldsymbol{u}^* = \text{PreSelection}(\{\boldsymbol{u}_1, ..., \boldsymbol{u}_k\}, \mathfrak{M})$;
9:    **if** $f(\boldsymbol{u}^*) \geq f(\boldsymbol{x})$ **then**
10:      $\boldsymbol{x} := \boldsymbol{u}^*$
11:    **end if**
12: **until** the termination condition is met

In the preselection procedure, we first generate k candidate offspring solutions, and then select a solution $\boldsymbol{u}^*$ which has the maximal predicted fitness value. That is, line 8 of Algorithm 2 changes to

$$\boldsymbol{u}^* = \arg\max_{u \in \{u_1, u_2, ..., u_k\}} \text{Predict}(u, \mathfrak{M}),$$

where $\text{Predict}(\boldsymbol{u}, \mathfrak{M})$ denotes the fitness of the candidate solution $\boldsymbol{u}$ predicted by regression model $\mathfrak{M}$.

We introduce a concept of (k, δ)-RPS surrogate as presented in Definition 1, which will be used in our analysis. It specifies the number of solutions generated in lines 5–7 of Algorithm 2, as well as the accuracy of the surrogate model. That is, we omit the specific training methods, and only assume that the obtained preselection model can predict the fitness of a solution approximately correctly with some probability. Note that for the pseudo-Boolean functions considered in this paper, the acceptable threshold is set to 0.5, while for general problems, 0.5 can be replaced by a parameter ϵ.

Definition 1. *A (k, δ)-RPS surrogate is a regression model-based preselection surrogate such that*
(1) k offspring solutions are generated before the real fitness evaluation,
(2) the prediction error exceeds the acceptable threshold 0.5 with probability δ, i.e., $\text{P}(|f(\boldsymbol{x}) - f'(\boldsymbol{x})| \geq 0.5) = \delta$, where $f'(\boldsymbol{x})$ denotes the predicted fitness of $\boldsymbol{x}$ by the surrogate.

RCPS Surrogate tries to learn a classifier which predicts whether a solution is better than another. Specifically, we first sample a set $\{\boldsymbol{x}_1, \boldsymbol{x}_2, ..., \boldsymbol{x}_N\}$ of solutions from the solution space, and then employ Algorithm 3 to assign a label for each pair of solutions. That is, a pair $(\boldsymbol{x}, \boldsymbol{y})$ of solutions will be assigned a label 1 if $\boldsymbol{x}$ is better than $\boldsymbol{y}$ (i.e., $\boldsymbol{x}$ wins), and a label -1 otherwise. After that, we employ a classification learning method, e.g., decision tree, to learn the classifier $\mathfrak{M}$. That is, line 2 of Algorithm 2 changes to

$$\mathfrak{M} = \text{ClassifierTrain}(\{\langle(\boldsymbol{x}_i, \boldsymbol{x}_j), l\rangle \mid 1 \leq i, j \leq N, i \neq j\}).$$

Algorithm 3. Training Data Preparation

1: **for** $i = 1$ to N **do**
2:   **for** $j = 1$ to $i - 1$ **do**
3:     **if** $f(\boldsymbol{x}_i) \geq f(\boldsymbol{x}_j)$ **then**
4:       assign the pair $(\boldsymbol{x}_i, \boldsymbol{x}_j)$ a label $l = 1$
5:     **else**
6:       assign the pair $(\boldsymbol{x}_i, \boldsymbol{x}_j)$ a label $l = -1$
7:     **end if**
8:     assign the pair $(\boldsymbol{x}_j, \boldsymbol{x}_i)$ a label $-l$
9:   **end for**
10: **end for**

In the preselection procedure, we first generate k candidate offspring solutions, and then select a solution $\boldsymbol{u}^*$ which wins the most times in the pairwise competition, with ties broken uniformly. Note that for each pair of candidate offspring solutions, only one of them can win, i.e., $\forall i, j$, Predict$((\boldsymbol{x}_i, \boldsymbol{x}_j), \mathfrak{M}) = -$Predict$((\boldsymbol{x}_j, \boldsymbol{x}_i), \mathfrak{M})$. That is, line 8 of Algorithm 2 changes to

$$\boldsymbol{u}^* = \arg\max_{\boldsymbol{u} \in \{\boldsymbol{u}_1, \ldots, \boldsymbol{u}_k\}} \sum_{\boldsymbol{u}_i \in \{\boldsymbol{u}_1, \ldots, \boldsymbol{u}_k\} \setminus \{\boldsymbol{u}\}} \text{Predict}((\boldsymbol{u}, \boldsymbol{u}_i), \mathfrak{M}),$$

where Predict$((\boldsymbol{u}, \boldsymbol{u}_i), \mathfrak{M})$ denotes the label of the pair $(\boldsymbol{u}, \boldsymbol{u}_i)$ of solutions predicted by classification model $\mathfrak{M}$.

Similar to the (k, δ)-RPS surrogate, in our analysis, we will omit the specific training methods, and only assume that the obtained classification model can predict the relation between two solutions correctly with some probability, as presented in Definition 2.

Definition 2. *A (k, δ)-RCPS surrogate is a binary relation classification-based preselection surrogate such that*
   (1) k offspring solutions are generated before the real fitness evaluation,
   (2) the relation between any two solutions is predicted wrong with probability δ.

2.3   OneMax and LeadingOnes

In this section, we introduce two well-known pseudo-Boolean functions OneMax and LeadingOnes, which will be used in this paper. The OneMax problem as presented in Definition 3 aims to maximize the number of 1-bits of a solution. Its optimal solution is 11...1 (briefly denoted as 1^n) with the function value n. It has been shown that the expected running time of the (1+1)-EA on OneMax is $\Theta(n \log n)$ [8]. For a Boolean solution $\boldsymbol{x}$, let x_i denote its i-th bit.

Definition 3 (Onemax). *The OneMax Problem of size n is to find an n bits binary string $\boldsymbol{x}^*$ such that $\boldsymbol{x}^* = \arg\max_{\boldsymbol{x} \in \{0,1\}^n} \sum_{i=1}^{n} x_i$.*

The LeadingOnes problem as presented in Definition 4 aims to maximize the number of consecutive 1-bits counting from the left of a solution. Its optimal solution is 1^n with the function value n. It has been proved that the expected running time of the (1+1)-EA on LeadingOnes is $\Theta(n^2)$ [8].

Definition 4 (LeadingOnes). *The LeadingOnes Problem of size n is to find an n bits binary string x^* such that $x^* = \arg\max_{x \in \{0,1\}^n} \sum_{i=1}^{n} \prod_{j=1}^{i} x_i$.*

2.4 Analysis Tools

Because an evolution process usually goes forward only based on the current population, an EA can be modeled as a Markov chain $\{\xi_t\}_{t=0}^{+\infty}$ [11,25]. The state space of the chain (denote as $\mathcal{X}$) is exactly the population space of the EA. The target state space $\mathcal{X}^*$ is the set of all optimal populations, where an "optimal" population implies containing an optimal solution. Note that we consider the discrete state space (i.e., $\mathcal{X}$ is discrete) in this paper.

Given a Markov chain $\{\xi_t\}_{t=0}^{+\infty}$ and $\xi_0 = x$, we define its *first hitting time* (FHT) as a random variable τ such that $\tau = \min\{t | \xi_t \in \mathcal{X}^*, t \geq 0\}$. That is, τ is the number of generations required to reach the optimal state space $\mathcal{X}^*$ from $\xi_0 = x$ for the first time. Then, we define the chain's *expected first hitting time* (EFHT) as the mathematical expectation of τ, i.e., $\mathbb{E}[\tau | \xi_0] = \sum_{i=0}^{+\infty} i \cdot \mathrm{P}(\tau = i)$.

In the following, we introduce two drift theorems which will be used to derive the EFHT of Markov chains in the paper. Drift analysis was first introduced to the running time analysis of EAs by He and Yao [11], and has become a popular tool with many variants [5,6]. We will use its additive (i.e., Lemma 1) as well as multiplicative (i.e., Lemma 2) version. To use drift analysis, we first need to construct a distance function $V(x)$ to measure the distance of a state x to the optimal state space $\mathcal{X}^*$, where $V(x)$ satisfies that $V(x \in \mathcal{X}^*) = 0$ and $V(x \notin \mathcal{X}^*) > 0$. Then, we need to investigate the progress on the distance to $\mathcal{X}^*$ in each step, i.e., $\mathbb{E}[V(\xi_t) - V(\xi_{t+1}) | \xi_t]$. For additive drift analysis in Lemma 1, an upper bound on the EFHT can be derived through dividing the initial distance by a lower bound on the progress. Multiplicative drift analysis in Lemma 2 is much easier to use when the progress is roughly proportional to the current distance to the optimum.

Lemma 1 (Additive Drift [11]). *Given a Markov chain $\{\xi_t\}_{t=0}^{+\infty}$ and a distance function $V(x)$, if for any $t \geq 0$ and any ξ_t with $V(\xi_t) > 0$, there exists a real number $c > 0$ such that $\mathbb{E}[V(\xi_t) - V(\xi_{t+1}) | \xi_t] \geq c$, then the EFHT satisfies that $\mathbb{E}[\tau | \xi_0] \leq V(\xi_0)/c$.*

Lemma 2 (Multiplicative Drift [6]). *Given a Markov chain $\{\xi_t\}_{t=0}^{+\infty}$ and a distance function $V(x)$, if for any $t \geq 0$ and any ξ_t with $V(\xi_t) > 0$, there exists a real number $c > 0$ such that $\mathbb{E}[V(\xi_t) - V(\xi_{t+1}) | \xi_t] \geq c \cdot V(\xi_t)$, then the EFHT satisfies that $\mathbb{E}[\tau | \xi_0] \leq (1 + \ln(V(\xi_0)/V_{\min}))/c$, where $V_{\min}$ denotes the minimum among all possible positive values of V.*

3 Analysis of the (1+1)-EA Using the RPS Surrogate

In this section, we analyze the expected running time of the (1+1)-EA using the (k, δ)-RPS surrogate on OneMax and LeadingOnes, respectively. Note that the acceptable threshold in Definition 1 is set to 0.5 on OneMax and LeadingOnes. Under such setting, the condition (2) in Definition 1 implies that for any two solutions x and y with $f(x) \geq f(y)$, x will be predicted better than y if the prediction error doesn't exceed the acceptable threshold.

We prove in Theorem 1 that when $\delta < 1/2$ and $k = c/\delta$, the expected running time of the (1+1)-EA using the (k, δ)-RPS surrogate on the OneMax problem is $O(n + \delta n \log n)$. Note that without surrogate model, the expected running time of the (1+1)-EA on the OneMax problem is $\Theta(n \log n)$ [8]. Therefore, if $\delta = O(1/\log n)$, the expected running time can be improved from $\Theta(n \log n)$ to $O(n)$. Intuitively, the results show that the running time can be significantly improved when the surrogate model is accurate enough and properly used.

The main proof idea can be summarized as follows. Since the comparison of the parent solution x and the preselected offspring solution u^* is under the real fitness, the distance function used in drift analysis does not increase. Furthermore, when at least one of the k offspring solutions is better than the parent, and all the k offspring solutions are "correctly" evaluated by the surrogate model, i.e., the prediction error doesn't exceed the acceptable threshold, there will be a positive progress on the distance function.

Theorem 1. *For the (1+1)-EA using the (k, δ)-RPS surrogate on the OneMax problem, the expected running time is $O(n + \delta n \log n)$ if $\delta < 1/2$ and $k = c/\delta$ (where c is a positive constant). Particularly, it is $O(n)$ if $\delta = O(1/\log n)$.*

Proof. We use additive and multiplicative drift analysis to prove this theorem. Let the distance function $V(x) = |x|_0$ be the number of 0-bits of a solution x. It is easy to verify that $V(x \in \mathcal{X}^* = \{1^n\}) = 0$ and $V(x \notin \mathcal{X}^*) > 0$.

Suppose that the current solution x has i 0-bits, i.e., $|x|_0 = i$. Then, we examine the expected progress $\mathbb{E}[V(\xi_t) - V(\xi_{t+1})|\xi_t = x]$. We decompose the progress into two parts, i.e., $\mathbb{E}[V(\xi_t) - V(\xi_{t+1})|\xi_t = x] = E^+ - E^-$, where

$$E^+ = \sum\nolimits_{\xi_{t+1}:V(\xi_{t+1})<i} P(\xi_{t+1}|\xi_t = x)(i - V(\xi_{t+1})),$$

$$E^- = \sum\nolimits_{\xi_{t+1}:V(\xi_{t+1})>i} P(\xi_{t+1}|\xi_t = x)(V(\xi_{t+1}) - i).$$

That is, E^+ and E^- denote the positive and negative drift towards the optimal state, respectively. Since the comparison of the parent solution and the preselected offspring solution is under the real fitness, the fitness of the solution will never decrease. Thus, the distance function will not increase, implying $E^- = 0$. To analyze the positive drift E^+, we consider the probability that one offspring solution x' is better than the parent solution x. We have

$$P(f(x') > f(x)) \geq (i/n) \cdot (1 - 1/n)^{n-1} \geq i/(en), \tag{1}$$

where the first inequality holds because it is sufficient to flip one of the i 0-bits of x by mutation and keep the other bits unchanged, and the second inequality is by $(1 - 1/n)^{n-1} \geq 1/e$. Then, we can derive a lower bound on the probability of generating at least one offspring solution which is better than the parent solution, i.e.,

$$P(\exists \boldsymbol{u}^* \in \{\boldsymbol{u}_j\}_{j=1}^k, f(\boldsymbol{u}^*) > f(\boldsymbol{x})) \geq 1 - (1 - i/(en))^k \geq 1 - e^{-ki/(en)}$$

$$\geq 1 - \frac{1}{1 + ki/(en)} = ki/(ki + en),$$

where the last two inequalities are both by $1 + a \leq e^a$. When all the k offspring solutions are correctly evaluated by the surrogate model, whose probability is $(1 - \delta)^k$, the best one will be chosen. Thus, we have

$$P(V(\xi_{t+1}) < i | \xi_t = \boldsymbol{x}) \geq (ki/(ki + en)) \cdot (1 - \delta)^k,$$

implying that

$$\mathbb{E}[V(\xi_t) - V(\xi_{t+1}) | \xi_t = \boldsymbol{x}] \geq P(V(\xi_{t+1}) < i | \xi_t = \boldsymbol{x}) \cdot 1$$
$$\geq (ki/(ki + en)) \cdot (1 - \delta)^k. \tag{2}$$

To derive the expected running time for finding the optimal solution, we divide the evolution process into two phases. The first phase starts from the initial solution and ends when $|\boldsymbol{x}|_0 \leq en/k$, and the second phase starts after the first phase finishes and ends when the optimal solution is found. Let τ_1 and τ_2 denote the running time of these two phases, respectively. For the first phase, i.e., $en/k \leq i \leq n$, because $ki/(ki + en) \geq ki/(ki + ki) = 1/2$, we have

$$\mathbb{E}[V(\xi_t) - V(\xi_{t+1}) | \xi_t = \boldsymbol{x}] \geq (1 - \delta)^k/2.$$

Thus, by Lemma 1, we get

$$\mathbb{E}[\tau_1 | \xi_0] \leq 2n/(1 - \delta)^k.$$

For the second phase, i.e., $i < en/k$, because $ki/(ki + en) \geq ki/(en + en) = ki/(2en)$, we have

$$\mathbb{E}[V(\xi_t) - V(\xi_{t+1}) | \xi_t = \boldsymbol{x}] \geq ki (1 - \delta)^k /(2en),$$

Thus, by Lemma 2, we get

$$\mathbb{E}[\tau_2 | \xi_0] \leq \frac{1 + \ln \frac{en}{k}}{\frac{k(1-\delta)^k}{2en}} = \frac{2e(2 - \ln k)}{k(1 - \delta)^k} n + \frac{2e}{k(1 - \delta)^k} n \ln n.$$

Combining the analysis of the two phases, we have

$$\mathbb{E}[\tau | \xi_0] = \mathbb{E}[\tau_1 | \xi_0] + \mathbb{E}[\tau_2 | \xi_0] \leq \frac{1}{\left((1 - \delta)^{\frac{1}{\delta}}\right)^c} \cdot \left(2n + \frac{4e}{c}\delta n + \frac{2e}{c}\delta n \ln n\right),$$

where the last inequality is by $k = c/\delta$. Note that

$$
\frac{1}{\left((1-\delta)^{\frac{1}{\delta}}\right)^c} = \left(\left(1 + \frac{\delta}{1-\delta}\right)^{\frac{1}{\delta}-1}\right)^c \cdot \frac{1}{(1-\delta)^c}
$$

$$
\leq e^{\frac{\delta}{1-\delta} \cdot (\frac{1}{\delta}-1) \cdot c} \cdot \frac{1}{(1-\delta)^c} = \left(\frac{e}{1-\delta}\right)^c < (2e)^c,
$$

(3)

where the first inequality is by $1 + a \leq e^a$, and the second inequality is by $\delta < 1/2$. Furthermore, as c is a constant, we get $\mathbb{E}[\tau|\xi_0] = O(n + \delta n \log n)$. Thus, the theorem holds. □

We prove in Theorem 2 that when $\delta < 1/2$ and $k = c/\delta$, the expected running time of the (1+1)-EA using the (k,δ)-RPS surrogate on the LeadingOnes problem is $O(\max\{n, \delta n^2\})$. Note that without surrogate model, the expected running time of the (1+1)-EA on the LeadingOnes problem is $\Theta(n^2)$ [8]. Therefore, if $\delta = O(1/n)$, the expected running time can be improved from $\Theta(n^2)$ to $O(n)$. The main proof idea is similar to that of Theorem 1. That is, the distance function used in drift analysis does not increase; meanwhile, it can decrease if at least one of the offspring solutions is better than the parent solution and all the offspring solutions are correctly evaluated by the surrogate model.

Theorem 2. *For the (1+1)-EA using the (k,δ)-RPS surrogate on the LeadingOnes problem, the expected running time is $O(\max\{n, \delta n^2\})$ if $\delta < 1/2$ and $k = c/\delta$ (where c is a positive constant). Particularly, it is $O(n)$ if $\delta = O(1/n)$.*

Proof. We use additive drift analysis to prove this theorem. Let the distance function $V(\boldsymbol{x}) = n - LO(\boldsymbol{x})$, where $LO(\boldsymbol{x})$ is the number of leading 1-bits of $\boldsymbol{x}$. It is easy to verify that $V(\boldsymbol{x} \in \mathcal{X}^* = \{1^n\}) = 0$ and $V(\boldsymbol{x} \notin \mathcal{X}^*) > 0$. Suppose that the current solution $\boldsymbol{x}$ has i leading 1-bits, i.e., $LO(\boldsymbol{x}) = i < n$. Then, Eq. (1) becomes

$$
P(f(\boldsymbol{x}') > f(\boldsymbol{x})) \geq (1/n) \cdot (1 - 1/n)^i \geq 1/(en),
$$

(4)

since it is sufficient to flip the first 0-bit and keep the i leading 1-bits unchanged. Equation (2) becomes

$$
\mathbb{E}[V(\xi_t) - V(\xi_{t+1})|\xi_t = \boldsymbol{x}] \geq P(V(\xi_{t+1}) < i|\xi_t = \boldsymbol{x}) \cdot 1 \geq (k/(k+en)) \cdot (1-\delta)^k.
$$

We consider two cases for k. If $k \geq en$, we have $k/(k+en) \geq 1/2$. Then, we get $\mathbb{E}[V(\xi_t) - V(\xi_{t+1})|\xi_t = \boldsymbol{x}] \geq (1-\delta)^k/2$. By Lemma 1, we get

$$
\mathbb{E}[\tau|\xi_0] \leq \frac{2n}{(1-\delta)^k} = \frac{1}{\left((1-\delta)^{\frac{1}{\delta}}\right)^c} \cdot 2n = O(n),
$$

where the first equality is by $k = c/\delta$, and the second inequality holds by Eq. (3). If $k < en$, we have $k/(k+en) \geq k/(2en)$. Then, we get $\mathbb{E}[V(\xi_t) - V(\xi_{t+1})|\xi_t = \boldsymbol{x}] \geq k(1-\delta)^k/(2en)$. By Lemma 1, we get

$$
\mathbb{E}[\tau|\xi_0] \leq \frac{2en^2}{k(1-\delta)^k} = O(\delta n^2),
$$

Thus, the analysis of the above two cases leads to $\mathbb{E}[\tau|\xi_0] = O\left(\max\{n, \delta n^2\}\right)$, implying that the theorem holds. ▯

4   Analysis of the (1+1)-EA Using the RCPS Surrogate

In this section, we analyze the expected running time of the (1+1)-EA using the (k, δ)-RCPS surrogate on OneMax and LeadingOnes, respectively.

We prove in Theorem 3 that when $\delta < 1/2$ and $k = c/\delta$, the expected running time of the (1+1)-EA using the (k, δ) RCPS surrogate on the OneMax problem is $O(n + \delta n \log n)$. Thus, the expected running time can be reduced by a factor of $O(\log n)$ if $\delta = O(1/\log n)$, which also suggests the effectiveness of the surrogate model. The main proof idea can be summarized as follows. Similar to the proof of Theorem 1, the distance function does not increase. When some offspring solutions are better than the parent solution and one of these offspring solutions wins the competition with the other offspring solutions, the preselected offspring solution can be better than the parent solution, leading to a positive drift towards the optimal solution.

Theorem 3. *For the (1+1)-EA using the (k, δ)-RCPS surrogate on the OneMax problem, the expected running time is $O(n + \delta n \log n)$ if $\delta < 1/2$ and $k = c/\delta$ (where c is a positive constant). Particularly, it is $O(n)$ if $\delta = O(1/\log n)$.*

Proof. We use additive and multiplicative drift analysis to prove this theorem. Let the distance function $V(\boldsymbol{x}) = |\boldsymbol{x}|_0$ be the number of 0-bits of a solution $\boldsymbol{x}$. Suppose that the current solution $\boldsymbol{x}$ has i 0-bits, i.e., $|\boldsymbol{x}|_0 = i$.

Suppose that the offspring solutions $\boldsymbol{x}_1, ..., \boldsymbol{x}_m$ are better the parent solution $\boldsymbol{x}$ and $\boldsymbol{x}_{m+1}, ..., \boldsymbol{x}_k$ are worse than the parent solution (or have the same fitness as the parent solution). If $\exists j \in \{1, ..., m\}$, $\boldsymbol{x}_j$ wins the competitions with all the other offspring solutions, then $\boldsymbol{x}_j$ will be chosen and will bring progress as it is better than the parent solution. The probability of this event is

$$\sum_{j=1}^{m} \delta^{j-1}(1-\delta)^{k-j} = \frac{(1-\delta)^k}{1-2\delta}\left(1 - \left(\frac{\delta}{1-\delta}\right)^m\right).$$

Let p denote the probability that a better individual is produced by mutation, which is at least $i/(en)$ by Eq. (1). Then, we have

$$P(V(\xi_{t+1}) < i|\xi_t = \boldsymbol{x}) \geq \sum_{m=1}^{k} \binom{k}{m} p^m (1-p)^{k-m} \frac{(1-\delta)^k}{1-2\delta}\left(1 - \left(\frac{\delta}{1-\delta}\right)^m\right)$$

$$= \frac{(1-\delta)^k}{1-2\delta}\left(1 - \left(1 - \frac{1-2\delta}{1-\delta}p\right)^k\right)$$

$$\geq \frac{(1-\delta)^k}{1-2\delta}\left(1 - \frac{1}{1 + \frac{1-2\delta}{1-\delta}\frac{ki}{en}}\right).$$

Similar to the analysis of Theorem 1, we have

$$\mathbb{E}[V(\xi_t) - V(\xi_{t+1})|\xi_t = x] \geq \mathrm{P}(V(\xi_{t+1}) < i|\xi_t = x) \cdot 1$$

$$\geq \frac{(1-\delta)^k}{1-2\delta}\left(1 - \frac{1}{1 + \frac{1-2\delta}{1-\delta}\frac{ki}{en}}\right).$$

To derive the expected running time for finding the optimal solution, we divide the evolutionary process into two phases. The first phase starts from the initial solution and ends when $|x|_0 \leq \frac{1-\delta}{1-2\delta}\frac{en}{k}$, and the second phase starts after the first phase finishes and ends when the optimal solution is found. Let τ_1 and τ_2 denote the running time of these two phases, respectively. For the first phase, i.e., $\frac{1-\delta}{1-2\delta}\frac{en}{k} \leq i \leq n$, we have

$$\mathbb{E}[V(\xi_t) - V(\xi_{t+1})|\xi_t = x] \geq (1-\delta)^k/(2-4\delta),$$

By Lemma 1, we get

$$\mathbb{E}[\tau_1|\xi_0] \leq n(2-4\delta)/(1-\delta)^k \leq 2n/(1-\delta)^k.$$

For the second phase, i.e., $i < \frac{1-\delta}{1-2\delta}\frac{en}{k} \leq n$, we have

$$\mathbb{E}[V(\xi_t) - V(\xi_{t+1})|\xi_t = x] \geq ki(1-\delta)^{k-1}/(2en).$$

By Lemma 2, we get

$$\mathbb{E}[\tau_2|\xi_0] \leq \frac{1 + \ln\left(\frac{1-\delta}{1-2\delta}\frac{en}{k}\right)}{\frac{k(1-\delta)^{k-1}}{2en}} \leq \frac{2en(1+\ln n)}{k(1-\delta)^k}.$$

Combining the analysis of the two cases, we have

$$\mathbb{E}[\tau|\xi_0] = \mathbb{E}[\tau_1|\xi_0] + \mathbb{E}[\tau_2|\xi_0] \leq \frac{2n}{(1-\delta)^k} + \frac{2en(1+\ln n)}{k(1-\delta)^k} = O(n + \delta n \log n),$$

where the last equality is by $k = c/\delta$ and Eq. (3). Thus, the theorem holds. □

We prove in Theorem 4 that when $\delta < 1/2$ and $k = c/\delta$, the expected running time of the (1+1)-EA using the (k, δ)-RCPS surrogate on the LeadingOnes problem is $O(\max\{n, \delta n^2\})$. The results show that the expected running time can be reduced by a factor of $O(n)$ if $\delta = O(1/n)$. The main proof idea is similar to that of Theorem 3.

Theorem 4. *For the (1+1)-EA using the (k, δ)-RCPS surrogate on the LeadingOnes problem, the expected running time is $O(\max\{n, \delta n^2\})$ if $\delta < 1/2$ and $k = c/\delta$ (where c is a positive constant). Particularly, it is $O(n)$ if $\delta = O(1/n)$.*

Proof. We use additive drift analysis to prove this theorem. Let the distance function $V(\boldsymbol{x}) = n - LO(\boldsymbol{x})$. Suppose that the current solution $\boldsymbol{x}$ has i leading 1-bits, i.e., $LO(\boldsymbol{x}) = i < n$.

Similar to the analysis in Theorem 3, we have

$$
\begin{aligned}
\mathrm{P}(V(\xi_{t+1}) < i|\xi_t = \boldsymbol{x}) &\geq \frac{(1-\delta)^k}{1-2\delta}\left(1 - \frac{1}{1 + \frac{1-2\delta}{1-\delta}pk}\right) \\
&\geq \frac{(1-\delta)^{\frac{c}{\delta}}}{1-2\delta}\left(1 - \frac{1}{1 + \left(\frac{1}{\delta} - 2\right)\frac{c}{en}}\right),
\end{aligned}
$$

where the second inequality holds by $p \geq 1/(en)$ as shown in Eq. (4), $k = c/\delta$, and $\delta < 1/2$. We consider two cases for δ. If $\left(\frac{1}{\delta} - 2\right) \cdot \frac{c}{en} \geq 1$, i.e., $\delta \leq \frac{c}{en+2c}$, we have

$$
\mathbb{E}[V(\xi_t) - V(\xi_{t+1})|\xi_t = \boldsymbol{x}] \geq \mathrm{P}(V(\xi_{t+1}) < i|\xi_t = \boldsymbol{x}) \cdot 1 \geq (1-\delta)^{c/\delta}/2.
$$

By Lemma 1, we get

$$
\mathbb{E}[\tau|\xi_0] \leq 2n/(1-\delta)^{c/\delta} = O(n).
$$

If $\left(\frac{1}{\delta} - 2\right) \cdot \frac{c}{en} < 1$, i.e., $\delta > \frac{c}{en+2c}$, we have

$$
\mathbb{E}[V(\xi_t) - V(\xi_{t+1})|\xi_t = \boldsymbol{x}] \geq c\,(1-\delta)^{c/\delta}/(2e\delta n).
$$

By Lemma 1, we get

$$
\mathbb{E}[\tau|\xi_0] \leq \frac{2e\delta n^2}{c\,(1-\delta)^{c/\delta}} = O(\delta n^2).
$$

Thus, $\mathbb{E}[\tau|\xi_0] = O\left(\max\{n, \delta n^2\}\right)$, implying that the theorem holds. □

5   Conclusion and Discussion

In this paper, we conduct a preliminary study on the running time analysis of surrogate-assisted EAs, by considering the (1+1)-EA using the RPS and RCPS surrogates solving OneMax and LeadingOnes. We introduce the concept of the (k, δ)-RPS and (k, δ)-RCPS surrogates, and derive the parameter values that can make using the surrogate model accelerate the evolution process. The results imply that if the surrogate model is accurate enough and used properly, the running time can be significantly improved.

We hope this work can encourage more work on the running time analysis of EAs using surrogate models. In the future, the following two directions can be considered. On one hand, in this paper, we simply assume that the surrogate model is trained in advance before optimization, while in practical applications, the surrogate model is usually updated along with the optimization process.

It is interesting to theoretically study the impact of updating the surrogate model with newly obtained data. On the other hand, when analyzing the running time, we only consider the cost of true fitness evaluation during the evolutionary process, which is somewhat unfair. It is interesting to examine the total cost of surrogate-assisted EAs, i.e., the cost of training before evolution and the cost during the evolutionary process.

References

1. Auger, A., Doerr, B.: Theory of Randomized Search Heuristics: Foundations and Recent Developments. World Scientific, Singapore (2011)
2. Back, T.: Evolutionary Algorithms in Theory and Practice: Evolution Strategies, Evolutionary Programming, Genetic Algorithms. Oxford University Press, Oxford (1996)
3. Bian, C., Qian, C., Tang, K.: A general approach to running time analysis of multi-objective evolutionary algorithms. In: Proceedings of the 27th International Joint Conference on Artificial Intelligence (IJCAI 2018), Stockholm, Sweden, pp. 1405–1411 (2018)
4. Corus, D., Dang, D.C., Eremeev, A.V., Lehre, P.K.: Level-based analysis of genetic algorithms and other search processes. IEEE Trans. Evol. Comput. **22**(5), 707–719 (2017)
5. Doerr, B., Goldberg, L.A.: Adaptive drift analysis. Algorithmica **65**(1), 224–250 (2013)
6. Doerr, B., Johannsen, D., Winzen, C.: Multiplicative drift analysis. Algorithmica **64**(4), 673–697 (2012)
7. Doerr, B., Neumann, F.: Theory of Evolutionary Computation: Recent Developments in Discrete Optimization. Springer, Cham (2020). https://doi.org/10.1007/978-3-030-29414-4
8. Droste, S., Jansen, T., Wegener, I.: On the analysis of the (1+1) evolutionary algorithm. Theor. Comput. Sci. **276**(1–2), 51–81 (2002)
9. Hao, H., Zhang, J., Lu, X., Zhou, A.: Binary relation learning and classifying for preselection in evolutionary algorithms. IEEE Trans. Evol. Comput. **24**(6), 1125–1139 (2020)
10. Hao, H., Zhang, J., Zhou, A.: A comparison study of surrogate model based preselection in evolutionary optimization. In: Huang, D.-S., Jo, K.-H., Zhang, X.-L. (eds.) ICIC 2018. LNCS, vol. 10955, pp. 717–728. Springer, Cham (2018). https://doi.org/10.1007/978-3-319-95933-7_80
11. He, J., Yao, X.: Drift analysis and average time complexity of evolutionary algorithms. Artif. Intell. **127**(1), 57–85 (2001)
12. Jin, Y.: A comprehensive survey of fitness approximation in evolutionary computation. Soft. Comput. **9**(1), 3–12 (2005)
13. Jin, Y.: Surrogate-assisted evolutionary computation: recent advances and future challenges. Swarm Evol. Comput. **1**(2), 61–70 (2011)
14. Jin, Y., Olhofer, M., Sendhoff, B.: A framework for evolutionary optimization with approximate fitness functions. IEEE Trans. Evol. Comput. **6**(5), 481–494 (2002)
15. Jin, Y., Wang, H., Sun, C.: Data-Driven Evolutionary Optimization. SCI, vol. 975. Springer, Cham (2021). https://doi.org/10.1007/978-3-030-74640-7
16. Mockus, J.: Application of Bayesian approach to numerical methods of global and stochastic optimization. J. Global Optim. **4**(4), 347–365 (1994)

17. Neumann, F., Witt, C.: Bioinspired Computation in Combinatorial Optimization: Algorithms and Their Computational Complexity. Springer, Heidelberg (2010). https://doi.org/10.1007/978-3-642-16544-3
18. Oliveto, P.S., Witt, C.: Simplified drift analysis for proving lower bounds in evolutionary computation. Algorithmica **59**(3), 369–386 (2011)
19. Qian, C., Xiong, H., Xue, K.: Bayesian optimization using pseudo-points. In: Proceedings of the 29th International Joint Conference on Artificial Intelligence (IJCAI 2020), Yokohama, Japan, pp. 3044–3050 (2020)
20. Sudholt, D.: A new method for lower bounds on the running time of evolutionary algorithms. IEEE Trans. Evol. Comput. **17**(3), 418–435 (2012)
21. Wegener, I.: Methods for the analysis of evolutionary algorithms on pseudo-Boolean functions. In: Evolutionary Optimization, pp. 349–369. Kluwer, Norwell (2002)
22. Yu, Y., Qian, C.: Running time analysis: convergence-based analysis reduces to switch analysis. In: Proceedings of the IEEE Congress on Evolutionary Computation (CEC), Sendai, Japan, pp. 2603–2610 (2015)
23. Yu, Y., Qian, C., Zhou, Z.H.: Switch analysis for running time analysis of evolutionary algorithms. IEEE Trans. Evol. Comput. **19**(6), 777–792 (2014)
24. Zhang, J., Zhou, A., Tang, K., Zhang, G.: Preselection via classification: a case study on evolutionary multiobjective optimization. Inf. Sci. **465**, 388–403 (2018)
25. Zhou, Z.H., Yu, Y., Qian, C.: Evolutionary Learning: Advances in Theories and Algorithms. Springer, Singapore (2019). https://doi.org/10.1007/978-981-13-5956-9

Runtime Analysis of Simple Evolutionary Algorithms for the Chance-Constrained Makespan Scheduling Problem

Feng Shi[1]([✉])[ID], Xiankun Yan[2][ID], and Frank Neumann[2][ID]

[1] School of Computer Science and Engineering, Central South University,
Changsha 410083, People's Republic of China
fengshi@csu.edu.cn
[2] Optimisation and Logistics, School of Computer Science,
The University of Adelaide, Adelaide, Australia

Abstract. The Makespan Scheduling problem is an extensively studied NP-hard problem, and its simplest version looks for an allocation approach for a set of jobs with deterministic processing times to two identical machines such that the makespan is minimized. However, in real life scenarios, the actual processing time of each job may be stochastic around an expected value with a variance under the influence of external factors, and these actual processing times may be correlated with covariances. Thus within this paper, we propose a chance-constrained version of the Makespan Scheduling problem and investigate the performance of Randomized Local Search and $(1 + 1)$ EA for it. More specifically, we study two variants of the Chance-constrained Makespan Scheduling problem and analyze the expected runtime of the two algorithms to obtain an optimal or almost optimal solution to the instances of the two variants.

Keywords: Chance-constraint · Makespan scheduling problem · RLS · $(1 + 1)$ EA

1 Introduction

To discover the reasons behind the successful applications of evolutionary algorithms in various areas including engineering and economics, lots of researchers made efforts to study the theoretical performance of evolutionary algorithms for classical combinatorial optimization problems. But most of these studied problems are deterministic (such as Vertex Cover problem [4,5,13,15,24,25,27–29,40] and Minimum Spanning Tree problem [3,14,21,22,35,37]), and the optimization problems in real-world are often stochastic and have dynamic components. Hence in the past few years, the related researchers paid attentions to the theoretical

This work has been supported by the National Natural Science Foundation of China under Grants 62072476 and 61872048, the Hunan Provincial Natural Science Foundation of China under Grant 2021JJ40791, and the Australian Research Council (ARC) through grant FT200100536.

performance of evolutionary algorithms for dynamic and stochastic combinatorial optimization problems [9,16,19,30,32,33] and obtained a series of theoretical results that further advance the understanding of evolutionary algorithms.

Chance-constrained optimization problems is an important class of stochastic optimization problems. They consider that the constraints may be influenced by the noise of stochastic components, thus their goal is to optimize the given objective function under that the constraints can be violated up to certain probability levels [2,12,17,26]. The basic technique for solving chance-constrained optimization problems is to convert the stochastic constraints to their respective deterministic equivalents according to the predetermined confidence level. Recently, researchers began to focus on the *chance-constrained* optimization problems and analyze the theoretical performance of evolutionary algorithms for them.

The classical Makespan Scheduling problem (abbr. MSP) [1] considers two identical machines and a set of jobs with deterministic processing times, and its aim is to allocate the jobs to the machines such that the makespan is minimized (we only consider its simplest version, please refer to [6,11,31] for its approximation algorithms). In real life scenarios, the actual processing time of each job may be stochastic around an expected value with a variance, and the actual processing times of the jobs may be correlated with covariances. Thus a chance-constrained version of MSP, named *Chance-constrained Makespan Scheduling Problem* (abbr. CCMSP), is proposed in the paper. CCMSP considers two identical machines and several groups of jobs, where the jobs have the same expected processing time and variance if they are in the same group, and their actual processing times are correlated by a covariance if they are in the same group and allocated to the same machine. The goal of CCMSP is to minimize a deterministic makespan value and subject to the probability that the actual makespan exceeds the deterministic makespan is no more than an acceptable threshold.

A few theoretical results have been obtained about the performance of evolutionary algorithms for MSP and chance-constrained problems. Witt [36] carried out the runtime analysis of evolutionary algorithms for MSP with two machines. Later Gunia [7] extended the results to MSP with a constant number of machines. Sutton et al. [34] gave the parameterized runtime analysis of RLS and $(1 + 1)$ EA for MSP with two machines. Neumann et al. [23] proposed the dynamic version of MSP with two machines and analyzed the performance of RLS and $(1 + 1)$ EA. Xie et al. [38] studied the single- and multi-objective evolutionary algorithms for the Chance-constrained Knapsack problem, where they used the Chebyshev inequality and Chernoff bounds to estimate the constraint violation probability of a given solution. Then Neumann et al. [20] followed the work of Xie et al. [38] and analyzed special cases of this problem. Note that the Chance-constrained Knapsack problem studied in the above two work does not consider the correlationship among the weights of items. Thus recently Xie et al. [39] analyzed the expected optimization time of RLS and $(1 + 1)$ EA for the Chance-constrained Knapsack Problem with correlated uniform weights. Neumann et al. [18] presented the first runtime analysis of multi-objective evolutionary algorithms for chance-constrained submodular functions.

Within this paper, we investigate the expected runtime of RLS and $(1 + 1)$ EA for CCMSP. More specifically, we consider two special variants of CCMSP: (1). CCMSP-1, all jobs have the same expected processing time and variance, and all groups have the same covariance and *even* size; (2). CCMSP-2, the difference from CCMSP-1 is that the groups have different sizes. For CCMSP-1, we prove that CCMSP-1 is polynomial-time solvable by showing that RLS and $(1 + 1)$ EA can obtain an optimal solution to any instance I_1 of it in expected runtime $O(n^2/m)$ and $O((k + m)n^2)$, respectively, where n and k are the numbers of jobs and groups considered in I_1, and $m = n/k$. For CCMSP-2, the size difference among groups makes the discussion complicated, thus a simplified variant of CCMSP-2 named CCMSP-2$^+$ is proposed: The sum of the variances and covariances of the jobs allocated to the same machine cannot be over the expected processing time of a job, no matter how many jobs are allocated to the machine. We prove that CCMSP-2$^+$ is NP-hard and that RLS can get an optimal solution to the instance I_2^+ of CCMSP-2$^+$ in expected polynomial-runtime if the total number of jobs is odd; otherwise, an almost optimal solution to I_2^+.

2   Preliminaries

Consider two identical machines M_0 and M_1, and k groups of jobs, where each group G_i has m_i many jobs (i.e., there are $n = \sum_{i=1}^k m_i$ many jobs in total). W.l.o.g., assume $m_1 \leq m_2 \leq \ldots \leq m_k$. The j-th job in group G_i ($j \in [1, m_i]$, where the notation $[x, y]$ denotes the set containing all integers ranging from x to y), denoted by b_{ij}, has actual processing time p_{ij} with expect value $E[p_{ij}] = a_{ij} > 0$ and variance $\sigma_{ij}^2 > 0$. Additionally, for any two jobs of the same group G_i, if they are allocated to the same machine, then their actual processing times are correlated with each other by a covariance $c_i > 0$; otherwise, independent.

The *Chance-constrained Makespan Scheduling Problem* (abbr. CCMSP) studied in the paper looks for an allocation of the n jobs to the two machines that minimizes the makespan M such that the probabilities of the *loads* on M_0 and M_1 exceeding M are no more than a threshold $0 < \gamma < 1$, where the load on M_t ($t \in [0, 1]$) is the sum of the actual processing times of the jobs allocated to M_t.

An allocation (or simply called solution) x to an instance of CCMSP, is represented as a bit-string with length n, $x = x_{11} \cdots x_{ij} \cdots x_{km_k} \in \{0, 1\}^n$, where the job b_{ij} is allocated to M_0 if $x_{ij} = 0$; otherwise, M_1 (in the remaining text, we simply say that a bit is of G_i if its corresponding job is of G_i). Denote by $M_0(x)$ and $M_1(x)$ the sets of jobs allocated to M_0 and M_1, respectively, w.r.t. x. Denote by $l_t(x) = \sum_{b_{ij} \in M_t(x)} p_{ij}$ the *load* on M_t ($t \in [0, 1]$). Let $\alpha_i(x) = |M_0(x) \cap G_i|$ and $\beta_i(x) = |M_1(x) \cap G_i|$ for all $i \in [1, k]$. The CCMSP can be formulated as:

Minimize M

Subject to $\Pr(l_t(x) > M) \leq \gamma$ for all $t \in [0, 1]$.

Observe that the expected value of $l_t(x)$ is $E[l_t(x)] = \sum_{b_{ij} \in M_t(x)} a_{ij}$. Considering the variance σ_{ij}^2 of each job b_{ij} and the covariance among the jobs of the same group that are allocated to the same machine, the variance of $l_t(x)$ is $Var[l_t(x)] = \sum_{b_{ij} \in M_t(x)} \sigma_{ij}^2 + cov[l_t(x)]$, where $cov[l_t(x)] = \sum_{i=1}^{k} 2c_i \binom{|M_t(x) \cap G_i|}{2}$. Note that $\binom{|M_t(x) \cap G_i|}{2} = 0$ if $0 \leq |M_t(x) \cap G_i| \leq 1$. For the probability $Pr(l_t(x) > M)$ with $t \in [0,1]$, as the work [38,39], we use the one-sided Chebyshev's inequality (cf. Theorem 1) to construct a usable surrogate of the chance-constraint.

Theorem 1. *(One-sided Chebyshev's inequality). Let X be a random variable with expected value $E[X]$ and variance $Var[X]$. Then for any $\Delta \in \mathbb{R}^+$, $Pr(X > E[X] + \Delta) \leq \frac{Var[X]}{Var[X]+\Delta^2}$.*

By the One-sided Chebyshev's inequality, upper bounding the probability of the actual makespan exceeding M by γ indicates that for all $t \in [0,1]$,

$$\Pr(l_t(x) > M) \leq \frac{Var[l_t(x)]}{Var[l_t(x)] + (M - E[l_t(x)])^2} \leq \gamma$$

$$\iff \sqrt{\frac{(1-\gamma)}{\gamma} Var[l_t(x)]} + E[l_t(x)] = l_t'(x) \leq M.$$

Thus $\max\{\Pr(l_0(x) > M), \Pr(l_1(x) > M)\} \leq \gamma$ hold iff $L(x) = \max\{l_0'(x), l_1'(x)\} \leq M$. In other words, $L(x)$ is the tight lower bound for the value of M, if using the surrogate of the chance-constraint by the One-sided Chebyshev's inequality. Therefore, $l_t'(x)$ can be treated as a *new* measure for the load on M_t, and the goal of CCMSP is simplified to minimize $L(x)$. Let $t(x) = \arg\max_t\{l_0'(x), l_1'(x)\}$.

It is not hard to derive that CCMSP is NP-hard as MSP is NP-hard. Within the paper, we study the two specific variants of CCMSP given below.

CCMSP-1. All the n jobs have the same expected processing time $a_{ij} = a > 0$ and variance $\sigma_{ij}^2 = d > 0$, and the k groups have the same covariance $c > 0$ and size $m > 0$. Moreover, m is even.

CCMSP-2. All the n jobs have the same expected processing time $a_{ij} = a > 0$ and variance $\sigma_{ij}^2 = d > 0$, and the k groups have the same covariances $c > 0$. However, the k groups may have different sizes (may be even or odd).

Given an instance I of CCMSP-1 or CCMSP-2 and a solution x to I, if $||M_0(x)| - |M_1(x)|| \leq 1$ (i.e., $|M_0(x)| = |M_1(x)|$ if n is even), then x is an *equal-solution*; if $||M_0(x)| - |M_1(x)|| \leq 1$, and $|\alpha_i(x) - \beta_i(x)| \leq 1$ for all $i \in [1,k]$ (i.e., $\alpha_i = \beta_i$ if m_i is even), then x is a *balanced-solution*.

3   Algorithms

We study the performance of Randomized Local Search (abbr. RLS, given as Algorithm 1) and $(1 + 1)$ EA (given as Algorithm 2) for the two variants of

CCMSP. The two algorithms run in a similar way, randomly generating an offspring based on the maintained solution and replacing it if the offspring is not worse than it regarding their fitness. The difference between the two algorithms is the way to generate offspring: With probability $1/2$, RLS chooses one bit of the maintained solution uniformly at random and flips it, and $1/2$ chooses two bits of the maintained solution uniformly at random and flips them; $(1 + 1)$ EA flips each bit of the maintained solution with probability $1/n$. The fitness function considered in the two algorithms is the natural one, $f(x) = L(x) = \max\{l'_0(x), l'_1(x)\}$.

Algorithm 1: RLS

---

1 choose $x \in \{0, 1\}^n$ uniformly at random;
2 **while** *stopping criterion not met* **do**
3   choose $b \in \{0, 1\}$ uniformly at random;
4   **if** $b = 0$ **then**
5          $y \leftarrow$ flip one bit of x chosen uniformly at random;
6   **else**
7   choose $(i, j) \in \{(k, l) | 1 \leq k < l \leq n\}$ uniformly at random;
8          $y \leftarrow$ flip the i-th and j-th bits of x;
9   **if** $f(y) \leq f(x)$ **then**
10         $x \leftarrow y$;

Algorithm 2: (1+1) EA

---

1 choose $x \in \{0, 1\}^n$ uniformly at random;
2 **while** *stopping criterion not met* **do**
3      $y \leftarrow$ flip each bit of x independently with probability $1/n$;
4   **if** $f(y) \leq f(x)$ **then**
5          $x \leftarrow y$;

---

4 Performance for CCMSP-1

The section starts with an observation that will be used throughout the paper.

Observation 1. $\binom{\lfloor \frac{x+y}{2} \rfloor}{2} + \binom{\lceil \frac{x+y}{2} \rceil}{2} \leq \binom{x}{2} + \binom{y}{2} \leq \binom{x+y}{2}$ *holds for any two natural numbers x and y.*

Consider an instance $I_1 = (a, c, d, \gamma, k, m)$ of CCMSP-1 and a solution x to I_1. As the groups considered in I_1 have the same size m, there is a variable $\delta_i(x)$ such that $\alpha_i(x) = \frac{m}{2} + \delta_i(x)$ and $\beta_i(x) = \frac{m}{2} - \delta_i(x)$ for any $i \in [1, k]$. Thus,

$$cov[l_0(x)] - cov[l_1(x)] = 2c \sum_{i=1}^{k} \left(\binom{\alpha_i}{2} - \binom{\beta_i}{2} \right) = 2c(m-1) \sum_{i=1}^{k} \delta_i$$

$$= c(m-1) \left(\sum_{i=1}^{k} \alpha_i(x) - \sum_{i=1}^{k} \beta_i(x) \right) = c(m-1) \left(|M_0(x)| - |M_1(x)| \right).$$

Based on the conclusion, it is not hard to derive the following two lemmata.

Lemma 1. *For any solution x to the instance $I_1 = (a, c, d, \gamma, k, m)$ of CCMSP-1, if $|M_0(x)| > |M_1(x)|$ (resp., $|M_1(x)| > |M_0(x)|$) then $l_0'(x) > l_1'(x)$ (resp., $l_1'(x) > l_0'(x)$); if $|M_0(x)| = |M_1(x)|$ then $l_0'(x) = l_1'(x)$.*

Lemma 2. *For any solution x to the instance $I_1 = (a, c, d, \gamma, k, m)$ of CCMSP-1, if x is a balanced-solution then $L(x) = l_0'(x) = l_1'(x)$ gets the minimum value; more specifically, x is an optimal solution to I_1 iff x is a balanced-solution to I_1.*

Theorem 2. *The expected runtime of RLS to obtain an optimal solution to the instance $I_1 = (a, c, d, \gamma, k, m)$ of CCMSP-1 is $O(n^2/m) = O(kn)$.*

Proof. Let x_0 be the initial solution maintained by RLS. Assume that $|M_0(x_0)| > |M_1(x_0)|$. Thus $L(x) = l_0'(x_0) > l_1'(x_0)$ by Lemma 1 and $|M_0(x_0)| - |M_1(x_0)| \geq 2$ as $n = mk$ is even. The following discussion first analyzes the process of RLS to obtain the first equal-solution x_1 based on x_0. Five possible cases for the mutation of RLS on x_0 are listed as follows, obtaining an offspring x_0' of x_0.

Case (1). Flipping a 0-bit in x_0 (i.e., $|M_0(x_0')| = |M_0(x_0)| - 1$). Observe that $L(x_0) = l_0'(x_0) > l_0'(x_0')$. As $|M_0(x_0)| - |M_1(x_0)| \geq 2$, $|M_0(x_0')| \geq |M_1(x_0')|$ and $L(x_0') = l_0'(x_0')$ by Lemma 1. Thus $L(x_0') < L(x_0)$ and x_0' can be accepted.

Case (2). Flipping a 1-bit in x_0 (i.e., $|M_0(x_0')| = |M_0(x_0)| + 1$). Observe that $L(x_0') = l_0'(x_0') > l_0'(x_0) = L(x_0)$, thus x_0' cannot be accepted.

Case (3). Flipping two 0-bits in x_0 (i.e., $|M_0(x_0')| = |M_0(x_0)| - 2$). If $|M_0(x_0')| \geq |M_1(x_0')|$, then using the reasoning for Case (1) gets that $L(x_0') \leq L(x_0)$ and x_0' can be accepted. If $|M_0(x_0')| < |M_1(x_0')|$ then $|M_0(x_0)| = |M_1(x_0')|$ as n is even. By Lemma 1, $L(x_0) - L(x_0') = \sqrt{\frac{1-\gamma}{\gamma}} \left(\sqrt{Var[l_0(x_0)]} - \sqrt{Var[l_1(x_0')]} \right)$. As $Var[l_0(x_0)] \geq Var[l_1(x_0')] \iff cov[l_0(x_0)] \geq cov[l_1(x_0')]$, x_0' can be accepted iff $cov[l_0(x_0)] \geq cov[l_1(x_0')]$.

Case (4). Flipping a 0-bit and a 1-bit in x (i.e., $|M_0(x_0')| = |M_0(x_0)|$). Using the reasoning similar to that for Case (3), we have that x_0' can be accepted iff $cov[l_0(x_0)] \geq cov[l_0(x_0')]$.

Case (5). Flipping two 1-bits in x_0 (i.e., $|M_0(x_0')| = |M_0(x_0)| + 2$). Using the reasoning similar to that for Case (2), we have that $L(x_0') > L(x_0)$ and x_0' cannot be accepted.

Summarizing the above analysis gets that if x_0' is accepted by RLS, then it satisfies one of the following two conditions: (1). $|M_{t(x_0')}(x_0')| < |M_0(x_0)|$ and $cov[l_{t(x_0')}(x_0')] < cov[l_0(x_0)]$; (2). $|M_{t(x_0')}(x_0')| = |M_0(x_0)|$ and $cov[l_{t(x_0')}(x_0')] \leq cov[l_0(x_0)]$. That is, the gap between the numbers of jobs in the two machines cannot increase during the optimization process. The mutation considered in Case (1) can be generated by RLS with probability $\Omega(1/4)$ that decreases the gap between the numbers of jobs in the two machines by 2. As $||M_0(x_0)| - |M_1(x_0)|| \leq n$, using the Additive Drift analysis [10] gets that RLS takes expected runtime $O(n)$ to obtain the first equal-solution x_1 based on x_0.

Now we consider the expected runtime of RLS to obtain an optimal solution x^* based on x_1. Let $p(x) = \sum_{i=1}^{k} |\alpha_i(x) - \beta_i(x)| = \sum_{i=1}^{k} |2\alpha_i(x) - m|$ be the potential of the solution x maintained during the process, and we show that during the optimization process the potential value cannot increase. Note that once the first equal-solution x_1 is obtained, then all solutions subsequently accepted by RLS are equal-ones, thus only the mutations flipping a 0-bit and a 1-bit of x_1 are considered below. Assume that the mutation flips a 0-bit of G_i and a 1-bit of G_j in x_1, and denoted by x_1' the solution obtained. The potential change is

$$\Delta_p = p(x_1) - p(x_1') = |2\alpha_i(x_1) - m| + |2\alpha_j(x_1) - m| - (|2\alpha_i(x_1') - m| + |2\alpha_j(x_1') - m|),$$

where $\alpha_i(x_1') = \alpha_i(x_1) - 1$ and $\alpha_j(x_1') = \alpha_j(x_1) + 1$. The above discussion shows that x_1' can be accepted by RLS iff $\Delta_{cov} = cov[l_0(x_1)] - cov[l_0(x_1')] \geq 0$, where

$$\Delta_{cov}/2c = (cov[l_0(x_1)] - cov[l_0(x_1')])/2c = \alpha_i(x_1) - 1 - \alpha_j(x_1).$$

We divide the analysis for the values of Δ_p and Δ_{Var} into four cases.

Case (I). $\alpha_i(x_1) > \frac{m}{2}$ and $\alpha_j(x_1) \geq \frac{m}{2}$. Observe that $\Delta_p = 0$, but the value of Δ_{Var} depends on the relationship between $\alpha_i(x_1)$ and $\alpha_j(x_1)$.

Case (II). $\alpha_i(x_1) \leq \frac{m}{2}$ and $\alpha_j(x_1) \geq \frac{m}{2}$. Observe that $\Delta_p = -4$, but $\Delta_{Var} < 0$, implying that x_1' cannot be accepted by RLS.

Case (III). $\alpha_i(x_1) > \frac{m}{2}$ and $\alpha_j(x_1) < \frac{m}{2}$. Observe that $\Delta_p = 4$ and $\Delta_{Var} > 0$, implying that x_1' can be accepted by RLS.

Case (IV). $\alpha_i(x_1) \leq \frac{m}{2}$ and $\alpha_j(x_1) < \frac{m}{2}$. Observe that $\Delta_p = 0$, but the value of Δ_{Var} depends on the relationship between $\alpha_i(x_1)$ and $\alpha_j(x_1)$.

Summarizing the analysis of the four cases gets that during the optimization process, the potential value cannot increase. Observe that there exist $i, j \in [1, k]$ such that $\alpha_i(x_1) = |M_0(x_1) \cap G_i| > \frac{m}{2}$ and $\alpha_j(x_1) = |M_0(x_1) \cap G_j| < \frac{m}{2}$ (i.e., Case (III) holds), and the offspring obtained by the mutation flipping a 0-bit of

G_i and a 1-bit of G_j in x_1 can be accepted. Now we consider the probability to generate such a mutation. Let $S_0 \subset [1, k]$ (resp., $S_1 \subset [1, k]$) such that for any $i \in S_0$, $\alpha_i(x_1) > \beta_i(x_1)$ (resp., $\alpha_i(x_1) < \beta_i(x_1)$). Since x_1 is an equal-solution,

$$\sum_{i \in S_0} \alpha_i(x_1) - \beta_i(x_1) = \sum_{i \in S_1} \beta_i(x_1) - \alpha_i(x_1) = p(x_1)/2. \tag{1}$$

Combining Equality (1) with $\sum_{i \in S_0} \alpha_i(x_1) + \beta_i(x_1) = |S_0|m$ and $\sum_{i \in S_1} \alpha_i(x_1) + \beta_i(x_1) = |S_1|m$ gets $\sum_{i \in S_0} \alpha_i(x_1) = \frac{p(x_1)}{4} + \frac{|S_0|m}{2} \geq \frac{p(x_1)}{4} + \frac{m}{2}$ and $\sum_{i \in S_1} \beta_i(x_1) = \frac{p(x_1)}{4} + \frac{|S_1|m}{2} \geq \frac{p(x_1)}{4} + \frac{m}{2}$. Thus there are $\frac{p(x_1)}{4} + \frac{m}{2}$ 0-bits, each of which is in a group G_u with $\alpha_u(x_1) > \frac{m}{2}$, and $\frac{p(x_1)}{4} + \frac{m}{2}$ 1-bits, each of which is in a group G_v with $\alpha_v(x_1) < \frac{m}{2}$. That is, RLS generates such a mutation with probability $\Omega((\frac{2m + p(x_1)}{4n})^2)$ and takes expected runtime $O((\frac{n}{2m + p(x_1)})^2)$ to obtain an offspring x_1' with $p(x_1') = p(x_1) - 4$. Considering all possible values for the potential of the maintained solution (note that $1 \leq p(x_1) \leq n$), the total expected runtime of RLS to obtain x^* based on x_1 can be upper bounded by

$$\sum_{t=1}^{n} O(\frac{n^2}{(t + 2m)^2}) = O(n^2) \sum_{t=1}^{n} (t + 2m)^{-2} = O(n^2) \int_{1}^{n} (t + 2m)^{-2} dt = O(n^2/m).$$

In summary, RLS takes expected runtime $O(n^2/m) = O(kn)$ to obtain an optimal solution to I_1 based on the initial solution x_0. □

Theorem 3. *The expected runtime of $(1 + 1)$ EA to obtain an optimal solution to the instance $I_1 = (a, c, d, \gamma, k, m)$ of CCMSP-1 is $O((k + m)n^2)$.*

Proof. As the mutation of $(1 + 1)$ EA may flip more than two bits simultaneously, the reasoning given in Theorem 2 cannot be directly applied for the performance of $(1 + 1)$ EA. We first consider the expected runtime of $(1 + 1)$ EA to get the first equal-solution x_1 based on the initial solution x_0 that is assumed to have $|M_0(x_0)| > |M_1(x_0)|$. A vector function $v(x) = (|M_{t(x)}(x)|, b(x))$ is designed for the solutions x obtained during the process, where $b(x) = \sum_{i=1}^{k} \binom{|M_{t(x)}(x) \cap G_i|}{2}$.

For ease of notation, let $|M_{t(x)}(x)| = \ell$, where $\ell \in [\frac{n}{2}, n]$ (as $M_{t(x)}(x)$ is the fuller machine by Lemma 1). Then $0 < b(x) \leq \lfloor \frac{\ell}{m} \rfloor \binom{m}{2} + \binom{\ell\%m}{2} \leq (\frac{\ell}{m} + 1)\binom{m}{2}$, where the first $\leq$ holds by Observation 1. Hence the number of possible values of $v(x)$ can be upper bounded by $\sum_{\ell=\frac{n}{2}+1}^{n} (\frac{\ell}{m} + 1)\binom{m}{2} = O(mn^2)$. Observe that for any two solutions x and x', if $v(x) = v(x')$ then $L(x) = L(x')$. Thus the number of possible values of $L(x)$ can be upper bounded by $O(mn^2)$ as well.

Consider a mutation flipping a $t(x)$-bit on x (i.e., if $t(x) = 0$ then flipping a 0-bit; otherwise, a 1-bit). By the discussion for Case (1) given in Theorem 2, the solution x' obtained by the mutation has $L(x') < L(x)$ and can be accepted. The probability of $(1 + 1)$ EA to generate such a mutation is $\Omega(1/2)$. Thus combining the probability and the number of possible values of $L(x)$ gives that $(1 + 1)$ EA takes expected runtime $O(mn^2)$ to get the first equal-solution x_1 based on x_0.

Now we consider the runtime of $(1 + 1)$ EA to obtain an optimal solution based on x_1. As all solutions accepted subsequently are equal-ones, we take $b(x)$

as the potential function, where the number of possible values of $b(x)$ can be bounded by $O(km^2)$. By the reasoning given in Theorem 2 a mutation flipping a 0-bit and a 1-bit that can obtain an improved solution can be generated with probability $\Omega((\frac{m}{2n})^2)$. Consequently, $(1 + 1)$ EA takes expected runtime $O(kn^2)$ to obtain an optimal solution based on x_1. In summary, $(1 + 1)$ EA takes expected runtime $O((k + m)n^2)$ to obtain an optimal solution to I_1.   □

5   Performance for CCMSP-2

The section starts with a lemma to show that the discussion for CCMSP-2 would be more complicated than that for CCMSP-1.

Lemma 3. *Given a solution x to an instance $I_2 = (a, c, d, \gamma, k, \{m_i | i \in [1, k]\})$ of CCMSP-2, whether $l'_0(x) > l'_1(x)$ holds is unknown even if $|M_0(x)| > |M_1(x)|$.*

Proof. Recall that the group G_i has size m_i, and there is a variable $\delta_i(x)$ such that $\alpha_i(x) = m_i/2 + \delta_i(x)$ and $\beta_i(x) = m_i/2 - \delta_i(x)$ for any $i \in [1, k]$. Thus

$$cov[l_0(x)] - cov[l_1(x)] = 2c \sum_{i=1}^{k} \left(\binom{\alpha_i}{2} - \binom{\beta_i}{2} \right) = 2c \sum_{i=1}^{k} (m_i - 1)\delta_i.$$

Observe that $2c \sum_{i=1}^{k} (m_i - 1)\delta_i$ can be treated as a weighted version of $\sum_{i=1}^{k} \delta_i(x)$, where $\sum_{i=1}^{k} \delta_i(x) > 0$ due to $|M_0(x)| > |M_1(x)|$, but it is impossible to decide whether $2c \sum_{i=1}^{k} (m_i - 1)\delta_i$ is greater than 0. Furthermore, the relationship among the values of a, c and d are unrestricted. Consequently, it is also impossible to decide whether or not $l'_0(x) > l'_1(x)$ holds.   □

For ease of analysis, we set an *extra constraint* on the values of a, c and d considered in the instances of CCMSP-2:

$$\sqrt{\frac{(1 - \gamma)}{\gamma} \left(nd + 2c \sum_{i=1}^{k} \binom{m_i}{2} \right)} < a. \tag{2}$$

That is, for any solution x to any instance of CCMSP-2 and any $t \in [0, 1]$, $E[l_t(x)]$ contributes much more than $\sqrt{\frac{(1-\gamma)}{\gamma} Var[l_t(x)]}$ to $l'_t(x)$ under the extra constraint, because $\sqrt{\frac{(1-\gamma)}{\gamma} Var[l_t(x)]} \leq \sqrt{\frac{(1-\gamma)}{\gamma} \left(nd + 2c \sum_{i=1}^{k} \binom{m_i}{2} \right)} < a$. The new variant of CCMSP-2 is called CCMSP-2$^+$ in the remaining text.

Due to the extra constraint of CCMSP-2$^+$, for any solution x to I_2^+, if $|M_0(x)| > |M_1(x)|$ (resp., $|M_0(x)| < |M_1(x)|$) then $l'_0(x) > l'_1(x)$ (resp., $l'_0(x) < l'_1(x)$). Thus it is easy to derive the following lemma.

Lemma 4. *Given an instance $I_2^+ = (a, c, d, \gamma, k, \{m_i | i \in [1, k]\})$ of CCMSP-2$^+$, any optimal solution to I_2^+ is an equal-solution.*

Lemma 5. *CCMSP-2$^+$ is NP-hard.*

Proof. For the computational hardness of CCMSP-2$^+$, the discussion is divided based on the number of jobs considered in the instances of CCMSP-2$^+$.

Case 1. The instances of CCMSP-2^+ that consider odd many jobs.

Let $I_2^+ = (a, c, d, \gamma, k, \{m_i | i \in [1, k]\})$ be an instance of CCMSP-2^+, where $n = \sum_{i=1}^{k} m_i$ is odd. We construct an optimal solution x^* to I_2^+ as follows. By Lemma 4, x^* is an equal-solution. Assume $|M_0(x^*)| = |M_1(x^*)| + 1 = \frac{n+1}{2}$. By the extra constraint of CCMSP-2^+, $l_0'(x^*) > l_1'(x^*)$. Thus we only need to analyze the optimal allocation approach of $\frac{n+1}{2}$ many jobs on M_0 w.r.t. x^* such that $cov[l_0(x^*)]$ is minimized. By Observation 1, $k(\frac{n+1}{2k}) \leq cov[l_0(x^*)]$ (i.e., each group allocates $\frac{n+1}{2k}$ many jobs to M_0), but $\frac{n+1}{2k}$ may be not an integer. Fortunately, by Observation 1, it is easy to get that the optimal allocation approach of $\frac{n+1}{2}$ many jobs on M_0 w.r.t. x^* can be obtained as: For each $1 \leq i \leq k$ (assume that the values of $\alpha_j(x^*)$ for all $1 \leq j < i$ have been specified), if

$$m_i < (\frac{n+1}{2} - \sum_{j=1}^{i-1} \alpha_j(x^*))/(k+1-i),$$

then let $\alpha_i(x^*) = m_i$; otherwise, let $\alpha_i(x^*) = \lceil (\frac{n+1}{2} - \sum_{j=1}^{i-1} \alpha_j(x^*))/(k+1-i) \rceil$.

Observe that once $\alpha_i(x^*)$ is set as $\lceil (\frac{n+1}{2} - \sum_{j=1}^{i-1} \alpha_j(x^*))/(k+1-i) \rceil$, then for all $i < j \leq k$, $|\alpha_j(x^*) - \alpha_i(x^*)| \leq 1$ (as $m_1 \leq m_2 \leq \ldots \leq m_k$). In a word, the optimal solution x^* to I_2^+ satisfies the following property.

Property-Odd: For any $i \in [1, k]$, either $\alpha_i(x^*) = m_i$ or $0 \leq \alpha_{max}(x^*) - \alpha_i(x^*) \leq 1$, where $\alpha_{max}(x^*) = \max\{\alpha_1(x^*), \ldots, \alpha_k(x^*)\}$.

Case 2. The instances of CCMSP-2^+ that consider even many jobs.

The definition of the *Two-way Balanced Partition problem* is: Given a multiset S that contains non-negative integers such that both $|S|$ and $\sum_{e \in S} e$ are even, can S be partitioned into two subsets S_1 and S_2 such that $|S_1| = |S_2|$ and $\sum_{a \in S_1} a = \sum_{b \in S_2} b$? The NP-hardness of the Two-way Balanced Partition problem can be shown by reducing the well-known Partition problem [8] to it. It can be shown that any instance of the Two-way Balanced Partition problem can be polynomial-time reduced to an instance I_2^+ of CCMSP-2^+ such that I_2^+ has even many groups and each group has odd size. Due to the page limit, the detailed discussion will be given in a complete version.   □

Corollary 1. *CCMSP-2 is NP-hard.*

5.1   Performance for CCMSP-2^+

Theorem 4. *Given an instance $I_2^+ = (a, c, d, \gamma, k, \{m_i | i \in [1, k]\})$ of CCMSP-2^+ that considers odd many jobs (i.e., $n = \sum_{i=1}^{k} m_i$ is odd), RLS takes expected runtime $O(\sqrt{k} n^3)$ to obtain an optimal solution to I_2^+.*

Proof. Let x_0 be the initial solution maintained by RLS. The optimization process of RLS for x_0 discussed below is divided into two phases.

Phase-1. Obtaining the first equal-solution x_1 based on x_0.

Let $p_1(x) = ||M_0(x)| - |M_1(x)||$ be the potential of the solution x maintained during Phase-1. Observe that $1 \le p_1(x) \le n$, and the extra constraint of CCMSP-2$^+$ indicates that for any two solutions x' and x'' to I_2^+, if $p_1(x') < p_1(x'')$ then $L(x') < L(x'')$. The mutation of RLS flipping exactly one bit in x whose corresponding job is allocated to the fuller machine w.r.t. x, can be generated by RLS with probability $\Omega(1/4)$, and the obtained solution x' has potential value $p_1(x') = p_1(x) - 2$. Combining $p_1(x') = p_1(x) - 2$ with the conclusion given above gets $L(x') < L(x)$, and x' can be accepted by RLS. Then using the Additive Drift analysis [10], we can derive that Phase-1 takes expected runtime $O(n)$. Note that after the acceptance of x_1, any non-equal-solution cannot be accepted. W.l.o.g., assume that $|M_0(x_1)| = |M_1(x_1)| + 1$.

Phase-2. Obtaining the first optimal solution based on x_1.

Case (1). $cov[l_0(x_1)] < cov[l_1(x_1)]$.

First of all, it is not hard to get that any mutation flipping exactly one bit of x_1 cannot get an improved solution under Case (1). Thus the following discussion only considers the mutations flipping a 0-bit of G_i and a 1-bit of G_j in x_1 (note that the other kinds of mutations flipping two bits cannot get equal-solutions). Denote by x_1' the obtained solution. Hence $|M_0(x_1')| = |M_0(x_1)| = |M_1(x_1')| + 1 = |M_1(x_1)| + 1$ and $cov[l_0(x_1)] - cov[l_0(x_1')] = 2c(\alpha_i(x_1) - 1 - \alpha_j(x_1))$. If $\alpha_i(x_1) - \alpha_j(x_1) \ge 1$ then $cov[l_0(x_1')] \le cov[l_0(x_1)]$, and x_1' can be accepted.

Assume that RLS obtains a solution x_1^* based on x_1, on which all possible mutations flipping exactly a 0-bit and a 1-bit of x_1^* cannot get an improved solution, where the 0-bit and 1-bit are of G_i and G_j, respectively. Then x_1^* satisfies the property: For any $1 \le i \ne j \le k$, if $\alpha_i(x_1^*) - \alpha_j(x_1^*) \ge 2$, then all jobs of G_j are allocated to M_0 w.r.t. x_1^*, i.e., $\alpha_j(x_1^*) = m_j$ and $\beta_j(x_1^*) = 0$. In other words, for any $1 \le j \le k$, either $\alpha_j(x_1^*) = m_j$ or $0 \le \alpha_{max}(x_1^*) - \alpha_j(x_1^*) \le 1$, where $\alpha_{max}(x_1^*) = \max\{\alpha_1(x_1^*), \dots, \alpha_k(x_1^*)\}$. Thus x_1^* satisfies Property-Odd given in the proof of Lemma 5, and x_1^* is an optimal solution to I_2^+.

For the expected runtime of RLS for Phase-2, let $p_{21}(x) = cov[l_0(x)]/2c$ be the potential of the solution x maintained during Phase-2. The above discussion shows that $|M_0(x)| = |M_1(x)| + 1$. Let $i_{max} = \arg\max\{\alpha_1(x), \dots, \alpha_k(x)\}$. Then $\binom{\alpha_{i_{max}}(x)}{2} \ge \frac{p_{21}(x)}{k}$, implying that $\alpha_{i_{max}}(x) \ge (\sqrt{1 + \frac{8p_{21}(x)}{k}} + 1)/2$. Since x does not satisfy Property-Odd, there exists a $1 \le j' \ne i_{max} \le k$ such that $\alpha_{i_{max}}(x) - \alpha_{j'}(x) \ge 2$ but $\alpha_{j'}(x) < m_{j'}$. Thus $\beta_{j'}(x) \ge 1$. The mutation flipping a 0-bit of $G_{i_{max}}$ and a 1-bit of $G_{j'}$ in x can be generated by RLS with probability $\Omega(\frac{\alpha_{i_{max}}(x) \cdot \beta_{j'}(x)}{n^2}) = \Omega(\frac{\alpha_{i_{max}}(x)}{n^2}) = \Omega(\frac{1}{n^2}\sqrt{\frac{p_{21}(x)}{k}})$, and the potential value of the obtained solution is decreased by at least 1 compared to $p_{21}(x)$. Observe that the upper bound of $p_{21}(x_1)$ and lower bound of $p_{21}(x_1^*)$ are $\binom{\frac{n+1}{2}}{2}$ and $k\binom{\frac{n+1}{2k}}{2}$, respectively. Considering all possible potential values of x, we have that the expected runtime of RLS for Phase-2 can be bounded by

$$\sum_{t=k(\frac{n+1}{2k})}^{(\frac{n+1}{2})} O(\frac{\sqrt{k}n^2}{\sqrt{t}}) = O(\sqrt{k}n^2) \int_{k(\frac{n+1}{2k})}^{(\frac{n+1}{2})} t^{-\frac{1}{2}} dt = O(\sqrt{k}n^3).$$

Case (2). $cov[l_0(x_1)] \geq cov[l_1(x_1)]$.

The main difference between the discussion for Case (2) and that for Case (1) is that the mutation flipping one bit may generate an improved solution, implying that the fuller machine may be M_0 or M_1. However, no matter which one is the fuller machine, the value $cov[l_{t(x)}(x)]$ cannot increase during Phase-2, where x is a solution maintained by RLS during Phase-2. By the reasoning given for Case (1), for a mutation flipping exactly one 0-bit of G_i and one 1-bit of G_j in x, if $t(x) = 0$ and $\alpha_i(x) - \alpha_j(x) \geq 2$, or $t(x) = 1$ and $\beta_j(x) - \beta_i(x) \geq 2$, then $cov[l_{t(x')}(x')] < cov[l_{t(x)}(x)]$ for the obtained solution x', and x' can be accepted.

Let $p_{22}(x) = cov[l_{t(x)}(x)]/2c$ be the potential of the solution x. Using the reasoning similar to that given for Case (1), we can get that RLS takes expected runtime $O(\sqrt{k}n^3)$ to obtain an optimal solution to I_2^+ under Case (2).   □

Theorem 5. *Given an instance* $I_2^+ = (a, c, d, \gamma, k, \{m_i | i \in [1, k]\})$ *of CCMSP-2^+ that considers even many jobs (i.e., $n = \sum_{i=1}^{k} m_i$ is even), RLS takes expected runtime $O(n^4)$ to obtain an equal-solution x^* such that either $|cov[l_0(x^*)] - cov[l_1(x^*)]| \leq 2c(m_k - m_1 - 1)$ or $cov[l_{t(x^*)}(x^*)] \leq \frac{c}{4}(\frac{n^2}{k} - 2n + k)$.*

Proof. Let x_0 be the initial solution maintained by RLS. The proof runs in a similar way to that of Theorem 4, dividing the optimization process into two phases: **Phase-1**, obtaining the first equal-solution x_1 based on x_0; **Phase-2**, optimizing the solution x_1. Moreover, the analysis for Phase-1 is the same as that given in the proof of Theorem 4, i.e., Phase-1 takes expected runtime $O(n)$. Now we consider Phase-2, where the solution x_1 is assumed to have $cov[l_0(x_1)] > cov[l_1(x_1)]$. Let $\Delta(x_1) = cov[l_0(x_1)] - cov[l_1(x_1)]$. The following discussion only considers the mutations flipping a 0-bit of G_i and a 1-bit of G_j in x_1. Denote by x_1' the obtained solution. We have

$$cov[l_0(x_1')]$$

$$= cov[l_0(x_1)] - 2c\left[\left(\binom{\alpha_i(x_1)}{2} + \binom{\alpha_j(x_1)}{2}\right) - \left(\binom{\alpha_i(x_1) - 1}{2} + \binom{\alpha_j(x_1) + 1}{2}\right)\right]$$

$$= cov[l_0(x_1)] + 2c(\alpha_j(x_1) - \alpha_i(x_1) + 1)$$

and $cov[l_1(x_1')] = cov[l_1(x_1)] + 2c(\beta_i(x_1) - \beta_j(x_1) + 1)$ similarly.

If $\alpha_j(x_1) \leq \alpha_i(x_1) - 1$ (i.e., $cov[l_0(x_1')] \leq cov[l_0(x_1)]$) and $\beta_i(x_1) - \beta_j(x_1) + 1 \leq \Delta(x_1)/2c$ (i.e., $cov[l_1(x_1')] \leq cov[l_0(x_1)]$), then $L(x_1') \leq L(x_1)$ and x_1' can be accepted by RLS, and $cov[l_0(x_1')] - cov[l_1(x_1')] = \Delta(x_1) + 2c(m_j - m_i)$.

Now we assume that RLS obtains a solution x_1^* based on x_1 such that any mutation flipping a 0-bit and a 1-bit of x_1^* cannot get an improved solution, and $cov[l_0(x_1^*)] \geq cov[l_1(x_1^*)]$. Let $i_{max} = \arg \max\{\alpha_1(x_1^*), \alpha_2(x_1^*), \ldots, \alpha_k(x_1^*)\}$. Then the above discussion shows that for any $j \in [1, k]$, if $\alpha_j(x_1^*) < \alpha_{i_{max}}(x_1^*) - 1$ then $\beta_{i_{max}}(x_1^*) - \beta_j(x_1^*) + 1 \geq \Delta(x_1^*)/2c$, i.e.,

$$(m_{i_{max}} - \alpha_{i_{max}}(x_1^*)) - (m_j - \alpha_j(x_1^*)) \geq \Delta(x_1^*)/2c - 1,$$

implying that (recall that $m_1 \leq m_2 \leq \ldots \leq m_k$)

$$\Delta(x_1^*)/2c + 1 \leq \Delta(x_1^*)/2c - 1 + (\alpha_{i_{max}}(x_1^*) - \alpha_j(x_1^*)) \leq m_{i_{max}} - m_j \leq m_k - m_1.$$

In other words, for x_1^*, if there is a $j \in [1, k]$ with $\alpha_j(x_1^*) < \alpha_{i_{max}}(x_1^*) - 1$, then $\Delta(x_1^*)/2c \leq m_k - m_1 - 1$. If there is no $j \in [1, k]$ with $\alpha_j(x_1^*) < \alpha_{i_{max}}(x_1^*) - 1$, then for each $j \in [1, k]$, $0 \leq \alpha_{i_{max}}(x_1^*) - \alpha_j(x_1^*) \leq 1$. Now we bound the value of $cov[l_0(x_1^*)]$. Let $\tau = |\{1 \leq j \leq k | \alpha_j(x_1^*) = \alpha_{i_{max}}(x_1^*) - 1\}|$. Then $(k - \tau)\alpha_{i_{max}}(x_1^*) + \tau(\alpha_{i_{max}}(x_1^*) - 1) = n/2$ implies that $\alpha_{i_{max}}(x_1^*) = \frac{n}{2k} + \frac{\tau}{k}$, and

$$cov[l_0(x_1^*)]/2c = (k - \tau)\binom{\alpha_{i_{max}}(x_1^*)}{2} + \tau\binom{\alpha_{i_{max}}(x_1^*) - 1}{2}$$

$$= \frac{n^2}{8k} - \frac{n}{4} - (\frac{\tau^2}{2k} - \frac{\tau}{2}) \leq \frac{n^2}{8k} - \frac{n}{4} + \frac{k}{8},$$

where $\frac{\tau^2}{2k} - \frac{\tau}{2}$ gets its minimum value $-\frac{k}{8}$ when $\tau = \frac{k}{2}$.

For the expected runtime of RLS to get x_1^* based on x_1, let $p(x) = cov[l_{t(x)}(x)]/2c$ be the potential of x that is a solution maintained by RLS during the process. Observe that $p(x)$ cannot increase during the process. The probability of RLS to generate such a mutation mentioned above is $\Omega(1/n^2)$, and the potential value decreases by at least 1. As $p(x_1)$ can be upper bounded by $O(n^2)$, using the Additive Drift analysis [10] gets that RLS takes expected runtime $O(n^4)$ to obtain x_1^* based on x_1. In summary, RLS takes expected runtime $O(n^4)$ to obtain an equal-solution x_1^* satisfying the claimed condition based on x_0. □

6 Conclusion

In the paper, we studied a chance-constrained version of the Makespan Scheduling problem and investigated the performance of RLS and $(1 + 1)$ EA for it. More specifically, we studied two simple variants of the problem (namely, CCMSP-1 and CCMSP-2^+) and obtained a series of results: CCMSP-1 was shown to be polynomial-time solvable by giving the expected runtime of RLS and $(1 + 1)$ EA separately to obtain an optimal solution to the given instance of CCMP-1; CCMSP-2^+ was shown to be NP-hard by reducing the Two-way Balanced Partition problem to it, but any instance of CCMSP-2^+ that considers odd many jobs was shown to be polynomial-time solvable by giving the expected runtime of RLS to obtain an optimal solution to it.

Future work on the Chance-constrained Makespan Scheduling problem or the chance-constrained version of other classical combinatorial optimization problems would be interesting, and these related results would further advance and broaden the understanding of evolutionary algorithms.

References

1. Blazewicz, J., Lenstra, J., Kan, A.: Scheduling subject to resource constraints: classification and complexity. Discret. Appl. Math. **5**(1), 11–24 (1983)

2. Charnes, A., Cooper, W.W.: Chance-constrained programming. Manage. Sci. **6**(1), 73–79 (1959)
3. Corus, D., Lehre, P.K., Neumann, F.: The generalized minimum spanning tree problem: a parameterized complexity analysis of bi-level optimisation. In: Proceedings of the Genetic and Evolutionary Computation Conference (GECCO), pp. 519–526. ACM (2013)
4. Friedrich, T., He, J., Hebbinghaus, N., Neumann, F., Witt, C.: Analyses of simple hybrid algorithms for the vertex cover problem. Evol. Comput. **17**(1), 3–19 (2009)
5. Friedrich, T., He, J., Hebbinghaus, N., Neumann, F., Witt, C.: Approximating covering problems by randomized search heuristics using multi-objective models. Evol. Comput. **18**(4), 617–633 (2010)
6. Graham, R.L.: Bounds for certain multiprocessing anomalies. Bell Syst. Tech. J. **45**(9), 1563–1581 (1966)
7. Gunia, C.: On the analysis of the approximation capability of simple evolutionary algorithms for scheduling problems. In: Proceedings of the Genetic and Evolutionary Computation Conference (GECCO), pp. 571–578. ACM (2005)
8. Hayes, B.: Computing science: the easiest hard problem. Am. Sci. **90**(2), 113–117 (2002)
9. He, J., Mitavskiy, B., Zhou, Y.: A theoretical assessment of solution quality in evolutionary algorithms for the knapsack problem. In: Proceedings of the IEEE Congress on Evolutionary Computation (CEC), pp. 141–148. IEEE (2014)
10. He, J., Yao, X.: A study of drift analysis for estimating computation time of evolutionary algorithms. Nat. Comput. **3**, 21–35 (2004)
11. Hochbaum, D.S., Shmoys, D.B.: Using dual approximation algorithms for scheduling problems theoretical and practical results. J. ACM **34**(1), 144–162 (1987)
12. Iwamura, K., Liu, B.: A genetic algorithm for chance constrained programming. J. Inf. Optim. Sci. **17**(2), 409–422 (1996)
13. Jansen, T., Oliveto, P.S., Zarges, C.: Approximating vertex cover using edge-based representations. In: Proceedings of the Workshop on Foundations of Genetic Algorithms (FOGA), pp. 87–96. ACM (2013)
14. Kratsch, S., Lehre, P.K., Neumann, F., Oliveto, P.S.: Fixed parameter evolutionary algorithms and maximum leaf spanning trees: a matter of mutation. In: Schaefer, R., Cotta, C., Kołodziej, J., Rudolph, G. (eds.) PPSN 2010. LNCS, vol. 6238, pp. 204–213. Springer, Heidelberg (2010). https://doi.org/10.1007/978-3-642-15844-5_21
15. Kratsch, S., Neumann, F.: Fixed-parameter evolutionary algorithms and the vertex cover problem. Algorithmica **65**(4), 754–771 (2013)
16. Lissovoi, A., Witt, C.: A runtime analysis of parallel evolutionary algorithms in dynamic optimization. Algorithmica **78**(2), 641–659 (2017)
17. Miller, B.L., Wagner, H.M.: Chance constrained programming with joint constraints. Oper. Res. **13**(6), 930–945 (1965)
18. Neumann, A., Neumann, F.: Optimising monotone chance-constrained submodular functions using evolutionary multi-objective algorithms. In: Bäck, T., et al. (eds.) PPSN 2020. LNCS, vol. 12269, pp. 404–417. Springer, Cham (2020). https://doi.org/10.1007/978-3-030-58112-1_28
19. Neumann, F., Pourhassan, M., Roostapour, V.: Analysis of evolutionary algorithms in dynamic and stochastic environments. In: Theory of Evolutionary Computation. NCS, pp. 323–357. Springer, Cham (2020). https://doi.org/10.1007/978-3-030-29414-4_7

20. Neumann, F., Sutton, A.M.: Runtime analysis of the (1+1) evolutionary algorithm for the chance-constrained knapsack problem. In: Proceedings of the Workshop on on Foundations of Genetic Algorithms (FOGA), pp. 147–153. ACM (2019)
21. Neumann, F., Wegener, I.: Minimum spanning trees made easier via multi-objective optimization. Nat. Comput. **5**, 305–319 (2006)
22. Neumann, F., Wegener, I.: Randomized local search, evolutionary algorithms, and the minimum spanning tree problem. Theoret. Comput. Sci. **378**(1), 32–40 (2007)
23. Neumann, F., Witt, C.: On the runtime of randomized local search and simple evolutionary algorithms for dynamic makespan scheduling. In: Proceedings of the International Joint Conference on Artificial Intelligence (IJCAI), pp. 3742–3748 (2015)
24. Oliveto, P.S., He, J., Yao, X.: Analysis of population-based evolutionary algorithms for the vertex cover problem. In: Proceedings of the IEEE Congress on Evolutionary Computation (IEEE World Congress on Computational Intelligence), pp. 1563–1570. IEEE (2008)
25. Oliveto, P.S., He, J., Yao, X.: Analysis of the (1+1) EA for finding approximate solutions to vertex cover problems. IEEE Trans. Evol. Comput. **13**(5), 1006–1029 (2009)
26. Poojari, C.A., Varghese, B.: Genetic algorithm based technique for solving chance constrained problems. Eur. J. Oper. Res. **185**(3), 1128–1154 (2008)
27. Pourhassan, M., Friedrich, T., Neumann, F.: On the use of the dual formulation for minimum weighted vertex cover in evolutionary algorithms. In: Proceedings of the Workshop on Foundations of Genetic Algorithms (FOGA), pp. 37–44. ACM (2017)
28. Pourhassan, M., Gao, W., Neumann, F.: Maintaining 2-approximations for the dynamic vertex cover problem using evolutionary algorithms. In: Proceedings of the Genetic and Evolutionary Computation Conference (GECCO), pp. 513–518. ACM (2015)
29. Pourhassan, M., Shi, F., Neumann, F.: Parameterized analysis of multi-objective evolutionary algorithms and the weighted vertex cover problem. In: Handl, J., Hart, E., Lewis, P.R., López-Ibáñez, M., Ochoa, G., Paechter, B. (eds.) PPSN 2016. LNCS, vol. 9921, pp. 729–739. Springer, Cham (2016). https://doi.org/10.1007/978-3-319-45823-6_68
30. Roostapour, V., Neumann, A., Neumann, F., Friedrich, T.: Pareto optimization for subset selection with dynamic cost constraints. Artif. Intell. **302**, 103597 (2022)
31. Sahni, S.K.: Algorithms for scheduling independent tasks. J. ACM **23**(1), 116–127 (1976)
32. Shi, F., Neumann, F., Wang, J.: Runtime performances of randomized search heuristics for the dynamic weighted vertex cover problem. Algorithmica **83**(4), 906–939 (2021)
33. Shi, F., Schirneck, M., Friedrich, T., Kötzing, T., Neumann, F.: Reoptimization time analysis of evolutionary algorithms on linear functions under dynamic uniform constraints. Algorithmica **81**(2), 828–857 (2019)
34. Sutton, A.M., Neumann, F.: A parameterized runtime analysis of simple evolutionary algorithms for makespan scheduling. In: Coello, C.A.C., Cutello, V., Deb, K., Forrest, S., Nicosia, G., Pavone, M. (eds.) PPSN 2012. LNCS, vol. 7491, pp. 52–61. Springer, Heidelberg (2012). https://doi.org/10.1007/978-3-642-32937-1_6
35. Roostapour, V., Bossek, J., Neumann, F.: Runtime analysis of evolutionary algorithms with biased mutation for the multi-objective minimum spanning tree problem. In: Proceedings of the Genetic and Evolutionary Computation Conference (GECCO), pp. 551–559. ACM (2020)

36. Witt, C.: Worst-case and average-case approximations by simple randomized search heuristics. In: Diekert, V., Durand, B. (eds.) STACS 2005. LNCS, vol. 3404, pp. 44–56. Springer, Heidelberg (2005). https://doi.org/10.1007/978-3-540-31856-9_4
37. Witt, C.: Revised analysis of the $(1+1)$ EA for the minimum spanning tree problem. In: Proceedings of the Genetic and Evolutionary Computation Conference (GECCO), pp. 509–516. ACM (2014)
38. Xie, Y., Harper, O., Assimi, H., Neumann, A., Neumann, F.: Evolutionary algorithms for the chance-constrained knapsack problem. In: Proceedings of the Genetic and Evolutionary Computation Conference (GECCO), pp. 338–346. ACM (2019)
39. Xie, Y., Neumann, A., Neumann, F., Sutton, A.M.: Runtime analysis of RLS and the $(1+1)$ EA for the chance-constrained knapsack problem with correlated uniform weights. In: Proceedings of the Genetic and Evolutionary Computation Conference (GECCO), pp. 1187–1194. ACM (2021)
40. Yu, Y., Yao, X., Zhou, Z.H.: On the approximation ability of evolutionary optimization with application to minimum set cover. Artif. Intell. **180**, 20–33 (2012)

Runtime Analysis of the (1+1) EA on Weighted Sums of Transformed Linear Functions

Frank Neumann[1(✉)] and Carsten Witt[2]

[1] Optimisation and Logistics, School of Computer Science,
The University of Adelaide, Adelaide, Australia
frank.neumann@adelaide.edu.au
[2] DTU Compute, Technical University of Denmark, Kongens Lyngby, Denmark
cawi@imm.dtu.dk

Abstract. Linear functions play a key role in the runtime analysis of evolutionary algorithms and studies have provided a wide range of new insights and techniques for analyzing evolutionary computation methods. Motivated by studies on separable functions and the optimization behaviour of evolutionary algorithms as well as objective functions from the area of chance constrained optimization, we study the class of objective functions that are weighted sums of two transformed linear functions. Our results show that the (1+1) EA, with a mutation rate depending on the number of overlapping bits of the functions, obtains an optimal solution for these functions in expected time $O(n \log n)$, thereby generalizing a well-known result for linear functions to a much wider range of problems.

1 Introduction

Runtime analysis is one of the major theoretical tools to provide rigorous insights into the working behavior of evolutionary algorithms and other randomized search heuristics [12,20,27]. The class of pseudo-Boolean linear functions plays a key role in the area of runtime analysis. Starting with the simplest linear functions called OneMax for which the first runtime analysis has been carried out, a wide range of results have been obtained for the general class of linear functions. This includes the study of Droste, Jansen and Wegener [14] who were the first to obtain an upper bound of $O(n \log n)$ for the (1+1) EA on the general class of pseudo-Boolean linear functions. This groundbreaking result has been based on a very lengthy proof and subsequently a wide range of improvements have been made in terms of the development of new techniques for the analysis as well as the precision of the results. The proof has been simplified significantly using the analytic framework of drift analysis [15] by He and Yao [16]. Jägersküpper [18,19] provided the first analysis of the leading coefficient in the bound $O(n \log n)$ on the optimisation time for the problem. Furthermore, advances to simplify proofs and getting precise results have been made using the framework of multiplicative drift [10]. Doerr, Johannsen and Winzen improved the upper bound result to $(1.39 + o(1))en \ln n$ [9]. Finally, Witt [28] improved this bound to $en \ln n + O(n)$

G. Rudolph et al. (Eds.): PPSN 2022, LNCS 13399, pp. 542–554, 2022.
https://doi.org/10.1007/978-3-031-14721-0_38

by using adaptive drift analysis [6,7]. We expand such investigations for the
(1+1) EA into a wider class of problems that are modelled by two transformed
linear functions. This includes classes of separable functions and chance constrained optimization problems.

1.1   Separable Functions

As an example, consider the separable objective function

$$f(x) = \left(\sum_{i=1}^{n/2} w_i x_i \right)^2 + \sqrt{\sum_{i=n/2+1}^{n} w_i x_i} \tag{1}$$

where $w_i \in \mathbb{Z}^+$, $1 \le i \le n$, and $x = (x_1, \ldots, x_n) \in \{0,1\}^n$. The function f
consists of two objective functions

$$f_1(x_1, \ldots, x_{n/2}) = \left(\sum_{i=1}^{n/2} w_i x_i \right)^2 \text{ and } f_2(x_{n/2+1}, \ldots, x_n) = \sqrt{\sum_{i=n/2+1}^{n} w_i x_i}.$$

Here f_1 is the square of a function linear in the first half of variables and f_2
is the square root of a linear function in the remaining variables. Some investigations on how evolutionary algorithms optimize separable fitness functions have
been carried out in [13]. It has been shown that if the different functions only
have a small range, then the (1+1) EA optimizes separable functions efficiently if
the different separable functions themselves are easy to be optimized. However,
in our example above the two separable functions may take on exponentially
many values but both functions on their own are optimized by the (1+1) EA
in time $O(n \log n)$ using the results for the (1+1) EA on linear functions. This
holds as the transformation applying the square in f_1 or the square root in f_2
does not change the behavior of the (1+1) EA. The questions arises whether the
$O(n \log n)$ bounds also holds for the function f which combines f_1 and f_2. We
investigate this setting of separable functions for the more general case where
the objective function is given as a weighted sum of two separable transformed
linear functions. For technical reasons, we consider a (1+1) EA with potentially
reduced mutation probability depending on the number of overlapping bits of
the two functions.

1.2   Chance Constrained Problems

Another motivation for our work comes from problems from the area of chance
constrained optimization [2] and considers the case where the two functions are
overlapping or are even defined on the same set of variables. Recently evolutionary algorithms have been used for chance constrained problems which motivates
our investigations. In a chance constrained setting the input involves stochastic
components and the goal is to optimize a given objective function under the condition that constraints are met with high probability or that function values are

guaranteed with a high probability. Evolutionary algorithms have been designed for the chance constrained knapsack problem [1,29,30], chance constrained stock pile blending problems [31], and chance constrained submodular functions [24].

Runtime analysis results have been obtained for restricted settings of the knapsack problem [26,32] where the weights are stochastic and the constraint bound has to be met with high probability. The analysis for the case of stochastic constraints and the class of submodular function [5] and the knapsack problem [29] already reveal constraint functions that are a linear combination of the expected weight and the standard deviation of a solution when using Chebyshev's inequality for constraint evaluation. Such functions are the subject of our investigations.

To make the type of problems that we are interested in clear, we state the following problem. Given a set of m items $E = \{e_1, \ldots, e_m\}$ with random weights w_i, $1 \leq i \leq m$. We assume that the weights are independent and each w_i is distributed according to a normal distribution $N(\mu_i, \sigma_i^2)$, $1 \leq i \leq m$. We assume $\mu_i \geq 0$ and $\sigma_i \geq 0$, $1 \leq i \leq m$. Our goal is to

$$\min W \text{ subject to } \Pr(w(x) \leq W) \geq \alpha \tag{2}$$

where $w(x) = \sum_{i=1}^{n} w_i x_i$, $x \in \{0,1\}^m$, and $\alpha \in \,]0,1[$. The problem given in Equation (2) is usually considered under additional constraints, e.g. spanning tree constraints in [17], which we do not consider in this paper.

According to [17] the problem given in Eq. 2 is equivalent to minimizing the fitness function

$$g(x) = \sum_{i=1}^{m} \mu_i x_i + K_\alpha \left(\sum_{i=1}^{m} \sigma_i^2 x_i \right)^{1/2} \tag{3}$$

where K_α is the α-fractile point of the standard normal distribution.

The fitness function g is a linear combination of the expected value of a solution which is a linear function and the square root of its variance where the variance is again a linear function. In order to understand the behaviour of evolutionary algorithms on fitness functions obtained for chance constrained optimization problems, our runtime analysis for the (1+1) EA covers such fitness functions if we assume the reduced mutation probability mentioned above.

1.3   Transformed Linear Functions

In our investigations, we consider the much wider class of problems where a given fitness function is obtained by the linear combination of two transformed linear functions. The transformations applied to the linear functions only have to be monotonically increasing in terms of the functions values of the linear functions. This includes the setting of separable functions and chance constrained problems described previously. Furthermore, we do not require that the two linear functions are defined on the same number of bits.

The main result of our paper is an $O(n \log n)$ upper bound for the (1+1) EA with mutation probability $1/(n + s)$ on the class of sums of two transformed linear functions where s is the number of bits for which the two linear functions overlap. This directly transfers to the separable problem type given in Eq. 1 with standard bit mutation probability $1/n$ and to the chance constraint formulation given in Eq. 3 when using mutation probability $1/(2n)$.

The outline of the paper is as follows. In Sect. 2, we formally introduce the problem formulation for which we analyze the (1+1) EA in this paper. We discuss the exclusion of negative weights in our setup in Sect. 3 and present the $O(n \log n)$ bound in Sect. 4. Finally, we finish with some discussion and conclusions.

2   Preliminaries

The (1+1) EA shown in Algorithm 1 (generalized with a parameter s discussed below; classically $s = 0$ is assumed) is a simple evolutionary algorithm using independent bit flips and elitist selection. It is very well studied in the theory of evolutionary computation [3] and serves as a stepping stone towards the analysis of more complicated evolutionary algorithms. As common, in the area of runtime analysis, we measure the run time of the (1+1) EA by the number of iterations of the repeat loop. The optimization time refers to the number of fitness evaluations until an optimal solution has been obtained for the first time, and the expected optimization time refers to the expectation of this value.

2.1   Sums of Two Transformed Linear Functions Without Constraints

We will study the (1+1) EA on the scenario given in (1) and (3), assuming no additional constraints. In fact, we will generalize the scenario to the sum of two transformed pseudo-Boolean linear functions which may be (partially) overlapping. Note, that in (1) there is no overlap on the domains of the two linear functions and the transformations are the square and the square root, whereas in (3) there is complete overlap on the domains and the transformations are the identity function and the square root.

The crucial observation in our analysis is that the scenario considered here extends the linear function problem [28] that is heavily investigated in the theory of evolutionary algorithms. Despite the simple structure of the problem, there is no clear fitness-distance correlation in the linear function problem, which makes the analysis of the global search operator of the (1+1) EA difficult. If only local mutations are used, leading to the well known randomized local search (RLS) algorithm [4], then both the linear function problem and the generalized scenario considered here are very easy to analyze using standard coupon collector arguments [23], leading to $O(n \log n)$ expected optimization time. For the globally searching (1+1) EA, we will obtain the same bound, proving that the problem is easy to solve for it; however, we need advanced drift analysis methods to prove this.

We note that the class of functions we consider falls within the more general class of so-called monotone functions. Such functions can be difficult to optimize

---

Algorithm 1: (1+1) EA for minimization of a pseudo-Boolean function $f\colon \{0,1\}^{n-s} \to \mathbb{R}$, where $s \in \{0, \ldots, n/2\}$

---

1 Choose $x \in \{0,1\}^{n-s}$ uniformly at random;
2 **repeat**
3 $\quad$ Create y by flipping each bit x_i of x with probability $p = \frac{1}{n}$;
4 $\quad$ **if** $f(y) \leq f(x)$ **then**
5 $\quad\quad \lfloor\ x \leftarrow y;$

6 **until** *stop*;

---

with a (1+1) EA using mutation probabilities larger than $1/n$ [8]; however, it is also known that the standard (1+1) EA with mutation probability $1/n$ as considered here optimizes all monotone functions in expected time $O(n \log^2 n)$ [22]. Our bound is by an asymptotic factor of $\log n$ better if $s = o(n)$. However, it should be noted that for $s = \Omega(n)$, the bound $O(n \log n)$ already follows directly from [8] since it corresponds to a mutation probability of c/n for a constant $c < 1$. In fact, the fitness function $g(x)$ arising from the chance-constrained scenario presented in (3) above would fall into the case $s = n/2$.

Set-Up. We will investigate a general optimization scenario involving two linear pseudo-Boolean functions in an unconstrained search space. The objective function is an arbitrarily weighted sum of monotone transformations of two linear functions defined on (possibly overlapping) subspaces of $\{0,1\}^{n-s}$ for some $s \geq 0$, where s denotes the number of shared bits. Note that the introduction of this paper mentions a search space of dimension n and a mutation probability of $p = 1/(n+s)$ for the (1+1) EA. While the former perspective is more natural to present, from now on, we consider the asymptotically equivalent setting of search space dimension $n - s$ and mutation probability $p = 1/n$, which eases notation in the upcoming calculations.

Let α be a constant such that $1/2 \leq \alpha \leq \ln(2 - \epsilon) \approx 0.693 - \epsilon/2$ for some constant $\epsilon > 0$ and assume that αn is an integer. We allow the subfunctions to depend on a number of bits in $[(1 - \alpha)n, \alpha n]$, including the balanced case that both subfunctions depend on exactly $n/2$ bits. Formally, we have

– linear functions

$$\ell_1\colon \{0,1\}^{\alpha n} \to \mathbb{R} \text{ and } \ell_2\colon \{0,1\}^{(1-\alpha)n} \to \mathbb{R},$$

where $\ell_1(y_1, \ldots, y_{\alpha n}) = \sum_{i=1}^{\alpha n} w_i^{(1)} y_i$, and similarly $\ell_2(z_1, \ldots, z_{(1-\alpha)n}) = \sum_{i=1}^{(1-\alpha)n} w_i^{(2)} z_i$ with non-negative weights $w_i^{(1)}$ and $w_i^{(2)}$.
– $B_1 \subseteq \{1, \ldots, n\}$ and $B_2 \subseteq \{1, \ldots, n\}$, denoting the bit positions that ℓ_1 resp. ℓ_2 are defined on in the actual objective function $f\colon \{0,1\}^{n-s} \to \mathbb{R}$.
– The overlap count $s := |B_1 \cap B_2|$, where $s \leq \min\{(1 - \alpha)n, \alpha n\} = (1 - \alpha)n \leq n/2$

- the linear functions with extended domain $\ell_1^*(x_1, \ldots, x_{n-s}) = \sum_{i \in B_1} w_{r^{(1)}(i)}^{(1)} x_i$ where $r^{(1)}(i)$ is the rank of i in B_1 (with the smallest number receiving rank number 1); and analogously $\ell_2^*(x_1, \ldots, x_{n-s}) = \sum_{i \in B_2} w_{r^{(2)}(i)}^{(2)} x_i$; note that ℓ_1^* and ℓ_2^* only depend essentially on αn and $(1 - \alpha)n$ bits, respectively.
- monotone increasing functions $h_1 : \mathbb{R} \to \mathbb{R}$ and $h_2 : \mathbb{R} \to \mathbb{R}$.

Then the objective function $f : \{0, 1\}^{n-s} \to \mathbb{R}$, which w.l.o.g. is to be minimized, is given by

$$f(x_1, \ldots, x_{n-s}) = h_1(\ell_1^*(x_1, \ldots, x_{n-s})) + h_2(\ell_2^*(x_1, \ldots, x_{n-s})).$$

For $s = 0$, h_1 being the square function, and h_2 being the square root function, this matches the setting of separable functions given in Eq. 1. This set-up also includes the case that

$$f(x_1, \ldots, x_m) = \ell_1(x_1, \ldots, x_m) + R\sqrt{\ell_2(x_1, \ldots, x_m)}$$

for two m-dimensional, completely overlapping linear functions ℓ_1 and ℓ_2 and an arbitrary factor $R \geq 0$, as motivated and given in Eq. 3. Note that this matches our set-up with $n = 2\,m$ and $s = n$.

For our analysis we will make use of the multiplicative drift theorem (Theorem 1) that has been introduced in [11] and was enhanced with tail bounds by [7]. We use a slightly generalised presentation that can be found in [21].

Theorem 1 (Multiplicative Drift, cf. [7,11,21]). *Let $(X_t)_{t \geq 0}$, be a stochastic process, adapted to a filtration $\mathcal{F}_t$, over some state space $S \subseteq \{0\} \cup [s_{\min}, s_{\max}]$, where $0 \in S$ and $s_{\min} > 0$. Suppose that there exists a $\delta > 0$ such that for all $t \geq 0$*

$$E(X_t - X_{t+1} \mid \mathcal{F}_t) \geq \delta X_t.$$

Then it holds for the first hitting time $T := \min\{t \mid X_t = 0\}$ that

$$E(T \mid \mathcal{F}_0) \leq \frac{\ln(X_0/s_{\min}) + 1}{\delta}.$$

Moreover, $\Pr(T > (\ln(X_0/s_{\min}) + r)/\delta) \leq e^{-r}$ for any $r > 0$.

3 Negative Weights Allow for Multimodal Functions

We will now justify that the inclusion of negative weights in the underlying linear functions, along with overlapping domains, can lead to multimodal problems that cannot be optimized in expected time $O(n \log n)$ any longer. In the following example, the two linear functions depend essentially on all n bits.
Let

$$f(x_1, \ldots, x_n) = \underbrace{\left(\frac{x_1}{2} + \sum_{i=2}^{n} x_i \right)}_{h_1(\ell_1(x))} + \underbrace{\left(\frac{\sum_{i=1}^{n}(1 - x_i)}{n - 0.5} \right)^{n^2}}_{h_2(\ell_2(x))}$$

Basically, the first linear function $\ell_1(x) = x_1/2 + \sum_{i=2}^{n} x_i$ is a ONEMAX function except for the first bit that has a smaller weight than the rest. The second linear function $\ell_2(x)$ is linear in the number of zeros, i.e., corresponds to the ZEROMAX function that is equivalent to ONEMAX for the (1+1) EA due to symmetry reasons. The transformation h_2 that is applied to ZEROMAX divides the number of zero-bits by $n - 0.5$ and raises the result to a large power. Essentially, the value of $h_2(z)$ is $e^{\Theta(n)}$ if $z = n$ and $e^{-\Theta(n)}$ otherwise. This puts a constraint on the number of zero-bits. If $|x|_1 \geq 1$, then f is monotone increasing in ℓ_1, i.e., search points decreasing the ℓ_1-value also decrease the f-value. However, the all-zeros string has the largest f-value, i.e., is worst.

We can now see that all search points having one exactly one one-bit at one of the positions $2, \ldots, n$ are local optima. To create the global optimum from such a point, two bits have to flip simultaneously, leading to $\Omega(n^2)$ expected time to reach the optimum from the point. The situation is similar to the optimization of linear function under uniform constraints [25].

4   Upper Bound

The following theorem is the main result of this paper, showing that the (1+1) EA can optimize the generalized class of functions in asymptotically the same time as an ordinary linear function.

Theorem 2. *Let f be the sum of two transformed linear functions as defined in the set-up in Sect. 2.1. Then the expected optimization time of the (1+1) EA on f is $O(n \log n)$.*

The proof of Theorem 2 uses drift analysis with a carefully defined potential function, explained in the following.

Potential Function. We build upon the approach from [28] to construct a potential function $g^{(1)}$ for ℓ_1 and a potential function $g^{(2)}$ for ℓ_2, resulting in a combined potential function $\phi(x) = g^{(1)}(x) + g^{(2)}(x)$. The individual potential functions are obtained in the same way as if the (1+1) EA with mutation probability $1/n$ was only optimizing ℓ_1 and ℓ_2, respectively, on an αn-dimensional and $(1 - \alpha)n$-dimensional search space, respectively. The key idea is that accepted steps of the (1+1) EA on g must improve at least one of the two functions ℓ_1 and ℓ_2. This event leads to a high enough drift of the respective potential function that is still positive after pessimistically incorporating the potential loss due to flipping zero-bits that only the other linear function depends on.

We proceed with the definition of the potential functions $g^{(1)}$ and $g^{(2)}$ (similarly to Sect. 5 in [28]). For the two underlying linear functions we assume their arguments are reordered according to increasing weights. Note we cannot necessarily sort the set of all indices $1, \ldots, n - s$ of the function f so that both underlying linear functions have increasing coefficients; however, as we analyze the underlying functions separately, we can each time use the required sorting in these separate considerations.

Definition 1. *Given a linear function $\sum_{i=1}^{k} w_i x_i$, where $w_1 \leq \cdots \leq w_k$, we define the potential function $g(x_1, \ldots, x_k) = \sum_{i=1}^{k} g_i x_i$ by*

$$g_i = \left(1 + \frac{1}{n}\right)^{\min\{j \leq i \mid w_j = w_i\} - 1}.$$

In our scenario, $g^{(1)}(z)$ is the potential function obtained from applying this construction to the αn-dimensional linear function $\ell_1(z)$, and proceeding accordingly with the $(1-\alpha)n$-dimensional function $g^{(2)}(y)$ and $\ell_2(y)$. Finally, we define $\phi(x) = g^{(1)}(z) + g^{(2)}(y)$.

We can now give the proof of our main theorem.

Proof (Proof of Theorem 2). Using the potential function from Definition 1, we analyze the (1+1) EA on f, assume an arbitrary, non-optimal search point $x_t \in \{0,1\}^{n-s}$ and consider the expected change of g from time t to time $t+1$. We consider an accepted step where the offspring differs from the parent since this is necessary for g to change. That is, at least one 1-bit flips and f does not grow. Let A be the event that an offspring $x' \neq x_t$ is accepted. For A to occur, it is necessary that at least one of the two functions ℓ_1 and ℓ_2 does not grow. Since the two cases can be handled in an essentially symmetrically manner (they become perfectly symmetrical for $\alpha = n/2$), we only analyze the case that ℓ_2 does not grow and that at least one bit in B_2 is flipped from 1 to 0. Hence, we consider exactly the situation that the (1+1) EA with the linear function ℓ_2 as $(1-\alpha)n$-bit fitness function produces an offspring that is accepted and different from the parent.

Let $Y_t = g^{(2)}(y_t)$, where y_t is the restriction of the search point x_t at time t to the $(1-\alpha)n$ bits in B_2 that $g^{(2)}$ depends on, assuming the indices of x_t to be reordered with respect to increasing coefficients $w_1^{(2)}, \ldots, w_{(1-\alpha)n}^{(2)}$. To compute the drift of g, we distinguish between several cases and events in a way similar to the proof of Theorem 5.1 in [28]. Each of these cases first bounds the drift of Y_t sufficiently precisely and then adds a pessimistic estimate of the drift of $Z_t = g^{(1)}(z_t)$, which corresponds to the other linear function on bits from B_1, i.e., the function whose value may grow under the event A. Note that $g^{(1)}$ depends on at least as many bits as $g^{(2)}$ does.

Since the estimate of the drift of Z_t is always the same, we present it first. Let $\tilde{Z}_{t+1}$ denote the $g^{(1)}$-value of the mutated bit string x' (restricted to the bits in B_1). If x' is accepted, then $Z_{t+1} = \tilde{Z}_{t+1}$; otherwise $Z_{t+1} = Z_t$. If we pessimistically assume that each bit in z_t (i.e., the restriction of x_t to the bits in B_1) is a zero-bit that can flip to 1, we obtain the upper bound

$$\mathrm{E}\left(\tilde{Z}_{t+1} - Z_t \mid Z_t\right) \leq \frac{1}{n} \sum_{i=1}^{\alpha n} \left(1 + \frac{1}{n}\right)^{i-1} = \frac{1}{n} \frac{\left(1 + \frac{1}{n}\right)^{\alpha n - 1} - 1}{1/n}$$

$$\leq e^{\alpha} - 1 \leq e^{\ln(2-\epsilon)} - 1 \leq 1 - \epsilon, \tag{4}$$

where we use that $\alpha < \ln(2-\epsilon)$ for some constant $\epsilon > 0$. Also, since $Z_{t+1} = Z_t$ if the mutation is rejected and we only consider flipping zero-bits, we have under A (the event that x' is accepted) that

$$E(Z_{t+1} - Z_t \mid Z_t; A) \leq E\left(\tilde{Z}_{t+1} - Z_t \mid Z_t\right) \leq 1 - \epsilon. \tag{5}$$

Note that the estimations (4) and (5) include the case that s of the bits in z_t are shared with the input string y_t of the other linear function $g^{(2)}(y_t)$.

We next conduct the detailed drift analysis to bound $E(\phi(x_t) - \phi(x_{t+1}) \mid x_t)$, considering certain events necessary for A. Two different cases are considered.

Case 1: at least two one-bits in y_t flip (event S_1). Let $\tilde{Y}_{t+1}$ denote the $g^{(2)}$-value of the mutated bit string x', restricted to the bits in B_2, under event S_1 *before selection.* If x' is accepted, then $Y_{t+1} = \tilde{Y}_{t+1}$; otherwise $Y_{t+1} = Y_t$. Since $g_i \geq 1$ for all i, every zero-bit in y_t flips to one with probability at most $1/n$, and $(1 - \alpha) \leq \alpha$, we can re-use the estimations from (4). Bounding the contribution of the flipping one-bits from below by 2, we obtain

$$E\left(Y_t - \tilde{Y}_{t+1} \mid Y_t; S_1\right) \geq 2 - \frac{1}{n} \sum_{i=1}^{(1-\alpha)n} \left(1 + \frac{1}{n}\right)^{i-1}$$

$$\geq 2 - \frac{1}{n} \sum_{i=1}^{\alpha n} \left(1 + \frac{1}{n}\right)^{i-1} \geq 2 - (1 - \epsilon) = 1 + \epsilon.$$

Along with (4), we have

$$E(\phi(x_t) - \phi(x') \mid x_t; S_1) \geq E\left(Y_t - \tilde{Y}_{t+1} \mid Y_t; S_1\right) - E\left(Z_t - \tilde{Z}_{t+1} \mid Z_t; S_1\right)$$

$$\geq 2 - (1 - \epsilon) - (1 - \epsilon) > \epsilon.$$

Since the drift of ϕ is non-negative in Case 1, we estimate it from below by 0 regardless of whether A occurs or not and focus only on the event defined in the following case.

Case 2: exactly one one-bit in y_t flips (event S_2). Let i^* denote the random index of the flipping one-bit in y_t. Moreover, let the function $\beta(i) = \min\{j \leq i \mid w_j^{(2)} = w_i^{(2)}\}$ denote the smallest index at most i with the same weight as $w_i^{(2)}$, i.e., $\beta(i) - 1$ is the largest index of a strictly smaller weight; using our assumption that the weights are monotonically increasing with their index. If at least one zero-bit having the same or a larger weight than bit i^* flips, neither ℓ_2 nor g_2 change (because the offspring has the same function value or is rejected); hence, we now, without loss of generality, only consider the subevents of S_2 where all flipping zero-bits have an index of at most $\beta(i^*)$. (This reasoning is similar to the analysis of *Subcase 2.2.2* in the proof of Theorem 5 from [28].)

Redefining notation, let $\tilde{Y}_{t+1}$ denote the $g^{(2)}$-value of the mutated bit string x' (restricted to the bits in B_2) under event S_2 *before selection.* If x' is accepted, then $Y_{t+1} = \tilde{Y}_{t+1}$; otherwise $Y_{t+1} = Y_t$. Recalling that A is the event that the mutation x' is accepted, we have by the law of total probability

$$\mathrm{E}(Y_t - Y_{t+1} \mid Y_t; S_2) = \Pr(A \mid S_2) \cdot \mathrm{E}\Big(Y_t - \tilde{Y}_{t+1} \mid Y_t; A \cap S_2\Big)$$
$$\geq \Pr(A \mid S_2) \cdot \mathrm{E}\Big(Y_t - \tilde{Y}_{t+1} \mid Y_t; S_2\Big),$$

where the inequality holds since the our estimation of $\mathrm{E}\Big(Y_t - \tilde{Y}_{t+1} \mid Y_t; S_2\Big)$ below will consider exactly one one-bit to flip and assume all zero-bits to flip independently, even though already steps flipping two zero-bits right of $\beta(i^*)$ may be rejected.

Moreover, using the law of total probability and (5),

$$\mathrm{E}(Z_{t+1} - Z_t \mid Z_t; S_2) \leq \Pr(A \mid S_2) \cdot \mathrm{E}\Big(\tilde{Z}_{t+1} - Z_t \mid Z_t; S_2\Big)$$

and therefore

$$\mathrm{E}(\phi(x_t) - \phi(x_{t+1}) \mid x_t; S_2) = \Pr(A \mid S_2) \cdot \mathrm{E}(\phi(x_t) - \phi(x') \mid Y_t; A \cap S_2)$$
$$\geq \Pr(A \mid S_2)\mathrm{E}\Big((Y_t - \tilde{Y}_{t+1}) - (\tilde{Z}_{t+1} - Z_t) \mid x_t; S_2\Big) \tag{6}$$

It holds that $\Pr(A \mid S_2) \geq (1 - 1/n)^{n-1} \geq e^{-1}$ since the mutation flipping i^* is certainly accepted if no other bits flip. To bound the drift, we use that every zero-bit j right of $\beta(i^*)$ flips with probability $1/n$ and contributes $g_j^{(2)}$ to the difference $Y_t - \tilde{Y}_{t+1}$. Moreover, the flip of i^* contributes the term $g_{i^*}^{(2)}$ to the difference. Altogether,

$$\mathrm{E}\Big(Y_t - \tilde{Y}_{t+1} \mid Y_t; S_2\Big) = \Big(1 + \frac{1}{n}\Big)^{\beta(i^*)-1} - \frac{1}{n}\sum_{j=1}^{\beta(i^*)-1}\Big(1 + \frac{1}{n}\Big)^{j-1}$$

$$= \Big(1 + \frac{1}{n}\Big)^{\beta(i^*)-1} - \frac{1}{n}\left(\frac{(1 + \frac{1}{n})^{\beta(i^*)-1} - 1}{1/n}\right) = 1.$$

Combining this with (5), we have

$$\mathrm{E}\big(\phi(x_t) - \phi(x') \mid x_t; S_2\big) = \mathrm{E}((Y_t - \tilde{Y}_{t+1}) - (\tilde{Z}_{t+1} - Z_t) \mid x_t; S_2) \geq 1 - (1 - \epsilon) = \epsilon.$$

Altogether, using (6) and our lower bound $\Pr(A \mid S_2) \geq e^{-1}$, we have the following lower bound on the drift under S_2:

$$\mathrm{E}(\phi(x_t) - \phi(x_{t+1}) \mid x_t; S_2) \geq e^{-1}\epsilon.$$

Finally, we compute the total drift considering all possible one-bits that can flip under S_2. Let I be the set of one-bits in the whole bit string x_t. Since the analysis is analogous when considering an index $i \in I$, we still consider the situation that the corresponding linear function decreases or stays the same if $i \in B_2$, i.e., i belongs to y_t and remark that an analogous event A' with respect to the bits B_1 and the string z_t can be analyzed in the same way.

Now, for $i \in I$, let F_i denote the event that bit i is the only flipping one-bit in the considered part of the bit string and let F be the event that exactly one bit from I flips. We have for all $i \in I$ that

$$\mathrm{E}(\phi(x_t) - \phi(x_{t+1}) \mid x_t; F_i) \geq e^{-1}\epsilon.$$

and therefore also $\mathrm{E}(\phi(x_t) - \phi(x_{t+1}) \mid x_t; F) \geq e^{-1}\epsilon$. It is sufficient to flip one of the $|I|$ one-bits and no other bit to have an accepted mutation, which has probability at least $(|I|/n)(1-1/n)^{n-1} \geq \frac{|I|}{en}$. We obtain the unconditional drift

$$\mathrm{E}(\phi(x_t) - \phi(x_{t+1}) \mid x_t) \geq \frac{|I|}{en}\mathrm{E}(\phi(x_t) - \phi(x_{t+1}) \mid x_t; F_i) \geq \frac{|I|e^{-2}}{n}\epsilon,$$

recalling that we estimated the drift from below by 0 if at least two one-bits flip. To conclude the proof, we relate the last bound to $\phi(x_t)$. Clearly, since $g_i \leq (1+1/n)^{n-1} \leq e$ for all $i \in \{1, \ldots, \alpha n\}$ and since each one-bit can contribute to both $g^{(1)}(x_t)$ and $g^{(2)}(y_t)$, we have $\phi(x_t) \leq 2e|I|$ so that

$$\mathrm{E}(\phi(x_t) - \phi(x_{t+1}) \mid x_t) \geq \left(\frac{e^{-3}\epsilon}{2n}\right)\phi(x_t).$$

Hence, we have established a multiplicative drift of the potential ϕ with a factor of $\delta = (e^{-3}\epsilon)/(2n)$ and we obtain the claimed $O(n \log n)$ bound on the expected optimization time via the multiplicative drift theorem (Theorem 1), using $X_0 \leq n(1 + 1/(n-1))^n = O(n)$ and $s_{\min} = 1$.   □

We remark that the drift factor $(e^{-3}\epsilon)/n$ from the previous proof can be improved by constant factors using a more detailed case analysis; however, since ϵ can be arbitrarily small and the final bound is in O-notation, this does not seem worth the effort.

5   Discussion and Conclusions

Motivated by studies on separable functions and objective functions for chance constrained problems based on the expected value and variance of solutions, we investigated the quite general setting of the sum of two transformed linear functions and established an $O(n \log n)$ bound for the (1+1) EA.

We now would like to point out some topics for further investigations. Our result from Theorem 2 has some limitations. First of all, the domains of the two linear functions may not differ very much in size; more precisely they must be within a factor of $\alpha/(1 - \alpha) \leq (\ln(2))/(1 - \ln(2)) \approx 2.26$. With the current pessimistic assumption that an improving mutation only improves one of the two linear functions and simultaneously may flip any bit in the other function to 1 without the mutation being rejected, we cannot improve this to larger size differences for the domain. For the same reason, the result cannot easily be generalized to mutation probabilities c/n for arbitrary constants $c > 0$ as shown for the original case of simple linear functions in [28]. Although that paper also suggests a different, more powerful class of potential functions to handle high mutation probabilities, it seems difficult to apply these more powerful potential functions in the presence of our pessimistic assumptions. With stronger conditions on α, it may be possible to extend the present results to mutation probabilities up to $(1 + \epsilon)/(n + s)$ for a positive constant ϵ depending on α. However, it would be more interesting to see whether the $O(n \log n)$ bound would also hold for mutation probability $1/n$ for all $s \geq 1$, which would include the function $g(x)$ from the chance-constrained scenario in (3) for the usual mutation probability.

Acknowledgments. This work has been supported by the Australian Research Council (ARC) through grant FT200100536 and by the Independent Research Fund Denmark through grant DFF-FNU 8021-00260B.

References

1. Assimi, H., Harper, O., Xie, Y., Neumann, A., Neumann, F.: Evolutionary bi-objective optimization for the dynamic chance-constrained knapsack problem based on tail bound objectives. In: ECAI, vol. 325, pp. 307–314. IOS Press (2020)
2. Charnes, A., Cooper, W.W.: Chance-constrained programming. Manage. Sci. **6**(1), 73–79 (1959)
3. Doerr, B.: Probabilistic tools for the analysis of randomized optimization heuristics. In: Theory of Evolutionary Computation. NCS, pp. 1–87. Springer, Cham (2020)
4. Doerr, B., Doerr, C.: The impact of random initialization on the runtime of randomized search heuristics. Algorithmica **75**(3), 529–553 (2016)
5. Doerr, B., Doerr, C., Neumann, A., Neumann, F., Sutton, A.M.: Optimization of chance-constrained submodular functions. In: AAAI, pp. 1460–1467. AAAI Press (2020)
6. Doerr, B., Goldberg, L.A.: Adaptive drift analysis. In: Schaefer, R., Cotta, C., Kołodziej, J., Rudolph, G. (eds.) PPSN 2010. LNCS, vol. 6238, pp. 32–41. Springer, Heidelberg (2010)
7. Doerr, B., Goldberg, L.A.: Adaptive drift analysis. Algorithmica **65**(1), 224–250 (2013)
8. Doerr, B., Jansen, T., Sudholt, D., Winzen, C., Zarges, C.: Mutation rate matters even when optimizing monotonic functions. Evol. Comput. **21**(1), 1–27 (2013)
9. Doerr, B., Johannsen, D., Winzen, C.: Drift analysis and linear functions revisited. In: Proceedings of CEC 2010, pp. 1–8. IEEE Press (2010)
10. Doerr, B., Johannsen, D., Winzen, C.: Multiplicative drift analysis. In: Proceedings of GECCO 2010, pp. 1449–1456. ACM Press (2010)
11. Doerr, B., Johannsen, D., Winzen, C.: Multiplicative drift analysis. Algorithmica **64**(4), 673–697 (2012)
12. Doerr, B., Neumann, F. (eds.): Theory of Evolutionary Computation-Recent Developments in Discrete Optimization. Springer, Cham (2020)
13. Doerr, B., Sudholt, D., Witt, C.: When do evolutionary algorithms optimize separable functions in parallel? In: FOGA, pp. 51–64. ACM (2013)
14. Droste, S., Jansen, T., Wegener, I.: On the analysis of the (1+1) evolutionary algorithm. Theoret. Comput. Sci. **276**, 51–81 (2002)
15. Hajek, B.: Hitting-time and occupation-time bounds implied by drift analysis with applications. Adv. Appl. Probab. **13**(3), 502–525 (1982)

16. He, J., Yao, X.: A study of drift analysis for estimating computation time of evolutionary algorithms. Nat. Comput. **3**(1), 21–35 (2004)
17. Ishii, H., Shiode, S., Nishida, T., Namasuya, Y.: Stochastic spanning tree problem. Discret. Appl. Math. **3**(4), 263–273 (1981)
18. Jägersküpper, J.: A blend of Markov-chain and drift analysis. In: Rudolph, G., Jansen, T., Beume, N., Lucas, S., Poloni, C. (eds.) PPSN 2008. LNCS, vol. 5199, pp. 41–51. Springer, Heidelberg (2008)
19. Jägersküpper, J.: Combining Markov-chain analysis and drift analysis. Algorithmica **59**(3), 409–424 (2011)
20. Jansen, T.: Analyzing Evolutionary Algorithms - The Computer Science Perspective. Springer, Cham (2013)
21. Lehre, P.K., Witt, C.: Tail bounds on hitting times of randomized search heuristics using variable drift analysis. Combinatorics Probab. Comput. **30**(4), 550–569 (2021)
22. Lengler, J., Martinsson, A., Steger, A.: When does hillclimbing fail on monotone functions: an entropy compression argument. In: ANALCO, pp. 94–102. SIAM (2019)
23. Motwani, R., Raghavan, P.: Randomized Algorithms. Cambridge University Press, Cambridge (1995)
24. Neumann, A., Neumann, F.: Optimising monotone chance-constrained submodular functions using evolutionary multi-objective algorithms. In: Bäck, T., et al. (eds.) PPSN 2020. LNCS, vol. 12269, pp. 404–417. Springer, Cham (2020)
25. Neumann, F., Pourhassan, M., Witt, C.: Improved runtime results for simple randomised search heuristics on linear functions with a uniform constraint. Algorithmica **83**(10), 3209–3237 (2021)
26. Neumann, F., Sutton, A.M.: Runtime analysis of the (1+1) evolutionary algorithm for the chance-constrained knapsack problem. In: FOGA, pp. 147–153. ACM (2019)
27. Neumann, F., Witt, C.: Bioinspired Computation in Combinatorial Optimization - Algorithms and Their Computational Complexity. Springer, Cham (2010)
28. Witt, C.: Tight bounds on the optimization time of a randomized search heuristic on linear functions. Combinatorics Probab. Comput. **22**(2), 294–318 (2013)
29. Xie, Y., Harper, O., Assimi, H., Neumann, A., Neumann, F.: Evolutionary algorithms for the chance-constrained knapsack problem. In: GECCO, pp. 338–346. ACM (2019)
30. Xie, Y., Neumann, A., Neumann, F.: Specific single- and multi-objective evolutionary algorithms for the chance-constrained knapsack problem. In: GECCO, pp. 271–279. ACM (2020)
31. Xie, Y., Neumann, A., Neumann, F.: Heuristic strategies for solving complex interacting stockpile blending problem with chance constraints. In: GECCO, pp. 1079–1087. ACM (2021)
32. Xie, Y., Neumann, A., Neumann, F., Sutton, A.M.: Runtime analysis of RLS and the (1+1) EA for the chance-constrained knapsack problem with correlated uniform weights. In: GECCO, pp. 1187–1194. ACM (2021)

Runtime Analysis of Unbalanced Block-Parallel Evolutionary Algorithms

Brahim Aboutaib[1,2] and Andrew M. Sutton[3(✉)]

[1] Université du Littoral Côte d'Opale, LISIC, 62100 Calais, France
[2] Faculty of Science, LRIT, Mohammed V University in Rabat, Rabat, Morocco
[3] Department of Computer Science, University of Minnesota Duluth,
55812 Duluth, MN, USA
amsutton@d.umn.edu

Abstract. We revisit the analysis of the (1+λ) EA in a parallel setting when the offspring population size is significantly larger than the number of processors available. If the workload is not balanced across the processors, existing runtime results do not transfer directly. We therefore consider two new scenarios that produce unbalanced processors: (1) when the computation time of the fitness function is variable and depends on the structure of the individual, and (2) when processing is interrupted as soon as a viable offspring is found on one of the machines. We derive parallel execution times for both these models as a function of both the population size and the number of parallel machines. We discuss the potential trade-off between communication overhead and execution time, and we conduct some experiments.

Keywords: Parallel evolutionary algorithms · Runtime analysis

1 Introduction

The (1+λ) EA is a simple population-based evolutionary algorithm that maintains a single parent $x^{(t)}$ in generation t and produces an offspring population of λ individuals using mutation. A new parent individual $x^{(t+1)}$ for the following generation $t + 1$ is obtained by finding the fittest individual among the λ offspring and parent $x^{(t)}$. In this sense, the (1+λ) EA is a mutation-only evolutionary algorithm that employs elitism via truncation survival selection.

The first theoretical analysis of the (1+λ) EA considered, along with the total number of fitness evaluations, the so-called *parallel execution time* [10], which assumes there are exactly λ parallel processors available, the time to compute the fitness is uniform for all individuals, and the master process must wait until every worker task has finished.

The goal of this paper is to extend this analysis to broader settings in which the number of processors might be far smaller than λ so that large blocks of individuals must be distributed among the processors. If computational effort is uniform across all individuals, or if the parallelism is not additionally leveraged

G. Rudolph et al. (Eds.): PPSN 2022, LNCS 13399, pp. 555–568, 2022.
https://doi.org/10.1007/978-3-031-14721-0_39

to balance the load, the analysis is somewhat trivial: one needs only to divide the sequential running time by the number of processors available. However, if the processors' workload is unbalanced due to these factors, it becomes necessary to carry out additional analyses. In this paper, we consider two new scenarios for parallelizing the $(1+\lambda)$ EA: (1) when the computational effort is nonuniform and depends on the structure of the solutions generated or, (2) when the parallel architecture is leveraged to abort additional fitness computations as soon as a viable offspring is found on any processor. The latter case might be especially helpful in situations where the population size is set too high. In many optimization problems, one may have to deal with non-uniform fitness evaluation time. For example, the evaluation time of the LEADINGONES function can vary between $O(1)$ and $O(n)$. This can also be the case in Genetic Programming where individuals represent programs of different length and complexity. Another case is the so-called automated algorithm configuration, where different configurations (e.g., parameters' values) are evaluated. Besides these algorithmic motivations, there is also a hardware motivation: when running parallel algorithms on a heterogeneous cluster, it is not uncommon for processors to have different clock speeds. This would lead to significantly different execution times.

1.1   Background

Cantú-Paz [2] remarked that the easiest way to parallelize an evolutionary algorithm is to distribute fitness evaluation among a pool of worker processes, while a single master process handles genetic operators such as mutation and selection. This technique is especially beneficial when the evaluation of the fitness function is the main bottleneck. In this approach, which he called the "master-slave model"[1], the master process produces a population of individuals, and sends a fraction of the population to each of the other processes to offload the effort of fitness evaluation. Communication occurs in every generation as the master process distributes and collects data to and from the so-called *slave* processes. This process is depicted in Fig. 1.

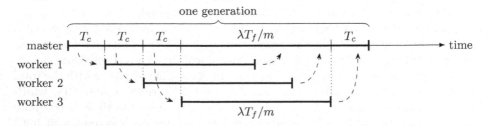

Fig. 1. A single generation of a parallel EA with communication time T_c and task execution time $\lambda T_f/m$. Dashed arrows represent communication between tasks.

---

[1] In this work, we will instead adopt the terminology *master-worker*.

This master-worker model is far simpler than other traditional parallelization techniques such as island models, cellular EAs or hybrid models [13]. In the absence of other factors (such as the ones we consider in this paper), runtime analyses would directly translate to this model because the parallelized variant visits the search space in the same way as the sequential variant [2].

Given a population size of λ and m processors, if T_f is the time to evaluate the fitness of a single individual, then if each of the m processors evaluates a λ/m fraction of the population, Cantú-Paz [2] estimates the elapsed time for a single generation as $mT_c + \frac{\lambda T_f}{m}$, where T_c is the inter-task communication time (see Fig. 1). He also derives the optimal number of parallel processors that minimizes the cost incurred from communication/parallelization trade-off, which, stated in our notation, is $m^* = \sqrt{\lambda T_f / T_c}$.

Using this estimate, one can easily translate existing results on the $(1+\lambda)$ EA (e.g., [3,4,6,7,10]) by simply multiplying by this factor once T_c and T_f are known. Implicit in this estimate is the assumption that the cost to compute a λ/m fraction of the population is static.

A detailed mathematical model of a concrete implementation of the master-worker model was considered by Dubreuil, Gagné and Parizeau [5]. The authors derived an expression for predicted speed-up as a function of the number of processors, fitness evaluation time, communication time and latency. They also validated this model via simulation. For a more comprehensive treatment of parallel evolutionary algorithms, we refer the reader to the book chapter of Sudholt [13] and the survey by Alba and Troya [1].

2   Block-Parallel $(1+\lambda)$ EA

Block mapping is a simple data distribution scheme for parallel processing that assigns contiguous blocks of data to parallel tasks [8]. In this scheme, a d-dimensional array is distributed among m processes so that each process receives a contiguous block of the array along a subset of array dimensions. In our setting, a set of λ length-n binary strings can be represented by a $\lambda \times n$ two-dimensional bit array. Selecting the first dimension, block mapping partitions the array into m blocks so that the k-th block contains rows $k\lambda/m$ to $(k+1)\lambda/m - 1$.

The Block-Parallel $(1+\lambda)$ parEA is listed in Algorithm 1 and is otherwise identical to the standard $(1+\lambda)$ EA, except that first an array P of λ individuals (together with a location to store their fitness value) is computed via standard mutation by the master process and then a one-to-many *scatter* communication is executed with P so each of the m processors receive a block P_{local} of λ/m individuals. Each processor sequentially computes the fitness of its individuals, storing its result in P_{local}. The master process waits until all processes have completed, and then executes a many-to-one *gather* communication to retrieve the λ fitness values stored in P. It then finds the maximum fitness of P (breaking ties arbitrarily) and replaces x if the fittest of P is at least as fit as x.

Algorithm 1: Block-Parallel $(1+\lambda)$ parEA

---

1   choose x from $\{0,1\}^n$ uniformly at random;
2   compute $f(x)$;
3   **while** *termination criterion not met* **do**
4      $P \leftarrow$ an array of λ length-n string, value pairs;
5   **for** $i \leftarrow 1$ **to** λ **do**
6   Create an offspring y by flipping each bit of x with probability $1/n$;
7         $P[i] \leftarrow (y, -\infty)$;
8   Scatter P to m processors;   *parallel section*
9   **do (in parallel)**
10   **for** $(y, f_y) \in P_{local}$ **do**
11   compute $f(y)$ and set f_y to $f(y)$;
12   Gather P from processors;
13   Let $(y_{max}, f_{max}) \in P$ be the pair with maximal fitness;
14   **if** $f_{max} \geq f(x)$ **then**
15         $x \leftarrow y_{max}$

Runtime analysis on sequential evolutionary algorithms typically characterize the runtime of an algorithm $\mathcal{A}$ on a function f as a random variable $T_{\mathcal{A},f}$ that corresponds to the number of fitness evaluations executed until an optimal solution is first generated. Obviously, if fitness evaluation costs are all equal, then the block-parallel runtime is proportional to $T_{\mathcal{A},f}/m$ on m machines, because the workload is more or less balanced. Workload imbalance can arise when the cost of fitness evaluation is variable, and thus some individuals may require significantly more effort than others. Discrepancies that arise from the imbalance may be small within a single generation, but their effects are additive, and would likely accumulate over the entire run. We visit a simple model of this scenario in the next section in which fitness evaluation depends on solution structure. Later, in Sect. 4, we will investigate imbalance arising from early stopping.

3   Heterogeneous Fitness Evaluation

Workload imbalance in parallel evolutionary algorithms can arise from nonuniformity in the computational effort required by the fitness function. We begin by considering the setting in which the cost of fitness evaluation of an individual depends on the structure of the individual. Given a bit string $x \in \{0,1\}^n$, we denote as $\text{cost}(x)$ the computational cost for evaluating $f(x)$, which is defined as follows. Fix $\alpha_1, \alpha_2 > 0$, then

$$\text{cost}(x) = \alpha_1 |x|_1 + \alpha_0 |x|_0. \tag{1}$$

In order to understand how processing a block of individuals affects the running time of the algorithm, it is useful to characterize the moment generating function of the random variable associated with the cost of evaluating the fitness of an offspring. This is captured in the following lemma.

Lemma 1. *Let y be an offspring generated from x by standard uniform random mutation and let $Z = cost(y)$ be the random variable associated with the cost of computing $f(y)$. Then the moment generating function of Z is*

$$M_Z(t) = n^{-n} \left(e^{t\alpha_0}(n-1) + e^{t\alpha_1}\right)^{|x|_1} \left(e^{t\alpha_1}(n-1) + e^{\alpha_0}\right)^{n-|x|_1}.$$

Proof. To produce y, each bit of x flips independently with probability $1/n$, so we can write Z as the sum of n independent random variables $Z = \sum_{i:x_i=1} Z_{i,1} + \sum_{i:x_i=0} Z_{i,0}$, where

$$Z_{i,1} = \begin{cases} \alpha_1 & \text{w/ prob. } (1-1/n), \\ \alpha_0 & \text{w/ prob. } 1/n, \end{cases} \quad \text{and,} \quad Z_{i,0} = \begin{cases} \alpha_0 & \text{w/ prob. } (1-1/n), \\ \alpha_1 & \text{w/ prob/ } 1/n. \end{cases}$$

Thus we have

$$M_Z(t) = \mathbb{E}[e^{tZ}] = \left(\prod_{i:x_i=1} \mathbb{E}[e^{Z_{i,1}}]\right) \left(\prod_{i:x_i=1} \mathbb{E}[e^{Z_{i,0}}]\right)$$

$$= \left(\prod_{i:x_i=1} \left(1-\frac{1}{n}\right) e^{t\alpha_1} + e^{t\alpha_0}\left(\frac{1}{n}\right)\right) \left(\prod_{i:x_i=0} \left(1-\frac{1}{n}\right) e^{t\alpha_0} + e^{t\alpha_1}\left(\frac{1}{n}\right)\right)$$

$$= n^{-n} \left(e^{t\alpha_0}(n-1) + e^{t\alpha_1}\right)^{|x|_1} \left(e^{t\alpha_1}(n-1) + e^{\alpha_0}\right)^{n-|x|_1},$$

where we have used the independence of the $Z_{i,j}$s in the first line.   □

We now state a useful result that, ignoring delays caused by task communication, relates α_0, α_1, λ, m and n to the expected cost of evaluating the offspring population of an individual.

Theorem 1. *The expected time to compute the λ offspring of a parent individual x in a single generation of the Block-Parallel $(1+\lambda)$ parEA on m machines is bounded above by*

$$\max\{\alpha_0, \alpha_1\} \left(\ln m + \frac{\lambda}{m}(|x|_1 \ln n + n)\right)$$

Proof. Denote the time on machine j as the random variable B_j. Assuming negligible communication and spin-up time, the time to compute the fitness of all offspring is bounded by the bottleneck of the largest value of B_j. Thus we define $Y = \max\{B_1, B_2, \ldots, B_m\}$ and seek to bound the expectation $\mathbb{E}[Y]$. Note that for any $t > 0$, we have

$$\exp\left(t\,\mathbb{E}[Y]\right) \le \mathbb{E}\left[\exp(tY)\right] = \mathbb{E}\left[\max_j\{\exp(tB_j)\}\right] \tag{2}$$

$$\le \sum_{j=1}^{m} \mathbb{E}\left[\exp(tB_j)\right] \tag{3}$$

$$= \sum_{j=1}^{m} \mathbb{E}\left[\exp\left(t \sum_{y\in P_{\text{local},j}} \text{cost}(y)\right)\right] \tag{4}$$

$$= \sum_{j=1}^{m} \prod_{y\in P_{\text{local},j}} \mathbb{E}\left[\exp\left(t\,\text{cost}(y)\right)\right] \tag{5}$$

$$= m\left(M_Z(t)\right)^{\lambda/m}. \tag{6}$$

The inequality in (2) is obtained by Jensen's inequality, and the inequality in (3) is obtained by a union bound. In (5) we use the fact that each of the λ/m offspring in a block are generated independently.

Putting together the above with Lemma 1, we have the inequality

$$\exp\left(t\,\mathbb{E}[Y]\right) \le m\left(n^{-n}\left(e^{t\alpha_0}(n-1)+e^{t\alpha_1}\right)^{|x|_1}\left(e^{t\alpha_1}(n-1)+e^{\alpha_0}\right)^{n-|x|_1}\right)^{\lambda/m}.$$

Taking the natural log of both sides of this inequality, we have

$$t\,\mathbb{E}[Y] \le \ln m + \frac{\lambda}{m}\left(|x|_1\ln\left(\frac{e^{t\alpha_0}(n-1)+e^{t\alpha_1}}{e^{t\alpha_1}(n-1)+e^{t\alpha_0}}\right) + n\ln(e^{t\alpha_1}(n-1)+e^{t\alpha_0}) - n\ln n\right),$$

and setting $t := 1/\max\{\alpha_0,\alpha_1\}$,

$$\mathbb{E}[Y] \le \max\{\alpha_0,\alpha_1\}\left(\ln m + \frac{\lambda}{m}(|x|_1\ln n + n)\right),$$

since $e^{t\alpha_0}(n-1)+e^{t\alpha_1} \le en$ and $e \le e^{t\alpha_1}(n-1)+e^{t\alpha_0} \le en$. $\qquad\square$

Lemma 2 (Jansen, De Jong and Wegener [10]). *Let $x \in \{0,1\}^n$ with $\text{OneMax}(x) = i < n$, and let $P = \{y^{(1)}, y^{(2)}, \ldots, y^{(\lambda)}\}$ be a set of λ offspring generated independently by standard uniform mutation. The probability that there exists a $y \in P$ with $\text{OneMax}(y) > \text{OneMax}(x)$ is at least $(\lambda(n-i))/(\lambda(n-i)+en)$.*

Theorem 2. *The expected time for the Block-Parallel $(1+\lambda)$ parEA with $m < \lambda$ machines on OneMax is*

$$O\left(\max\{\alpha_0,\alpha_1\}\left(\frac{\lambda n^2\log n}{m}\right)\right).$$

Proof. For $i \in \{0,1,\ldots,n-1\}$, denote as G_i the number of generations the algorithm spends on fitness level i, that is, the number of generation in which the parent individual has a OneMax value of i. Denote as T_i the total time spent on fitness level i. Note that G_i is distributed geometrically, and by Lemma 2, $\mathbb{E}[G_i] \leq 1 + \frac{en}{\lambda(n-i)}$. By Theorem 1,

$$\mathbb{E}[T_i \mid G_i = k] \leq k \max\{\alpha_0, \alpha_1\} \left(\ln m + \frac{\lambda}{m}(i \ln n + n) \right).$$

For visual clarity we set $\alpha := \max\{\alpha_0, \alpha_1\}$ and write

$$\mathbb{E}[T_i] \leq \sum_{k=1}^{\infty} \mathbb{E}[T_i \mid G_i = k] \Pr(G_i = k)$$

$$\leq \alpha \cdot \left(\ln m + \frac{\lambda}{m}(i \ln n + n) \right) \sum_{k=1}^{\infty} k \Pr(G_i = k)$$

$$\leq \alpha \cdot \left(\ln m + \frac{\lambda}{m}(i \ln n + n) \right) \left(1 + \frac{en}{\lambda(n-i)} \right)$$

$$= \alpha \cdot \left(\frac{en^2}{m(n-i)} + \frac{en \ln m}{\lambda(n-i)} + \frac{\lambda i \ln n}{m} + \frac{ei \ln n}{m(n-i)} + \frac{\lambda n}{m} + \ln m \right).$$

The total expected time is bounded by

$$\mathbb{E}\left[\sum_{i=0}^{n-1} T_i \right] = \sum_{i=0}^{n-1} \mathbb{E}[T_i]$$

$$\leq \alpha \cdot \left[\left(\frac{en^2}{m} + \frac{en \ln m}{\lambda} \right) \sum_{i=1}^{n} \frac{1}{i} + \frac{\lambda n(n-1) \ln n}{2m} + \frac{e \ln n}{m} \sum_{i=0}^{n-1} \frac{i}{n-i} + \frac{n^2 \lambda}{m} + n \ln m \right]$$

$$\leq \alpha \cdot \left[\left(\frac{en^2}{m} + \frac{en \ln m}{\lambda} \right) \ln(en) + \frac{\lambda n(n-1) \ln n}{2m} + \frac{en \ln^2 n}{m} + \frac{n^2 \lambda}{m} + n \ln m \right],$$

where we have used the bound

$$\sum_{i=0}^{n-1} \frac{i}{n-i} = (n-1) + \sum_{i=0}^{n-2} \frac{i}{n-i} \leq (n-1) + \int_0^{n-1} \frac{x}{n-x} \, dx = n \ln n.$$

The asymptotic bound is obtained by observing $m < \lambda$ and discarding lower order terms. □

We can take a similar approach to bound the run time on LeadingOnes.

Theorem 3. *The expected time for the Block-Parallel $(1+\lambda)$ parEA with $m < \lambda$ machines on LeadingOnes is*

$$O\left(\max\{\alpha_0, \alpha_1\} \left(\frac{n^2 \lambda \log n}{m} + \frac{n^3 \log n}{m} \right) \right).$$

Proof. Let $x \in \{0,1\}^n$ where LeadingOnes$(x) = i < n$. The probability that a strictly improving string occurs in the offspring population is at least

$$1 - \left(1 - \frac{1}{en} \right)^{\lambda} \geq 1 - e^{-\lambda/(en)} \geq 1 - \frac{en}{\lambda + en} = \frac{\lambda}{\lambda + en}.$$

The number of generations on fitness level i is geometrically distributed with probability bounded as above, and we can bound the expected number of generations on level i as $(\lambda + en)/\lambda$. As in the proof of Theorem 2, we bound the total time on fitness level i as

$$
\mathbb{E}[T_i] \leq \max\{\alpha_0, \alpha_1\} \left(\ln m + \frac{\lambda}{m}(i \ln n + n) \right) \left(\frac{en + \lambda}{\lambda} \right)
$$

$$
= \max\{\alpha_0, \alpha_1\} \left(\frac{en^2 + \lambda n}{m} + \frac{en \ln m}{\lambda} + \frac{i \ln n(en + \lambda)}{m} + \ln m \right)
$$

using Theorem 1. The bound on the expected total time is obtained by summing over all fitness levels $i \in \{0, 1, \ldots, n-1\}$:

$$
\max\{\alpha_0, \alpha_1\} \left(\frac{en^3 + \lambda n^2}{m} + \frac{en^2 \ln m}{\lambda} + \frac{n(n-1) \ln n(en + \lambda)}{2m} + n \ln m \right).
$$

The asymptotic bound arises from the fact that $m < \lambda$ and dropping lower order terms. □

4 First-Improving Search Using Task Abortion

The Block-Parallel $(1+\lambda)$ parEA can be classified as a variant of *single-walk, single-step* parallel local search [14] in which the search neighborhood of a candidate solution is partitioned up and processed in parallel. In such scenarios, there can be a dramatic difference between accepting the best neighbor found vs. accepting the first improving neighbor found. Indeed, it has been shown that this choice has a significant impact on search behavior and landscape structure [9,11].

Roussel-Ragot and Dreyfus [12] considered a master-worker parallel implementation of simulated annealing for which, in the low temperature regime, m worker processors attempt moves on their own asynchronously until one accepts a move after which the entire process synchronizes and the current solution is updated. This inspires us to modify the Block-Parallel $(1+\lambda)$ parEA in a similar way: instead of waiting until all λ offspring have been processed, we abort all tasks as soon as one of the machines has found an acceptable offspring, effectively transforming the process into a first-improving search.

We also change our perspective to homogeneous execution time, that is, the fitness evaluation cost is uniform for all processors and all individuals. Algorithm 2 lists the Abortive Block-Parallel $(1+\lambda)$ parEA, which is identical to the Block-Parallel $(1+\lambda)$ parEA (Algorithm 1), except that a process can broadcast a message to all parallel tasks once it finds a suitable parent for the next generation. This message interrupts all offspring processing for the current generation, and processing for the next generation can already start.

The following lemma shows that the minimum of m geometrically distributed random variables is geometrically distributed.

Algorithm 2: Abortive Block-Parallel $(1+\lambda)$ parEA

---

1   choose x from $\{0,1\}^n$ uniformly at random;
2   compute $f(x)$;
3   **while** *termination criterion not met* **do**
4 | $P \leftarrow$ an array of λ length-n string, value pairs;
5 | **for** $i \leftarrow 1$ **to** λ **do**
6 | | Create an offspring y by flipping each bit of x with probability $1/n$;
7 | | $P[i] \leftarrow (y, -\infty)$;
8 | Scatter P to m processors;   ⟵ *parallel section*
9 | **do (in parallel)**
10 | | **for** $(y, f_y) \in P_{local}$ **do**
11 | | | compute $f(y)$ and set f_y to $f(y)$;
12 | | | **if** $f_y \geq f(x)$ **then**
13 | | | | break all processes out of loop
14 | Gather P from processors;
15 | Let $(y_{\max}, f_{\max}) \in P$ be the pair with maximal fitness;
16 | **if** $f_{\max} \geq f(x)$ **then**
17 | | $x \leftarrow y_{\max}$;

Lemma 3. *Let $(X_1, X_2, \ldots, X_m)$ be a sequence of iid random variables with each $X_j \sim \text{Geom}(p)$. Define $Y = \min\{X_1, X_2, \ldots, X_m\}$. Then $Y \sim \text{Geom}(1 - (1-p)^m)$.*

Proof. The event $\{Y = k\}$ occurs when, for some nonempty subsequence $I \subseteq [m]$, $X_j = k$ for $j \in I$ and $X_j > k$ for $j \in [m] \setminus I$. Thus,

$$\Pr(Y = k) = \sum_{I \subseteq [m]: I \neq \emptyset} \Pr(X_j = k : j \in I) \Pr(X_{j'} > k : j' \in [m] \setminus I)$$

$$= \sum_{j=1}^{m} \binom{m}{j} \left((1-p)^{k-1} p \right)^j (1-p)^{k(m-j)}$$

$$= (1-p)^{m(k-1)} \left(\sum_{j=1}^{m} \binom{m}{j} p^j (1-p)^{m-j} \right)$$

$$= (1 - (1 - (1-p)^m))^{k-1} (1 - (1-p)^m).$$

Thus Y is distributed as $\text{Geom}(1 - (1-p)^m)$.   □

Lemma 3 provides a mechanism to capture the running time of the first task to find a viable offspring in its block. The following theorem uses this result to bound above and below the execution time of the Abortive Block-Parallel $(1+\lambda)$ parEA on ONEMAX.

Theorem 4. *Let T be the time for the Abortive Block-Parallel $(1+\lambda)$ parEA with $m < \lambda$ machines to solve* ONEMAX. *Then*

$$\frac{n}{m}\ln n + \Omega\left(\frac{n}{m}\right) \leq \mathbb{E}[T] \leq \frac{n}{m}\ln\left(\frac{n}{m}\right) + \frac{n}{2e} + O\left(\frac{n}{m}\right).$$

Proof. For simplicity, we assume m divides λ, but the proof is easily adapted to the general case. We divide the running time of the Block-Parallel $(1+\lambda)$ parEA into segments delineated by the fitness of the search point on the master process. In this setting, the time can be characterized a random variable $T = \sum_{i=1}^{n} X_i$, where X_i denotes the total time spent in which $f(x) = n-i$. This is the waiting time until the first machine processes a strictly improving offspring y of the parent individual x where $f(y) > f(x)$.

Suppose x is the current parent individual. A given machine continually processes the offspring in its local block P_{local} until one of the following events occur: (1) it generates an offspring y with $f(y) > f(x)$, (2) it generates an offspring y with $f(y) = f(x)$, (3) it is aborted by another machine that has found an offspring y with $f(y) > f(x)$, (4) it is aborted by another machine that has found an offspring y with $f(y) = f(x)$, or (5) it finishes processing offspring in its local block.

The Block-Parallel $(1+\lambda)$ parEA leaves fitness level $n-i$ only when the above events (1) or (3) occur. Otherwise, the next offspring the machine will process is an offspring of an individual on the same fitness level as x (perhaps even x itself, as in the case of (5)).

In the absence of abortion by another machine, the time until a machine finds an improving offspring is distributed geometrically with success probability $\frac{i}{n}(1 - \frac{1}{n})^{n-i}$. Since abortion only happens when the first machine succeeds in processing a strictly improving offspring, the waiting time for the Block-Parallel $(1+\lambda)$ parEA to leave fitness level $n-i$ is the minimum over m geometric random variables. By Lemma 3,

$$\mathbb{E}[X_i] = \frac{1}{1 - \left(1 - \frac{i}{n}\left(1 - \frac{i}{n}\right)^{n-i}\right)^m} \leq \frac{1}{1 - (1 - \frac{i}{en})^m} \leq \frac{1}{1 - \exp(-\frac{im}{en})}.$$

We may bound $\mathbb{E}[X]$ from above by

$$\mathbb{E}\left[\sum_{i=1}^{n} X_i\right] \leq \sum_{i=1}^{n} \frac{1}{1 - \exp\left(-\frac{im}{en}\right)} \leq \int_1^n \frac{dx}{1 - \exp\left(-\frac{xm}{en}\right)} + \frac{1}{1 - \exp\left(-\frac{m}{en}\right)}$$

$$= \frac{en}{m}\ln\left(1 - e^{(mx)/(en)}\right)\Big|_1^n + \frac{e^{m/(en)}}{e^{m/(en)} - 1}$$

$$= \frac{en}{m}\ln\left(\frac{e^{m/e} - 1}{e^{m/(en)} - 1}\right) + O(n/m).$$

Note that $\ln\left(\frac{q-1}{q^x-1}\right) = \ln\left(\frac{q-1}{x\ln q}\right) - \frac{1}{2}x\ln q + O(x^2)$ as $x \to 0$, so setting $x = 1/n$ we can estimate

$$\ln\left(\frac{e^{m/e} - 1}{e^{m/(en)} - 1}\right) = \ln\left(\frac{e^{m/e} - 1}{(1/n)\ln e^{m/e}}\right) - \frac{1}{2n}\ln e^{m/e} + O(1/n^2)$$
$$\leq \ln(n/m) + m/(2e) + O(1).$$

For the lower bound, the probability of an improving offspring is at most i/n, thus Bernoulli's inequality yields

$$\mathbb{E}[X_i] \geq \frac{1}{1 - (1 - i/n)^m} \geq \frac{1}{1 - (1 - im/n)} = \frac{n}{im},$$

and by linearity of expectation, $\mathbb{E}[T] \geq (n/m)(\ln n + \Omega(1))$ □

Aside from the fact that we require $m \leq \lambda$, the bound of Theorem 4 does not depend on λ. One consequence is that the Abortive Block-Parallel $(1+\lambda)$ parEA can potentially overcome situations in which the offspring population is set too large. However, our analysis so far has not taken into account communication effort, which would indeed depend on λ. Analyzing the trade-off between these two artifacts is a direction for future work.

5   Experimental Analysis

5.1   Experimental Settings

We consider the following experiments with the Block-Parallel EAs. We consider ONEMAX instances of size $n \in \{128, 256, 512, 1024, 2048\}$, and $\lambda \in \{64, 128, 256, 512, 1024\}$. We experiment with different numbers of workers $m \in \{2, 4, 8, 16, 32, 64\}$. The stop condition is either finding the optimum solution, the $\{1\}^n$ string, or exhausting a budget of n^2 function evaluations. We report the mean optimization time, in seconds, and its standard deviation averaged over 100 runs, as a function of different parameters.

5.2   The Block-Parallel $(1 + \lambda)$parEA Analysis

We first start by reporting the impact of considering a number of offsprings greater than the available number of processors. Figure 2a reports the mean optimization time of the Block-Parallel $(1+\lambda)$parEA as a function of λ optimizing the ONEMAX function of a size $n = 512$. This figure characterizes the overall cost of communicating batches of larger offsprings $\frac{\lambda}{m}$ to each worker on the optimization time. That is the cost of sending these batches and the waiting time they take to be evaluated by all the m workers. Though one would expect a better runtime for larger λ values, one has also to consider the time these solutions will take to be evaluated and that the Block-Parallel $(1 + \lambda)$parEA algorithm is synchronous, which means that it waits until all the batches are evaluated before considering a new generation.

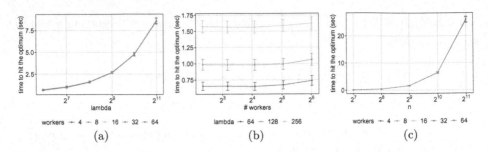

Fig. 2. Mean optimization time in seconds as a function of λ (a), number of workers m (b), and the problem size n (c).

To assess the impact of the number of workers, we plot in Fig. 2b the mean optimization time as a function of the m. This figure suggests that considering a number of workers smaller than the number of offspring may be slightly better than considering $m = \lambda$. This relative improvement could be due to the reduction of the communication overhead the algorithm incurs when the number of workers is lower compared to a large number of workers. Notice also that this improvement is stable relative to the number of offspring.

Figure 2c reports the wall-clock time as a function of the problem size n for $\lambda = 256$ and reveals a faster growth of the optimization time as a function of n.

5.3   Simulations on the Abortive Block-Parallel (1+λ) parEA

For the Abortive Block-Parallel (1+λ) parEA on ONEMAX we simulated the environment by dividing the offspring population into blocks of size λ/m and measuring in each block, how many offspring were processed until a viable one

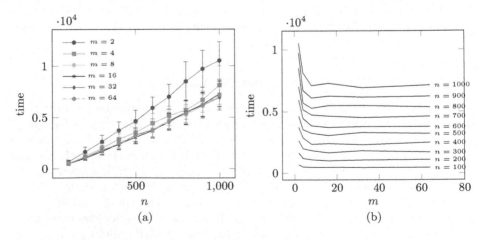

Fig. 3. Simulated wall-clock time of Abortive Block-Parallel (1+λ) parEA on ONEMAX for $\lambda = 100$ as a function of n (a) and m (b).

was found. The minimum of this count was added to the master clock, which was stopped as soon as an optimal string was found. We report the results from this simulation in Fig. 3. The advantage increases with m up to a point (roughly around $\ln n$) and then returns diminish, as predicted by the bounds in Theorem 4. Note that the simulation does not account for communication costs.

6 Conclusion

In this paper, we have considered a parallel setting of the $(1+\lambda)$ EA in which the offspring population size is larger than the number of parallel processors available and the computational effort is unbalanced due to variable fitness computation times or early stopping when viable offspring are found. For these cases we have derived detailed runtime bounds for ONEMAX and LEADINGONES. We have also presented experiments for this scenario that help clarify the picture for ONEMAX.

A promising avenue for future work is to explore the trade-offs between execution time and communication overhead by investigating the interplay between population size and processor pool size.

References

1. Alba, E., Troya, J.M.: A survey of parallel distributed genetic algorithms. Complex. **4**(4), 31–52 (1999)
2. Cantú-Paz, E.: Master-slave parallel genetic algorithms. In: Efficient and Accurate Parallel Genetic Algorithms, pp. 33–48. Springer, Boston (2001). https://doi.org/10.1007/978-1-4615-4369-5_3
3. Doerr, B., Künnemann, M.: How the $(1+\lambda)$ evolutionary algorithm optimizes linear functions. In: Blum, C., Alba, E. (eds.) Genetic and Evolutionary Computation Conference, GECCO '13, Amsterdam, The Netherlands, 6–10 July 2013, pp. 1589–1596. ACM (2013). https://doi.org/10.1145/2463372.2463569
4. Doerr, B., Künnemann, M.: Royal road functions and the $(1 + \lambda)$ evolutionary algorithm: almost no speed-up from larger offspring populations. In: Proceedings of the IEEE Congress on Evolutionary Computation, CEC 2013, Cancun, Mexico, 20–23 June 2013, pp. 424–431. IEEE (2013). https://doi.org/10.1109/CEC.2013.6557600
5. Dubreuil, M., Gagné, C., Parizeau, M.: Analysis of a master-slave architecture for distributed evolutionary computations. IEEE Trans. Syst. Man Cybern. Part B (Cybernetics) **36**(1), 229–235 (2006). https://doi.org/10.1109/TSMCB.2005.856724
6. Gießen, C., Witt, C.: The interplay of population size and mutation probability in the $(1 + \lambda)$ EA on OneMax. Algorithmica **78**(2), 587–609 (2016). https://doi.org/10.1007/s00453-016-0214-z
7. Gießen, C., Witt, C.: Optimal mutation rates for the $(1+\lambda)$ EA on OneMax through asymptotically tight drift analysis. Algorithmica **80**(5), 1710–1731 (2017). https://doi.org/10.1007/s00453-017-0360-y
8. Grama, A., Gupta, A., Karypis, G., Kumar, V.: Introduction to Parallel Computing. Pearson Education Limited (2003)

9. Hansen, P., Mladenović, N.: First vs. best improvement: an empirical study. Discr. Appl. Math. **154**(5), 802–817 (2006). https://doi.org/10.1016/j.dam.2005.05.020

10. Jansen, T., De Jong, K.A., Wegener, I.: On the choice of the offspring population size in evolutionary algorithms. Evolution. Comput. **13**(4), 413–440 (2005). https://doi.org/10.1162/106365605774666921

11. Ochoa, G., Verel, S., Tomassini, M.: First-improvement vs. best-improvement local optima networks of NK landscapes. In: Schaefer, R., Cotta, C., Kołodziej, J., Rudolph, G. (eds.) PPSN 2010. LNCS, vol. 6238, pp. 104–113. Springer, Heidelberg (2010). https://doi.org/10.1007/978-3-642-15844-5_11

12. Roussel-Ragot, P., Dreyfus, G.: A problem independent parallel implementation of simulated annealing: models and experiments. IEEE Trans. Comput. Aided Design Integrat. Circuits Syst. **9**(8), 827–835 (1990). https://doi.org/10.1109/43.57790

13. Sudholt, D.: Parallel evolutionary algorithms. In: Kacprzyk, J., Pedrycz, W. (eds.) Springer Handbook of Computational Intelligence, pp. 929–959. Springer, Heidelberg (2015). https://doi.org/10.1007/978-3-662-43505-2_46

14. Verhoeven, M.G.A., Aarts, E.H.L.: Parallel local search. J. Heurist. **1**(1), 43–65 (1995). https://doi.org/10.1007/BF02430365

Self-adjusting Population Sizes
for the $(1, \lambda)$-EA on Monotone Functions

Marc Kaufmann, Maxime Larcher, Johannes Lengler, and Xun Zou[✉]

Department of Computer Science, ETH Zürich, Zürich, Switzerland
{marc.kaufmann,larcherm,johannes.lengler,xun.zou}@inf.ethz.ch

Abstract. We study the $(1, \lambda)$-EA with mutation rate c/n for $c \leq 1$, where the population size is adaptively controlled with the $(1 : s + 1)$-success rule. Recently, Hevia Fajardo and Sudholt have shown that this setup with $c = 1$ is efficient on ONEMAX for $s < 1$, but inefficient if $s \geq 18$. Surprisingly, the hardest part is not close to the optimum, but rather at linear distance. We show that this behavior is not specific to ONEMAX. If s is small, then the algorithm is efficient on all monotone functions, and if s is large, then it needs superpolynomial time on all monotone functions. For small s and $c < 1$ we show a $O(n)$ upper bound for the number of generations and $O(n \log n)$ for the number of function evaluations; for small s and $c = 1$ we show the optimum is reached in $O(n \log n)$ generations and $O(n^2 \log \log n)$ evaluations. We also show formally that optimization is always fast, regardless of s, if the algorithm starts in proximity of the optimum. All results also hold in a dynamic environment where the fitness function changes in each generation.

Keywords: Parameter control · Self-adaptation · $(1, \lambda)$-EA · One-fifth rule · Monotone functions · Dynamic environments · Evolutionary algorithm

1 Introduction

Randomized Optimization Heuristics (ROHs) like Evolutionary Algorithms (EAs) are simple general-purpose optimizers. One of their strengths is that they can often be applied with little adaptation to the problem at hand. However, ROHs usually come with parameters, and their efficiency often depends on the parameter settings. Therefore, *parameter control* is a classical topic in the design and analysis of ROHs [23]. It aims at providing methods to automatically tune parameters over the course of optimization. It is not the goal to remove parameters altogether, as the parameter control mechanisms introduce new meta-parameters. However, the following two objectives can sometimes be achieved.

Extended Abstract. The proofs and further details are available on arxiv [32].
M. Kaufmann—Supported by the Swiss National Science Foundation [grant number 192079].
X. Zou—Supported by the Swiss National Science Foundation [grant number CR-SII5_173721].

Firstly, some ROHs are rather sensitive to small changes in the parameters, and inadequate setting can slow down or even prevent success. Two examples relevant for this paper are the $(1, \lambda)$-EA, which fails to optimize even the ONE-MAX benchmark if λ is too small [2,28,53], and the $(1 + 1)$-EA, which fails on monotone functions if the mutation rate is too large [17,18,43]. In both cases, changing the parameters just by a constant factor makes all the difference between finding the optimum in time $O(n \log n)$, and not even finding an ε-approximation of the optimal solution in polynomial time. So these algorithms are extremely sensitive to small changes of parameters. In such cases, one hopes that performance is *more robust* with respect to the meta-parameters, i.e., that the parameter control mechanism finds a decent parameter setting regardless of its meta-parameters.

Secondly, often there is no single parameter setting that is optimal throughout the course of optimization. Instead, different phases of the optimization process profit from different parameter settings, and the overall performance with *dynamically adapted* parameters is better than for any static parameters [4,5,11,15,22]. This topic, which has always been studied in continuous optimization, has taken longer to gain traction in discrete domains [1,30] but has attracted increasing interest over the last years [6,13,14,16,21,25,35,45–50,52]. Instead of a detailed discussion we refer the reader to the book chapter [10] for an overview of theoretical results, and to [26] for a discussion of some recent development.

One of the most traditional and influential methods for parameter control is the $(1 : s + 1)$-success rule [33], independently developed several times [8,51,54] and traditionally used with $s = 4$ as one-fifth rule in continuous domains, e.g. [3]. This rule has been transferred to the $(1, \lambda)$-EA [26,27], which yields the so-called *self-adjusting* $(1, \lambda)$-*EA* or *SA*-$(1, \lambda)$-*EA*, also called $(1, \{\lambda/F, F^{1/s}\lambda\})$-EA. As in the basic $(1, \lambda)$-EA, in each generation the algorithm produces λ offspring, and selects the fittest of them as the unique parent for the next generation. The difference to the basic $(1, \lambda)$-EA is that the parameter λ is replaced by λ/F if the fittest offspring is fitter than the parent, and by $\lambda \cdot F^{1/s}$ otherwise. Thus the $(1 : s + 1)$-success rule replaces the parameter λ by two parameters s and F. As outlined above, there are two hopes associated with this scheme:

(i) that the performance is *more robust* with respect to F and s than with respect to λ;
(ii) that the scheme can adaptively find the *locally optimal* value of λ throughout the course of optimization.

Recently, Hevia Fajardo and Sudholt have investigated both hypotheses on the ONEMAX benchmark [26,27]. They found a negative result for (i), and a (partial) positive result for (ii). The negative result says that performance is at least as fragile with respect to the parameters as before: if $s < 1$, then the SA-$(1, \lambda)$-EA finds the optimum of ONEMAX in $O(n)$ generations, but if $s \geq 18$ and $F \leq 1.5$ the runtime becomes exponential with overwhelming probability. Experimentally, they find that the range of bad parameter values even seems to include the standard choice $s = 4$, which corresponds to the 1/5-rule. On the other hand, they show that for $s < 1$, the algorithm successfully achieves (ii): they show that the expected number of function evaluations is $O(n \log n)$, which is optimal among all unary unbiased black-box algorithms [15,36]. Moreover, they show that the algorithm makes steady progress over the course of optimization, needing $O(b - a)$ generations to increase the fitness from a to b whenever $b - a \geq C \log n$ for a suitable constant C. The crucial point is that this is independent of a and b, so independent of the current state of the algorithm. It implies that the algorithm chooses $\lambda = O(1)$ in early stages when progress is easy, and (almost) linear values $\lambda = \Omega(n)$ in the end when progress is hard. Thus it achieves (ii) conditional on having appropriate parameter settings.

Interestingly, it is shown in [26] that for $s \geq 18$, the SA-$(1, \lambda)$-EA fails in a region far away from the optimum, more precisely in the region with 85% one-bits. Consequently, it also fails for every other function that is identical with ONEMAX in the range of $[0.84n, 0.85n]$ one-bits, which includes other classical benchmarks like JUMP, CLIFF, and RIDGE. It is implicit that the algorithm would be efficient in regions that are closer to the optimum. This is remarkable, since usually optimization is harder close to the optimum. Such a reversed failure profile has previously only been observed in very few situations. One is the $(\mu + 1)$-EA with mutation rate c/n for an arbitrary constant $0 < c \leq 1$ on certain monotone functions. This algorithm is efficient close to the optimum, but fails to cross some region at linear distance of the optimum if $\mu > \mu_0$ for some μ_0 that depends on c [44]. A similar phenomenon has been shown for $\mu = 2$ and a specific value of c in the dynamic environment DYNAMIC BINVAL [40,41]. These are the only examples for this phenomenon that the authors are aware of.

A limitation of [26] is that it studies only a single benchmark, the ONEMAX function. Although the negative result also holds for functions that are identical to ONEMAX in some range, the agreement with ONEMAX in this range must be perfect, and the positive result does not extend to other functions in such a way. This leaves the question on what happens for larger classes of benchmarks:

(a) Is there a safe choice for s that makes the algorithm efficient for a whole class of functions?
(b) Does the positive result (ii) extend to other benchmarks beyond ONEMAX?

In this paper, we answer both questions with *Yes* for the set of all (strictly) monotone pseudo-Boolean functions. This is a very large class; for example, it contains all linear functions. In fact, all our results hold in an even more general *dynamic* setting: the fitness function may be different in each generation, but it is a monotone function every time and therefore shares the same global optimum $(1, \ldots, 1)$. We show an upper bound of $O(n)$ generations and $O(n \log n)$ function evaluations if the mutation rate is c/n for some $c < 1$, which is a very natural assumption for monotone functions as many algorithms become inefficient for large values of c [18,37,39,43]. Those results are as strong as the positive results in [26], except that we replace the constant "1" in the condition $s < 1$ by a different constant that depends on c. For $c = 1$ we still show a bound of $O(n \log n)$ generations and $O(n^2 \log \log n)$ evaluations. This polynomial bound is in line with the general frameworks for elitist algorithms [18,29], although we believe the true answer should be smaller; see discussion after Theorem 4 for details.

Both parts of the answer are encouraging news for the SA-$(1, \lambda)$-EA. It means that, at least for this class of benchmarks, there is a universal parameter setting that works in all situations. This resembles the role of the mutation rate c/n for the $(1 + 1)$-EA on monotone functions: If $c < 1 + \varepsilon$, then the $(1 + 1)$-EA is quasilinear on all monotone functions [18,39,43], and for $c < 1$ this is known for many other algorithms as well [37]. On the other hand, the $(\mu + 1)$-EA is an example where such a safe parameter choice for c does not exist: for any $c > 0$ there is μ such that the $(\mu+1)$-EA with mutation rate c/n needs superpolynomial time to find the optimum of some monotone functions.

We do not just strengthen the positive result, but we show that the negative result generalizes in a surprisingly strong sense, too: for an arbitrary mutation rate c/n where $c < 1$, if s is sufficiently large, then the SA-$(1, \lambda)$-EA needs exponential time on *every* monotone function. Thus the failure mode for large s is not specific to ONEMAX. On the other hand, we also generalize the result (implicit in [26]) that the only hard region is at linear distance from the optimum: for any value of s, if the algorithm starts close enough to the optimum (but still at linear distance), then with high probability it optimizes every monotone function efficiently. Finally, we complement the theoretical analysis with simulations, which show another interesting aspect: in a 'middle region' of s, it seems to depend on F whether the algorithm is efficient or not, similarly to the SA $(1 + (\lambda, \lambda))$-GA [9, Sect. 6.5]. However, there the reason was that the success probability was universally bounded by $\approx 0.31 < 1$, independent of λ. In our setting, the success probability approaches one as λ grows, so the underlying reason seems different.

Our proofs re-use ideas from [26]. In particular, we use a potential function of the form $g(x^t, \lambda^t) = \text{ZM}(x^t) + h(\lambda^t)$, where $\text{ZM}(x^t)$ is the number of zero-bits in x^t and $h(\lambda^t)$ is a penalty term for small values of λ^t. Similar decompositions have been used before [12]. The exact form of h depends on the situation; sometimes it is very similar to the choices in [26] (h_1, h_4 below), but some cases are completely different (h_2, h_3). With these potential functions, we obtain a positive or negative drift, depending on the situation. Once the drift is established, the positive and negative statements about generations follow from standard drift analysis [34,38]. For the number of function evaluations, while some themes from [26] reappear, the proof strategy is different, see the discussion after Corollary 6 for the reasons.

Due to space restrictions, this extended abstract only contains the results, and only sketches of the proofs. The complete proofs may all be found in [32].

2   Preliminaries and Definitions

The Algorithm: SA-$(1, \lambda)$-EA. We will consider the self-adjusting $(1, \lambda)$-EA with $(1 : s + 1)$-success rate, with mutation rate c/n, success ratio s and update strength F, and we denote this algorithm by SA-$(1, \lambda)$-EA. It is given by the pseudocode in Algorithm 1. Note that the parameter λ may take non-integral values during the execution of the algorithm, however the number of children generated at each step is chosen as the closest integer $\lfloor \lambda \rceil$ to λ.

Let us give a short explanation of the concept of the $(1 : s + 1)$-success rule, or $(1 : s + 1)$-rule for short. For given λ and given position x in the search space (we omit the index t), the algorithm has some *success probability* p, where success means that $f(y) > f(x)$ for the fittest of λ offspring y of x. To keep this explanation simple we now make a few assumptions, which we lift in the formal analysis: we will ignore the rounding effect of λ and we will assume $p(\lambda) \le 1/(s+1)$ for $\lambda = 1$, and that $0 < p < 1$. The success probability $p = p(\lambda)$ is an increasing function in λ, and strictly increasing due to $0 < p < 1$. Hence there is a value λ^* such that $p(\lambda) < 1/(s+1)$ for $\lambda < \lambda^*$ and $p(\lambda) > 1/(s+1)$ for $\lambda > \lambda^*$. Now consider the potential $\log_F \lambda$. This potential decreases by 1 with probability p and increases by $1/s$ with probability $1 - p$. So in expectation it changes by $-p + (1 - p)/s = (1 - (s + 1)p)/s$. Hence, the expected change is positive if $\lambda < \lambda^*$ and negative if $\lambda > \lambda^*$. Therefore, λ has a drift towards λ^* from both sides (in a logarithmic scaling). So the rule implicitly has a *target population size* λ^*, and this population size λ^* corresponds to the *target success rate* $p = 1/(s + 1)$.

Note that a drift towards λ^* does not necessarily imply that λ always stays close to λ^*. Firstly, p depends on the current state x of the algorithm, and might vary rapidly as the algorithm progresses (though this does not seem to be a very typical situation). In this case, the target value λ^* also varies. Secondly, even if λ^* remains constant, there may be random fluctuations around this value, see [34, 38] for treatments on when drift towards a target guarantees concentration. However, we note that the $(1 : s + 1)$-rule for controlling λ gives stronger guarantees than the same rule for controlling other parameters like step size or mutation rate. The difference is that other parameters do not necessarily influence p in a monotone way, and therefore we can not generally guarantee that there is a drift towards success probability $1/(s + 1)$ when the $(1 : s + 1)$-rule is used to control them. Only when controlling λ we are guaranteed a drift in the right direction.

Algorithm 1. SA-$(1, \lambda)$-EA with success rate s, update strength F and mutation rate c/n for maximizing a fitness function $f : \{0, 1\}^n \to \mathbb{R}$.

Initialization: Choose $x^0 \in \{0, 1\}^n$ uniformly at random (u.a.r.) and $\lambda^0 := 1$
Optimization: for $t = 0, 1, \dots$ do
    Mutation: for $j \in \{1, \dots, \lfloor \lambda^t \rceil\}$ do
        $y^{t,j} \leftarrow$ mutate(x^t) by flipping each bit of x^t independently with prob. c/n
    Selection: Choose $y^t = \arg\max\{f(y^{t,1}), \dots, f(y^{t, \lfloor \lambda \rceil})\}$, breaking ties randomly
    Update:
    if $f(y^t) > f(x^t)$ then $\lambda^{t+1} \leftarrow \max\{1, \lambda^t/F\}$;   else $\lambda^{t+1} \leftarrow F^{1/s}\lambda^t$;
    $x^{t+1} \leftarrow y^t$;

The Benchmark: Dynamic Monotone Functions. Whenever we speak of "monotone" functions in this paper, we mean strictly monotone pseudo-Boolean functions, defined as follows.

Definition 1. *We call* $f : \{0,1\}^n \to \mathbb{R}$ *monotone if* $f(x) > f(y)$ *for every pair* $x, y \in \{0,1\}^n$ *with* $x \neq y$ *and* $x_i \geq y_i$ *for all* $1 \leq i \leq n$.

In this paper we will consider the following set of benchmarks. For each $t \in \mathbb{N}$, let $f^t : \{0,1\}^n \to \mathbb{R}$ be a monotone function that may change at each step depending on x^t. Then the selection step in the t-th generation of Algorithm 1 is performed with respect to f^t. By slight abuse of language we will still speak of *a* dynamic monotone function f.

All our results (positive and negative) hold in this dynamic setup. This set of benchmarks is quite general. Of course, it contains the static setup that we only have a single monotone function to optimize, which includes linear functions and ONEMAX as special cases. It also contains the setup of Dynamic Linear Functions (originally introduced as Noisy Linear Functions in [42]) and DYNAMIC BINVAL [40,41]. On the other hand, all monotone functions share the same global optimum $(1 \ldots 1)$, have no local optima, and flipping a zero-bit into a one-bit strictly improves the fitness. In the dynamic setup, these properties still hold "locally", within each selection step. Thus the setup falls into the general framework by Jansen [29], which was extended to the *partially ordered EA* (PO-EA) by Colin, Doerr, Férey [7]. This implies that the $(1+1)$-EA with mutation rate c/n finds the optimum of every such Dynamic Monotone Function in expected time $O(n \log n)$ if $c < 1$, and in time $O(n^{3/2})$ if $c = 1$.

Potential Functions and Methods. Drift analysis is a key instrument in the theory of EAs. To apply it, we must define a *potential function* and compute the expected change of this potential. A common potential for simple problems are the ONEMAX and ZEROMAX potentials $\text{OM}(x^t) = \sum_i x_i^t$ and $\text{ZM}(x^t) = n - \text{OM}(x^t)$, which respectively count the number of one-bits and zero-bits of the current state x^t. We will write $Z^t := \text{ZM}(x^t)$ throughout the paper.

As mentioned before, this potential function will not be sufficient, so as in [26] we use a composite potential function of the form $g(x^t, \lambda^t) = \text{ZM}(x^t) + h(\lambda^t)$, where $h(\lambda^t)$ varies from application to application. Since the potential always contains Z^t as additive term, the drift of Z^t enters the drift of the potential in all cases. Thus we compute this drift in Lemma 3 below. We omit all proofs, and only give some key intermediate result. For all times t we define $A^{t,j}$ as the event that the j-th offspring at time t does not flip any one-bit of the parent, and A^t is the event that such a child exists at time t. We also define $B^{t,j}$ as the event that the j-th child *does* flip a *zero*-bit of the parent and B^t the event that such a child exists. We drop the superscript t when the time is clear from context.

Lemma 2. *There exist constants* $b_1, b_2, b_3 > 0$ *depending only on* c *such that at all times* t *with* $Z^t \geq 1$ *we have*

$$\Pr[\bar{A}] \leq e^{-b_1 \lambda} \qquad \text{and} \qquad e^{-b_2 \lambda Z^t / n} \leq \Pr[\bar{B}] \leq e^{-b_3 \lambda Z^t / n}.$$

The key ingredient in our analysis is the following lemma.

Lemma 3. *Consider the SA-$(1, \lambda)$-EA with mutation rate $0 < c \le 1$ and update strength $1 < F$. There exist constants $a_1, a_2, b > 0$ depending only on c such that at all times t with $Z^t > 0$ we have*

$$\mathbf{E}[Z^t - Z^{t+1} \mid x^t, \lambda^t] \ge \Pr[B] \cdot a_1 \left(1 - c(1 - Z^t/n)\right) - a_2 e^{-b\lambda^t}.$$

We give two key steps in proving Lemma 3.

Claim 1. *At all times $t \ge 0$ with $Z^t > 0$ we have*

$$\mathbf{E}[Z^t - Z^{t+1} \mid A, B] \ge e^{-c}(1 - c(1 - Z^t/n)).$$

Claim 2. *At all times $t \ge 0$ with $Z^t > 0$ we have*

$$\mathbf{E}[Z^t - Z^{t+1} \mid \bar{A}] \ge -c/(1 - e^{-c}).$$

We omit the proofs. Both claims are rather easy for $\lambda = 1$. The main part of the proof is showing that the selection step influences the expectation in the right direction, for which we use the FKG inequality [24]. Details can be found in [32].

We remark that Lemma 3 and both claims also hold if we replace $Z^t - Z^{t+1}$ by $\min\{1, Z^t - Z^{t+1}\}$, which is helpful for concentration results as the latter term has exponentially decaying tail probabilities.

Proof (of Lemma 3). The drift of $Z^t = Z_M(x^t)$ may be decomposed into,

$$\mathbf{E}[Z^t - Z^{t+1}] = \Pr[\bar{A}] \cdot \mathbf{E}[Z^t - Z^{t+1} \mid \bar{A}] + \Pr[A, B] \cdot \mathbf{E}[Z^t - Z^{t+1} \mid A, B]$$
$$+ \Pr[A, \bar{B}] \cdot \mathbf{E}[Z^t - Z^{t+1} \mid A, \bar{B}],$$

where we omitted the conditioning on x, λ for brevity. The events A, B are independent so we get $\Pr[A, B] = \Pr[A] \Pr[B]$. Also, the third conditional expectation is 0: if $\bar{B}$ holds then no child is a strict improvement of the parent, but A guarantees that some child is at least as good. Hence, if $A, \bar{B}$ hold then $x^t = x^{t+1}$. Combining those remarks with the bounds of Claims 1 and 2 gives

$$\mathbf{E}[Z^t - Z^{t+1} \mid x, \lambda] \ge \Pr[A] \Pr[B] \cdot \tfrac{1 - c(1 - Z^t/n)}{e^c} - \Pr[\bar{A}] \cdot \tfrac{c}{1 - e^{-c}}.$$

Lemma 2 guarantees that $\Pr[A] \ge C = \Omega(1)$ since $\lambda \ge 1$ and that $\Pr[\bar{A}] \le e^{-b_1 \lambda}$. Choosing $a_1 = Ce^{-c}$, $a_2 = c/(1 - e^{-c})$ and $b = b_1$ gives the result. $\qquad\square$

3 Monotone Functions Are Efficient for Large Success Rates

In this section we give the positive results, which hold if s is small, or if $s > 0$ is arbitrary and we start sufficiently close to the threshold. We show that in these cases, for *any* strictly monotone fitness function the optimum is found efficiently both in the number of generations and evaluations. For brevity, we present all those results in a single theorem.

Theorem 4. *Let $0 < c \leq 1 < F$ and $0 < s$ be the parameters of the EA. There exist constants $C(c, F, s), s_0(c) > 0$, $F_0 > 1$ and $\varepsilon(c, F, s) \in (0, 1)$ such that the following holds. For every dynamic monotone function and any starting values $x^{\mathrm{init}}, \lambda^{\mathrm{init}}$, the number of generations G (resp. number of evaluations E) of the SA-$(1, \lambda)$-EA satisfies the following:*

(i) *if $c < 1$ and $s < s_0$, then $\mathbf{E}[G] \leq Cn$ and $\mathbf{E}[E] \leq C(\lambda^{\mathrm{init}} + n \log n)$;*

(ii) *if $c = 1$, $s < s_0$ and $F < F_0$, then $\mathbf{E}[G] \leq Cn \log n$ and $\mathbf{E}[E] \leq C(\lambda^{\mathrm{init}} + n^2 \log \log)$;*

(iii) *if $c < 1$ and $\mathrm{ZM}(x^{\mathrm{init}}) \leq \varepsilon n$ then $G \leq Cn$ and $E \leq C(\lambda^{\mathrm{init}} + n \log n)$ with high probability.*

We remark that the number of evaluations is tight for $c < 1$ since any unary unbiased algorithm needs at least $\Omega(n \log n)$ function evaluations to optimize ONEMAX [36]. For $c = 1$, we suspect that even the main order n^2 is not tight, since the $(1 + 1)$-EA with $c = 1$ is known to need time $O(n^{3/2})$ even in the pessimistic PO-EA model [7], which includes every dynamic monotone function. The order $n^{3/2}$ is tight for the PO-EA, but a stronger bound of $O(n \log^2 n)$ is known for all static monotone functions [39]. Thus we conjecture that the number of function evaluations is quasi-linear even for $c = 1$.

The key ingredient for the proof is an appropriate potential function of the form $g(x, \lambda) = \mathrm{ZM}(x) + h(\lambda)$. We will use a different h in all three cases.

Definition 5 (Potential functions for positive result). *Let*

$$h_1(\lambda) := -K_1 \cdot \min\{0, \log_F(\lambda/\lambda_{\max})\},$$

$$h_2(\lambda) := -K_2 \cdot \min\{0, \tfrac{1}{\lambda_{\max}} - \tfrac{1}{\lambda}\},$$

$$h_3(\lambda) := -K_3 \cdot \min\left\{0, \log_F \lambda/\lambda_{\max}\right\} + K_4 e^{-K_5\lambda},$$

where $\lambda_{\max} := F^{1/s}n$ and the $K_j > 0$ are constants to be chosen later. Then for $i = 1, 2, 3$, for all $x \in \{0, 1\}^n$ and $\lambda \in [1, \infty)$ we define

$$g_i(x, \lambda) := \mathrm{ZM}(x) + h_i(\lambda),$$

and we denote $Z^t := \mathrm{ZM}(x^t)$, $H_i^t := h_i(\lambda^t)$ and $G_i^t := g_i(x^t, \lambda^t)$.

Very roughly speaking, there are two reasons why we choose those functions: first $|h|$ is rather small (e.g. $0 \leq H_1^t \leq K_1 \log_F(nF^{1/s})$) meaning that $g = z + h$ is a good approximation of z. The second reason is that we want g to consistently have negative drift towards 0 to be able to conclude with standard drift theorems. The number of zero-bits z has negative drift when λ is large, but positive drift when λ is small; the functions H_i above are chosen to have sufficiently large negative drift when λ is small, and negligible drift (compared to that of z) when λ is large. A technical, but easy computation gives the drift of H_i^t:

Claim 3. *At all times $t \geq 0$ with $Z^t > 0$ we have*

$$\mathbf{E}[H_1^t - H_1^{t+1} \mid x, \lambda] \geq -K_1 \cdot \Pr[B] + \frac{K_1}{s} \cdot \Pr[\bar{B}] \cdot \mathbf{1}_{\lambda < n},$$

$$\mathbf{E}[H_2^t - H_2^{t+1} \mid x, \lambda] \geq -\frac{K_2}{\lambda}(F - 1) \cdot \Pr[B] + \frac{K_2}{\lambda}(1 - F^{-1/s}) \cdot \Pr[\bar{B}] \cdot \mathbf{1}_{\lambda < n},$$

$$\mathbf{E}[H_3^t - H_3^{t+1} \mid x, \lambda] \geq \mathbf{1}_{\lambda < n} \Pr[\bar{B}]\big(K_3/s + K_4(1 - e^{-K_5(F^{1/s} - 1)})e^{-K_5\lambda}\big)$$
$$- \Pr[B] \cdot \big(K_3 + K_4 e^{-K_5\lambda/F}\big).$$

With a bit of case distinction, Claims 1, 2 and 3 may now be combined to obtain the following drifts of G_1^t, G_2^t, and G_3^t.

Corollary 6. *There exist constants $s_0 > 0$ and $F_0 > 1$ such that for all $0 < s \leq s_0$ there are $\delta, \varepsilon > 0$ and a choice of $K_1, \ldots, K_5$ such that for all t with $Z^t > 0$,*

$$\mathbf{E}[G_1^t - G_1^{t+1} \mid x, \lambda] \geq \delta \qquad \text{if } c < 1,$$

$$\mathbf{E}[G_2^t - G_2^{t+1} \mid x, \lambda] \geq \delta G_2^t/n \qquad \text{if } c = 1 \text{ and } F < F_0,$$

$$\mathbf{E}[G_3^t - G_3^{t+1} \mid x, \lambda] \geq \delta \qquad \text{if } c < 1 \text{ and } \text{ZM}(x)/n \leq 2\varepsilon.$$

Proof (Sketch for G_1^t). Combining Lemma 3 and Claim 3, the drift of G_1 is

$$\mathbf{E}[G_1^t - G_1^{t+1} \mid x, \lambda] \geq \Pr[B](\alpha_1 - K_1) + \mathbf{1}_{\lambda < n} \Pr[\bar{B}]K_1/s - \alpha_2 e^{-\beta\lambda}.$$

We choose $K_1 = \alpha_1/2$ so that $\alpha_1 - K_1 \geq K_1$, and we may assume $s \leq 1$. Now we make a case distinction. If $\lambda < n$, then one can check

$$\mathbf{E}[G_1^t - G_1^{t+1} \mid x, \lambda] \geq K_1 + \Pr[\bar{B}]K_1(1 - s)/s - \alpha_2 e^{-\beta\lambda}.$$

There is $\lambda_0 = \lambda_0(\alpha_2, \beta, K_1)$ such that for $\lambda \geq \lambda_0$ the last term can be bounded as $\alpha_2 e^{-\beta\lambda} \leq K_1/2$, in which case the drift is at least $K_1/2$. For $\lambda < \lambda_0$, the drift is also $\Omega(1)$ for small enough s since $\Pr[\bar{B}] = \Omega(1)$ by Lemma 2.

If $\lambda \geq n$, the drift is $\mathbf{E}[G_1^t - G_1^{t+1} \mid x, \lambda] \geq \Pr[B]K_1 - \alpha_2 e^{-\beta n}$. The first term is at least $\Pr[B]K_1 = \Omega(1)$ since K_1 is a constant and so is $\Pr[B]$ by Lemma 2, while the second is $e^{-\Omega(n)} = o(1)$; this implies that the drift is $\Omega(1)$. □

From Corollary 6, it is not hard to prove the claim on the number of generations. For the number of evaluations, we use the best-so-far ZeroMax value $Z_*^t := \min_{\tau \leq t}(Z^\tau)$ as in [26,27], but otherwise our proof is different. In fact, we believe that the proof in these papers is not fully correct. In [26, Theorem 3.5], the authors bound the number of evaluations per generation by identically distributed random variables, and use Wald's equation to bound the total number of evaluations. However, Wald's equation is only true for the sum of *independent* random variables (or similar conditions, e.g. [20]), and such a condition is not satisfied in our case. Thus we need to use a different approach based on concentration of hitting times. We only give the key steps for the case $c < 1$ without proofs. For the following lemmas, we consider the SA-$(1, \lambda)$-EA as in Theorem 4, with an arbitrary initial search point x^{init} and an initial value of $\lambda = \lambda^{\text{init}}$.

Lemma 7 (Fajardo, Sudholt). *There exists a constant $C > 0$ such that at all times $t \geq 0$, $\mathbf{E}\left[\lambda^t \cdot 1_{Z^t_* \geq z}\right] \leq \lambda^{\mathrm{init}}/F^t + Cn/z$.*

Lemma 8. *Let T denote the first time t at which $\lambda^t \leq 8en\log n/Z^t$. There exists an absolute constant $C > 0$ such that $\mathbf{E}[\sum_{t=1}^{T}\lambda^t] \leq C\lambda^{\mathrm{init}}$.*

Lemma 9. *Let $c < 1$ and (a, b) be an interval of length $b - a = \log n$. Assume $\mathrm{ZM}(x^{\mathrm{init}}) \leq b$, and let T be first time t at which $Z^t \leq a$. Then there exists an absolute constant $D > 0$ such that $T \leq D\log n$ with probability at least $1 - n^{-4}$.*

To prove the statement about the number of evaluations, we divide a run of the algorithm into *blocks* and *phases* as follows. A block starts with an initialisation phase which lasts until the condition $\lambda^t \leq F^{1/s}8en\log n/Z^t$ is met. Once this phase is over, the block runs for $n/\log n$ phases of length $D\log n$, with D the constant of Lemma 9. During the i-th such phase the process attempts to improve Z^t from $n - i\log n$ to $n - (i+1)\log n$. If such an improvement is made before the $D\log n$ steps are over, then the process remains idle during the remaining steps of that phase. If a phase fails to make the correct improvement in $D\log n$ generations, or if $\lambda^t \geq F^{1/s}8en\log n/Z^t_*$ at any point after the initialisation phase is over, then the whole block is considered a failure, and the next block starts. This trick ensures that no block starts with too large λ^{init}. Then the proof relies on the following two facts, stated without proof:

(i) every block finds the optimum whp;
(ii) consider a block starting with $\lambda = \lambda^{\mathrm{init}}$, then the expected number of function evaluations during this block is $O(\lambda^{\mathrm{init}} + n\log n)$.

For $c = 1$, we use the same idea, but we use $\log n/\log\log n$ phases, where the i-th phase attempts to improve Z^t from $n/\log^{i-1}n$ to $n/\log^i n$ in $Dn\log\log n$ steps for a large constant D. Then the probability that a phase fails to improve Z^t is at most $1/\log^2 n$, and the total number of function evaluations per block after the initialization phase is $O(\sum_{i=0}^{\log n/\log\log n} n\log\log n \cdot \log^i n) = O(n^2\log\log n)$.

4   Small Success Rates Yield Exponential Runtimes

For large s, that is, for a small enough success rate, we show that the SA-$(1, \lambda)$-EA needs super-polynomial time to find the optimum of any dynamic monotone function. The reason is that the algorithm has negative drift in a region that is still far away from the optimum, in linear distance. In fact, we have shown before that the drift is *positive* close to the optimum. Thus the hardest region for the SA-$(1, \lambda)$-EA is not around the optimum. This surprising phenomenon was discovered for ONEMAX in [27]. We show that it is not caused by any specific property of ONEMAX, but that it occurs for *every* dynamic monotone function. Even in the ONEMAX case, our result is slightly stronger than [26], since they show their result only for $1 < F < 1.5$, while ours holds for all $F > 1$. On the other hand, they give an explicit constant $s_1 = 18$ for ONEMAX.

Theorem 10. *Let $0 < c \leq 1 < F$. For every $\varepsilon > 0$, there exists $s_1 > 0$ such that for all $s \geq s_1$, for every dynamic monotone function and every initial search point x^{init} satisfying $\text{ZM}(x^{\text{init}}) \geq \varepsilon n$ the number of generations of the SA-$(1, \lambda)$-EA with parameters s, F, c is $e^{\Omega(n/\log^2 n)}$ with high probability.*

The proof is quite similar to the other cases, except that we prove a negative drift. Similar to [26], we use the potential function

$$h_4(\lambda) := -K_6 \log_F^2(\lambda F) = -K_6 \cdot (\log_F(\lambda) + 1)^2,$$

and we define $g_4(x, \lambda) = \text{ZM}(x) + h_4(\lambda)$, $G_4^t := g_4(x^t, \lambda^t)$, $H_4^t := h_4(\lambda^t)$. Then we use the following key steps. Note the switched order of $t + 1$ and t.

Lemma 11. *There exists a constant $\alpha_1 > 0$ depending only on c such that at all times t we have $\mathbf{E}[Z^{t+1} - Z^t \mid x, \lambda] \geq -\Pr[B]\alpha_1(1 + \log \lambda)$.*

Lemma 12. *There exist constants $\varepsilon, \alpha_2 > 0$ depending only on c, F such that if $Z^t \leq \varepsilon n$ and $\lambda \leq F$, then $\mathbf{E}[Z^{t+1} - Z^t \mid x, \lambda] \geq \alpha_2$.*

Lemma 13. *Assume that $s \geq 1$. At all times t with $Z^t > 0$ we have*

$$\mathbf{E}[H_4^{t+1} - H_4^t \mid x, \lambda] \geq \tfrac{1}{3} \Pr[B]K_4(1 + \log_F \lambda)\mathbf{1}_{\lambda \geq F} - \tfrac{3}{s}K_4(1 + \log_F \lambda).$$

Corollary 14. *For all $0 < c \leq 1 < F$ and every sufficiently small $\varepsilon > 0$ there exists $s_1 > 0$ such that for all $s \geq s_1$, the following holds. There exists a constant $\delta > 0$ such that if $\varepsilon n/2 \leq Z^t \leq \varepsilon n$ then $\mathbf{E}[G_4^{t+1} - G_4^t \mid x, \lambda] \geq \delta$.*

5   Simulations

In this section, we provide simulations that complement our theoretic analysis. The functions that we optimize in our simulations include ONEMAX, BINARY, HOTTOPIC [37], BINARYVALUE, and DYNAMIC BINVAL [40], where BINARY is defined as $f(x) = \sum_{i=1}^{\lfloor n/2 \rfloor} x_i n + \sum_{i=\lfloor n/2 \rfloor+1}^{n} x_i$, and BINARYVALUE is defined as $f(x) = \sum_{i=1}^{n} 2^{i-1}x_i$. The definition of HOTTOPIC can be found in [37], and we set the parameters to $L = 100$, $\alpha = 0.25$, $\beta = 0.05$, and $\varepsilon = 0.05$. DYNAMIC BINVAL is the dynamic environment which applies the BINARYVALUE function to a random permutation of the n bit positions, see [40] for its formal definition. In all experiments, we start the SA-$(1, \lambda)$-EA with a randomly sampled search point and an initial offspring size of $\lambda^{\text{init}} = 1$. The algorithm terminates when the optimum is found or after $500n$ generations. The code for the simulations can be found at https://github.com/zuxu/OneLambdaEA.

Figure 1 follows the same setup with $F = 1.5$ as in [26], but for a larger set of functions. We recover the same threshold $s = 3.4$ for ONEMAX. For the other monotone functions, the threshold is smaller than $s = 3.4$. This opposes a conjecture in [26], which we formally disprove in a companion paper [31].

Effect of F. We have shown that
the SA-$(1, \lambda)$-EA with $c < 1$ optimizes every dynamic monotone function efficiently when s is small and is
inefficient when s is too large. Both
results hold for arbitrary F. It is natural to assume that there is a threshold s_0 between the efficient and inefficient regime. However, Fig. 2 shows
that the situation might be more complex. For this plot, we have fixed s
slightly below the threshold for $F =$
1.5 on DYNAMIC BINVAL (empirically
determined from Fig. 1) and systematically varied the value of F. For this
intermediate value of s, we see a phase
transition in terms of F.

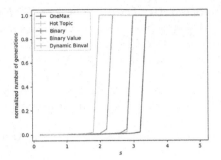

Fig. 1. Average number of generations of
the self-adjusting $(1, \lambda)$-EA with $F = 1.5$
and $c = 1$ in 10 runs when optimizing
monotone functions with $n = 10000$, normalised and capped at $500n$ generations.
Curves for HOTTOPIC and BINARY mostly
overlap. The evaluated values of s range
from 0.2 to 5 with a step size of 0.2 for
all functions except that DYNAMIC BINVAL
was not evaluated for $3.2 \leq s \leq 5$ due to
performance issues.

Hence, we conjecture that there
is no threshold s_0 such that the
SA-$(1, \lambda)$-EA is efficient for all $s < s_0$
and all $F > 1$, and inefficient for all
$s > s_0$ and all $F > 1$. Rather, we con-
jecture that there is 'middle range' of values of s for which it depends on F
whether the SA-$(1, \lambda)$-EA is efficient. We know from this paper that this phenomenon can *only* occur for a 'middle range': both for sufficiently small s and
for sufficiently large s, the value of F does not play a role.

In general, smaller values of F seem to be beneficial. However, the correlation
is not perfect, see for example the dip for $c = 0.98$ and $F = 5.5$ in the left subplot
of Fig. 2. These dips also happen for some other combinations of s, F and c (not
shown), and they seem to be consistent, i.e., they do not disappear with a larger
number of runs or larger values of n up to $n = 5000$. To test whether this is due
to the rounding scheme, we checked whether the effect disappears if we round λ
in each generation stochastically to the next integer; e.g., $\lambda^t = 2.6$ means that
in generation t we create two offspring with probability 40% and three offspring
with probability 60%. The effect remains, and the runtime still seems to depend
on F in a non-monotone fashion, see the right subplot of Fig. 2.

The impact of F is visible for all ranges $c < 1$, $c = 1$ and $c > 1$. For $c = 1$ we
have only proven efficiency for sufficiently small F. However, we conjecture that
there is no real phase transition at $c = 1$, and the 'only' difference is that our
proof methods break down at this point. For the fixed s, with increasing c the
range of F becomes narrower and restricts to smaller values while larger values
of c admit a larger range of values for F.

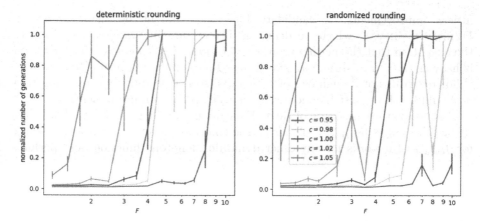

Fig. 2. Average number of generations of the self-adjusting $(1, \lambda)$-EA with $s = 1.8$ and in 50 runs when optimizing DYNAMIC BINVAL with $n = 1000$, normalised and capped at $500n$ generations. The left and right subplots correspond to the deterministic and randomized rounding schemes respectively. The vertical bars indicate standard deviation.

6 Conclusion

We have studied the SA-$(1, \lambda)$-EA on dynamic monotone functions. Hevia Fajardo and Sudholt had shown an extremely strong dependency of the performance on the success rate s for the ONEMAX benchmark. We have shown that there is nothing specific to ONEMAX about the situation. The same effect happens for any (static or dynamic) monotone fitness function: for small values of s, the SA-$(1, \lambda)$-EA is efficient on all dynamic monotone functions, while for large values of s, the SA-$(1, \lambda)$-EA is inefficient on every dynamic monotone function. In the latter case, the bottleneck is not around the optimum, but rather in some area of linear distance from the optimum. Thus the SA-$(1, \lambda)$-EA is one of the surprising examples showing that some algorithms may fail in easy fitness landscapes, but succeed in hard fitness landscapes.

Hevia Fajardo and Sudholt have conjectured that the problem becomes worse the easier the fitness landscape is. Concretely, they conjectured that any parameter choice that works for ONEMAX should also give good result for any other landscape [27]. In a companion paper [31], we disprove this conjecture, but for an unexpected reason: there are different ways to measure 'easiness' of a fitness landscape. While it is theoretically proven that ONEMAX is the easiest fitness function with respect to decreasing the distance from the optimum [19], this is not the aspect that matters for the SA-$(1, \lambda)$-EA. Here, the important aspect is how easy it is to find a fitness improvement, since this may induce too small target population sizes in the SA-$(1, \lambda)$-EA. For finding fitness improvements, there are easier functions than ONEMAX, for example the DYNAMIC BINVAL function [40] or HOTTOPIC functions [37], see [31] for details. It remains open to determine the easiest dynamic monotone function f_{easiest} with respect to fitness improvements. A candidate for $f_{easiest}$ might be the 'adversarial' DYNAMIC BINVAL, which we define as DYNAMIC BINVAL (see Sect. 5) with the exception that the permutation is not random but chosen so that any 0-bit is heavier than any 1-bit. With this fitness function, any 0-bit flip gives a fitter child, regardless of the number of 1-bit flips, so it is intuitively convincing that it should be the easiest function with respect to fitness improvement.

Moreover, the conjecture of Hevia Fajardo and Sudholt might still hold if we replace ONEMAX by $f_{easiest}$. I.e., is it true that any parameter choice that works for $f_{easiest}$ also works for any other dynamic monotone function, and perhaps even in yet more general settings?

Apart from that, the most puzzling part of the picture is the experimental finding that in a 'middle regime' of success rates, the update strength F seems to play a role in a non-monotone way (for fixed success rate s). It is open to prove theoretically that there is indeed such a 'middle regime' where F plays a role at all. For why this effect is non-monotone in F, we do not even have a good hypothesis. As outlined in Sect. 5, it does not seem to be a rounding effect. This shows that we are still missing important parts of the overall picture.

Acknowledgements. We thank Dirk Sudholt for helpful discussions during the Dagstuhl seminar 22081 "Theory of Randomized Optimization Heuristics". We also thank the reviewers for their helpful comments and suggestions.

References

1. Aleti, A., Moser, I.: A systematic literature review of adaptive parameter control methods for evolutionary algorithms. ACM Comput. Surv. (CSUR) **49**(3), 1–35 (2016)
2. Antipov, D., Doerr, B., Yang, Q.: The efficiency threshold for the offspring population size of the (μ, λ) EA. In: Genetic and Evolutionary Computation Conference (GECCO), pp. 1461–1469 (2019)
3. Auger, A.: Benchmarking the (1+ 1) evolution strategy with one-fifth success rule on the BBOB-2009 function testbed. In: Genetic and Evolutionary Computation Conference (GECCO), pp. 2447–2452 (2009)
4. Badkobeh, G., Lehre, P.K., Sudholt, D.: Unbiased black-box complexity of parallel search. In: Bartz-Beielstein, T., Branke, J., Filipič, B., Smith, J. (eds.) PPSN 2014. LNCS, vol. 8672, pp. 892–901. Springer, Cham (2014). https://doi.org/10.1007/978-3-319-10762-2_88
5. Böttcher, S., Doerr, B., Neumann, F.: Optimal Fixed and adaptive mutation rates for the leadingones problem. In: Schaefer, R., Cotta, C., Kołodziej, J., Rudolph, G. (eds.) PPSN 2010. LNCS, vol. 6238, pp. 1–10. Springer, Heidelberg (2010). https://doi.org/10.1007/978-3-642-15844-5_1
6. Case, B., Lehre, P.K.: Self-adaptation in nonelitist evolutionary algorithms on discrete problems with unknown structure. IEEE Trans. Evol. Comput. **24**(4), 650–663 (2020)
7. Colin, S., Doerr, B., Férey, G.: Monotonic functions in EC: anything but monotone! In: Genetic and Evolutionary Computation Conference (GECCO), pp. 753–760 (2014)

8. Devroye, L.: The compound random search. Ph.D. dissertation, Purdue Univ., West Lafayette, IN (1972)
9. Doerr, B., Doerr, C.: Optimal static and self-adjusting parameter choices for the $(1 + (\lambda, \lambda))$ Genetic Algorithm. Algorithmica **80**(5), 1658–1709 (2018)
10. Doerr, B., Doerr, C.: Theory of Parameter control for discrete black-box optimization: provable performance gains through dynamic parameter choices. In: Theory of Evolutionary Computation. NCS, pp. 271–321. Springer, Cham (2020). https://doi.org/10.1007/978-3-030-29414-4_6
11. Doerr, B., Doerr, C., Ebel, F.: From black-box complexity to designing new genetic algorithms. Theoret. Comput. Sci. **567**, 87–104 (2015)
12. Doerr, B., Doerr, C., Kötzing, T.: Provably optimal self-adjusting step sizes for multi-valued decision variables. In: Handl, J., Hart, E., Lewis, P.R., López-Ibáñez, M., Ochoa, G., Paechter, B. (eds.) PPSN 2016. LNCS, vol. 9921, pp. 782–791. Springer, Cham (2016). https://doi.org/10.1007/978-3-319-45823-6_73
13. Doerr, B., Doerr, C., Kötzing, T.: Static and self-adjusting mutation strengths for multi-valued decision variables. Algorithmica **80**(5), 1732–1768 (2018)
14. Doerr, B., Doerr, C., Lengler, J.: Self-adjusting mutation rates with provably optimal success rules. Algorithmica **83**(10), 3108–3147 (2021)
15. Doerr, B., Doerr, C., Yang, J.: Optimal parameter choices via precise black-box analysis. Theoret. Comput. Sci. **801**, 1–34 (2020)
16. Doerr, B., Gießen, C., Witt, C., Yang, J.: The $(1 + \lambda)$ evolutionary algorithm with self-adjusting mutation rate. Algorithmica **81**(2), 593–631 (2019)
17. Doerr, B., Jansen, T., Sudholt, D., Winzen, C., Zarges, C.: Optimizing monotone functions can be difficult. In: Schaefer, R., Cotta, C., Kołodziej, J., Rudolph, G. (eds.) PPSN 2010. LNCS, vol. 6238, pp. 42–51. Springer, Heidelberg (2010). https://doi.org/10.1007/978-3-642-15844-5_5
18. Doerr, B., Jansen, T., Sudholt, D., Winzen, C., Zarges, C.: Mutation rate matters even when optimizing monotonic functions. Evol. Comput. **21**(1), 1–27 (2013)
19. Doerr, B., Johannsen, D., Winzen, C.: Multiplicative drift analysis. Algorithmica **64**, 673–697 (2012)
20. Doerr, B., Künnemann, M.: Optimizing linear functions with the $(1+ \lambda)$ evolutionary algorithm-different asymptotic runtimes for different instances. Theoret. Comput. Sci. **561**, 3–23 (2015)
21. Doerr, B., Lissovoi, A., Oliveto, P.S., Warwicker, J.A.: On the runtime analysis of selection hyper-heuristics with adaptive learning periods. In: Genetic and Evolutionary Computation Conference (GECCO), pp. 1015–1022 (2018)
22. Doerr, B., Witt, C., Yang, J.: Runtime analysis for self-adaptive mutation rates. Algorithmica **83**(4), 1012–1053 (2021)
23. Eiben, A.E., Hinterding, R., Michalewicz, Z.: Parameter control in evolutionary algorithms. IEEE Trans. Evol. Comput. **3**, 124–141 (1999)
24. Grimmett, G.R., et al.: Percolation, vol. 321. Springer, Heidelberg (1999). https://doi.org/10.1007/978-3-662-03981-6
25. Hevia Fajardo, M.A., Sudholt, D.: On the choice of the parameter control mechanism in the $(1+(\lambda, \lambda))$ genetic algorithm. In: Genetic and Evolutionary Computation Conference (GECCO), pp. 832–840 (2020)
26. Hevia Fajardo, M.A., Sudholt, D.: Self-adjusting population sizes for non-elitist evolutionary algorithms: why success rates matter. arXiv preprint arXiv:2104.05624 (2021)
27. Hevia Fajardo, M.A., Sudholt, D.: Self-adjusting population sizes for non-elitist evolutionary algorithms: why success rates matter. In: Genetic and Evolutionary Computation Conference (GECCO), pp. 1151–1159 (2021)

28. Jagerskupper, J., Storch, T.: When the plus strategy outperforms the comma strategy and when not. In: 2007 IEEE Symposium on Foundations of Computational Intelligence, pp. 25–32. IEEE (2007)
29. Jansen, T.: On the brittleness of evolutionary algorithms. In: Stephens, C.R., Toussaint, M., Whitley, D., Stadler, P.F. (eds.) FOGA 2007. LNCS, vol. 4436, pp. 54–69. Springer, Heidelberg (2007). https://doi.org/10.1007/978-3-540-73482-6_4
30. Karafotias, G., Hoogendoorn, M., Eiben, Á.E.: Parameter control in evolutionary algorithms: trends and challenges. IEEE Trans. Evol. Comput. **19**(2), 167–187 (2014)
31. Kaufmann, M., Larcher, M., Lengler, J., Zou, X.: OneMax is not the easiest function for fitness improvements (2022). https://arxiv.org/abs/2204.07017
32. Kaufmann, M., Larcher, M., Lengler, J., Zou, X.: Self-adjusting population sizes for the $(1, \lambda)$-EA on monotone functions (2022). https://arxiv.org/abs/2204.00531
33. Kern, S., Müller, S.D., Hansen, N., Büche, D., Ocenasek, J., Koumoutsakos, P.: Learning probability distributions in continuous evolutionary algorithms-a comparative review. Nat. Comput. **3**(1), 77–112 (2004)
34. Kötzing, T.: Concentration of first hitting times under additive drift. Algorithmica **75**(3), 490–506 (2016)
35. Lässig, J., Sudholt, D.: Adaptive population models for offspring populations and parallel evolutionary algorithms. In: Foundations of Genetic Algorithms (FOGA), pp. 181–192 (2011)
36. Lehre, P.K., Witt, C.: Black-box search by unbiased variation. Algorithmica **64**, 623–642 (2012)
37. Lengler, J.: A general dichotomy of evolutionary algorithms on monotone functions. IEEE Trans. Evol. Comput. **24**(6), 995–1009 (2019)
38. Lengler, J.: Drift analysis. In: Theory of Evolutionary Computation. NCS, pp. 89–131. Springer, Cham (2020). https://doi.org/10.1007/978-3-030-29414-4_2
39. Lengler, J., Martinsson, A., Steger, A.: When does hillclimbing fail on monotone functions: an entropy compression argument. In: Analytic Algorithmics and Combinatorics (ANALCO), pp. 94–102. SIAM (2019)
40. Lengler, J., Meier, J.: Large population sizes and crossover help in dynamic environments. In: Bäck, T., et al. (eds.) PPSN 2020. LNCS, vol. 12269, pp. 610–622. Springer, Cham (2020). https://doi.org/10.1007/978-3-030-58112-1_42
41. Lengler, J., Riedi, S.: Runtime analysis of the $(\mu + 1)$-EA on the dynamic BinVal function. In: Zarges, C., Verel, S. (eds.) EvoCOP 2021. LNCS, vol. 12692, pp. 84–99. Springer, Cham (2021). https://doi.org/10.1007/978-3-030-72904-2_6
42. Lengler, J., Schaller, U.: The $(1 + 1)$-EA on noisy linear functions with random positive weights. In: Symposium Series on Computational Intelligence (SSCI), pp. 712–719. IEEE (2018)
43. Lengler, J., Steger, A.: Drift analysis and evolutionary algorithms revisited. Comb. Probab. Comput. **27**(4), 643–666 (2018)
44. Lengler, J., Zou, X.: Exponential slowdown for larger populations: the $(\mu + 1)$-EA on monotone functions. Theoret. Comput. Sci. **875**, 28–51 (2021)
45. Lissovoi, A., Oliveto, P., Warwicker, J.A.: How the duration of the learning period affects the performance of random gradient selection hyper-heuristics. In: AAAI Conference on Artificial Intelligence (AAAI), vol. 34, no. 3, pp. 2376–2383 (2020)
46. Lissovoi, A., Oliveto, P.S., Warwicker, J.A.: On the time complexity of algorithm selection hyper-heuristics for multimodal optimisation. In: AAAI Conference on Artificial Intelligence (AAAI), vol. 33, no. 1, pp. 2322–2329 (2019)

47. Lissovoi, A., Oliveto, P.S., Warwicker, J.A.: Simple hyper-heuristics control the neighbourhood size of randomised local search optimally for LeadingOnes. Evol. Comput. **28**(3), 437–461 (2020)
48. Mambrini, A., Sudholt, D.: Design and analysis of schemes for adapting migration intervals in parallel evolutionary algorithms. Evol. Comput. **23**(4), 559–582 (2015)
49. Rajabi, A., Witt, C.: Evolutionary algorithms with self-adjusting asymmetric mutation. In: Bäck, T., et al. (eds.) PPSN 2020. LNCS, vol. 12269, pp. 664–677. Springer, Cham (2020). https://doi.org/10.1007/978-3-030-58112-1_46
50. Rajabi, A., Witt, C.: Self-adjusting evolutionary algorithms for multimodal optimization. In: Genetic and Evolutionary Computation Conference (GECCO), pp. 1314–1322 (2020)
51. Rechenberg, I.: Evolutionsstrategien. In: Schneider, B., Ranft, U. (eds.) Simulationsmethoden in der Medizin und Biologie, pp. 83–114. Springer, Heidelberg (1978). https://doi.org/10.1007/978-3-642-81283-5_8
52. Rodionova, A., Antonov, K., Buzdalova, A., Doerr, C.: Offspring population size matters when comparing evolutionary algorithms with self-adjusting mutation rates. In: Genetic and Evolutionary Computation Conference (GECCO), pp. 855–863 (2019)
53. Rowe, J.E., Sudholt, D.: The choice of the offspring population size in the $(1, \lambda)$ evolutionary algorithm. Theoret. Comput. Sci. **545**, 20–38 (2014)
54. Schumer, M., Steiglitz, K.: Adaptive step size random search. IEEE Trans. Autom. Control **13**(3), 270–276 (1968)

Theoretical Study of Optimizing Rugged Landscapes with the cGA

Tobias Friedrich[1], Timo Kötzing[1], Frank Neumann[2],
and Aishwarya Radhakrishnan[1(✉)]

[1] Hasso Plattner Institute, University of Potsdam, Potsdam, Germany
{friedrich,timo.koetzing,aishwarya.radhakrishnan}@hpi.de
[2] School of Computer Science, The University of Adelaide, Adelaide, Australia
frank.neumann@adelaide.edu.au

Abstract. Estimation of distribution algorithms (EDAs) provide a distribution-based approach for optimization which adapts its probability distribution during the run of the algorithm. We contribute to the theoretical understanding of EDAs and point out that their distribution approach makes them more suitable to deal with rugged fitness landscapes than classical local search algorithms.

Concretely, we make the ONEMAX function rugged by adding noise to each fitness value. The cGA can nevertheless find solutions with $n(1 - \varepsilon)$ many 1s, even for high variance of noise. In contrast to this, RLS and the (1+1) EA, with high probability, only find solutions with $n(1/2 + o(1))$ many 1s, even for noise with small variance.

Keywords: Estimation-of-distribution algorithms · Compact genetic algorithm · Random local search · Evolutionary algorithms · Run time analysis · Theory

1 Introduction

Local search [1], evolutionary algorithms [9] and other types of search heuristics have found applications in solving classical combinatorial optimization problems as well as challenging real-world optimization problems arising in areas such as mine planning and scheduling [18,20] and renewable energy [19,24].

Local search techniques perform well if the algorithm can achieve improvement through local steps whereas other more complex approaches such as evolutionary algorithms evolving a set of search points deal with potential local optima by diversifying their search and allowing to change the current solutions through operators such as mutation and crossover. Other types of search heuristics, such as estimation of distribution algorithms [21] and ant colony optimization [8], sample their solutions in each iteration from a probability distribution that is adapted based on the experience made during the run of the algorithm. One of the key questions that arises is when to favour distribution-based algorithms over search point-based methods. We will investigate this in the context of rugged landscapes that are obtained by stochastic perturbation.

G. Rudolph et al. (Eds.): PPSN 2022, LNCS 13399, pp. 586–599, 2022.
https://doi.org/10.1007/978-3-031-14721-0_41

Real-world optimization problems are often rugged with many local optima and quantifying and handling rugged landscapes is an important topic when using search heuristics [3,17,22]. Small details about the chosen search point can lead to a rugged fitness landscapes even if the underlying problem has a clear fitness structure which, by itself, would allow local search techniques to find high quality solution very quickly.

In this paper we model a rugged landscape with underlying fitness structure via ONEMAX,[1] where each search point is perturbed by adding an independent sample from some given distribution D. We denote the resulting (random) fitness function OM_D.

Note that this setting of D-rugged ONEMAX is different from noisy ONEMAX (with so-called additive posterior noise) [2,11–13,23] in that evaluating a search point multiple times does not lead to different fitness values (which then could be averaged, implicitly or explicitly) to get a clearer view of the underlying fitness signal. Note that another setting without reevaluation (but on a combinatorial path problem and for an ant colony algorithm, with underlying non-noisy ground truth) was analyzed in [14].

In this paper we consider as distribution the normal distribution as well as the geometric distribution. Since all search points get a sample from the same distribution added, the mean value of the distribution is of no importance (no algorithm we consider makes use of the absolute fitness value, only of relative values). Mostly important is the steepness of the tail, and in this respect the two distributions are very similar.

An important related work, [10] discusses what impact the *shape* of the chosen noise model has on optimization of noisy ONEMAX. They find that steep tails behave very differently from uniform tails, so for this first study we focus on two distributions with steep tails, which we find to behave similarly.

As a first algorithm, which was found to perform very well under noise [2], we consider the *compact genetic algorithm* (cGA), an estimation of distribution algorithm which has been subject to a wide range of studies in the theoretical analysis of estimation of distribution algorithms [5,6,12,16]. See Sect. 2 for an exposition of the algorithm.

In Theorem 4 we show the cGA to be efficient on $\mathcal{N}(0, \sigma^2)$-rugged ONEMAX, even for arbitrarily large values of σ^2 (at the cost of a larger run time bound). Note that, since the optimum is no longer guaranteed to be at the all-1s string (and the global optimum will be just a rather random search point with a lot of 1s, we only consider the time until the cGA reaches $n(1 - \varepsilon)$ many 1s (similar to [23]). The idea of the proof is to show that, with sufficiently high probability, no search point is evaluated twice in a run of the cGA; then the setting is identical to ONEMAX with additive posterior noise and we can cite the corresponding theorem from the literature [12]. Thus, working with a distribution over the search space and adapting it during the search process leads to a less coarse-

---

[1] ONEMAX is the well-studied pseudo-Boolean test function mapping $x \in \{0, 1\}^n$ to $\sum_{i=1}^{n} x_i$.

grained optimization process and presents a way to deal effectively with rugged fitness landscapes.

We contrast this positive result with negative results on search point based methods, namely random local search (RLS, maintaining a single individual and flipping a single bit each iteration, discarding the change if it worsened the fitness) and the so-called (1+1) EA (which operates like RLS, but instead of flipping a single bit, each bit is flipped with probability $1/n$), as well as with Random Search (RS, choosing a uniformly random bit string each iteration).

We first consider random local search on $\mathcal{N}(0, \sigma^2)$-rugged ONEMAX. Theorem 5 shows that, for noise in $\Omega(\sqrt{\log n})$, RLS will not make it further than half way from the random start to the ONEMAX-optimum and instead get stuck in a local optimum. The proof computes, in a rather lengthy technical lemma, the probability that a given search point has higher fitness than all of its neighbors.

Going a bit more into detail about what happens during the search, we consider the geometric distribution. In Theorem 6 we prove that, even for constant variance, with high probability the algorithm is stuck after transitioning to a new search point at most $\log^2(n)$ many times. This means that essentially no progress is made over the initial random search point, even for small noise values! The proof proceeds by showing that successive accepted search points have higher and higher fitness values; in fact, the fitness values grow faster than the number of 1s in the underlying bit string. Since every new search point has to have higher fitness than the previous, it quickly is unfeasible to find even better search points (without going significantly higher in the 1s of the bit string).

In Theorem 7 we translate the result for RLS to the (1+1) EA. We require small but non-constant variance of D to make up for the possibility of the (1+1) EA to jump further, but otherwise get the result with an analogous proof. To round off these findings, we show in Theorem 8 that Random Search has a bound of $O(\sqrt{n \log n})$ for the number of 1s found within a polynomial number of iterations.

In Sect. 7 we give an experimental impression of the negative results. We depict that, within n^2 iterations, the proportion of 1s in the best string found is decreasing in n for all algorithms RLS, (1+1) EA and RS, where RS is significantly better than the (1+1) EA, RLS being the worst.

This paper proceeds with some preliminaries on the technical details regarding the algorithms and problems considered. After the performance analyses of the different algorithms in Sect. 3 through 6 and the experimental evaluation in Sect. 7, we conclude in Sect. 8.

2   Algorithms and Problem Setting

In this section we define the D-rugged ONEMAX problem and describe all the algorithms which we are analyzing in this paper. Random local search (RLS) on a fitness function f is given in Algorithm 1. RLS samples a point uniformly at random and at each step creates an offspring by randomly flipping a bit. At the end of each iteration it retains the best bit string available.

Algorithm 1: RLS on fitness function f

---

1 Choose $x \in \{0,1\}^n$ uniformly at random;
2 **while** *stopping criterion not met* **do**
3 $\quad$ $y \leftarrow$ flip one bit of x chosen uniformly at random;
4 $\quad$ **if** $f(y) \geq f(x)$ **then** $x \leftarrow y$

The (1+1) EA on a fitness function f is given in Algorithm 2. The difference between RLS and (1+1) EA is that (1+1) EA creates an offspring by flipping each bit with probability $1/n$.

Algorithm 2: (1+1) EA on fitness function f

---

1 Choose $x \in \{0,1\}^n$ uniformly at random;
2 **while** *stopping criterion not met* **do**
3 $\quad$ $y \leftarrow$ flip each bit of x independently with probability $1/n$;
4 $\quad$ **if** $f(y) \geq f(x)$ **then** $x \leftarrow y$

The cGA on a fitness function f is given in Algorithm 3. This algorithm starts with two bit strings which have the probability of $1/2$ for each of their bit to be 1. After each step this probability is updated based on the best bit string encountered.

Algorithm 3: The compact GA on fitness function f

---

1 $t \leftarrow 0$, $K \leftarrow$ initialize;
2 $p_{1,t} \leftarrow p_{2,t} \leftarrow \cdots \leftarrow p_{n,t} \leftarrow 1/2$;
3 **while** *termination criterion not met* **do**
4 $\quad$ **for** $i \in \{1, \ldots, n\}$ **do**
5 $\quad\quad$ $x_i \leftarrow 1$ with probability $p_{i,t}$, $x_i \leftarrow 0$ else;
6 $\quad$ **for** $i \in \{1, \ldots, n\}$ **do**
7 $\quad\quad$ $y_i \leftarrow 1$ with probability $p_{i,t}$, $y_i \leftarrow 0$ else;
8 $\quad$ **if** $f(x) < f(y)$ **then** swap x and y for $i \in \{1, \ldots, n\}$ **do**
9 $\quad\quad$ **if** $x_i > y_i$ **then** $p_{i,t+1} \leftarrow p_{i,t} + 1/K$ **if** $x_i < y_i$ **then** $p_{i,t+1} \leftarrow p_{i,t} - 1/K$
$\quad\quad$ **if** $x_i = y_i$ **then** $p_{i,t+1} \leftarrow p_{i,t}$
10 $\quad$ $t \leftarrow t + 1$;

2.1 D-Rugged OneMax

To give a simple model for a rugged landscape with underlying gradient, we use a randomly perturbed version of the well-studied OneMax test function. We fix a dimension $n \in \mathbb{N}$ and a random distribution D. Then we choose, for every $x \in \{0,1\}^n$, a random distortion y_x from the distribution D. We define a D-rugged OneMax function as $\mathrm{OM}_D \colon \{0,1\}^n \to \mathbb{R} := x \mapsto \|x\|_1 + y_x$ where $\|x\|_1 := |\{i \mid x_i = 1\}|$ is the number of 1s in x.

In the following sections we show that the cGA optimizes even very rugged distortions of OM_D efficiently, while RLS will get stuck in a local optimum.

3   Performance of the cGA

Let $D \sim N(0, \sigma^2)$. The following is Lemma 5 from [12], which shows that, while optimizing OM_D, the probability that marginal probabilities falls a constant less than $1/2$ is superpolynomially small in n. Note that here and in all other places in this paper, we give probabilities that range over the random choices of the algorithm as well as the random landscape of the instance.

Lemma 1. *Let $\varepsilon \in (0, 1)$ and define*

$$M_\varepsilon = \left\{ p \in [0.25, 1]^n \;\middle|\; \sum_{i=1}^{n} p_i \leq n(1 - \varepsilon) \right\}.$$

Then

$$\max_{p, q \in M_\varepsilon} \sum_{i=1}^{n} (2p_i q_i - p_i - q_i) \leq -n\varepsilon/2.$$

Lemma 2 ([7,12]). *Let $D \sim N(0, \sigma^2)$. Consider the cGA optimizing OM_D with $\sigma^2 > 0$. Let $0 < a < 1/2$ be an arbitrary constant and $T' = \min\{t \geq 0 : \exists i \in [n], p_{i,t} \leq a\}$. If $K = \omega(\sigma^2 \sqrt{n} \log n)$, then for every polynomial $\mathrm{poly}(n)$, n sufficiently large, $\Pr(T' < \mathrm{poly}(n))$ is superpolynomially small.*

Theorem 3. *Let $\varepsilon \in (0, 1)$ and define*

$$M_\varepsilon = \left\{ p \in [0.25, 1]^n \;\middle|\; \sum_{i=1}^{n} p_i \leq n(1 - \varepsilon) \right\}.$$

Let $S \subseteq \{0, 1\}^n$ be a random multi set of polynomial size (in n), where each member is drawn independently according to some marginal probabilities from M_ε. Then the probability that the multi set contains two identical bit strings is $2^{-\Omega(n\varepsilon)}$.

Proof. Let $x, y \in S$ be given, based on marginal probabilities $p, q \in M_\varepsilon$, respectively. Using Lemma 1 in the last step, we compute

$$P(x = y) = \prod_{i=1}^{n} P(x_i = y_i)$$

$$= \prod_{i=1}^{n} (p_i q_i + (1 - p_i)(1 - q_i))$$

$$= \prod_{i=1}^{n} (1 - p_i - q_i + 2p_i q_i)$$

$$\leq \prod_{i=1}^{n} \exp(-p_i - q_i + 2p_i q_i)$$

$$\leq \exp\left(\sum_{i=1}^{n} -p_i - q_i + 2p_i q_i\right)$$

$$\underset{\text{Lemma 1}}{\leq} \exp(-n\varepsilon/2).$$

Since there are only polynomially many pairs of elements from S, we get the claim by an application of the union bound.

$\square$

From the preceding theorem we can now assume that the cGA always samples previously unseen search points. We can use [12, Lemma 4] which gives an additive drift of $O(\sqrt{n}/(K\sigma^2))$ in our setting.

Now we can put all ingredients together and show the main theorem of this section.

Theorem 4. *Let $D \sim N(0, \sigma^2)$, $\sigma^2 > 0$ and let ε be some constant. Then the cGA with $K = \omega(\sigma^2\sqrt{n}\log n)$ optimizes OM_D up to a fitness of $n(1 - \varepsilon)$ within an expected number of $O(K\sqrt{n}\sigma^2)$ iterations.*

Proof. We let $X_t = \sum_{i=1}^{n} p_{i,t}$ be the sum of all the marginal probabilities at time step t of the cGA. Using Lemma 2, we can assume that (for polynomially many time steps) the cGA has marginal probabilities of at least 0.25. Now we can employ Lemma 3 to see that the cGA does not sample the same search point twice in a polynomial number of steps. Thus, as mentioned, we can use [12, Lemma 4] to get an additive drift of $O(\sqrt{n}/(K\sigma^2))$ as long as X_t does not reach a value of $n(1 - \varepsilon)$.

The maximal value of X_t is n, so we can use an additive drift theorem that allows for overshooting [15, Theorem 3.7] to show the claim. $\square$

4   Performance of RLS

In this section we show that RLS cannot optimize rugged landscapes even for small values of ruggedness. We show that RLS will not find a solution with more than $3n/4$ ones with high probability. This implies that RLS will get stuck in a local optimum with a high probability because there are exponentially many points with number of ones more than $3n/4$ and the probability that none of this points have noise more than the noise associated with the best solution found by RLS is very low.

Theorem 5. *Let $D \sim N(0, \sigma^2)$. Let $\sigma^2 \geq 4\sqrt{2\ln(n + 1)}$. Then there is a constant $c < 1$ such that RLS optimizing OM_D will reach a solution of more than $3n/4$ many 1s (within any finite time) will have a probability of at most c.*

Proof. We consider the event A_0 that RLS successfully reached a fitness of $3n/4$ starting from an individual with at most $n/2$ many 1s.

With probability at least $1/2$ the initial search point of *RLS* has at most $n/2$ many 1s.

We define, for each level $i \in \{n/2, n/2 + 3, n/2 + 6, \ldots, 3n/4 - 3\}$, the first *accepted* individual x_i which RLS found on that level. For the event A_0 to hold, it must be the case that all x_i are *not* local optima. Any search points in the neighborhood of x_i sampled previous to the encounter of x_i will have a value less than x_i (since x_i is accepted) and the decision of whether x_i is a local optimum depends only on the $k < n$ so far not sampled neighbors. Since two different x_i have a Hamming distance of at least 3, these neighbors are disjoint sets (for different i) and their noises are independent.

For any point x to be a local optimum, it needs to have a higher fitness than any of its at most n neighbors. We assume pessimistically that all neighbors have one more 1 in the bit string and compute a lower bound on the probability that the random fitness of x is higher than the random fitness of any neighbor by bounding the probability that a Gaussian random variable is larger than the largest of n Gaussian random variables plus 1. By scaling, this is the probability that some $\mathcal{N}(0,1)$-distributed random variable is higher than the maximum of n independent $\mathcal{N}(1/\sigma^2, 1)$-distributed random variables.

Using the symmetry of the normal distribution, this is equivalent to the probability that some $\mathcal{N}(1/\sigma^2, 1)$-distributed random variable is *less* than the *minimum* of n $\mathcal{N}(0,1)$-distributed random variables. This is exactly the setting of Lemma 1, where we pick $c = 1/\sigma^2$. Plugging in our bound for σ^2, we get a probability of $\Omega(1/n)$ that an arbitrary point in our landscape is a local optimum.

Thus we get with constant probability that one of the $\Theta(n)$ many x_i is a local optimum. With this constant probability, event A_0 cannot occur, as desired. $\square$

4.1   Performance of RLS – A Detailed Look

We now want to give a tighter analysis of RLS on rugged ONEMAX by showing that, in expectation, the noise of new accepted search points is growing. For the analysis, we will switch to a different noise model: We now assume our noise to be $\text{Geo}(p)$-distributed, for some $p \leq 1/2$. We believe that a similar analysis is also possible for normal-distributed noise, but in particular with much more complicated Chernoff bounds.

Theorem 6. *Let $p \leq 1/2$ and let $D \sim \text{Geo}(p)$. Then, for all c, the probability that RLS optimizing D-rugged* ONEMAX *will transition to a better search point more than $\log^2(n)$ times is $O(n^{-c})$.*

In particular, in this case, RLS does not make progress of $\Omega(n)$ over the initial solution and does not find the optimum.

Proof. We consider the run of RLS optimizing D-rugged ONEMAX and denote with X_t the *noise* of the t-th *accepted* search point. We know that $X_0 \sim D$ and each next point has to be larger than at least the previous search point minus

1: in each iteration either a 0 bit or a 1 bit is flipped and RLS accepts the new search point only if its fitness value is greater than or equal to the previous search point.

We will show that X_t is, in expectation, growing. Furthermore we will show that, with high probability, for $t = \log^2(n)$ we have that $X_t \geq \log^2(n)/2$. We finish the proof by showing that, with a search point with such a noise value, it is very unlikely to find a better search point.

Note that, for all t, we have that the distribution of X_{t+1} is the distribution of D conditional on being at least $X_t - 1$ if a 0 was flipped to a 1 and $X_t + 1$ otherwise. Pessimistically assuming the first case and since D is memory-less, we get $X_{t+1} \sim X_t - 1 + D$. In particular, since $E[D] = \frac{1}{p}$, we have $E[X_{t+1}] \geq E[X_t] + 1/p - 1$. Inductively we get

$$E[X_t] \geq E[X_0] + \frac{t}{p} - t. \tag{1}$$

Let the geometric random variable attached with each X_t be D_t and we have $X_t \sim X_{t-1} - 1 + D_t$, therefore $X_t \sim \sum_{i=0}^{t} D_i - t$. This implies $P(X_t \leq (\frac{t+1}{p} - t)/2)$ is nothing but $P(\sum_{i=0}^{t} D_i \leq (\frac{t+1}{p} + t)/2)$. By using Chernoff bounds for the sum of independent geometric random variables [4, Theorem 1.10.32] and by letting $\delta = \frac{1}{2} - \frac{tp}{2(t+1)}$ we have,

$$P\left(X_t \leq \left(\frac{t+1}{p} - t\right)/2\right) = P\left(\sum_{i=0}^{t} D_i \leq \left(\frac{t+1}{p} + t\right)/2\right)$$

$$= P\left(\sum_{i=0}^{t} D_i \leq (1-\delta)\frac{t+1}{p}\right)$$

$$\leq \exp\left(-\frac{\delta^2(t+1)}{2 - \frac{4\delta}{3}}\right).$$

Since $p \leq \frac{1}{2}$, we have $\delta \geq \frac{1}{4}$. Therefore,

$$P\left(X_t \leq \left(\frac{t+1}{p} - t\right)/2\right) \leq \exp\left(-\frac{\delta^2(t+1)}{2 - \frac{4\delta}{3}}\right) \leq \exp\left(-\frac{3t}{80}\right).$$

When $t = \log^2(n)$,

$$P\left(X_t \leq \left(\frac{t+1}{p} - t\right)/2\right) \leq \exp\left(-\frac{3t}{80}\right) = n^{-\frac{3}{80}\log(n)}.$$

Assume that we sampled a search point with noise at least $m = \frac{t+1}{2p} - \frac{t}{2}$, where $t = \log^2(n)$. For a neighbor of the current search point to have higher fitness it should have at least $m - 1$ or $m + 1$ noise, depending on whether it has an extra 1 bit or an extra zero bit. The probability for this to happen is,

$$P(D \geq m + 1) \leq P(D \geq m - 1)$$

$$= p(1-p)^{-2}(1-p)^m$$
$$\leq e^{\frac{1}{2}} n^{-\frac{1}{4}\log(n)}.$$

Using this, we will show that once a search point with noise at least m is sampled, the probability that at least one of the neighbours is of higher fitness is $O\left(n^{1-\frac{\log(n)}{4}}\right)$. Let $D_{m_1}, \ldots, D_{m_n}$ denote the random geometric noise associated with the neighbors of the current search point with at least noise m. Then probability that at least one of the neighbours is of higher fitness is

$$\leq P(D_{m_1} \geq m - 1 \cup \cdots \cup D_{m_n} \geq m - 1)$$
$$\leq \sum_{i=1}^{n} P(D_{m_i} \geq m - 1)$$
$$\leq e^{\frac{1}{2}} n^{1-\frac{\log(n)}{4}}.$$

$\square$

5   Performance of the (1+1) EA

In this section we extend the analysis given for RLS in Theorem 6 to the (1+1) EA.

Theorem 7. *Let $p \leq 1/(2\log(n))$ and let $D \sim \text{Geo}(p)$. Then, for all $c > 0$ and $k \in \mathbb{N}$, the probability that the (1+1) EA optimizing D-rugged ONEMAX will transition to a better search point more than $\log^2(n)$ times within n^k steps is $O(n^{-c})$.*

In particular, the probability that the (1+1) EA makes progress of $\Omega(n)$ over the initial solution within n^k steps is $O(n^{-c})$ and thus does not find the optimum.

Proof. We first show that, for $c > 0$, in any iteration the new accepted search point by (1+1) EA does not have more than $ck\log(n) - 1$ ones than the previous accepted point with probability at least $1 - O(n^{-c})$. Then we assume the worst case scenario that the new search point has $ck\log(n) - 1$ more ones than the previous search point to proceed with the proof similar to Theorem 6 for RLS.

Let X denote the number of bit flips happened to get the current search point. Then $X \sim \text{Bin}(n, 1/n)$ (which has an expectation of 1). If the current search point has $ck\log(n) - 1$ more ones than the previous search point then X has to be at least $ck\log(n) - 1$. By a multiplicative Chernoff bound for the sum of independent Bernoulli trails, if $\delta = ck\log(n) - 2$,

$$P(X \geq ck\log(n) - 1) = P(X \geq 1 + \delta)$$
$$\leq \exp\left(-\frac{(ck\log(n) - 2)^2}{ck\log(n)}\right)$$
$$\leq \frac{e^4}{n^{ck}}.$$

A union bound over n^k iterations gives that the probability that within any of these iterations the (1+1) EA jumps further than $ck \log(n) - 1$ is $O(n^{-c})$.

Consider now the run of the (1+1) EA optimizing D−rugged ONEMAX and let X_t denote the noise associated with the t−th accepted search point. Similar to Theorem 6, we now show the following. (1) X_t grows in expectation. (2) For $t = \log^2(n)$, with high probability $X_t \geq \log^2(n)/2$. (3) If we have a search point with noise greater than $\log^2(n)/2$, the probability to find a better search point within n^k steps is very low. We pessimistically assume that the t-th accepted search point has $\log(n) - 1$ ones more than the previous search point. Since D is memory-less, we get $X_t \sim X_{t-1} - \log(n) + 1 + D$. This inductively along with the fact that $X_0 = D$ implies,

$$E[X_t] \geq \frac{t+1}{p} - t \log(n) + t.$$

Let D_t be the geometric random variable associated with X_t. Then we have $X_t \sim X_{t-1} - \log(n) + 1 + D_t$, which is $X_t \sim \sum_{i=0}^{t} D_i - t \log(n) + t$. If we let $\delta = \frac{1}{2} + \frac{tp}{2(t+1)} - \frac{tp \log(n)}{2(t+1)}$, $a_t = \left(\frac{t+1}{p} - t \log(n) + t \right)/2$ and use Chernoff bounds for the sum of independent geometric random variables [4, Theorem 1.10.32] we have,

$$P(X_t \leq a_t) = P\left(\sum_{i=0}^{t} D_i \leq \frac{t+1}{2p} + \frac{t \log(n)}{2} - \frac{t}{2} \right)$$

$$= P\left(\sum_{i=0}^{t} D_i \leq (1 - \delta)\frac{t+1}{p} \right)$$

$$\leq \exp\left(-\frac{\delta^2(t+1)}{2 - \frac{4\delta}{3}} \right).$$

Since $p \leq \frac{1}{2\log(n)}$, we have $\delta \geq \frac{1}{4}$. Therefore,

$$P\left(X_t \leq \left(\frac{t+1}{p} - t \log(n) + t \right)/2 \right) \leq \exp\left(-\frac{\delta^2(t+1)}{2 - \frac{4\delta}{3}} \right) \leq \exp\left(-\frac{3t}{80} \right).$$

When $t = \log^2(n)$,

$$P\left(X_t \leq \left(\frac{t+1}{p} - t \log(n) + t \right)/2 \right) \leq \exp\left(-\frac{3t}{80} \right)$$

$$= n^{-\frac{3}{80} \log(n)}.$$

Now assume that we sampled a search point with noise at least $m = \frac{t+1}{2p} - \frac{t \log(n)}{2} + \frac{t}{2}$, where $t = \log^2(n)$. As we have seen at any given iteration the probability that the standard mutation operator flips more than $ck \log(n) - 1$ bits is very low, we will again analyze the worst case scenario. For a neighbor of the current search point with at most $ck \log(n) - 1$ to have higher fitness it should have at least $m - ck \log(n)$ noise. The probability for this to happen is,

$$P(D \geq m - ck \log n + 1) = p(1-p)^{m-ck \log(n)} \leq e^{-p(m-ck \log(n))} = e^{\frac{ck}{2}} n^{-\frac{1}{4} \log n}$$

For a given k, let $D_{m_1}, \ldots, D_{m_{n^k}}$ denote the random geometric noise associated with n^k neighbors of the current search point with at least noise m. Then probability that within n^k steps at least one of the neighbours will be of higher fitness is at most

$$P\left(\bigcup_{i=1}^{n}(D_{m_i} \geq m - ck \log(n) + 1)\right) \leq \sum_{i=1}^{n^k} P(D_{m_i} \geq m - ck \log(n) + 1)$$

$$\leq e^{\frac{ck}{2}} n^{k - \frac{\log(n)}{4}}.$$

$\square$

6   Performance of Random Search

For comparison with the performance of RLS and the (1+1) EA, we briefly consider Random Search (RS) in this section. We state the theorem that Random Search has a bound of $O(\sqrt{n \log n})$ for the number of 1s found within a polynomial number of iterations and can be proved by Chernoff bounds.

Theorem 8. *Let $c > 0$ be given and $t \geq 1$. Then the bit string with the most number of 1s found by Random Search within $t \leq n^k$ iterations, choosing a uniformly random bit string each iteration, has at most $n/2 + O(\sqrt{n \log n})$ 1s with probability $1 - O(n^{-c})$.*

7   Experimental Evaluation

In this section we empirically analyze the performance of the cGA, the RLS, the (1+1) EA and the Random Search algorithms on the rugged ONEMAX with two different noise models. We considered noise sampled from the normal distribution with mean zero and variance 5 and another noise model sampled from the geometric distribution with variance 5. From the results we can see that after n^2 iterations, where n is length of the bit string, the RLS and the (1+1) EA does not sample a search point with more than 60% of 1s but the cGA with $K = \sqrt{n} \log(n)$ always finds the optimum. The plot in Fig. 1 is mean of 100 independent runs of each algorithm for bit string lengths 100 to 1000 with step size 100. To have closer look at the performance of the other algorithms, performance of the cGA(straight line at 100) is removed from the plot.

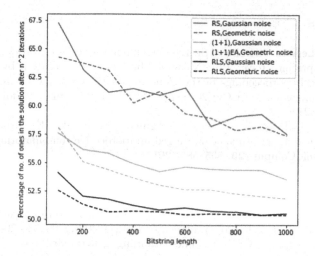

Fig. 1. Percentage of ones in the sampled search point after n^2 iterations of Random Search, (1+1) EA and RLS on optimizing rugged OneMax function with Normal and Geometric noise both having variance of 5. Performance of cGA which is a straight line at 100% is omitted.

Each time the first search point was set to have 50% of 1s. As n increases the percentage of ones in the search point sampled by RLS and (1+1) EA after n^2 iteration tends to 50% and it tends to less than 60% in case of random search.

8   Conclusion

Rugged fitness landscapes appear in many real-world problems and may lead to algorithms getting stuck in local optima. We investigated this problem in this paper for the rugged ONEMAX problem which is obtained through a noisy ONEMAX fitness function where each search point is only evaluated once. We have shown that RLS and the (1+1) EA can only achieve small improvements on an initial solution chosen uniformly at random. In contrast to this the cGA is able to improve solution quality significantly until it is (almost) optimal. Our experimental investigations show this behaviour for realistic input sizes and also point out that RLS and the (1+1) EA perform significantly worse than random search on the rugged ONEMAX problem.

Acknowledgements. This work was supported by the Australian Research Council through grant FT200100536.

References

1. Aarts, E., Lenstra, J.K.: Local Search in Combinatorial Optimization. Wiley (1997)
2. Aishwaryaprajna, Rowe, J.E.: Noisy combinatorial optimisation by evolutionary algorithms. In: Proceedings of the Genetic and Evolutionary Computation Conference Companion, GECCO 2019, pp. 139–140 (2019). https://doi.org/10.1145/3319619.3321955
3. Daolio, F., Liefooghe, A., Verel, S., Aguirre, H., Tanaka, K.: Problem features versus algorithm performance on rugged multiobjective combinatorial fitness landscapes. Evol. Comput. **25**, 555–585 (2017)
4. Doerr, B.: Probabilistic tools for the analysis of randomized optimization heuristics. In: Theory of Evolutionary Computation. NCS, pp. 1–87. Springer, Cham (2020). https://doi.org/10.1007/978-3-030-29414-4_1
5. Doerr, B.: The runtime of the compact genetic algorithm on jump functions. Algorithmica **83**, 3059–3107 (2021). https://doi.org/10.1007/s00453-020-00780-w
6. Doerr, B., Zheng, W.: From understanding genetic drift to a smart-restart parameter-less compact genetic algorithm. In: Coello, C.A.C. (ed.) Genetic and Evolutionary Computation Conference, GECCO 2020, pp. 805–813 (2020). https://doi.org/10.1145/3377930.3390163
7. Doerr, B., Zheng, W.: Sharp bounds for genetic drift in estimation of distribution algorithms. IEEE Trans. Evol. Comput. **24**(6), 1140–1149 (2020). https://doi.org/10.1109/TEVC.2020.2987361
8. Dorigo, M., Stützle, T.: Ant Colony Optimization. MIT Press (2004)
9. Eiben, A.E., Smith, J.E.: Introduction to Evolutionary Computing, 2nd edn. Natural Computing Series. Springer, Heidelberg (2015). https://doi.org/10.1007/978-3-662-44874-8
10. Friedrich, T., Kötzing, T., Krejca, M.S., Sutton, A.M.: Graceful scaling on uniform versus steep-tailed noise. In: Handl, J., Hart, E., Lewis, P.R., López-Ibáñez, M., Ochoa, G., Paechter, B. (eds.) PPSN 2016. LNCS, vol. 9921, pp. 761–770. Springer, Cham (2016). https://doi.org/10.1007/978-3-319-45823-6_71
11. Friedrich, T., Kötzing, T., Krejca, M., Sutton, A.M.: The benefit of sex in noisy evolutionary search (2015)
12. Friedrich, T., Kötzing, T., Krejca, M.S., Sutton, A.M.: The compact genetic algorithm is efficient under extreme gaussian noise. Trans. Evol. Comput. **21**, 477–490 (2017). https://doi.org/10.1109/TEVC.2016.2613739
13. Gießen, C., Kötzing, T.: Robustness of populations in stochastic environments. Algorithmica **75**, 462–489 (2016). https://doi.org/10.1007/s00453-015-0072-0
14. Horoba, C., Sudholt, D.: Ant colony optimization for stochastic shortest path problems. In: Proceedings of the 12th Annual Conference on Genetic and Evolutionary Computation, GECCO 2010, pp. 1465–1472 (2010). https://doi.org/10.1145/1830483.1830750
15. Krejca, M.S.: Theoretical analyses of univariate estimation-of-distribution algorithms. Doctoral thesis, Universität Potsdam (2019). https://doi.org/10.25932/publishup-43487
16. Lengler, J., Sudholt, D., Witt, C.: The complex parameter landscape of the compact genetic algorithm. Algorithmica **83**, 1096–1137 (2021). https://doi.org/10.1007/s00453-020-00778-4
17. Malan, K.M., Engelbrecht, A.P.: Quantifying ruggedness of continuous landscapes using entropy. In: IEEE Congress on Evolutionary Computation, pp. 1440–1447 (2009)

18. Myburgh, C., Deb, K.: Evolutionary algorithms in large-scale open pit mine scheduling. In: GECCO, pp. 1155–1162. ACM (2010)

19. Neshat, M., Alexander, B., Wagner, M.: A hybrid cooperative co-evolution algorithm framework for optimising power take off and placements of wave energy converters. Inf. Sci. **534**, 218–244 (2020)

20. Osada, Y., While, R.L., Barone, L., Michalewicz, Z.: Multi-mine planning using a multi-objective evolutionary algorithm. In: IEEE Congress on Evolutionary Computation, pp. 2902–2909 (2013)

21. Pelikan, M., Hauschild, M., Lobo, F.G.: Estimation of distribution algorithms. In: Kacprzyk, J., Pedrycz, W. (eds.) Handbook of Computational Intelligence, pp. 899–928 (2015)

22. Poursoltan, S., Neumann, F.: Ruggedness quantifying for constrained continuous fitness landscapes. In: Datta, R., Deb, K. (eds.) Evolutionary Constrained Optimization. ISFS, pp. 29–50. Springer, New Delhi (2015). https://doi.org/10.1007/978-81-322-2184-5_2

23. Prugel-Bennett, A., Rowe, J., Shapiro, J.: Run-time analysis of population-based evolutionary algorithm in noisy environments. In: Proceedings of the 2015 8th Conference on Foundations of Genetic Algorithms, FOGA 2015, pp. 69–75 (2015). https://doi.org/10.1145/2725494.2725498

24. Tran, R., Wu, J., Denison, C., Ackling, T., Wagner, M., Neumann, F.: Fast and effective multi-objective optimisation of wind turbine placement. In: GECCO, pp. 1381–1388. ACM (2013)

Towards Fixed-Target Black-Box Complexity Analysis

Dmitry Vinokurov[ID] and Maxim Buzdalov[(✉)][ID]

ITMO University, Saint Petersburg, Russia
mbuzdalov@gmail.com

Abstract. Recently, fine-grained measures of performance of randomized search heuristics received attention in the theoretical community. In particular, some results were proven specifically for fixed-target runtime analysis. However, this research domain still lacks an important counterpart, namely, the (black-box) complexity analysis, which shall augment runtime analyses of particular algorithms with the bounds on what can be achieved with the best possible algorithms.

This paper makes few first steps in this direction. We prove upper and lower bounds on the fixed-target black-box complexity of the standard benchmark function ONEMAX given the problem size n and the target fitness k that we want to achieve. On the way to these bounds, we prove a general lower bound theorem suitable to derive bounds not only in fixed-target settings, but also in settings where a problem instance may have multiple optima.

Keywords: Black-box complexity · Fixed-target analysis · OneMax

1 Introduction

Fixed-target analysis, which measures how much work an evolutionary algorithm shall perform to find a solution with the predefined quality, has been recently put in the context of rigorous runtime analysis [4]. The discipline of fixed-target runtime analysis studies mainly the expected running times needed for a given evolutionary algorithm solving a given optimization problem to obtain a solution with the fitness at least k (assuming maximization). One typically proves theorems that bound these expected running times from above and from below by certain expressions that depend on the target fitness k, as well as on the problem size, the parameters of the algorithm and so on. By comparing these expressions for different algorithms, one may find out which algorithm is better suited for a particular application. By looking at how they change as k increases, one may derive some predictions about the dynamics of the algorithms, including questions of switching from one algorithm to another one when the fitness reaches a certain threshold.

In the classical analysis of evolutionary algorithms, the runtime analysis is complemented by the black-box complexity analysis, which studies how well any

G. Rudolph et al. (Eds.): PPSN 2022, LNCS 13399, pp. 600–611, 2022.
https://doi.org/10.1007/978-3-031-14721-0_42

black-box search algorithm that belongs to a certain class may solve the given problem. A similar duality exists in the classical algorithm analysis. Comparisons between the running times of existing evolutionary algorithms and the corresponding black-box complexities raise and help to answer important questions [15] and also assist in designing new, more efficient algorithms [6].

The kind of black-box complexity analysis that would be a counterpart to the fixed-target runtime analysis is currently missing. In this paper, we aim to initiate the research on the *fixed-target black-box complexity analysis* that is to answer how fast any black-box search algorithm (that belongs to a certain class) may reach the given target fitness value of an optimization problem. Since this is the very first paper on this topic, we limit ourselves with the unrestricted complexity (that is, without any constraints on which algorithms we may consider, e.g. whether the crossover may be used, or whether the selection shall be elitist) and with one particular optimization problem, the famous ONEMAX.

Some of the presented results extend beyond the fixed-target setting. In particular, in Sect. 4 we present the first theorem suitable for proving general lower bounds on black-box complexities without assuming that problem instances and search space elements have a bijection on them. This theorem covers functions with multiple optima, of which the fixed-target setting is a special case.

Other sections of this paper are as follows. Section 2 covers the closest related work, while Sect. 3 introduces some notation, the formal problem setting, important existing results and even the trivial cases of the claimed bounds. Section 5 proves the lower bound on the unrestricted fixed-target black-box complexity of ONEMAX, whereas Sect. 6 proves the corresponding upper bound. While these bounds do not provide an exact asymptotic match for the whole range of targets, they do when the target is large enough.

2 Related Work

Reporting something more than just the average time required to find the optimum of a problem has been well-recognized in the area of benchmarking for quite a long time [10,13]. Apart from the fixed-target analysis, other perspectives include the fixed-budget analysis [7,16,18] and unlimited budget analysis [14]. The particular focus on the fixed-target runtime analysis is motivated and popularized in [4], where many examples are given for when it is easy to derive the fixed-target results based on the already existing proofs of complete running times, and several tools are presented to ease this kind of runtime analysis.

In this paper, we focus on developing the basics for the counterpart of the fixed-target runtime analysis, and it is natural to begin with the famous ONE-MAX problem. Erdős and Rényi prove what appears to be the upper and lower bounds on ONEMAX [12], although under a very different name. In [2], an upper bound has been proven independently, but not without some flaws in proofs. A follow-up paper [8] did the accurate proof, which includes the constant factors that are still the tightest possible ones as of now.

The seminal paper [11] introduces the notion of the black-box complexity, as well as first upper and lower bounds. It also contains the now-classic theorem for proving lower bounds under very few assumptions. The generality of the latter made it hard to use to prove sharper bounds, for which reason one of the subsequent works [5] formulates and proves the so-called matrix theorem, which can be used to prove sharper lower bounds given more information about the properties of the problem being analyzed. To date, these results remain the only ones with a very wide range of applicability, although more black-box complexity results exist for certain specific settings (e.g. [1,9] for unrestricted and elitist complexities of a problem different from ONEMAX).

3   Preliminaries

Basic Notation. We use the notation $[a...b]$ to denote a set of integers $\{a, a + 1, ..., b - 1, b\}$, and we denote the set $[1...n]$ as $[n]$. We use $\log n$ to denote the logarithm of n to base 2. The Hamming distance $\mathcal{H}(x, y)$ between two bit strings x and y of the same size n is the number of bit positions they differ at, that is, $\mathcal{H}(x, y) := |\{i \in [n] \mid x_i \neq y_i\}|$.

The factorial of n is denoted by $n!$ and is equal to $n! = 1 \cdot 2 \cdot ... \cdot n$, where $0! = 1$ for convenience. The binomial coefficient $\binom{n}{k}$ is the value $n!/(k!(n - k)!)$. We denote as $V(n, k)$ the sum of the first $k + 1$ binomial coefficients with the fixed n: $V(n, k) := \sum_{i=0}^{k} \binom{n}{k}$; this is the number of bit strings of length n at a distance of at most k from the given bit string. If $k \leq n/2$, one particularly convenient upper bound is known [3, Lemma 4.7.2]:

$$V(n, k) = \sum_{i=0}^{k} \binom{n}{i} \leq 2^{nH\left(\frac{k}{n}\right)}, \tag{1}$$

where $H(p) = -p \log_2 p - (1 - p) \log_2 (1 - p)$ is the binary entropy function.

Optimization Problem. The object of our research is the ONEMAX problem defined on bit strings of some fixed length n, which is called the *problem size*. We use n for the problem size throughout the entire paper. The problem consists of 2^n problem instances parameterized by a hidden optimum $z \in \{0, 1\}^n$, which are maximization problems with the following objective functions:

$$\mathrm{OM}_z : \{0, 1\}^n \to \mathbb{N}; \quad x \mapsto |\{i \in [n] \mid x_i = z_i\}|.$$

One may also note that $\mathrm{OM}_z(x) = n - \mathcal{H}(x, z)$.

Fixed-Target Problem Setting. When we maximize some function $f(x)$ with the fixed target, we aim at finding a point x^* with fitness equal to, or exceeding, some predefined threshold k, which is called the *target*. Our setting in this paper is as follows. We are given a problem instance OM_z of the known size n, but an unknown hidden optimum z, and an integer target $k \in [0...n]$. Our aim is, by querying search points x_i for $i = 1, 2, ..., t, ...$ and analyzing the answers to these queries $f(x_i)$, to obtain an x^* such that $\mathrm{OM}_z(x^*) \geq k$. Note that if $k = n$, this is the same as finding the maximum of OM_z, but we investigate the general case.

Black-Box Complexity. Given an algorithm $\mathcal{A}$ that generates the search points x_i, its running time on the ONEMAX problem is the maximum over all its instances: $T_{\mathcal{A}}(n,k) = \max_z \min\{i \mid \mathrm{OM}_z(x_i) \geq k\}$. As $\mathcal{A}$ is allowed to be randomized, $T_{\mathcal{A}}(n,k)$ is a random variable, and we are interested in minimizing its expected value $E[T_{\mathcal{A}}(n,k)]$. The *fixed-target black-box complexity* of OM_z of problem size n with respect to the target k is $BBC(n,k) = \min_{\mathcal{A}} E[T_{\mathcal{A}}(n,k)]$.

Since it is very hard, and not actually really needed, to determine the exact value of $BBC(n,k)$, we aim at proving upper and lower bounds on that value that are reasonably close to each other, similarly to what most of the black-box complexity papers do. The upper bound is proven by creating some algorithm $\mathcal{A}$ and proving an upper bound on its running time. Since any reasonable algorithm, including the one we propose, is unbiased in the sense of [17,19], the expression for its running time $T_{\mathcal{A}}(n,k)$ does not depend on the hidden optimum z and can be simplified by choosing an arbitrary z, without loss of generality $z = 1^n$. The lower bound is proven in this paper by specialized information theory arguments similar to those in [5,11,12].

Trivial Case. If $k \leq n/2$, the target is too easy, so we can immediately formulate and prove the following lemma.

Lemma 1. *The fixed-target black-box complexity of* ONEMAX *of size* n *and target* $k \leq n/2$ *is* $1 + 2^{-n} \cdot \sum_{t=0}^{k-1} \binom{n}{t} \leq \frac{3}{2}$.

Proof. The upper bound is as follows. If the first query is done by uniformly choosing a random bitstring, it already hits the target with the probability of $p_1 = 2^{-n} \cdot \sum_{t=k}^{n} \binom{n}{t} \geq \frac{1}{2}$. If it does not, the query to the opposite point, which is performed by flipping all the bits, definitely does. The resulting expected running time is $1 + (1 - p_1)$, which matches the theorem statement.

The lower bound follows from the Yao's minimax principle [21] by choosing a uniform distribution over the problem instances and using a deterministic algorithm that samples the all-zeros point 0^n first, and the all-ones point 1^n second. Its average running time is the same as above by the same argument, and it is easy to see that nothing better is possible. $\square$

For this reason, in the rest of the paper we assume that $k > n/2$.

More Notation. For convenience, we denote $\bar{k} := n - k$, which will make some of our equations shorter and symmetric. The meaning of $\bar{k}$ can be seen as the distance between the easiest target to the true optimum. Obviously, if $k > n/2$ then $\bar{k} < k$. We also frequently use a symbol $\delta := k - n/2$. In our proofs, all the symbols k, $\bar{k}$ and δ can be used with these relations implied. We take some precautions, however, to remind the reader about them.

4   Generic Lower Bounds

We start with a generic theorem useful for proving lower bounds on unrestricted fixed-target black-box complexities. It is based on [11, Theorem 2] and integrates one of the improvements from its extension, the matrix theorem [5]. Our theorem explicitly allows multiple optima for each function instance: more precisely, it takes the view from the other side and considers that each query may be an optimum of several function instances. To apply this theorem to the fixed-target setting, one counts hitting a target as hitting one of the optima of the function being optimized. This theorem is quite complicated, but it allows exploitation of the problem structure to gain better precision.

Theorem 1. *Let S be the finite search space of an optimization problem. Let F be a set of problem instances $\{f : S \to \mathbb{R}\}$, where each problem instance may have more than one optimum, and $\mathcal{O}_f$ be the set of optima of a problem instance f. For each $f \in F$, assume that each call to $f(x)$, $x \in S$, results in:*

- *at least one answer indicating that $x \in \mathcal{O}_f$;*
- *at most $b \geq 2$ answers indicating that $x \notin \mathcal{O}_f$.*

Define a covering subset $S' \subseteq S$ as a subset of the search space that contains at least one optimum of each problem instance, that is, $\forall f \in F$, $\mathcal{O}_f \cap S' \neq \emptyset$. Let $\mathcal{S}$ be the set of all covering subsets.

Let π be an ordering of a covering subset S', such that $\pi(S', i)$ is the i-th element of S' in a sequence. Let $\Pi(S')$ be the set of all such orderings. Then the black-box complexity of this optimization problem is at least:

$$1 + \frac{1}{|F|} \cdot \min_{S' \in \mathcal{S}} \min_{\pi \in \Pi(S')} \sum_{f \in F} \min\{\lceil \log_b i \rceil \mid 1 \leq i \leq |S'|, \pi(S', i) \in \mathcal{O}_f\}.$$

Proof. Similarly to [11, Theorem 2], we reduce our setting to the expected runtime of the best deterministic algorithm following the Yao's minimax principle [21]. Each deterministic algorithm is, in turn, represented as a decision tree, where the nodes correspond to queries to the optimized function, and the edges correspond to different answers. Similar to [5], we consider only edges corresponding to receiving non-optimal answers, as otherwise the algorithm finds an optimum and terminates. Whenever a query results in finding an optimum of some problem instances from F, we write down these instances to the corresponding node.

Let $d(f)$ be the smallest depth of any node (the root has the depth of 1) where the problem instance $f \in F$ has been written down. The average runtime of the algorithm is the average depth of all occurrences of f, that is, $\frac{1}{|F|} \cdot \sum_{f \in F} d(f)$.

Now we need to arrange the queries in such a way that this *total average depth* is not larger than the average depth of the best possible algorithm. This arrangement need not correspond to any particular algorithm, or even to a correct algorithm at all, but it still needs to be a valid arrangement. For this reason, we may only consider the covering subsets of queries to be assigned to nodes of the decision tree, since otherwise some problem instances will remain unsolved. Surely, if a node has a query assigned, then its parent node also has one. However, if a node has a query assigned but some node at a smaller depth does not, we can move that query to the free node. The total average depth will not decrease due to this move. Hence, similarly to [11, Theorem 2], we can limit ourselves to greedy assignments, where the depth of the i-th assigned node is $1 + \lceil \log_b i \rceil$. Here, it is also assumed that each node has the maximum outdegree of b whenever possible, which again does not increase the answer.

Unlike [11, Theorem 2], the order of assigned queries matters, since each query may have more than one associated problem instance. Without any assumptions on how $\mathcal{O}_f$ interact, the only safe way to obtain the minimum total average depth is:

- to iterate over all covering subsets of the search space;
- for each of them to iterate over all possible orderings of its elements,
- to assign these elements to the nodes of the decision tree in a greedy way,
- and finally to compute the total average depth and to update the minimum.

This is precisely what the theorem statement reflects, remembering that the depth of the i-th assigned node is $1 + \lceil \log_b i \rceil$. The only difference is that the leading $1+$ is moved outside the minimum clauses.   □

Theorem 1 is arguably hard to use. The following simplified theorem may instead be applied to trade some precision for the ease of use.

Theorem 2. *Let S be the finite search space of an optimization problem. Let F be a set of problem instances $\{f : S \to \mathbb{R}\}$, where each problem instance may have more than one optimum, and $\mathcal{O}_f$ be the set of optima of a problem instance f. For each $f \in F$, assume that each call to $f(x)$, $x \in S$, results in:*

- *at least one answer indicating that $x \in \mathcal{O}_f$;*
- *at most $b \geq 2$ answers indicating that $x \notin \mathcal{O}_f$.*

Assume each search point is an optimum of at most m different problem instances, that is, $\forall x \in S$, $|\{f \in F \mid x \in \mathcal{O}_f\}| \leq m$, with an obvious restriction that $|S| \cdot m \geq |F|$. The black-box complexity of this optimization problem is at least

$$\left\lfloor \log_b \left(1 + (b-1) \cdot \left\lfloor \frac{|F|}{m} \right\rfloor \right) \right\rfloor - \frac{1}{b-1} \cdot \frac{m}{|F|} \cdot \left\lfloor \frac{|F|}{m} \right\rfloor.$$

Proof. We use the ideas of Theorem 1 while safely assuming that all the sets of problem instances assigned to nodes with depths smaller than the maximum depth do not intersect, and at the maximum depth only at most one node may have less than m problem instances assigned. This way, one can re-use [5, Theorem 6] with the "new" search space size of $\lfloor \frac{|F|}{m} \rfloor$ and obtain the following bound:

$$\left\lfloor \log_b \left(1 + (b-1) \cdot \left\lfloor \frac{|F|}{m} \right\rfloor \right) \right\rfloor - \frac{1}{b-1}.$$

If $|F|$ does not divide evenly by m, one can account for the remainder by taking into account from [5, proof of Theorem 6] that the logarithm rounded down is actually the depth of the remaining node. By assuming D to be this depth, one can see that

$$\frac{(D - \frac{1}{b-1}) \cdot m \cdot \lfloor \frac{|F|}{m} \rfloor + D \cdot (|F| - m \cdot \lfloor \frac{|F|}{m} \rfloor)}{|F|} = D - \frac{1}{b-1} \cdot \frac{m}{|F|} \cdot \left\lfloor \frac{|F|}{m} \right\rfloor.$$

If $|F|$ divides evenly by m, this equation also holds. This proves the theorem. □

5   Lower Bound for OneMax

We now prove the lower bound on the unrestricted fixed-target black-box complexity of ONEMAX. We will use Theorem 2 together with one particular feature of ONEMAX. When the target is k, and the answer to one of the queries is $\overline{k} = n - k$ or smaller, we may immediately hit the target in the next query by just inverting all the bits. For the sake of lower bounds, this one query may be simply neglected in our bounds. As a result, we can obtain a much sharper bound by considering only fitness values in $(\overline{k}, k)$ as non-terminating ones.

We prove two versions of the lower bound, the sharper one (Theorem 3) and the one in the closed form (Theorem 4).

Theorem 3. *The unrestricted fixed-target black-box complexity of* ONEMAX *with problem size n and target $k = n/2 + \delta$, where $\delta \geq 2$, is at least:*

$$BBC(n, k) \geq \left\lfloor \log_{2\delta-2} \left(1 + (2\delta - 3) \left\lfloor \frac{2^n}{V(n, n-k)} \right\rfloor \right) \right\rfloor - \frac{1}{2\delta - 3}.$$

Proof. The requirements of Theorem 2 are the number of possible problem instances $|F| = 2^n$, the number of non-terminating answers $b = (k - 1) - (n - k + 1) = 2k - n - 2 = 2\delta - 2$ and the upper limit m on the number of problem instances covered by one point of the search space. The latter is equal to $V(n, \overline{k})$ for all points, so the bound itself can be written as follows:

$$BBC(n, k) = \left\lfloor \log_{2\delta-2} \left(1 + (2\delta - 3) \left\lfloor \frac{2^n}{V(n, \overline{k})} \right\rfloor \right) \right\rfloor - \frac{1}{2\delta - 3} \frac{V(n, \overline{k})}{2^n} \left\lfloor \frac{2^n}{V(n, \overline{k})} \right\rfloor$$

$$\geq \left\lfloor \log_{2\delta-2} \left(1 + (2\delta - 3) \left\lfloor \frac{2^n}{V(n, \overline{k})} \right\rfloor \right) \right\rfloor - \frac{1}{2\delta - 3},$$

which proves the theorem.   □

Theorem 4. *The unrestricted fixed-target black-box complexity of* ONEMAX *with problem size n and target $k = n/2 + \delta$, where $\delta \geq 2$, is at least:*

$$BBC(n, k) \geq \frac{\left(\frac{n}{2} + \delta\right) \log_2 \left(1 + \frac{2\delta}{n}\right) + \left(\frac{n}{2} - \delta\right) \log_2 \left(1 - \frac{2\delta}{n}\right)}{\log_2(2\delta - 2)} - 2.$$

Proof. We start with the result of Theorem 3. Since $\delta \geq 2$, we simplify the argument of the logarithm as follows:

$$1 + (2\delta - 3) \left\lfloor \frac{2^n}{V(n, \overline{k})} \right\rfloor \geq 1 + \left\lfloor \frac{2^n}{V(n, \overline{k})} \right\rfloor \geq 1 + \left(\frac{2^n}{V(n, \overline{k})} - 1 \right) = \frac{2^n}{V(n, \overline{k})}.$$

We now use the upper bound on $V(n, \overline{k})$ given in Eq. (1):

$$V(n, \overline{k}) \leq 2^{-n\left(\frac{k}{n} \log_2 \frac{k}{n} + \frac{\overline{k}}{n} \log_2 \frac{\overline{k}}{n}\right)} = 2^{n \log_2 n - k \log_2 k - \overline{k} \log_2 \overline{k}},$$

which allows to simplify further

$$\log_{2\delta-2}\frac{2^n}{V(n,\overline{k})} \geq \frac{k\log_2 k + \overline{k}\log_2\overline{k} - n\log_2 n + n}{\log_2(2\delta - 2)}$$

$$= \frac{\left(\frac{n}{2}+\delta\right)\log_2\left(1+\frac{2\delta}{n}\right)+\left(\frac{n}{2}-\delta\right)\log_2\left(1-\frac{2\delta}{n}\right)}{\log_2(2\delta-2)}.$$

By applying also inequalities $\lfloor x\rfloor \geq x - 1$ and $2\delta - 3 \geq 1$ to the result of Theorem 3, we obtain the required bound. $\qquad\square$

To understand the asymptotic behaviour of the bound above, we consider two cases, $\delta = o(n)$ and $\delta = \Theta(n)$. For the first one, we use the Taylor series $\ln(1+x) = x - x^2/2 + O(x^3)$ and rewrite the bound as:

$$BBC(n,k) \geq \frac{\left(\frac{n}{2}+\delta\right)\log_2\left(1+\frac{2\delta}{n}\right)+\left(\frac{n}{2}-\delta\right)\log_2\left(1-\frac{2\delta}{n}\right)}{\log_2(2\delta-2)} - 2$$

$$= \frac{\left(\frac{n}{2}+\delta\right)\ln\left(1+\frac{2\delta}{n}\right)+\left(\frac{n}{2}-\delta\right)\ln\left(1-\frac{2\delta}{n}\right)}{\ln(2\delta-2)} - 2$$

$$= \frac{\left(\frac{n}{2}+\delta\right)\cdot\left(\frac{2\delta}{n}-\frac{2\delta^2}{n^2}+O\left(\left(\frac{\delta}{n}\right)^3\right)\right)+\left(\frac{n}{2}-\delta\right)\cdot\left(-\frac{2\delta}{n}-\frac{2\delta^2}{n^2}-O\left(\left(\frac{\delta}{n}\right)^3\right)\right)}{\ln(2\delta-2)} - 2$$

$$\geq \frac{\frac{2\delta^2}{n}\pm O\left(\left(\frac{\delta^3}{n^3}\right)\right)}{\ln(2\delta-2)} - 2 = \frac{(2-O(\frac{\delta}{n}))\cdot\delta^2}{n\ln(2\delta-2)} - 2.$$

This results in positive lower bounds for $\delta \geq C_1\sqrt{n\ln n}$ with a large enough constant C_1. If $\delta = \Theta(n)$, the bound amounts to $\Theta(n/\log n)$, where the leading constant is influenced by the particular δ, e.g. if $\delta = C_2 n$ for a constant C_2, the result would be

$$\frac{n}{\log_2 n}\cdot\left(\left(\frac{1}{2}+C_2\right)\log_2(1+2C_2)+\left(\frac{1}{2}-C_2\right)\log_2(1-2C_2)\right) - 2.$$

Note that the result remains correct even if $C_2 = 1/2$, since the second addend tends to zero (e.g. $0\log_2 0$ is taken to be zero, which matches the usage pattern common to applications of the binary entropy function in information theory). In this respect, it preserves compatibility with the existing results on the black-box complexity of complete optimization.

6   Upper Bound for OneMax

For an upper bound, we employ the trick similar to the one used in [5] to prove refined bounds for the JUMP functions. Namely, we reduce the problem of hitting the target k in a ONEMAX problem of size n to finding the optimum of ONEMAX of size $\overline{k}$ on a subset of its bits.

Theorem 5. *The unrestricted fixed-target black-box complexity of* ONEMAX *with problem size n and target $k = n/2 + \delta$, where $\delta \geq 1$, is at most*

$$\frac{(4 + o_\delta(1))\delta}{\log_2(2\delta)} + \Theta_n(\sqrt{n}),$$

where the asymptotic notation is with respect to the variables given in subscripts.

Proof. We employ the following algorithm:

- We find a search point x_1 with fitness $\lceil n/2 \rceil$ by repeated random sampling.
- Next, we find a search point x_2 with the same fitness and $\mathcal{H}(x_1, x_2) = 2\bar{k}$, by repeated random sampling of points at the specified Hamming distance.
- Finally, we fix the bits differing between x_1 and x_2 and find the optimum of ONEMAX defined on the remaining bits, paying attention that the original function returns the value that is by $\bar{k}$ larger.

The first step requires $2^n / \binom{n}{\lceil n/2 \rceil} = \Theta_n(\sqrt{n})$ queries in expectation. Similarly, the second step needs $\Theta_{\bar{k}}(\sqrt{\bar{k}}) = O_n(\sqrt{n})$ queries in expectation. The final step needs $\frac{(2 + o_\delta(1)) \cdot (2\delta)}{\log_2(2\delta)}$ steps using the algorithm from [8]. The optimum returned by that algorithm corresponds to the string with the fitness of exactly k, which hits the target. □

We can see that while our lower bound has an order of $\Theta(\frac{\delta^2}{n \log \delta})$, our upper bound is rather $\Theta(\frac{\delta}{\log \delta})$. Hence, as long as $\delta = \Theta(n)$, the bounds match asymptotically, however, they actually differ by a factor of $\Theta(\frac{\delta}{n})$, which becomes significant if the target is very close to $n/2$.

To obtain a clearer picture, in Fig. 1 we provide the plots of the bounds for a fixed problem size $n = 10^3$ and varying targets. Here, we plot the bounds obtained from the respective theorems, as well as the slightly sharper lower bound given by Theorem 3 that does not undergo simplifications. We also plot the version of the upper bound that takes care of occasional hitting the target while searching for the point with the fitness of $n/2$, which has a visible impact for small targets. From Fig. 1 one can see that the simpler version of the lower bound does not lose much compared to the more sophisticated one: the differences become significant when the bounds approach the value of 1. However, the gap between the lower and the upper bounds is noticeable: the ratio of the bounds reaches approximately 50 at $k \approx 2^5$. This comes at least partially from the asymptotic ratio of the bounds, and can be also influenced by the additional $\Theta(\sqrt{n})$ that is present in the upper bound but not in the lower bound.

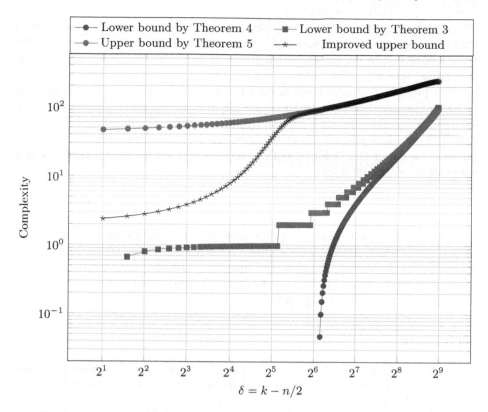

Fig. 1. Visual comparison of the obtained bounds for $n = 10^3$. Asymptotically lower terms, which are positive and are present only in upper bounds, are evaluated exactly for the purpose of these plots.

7   Conclusion

We have presented the first results that can be classified as the unrestricted fixed-target black-box complexity bounds for the ONEMAX problem. Our lower bound is $\Theta(\frac{\delta^2}{n \log \delta})$ and our upper bound is $\Theta(\frac{\delta}{\log \delta})$, both terms simplified, where n is the problem size and $\delta = k - n/2$ is the difference between the target fitness k and half the problem size. These bounds agree well with the existing (complete optimization) black-box complexity of ONEMAX, however start to diverge slightly for smaller targets. In particular, the lower bound is non-trivial only for $\delta = \Omega(\sqrt{n \log n})$, and the upper bound is obviously loose for $\delta = O(\sqrt{n})$, for which the true bound is already constant.

It is unclear as of now which of these bounds is closer to the true fixed-target black-box complexity of ONEMAX. While we admit that the lower bound may be too loose due to optimistic assumptions, we still do not know whether the algorithm provided for the upper bound exploits the problem features, including the requirement to reach the fitness of only at least k, strongly enough. The alternative strategy we tried to consider, and which might be more fruitful, is to query search points only from some "good" subset of points, such as those that form a sphere covering of the search space. It seems to be much more difficult to analyse, though.

The presented generic lower bound theorem may also cover problems well outside of the fixed-target settings, including multimodal problems and probably even multiobjective problems (in a different way from, e.g., linear multiobjective drift [20]). It may possibly be generalized similarly to the matrix theorem [5], however we admit that a lot of work is necessary to derive such a theorem in a convenient and usable form.

The fixed-target black-box complexity analysis may also be conducted for limited algorithm classes, such as mutation-only algorithms, fixed-arity algorithms and elitist algorithms. Although everyone knows that complexity analysis is hard, we nevertheless encourage the community to undertake this direction: as the fixed-target setting is inherently more practice-oriented than searching for the optimal solution, the obtained insights—and the way they are different from what is recommended for complete optimization—may appear to be useful for solving hard problems in practice.

References

1. Afshani, P., Agrawal, M., Doerr, B., Doerr, C., Larsen, K.G., Mehlhorn, K.: The query complexity of a permutation-based variant of Mastermind. Discret. Appl. Math. **260**, 28–50 (2019)
2. Anil, G., Wiegand, R.P.: Black-box search by elimination of fitness functions. In: Proceedings of Foundations of Genetic Algorithms, pp. 67–78 (2009)
3. Ash, R.B.: Information Theory. Dover Publications (1990)
4. Buzdalov, M., Doerr, B., Doerr, C., Vinokurov, D.: Fixed-target runtime analysis. Algorithmica **84**(6), 1762–1793 (2022)
5. Buzdalov, M., Doerr, B., Kever, M.: The unrestricted black-box complexity of jump functions. Evol. Comput. **24**(4), 719–744 (2016)
6. Doerr, B., Doerr, C., Ebel, F.: From black-box complexity to designing new genetic algorithms. Theoret. Comput. Sci. **567**, 87–104 (2015)
7. Doerr, B., Jansen, T., Witt, C., Zarges, C.: A method to derive fixed budget results from expected optimisation times. In: Proceedings of Genetic and Evolutionary Computation Conference, pp. 1581–1588 (2013)
8. Doerr, B., Johannsen, D., Kötzing, T., Lehre, P.K., Wagner, M., Winzen, C.: Faster black-box algorithms through higher arity operators. In: Proceedings of Foundations of Genetic Algorithms, pp. 163–172 (2011)
9. Doerr, C., Lengler, J.: The (1+1) elitist black-box complexity of LeadingOnes. Algorithmica **80**(5), 1579–1603 (2018)
10. Doerr, C., Ye, F., Horesh, N., Wang, H., Shir, O.M., Bäck, T.: Benchmarking discrete optimization heuristics with IOHprofiler. Appl. Soft Comput. **88**, 106027 (2020)
11. Droste, S., Jansen, T., Wegener, I.: Upper and lower bounds for randomized search heuristics in black-box optimization. Theor. Comput. Syst. **39**(4), 525–544 (2006)
12. Erdős, P., Rényi, A.: On two problems of information theory. Magyar Tudományos Akadémia Matematikai Kutató Intézet Közleményei **8**, 229–243 (1963)

13. Hansen, N., Auger, A., Ros, R., Mersmann, O., Tusar, T., Brockhoff, D.: COCO: a platform for comparing continuous optimizers in a black-box setting. Optim. Meth. Softw. **36**(1), 114–144 (2021)
14. He, J., Jansen, T., Zarges, C.: Unlimited budget analysis. In: Proceedings of Genetic and Evolutionary Computation Conference Companion, pp. 427–428 (2019)
15. Jansen, T., Wegener, I.: The analysis of evolutionary algorithms–a proof that crossover really can help. Algorithmica **34**, 47–66 (2002)
16. Jansen, T., Zarges, C.: Performance analysis of randomised search heuristics operating with a fixed budget. Theoret. Comput. Sci. **545**, 39–58 (2014)
17. Lehre, P.K., Witt, C.: Black-box search by unbiased variation. Algorithmica **64**, 623–642 (2012)
18. Lengler, J., Spooner, N.: Fixed budget performance of the (1+1) EA on linear functions. In: Foundations of Genetic Algorithms XIII, pp. 52–61 (2015)
19. Rowe, J., Vose, M.: Unbiased black box search algorithms. In: Proceedings of Genetic and Evolutionary Computation Conference, pp. 2035–2042 (2011)
20. Rowe, J.E.: Linear multi-objective drift analysis. Theoret. Comput. Sci. **736**, 25–40 (2018)
21. Yao, A.C.C.: Probabilistic computations: toward a unified measure of complexity. In: 18th Annual Symposium on Foundations of Computer Science, pp. 222–227 (1977)

Two-Dimensional Drift Analysis:
Optimizing Two Functions Simultaneously Can Be Hard

Duri Janett and Johannes Lengler[✉]

Department of Computer Science, ETH Zürich, Zürich, Switzerland
johannes.lengler@inf.ethz.ch

Abstract. In this paper we show how to use drift analysis in the case of two random variables X_1, X_2, when the drift is approximatively given by $A \cdot (X_1, X_2)^T$ for a matrix A. The non-trivial case is that X_1 and X_2 impede each other's progress, and we give a full characterization of this case. As application, we develop and analyze a minimal example TwoLinof a dynamic environment that can be hard. The environment consists of two linear function f_1 and f_2 with positive weights 1 and n, and in each generation selection is based on one of them at random. They only differ in the set of positions that have weight 1 and n. We show that the $(1 + 1)$-EAwith mutation rate χ/n is efficient for small constant χ on TwoLin, but does not find the shared optimum in polynomial time for large constant χ.

1 Introduction

1.1 The Application: TwoLin

Randomized Search Heuristics (RSHs) like Evolutionary Algorithms (EAs), are general-purpose optimizers that are applied in a wide range of applications. To apply them more efficiently, it is important to understand failure modes that are specific to some EAs or some parameter settings, so that one can avoid employing an algorithm in inadequate situations. This line of research started with the seminal work of Doerr et al. [7], in which they showed that the $(1 + 1)$-EA with mutation rate χ/n is inefficient on some monotone functions if $\chi > 16$, while it is efficient on all strictly monotone functions if $\chi < 1$. Note that the transition happens for constant χ, which is the most reasonable parameter regime. In particular, the failure mode is not related to the trivial problems that occur for extremely large mutation rates, $\chi \gg \log n$, where the algorithm fails to produce neighbours in Hamming distance one [31]. Subsequently these results on how the mutation rate and related parameters affect optimiziation of monotone functions have been refined [2, 19, 23] and extended to a large collection of other EAs [17, 25]. To highlight just one result, the $(\mu + 1)$-EA with standard mutation rate $1/n$ fails on some monotone functions if the population size μ is too large [25]. Other algorithm-specific failure modes include

---

Extended Abstract. All proofs and further details are available on arxiv [11].

G. Rudolph et al. (Eds.): PPSN 2022, LNCS 13399, pp. 612–625, 2022.
https://doi.org/10.1007/978-3-031-14721-0_43

(i) non-elitist selection strategies with too small offspring reproductive rate (e.g., comma strategies with small offspring population size) [1,5,6,10,16];

(ii) elitist algorithms (and also some non-elitist strategies) in certain landscapes with deceptive local optima [3,4];

(iii) the self-adjusting $(1, \lambda)$-EA if the target success probability is too large [9, 13,14];

(iv) Min-Max Ant Systems with too few ants [26]; probably the compact Genetic Algorithm and the Univariate Marginal Distribution Algorithms have a similar failure mode for too large step sizes [24,30].

It is important to note that all aforementioned failure modes are specific to the algorithm, not to the problem. Of course, there are many problems which are intrinsically hard, and where algorithms fail due to the hardness of the problem. However, in the above examples there is a large variety of other RSHs which can solve the problems easily. Except for (ii), the failure modes above even happen on the OneMax problem, which is traditionally the easiest benchmark for RSHs. Since failure modes can occur even in simple situations, it is important to understand them in order to help practitioners avoiding them.

Unfortunately, some other benchmarks for failure modes are rather technical, in particular in the context of monotone functions. Recently, it was discovered that the same failure modes as for monotone functions could be observed by studying certain *dynamic environments*, more concretely *Dynamic Linear Functions* and the *Dynamic Binary Value* function DynBV [20–22]. These environments are very simple, so they allow to study failure modes in greater detail. Crucially, failure in such environments is due to the algorithms, not the problems: the environments all fall within a general class of problems introduced by Jansen [12] and called *partially-ordered EA* (PO-EA) by Colin, Doerr and Ferey [2], for which the $(1 + 1)$-EA with mutation rate χ/n is known to have optimization time $O(n \log n)$ for $\chi < 1$, and $O(n^{3/2})$ for $\chi = 1$.[1] Hence, they are not intrinsically hard. Both dynamic environments define a set of linear functions with positive weights, and redraw the fitness function in each generation from this set.

A potential counterargument against these dynamic environments is that the set of fitness functions is very large, for dynamic linear functions even infinite. Thus, during optimization the algorithm may never encounter the same environment twice. In applications, it seems more reasonable that the setup switches between a small set of different environments. Such as a chess engine which is trained against several, but not arbitrarily varying number of opponents, or a robot which is trained in a few training environments. Thus, in this paper we propose a *minimal example* of a dynamic environment, TwoLin, in which EAs may exhibit failure modes. For $0 \le \ell \le 1$, we define two functions via

---

[1] The statement for $\chi = 1$ is contained in [12], but the proof was wrong. It was later proven in [2].

$$f_1(x) := f_1^\ell(x) := \sum\nolimits_{i=1}^{\lfloor \ell n \rfloor} nx_i + \sum\nolimits_{i=\lfloor \ell n \rfloor+1}^{n} nx_i,$$

$$f_2(x) := f_2^\ell(x) := \sum\nolimits_{i=1}^{\lfloor \ell n \rfloor} nx_i + \sum\nolimits_{i=\lfloor \ell n \rfloor+1}^{n} nx_i. \tag{1}$$

Then for $0 \le \rho \le 1$, TwoLin$^{\rho,\ell}$ is the probability distribution over $\{f_1^\ell, f_2^\ell\}$ that chooses f_1 with probability ρ and f_2 with probability $1-\rho$. In each generation t, a random function $f^t \in \{f_1, f_2\}$ is chosen according to TwoLin$^{\rho,\ell}$, and selection of the next population is based on this fitness function f^t. Note that, similar to monotone functions, f_1 and f_2 share the global optimum at $(1 \ldots 1)$ (which is crucial for benchmarks in dynamic optimization), have no local optima, and flipping a zero-bit into a one-bit always increases the fitness. This is why it falls into the framework of PO-EA and is hence optimized in time $O(n \log n)$ by the $(1+1)$-EA with mutation rate c/n, for any constant $c < 1$.

Results. We show that even for the simple setting of TwoLin$^{\rho,\ell}$, the $(1+1)$-EA has a failure mode for too large mutation rates χ/n. For all constant values $\rho, \ell \in (0,1)$, we show that for sufficiently small χ the algorithm finds the optimum of TwoLin$^{\rho,\ell}$ in time $O(n \log n)$ if started with $o(n)$ zero-bits, but it takes superpolynomial time for large values of χ. For the symmetric case $\rho = \ell = .5$ the treshold between the two regimes is at $\chi_0 \approx 2.557$, which is only slightly larger than the best known thresholds for the $(1+1)$-EA on monotone functions ($\chi_0 \approx 2.13$ [17,23]) and for general Dynamic Linear Functions and DynBV ($\chi_0 \approx 1.59$ [20–22]). Thus, we successfully identify a minimal example in which the same failure mode of the $(1+1)$-EA shows as for monotone functions and for the general dynamic settings, and it shows almost as early as in those settings.

It remains open whether the positive result also holds when the algorithm starts with $\Omega(n)$ zero-bits. However, we provide an interesting *Domination Lemma* that sheds some light on this question. We call $\Delta_{1,0}$ the drift conditional on flipping one zero-bit in the left part of the bit string and no zero-bit in the right part, and conversely for $\Delta_{0,1}$. Those two terms dominate the drift close to the optimum. We prove that (throughout the search space, not just close to the optimum) whenever *both* $\Delta_{1,0}$ and $\Delta_{0,1}$ are positive, then the total drift is also positive. This is enough to remove the starting condition for the symmetric case $\rho = \ell = 1/2$. However, in general the drift close to the optimum is a weighted sum of those two terms, which may be positive without both terms individually being positive.

1.2 The Method: Two-Dimensional Multiplicative Drift Analysis

Although we believe that the subject and results of the paper are well-motivated, the main contribution lies in the method. The main workhorse in the runtime analysis of evolutionary algorithms is drift analysis. For the simple symmetric case $p = 1/2$ and $\ell = 1/2$, we will see that it suffices to track the number Z of zero-bits, which leads to a fairly standard application of common drift theorems, in particular of multiplicative and negative drift.

However, the situation changes completely when we turn to the cases $\rho \neq 1/2$ and $\ell \neq 1/2$. Here we don't have symmetry, and the drift is no longer a function of Z. This is inherent to the problem. The state of the algorithm is insufficiently characterized by a single quantity. Instead, a natural characterization of the state needs to specify *two* quantities: the number of zero-bits in the left and right part of the string, denoted by X_L and X_R, respectively. Then we obtain a multiplicative drift *in two dimensions*, i.e., there is a 2×2-matrix A such that the drift of the column vector $X = (X_L, X_R)$ is approximatively $A \cdot X$. This can also be called *linear* drift, but it is traditionally called multiplicative drift in the one-dimensional case. Crucially, X_L and X_R contribute *negatively* to each other, i.e., the non-diagonal entries of A are negative.[2] In this case, there are values for X for which the distance from the optimum increases in expectation.

The main contributions of the paper is that we give a general solution for two-dimensional processes that follow such a multiplicative drift. Translated to the algorithm, our result shows that the $(1+1)$-EA is efficient *if and only if the matrix A has a positive eigenvalue*, except for the threshold cases. The formal statement is in Theorem 2. The only restriction that we make is that the two random variables X_L, X_R *individually* have positive drift (towards the optimum), but impose a negative drift *on each other*. This is the non-trivial case. For example, if X_L has negative drift by itself, then it is not hard to see that the optimum will not be reached efficiently from a situation where X_L is much larger than X_R. On the other hand, if X_L and X_R both have positive drift individually, and X_L contributes positively to the drift of X_R, then it is easy to argue that X_R goes to zero regardless of X_L, and afterwards X_L also goes to zero. Thus the hard case is covered and settled by Theorem 2.

Our setting has some similarity with the breakthrough result by Jonathan Rowe on linear multi-objective drift analysis [28], one of the most underrated papers of the field in the last years. Our positive result overlaps with the result of [28].[3] In fact, [28] is more general in the sense that it gives a sufficient criterion for fast convergence in arbitrary constant dimension, also in terms of the eigenvalues and eigenvectors of matrix A. (It gives a very nice collection of examples, too!) However, it is unclear whether the criterion in [28] is also sufficient in general, while our result contains matching positive and negative results. Moreover, while [28] assumes that the drift is *exactly* given by $A \cdot X$ (or lower bounded by that), we show our result even if the drift is only *approximately* given by $A \cdot X$. This extension is non-trivial, and indeed the largest portion of the proof goes into showing that the result is robust under such error terms. The vast majority of applications have minor order error terms in the drift (and so do the applications in this paper), so we believe that this is quite valuable.

---

[2] We follow the convention that we always call drift towards the optimum *positive*, and drift away from the optimum *negative*. Since the optimum in our case is at 0, this means that we consider the difference $X^t - X^{t+1}$ for the drift, not vice versa.

[3] For direct comparison it is important to note that [28] works with the matrix $I - A$ instead of A, where I is the identity matrix.

2   Preliminaries and Definitions

Throughout the paper, $\chi > 0$ and $\rho, \ell \in [0,1]$ are constants, independent of n, and all Landau notation is with respect to $n \to \infty$. We say that an event $\mathcal{E}_n$ holds *with high probability* or *whp* if $\Pr[\mathcal{E}_n] \to 1$ for $n \to \infty$. The environment TwoLin$^{\rho,\ell}$ is the probability distribution on $\{f_1^\ell, f_2^\ell\}$, which assigns probability ρ and $1 - \rho$ to f_1^ℓ and f_2^ℓ, respectively, where f_1^ℓ and f_2^ℓ are given by (1). The *left* and *right* part of a string $x \in \{0,1\}^n$, denoted by x_L and x_R, refers to the first $\lfloor \ell n \rfloor$ bits of x and to the remainder of the string, respectively. We will consider the $(1 + 1)$-EAwith mutation rate χ/n for maximization on $\mathcal{D} :=$ TwoLin$^{\rho,\ell}$, which is given in Algorithm 1. The *runtime* of an algorithm refers

---

Algorithm 1. The $(1 + 1)$-EA with mutation rate χ/n in environment $\mathcal{D}$.

Sample x^0 from $\{0,1\}^n$ uniformly at random (or start with pre-specified x^0).
for $t = 0, 1, 2, 3, \ldots$ **do**
    Draw f^t from $\mathcal{D}$.
    Create y^t by flipping each bit of x^t independently with probability χ/n.
    Set $x^{t+1} = \arg\max\{f^t(x) \mid x \in \{x^t, y^t\}\}$.

---

to the number of function evaluations before the algorithm evaluates the shared global maximum of TwoLin$^{\rho,\ell}$ for the first time. For typesetting reasons we will write column vectors in horizontal form in inline text, e.g. (X_L, X_R). We denote by $\|x\| := \max_i\{|x_i|\}$ the ∞-norm of a vector x, and similarly for matrices.

3   Two-Dimensional Multiplicative Drift

This section contains our main result in terms of presenting a tool, Theorem 2 below. We first give a lemma which states some basic facts about the matrices that we are interested in. All vectors in this section are column vectors.

Lemma 1. *Let $a, d > 0$ and $b, c < 0$, and consider the real 2×2-matrix $A = \begin{pmatrix} a & b \\ c & d \end{pmatrix} \in \mathbb{R}^{2\times2}$. Then the equation*

$$c\gamma^2 + d\gamma = a\gamma + b \tag{2}$$

has a unique positive root $\gamma_0 > 0$. The vector $e_1 := (\gamma_0, 1)$ is an eigenvector of A for the eigenvalue $\lambda_1 := c\gamma_0 + d$. The other eigenvalue is $\lambda_2 := a - c\gamma_0$ and has eigenvector $e_2 := (b, -c\gamma_0)$. The vector e_1 has two positive real entries, while the vector e_2 has a positive and a negative entry. Moreover,

- *if $a\gamma_0 + b > 0$, or equivalently $c\gamma_0 + d > 0$, then $\lambda_2 > \lambda_1 > 0$;*
- *if $a\gamma_0 + b < 0$, or equivalently $c\gamma_0 + d < 0$, then $\lambda_2 > 0 > \lambda_1$.*

The values $\gamma_0, \lambda_1, \lambda_2$ and the entries of e_1 and e_2 are smooth in a, b, c, d.

Proof. The roots of Eq. (2) are the roots of the quadratic polynomial $c\gamma^2 + (d - a)\gamma - b$. Its discriminant $(d - a)^2 + 4bc > 0$ is positive, so it has two real roots. Since the product of the roots is $-b/c < 0$ by Vieta's rule, they must have different signs. This proves existence and uniqueness of γ_0.

For the eigenvalues and eigenvectors, we check:

$$Ae_1 = \begin{pmatrix} a\gamma_0 + b \\ c\gamma_0 + d \end{pmatrix} \overset{(2)}{=} \begin{pmatrix} \gamma_0(c\gamma_0 + d) \\ c\gamma_0 + d \end{pmatrix} = \lambda_1 e_1, \tag{3}$$

and

$$Ae_2 = \begin{pmatrix} ab - bc\gamma_0 \\ bc - cd\gamma_0 \end{pmatrix} \overset{(2)}{=} \begin{pmatrix} b(a - c\gamma_0) \\ c(c\gamma_0^2 - a\gamma_0) \end{pmatrix} = \lambda_2 e_2. \tag{4}$$

Recalling $a, d > 0$ and $b, c < 0$, it is trivial to see that $\lambda_2 > 0$ in all cases, e_1 has two positive entries, and e_2 has a positive and a negative entry. Since $\lambda_1 = c\gamma_0 + d$, it has the same sign as $a\gamma_0 + b$ by (2). Finally, by (2), $\gamma_0\lambda_1 = a\gamma_0 + b < a\gamma_0 < a\gamma_0 - c\gamma_0^2 = \gamma_0\lambda_2$, which implies $\lambda_1 < \lambda_2$.

For smoothness, it suffices to observe that γ_0 can be written in the form $x + \sqrt{y}$, where x, y depend smoothly on the parameters of (2), and $y > 0$. Hence, γ_0 is smooth, and the remaining values depend smoothly on γ_0 and a, b, c, d. □

The following theorem is our main result. It is concerned with the situation that we have two-dimensional state vectors, and the drift in state $x \in \mathbb{R}^2$ is approximatively given by Ax. There are a few complications that we need to deal with. Firstly, in our application the drift does not look *exactly* like that, but only approximatively, up to $(1 \pm o(1))$ factors. Second, even in the positive case, in our application we will only compute the drift if the number of one-bits is $o(n)$, since the computations would get much more complicated otherwise. These complications are reflected in the theorem.

In fact, these complications are rather typical. At least in the negative case $a\gamma_0 + b < 0$, it is impossible for *any* random process on a finite domain that the drift is given *exactly* by Ax everywhere. This is the same as in one dimension: it is impossible to have a negative drift throughout the whole of a finite domain; it can hold in a subset of the domain, but not everywhere.

Theorem 2 (Two-Dimensional Multiplicative Drift). *Assume that for each $n \in \mathbb{N}$ we have a two-dimensional real Markov chain $(X^t)_{t\geq 0} = (X^t(n))_{t\geq 0}$ on $D \subseteq [0, n] \times [0, n]$, i.e., $X^t = (X_1^t, X_2^t)$ as column vector, where $X_1^t, X_2^t \in [0, n]$. Assume further that there is a (constant) real 2×2-matrix $A = \begin{pmatrix} a & b \\ c & d \end{pmatrix} \in \mathbb{R}^{2\times 2}$ with $a, d > 0$ and $b, c < 0$, and that there are constants $\kappa, r > 0$ and a function $\sigma = \sigma(n) = \omega(\sqrt{\log n / n})$ such that X^t satisfies the following conditions.*

A. *Two-dimensional linear drift. For all $t \geq 0$ and all x with $\|x\| \leq \sigma n$, the drift at $X^t = x$ is $(1 \pm o(1))A \cdot x/n$, by which we mean*

$$\mathbb{E}\left[X^t - X^{t+1} \mid X^t = x\right] = \begin{pmatrix} (1 \pm o(1))a\frac{x_1}{n} + (1 \pm o(1))b\frac{x_2}{n} \\ (1 \pm o(1))c\frac{x_1}{n} + (1 \pm o(1))d\frac{x_2}{n} \end{pmatrix}, \tag{5}$$

where the $o(1)$ terms are uniform over all t and x.

B. Tail bound on step size. *For all $i \geq 1$, $t \geq 0$, and for all $x = (x_1, x_2) \in D$,*

$$\Pr[\|X^t - X^{t+1}\| \geq i \mid X^t = x] \leq \frac{\kappa}{(1+r)^i}. \tag{6}$$

Let γ_0 be the unique positive root of Eq. (2), and let T be the hitting time of $(0,0)$, i.e., the first point in time when $X^T = (0,0)$.

(a) If $a\gamma_0 + b > 0$ and $\|X^0\| = o(\sigma n)$, then $T = O(n \log n)$ with high probability.
(b) If $a\gamma_0 + b < 0$ and $\|X^0\| \geq \sigma n$, then $T = e^{\Omega(\sigma^2 n)}$ with high probability.

Remark 3. While we have included some necessary complications in the statement of Theorem 2, we have otherwise sacrificed generality to increase readability. Firstly, we restrict ourselves to the case that the drift is $(1 \pm o(1))A \cdot x/n$. The scaling factor $1/n$ is quite typical for applications in EAs, but the machinery would work for other factors as well. Secondly, in many applications including ours, the factors x_1/n and x_2/n reflect the probability of having any change at all, and *conditional on changing* the drift is of order $\Theta(1)$. In this situation, the Negative Drift Theorem used in the proof can be replaced by stronger versions (see [29, Section 2.2]), and we can replace the condition $\sigma = \omega(\sqrt{\log n/n})$ by the weaker condition $\sigma = \omega(1)$ for both parts (a) and (b). Moreover, part (b) then holds with a stronger bound of $e^{\Omega(\sigma n)}$ under the weaker condition $\|X^0\| = \omega(1)$. Finally, the step size condition B can be replaced by other conditions [15, 18].

Proof (of Theorem 2). (a). By Lemma 1, the matrix A has two positive eigenvalues λ_1, λ_2 with $\lambda_2 > \lambda_1 > 0$ with eigenvectors e_1 and e_2, where e_1 has two positive entries, while e_2 has a positive and a negative entry. In other words, the 2×2-matrix U whose columns are given by e_1 and e_2 satisfies $U^{-1}AU = \begin{pmatrix} \lambda_1 & 0 \\ 0 & \lambda_2 \end{pmatrix}$. Moreover, the eigenvalues and eigenvectors depend smoothly on a, b, c, d. Hence, changing a, b, c, d by some additive term β changes the eigenvalues and eigenvectors by $O(\beta)$ if β is small.[4] In particular, writing $\beta = \delta/n$, for every $\varepsilon > 0$ there exists $\delta > 0$ such that every matrix $\tilde{A}$ with $\|A - \tilde{A}\| < \delta/n$ is invertible and has two different positive eigenvalues $\tilde{\lambda}_1, \tilde{\lambda}_2$ with eigenvectors $\tilde{e}_1, \tilde{e}_2$ respectively, such that

i) $|\lambda_1 - \tilde{\lambda}_1| < \varepsilon/n$ and $|\lambda_2 - \tilde{\lambda}_2| < \varepsilon/n$ and $\tilde{\lambda}_2 > \tilde{\lambda}_1$.
ii) $\|e_1 - \tilde{e}_1\| < \varepsilon/n$ and $\|e_2 - \tilde{e}_2\| < \varepsilon/n$.
iii) $\|U^{-1} - \tilde{U}^{-1}\| < \varepsilon/(n \max\{\|U\|, \|\tilde{U}\|\})$, where $\tilde{U}$ is the matrix with columns $\tilde{e}_1$ and $\tilde{e}_2$.

For the last point, note that $\|\tilde{U}\|$ is uniformly bounded by an absolute constant (depending on a, b, c, d) if δ/n is sufficiently small.

---

[4] This is because for any matrix M of norm 1 we can write $\lambda_1(A + \beta M) = \lambda_1(A) + \beta D\lambda_1(A) \cdot M + O(\beta^2)$, where the total differential $D\lambda_1(A)$ has bounded norm, and analogously for the other eigenvalues and eigenvectors.

Let $x \in \mathbb{R}^2$, and let $\eta := (\eta_1, \eta_2) := U^{-1}x$ and $\tilde{\eta} := (\tilde{\eta}_1, \tilde{\eta}_2) := \tilde{U}^{-1}x$. In other words, we write x in the basis $\{e_1, e_2\}$ by decomposing $x = \eta_1 e_1 + \eta_2 e_2$, and analogously for the basis $\{\tilde{e}_1, \tilde{e}_2\}$. Then we claim that iii) implies

$$(1 - \tfrac{\varepsilon}{n})\|\eta\| \le \|\tilde{\eta}\| \le (1 + \tfrac{\varepsilon}{n})\|\eta\|. \tag{7}$$

To check this, first note that for the identity matrix $I \in \mathbb{R}^{2 \times 2}$,

$$\|\tilde{U}^{-1}U - I\| \le \|U\| \cdot \|\tilde{U}^{-1} - U^{-1}\| \le \|U\| \cdot \tfrac{\varepsilon}{n\|U\|} = \tfrac{\varepsilon}{n}. \tag{8}$$

For any vector v, this imples

$$\|\tilde{U}^{-1}Uv - v\| = \|(\tilde{U}^{-1}U - I)v\| \overset{(8)}{\le} \tfrac{\varepsilon}{n}\|v\|. \tag{9}$$

With $\tilde{\eta} = \tilde{U}^{-1}x = \tilde{U}^{-1}U\eta$, this implies that

$$\|\tilde{\eta} - \eta\| \le \tfrac{\varepsilon}{n}\|\eta\|, \tag{10}$$

and the right hand side of (7) follows from $\|\tilde{\eta}\| \le \|\eta\| + \|\tilde{\eta} - \eta\|$. Reversing the roles of η and $\tilde{\eta}$, we also have $\|\eta\| \le (1 + \varepsilon/n)\|\tilde{\eta}\|$, and multiplying with $(1 - \varepsilon/n)$ yields $(1 - \varepsilon/n)\|\eta\| \le (1 - (\varepsilon/n)^2)\|\tilde{\eta}\| \le \|\tilde{\eta}\|$, which is the left hand side of (7). Finally, we note for later reference that by an analogous computation, (8) and (9) also hold with U and $\tilde{U}$ reversed, so for all vectors v,

$$\|U^{-1}\tilde{U}v - v\| \le \tfrac{\varepsilon}{n}\|v\|. \tag{11}$$

Now we choose $\varepsilon > 0$ so small that $(1 - (\lambda_1 - \varepsilon)/n)(1 + \varepsilon/n)^2 \le 1 - \lambda_1/(2n)$ for all $n \in \mathbb{N}$, and a corresponding $\delta > 0$. We consider the potential function $f : [0, n] \times [0, n] \to \mathbb{R}_0^+; f(x) := \|\eta\| = \|U^{-1}x\|$. In other words, the potential of a search point x is the norm of the corresponding vector η in basis $\{e_1, e_2\}$. Let x be a search point with $\|x\| < \sigma n$, and assume that n is so large that all $o(1)$ terms are at most δ. Then by Condition A there is some matrix $\tilde{A}$ with $\|A - \tilde{A}\| < \delta$ such that the drift at x is $\tilde{A}x/n$. Using the same notation as above, in particular $\tilde{\lambda}_1, \tilde{\lambda}_2, \tilde{e}_1, \tilde{e}_2$ for the eigenvalues and eigenvectors of $\tilde{A}$, we can rewrite this as

$$\begin{aligned}
\mathbb{E}\left[X^{t+1} \mid X^t = x\right] &= x - \tilde{A}x/n = (I - \tilde{A}/n) \cdot (\tilde{\eta}_1 \tilde{e}_1 + \tilde{\eta}_2 \tilde{e}_2) \\
&= \left(1 - \tfrac{\tilde{\lambda}_1}{n}\right)\tilde{\eta}_1 \tilde{e}_1 + \left(1 - \tfrac{\tilde{\lambda}_2}{n}\right)\tilde{\eta}_2 \tilde{e}_2.
\end{aligned} \tag{12}$$

Recall that $\tilde{\lambda}_2 > \tilde{\lambda}_1 > \lambda_1 - \varepsilon$ (the latter by i)) and that the potential is given by $f(x) = \|\eta\|$. Moreover, the function $x \mapsto \eta = U^{-1}x$ is a linear function, so it commutes with expectations. Thus, the expected potential of X^{t+1} is

$$\begin{aligned}
\mathbb{E}\left[f(X^{t+1}) \mid X^t = x\right] &= \left\|U^{-1}\left(\left(1 - \tfrac{\tilde{\lambda}_1}{n}\right)\tilde{\eta}_1 \tilde{e}_1 + \left(1 - \tfrac{\tilde{\lambda}_2}{n}\right)\tilde{\eta}_2 \tilde{e}_2\right)\right\| \\
&= \left\|U^{-1}\tilde{U}\left(\left(1 - \tfrac{\tilde{\lambda}_1}{n}\right)\tilde{\eta}_1, \left(1 - \tfrac{\tilde{\lambda}_2}{n}\right)\tilde{\eta}_1\right)\right\| \\
&\overset{(11)}{\le} (1 + \tfrac{\varepsilon}{n})(1 - \tfrac{\lambda_1 - \varepsilon}{n})\|\tilde{\eta}\| \\
&\overset{(7)}{\le} (1 + \tfrac{\varepsilon}{n})^2(1 - \tfrac{\lambda_1 - \varepsilon}{n})\|\eta\| \le (1 - \tfrac{\lambda_1}{2n})f(X^t).
\end{aligned} \tag{13}$$

Hence, the potential has multiplicative drift towards zero, as long as $\|X^t\| \leq \sigma n$. Since $f(X^t) = \Theta(\|X^t\|)$, there is a constant $\nu > 0$ such that $f(X^t) \leq \nu\sigma n$ implies $\|X^t\| \leq \sigma n$. Thus (13) is applicable whenever $f(X^t) \leq \nu\sigma n$. In particular, in the interval $[\nu\sigma n/2, \nu\sigma n]$, the potential has a downwards drift of order at least $\Theta(\sigma)$. The starting potential is below this interval as $f(X^0) = \Theta(\|X^0\|) = o(\sigma n)$, so by the Negative Drift Theorem [27, Theorem 2], with high probability the potential does not reach the upper boundary of this interval for at least $e^{\Omega(\sigma^2 n)} = \omega(n \log n)$ steps. Thus with high probability the random process remains in the region where (13) holds, and by the Multiplicative Drift Theorem [8,18], with high probability it reaches the optimum in $O(n \log n)$ steps.

(b). We only give a sketch. Here we use a different potential function. As before, for a vector x, let $\eta = (\eta_1, \eta_2)$ be the corresponding vector in basis $\{e_1, e_2\}$. Then we define the potential of x as $f(x) := \eta_1$. To convince ourselves this choice makes sense, one first shows that

1. $f(x) > 0$ for all $x \in [0, n]^2 \setminus \{(0, 0)\}$;
2. $f((0, 0)) = 0$;
3. there is a constant $\kappa > 0$ such that $f(x) \geq \kappa\|x\|$ for all $x \in [0, n]^2 \setminus \{(0, 0)\}$.

For statements 1 and 3 to hold, it is crucial that we restrict x to the first quadrant, i.e., that x does not have negative entries. It is sufficient to consider this range because the random variable X^t is restricted to it. We omit the proof for space reasons.

Then by a similar calculation as in (a), one can show

$$\mathbb{E}\left[f(X^{t+1}) \mid X^t = x\right] \geq (1 + \tfrac{|\lambda_1|}{2n})f(X^t). \tag{14}$$

Thus we have a drift away from the optimum. The proof is then completed by a standard application of the Negative Drift Theorem [27, Theorem 2], which we omit.   □

3.1   Interpretation and Generalization

We give some intuition on how to interpret the equations occuring in Lemma 1 and Theorem 2 in the language of TwoLin. The following approach leads to the same equations in the case of multiplicative (linear) drift, but it is possible to apply it also in situations where the drift is not linear.

We are interested in the case that the two random variables X_1 and X_2 each have positive drift by themselves, but influence each other negatively. As outlined in the introduction, other cases are easier and thus less interesting. This means that when X_1 is much larger than X_2, then the drift coming from X_1 dominates the drift coming from X_2. In particular, X_1 has a drift towards zero in this case, while X_2 has a drift away from zero. This means in particular that X_1 and X_2 should approach each other. On the other hand, if X_2 is much larger than X_1, then the situation is reversed: X_1 increases in expectation, and X_2 decreases. But again, X_1 and X_2 approach each other. That means that, as the ratio X_1/X_2 decreases, the drift of X_1/X_2 changes from negative to positive.

This suggests the following approach. Let $\gamma := X_1/X_2$. We search for a value of γ that is self-stabilizing. To this end, we consider the random variable $Y := X_1 - \gamma X_2$, compute the drift of Y and then choose the value γ_0 for which the drift of Y at $Y = 0$ is zero. For multiplicative drift, this leads precisely to Eq. (2). By the considerations above, the drift of Y is decreasing in Y. Hence, if the drift at $Y = 0$ is zero, then the drift for positive Y points towards zero, and the drift for negative Y *also* points towards zero.

It is possible to turn this into an analysis as follows. We divide the optimization process in two phases. In the first phase, we only consider $Y = X_1 - \gamma_0 X_2$. We show that Y approaches 0 quickly (with the Multiplicative Drift Theorem, in case of multiplicative drift), and stays very close to zero for a long time (with the Negative Drift Theorem). In the second phase, we analyze the random variable $X := X_1 + X_2$. However, since we know that Y is close to zero, we do not need X to have positive drift in the whole search space; it suffices that it has positive drift in the subspace where $X_1 = \gamma_0 X_2$. This is a massive restriction, and in the linear case the drift is just reduced to $C \cdot X$ for some constant C under this restriction. (Actually, $C = a\gamma_0 + b$.) Thus we can show convergence of X with standard drift arguments.

While this approach is rather cumbersome, it works in large generality. We do not need that the drift is linear. Essentially, we just need that the drift of X_1 is increasing in X_1 and decreasing in X_2, and vice versa for X_2. This already guarantees a solution γ_0 (which may then depend on X) for the condition that the drift of $Y := X_1 - \gamma X_2$ at $Y = 0$ is zero, and it guarantees that Y then has a drift pushing it towards zero from both directions. Thus there is no fundamental obstacle for the machinery described above, though the details may become quite technical. Iterating this argument, it might even be possible to apply the process to dimensions larger than two, see the full version for more details [11].

4   The $(1+1)$-EA on $\text{TwoLin}^{\rho,\ell}$

In this section, we analyze the runtime of the $(1+1)$-EA on $\text{TwoLin}^{\rho,\ell}$. The main contribution of this section is Theorem 6. This is an application of the methods developed in the previous section. In Sect. 4.1 below, we will give more precise results for the special case $\rho = \ell = .5$, as well as the Domination Lemma that holds for arbitrary ρ and ℓ. For readability, we will assume in the following that $\ell n \in \mathbb{N}$. We denote by X_L^t and X_R^t the number of zero bits in the left and right part of the t-th search point x^t, respectively, and let $X^t := (X_L^t, X_R^t)$. It only remains to compute the two-dimensional drift of X^t, for which we give the following definition.

Definition 4. Let $\chi > 0$ and $\rho, \ell \in [0, 1]$. Let $A = \begin{pmatrix} a & b \\ c & d \end{pmatrix} \in \mathbb{R}^{2 \times 2}$, where

$$a = \rho\chi e^{-\ell\chi} + (1-\rho)\chi e^{-\chi}, \qquad b = -(1-\rho)\ell\chi^2 e^{-(1-\ell)\chi},$$
$$c = -\rho(1-\ell)\chi^2 e^{-\ell\chi}, \qquad d = (1-\rho)\chi e^{-(1-\ell)\chi} + \rho\chi e^{-\chi}. \tag{15}$$

The reason for the definition of a, b, c, d and A is the following proposition, which states that the two-dimensional drift is then given by $(1 \pm o(1))Ax/n$.

Proposition 5. *Let* $\chi > 0$ *and* $\rho, \ell \in [0, 1]$. *Let* $x_L \in [\ell n]$ *and* $x_R \in [(1 - \ell)n]$, *and* $x = (x_L, x_R)$ *as column vector. If* $x_L + x_R = o(n)$, *then*

$$\mathbb{E}[X^t - X^{t+1} \mid X^t = x] = \frac{1}{n} \begin{pmatrix} (1 \pm o(1)) \cdot a & (1 \pm o(1)) \cdot b \\ (1 \pm o(1)) \cdot c & (1 \pm o(1)) \cdot d \end{pmatrix} \cdot x =: \frac{1 \pm o(1)}{n} Ax,$$

where a, b, c, d *and* A *are given by Definition 4.*

The proof of Proposition 5 consists mostly of rather standard calculations, and we omit them due to space restrictions. Next we give our main result for the $(1+1)$-EA on TWOLIN$^{\rho, \ell}$, which now follows easily from Theorem 2. We note that, as explained in Remark 3, the bound $n^{1/2 + \Omega(1)}$ in the second case is not tight and could be strengthened.

Theorem 6. *Let* $\chi > 0$, $\rho, \ell \in [0, 1]$, *and consider the* $(1 + 1)$-EA *with mutation rate* χ/n *on* TWOLIN$^{\rho, \ell}$. *Let* a, b, c, d *as in Definition 4, and let* γ_0 *be the unique root of* $c\gamma^2 + d\gamma = a\gamma + b$ *as in (2). Let* T *be the hitting time of the optimum.*

(a) *Assume that* $a\gamma_0 + b > 0$ *and that the* $(1+1)$-EA *is started with* $o(n)$ *zero-bits. Then* $T = O(n \log n)$ *with high probability.*

(b) *Assume that* $a\gamma_0 + b < 0$ *and that the* $(1 + 1)$-EA *is started with* $n^{1/2 + \Omega(1)}$ *zero-bits. Then* T *is superpolynomial with high probability.*

Proof (of Theorem 6). We need to check that Theorem 2 is applicable. For (a), assume that the algorithm starts with $x = o(n)$ zero-bits. We choose a function $\sigma = \sigma(n)$ such that $x = o(\sigma n) = o(n)$. Moreover, we require $\sigma = \omega(\sqrt{\log n/n})$. For concreteness, we may set $\sigma := \max\{n^{-1/4}, \sqrt{x/n}\}$. Then condition A (two-dimensional linear drift) holds by Proposition 5. Condition B (tail bound on step size) holds since the number of bit flips per generation satisfies such a tail bound for any constant χ. Thus the claim of (a) follows. For (b), we choose $\sigma := \sqrt{x}n^{-3/4} = n^{-1/2 + \Omega(1)}$. Since $x \leq \sqrt{x}n^{1/2} = \omega(\sigma n)$, Theorem 2 applies and gives that with high probability $T = e^{-n^{\Omega(1)}}$. $\qquad\square$

Although it is not difficult to write down an explicit formula for γ_0 and for the expression $a\gamma_0 + b$, the formula is so complicated that we refrain from giving it here. We suspect that for all $\rho, \ell \in (0, 1)$ there is a threshold χ_0 such that $a\gamma_0 + b > 0$ for all $\chi < \chi_0$ and $a\gamma_0 + b < 0$ for all $\chi > \chi_0$, but we couldn't deduce it easily from the explicit formula. We will show this in Sect. 4.1 for the symmetric case $\rho = \ell = .5$. For the general case, we instead only give the following, slightly weaker corollary. We omit the proof due to space restrictions.

Corollary 7. *Consider the setting of Theorem 6. For all* $\rho, \ell \in (0, 1)$ *there are* $\chi_1, \chi_2 > 0$ *such that* $a\gamma_0 + b > 0$ *for all* $\chi < \chi_1$ *and* $a\gamma_0 + b < 0$ *for all* $\chi > \chi_2$.

4.1   Domination and the Symmetric Case $\rho = \ell = 1/2$

In this section, we will give the Domination Lemma 10, and we use it to study the symmetric case $\rho = \ell = 1/2$ in more detail. We will show two things for the $(1+1)$-EA on TwoLin$^{.5,.5}$ beyond the general statement in Theorem 6: for the positive result, we remove the condition that the algorithm must start with $o(n)$ zero-bits; and we show that there is a threshold $\chi_0 \approx 2.557$ such that the algorithm is efficient for $\chi < \chi_0$ and inefficient for $\chi > \chi_0$. We start by inspecting the threshold condition $a\gamma_0 + b > 0$.

Lemma 8. *Let a, b, c, d be as in Definition 4. For $\rho = \ell = 1/2$, we have*

$$a = d = \tfrac{\chi}{2}\left(e^{-\chi/2} + e^{-\chi}\right), \qquad b = c = -\tfrac{\chi^2}{4}e^{-\chi/2}, \qquad \gamma_0 = 1. \qquad (16)$$

Let $\chi_0 \approx 2.557$ be the unique positive root of $2 - \chi + 2e^{-\chi/2} = 0$. Then $a\gamma_0 + b > 0$ for $\chi < \chi_0$ and $a\gamma_0 + b < 0$ for $\chi > \chi_0$.

Proof. The formulas for a, b, c, d are simply obtained by plugging $\rho = \ell = 1/2$ into (15). Since $a = d$, the defining Eq. (2) for γ_0 simplifies to $c\gamma_0^2 = b$, which implies $\gamma_0 = 1$. For the critical expression $a\gamma_0 + b$ we obtain

$$a\gamma_0 + b = \tfrac{\chi}{2}\left(e^{-\chi/2} + e^{-\chi}\right) - \tfrac{\chi^2}{4}e^{-\chi/2} = \tfrac{\chi e^{-\chi/2}}{4}\left(2 - \chi + 2e^{-\chi/2}\right). \qquad (17)$$

The expression in the bracket is decreasing in χ, so it is positive for $\chi < \chi_0$ and negative for $\chi > \chi_0$. □

Now we are ready to prove a stronger version of Theorem 6 for the case $\rho = \ell = .5$. The main difference is that the threshold is explicit and that we may assume that the algorithm starts with an arbitrary search point. Finally, the results also hold in expectation.

Theorem 9. *Let $\chi > 0$, $\rho = \ell = .5$, and consider the $(1+1)$-EA with mutation rate χ/n on TwoLin$^{.5,.5}$, with uniformly random starting point. Let $\chi_0 \approx 2.557$ be the unique root of $2 - \chi + 2e^{-\chi/2} = 0$. Let T be the hitting time of the optimum.*

(a) If $\chi < \chi_0$, then $T = O(n \log n)$ in expectation and with high probability.
(b) If $\chi > \chi_0$, then T is superpolynomial in expectation and with high probability.

Proof. We only describe the key steps, without proving them. The negative statement (b) follows immediately from Theorem 6 and Lemma 8. Note that if the runtime is large whp, then it is also large in expectation. For (a), we consider the one-dimensional potential $Y^t = X_L^t + X_R^t$. Let $\mathcal{E}_{i,j}$ be the event of flipping i zero-bits in the left part, and j zero-bits in the right part, and let $\Delta_{i,j}(\boldsymbol{x})$ be the drift conditional on $\mathcal{E}_{i,j}$ and on $X^t = \boldsymbol{x}$. Then it can be shown that for all $\boldsymbol{x}$,

$$\Delta_{1,0}(\boldsymbol{x}) \geq (1 - o(1))\tfrac{e^{-\chi}}{4}(2 - \chi + 2e^{-\chi/2}),$$
$$\Delta_{0,1}(\boldsymbol{x}) \geq (1 - o(1))\tfrac{e^{-\chi}}{4}(2 - \chi + 2e^{-\chi/2}). \qquad (18)$$

By the Domination Lemma below, this implies that all other $\Delta_{i,j}$ are also positive (except for $\Delta_{0,0} = 0$), so the total drift is at least $\Pr[\mathcal{E}_{0,1}]\Delta_{0,1} + \Pr[\mathcal{E}_{0,1}]\Delta_{0,1}$, which is easily seen to be $\Omega(Y^t/n)$. The result then follows from the Multiplicative Drift Theorem [18]. $\qquad\qquad\qquad\qquad\qquad\qquad\qquad\qquad\qquad\qquad\qquad\qquad\qquad\square$

It remains to show the Domination Lemma. Interestingly, this lemma holds in larger generality, for arbitrary ℓ and ρ. However, this does not suffice to generalize Theorem 9 to arbitrary ρ and ℓ. The point where it breaks is that it is generally not true that $\Delta_{0,1}$ and $\Delta_{1,0}$ are both positive if Δ_1 is positive. The proof is purely algebraic, but also long and involved, and we omit it here.

Lemma 10 (Domination Lemma). *Let $\rho, \ell \in [0,1]$, $x_L \in [\ell n]$ and $x_R \in [(1-\ell)n]$. With the notation from above, if $\Delta_{0,1}(x_L, x_R) > 0$ and $\Delta_{1,0}(x_L, x_R) > 0$ then $\Delta_{i,j}(x_L, x_R) > 0$ for all $i, j \geq 0$ with $i + j > 0$.*

References

1. Antipov, D., Doerr, B., Yang, Q.: The efficiency threshold for the offspring population size of the (μ, λ) EA. In: Genetic and Evolutionary Computation Conference (GECCO), pp. 1461–1469 (2019)
2. Colin, S., Doerr, B., Férey, G.: Monotonic functions in EC: anything but monotone! In: Genetic and Evolutionary Computation Conference (GECCO), pp. 753–760 (2014)
3. Dang, D.C., Eremeev, A., Lehre, P.K.: Escaping local optima with non-elitist evolutionary algorithms. In: AAAI Conference on Artificial Intelligence, vol. 35, pp. 12275–12283 (2021)
4. Dang, D.C., Eremeev, A., Lehre, P.K.: Non-elitist evolutionary algorithms excel in fitness landscapes with sparse deceptive regions and dense valleys. In: Genetic and Evolutionary Computation Conference (GECCO), pp. 1133–1141 (2021)
5. Dang, D.-C., Lehre, P.K.: Self-adaptation of mutation rates in non-elitist populations. In: Handl, J., Hart, E., Lewis, P.R., López-Ibáñez, M., Ochoa, G., Paechter, B. (eds.) PPSN 2016. LNCS, vol. 9921, pp. 803–813. Springer, Cham (2016). https://doi.org/10.1007/978-3-319-45823-6_75
6. Doerr, B.: Lower bounds for non-elitist evolutionary algorithms via negative multiplicative drift. Evol. Comput. **29**(2), 305–329 (2021)
7. Doerr, B., Jansen, T., Sudholt, D., Winzen, C., Zarges, C.: Mutation rate matters even when optimizing monotonic functions. Evol. Comput. **21**(1), 1–27 (2013)
8. Doerr, B., Johannsen, D., Winzen, C.: Multiplicative drift analysis. Algorithmica **64**, 673–697 (2012)
9. Hevia Fajardo, M.A., Sudholt, D.: Self-adjusting population sizes for non-elitist evolutionary algorithms: why success rates matter. In: Genetic and Evolutionary Computation Conference (GECCO), pp. 1151–1159 (2021)
10. Jägersküpper, J., Storch, T.: When the plus strategy outperforms the comma strategyand when not. In: Foundations of Computational Intelligence (FOCI), pp. 25–32. IEEE (2007)
11. Janett, D., Lengler, J.: Two-dimensional drift analysis: optimizing two functions simultaneously can be hard (2022). https://arxiv.org/abs/2203.14547
12. Jansen, T.: On the brittleness of evolutionary algorithms. In: Stephens, C.R., Toussaint, M., Whitley, D., Stadler, P.F. (eds.) FOGA 2007. LNCS, vol. 4436, pp. 54–69. Springer, Heidelberg (2007). https://doi.org/10.1007/978-3-540-73482-6_4

13. Kaufmann, M., Larcher, M., Lengler, J., Zou, X.: OneMax is not the easiest function for fitness improvements (2022). https://arxiv.org/abs/2204.07017
14. Kaufmann, M., Larcher, M., Lengler, J., Zou, X.: Self-adjusting population sizes for the $(1, \lambda)$-EA on monotone functions. In: Parallel Problem Solving from Nature (PPSN). Springer (2022)
15. Kötzing, T.: Concentration of first hitting times under additive drift. Algorithmica **75**(3), 490–506 (2016)
16. Lehre, P.K.: Negative drift in populations. In: Schaefer, R., Cotta, C., Kołodziej, J., Rudolph, G. (eds.) PPSN 2010. LNCS, vol. 6238, pp. 244–253. Springer, Heidelberg (2010). https://doi.org/10.1007/978-3-642-15844-5_25
17. Lengler, J.: A general dichotomy of evolutionary algorithms on monotone functions. IEEE Trans. Evol. Comput. **24**(6), 995–1009 (2019)
18. Lengler, J.: Drift analysis. In: Theory of Evolutionary Computation. NCS, pp. 89–131. Springer, Cham (2020). https://doi.org/10.1007/978-3-030-29414-4_2
19. Lengler, J., Martinsson, A., Steger, A.: When does hillclimbing fail on monotone functions: An entropy compression argument. In: Analytic Algorithmics and Combinatorics (ANALCO), pp. 94–102. SIAM (2019)
20. Lengler, J., Meier, J.: Large population sizes and crossover help in dynamic environments. In: Bäck, T., et al. (eds.) PPSN 2020. LNCS, vol. 12269, pp. 610–622. Springer, Cham (2020). https://doi.org/10.1007/978-3-030-58112-1_42
21. Lengler, J., Riedi, S.: Runtime Analysis of the $(\mu + 1)$-EA on the Dynamic BinVal Function. In: Evolutionary Computation in Combinatorial Optimization (EvoCom). pp. 84–99. Springer, Heidelberg (2021)
22. Lengler, J., Schaller, U.: The $(1 + 1)$-EA on noisy linear functions with random positive weights. In: Symposium Series on Computational Intelligence (SSCI), pp. 712–719. IEEE (2018)
23. Lengler, J., Steger, A.: Drift analysis and evolutionary algorithms revisited. Comb. Probab. Comput. **27**(4), 643–666 (2018)
24. Lengler, J., Sudholt, D., Witt, C.: The complex parameter landscape of the compact genetic algorithm. Algorithmica **83**(4), 1096–1137 (2021)
25. Lengler, J., Zou, X.: Exponential slowdown for larger populations: the $(\mu + 1)$-EA on monotone functions. Theoret. Comput. Sci. **875**, 28–51 (2021)
26. Neumann, F., Sudholt, D., Witt, C.: A few ants are enough: ACO with iteration-best update. In: Genetic and Evolutionary Computation Conference (GECCO), pp. 63–70 (2010)
27. Oliveto, P.S., Witt, C.: Improved time complexity analysis of the simple genetic algorithm. Theoret. Comput. Sci. **605**, 21–41 (2015)
28. Rowe, J.E.: Linear multi-objective drift analysis. Theoret. Comput. Sci. **736**, 25–40 (2018)
29. Rowe, J.E., Sudholt, D.: The choice of the offspring population size in the $(1, \lambda)$ evolutionary algorithm. Theoret. Comput. Sci. **545**, 20–38 (2014)
30. Sudholt, D., Witt, C.: On the choice of the update strength in estimation-of-distribution algorithms and ant colony optimization. Algorithmica **81**(4), 1450–1489 (2019)
31. Witt, C.: Tight bounds on the optimization time of a randomized search heuristic on linear functions. Comb. Probab. Comput. **22**(2), 294–318 (2013)

Author Index

Printed in the United States
by Baker & Taylor Publisher Services